Student Study

Solutions Manual

to accompany

General, Organic, & Biological Chemistry

Third Edition

Janice Gorzynski Smith
University of Hawaii at Manoa

Prepared By
Janice Smith

STUDENT STUDY GUIDE/SOLUTIONS MANUAL TO ACCOMPANY
GENERAL, ORGANIC, & BIOLOGICAL CHEMISTRY, THIRD EDITION

Published by McGraw-Hill Education, 2 Penn Plaza, New York, NY 10121.

ISBN: 978-1-259-28974-3
MHID: 1-259-28974-5

www.mhhe.com

Table of Contents

Chapter 1 Matter and Measurement

Chapter Review

[1] Describe the three states of matter. (1.1, 1.2)

- Matter is anything that has mass and takes up volume. Matter has three common states:
    - The **solid state** is composed of highly organized particles that lie close together. A solid has a definite shape and volume.
    - The **liquid state** is composed of particles that lie close together but are less organized than the solid state. A liquid has a definite volume but not a definite shape.
    - The **gas state** is composed of highly disorganized particles that lie far apart. A gas has no definite shape or volume.

[2] How is matter classified? (1.3)

- Matter can be classified as a pure substance or a mixture:
    - A pure substance is composed of a single component with a constant composition. A pure substance is either an element, which cannot be broken down into simpler substances by a chemical reaction, or a compound, which is formed by combining two or more elements.
    - A mixture is composed of more than one substance and its composition can vary depending on the sample.

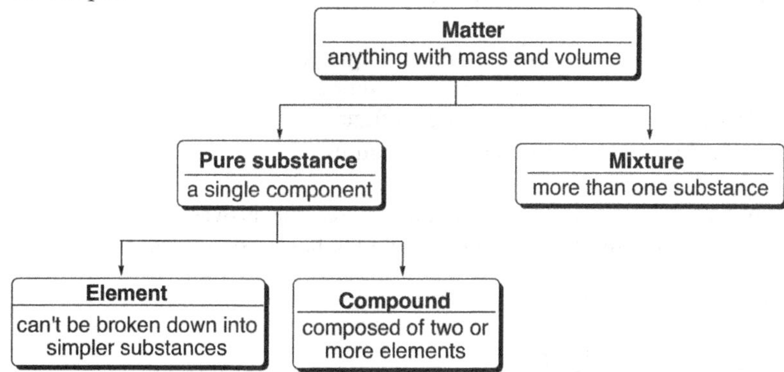

[3] What are the key features of the metric system of measurement? (1.4)

- In the metric system, each type of measurement has a base unit and all other units are related to the base unit by a prefix that indicates if the unit is larger or smaller than the base unit.
- The base units are meter (m) for length, gram (g) for mass, liter (L) for volume, and second (s) for time.

Common Prefixes Used for Metric Units

Prefix	Symbol	Meaning	Numerical Value	Scientific Notation
giga-	G	billion	1,000,000,000.	10^9
mega-	M	million	1,000,000.	10^6
kilo-	k	thousand	1,000.	10^3
deci-	d	tenth	0.1	10^{-1}
centi-	c	hundredth	0.01	10^{-2}
milli-	m	thousandth	0.001	10^{-3}
micro-	μ	millionth	0.000 001	10^{-6}
nano-	n	billionth	0.000 000 001	10^{-9}

[4] What are significant figures and how are they used in calculations? (1.5)

- Significant figures are all digits in a measured number including one estimated digit. All nonzero digits are significant. A zero is significant only if it occurs between two nonzero digits, or at the end of a number with a decimal point. A trailing zero in a number without a decimal point is not considered significant.

- All nonzero digits are always significant.

65.2 g	three significant figures
1,265 m	four significant figures
25 μL	two significant figures
255.345 g	six significant figures

- In multiplying and dividing with significant figures, the answer has the same number of significant figures as the original number with the fewest significant figures.

four significant figures

$$\text{miles per hour} = \frac{351.2 \text{ miles}}{5.5 \text{ hours}} = 63.854\ 545 \text{ miles per hour}$$

two significant figures

The answer must contain only **two** significant figures.

= **64 miles per hour**

- In adding or subtracting with significant figures, the answer has the same number of decimal places as the original number with the fewest decimal places.

weight at one year = 10.11 kg — 10.11 kg — two digits after the decimal point

weight at birth = 3.6 kg — − 3.6 kg — one digit after the decimal point

weight gain = 6.51 kg

last significant digit

- The answer can have only **one** digit after the decimal point.
- Round 6.51 to 6.5.
- The baby gained 6.5 kg during his first year of life.

[5] What is scientific notation? (1.6)

- Scientific notation is a method of writing a number as $y \times 10^x$, where y is a number between 1 and 10, and x is a positive or negative exponent.
- To convert a standard number to a number in scientific notation, move the decimal point to give a number between 1 and 10. Multiply the result by 10^x, where x is the number of places the decimal point was moved. When the decimal point is moved to the left, x is positive. When the decimal point is moved to the right, x is negative.
- To convert a number in scientific notation to a standard number, move the decimal point to the right for positive exponents, and to the left for negative exponents.

2.800×10^2 2.800 ---→ 280.0

Move the decimal point to the right two places.

7.4×10^{-11} 000 000 000 07.4 ---→ 0.000 000 000 074

Move the decimal point to the left 11 places.

[6] How are conversion factors used to convert one unit to another? (1.7, 1.8)

- A conversion factor is a term that converts a quantity in one unit to a quantity in another unit. To use conversion factors to solve a problem, set up the problem with any unwanted unit in the numerator of one term and the denominator of another term, so that unwanted units cancel.

conversion factor

$$130 \text{ lb} \times \frac{1 \text{ kg}}{2.20 \text{ lb}} = \boxed{59 \text{ kg}}$$ **answer in kilograms**

Pounds (lb) must be in the denominator to cancel the unwanted unit (lb) in the original quantity.

[7] What is temperature and how are the three temperature scales related? (1.9)

- Temperature is a measure of how hot or cold an object is. The Fahrenheit and Celsius temperature scales are divided into degrees. Both the size of the degree and the zero point of these scales differ. The Kelvin scale is divided into kelvins, and one kelvin is the same size as one degree Celsius.

To convert from Celsius to Fahrenheit:

$$T_F = 1.8(T_C) + 32$$

To convert from Fahrenheit to Celsius:

$$T_C = \frac{T_F - 32}{1.8}$$

To convert from Celsius to Kelvin:

$$T_K = T_C + 273$$

To convert from Kelvin to Celsius:

$$T_C = T_K - 273$$

[8] What are density and specific gravity? (1.10)

- **Density** is a physical property reported in g/mL or g/cc that relates the mass of an object to its volume. A less dense substance floats on top of a more dense liquid.
- **Specific gravity** is a unitless quantity that relates the density of a substance to the density of water. Since the density of water is 1.00 g/mL at 4 °C, the specific gravity of a substance equals its density but it contains no units.

$$\text{specific gravity} = \frac{\text{density of a substance (g/mL)}}{\text{density of water (g/mL)}}$$

Problem Solving

[1] Significant Figures (1.5)

Example 1.1 How many significant figures does each number contain?

a. 0.309 b. 72,000 c. 52.00 d. 94.03

Analysis
All nonzero digits are significant. A zero is significant only if it occurs between two nonzero digits, or at the end of a number with a decimal point.

Solution
Significant figures are shown in **bold.**

a. 0.**309** (three) b. **72**,000 (two) c. **52.00** (four) d. **94.03** (four)

Example 1.2 Round off each number to two significant figures.

a. 2.7983 b. 0.010 023 9 c. 42,980.0

Analysis
If the answer is to have *two* significant figures, look at the *third* number from the left. If this number is 4 or fewer, drop it and all remaining numbers to the right. If the third number from the left is 5 or greater, round the number up by adding one to the second digit.

Solution
a. 2.8 b. 0.010 c. 43,000 (Omit the decimal point after the 0. The number 43,000. has five significant figures.)

Example 1.3 Carry out each calculation and give the answer using the proper number of significant figures.

a. 4.29×0.023 b. $3,482 \div 52.7$

Analysis
Since these calculations involve multiplication and division, the answer must have the same number of significant figures as the original number with the fewest number of significant figures.

Solution

a. 4.29 × 0.023 = 0.098 67

- Since 0.023 has only two significant figures, round the answer to give it two significant figures.

  0.098 67 — Since this number is 6 (5 or greater), round the 8 to its left up by one.

- **Answer**: 0.099

---

b. 3,482 ÷ 52.7 = 66.072 106

- Since 52.7 has three significant figures, round the answer to give it three significant figures.

  66.072 106 — Since this number is 7 (5 or greater), round the 0 to its left up by one.

- **Answer**: 66.1

Example 1.4 On a trip to the market, you bought three packages of meat weighing 1.17 lb, 2.4 lb, and 0.97 lb. How much meat did you buy in all?

Analysis

Add up the number of pounds in each package to get the total pounds purchased. When adding, the answer has the same number of decimal places as the original number with the fewest decimal places.

Solution

```
 1.17 lb
 2.4  lb   one digit after the decimal point
 0.97 lb
--------
 4.54 lb  - - - - - ->  4.5 lb
           round off
  ↑
last significant digit
```

- Since 2.4 lb has only one digit after the decimal point, the answer can have only one digit after the decimal point.
- Round 4.54 to 4.5.
- Total meat purchased: 4.5 lb

---

[2] Scientific Notation (1.6)

Example 1.5 Write the numbers in scientific notation.

a. 62,000 b. 0.000 07

Analysis

Move the decimal point to give a number between 1 and 10. Multiply the number by 10^x, where x is the number of places the decimal point was moved. The exponent x is (+) when the decimal point moves to the left and (–) when the decimal point moves to the right.

Solution

a. 62000. = 6.2×10^4 ← the number of places the decimal point was moved to the left

Move the decimal point four places to the left.

- Write the coefficient as 6.2 (two significant figures), since 62,000 contains two significant figures.

b. 0.000 07 = 7×10^{-5} ← the number of places the decimal point was moved to the right

Move the decimal point five places to the right.

- Write the coefficient as 7 (one significant figure), since 0.000 07 contains one significant figure.

Example 1.6 Convert 1.83×10^4 to a standard number.

Analysis
The exponent in 10^x tells how many places to move the decimal point in the coefficient to generate a standard number. The decimal point goes to the right when x is positive and to the left when x is negative.

Solution

1.83×10^4 1.8300. ----→ **Answer:** 18,300

Move the decimal point to the right four places.

---

[3] Conversion Factors (1.7, 1.8)

Example 1.7 Write two conversion factors for each pair of units.

a. kilograms and pounds   b. liters and milliliters

Analysis
Use the equalities in Tables 1.3 and 1.4 to write a fraction that shows the relationship between the two units.

Solution

a. Conversion factors for kilograms and pounds:

$$\frac{2.20 \text{ lb}}{1 \text{ kg}} \quad \text{or} \quad \frac{1 \text{ kg}}{2.20 \text{ lb}}$$

b. Conversion factors for liters and milliliters:

$$\frac{1000 \text{ mL}}{1 \text{ L}} \quad \text{or} \quad \frac{1 \text{ L}}{1000 \text{ mL}}$$

Example 1.8 How many liters are in one gallon (1.0) of milk?

Analysis and Solution
[1] Identify the original quantity and the desired quantity.

1.0 gal ? L

original quantity desired quantity

[2] Write out the conversion factors.

- We have no conversion factor that directly relates gallons to liters. We do, however, know conversions for gallons to quarts and quarts to liters.

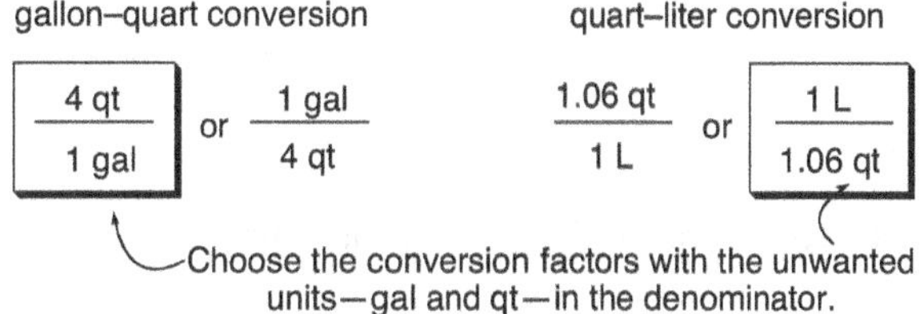

gallon–quart conversion
quart–liter conversion
4 qt / 1 gal or 1 gal / 4 qt
1.06 qt / 1 L or 1 L / 1.06 qt
Choose the conversion factors with the unwanted units—gal and qt—in the denominator.

[3] Solve the problem.

- To set up the problem so that unwanted units cancel, arrange each term so that the units in the numerator of one term cancel the units of the denominator of the adjacent term. In this problem we need to cancel both gallons and quarts to get liters.
- The single desired unit, liters, must be in the **numerator** of one term.

$$1.0\ \text{gal} \times \frac{4\ \text{qt}}{1\ \text{gal}} \times \frac{1\ \text{L}}{1.06\ \text{qt}} = 3.8\ \text{L}$$

Gallons cancel. Quarts cancel. Liters do not cancel.

[4] Check

- Since there are four quarts in a gallon and a quart is about the same size as a liter, one gallon should be about four liters. The answer, 3.8, is just about 4.
- Write the answer with two significant figures since the original quantity, 1.0 gal, has two significant figures.

[4] Temperature (1.9)

Example 1.9 An elderly patient who was brought in from his cold home in winter had a temperature of 92 °F. Convert this temperature to both T_C and T_K.

Analysis

First convert the Fahrenheit temperature to degrees Celsius using the equation $T_C = (T_F - 32)/1.8$. Then convert the Celsius temperature to kelvins by adding 273.

Solution

[1] Convert T_F to T_C:

$$T_C = \frac{T_F - 32}{1.8}$$

$$= \frac{92 - 32}{1.8} = 33\ ^oC$$

[2] Convert T_C to T_K:

$$T_K = T_C + 273$$

$$= 33 + 273 = 306\ K$$

[5] Density (1.10)

Example 1.10 Calculate the mass in grams of 20.0 mL of an antibiotic solution that has a density of 1.21 g/mL.

Analysis

Use density (g/mL) to interconvert the mass and volume of a liquid.

Solution

$$20.0\ \text{mL} \times \underbrace{\frac{1.21\ g}{1\ \text{mL}}}_{\text{density}} = 24.2\ \text{g of antibiotic solution}$$

Milliliters cancel.

The answer, 24.2 g, has three significant figures to match the number of significant figures in both factors in the problem.

Self-Test

[1] Fill in the blank with one of the terms listed below.

Chemical properties (1.2)	Liquid (1.2)	Physical properties (1.2)
Element (1.3)	Mass (1.4)	Solid (1.2)
Gas (1.2)	Matter (1.1)	Temperature (1.9)
Inexact number (1.5)	Mixture (1.3)	Weight (1.4)

1. ______ is the force that matter feels due to gravity.
2. An ______ is a pure substance that cannot be broken down into simpler substances by a chemical reaction.
3. ______ is anything that has mass and takes up volume.
4. A ______ has no definite shape or volume. The particles move randomly and are separated by a distance much larger than their size.
5. An ______ results from a measurement or observation and contains some uncertainty.
6. ______ are those properties that can be observed or measured without changing the composition of the material.
7. A ______ is composed of more than one substance. The composition can vary depending on the sample.

8. ______ is a measure of the amount of matter in an object.
9. A ______ has a definite volume, but takes on the shape of the container it occupies. The particles are close together but they can randomly move around.
10. ______ is a measurement of how hot or cold an object is.
11. A ______ has a definite shape and volume. The particles lie close together, and are arranged in a regular three-dimensional array.
12. ______ are those properties that determine how a substance can be converted to another substance by a chemical reaction.

[2] Fill in the blank with one of the terms listed below.

a. Kelvin b. Celsius c. Fahrenheit

13. The zero point on the ______ scale is called absolute zero.
14. On the ______ scale, water freezes at 32 ° and boils at 212 °.
15. On the ______ scale, water freezes at 0 ° and boils at 100 °.

[3] Pick the appropriate unit of measurement for each object.

a. meter
b. milliliter
c. centimeter
d. kilogram

16. The weight of a newborn infant
17. The amount of vaccine in a syringe
18. The length of a football field
19. The diameter of a pencil

[4] Match the standard number and the scientific notation.

a. 1.04×10^{-8} b. 1.04×10^{4} c. 1.04×10^{-2}

20. 10,400
21. 0.0104
22. 0.000 000 010 4

[5] How many significant figures are in each of the numbers below?

23. 54,800
24. 0.067 00
25. 119,000.0

Answers to Self-Test

1. Weight	6. Physical properties	11. solid	16. d	21. c
2. element	7. mixture	12. Chemical properties	17. b	22. a
3. Matter	8. Mass	13. a	18. a	23. 3
4. gas	9. liquid	14. c	19. c	24. 4
5. inexact number	10. Temperature	15. b	20. b	25. 7

Solutions to In-Chapter Problems

1.1 Naturally occurring: ice, blood. Synthetic: gloves, mask, plastic syringe, stainless steel needle.

1.2 *Physical properties* can be observed or measured without changing the composition of the material (a and d). *Chemical properties* determine how a substance can be converted to another substance by chemical reactions (b, c, and e).

1.3 This represents a chemical change because the "particles" on the left are different from the particles on the right. For example, on the left side there are particles containing only two red balls, while on the right there are none of these.

1.4 Representation (a) is a pure substance since each particle contains one red and two gray spheres. Representation (b) is a mixture since some of the particles are only red, and some are red and black.

1.5 A *pure substance* is composed of a single substance and has a constant composition regardless of the sample size (d). A *mixture* is composed of more than one component (a, b, and c). The composition of a mixture can vary depending on the sample.

1.6 An *element* is a pure substance that cannot be broken down into simpler substances by a chemical reaction (a). A *compound* is a pure substance formed by combining two or more elements together (b, c, and d).

1.7 Use Table 1.2 to determine the prefix for each unit.
a. a million liters = megaliter
b. a thousandth of a second = millisecond
c. a hundredth of a gram = centigram
d. a tenth of a liter = deciliter

1.8 One nanometer = 0.000 000 001 m (one billionth of a meter); therefore, 1 m = 1,000,000,000 nm.

1.9
a. 0.000 001 g = one microgram (1 μg)
b. 1,000,000,000 m = one gigameter (1 Gm)
c. 0.000 000 001 s = one nanosecond (1 ns)
d. 0.01 g = one centigram (1 cg)

1.10 Use Table 1.2 to determine which quantity is larger.
a. 3 cL
b. 1 μg
c. 5 km
d. 2 mL

1.11 All nonzero digits are significant. A zero is significant only if it occurs between two nonzero digits, or at the end of a number with a decimal point. The significant figures are in **bold.**

a. **23.45** 4 significant figures
b. **23.057** 5 significant figures
c. **23**0 2 significant figures
d. **231.0** 4 significant figures
e. 0.**202** 3 significant figures
f. 0.003 **60** 3 significant figures
g. **1,245,006** 7 significant figures
h. **1,2**00,000 2 significant figures
i. **10,04**0 4 significant figures
j. **10,040.** 5 significant figures

1.12 A zero is significant only if it occurs between two nonzero digits, or at the end of a number with a decimal point.

a. 0.003 04 — No Yes   b. 26,045 — Yes   c. 1,000,034 — Yes   d. 0.304 00 — No Yes

1.13 When the number to be rounded off is 4 or fewer, it and all other digits to the right are dropped. When the number is 5 or greater, 1 is added to the digit to its left.

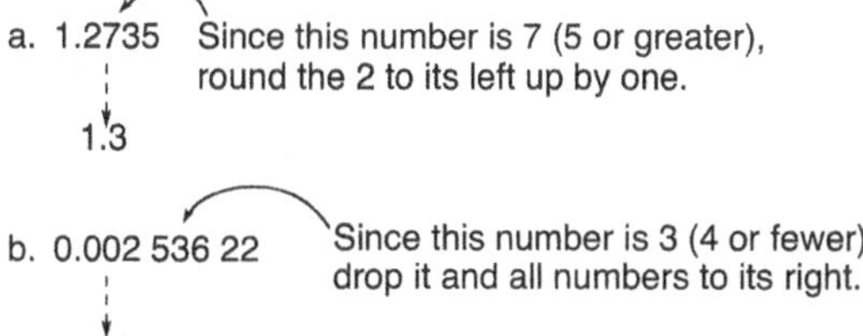
a. 1.2735
Since this number is 7 (5 or greater), round the 2 to its left up by one.
1.3
b. 0.002 536 22
Since this number is 3 (4 or fewer), drop it and all numbers to its right.
0.0025

c. 3,836.9 → 3,800
Since this number is 3 (4 or fewer), drop it and all numbers to its right.

1.14 The answers must have the same number of significant figures as the original number with the fewest number of significant figures.

a. 10.70 × 3.5 = 37.45
Since 3.5 has only two significant figures, round the answer to give it two significant figures.
37

b. 0.206 ÷ 25,993 = **0.000 007 93**
Since 0.206 has three significant figures, the answer has the appropriate number of significant figures.

c. 1,300 ÷ 41.2 = 31.553 398
Since 1,300 has only two significant figures, round the answer to give it two significant figures.
32

d. 120.5 × 26 = 3,133
Since 26 has only two significant figures, round the answer to give it two significant figures.
3,100

1.15 The answers must have the same number of decimal places as the original number with the fewest decimal places.

a. 27.8 cm + 0.246 cm = 28.046 cm
Since 27.8 has one digit after the decimal point, round the answer to one digit after the decimal point.
28.0 cm

b. 102.66 mL + 0.857 mL + 24.0 mL = 127.517 mL
Since 24.0 has one digit after the decimal point, round the answer to one digit after the decimal point.
127.5 mL

c. 54.6 mg – 25 mg = 29.6 mg
Since 25 has zero digits after the decimal point, round the answer to the nearest whole number.
30. mg

d. 2.35 s – 0.266 s = 2.084 s
Since 2.35 has two digits after the decimal point, round the answer to two digits after the decimal point.
2.08 s

1.16 To write a number in scientific notation:
[1] Move the decimal point to give a number between 1 and 10.
[2] Multiply the result by 10^x, where x is the number of places the decimal point was moved.

0.000 098 g/dL = 9.8×10^{-5} g/dL — the number of places the decimal point was moved to the right

Move the decimal point five places to the right.

1.17 To write a number in scientific notation:
[1] Move the decimal point to give a number between 1 and 10.
[2] Multiply the result by 10^x, where x is the number of places the decimal point was moved.

a. $93,200 = 9.32 \times 10^4$
The decimal point was moved four places to the left.

b. $0.000\,725 = 7.25 \times 10^{-4}$
The decimal point was moved four places to the right.

c. $6,780,000 = 6.78 \times 10^6$
The decimal point was moved six places to the left.

d. $0.000\,030 = 3.0 \times 10^{-5}$
The decimal point was moved five places to the right.

e. $4,520,000,000,000 = 4.52 \times 10^{12}$
The decimal point was moved 12 places to the left.

f. $0.000\,000\,000\,028 = 2.8 \times 10^{-11}$
The decimal point was moved 11 places to the right.

1.18 The exponent in 10^x tells how many places to move the decimal point in the coefficient to generate a standard number. The decimal point goes to the right when x is positive and to the left when x is negative.

a. $6.5 \times 10^3 = 6,500$
The decimal point was moved three places to the right.

b. $3.26 \times 10^{-5} = 0.000\,032\,6$
The decimal point was moved five places to the left.

c. $3.780 \times 10^{-2} = 0.037\,80$
The decimal point was moved two places to the left.

d. $1.04 \times 10^8 = 104,000,000$
The decimal point was moved eight places to the right.

e. $2.221 \times 10^6 = 2,221,000$
The decimal point was moved six placed to the right.

f. $4.5 \times 10^{-10} = 0.000\,000\,000\,45$
The decimal point was moved 10 places to the left.

1.19 Use the equalities in Tables 1.3 and 1.4 to write a fraction that shows the relationship between the two units.

a. $\frac{0.621 \text{ mi}}{1 \text{ km}}$ $\frac{1 \text{ km}}{0.621 \text{ mi}}$

c. $\frac{454 \text{ g}}{1 \text{ lb}}$ $\frac{1 \text{ lb}}{454 \text{ g}}$

b. $\frac{1000 \text{ mm}}{1 \text{ m}}$ $\frac{1 \text{ m}}{1000 \text{ mm}}$

d. $\frac{1000 \ \mu\text{g}}{1 \text{ mg}}$ $\frac{1 \text{ mg}}{1000 \ \mu\text{g}}$

1.20 To convert 4,120 km to miles:
[1] Identify the original quantity and the desired quantity, including units.
[2] Write out the conversion factor(s) needed to solve the problem.
[3] Set up and solve the problem.
[4] Write the answer using the correct number of significant figures and check by estimation.

[1] 4,120 km — original quantity; ? mi — desired quantity

[2] Two possible conversion factors: $\frac{1\ \text{km}}{0.621\ \text{mi}}$ or $\boxed{\frac{0.621\ \text{mi}}{1\ \text{km}}}$ Choose this factor to cancel the unwanted unit, km.

[3]

$$\underset{\text{original quantity}}{4{,}120\ \text{km}} \times \overset{\text{conversion factor}}{\frac{0.621\ \text{mi}}{1\ \text{km}}} = \underset{\text{desired quantity}}{2{,}558.52\ \text{mi}}$$

The number of km (unwanted unit) cancels.

[4] The initial number has three significant figures, so the final answer is rounded to 2,560 mi.

1.21 Use conversion factors to solve the problems.

a. $25\ \text{L} \times \frac{10\ \text{dL}}{1\ \text{L}} = 250\ \text{dL}$

b. $40.0\ \text{oz} \times \frac{28.3\ \text{g}}{1\ \text{oz}} = 1{,}132\ \text{g} = 1{,}130\ \text{g}$ rounded to 3 significant figures

c. $32\ \text{in.} \times \frac{2.54\ \text{cm}}{1\ \text{in.}} = 81.28\ \text{cm} = 81\ \text{cm}$ rounded to 2 significant figures

d. $10\ \text{cm} \times \frac{10\ \text{mm}}{1\ \text{cm}} = 100\ \text{mm}$

1.22

a. 0.46 mL b. $0.46\ \text{mL} \times \frac{1\ \text{L}}{1000\ \text{mL}} \times \frac{1000000\ \mu\text{L}}{1\ \text{L}} = 460\ \mu\text{L}$

1.23 Use conversion factors to solve the problems.

a. $6{,}250\ \text{ft} \times \frac{1\ \text{mi}}{5{,}280\ \text{ft}} \times \frac{1\ \text{km}}{0.621\ \text{mi}} = 1.91\ \text{km}$

Feet cancel. Miles cancel. Kilometers do not cancel.

b. $3\ \text{cups} \times \frac{1\ \text{qt}}{4\ \text{cups}} \times \frac{1\ \text{L}}{1.06\ \text{qt}} = 0.7\ \text{L}$

Cups cancel. Quarts cancel. Liters do not cancel.

c. $4.5\ \text{ft} \times \frac{12\ \text{in.}}{1\ \text{ft}} \times \frac{2.54\ \text{cm}}{1\ \text{in.}} = 140\ \text{cm}$

Feet cancel. Inches cancel. Centimeters do not cancel.

1.24 Use the conversion factors: 1 teaspoon = 5 mL and 80. mg acetaminophen/2.5 mL of Tylenol.

a. $2.5 \text{ tsp} \times \dfrac{5 \text{ mL}}{1 \text{ tsp}}$ = 12.5 mL rounded to 13 mL (2 significant figures)

b. $13 \text{ mL} \times \dfrac{80. \text{ mg}}{2.5 \text{ mL}}$ = 416 mg rounded to 420 mg (2 significant figures)

1.25

$$0.100 \text{ mg} \times \frac{1000 \ \mu\text{g}}{1 \text{ mg}} \times \frac{1 \text{ tablet}}{25 \ \mu\text{g}} = 4 \text{ tablets}$$

1.26

$$160 \text{ mg} \times \frac{5 \text{ mL}}{100 \text{ mg}} = 8 \text{ mL Children's Motrin}$$

1.27 Convert from T_C to T_F and T_K using the formulas listed in Section 1.9.

$T_F = 1.8(T_C) + 32$

$= 1.8(28.5) + 32 = 83.3 \text{ °F}$

$T_K = T_C + 273$

$= 28.5 + 273 = 302 \text{ K}$

1.28

a. $T_F = 1.8(T_C) + 32$

$= 1.8(20.) + 32 = 68 \text{ °F}$

b. $T_C = \dfrac{T_F - 32}{1.8}$

$= \dfrac{150. - 32}{1.8} = 66 \text{ °C}$

c. $T_C = T_K - 273$

$= 298 - 273 = 25 \text{ °C}$

$T_F = 1.8(T_C) + 32$

$= 1.8(25) + 32 = 77 \text{ °F}$

d. $T_K = T_C + 273$

$= 75 + 273 = 348 \text{ K}$

1.29
a. Since the densities of **A** and **B** are the same and there is a larger volume of **B,** the mass of **B** is greater than the mass of **A.**
b. The density of **A** is twice the density of **B,** but there is three times as much volume of **B** as **A,** so the mass of **B** is greater than the mass of **A.**
c. The density of **B** is greater than the density of **A** and there is a larger volume of **B,** so the mass of **B** is greater than the mass of **A.**

1.30 To convert volume (mL) to mass (g), multiply the volume by the density (g/mL).

$$10.0 \text{ mL} \times \frac{0.713 \text{ g}}{\text{mL}} = 7.13 \text{ g}$$

Milliliters cancel.

1.31 Convert pounds of lead to grams. Then use the density of lead (11.3 g/cc) to determine the volume.

$$5 \text{ weights} \times \frac{2.0 \text{ lb}}{1 \text{ weight}} \times \frac{454 \text{ g}}{1 \text{ lb}} \times \frac{1 \text{ cc}}{11.3 \text{ g}} = 402 \text{ cc}$$

4.0×10^2 cc lead

1.32

a. $$\text{specific gravity} = \frac{\text{density of a substance (g/mL)}}{\text{density of water (g/mL)}} = \frac{0.80 \text{ g/mL}}{1 \text{ g/mL}} = 0.80$$

b. $$2.3 = \frac{\text{density of a substance (g/mL)}}{1 \text{ g/mL}}$$ density = 2.3 g/mL

Solutions to Odd-Numbered End-of-Chapter Problems

1.33 Representation (a) is a pure element since each particle consists of a gray sphere. Representation (b) is a pure compound since each particle contains four gray spheres and one black sphere. Representation (c) is a mixture, because some of the particles contain two gray spheres, whereas others contain four gray spheres and one black sphere. Representation (d) is a mixture, because some of the particles are gray spheres, and some are blue.

1.35 Molecular art for an element shows spheres of one color only.
Elements: two blue spheres joined, two red spheres joined
Compounds: black sphere joined to two red spheres, red sphere joined to two gray spheres

1.37

Phase	a. Volume	b. Shape	c. Organization	d. Particle Proximity
Solid	Definite	Definite	Very organized	Very close
Liquid	Definite	Assumes shape of container	Less organized	Close
Gas	Not fixed	None	Disorganized	Far apart

1.39 A *chemical change* converts one substance to another substance by a chemical reaction (b). A *physical change* can be observed or measured without changing the composition of the material (a and c).

1.41 This is a physical change, because the compound CO_2 is unchanged in this transition. The same "particles" exist at the beginning and end of the process.

1.43 a, b: The temperature on the Fahrenheit thermometer is 76.5 °F, which has three significant figures.

c. $$T_C = \frac{T_F - 32}{1.8} = \frac{76.5 - 32}{1.8} = 24.7 \text{ °C}$$

1.45 An exact number results from counting objects or is part of a definition, such as having 20 people in a class. An inexact number results from a measurement or observation and contains some uncertainty, such as the distance from the earth to the sun, 9.3×10^7 miles.

1.47 Compare the measurements using Table 1.2. (< means *less than*; > means *greater than.*)

a. 5 mL < **5 dL** b. **10 mg** > 10 μg c. **5 cm** > 5 mm d. **10 Ms** > 10 ms

1.49 All nonzero digits are significant. A zero is significant only if it occurs between two nonzero digits, or at the end of a number with a decimal point. The significant figures are in **bold.**

a. **16.00**
4 significant figures

b. **160**
2 significant figures

c. 0.001 **60**
3 significant figures

d. **1,6**00,000
2 significant figures

e. **1.06**
3 significant figures

f. 0.**1600**
4 significant figures

g. $\mathbf{1.060} \times 10^{10}$
4 significant figures

h. $\mathbf{1.6} \times 10^{-6}$
2 significant figures

1.51 When the number to be rounded off is 4 or fewer, it and all other digits to the right are dropped. When the number is 5 or greater, 1 is added to the digit to its left.

a. 25,401 = 25,400
b. 1,248,486 = 1,250,000
c. 0.001 265 982 = 0.001 27
d. 0.123 456 = 0.123
e. 195.371 = 195
f. 196.814 = 197

1.53 The answers in problems with multiplication and division must have the same number of decimal places as the original number with the fewest decimal places. The answers in problems with addition and subtraction must have the same number of digits after the decimal point as the original number with the fewest digits after the decimal point.

a. $53.6 \times 0.41 = 21.976$
Since 0.41 has two significant figures, round the answer to two significant figures.
22

b. $25.825 - 3.86 = 21.965$
Since 3.86 has two digits after the decimal point, round the answer to two digits after the decimal point.
21.97

c. $65.2 \div 12 = 5.43333$
Since 12 has two significant figures, round the answer to two significant figures.
5.4

d. $41.0 + 9.135 = 50.135$
Since 41.0 has one digit after the decimal point, round the answer to one digit after the decimal point.
50.1

e. $694.2 \times 0.2 = 138.84$
Since 0.2 has one significant figure, round the answer to one significant figure.
100

f. $1{,}045 - 1.26 = 1{,}043.74$
Since 1,045 has no digits after the decimal point, round the answer to the closest whole number.
1,044

1.55 To write a number in scientific notation:
[1] Move the decimal point to give a number between 1 and 10.
[2] Multiply the result by 10^x, where x is the number of places the decimal point was moved.

a. 1,234 g = 1.234×10^3 g
The decimal point was moved three places to the left.
b. 0.000 016 2 m = 1.62×10^{-5} m
The decimal point was moved five places to the right.
c. 5,244,000 L = 5.244×10^6 L
The decimal point was moved six places to the left.
d. 0.005 62 g = 5.62×10^{-3} g
The decimal point was moved three places to the right.
e. 44,000 km = 4.4×10^4 km
The decimal point was moved four places to the left.

1.57 The exponent in 10^x tells how many places to move the decimal point in the coefficient to generate a standard number. The decimal point goes to the right when x is positive and to the left when x is negative.

a. 3.4×10^8 = 340,000,000
The decimal point was moved eight places to the right.
b. 5.822×10^{-5} = 0.000 058 22
The decimal point was moved five places to the left.
c. 3×10^2 = 300
The decimal point was moved two places to the right.
d. 6.86×10^{-8} = 0.000 000 068 6
The decimal point was moved eight places to the left.

1.59 Compare the two numbers. The number in **bold** is larger.

a. $\mathbf{4.44 \times 10^3}$ or 4.8×10^2
b. 5.6×10^{-6} or $\mathbf{5.6 \times 10^{-5}}$
c. $\mathbf{1.3 \times 10^8}$ or 52,300,000
d. $\mathbf{9.8 \times 10^{-4}}$ or 0.000 089

1.61 Write the number in scientific notation.

a. 0.000 400 g of folate = 4.00×10^{-4} g
The decimal point was moved four places to the right.
b. 0.002 g of copper = 2×10^{-3} g
The decimal point was moved three places to the right.
c. 0.000 080 g of vitamin K = 8.0×10^{-5} g
The decimal point was moved five places to the right.
d. 3,400 mg of chloride = 3.4×10^3 mg
The decimal point was moved three places to the left.

1.63 The scale shows the individual has a mass of 115 lb.

$$115\ \cancel{\text{lb}} \times \frac{1\ \text{kg}}{2.20\ \cancel{\text{lb}}} = 52.3\ \text{kg}$$

1.65 Use conversion factors to solve the problems.

a. $$1.5\ \cancel{\text{kg}} \times \frac{1000\ \text{g}}{1\ \cancel{\text{kg}}} = 1{,}500\ \text{g}$$

b. $$1.5\ \cancel{\text{kg}} \times \frac{2.20\ \text{lb}}{1\ \cancel{\text{kg}}} = 3.3\ \text{lb}$$

c. $$1{,}500\ \cancel{\text{g}} \times \frac{1\ \text{oz}}{28.3\ \cancel{\text{g}}} = 53\ \text{oz}$$

1.67 Use conversion factors to solve the problems.

a. $300\ \cancel{g} \times \frac{1000\ mg}{1\ \cancel{g}} = 300{,}000\ mg$

b. $2\ \cancel{L} \times \frac{1{,}000{,}000\ \mu L}{1\ \cancel{L}} = 2{,}000{,}000\ \mu L$

c. $5.0\ \cancel{cm} \times \frac{1\ m}{100\ \cancel{cm}} = 0.050\ m$

d. $300\ \cancel{g} \times \frac{1\ oz}{28.3\ \cancel{g}} = 10.60\ oz =$ **10 oz** rounded to one significant figure

e. $2\ \cancel{ft} \times \frac{12\ \cancel{in.}}{1\ \cancel{ft}} \times \frac{1\ m}{39.4\ \cancel{in.}} = 0.6091\ m =$ **0.6 m** rounded to one significant figure

f. $3.5\ \cancel{yd} \times \frac{3\ \cancel{ft}}{1\ \cancel{yd}} \times \frac{12\ \cancel{in.}}{1\ \cancel{ft}} \times \frac{1\ m}{39.4\ \cancel{in.}} = 3.198\ m =$ **3.2 m** rounded to two significant figures

1.69 Use conversion factors to solve the problems.

a. $50.\ \cancel{in.} \times \frac{2.54\ cm}{1\ \cancel{in.}} = 127\ cm =$ **130 cm** rounded to two significant figures

b. $3.0\ \cancel{pints} \times \frac{1\ \cancel{qt}}{2\ \cancel{pints}} \times \frac{1\ L}{1.06\ \cancel{qt}} = 1.415\ L =$ **1.4 L** rounded to two significant figures

c. $T_F = 1.8(T_C) + 32$

$= 1.8(37.7) + 32 = 99.9\ °F$

1.71 Use conversion factors to solve the problems.

a. $1\ \cancel{qt} \times \frac{946\ mL}{1\ \cancel{qt}} =$ **946 mL** rounded to three significant figures

b. $1\ \cancel{L} \times \frac{1.06\ \cancel{qt}}{1\ \cancel{L}} \times \frac{32\ fl\ oz}{1\ \cancel{qt}} = 33.92\ fl\ oz =$ **33.9 fl oz** rounded to three significant figures

1.73 Convert from T_C to T_F and T_K using the formulas listed in Section 1.9.

a. $T_F = 1.8(T_C) + 32$

$= 1.8(53) + 32 = 127\ °F$

$T_K = T_C + 273$

$= 53 + 273 = 326\ K$

b. $T_C = \frac{T_F - 32}{1.8}$

$= \frac{350. - 32}{1.8} = 177\ °C$

$T_K = T_C + 273$

$= 177 + 273 = 450.\ K$

1.75 Convert the temperatures to a common unit to compare.

a. $T_C = \dfrac{T_F - 32}{1.8} = \dfrac{10 - 32}{1.8} = -12\ °C < \mathbf{-10\ °C}$

($-12\ °C$ ← 10 °F; $-10\ °C$: higher temperature)

b. $T_C = \dfrac{T_F - 32}{1.8} = \dfrac{-50 - 32}{1.8} = \mathbf{-45\ °C} > -50\ °C$

($-45\ °C$ ← −50 °F; higher temperature)

1.77 a. Hexane is *less* dense than water, so 50 mL of hexane will be above the 100 mL of water.
b. Dichloromethane is *more* dense than water, so the 100 mL of water will be on top of the 50 mL of dichloromethane.

1.79

$$\frac{122\ \text{g}}{121\ \text{mL}} = 1.01\ \text{g/mL}$$

1.81

$$1\ \text{qt} \times \frac{946\ \text{mL}}{1\ \text{qt}} \times \frac{1.03\ \text{g}}{1\ \text{mL}} \times \frac{1\ \text{kg}}{1000\ \text{g}} = 0.974\ 38\ \text{kg} = \mathbf{0.974\ kg}$$

1.83 The density of a substance determines whether it floats or sinks in a liquid. The less dense liquid is the upper layer. The density of water is 1.0 g/mL.

a. heptane
(0.684 g/mL < 1.0 g/mL)

b. olive oil
(0.92 g/mL < 1.0 g/mL)

c. water
(1.0 g/mL < 1.49 g/mL)

d. water
(1.0 g/mL < 1.59 g/mL)

1.85 Use conversion factors to solve the problems.

a. $\dfrac{186\ \text{mg}}{\text{dL}} \times \dfrac{1\ \text{g}}{1000\ \text{mg}} = 0.186\ \text{g/dL}$

b. $\dfrac{186\ \text{mg}}{\text{dL}} \times \dfrac{10\ \text{dL}}{1\ \text{L}} = 1{,}860\ \text{mg/L}$

1.87

$$1.5\ \text{g} \times \frac{1000\ \text{mg}}{1\ \text{g}} \times \frac{1\ \text{tablet}}{500\ \text{mg}} = 3\ \text{tablets}$$

1.89

a. $2.0\ \text{L} \times \dfrac{1000\ \text{mL}}{1\ \text{L}} \times \dfrac{0.94\ \text{g}}{1\ \text{mL}} \times \dfrac{1\ \text{kg}}{1000\ \text{g}} = 1.9\ \text{kg}$

b. $1.9\ \text{kg} \times \dfrac{2.20\ \text{lb}}{1\ \text{kg}} = 4.2\ \text{lb}$

1.91

a. $\frac{20 \text{ mL}}{1 \text{ dose}} \times \frac{\$10.00}{300. \text{ mL}} = \frac{\$0.67}{\text{dose}}$

b. 2 tablespoons = 30. mL

$\frac{30 \text{ mL}}{1 \text{ dose}} \times \frac{\$10.00}{300. \text{ mL}} = \frac{\$1.00}{\text{dose}}$

1.93

2 tablets x 325 mg/tablet = 650. mg

$0.510 \text{ kg} \times \frac{1000 \text{ g}}{1 \text{ kg}} \times \frac{1000 \text{ mg}}{1 \text{ g}} \times \frac{1 \text{ dose}}{650. \text{ mg}} = 784.6 = 784 \text{ full doses}$

1.95

$4 \text{ times} \times \frac{2.0 \text{ g}}{\text{time}} \times \frac{1000 \text{ mg}}{1 \text{ g}} \times \frac{1 \text{ tablet}}{500. \text{ mg}} = 16 \text{ tablets}$

1.97

$\frac{2.0 \text{ mg}}{1 \text{ kg}} \times \frac{1 \text{ kg}}{2.20 \text{ lb}} \times 110 \text{ lb} = 1.0 \times 10^2 \text{ mg}$

1.99 Convert mass and height to kg and m, respectively. Use the formula, BMI = kg/m^2, to solve the problem.

$180 \text{ lb} \times \frac{1 \text{ kg}}{2.20 \text{ lb}} = 82 \text{ kg}$

6 ft 1 in. = 73 in.

$73 \text{ in.} \times \frac{1 \text{ m}}{39.4 \text{ in.}}$ 1.9 m

$\text{BMI} = \frac{\text{kg}}{\text{m}^2} = \frac{82}{(1.9)^2} = \frac{82}{3.61} = 23$

The BMI is in the normal range.

1.101

$42 \text{ lb} \times \frac{1 \text{ kg}}{2.20 \text{ lb}} \times \frac{10 \text{ mg}}{1 \text{ kg}} \times \frac{1 \text{ tablet}}{80 \text{ mg}} = 2.4 =$ **2 tablets**

1.103

$1.5 \text{ tsp} \times \frac{5.0 \text{ mL}}{1 \text{ tsp}} \times \frac{100. \text{ mg}}{5 \text{ mL}} \times \frac{1 \text{ g}}{1000 \text{ mg}} = 0.15 \text{ g}$

Chapter 2 Atoms and the Periodic Table

Chapter Review

[1] How is the name of an element abbreviated and how does the periodic table help to classify it as a metal, nonmetal, or metalloid? (2.1)

- An element is abbreviated by a one- or two-letter symbol. The periodic table contains a stepped line from boron to astatine. All metals are located to the left of the line. All nonmetals except hydrogen are located to the right of the line. The seven elements located along the line are metalloids.

[2] What are the basic components of an atom? (2.2)

- An atom is composed of two parts: a dense nucleus containing positively charged protons and neutral neutrons, and an electron cloud containing negatively charged electrons. Most of the mass of an atom resides in the nucleus, while the electron cloud contains most of its volume.
- The **atomic number (*Z*)** of a neutral atom tells the number of protons and the number of electrons.
- The **mass number (*A*)** is the sum of the number of protons and the number of neutrons.

[3] What are isotopes and how are they related to the atomic weight? (2.3)

- **Isotopes** are atoms that have the same number of protons but a different number of neutrons. The **atomic weight** is the weighted average of the mass of the naturally occurring isotopes of a particular element.

[4] What are the basic features of the periodic table? (2.4)

- The periodic table is a schematic of all known elements, arranged in rows (periods) and columns (groups), organized so that elements with similar properties are grouped together.
- The vertical columns are assigned group numbers using two different numbering schemes: 1–8 plus the letters A or B; or 1–18.
- The periodic table is divided into the main group elements (groups 1A–8A), the transition metals (groups 1B–8B), and the inner transition metals located in two rows below the main table.

[5] How are electrons arranged around an atom? (2.5)

- Electrons occupy discrete energy levels, organized into shells (numbered 1, 2, 3, and so on), subshells (*s*, *p*, *d*, and *f*), and orbitals.
- Each orbital holds two electrons.

Distribution of electrons in the first four shells

Shell	Number of electrons in a shell
4	32
3	18
2	8
lowest energy → 1	2

Increasing energy (upward arrow beside shells); Increasing number of electrons (upward arrow beside electron counts)

[6] What rules determine the electronic configuration of an atom? (2.6)

- To write the ground state electronic configuration of an atom, electrons are added to the lowest energy orbitals, giving each orbital two electrons. When two orbitals are equal in energy, one electron is added to each orbital until the orbitals are half-filled.
- Orbital diagrams that use boxes for orbitals and arrows for electrons indicate electronic configuration. Electron configuration can also be shown using superscripts to indicate how many electrons an orbital contains. For example, the electron configuration of the six electrons in a carbon atom is $1s^22s^22p^2$.

Orbital diagram for C: [↑↓] [↑↓] [↑][↑][ ]

1*s* 2*s* 2*p*

[7] How is the location of an element in the periodic table related to its electronic configuration? (2.6, 2.7)

- The periodic table is divided into four regions—the *s* block, *p* block, *d* block, and *f* block—based on the subshells that are filled with electrons last.
- Elements in the same group have the same number of valence electrons and similar electronic configurations.

[8] What is an electron-dot symbol? (2.7)

- An electron-dot symbol uses a dot to represent each valence electron around the symbol for an element.

	H	C	O	Cl
Number of valence electrons:	1	4	6	7
Electron-dot symbol:	H·	·C̣· (with dot above)	·Ö· (with pairs above and below)	·C̈l: (with pairs above and below)

[9] How are atomic size and ionization energy related to location in the periodic table? (2.8)

- The size of an atom decreases across a row and increases down a column.

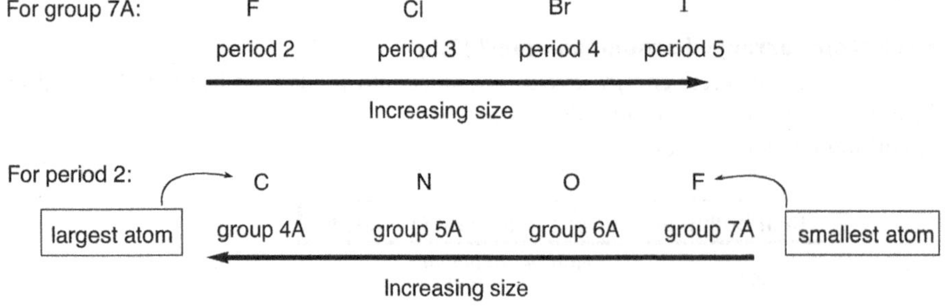

- **Ionization energy**—the energy needed to remove an electron from an atom—increases across a row and decreases down a column.

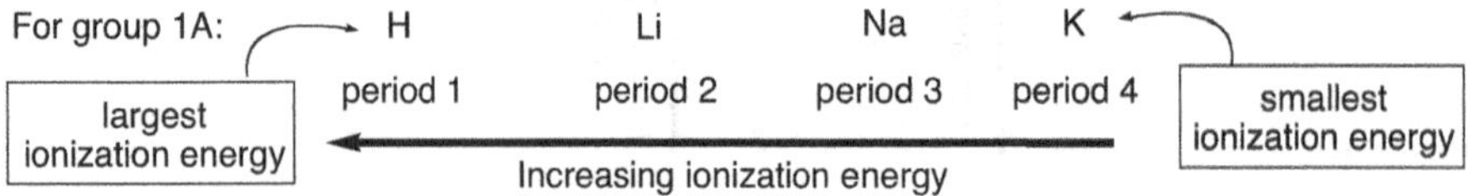

For period 2:	C	N	O	F
	group 4A	group 5A	group 6A	group 7A

→ Increasing ionization energy

Problem Solving

[1] Structure of the Atom (2.2)

Example 2.1 How many protons, neutrons, and electrons are contained in an atom of sodium, which has an atomic number of 11 and a mass number of 23?

Analysis

- In a neutral atom, the atomic number (Z) = the number of protons = the number of electrons.
- The mass number (A) = the number of protons + the number of neutrons.

Solution

The atomic number of 11 means that sodium has 11 protons and 11 electrons. To find the number of neutrons, subtract the atomic number (Z) from the mass number (A).

$$\begin{aligned} \text{number of neutrons} &= \text{mass number} - \text{atomic number} \\ &= 23 - 11 \\ &= 12 \text{ neutrons} \end{aligned}$$

---

[2] Isotopes (2.3)

Example 2.2 For each atom, give the following information: [1] the atomic number; [2] the mass number; [3] the number of protons; [4] the number of neutrons; [5] the number of electrons.

a. ${}^{109}_{47}Ag$ b. ${}^{81}_{35}Br$

Analysis

- The superscript gives the mass number and the subscript gives the atomic number for each element.
- The atomic number = the number of protons = the number of electrons.
- The mass number = the number of protons + the number of neutrons.

Solution

	Atomic Number	Mass Number	Number of Protons	Number of Neutrons	Number of Electrons
a. ${}^{109}_{47}Ag$	47	109	47	109 – 47 = 62	47
b. ${}^{81}_{35}Br$	35	81	35	81 – 35 = 46	35

Example 2.3 Determine the number of neutrons in each isotope.

a. sulfur-36 b. chlorine-37

Analysis

- The identity of the element tells us the atomic number.
- The number of neutrons = mass number – atomic number.

Solution

a. Sulfur's atomic number (*Z*) is 16. Sulfur-36 has a mass number (*A*) of 36.

$$\text{number of neutrons} = A - Z = 36 - 16 = 20 \text{ neutrons}$$

b. Chlorine's atomic number is 17 and the mass number of the given isotope is 37.

$$\text{number of neutrons} = A - Z = 37 - 17 = 20 \text{ neutrons}$$

---

[3] The Periodic Table (2.4)

Example 2.4 Give the period and group number for each element.
a. lead b. calcium

Analysis
Use the element symbol to locate an element in the periodic table. Count down the rows of elements to determine the period. The group number is located at the top of each column.

Solution
a. Lead (Pb) is located in the sixth row (period 6), and has group number 4A (or 14).
b. Calcium (Ca) is located in the fourth row (period 4), and has group number 2A (or 2).

---

[4] Electronic Configuration (2.6)

Example 2.5 Give the orbital diagram for the ground state electronic configuration of the element fluorine. Then, convert this orbital diagram to noble gas notation.

Analysis

- Use the atomic number to determine the number of electrons.
- Locate the element in the periodic table, and use Figure 2.7 to determine the order of orbitals filled with electrons. Read the table from left-to-right, row-by-row, beginning at the upper left corner and ending at the element in question. To fill orbitals of the same energy, place electrons one at a time in the orbitals until they are half-filled.
- To convert an orbital diagram to noble gas notation, replace the electronic configuration corresponding to the noble gas in the preceding row by the element symbol for the noble gas in brackets.

Solution

The atomic number of fluorine is 9, so nine electrons must be placed in orbitals. Four electrons are added in pairs to the 1*s* and 2*s* orbitals. The remaining five electrons are then added to the three 2*p* orbitals to give two pairs of electrons and one unpaired electron.

F
fluorine
9 electrons

[↑↓] [↑↓] [↑↓|↑↓|↑]
1*s* 2*s* 2*p*

Since fluorine is in the second period, use the noble gas helium in the preceding row to write the electronic configuration in noble gas notation. **Substitute [He] for the electrons in the first shell.**

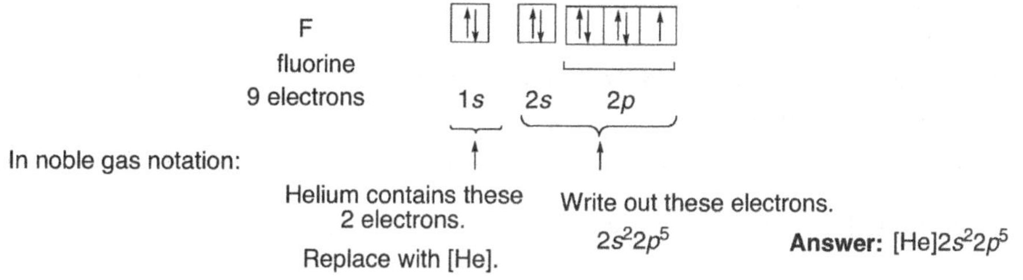

Example 2.6 Give the ground state electronic configuration of the element phosphorus. Convert the electronic configuration to noble gas notation.

Analysis

- Use the atomic number to determine the number of electrons.
- Locate the element in the periodic table and use Figure 2.7 to determine the order of orbitals filled with electrons.
- To convert the electronic configuration to noble gas notation, replace the electronic configuration corresponding to the noble gas in the preceding row by the element symbol for the noble gas in brackets.

Solution

The atomic number of phosphorus is 15, so 15 electrons must be placed in orbitals. Twelve electrons are added in pairs to the 1*s*, 2*s*, three 2*p*, and 3*s* orbitals. The remaining three electrons are added to the 3*p* orbitals. Since phosphorus is an element in period 3, use the noble gas neon in period 2 to write the noble gas configuration.

Electronic configuration for P (15 electrons) = $1s^2 2s^2 2p^6 3s^2 3p^3$ = $[Ne]3s^2 3p^3$

The noble gas neon contains these 10 electrons. Replace with [Ne]. — noble gas notation

---

[5] Valence Electrons (2.7)

Example 2.7 Determine the number of valence electrons and give the electronic configuration of the valence electrons of each element.

a. oxygen b. beryllium

Analysis

The group number of a main group element = the number of valence electrons. Use the general electronic configurations in Table 2.6 to write the configuration of the valence electrons.

Solution

a. Oxygen is located in group 6A, so it has six valence electrons. Since oxygen is a second period element, its valence electronic configuration is $2s^22p^4$.

b. Beryllium is located in group 2A, so it has two valence electrons. Since beryllium is a second period element, its valence electronic configuration is $2s^2$.

Example 2.8 Write an electron-dot symbol for each element.

a. calcium b. oxygen

Analysis

Write the element symbol for each element and use the group number to determine the number of valence electrons for a main group element. Represent each valence electron with a dot.

Solution

a. The element symbol for calcium is Ca. Ca is in group 2A and has two valence electrons. Electron-dot symbol:

Ca·

b. The element symbol for oxygen is O. O is in group 6A and has six valence electrons. Electron-dot symbol:

·Ö·

Self-Test

[1] Fill in the blank with one of the terms listed below.

Atomic number (2.2)	Ground state (2.6)	Metal (2.1)	Period (2.4)
Atomic weight (2.3)	Group (2.4)	Metalloid (2.1)	
Chemical formula (2.1)	Isotope (2.3)	Nucleus (2.2)	
Electron cloud (2.2)	Mass number (2.2)	Orbital (2.5)	

1. The _______ is the weighted average of the mass of the naturally occurring isotopes of a particular element reported in atomic mass units.
2. A _______ is a shiny material that is a good conductor of heat and electricity. All are solids except for mercury, which is a liquid.

3. The _______ is a dense core that contains protons and neutrons. Most of the mass of an atom resides here.
4. A row in the periodic table is called a _______. Elements in the same row are similar in size.
5. _______ are atoms of the same element having a different number of neutrons.
6. An _______ is a region of space where the probability of finding an electron is high. Each of these regions can hold *two* electrons.
7. The lowest energy arrangement of electrons is called the _______.
8. The _______ is the number of protons in the nucleus of an atom.
9. A _______ uses element symbols to show the identity of elements forming a compound and subscripts to show the ratio of atoms contained in the compound.
10. The _______ is the number of protons plus the number of neutrons.
11. A column in the periodic table is called a _______. Elements in the same column have similar electronic and chemical properties.
12. The _______, composed of electrons that move rapidly in the almost empty space surrounding the nucleus, comprises most of the volume of an atom.
13. A _______ has properties intermediate between metals and nonmetals. These include boron (B), silicon (Si), germanium (Ge), arsenic (As), antimony (Sb), tellurium (Te), and astatine (At).

[2] Fill in the blank with one of the terms listed below. You may use a term more than once.

a. alkali metals b. halogens c. noble gases

14. The _______ are especially stable as atoms, so they rarely combine with other elements to form compounds.
15. The _______ are located on the far left side of the periodic table.
16. The _______ and _______ are located on the far right side of the periodic table.

[3] Pick the statement that describes the type of orbital.

a. *s* orbital b. *p* orbital

17. An _______ has a sphere of electron density. It is lower in energy than other orbitals in the same shell because electrons are kept closer to the positively charged nucleus.
18. A _______ is an orbital that has a dumbbell shape.

[4] Fill in the blank with one of the terms listed below.

a. *s* block b. *p* block c. *d* block d. *f* block

19. The _______ consists of groups 3A–8A (except helium).
20. The _______ consists of the 10 columns of transition metals.
21. The _______ consists of groups 1A and 2A and the element helium.
22. The _______ consists of the two groups of 14 inner transition metals.

[5] Match the element symbol with the element name.

a. As b. Al c. Ar

23. Argon
24. Aluminum
25. Arsenic

Answers to Self-Test

1. atomic weight
2. metal
3. nucleus
4. period
5. Isotopes
6. orbital
7. ground state
8. atomic number
9. chemical formula
10. mass number
11. group
12. electron cloud
13. metalloid
14. c
15. a
16. b and c
17. a
18. b
19. b
20. c
21. a
22. d
23. c
24. b
25. a

Solutions to In-Chapter Problems

2.1 Each element is identified by a one- or two-letter symbol. Use the periodic table to find the symbol for each element.

a. Ca b. Rn c. N d. Au

2.2 Use the periodic table to find the element corresponding to each symbol.

a. neon
b. sulfur
c. iodine
d. silicon
e. boron
f. mercury
g. tin
h. antimony

2.3 *Metals* are shiny materials that are good conductors of heat and electricity. *Nonmetals* do not have a shiny appearance, and they are generally poor conductors of heat and electricity. *Metalloids* have properties intermediate between metals and nonmetals.

a, d, f, h: metals b, c, g: nonmetals e: metalloid

2.4 Use Figure 2.1 and the definitions in Answer 2.3 to determine if the micronutrients are metals, nonmetals, or metalloids.

As, B, Si: metalloids Cr, Co, Cu, Fe, Mn, Mo, Ni, Zn: metals F, I, Se: nonmetals

2.5 Use Figure 2.3 to determine which elements are represented in the molecular art.

a. 4 hydrogens, 1 carbon b. 3 hydrogens, 1 nitrogen c. 6 hydrogens, 2 carbons, 1 oxygen

2.6 The subscript tells how many atoms of a given element are in each chemical formula.

a. NaCN (sodium cyanide) = 1 sodium, 1 carbon, 1 nitrogen

b. H_2S (hydrogen sulfide) = 2 hydrogens, 1 sulfur

c. C_2H_6 (ethane) = 2 carbons, 6 hydrogens

d. SnF_2 (stannous fluoride) = 1 tin, 2 fluorines

e. CO (carbon monoxide) = 1 carbon, 1 oxygen

f. $C_3H_8O_3$ (glycerol) = 3 carbons, 8 hydrogens, 3 oxygens

2.7 Use Figure 2.3 to determine which elements are represented in the molecular art. Halothane contains 2 carbons, 1 hydrogen, 3 fluorines, 1 bromine, and 1 chlorine atom.

2.8

a. In a neutral atom, the number of protons = the number of electrons; 9 protons = 9 electrons.
b. The atomic number = the number of protons = 9.
c. This element is fluorine.

2.9 The atomic number is unique to an element and tells the number of protons in the nucleus and the number of electrons in the electron cloud.

Atomic Number	Element	Protons	Electrons
a. 2	Helium	2	2
b. 11	Sodium	11	11
c. 20	Calcium	20	20
d. 47	Silver	47	47
e. 78	Platinum	78	78

2.10 Answer the question as in Sample Problem 2.4.
a. There are 4 protons and 5 neutrons.
b. The atomic number = the number of protons = 4.
The mass number = the number of protons + the number of neutrons = 4 + 5 = 9.
c. The element is beryllium.

2.11 In a neutral atom, the atomic number (Z) = the number of protons = the number of electrons. The mass number (A) = the number of protons + the number of neutrons.

	Protons	Neutrons ($A - Z$)	Electrons
a.	17	18 (35 – 17)	17
b.	14	14 (28 – 14)	14
c.	92	146 (238 – 92)	92

2.12 The mass number (A) = the number of protons + the number of neutrons.

a. 42 protons, 42 electrons, 53 neutrons
42 + 53 = **95**

b. 24 protons, 24 electrons, 28 neutrons
24 + 28 = **52**

2.13 The superscript gives the mass number and the subscript gives the atomic number for each element. When the subscript is omitted, the identity of the element gives the atomic number. The atomic number = the number of protons = the number of electrons in a neutral atom. The mass number = the number of protons + the number of neutrons.

	Atomic Number	Mass Number	Protons	Neutrons	Electrons
a. $^{13}_{6}C$	6	13	6	7	6
b. ^{121}Sb	51	121	51	70	51

2.14 The identity of the element tells us the atomic number.
The mass number = the number of protons + the number of neutrons.

		Protons	Electrons	Atomic Number	Mass Number
With 12 neutrons:	$^{24}_{12}Mg$	12	12	12	12 + 12 = 24
With 13 neutrons:	$^{25}_{12}Mg$	12	12	12	12 + 13 = 25
With 14 neutrons:	$^{26}_{12}Mg$	12	12	12	12 + 14 = 26

2.15 a. The mass number is the sum of the number of protons and the number of neutrons. The highest mass number has the largest number of subatomic particles: [3] eight protons and nine neutrons.
b. In a neutral atom, the number of protons = the number of electrons; [1] eight electrons; [2] seven electrons; [3] eight electrons.

c. [1] $^{16}_{8}O$ [2] $^{16}_{7}N$ [3] $^{17}_{8}O$

2.16 Multiply the isotopic abundance by the mass of each isotope, and add up the products to give the atomic weight for the element.

a. Magnesium

Mass due to magnesium-24:	0.7899×23.99 amu	=	18.9497 amu
Mass due to magnesium-25:	0.1000×24.99 amu	=	2.499 amu
Mass due to magnesium-26:	0.1101×25.98 amu	=	2.8604 amu
	Atomic weight	=	24.3091 amu rounded to 24.31 amu **Answer**

b. Vanadium

Mass due to vanadium-50:	0.00250×49.95 amu	=	0.12488 amu
Mass due to vanadium-51:	0.99750×50.94 amu	=	50.8127 amu
	Atomic weight	=	50.93758 amu rounded to 50.94 amu **Answer**

2.17 a. There are 18 elements in period 4 (fourth row).
b. There are six elements in group 4A.

2.18 Use the element symbol to locate an element in the periodic table. Count down the rows of elements to determine the period. The group number is indicated at the top of each column.

Element	Period	Group
a. Oxygen	2	6A (or 16)
b. Calcium	4	2A (or 2)
c. Phosphorus	3	5A (or 15)
d. Platinum	6	8B (or 10)
e. Iodine	5	7A (or 17)

2.19 Use the definitions from Section 2.4 to identify the element fitting each description.

a. K
b. F
c. Ar
d. Sr
e. Zn
f. Nb

2.20

a. titanium, Ti, group 4B (or 4), period 4, transition metal
b. phosphorus, P, group 5A (or 15), period 3, main group element
c. dysprosium, Dy, no group number, period 6, inner transition element

2.21 Use Table 2.4 to tell how many electrons are present in each shell, subshell, or orbital.

a. a $2p$ orbital = 2 electrons
b. the $3d$ subshell = 10 electrons
c. a $3d$ orbital = 2 electrons
d. the third shell = 18 electrons

2.22 The **electronic configuration** of an individual atom is how the electrons are arranged in an atom's orbitals.

a. $1s^2 2s^2 2p^6 3s^2 3p^2$ = silicon
b. $1s^2 2s^2 2p^6 3s^2 3p^6 4s^2 3d^1$ = scandium
c. $[Ar]4s^2 3d^{10}$ = zinc

2.23 The **electronic configuration** of an individual atom shows how the electrons are arranged in an atom's orbitals.

a. lithium
b. beryllium, boron, carbon, nitrogen, oxygen, fluorine, neon
c. fluorine
d. oxygen

2.24 Use Example 2.5 to help draw the orbital diagram for each element.
[1] Use the atomic number to determine the number of electrons.
[2] Place electrons two at a time into the lowest energy orbitals, using Figure 2.7. When orbitals have the same energy, place electrons one at a time in the orbitals until they are half-filled.

a. magnesium [↑↓] [↑↓] [↑↓|↑↓|↑↓] [↑↓]
1s 2s 2p 3s

b. aluminum [↑↓] [↑↓] [↑↓|↑↓|↑↓] [↑↓] [↑| |]
1s 2s 2p 3s 3p

c. bromine [↑↓] [↑↓] [↑↓|↑↓|↑↓] [↑↓] [↑↓|↑↓|↑↓] [↑↓] [↑↓|↑↓|↑↓|↑↓|↑↓] [↑↓|↑↓|↑]
1s 2s 2p 3s 3p 4s 3d 4p

2.25 To convert the electronic configuration to noble gas notation, replace the electronic configuration corresponding to the noble gas in the preceding row by the element symbol for the noble gas in brackets.

a. sodium: $1s^22s^22p^63s^1$
$[Ne]3s^1$

b. silicon: $1s^22s^22p^63s^23p^2$
$[Ne]3s^23p^2$

c. iodine: $1s^22s^22p^63s^23p^64s^23d^{10}4p^65s^24d^{10}5p^5$
$[Kr]5s^24d^{10}5p^5$

2.26 To obtain the total number of electrons, add up the superscripts. This gives the atomic number and identifies the element. To determine the number of valence electrons, add up the number of electrons in the shell with the highest number.

a. $1s^22s^22p^63s^2$
12 electrons, 2 valence electrons in the $3s$ orbital, magnesium

b. $1s^22s^22p^63s^23p^3$
15 electrons, 5 valence electrons in the $3s$ and $3p$ orbitals, phosphorus

c. $1s^22s^22p^63s^23p^64s^23d^{10}4p^65s^24d^2$
40 electrons, 2 valence electrons in the $5s$ orbital, zirconium

d. $[Ar]4s^23d^6$
26 electrons, 2 valence electrons in the $4s$ orbital, iron

2.27 The group number of a main group element = the number of valence electrons. Use the general electronic configurations in Table 2.6 to write the configuration of the valence electrons.

a. fluorine = 7 valence electrons: $2s^22p^5$
b. krypton = 8 valence electrons: $4s^24p^6$
c. magnesium = 2 valence electrons: $3s^2$
d. germanium = 4 valence electrons: $4s^24p^2$

2.28

Se, selenium: $4s^24p^4$
Te, tellurium: $5s^25p^4$
Po, polonium: $6s^26p^4$

2.29 Write the symbol for each element and use the group number to determine the number of valence electrons for a main group element. Represent each valence electron with a dot.

a. :Br:
b. Li
c. Al·
d. ·S·
e. :Ne:

2.30 The atomic radius decreases left-to-right across a row and increases down a column of the periodic table.

a. Arsenic (As) has the larger atomic radius because it is located in the same group as nitrogen, but it is located farther down the column.
b. Silicon is farther to the left in the third row, so it has a larger atomic radius than sulfur.
c. Phosphorus has a larger atomic radius than fluorine because it is located farther to the left and farther down a column.

2.31 Ionization energy increases left-to-right across a row and decreases down a column of the periodic table.

a. Silicon has a higher ionization energy than sodium because silicon is located farther to the right in the same row.
b. Carbon has a higher ionization energy than silicon because carbon has a lower period number in the same group.
c. Fluorine has a higher ionization energy than sulfur because it is located farther to the right and closer to the top of a column of the periodic table.

2.32 a. Element [1] has the smaller atomic radius because atomic radius decreases left-to-right across a row and increases down a column of the periodic table.
b. Element [2] has the lower ionization energy because ionization energy increases left-to-right across a row and decreases down a column.
c. Element [1] is fluorine and element [2] is magnesium.

Solutions to Odd-Numbered End-of-Chapter Problems

2.33 Use Figure 2.3 to determine which elements are represented in the molecular art.
a. carbon (black) and oxygen (red) b. carbon (black), hydrogen (gray), and chlorine (green)

2.35 Use the periodic table to find the element corresponding to each symbol.

a. Au = gold, At = astatine, Ag = silver
b. N = nitrogen, Na = sodium, Ni = nickel
c. S = sulfur, Si = silicon, Sn = tin
d. Ca = calcium, Cr = chromium, Cl = chlorine
e. P = phosphorus, Pb = lead, Pt = platinum
f. Ti = titanium, Ta = tantalum, Tl = thallium

2.37 An *element* is a pure substance that cannot be broken down into simpler substances by a chemical reaction. A *compound* is a pure substance formed by combining two or more elements together.

a. H_2 = element
b. H_2O_2 = compound
c. S_8 = element
d. Na_2CO_3 = compound
e. C_{60} = element

2.39

a. cesium
b. ruthenium
c. chlorine
d. beryllium
e. fluorine
f. cerium

2.41

a. sodium: metal, alkali metal, main group element
b. silver: metal, transition metal
c. xenon: nonmetal, noble gas, main group element
d. platinum: metal, transition metal
e. uranium: metal, inner transition metal
f. tellurium: metalloid, main group element

2.43

a. 5 protons and 6 neutrons
b. The atomic number = the number of protons = 5.
c. The mass number = the number of protons + the number of neutrons = 5 + 6 = 11.
d. The number of electrons = the number of protons = 5.
e. element symbol: B

2.45

	Element Symbol	Atomic Number	Mass Number	Number of Protons	Number of Neutrons	Number of Electrons
a.	C	6	12	6	6	6
b.	P	15	31	15	16	15
c.	Zn	30	65	30	35	30
d.	Mg	12	24	12	12	12
e.	I	53	127	53	74	53
f.	Be	4	9	4	5	4
g.	Zr	40	91	40	51	40
h.	S	16	32	16	16	16

2.47

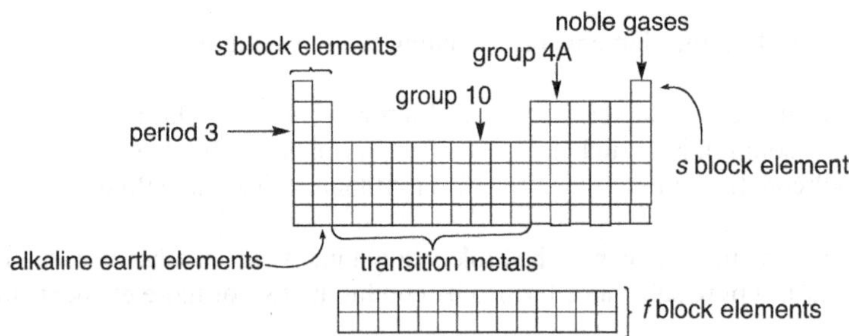

2.49 Hydrogen is located in group 1A but is not an alkali metal.

2.51 Use Figure 2.1 and the definitions in Answer 2.3 to classify each element in the fourth row of the periodic table as a metal, nonmetal, or metalloid.

K, Ca, Sc, Ti, V, Cr, Mn, Fe, Co, Ni, Cu, Zn, Ga: metals
Ge, As: metalloids
Se, Br, Kr: nonmetals

2.53 Group 8A (or 18) in the periodic table contains only nonmetals.

2.55 The atomic number = the number of protons = the number of electrons.
The mass number = the number of protons + the number of neutrons.

Mass Number	Protons	Neutrons	Electrons	Group	Symbol
16	8	8	8	6A (or 16)	$^{16}_{8}O$
17	8	9	8	6A (or 16)	$^{17}_{8}O$
18	8	10	8	6A (or 16)	$^{18}_{8}O$

2.57 The identity of the element tells us the atomic number.
The number of neutrons = mass number – atomic number.

	Symbol	Protons	Neutrons	Electrons
a.	$^{27}_{13}Al$	13	27 – 13 =14	13
b.	$^{35}_{17}Cl$	17	35 – 17 =18	17
c.	^{34}S	16	34 – 16 =18	16

2.59

a. $^{127}_{53}I$ b. $^{79}_{35}Br$ c. $^{107}_{47}Ag$

2.61 Multiply the isotopic abundance by the mass of each isotope, and add up the products to give the atomic weight for the element.

Silver

Mass due to silver-107:	0.5184 × 106.91 amu	= 55.4221 amu
Mass due to silver-109:	0.4816 × 108.90 amu	= 52.4462 amu
	Atomic weight	= 107.8683 amu rounded to 107.9 amu

Answer

2.63

a. first shell (n = 1) = 1 orbital (1s)

b. second shell (n = 2) = 4 orbitals (2s, three 2p)

c. third shell (n = 3) = 9 orbitals (3s, three 3p, five 3d)

d. fourth shell (n = 4) = 16 orbitals (4s, three 4p, five 4d, seven 4f)

2.65 Identify the element in the periodic table, and use Example 2.5 to help draw the orbital diagram for each element.

a. B [↑↓] [↑↓] [↑][ ][ ]

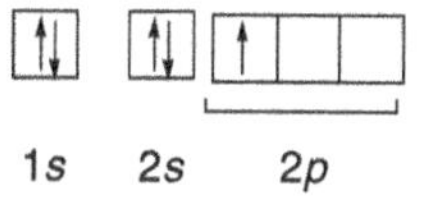

1s 2s 2p

b. K [↑↓] [↑↓] [↑↓][↑↓][↑↓] [↑↓] [↑↓][↑↓][↑↓] [↑]

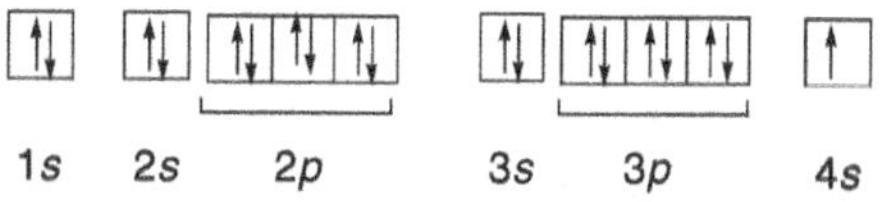

1s 2s 2p 3s 3p 4s

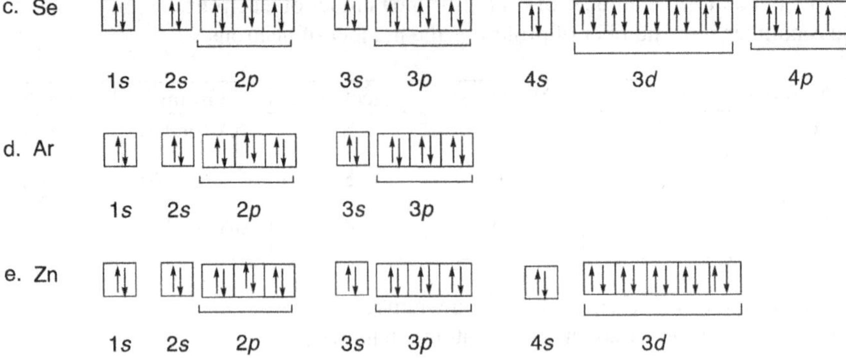

2.67 To convert the electronic configuration to noble gas notation, replace the electronic configuration corresponding to the noble gas in the preceding row by the symbol for the noble gas in brackets, as in Answer 2.25.

a. B: $1s^22s^22p^1$ or $[He]2s^22p^1$
b. K: $1s^22s^22p^63s^23p^64s^1$ or $[Ar]4s^1$
c. Se: $1s^22s^22p^63s^23p^64s^23d^{10}4p^4$ or $[Ar]4s^23d^{10}4p^4$
d. Ar: $1s^22s^22p^63s^23p^6$ or [Ar]
e. Zn: $1s^22s^22p^63s^23p^64s^23d^{10}$ or $[Ar]4s^23d^{10}$

2.69 To find the number of unpaired electrons, draw the orbital diagram for each element.

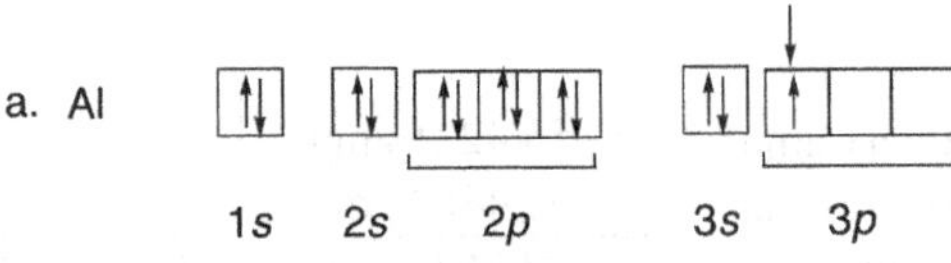

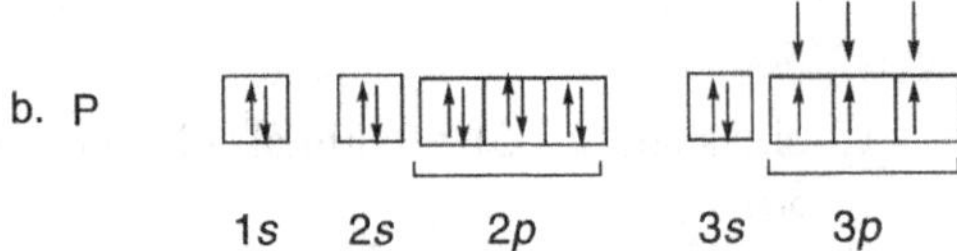

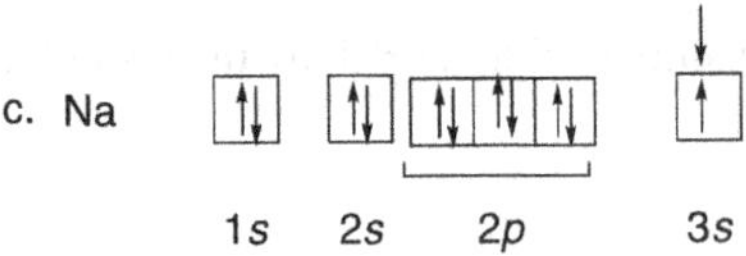

2.71

a. $1s^22s^22p^63s^23p^64s^23d^{10}4p^65s^2$ = 38 electrons, 2 valence electrons in the 5*s* orbital, strontium

b. $1s^22s^22p^63s^23p^4$= 16 electrons, 6 valence electrons in the 3*s* and the 3*p* orbitals, sulfur

c. $1s^22s^22p^63s^1$= 11 electrons, 1 valence electron in the 3*s* orbital, sodium

d. [Ne]$3s^23p^5$= 17 electrons, 7 valence electrons in the 3*s* and 3*p* orbitals, chlorine

2.73

	Electrons	Group Number	Valence Electrons	Period	Valence Shell
a. Carbon	6	4A (or 14)	4	2	2
b. Calcium	20	2A (or 2)	2	4	4
c. Krypton	36	8A (or 18)	8	4	4

2.75

a. carbon: $1s^22s^22p^2$; valence electrons $2s^22p^2$
b. calcium: $1s^22s^22p^63s^23p^64s^2$; valence electrons $4s^2$
c. krypton: $1s^22s^22p^63s^23p^64s^23d^{10}4p^6$; valence electrons $4s^24p^6$

2.77 The group number of a main group element = the number of valence electrons.

a. 2A = 2 valence electrons b. 4A = 4 valence electrons c. 7A = 7 valence electrons

2.79

a. sulfur: 6, $3s^23p^4$
b. chlorine: 7, $3s^23p^5$
c. barium: 2, $6s^2$
d. titanium: 2, $4s^2$
e. tin: 4, $5s^25p^2$

2.81 Write the element symbol for each element and use the group number to determine the number of valence electrons for a main group element. Represent each valence electron with a dot.

a. beryllium	b. silicon	c. iodine	d. magnesium	e. argon
Be· (2 dots)	·Si· (4 dots)	:I: (7 dots)	·Mg·	:Ar: (8 dots)

2.83 Elements [1] and [2] are in the same group (same column) of the periodic table.

a. Element [2] has the higher atomic number, because it has the larger atomic radius and atomic radius increases down a column.
b. Element [1] has the higher ionization energy, because ionization energy decreases down a column.
c. Element [2] gives up electrons more easily, because it has the lower ionization energy.

2.85 Use the size rules from Answer 2.30.

a. iodine b. carbon c. potassium d. selenium

2.87 Use the rules from Answer 2.31 to decide which has the higher ionization energy.

a. bromine b. nitrogen c. silicon d. chlorine

2.89 The size of atoms increases down a column of the periodic table, as the valence electrons are farther from the nucleus. The size of atoms decreases across a row of the periodic table as the number of protons in the nucleus increases.

a. neon < carbon < boron
b. beryllium < magnesium < calcium
c. sulfur < silicon < magnesium
d. neon < krypton < xenon
e. oxygen < sulfur < silicon
f. fluorine < sulfur < aluminum

2.91 Use the size rules from Answer 2.89 to rank the elements in order of increasing size.

fluorine < oxygen < sulfur < silicon < magnesium

2.93 Use the rules from Answer 2.31 to rank the elements in order of increasing ionization energy.

sodium < magnesium < phosphorus < nitrogen < fluorine

2.95 a. $C_{13}H_{18}ClNO$ (C–black; H–gray; Cl–green; N–blue; O–red)
b. chlorine

2.97

	a. Type	b. Block	c,d: Radius	e,f: Ionization Energy	g. Valence Electrons
Sodium	Metal	*s*			1
Potassium	Metal	*s*	Largest	Lowest	1
Chlorine	Nonmetal	*p*	Smallest	Highest	7

2.99

Carbon-11 has the same number of protons and electrons as carbon-12; that is, six. Carbon-11 has only five neutrons, whereas carbon-12 has six neutrons. The symbol for carbon-11 is ${}^{11}_{6}\text{C}$.

2.101 First, calculate how many gallons of gas are used. Then, use the number of grams of lead per gallon to determine the total amount of lead.

$$10{,}000.\ \cancel{\text{miles}} \times \frac{1\ \text{gal}}{25\ \cancel{\text{miles}}} = 4.0 \times 10^2\ \text{gal}$$

$$4.0 \times 10^2\ \cancel{\text{gal}} \times \frac{2.0\ \cancel{\text{g}}}{1\ \cancel{\text{gal}}} \times \frac{1\ \text{kg}}{1000\ \cancel{\text{g}}} = 0.80\ \text{kg lead}$$

Chapter 3 Ionic Compounds

Chapter Review

[1] What are the basic features of ionic and covalent bonds? (3.1)

- Both ionic bonding and covalent bonding follow one general rule: Elements gain, lose, or share electrons to attain the electronic configuration of the noble gas closest to them in the periodic table.
- **Ionic bonds** result from the transfer of electrons from one element to another. Ionic bonds form between a metal and a nonmetal. Ionic compounds consist of oppositely charged ions that feel a strong electrostatic attraction for each other.
- **Covalent bonds** result from the sharing of electrons between two atoms. Covalent bonds occur between two nonmetals, or when a metalloid combines with a nonmetal. Covalent bonding forms discrete molecules.

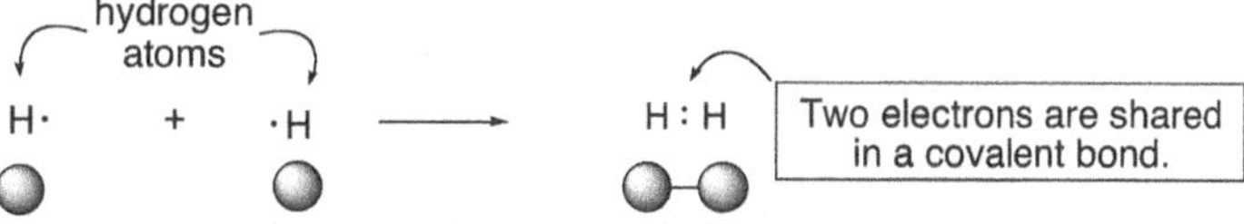

[2] How can the periodic table be used to determine whether an atom forms a cation or an anion, and its resulting ionic charge? (3.2)

- Metals form cations and nonmetals form anions.
- By gaining or losing one, two, or three electrons, an atom forms an ion with a completely filled outer shell of electrons.
- The charge on main group ions can be predicted from the position in the periodic table. For metals in groups 1A, 2A, and 3A, the group number = the charge on the cation. For nonmetals in groups 5A, 6A, and 7A, the charge on the anion = 8 – (the group number).
- **Ionic Charges of the Main Group Elements**

Group Number	Number of Valence Electrons	Number of Electrons Gained or Lost	General Structure of the Ion
1A (1)	1	1 e^- lost	M^+
2A (2)	2	2 e^- lost	M^{2+}
3A (3)	3	3 e^- lost	M^{3+}
5A (15)	5	3 e^- gained	X^{3-}
6A (16)	6	2 e^- gained	X^{2-}
7A (17)	7	1 e^- gained	X^-

[3] What is the octet rule? (3.2)

- Main group elements are especially stable when they possess an octet of electrons. Main group elements gain or lose one, two, or three electrons to form ions with eight outer shell electrons.

[4] What determines the formula of an ionic compound? (3.3)

- Cations and anions always form ionic compounds that have zero overall charge.
- Ionic compounds are written with the cation first, and then the anion, with subscripts to show how many of each are needed to have zero net charge.

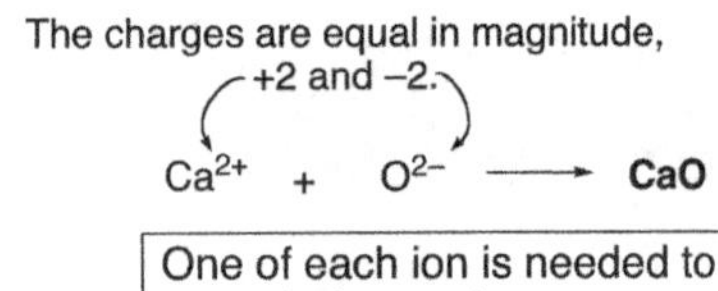

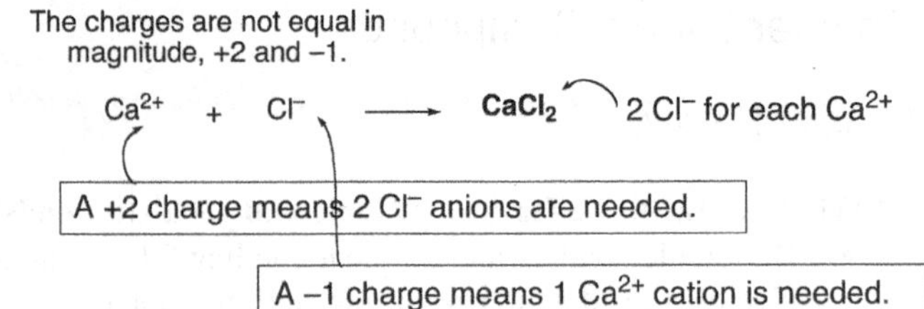

The charges are not equal in magnitude, +2 and −1.
Ca2+ + Cl− ⟶ CaCl2
2 Cl− for each Ca2+
A +2 charge means 2 Cl− anions are needed.
A −1 charge means 1 Ca2+ cation is needed.

[5] How are ionic compounds named? (3.4)

- Ionic compounds are always named with the name of the cation first.
- With cations having a fixed charge, the cation has the same name as its neutral element. The name of the anion usually ends in the suffix *-ide* if it is derived from a single atom or *-ate* (or *-ite*) if it is polyatomic.

NaF	$Ca_3(PO_4)_2$
sodium fluor*ide*	calcium phosph*ate*

- When the metal has a variable charge, use the overall anion charge to determine the charge on the cation. Then name the cation using a Roman numeral or the suffix *-ous* (for the ion with the lesser charge) or *-ic* (for the ion with the larger charge).

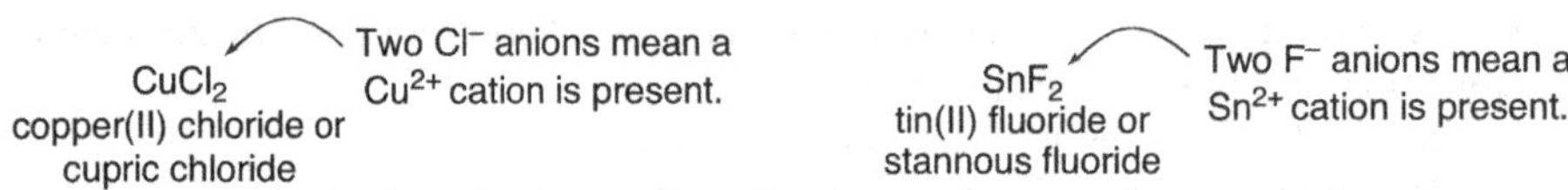

Two Cl− anions mean a Cu2+ cation is present.

tin(II) fluoride or

Two F− anions mean a Sn2+ cation is present.

[6] Describe the properties of ionic compounds. (3.5)

- Ionic compounds are crystalline solids with the ions arranged to maximize the interactions of the oppositely charged ions.
- Ionic compounds have high melting points and boiling points.
- Most ionic compounds are soluble in water and their aqueous solutions conduct an electric current.

[7] What are polyatomic ions and how are they named? (3.6)

- Polyatomic ions are charged species that are composed of more than one element.
- The names for polyatomic cations end in the suffix *-onium.*
- Many polyatomic anions have names that end in the suffix *-ate*. The suffix *-ite* is used for an anion that has one fewer oxygen atom than a similar anion named with the *-ate* ending. When two anions differ in the presence of a hydrogen, the word *hydrogen* or prefix *bi-* is added to the name of the anion.

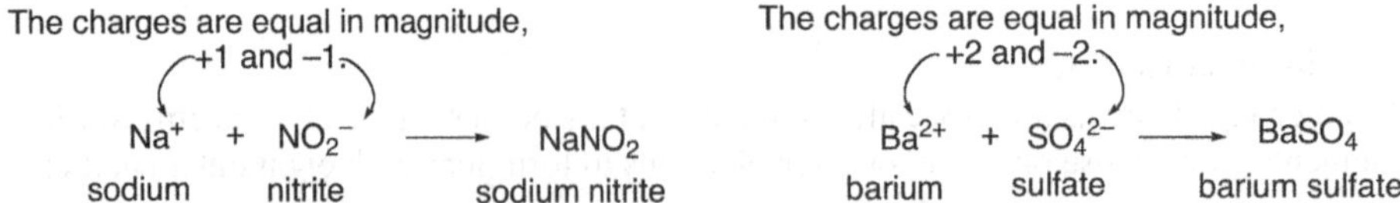

sodium nitrite

barium sulfate

[8] List useful consumer products and drugs that are composed of ionic compounds.

- Useful ionic compounds that contain alkali metal cations and halogen anions include KI (iodine supplement), NaF (source of fluoride in toothpaste), and KCl (potassium supplement). (3.3)

- Other products contain SnF_2 (fluoride source in toothpaste), Al_2O_3 (abrasive in toothpaste), and ZnO (sunblock agent). (3.4)
- Useful ionic compounds with polyatomic anions include $CaCO_3$ (antacid and calcium supplement), $Mg(OH)_2$ (antacid), and $FeSO_4$ (iron supplement). (3.6)

Problem Solving

[1] Introduction to Bonding (3.1)

Example 3.1 Predict whether the bonds in the following compounds are ionic or covalent.

a. $CaCl_2$ b. CCl_4

Analysis
The position of the elements in the periodic table determines the type of bonds they form. When a metal and nonmetal combine, the bond is ionic. When two nonmetals combine, or a metalloid bonds to a nonmetal, the bond is covalent.

Solution
a. Since Ca is a metal on the left side and Cl is a nonmetal on the right side of the periodic table, the bond must be ionic.
b. Since CCl_4 contains only the nonmetals carbon and chlorine, the bonds must be covalent.

---

[2] Ions (3.2)

Example 3.2 Write the ion symbol for an atom with 11 protons and 10 electrons.

Analysis
Since the number of protons equals the atomic number, this quantity identifies the element. The charge is determined by comparing the number of protons and electrons. If the number of electrons is greater than the number of protons, the charge is negative. If the number of protons is greater than the number of electrons, the charge is positive.

Solution
An element with 11 protons has an atomic number of 11, identifying it as sodium (Na). Since there is one more proton than electrons (11 vs. 10), the charge is +1. Thus, the answer is Na^+.

Example 3.3 How many protons and electrons are present in each ion?

a. Mg^{2+} b. Cl^-

Analysis
Use the identity of the element to determine the number of protons. The charge tells how many more or fewer electrons there are compared to the number of protons. A positive charge means more protons than electrons, while a negative charge means more electrons than protons.

Solution

a. Mg^{2+}: The element magnesium (Mg) has an atomic number of 12, so it has 12 protons. Since the charge is +2, there are two more protons than electrons, giving the ion 10 electrons.

b. Cl^{-}: The element chlorine (Cl) has an atomic number of 17, so it has 17 protons. Since the charge is –1, there is one more electron than protons, giving the ion 18 electrons.

Example 3.4 Use the group number to determine the charge on an ion derived from each element.

a. potassium b. selenium

Analysis

Locate the element in the periodic table. A metal in groups 1A, 2A, or 3A forms a cation equal in charge to the group number. A nonmetal in groups 5A, 6A, and 7A forms an anion whose charge equals 8 – (the group number).

Solution

a. Potassium (K) is located in group 1A, so it forms a cation with a +1 charge, K^{+}.

b. Selenium (Se) is located in group 6A, so it forms an anion with a charge of 8 – 6 = 2; Se^{2-}.

---

[3] Naming Ionic Compounds (3.4)

Example 3.5 Name each ionic compound.

a. KCl b. $CaBr_2$

Analysis

Name the cation and then the anion.

Solution

a. KCl: The cation is potassium and the anion is chloride (derived from chlorine); thus, the name is potassium chloride.

b. $CaBr_2$: The cation is calcium and the anion is bromide (derived from bromine); thus, the name is calcium bromide.

---

[4] Polyatomic Ions (3.6)

Example 3.6 Write a formula for an ionic compound formed from calcium and carbonate.

Analysis

- Identify the cation and anion and determine the charges.
- When ions of equal charge combine, one of each is needed. When ions of unequal charge combine, use the ionic charges to determine the relative number of each ion.
- Write the formula with the cation first and then the anion, omitting charges. Use parentheses around polyatomic ions when more than one appears in the formula, and use subscripts to indicate the number of each ion.

Solution

The cation from calcium has a +2 charge since calcium is located in group 2A. The carbonate anion has the formula CO_3^{2-}. Since the cation and anion have the same charge, there is one of each ion to give an overall charge of zero.

$$Ca^{2+} + CO_3^{2-} \longrightarrow CaCO_3$$

Answer: The formula is $CaCO_3$.

Example 3.7 Name each ionic compound.

a. $Fe_2(SO_4)_3$ b. $FeSO_4$

Analysis

First determine if the cation has fixed or variable charge. To name an ionic compound that contains a cation that always has the same charge, name the cation and then the anion. When the metal has variable charge, use the overall anion charge to determine the charge on the cation. Then name the cation (using a Roman numeral or the suffix *-ous* or *-ic*), followed by the anion.

Solution

a. $Fe_2(SO_4)_3$: Iron cations have different charges. This cation has a +3 charge = ferric or iron(III). The anion SO_4^{2-} is called sulfate.

Answer: iron(III) sulfate or ferric sulfate

b. $FeSO_4$: The iron cation has a charge of +2 = ferrous or iron(II). The anion SO_4^{2-} is called sulfate.

Answer: iron(II) sulfate or ferrous sulfate

Self-Test

[1] Fill in the blank with one of the terms listed below.

Anions (3.2)	Covalent bonds (3.1)	Nomenclature (3.4)
Bonding (3.1)	Ionic bonds (3.1)	Octet (3.2)
Cations (3.2)	Ionic compounds (3.1)	Polyatomic ion (3.6)

1. _______ result from the sharing of electrons between two atoms.
2. _______ are positively charged ions. This type of ion has fewer electrons than protons.
3. _______ are composed of cations and anions.
4. _______ is the joining of two atoms in a stable arrangement.
5. A main group element is especially stable when it possesses an _______ of electrons in its outer shell.
6. A _______ is a cation or anion that contains more than one atom.
7. Assigning an unambiguous name to each compound is called chemical _______.
8. _______ result from the transfer of electrons from one element to another.
9. _______ are negatively charged ions. This type of ion has more electrons than protons.

[2] Fill in the blank with one of the terms listed below.

a. cations
b. anions
c. ions
d. molecule

10. _______ are charged species in which the number of protons and electrons in an atom is *not* equal.
11. A _______ is a discrete group of atoms that share electrons.
12. Metals form _______.
13. Nonmetals form _______.

[3] Match the name of the anion with its symbol.

a. Br^-
b. Cl^-
c. F^-
d. I^-
e. O^{2-}
f. S^{2-}

14. bromide
15. fluoride
16. sulfide
17. chloride
18. iodide
19. oxide

[4] Match the polyatomic anion symbol to its name.

a. HCO_3^-
b. SO_3^{2-}
c. NO_3^-
d. $CH_3CO_2^-$
e. OH^-
f. HPO_4^{2-}

20. hydroxide
21. sulfite
22. hydrogen carbonate or bicarbonate
23. hydrogen phosphate
24. nitrate
25. acetate

[5] Write the molecular formula of the ionic compound derived from the cations on the left and each of the anions across the top.

	^-OH	SO_4^{2-}	PO_4^{3-}
K^+	KOH	27.	29.
Al^{3+}	26.	28.	30.

Answers to Self-Test

1. Covalent bonds
2. Cations
3. Ionic compounds
4. Bonding
5. octet
6. polyatomic ion
7. nomenclature
8. Ionic bonds
9. Anions
10. c
11. d
12. a
13. b
14. a
15. c
16. f
17. b
18. d
19. e
20. e
21. b
22. a
23. f
24. c
25. d
26. $Al(OH)_3$
27. K_2SO_4
28. $Al_2(SO_4)_3$
29. K_3PO_4
30. $AlPO_4$

Solutions to In-Chapter Problems

3.1 The position of the elements in the periodic table determines the type of bonds they form. When a metal and nonmetal combine, as in (b) and (c), the bond is ionic. When two nonmetals combine, or when a metalloid bonds to a nonmetal, the bond is covalent.

a. CO covalent
b. CaF_2 ionic
c. MgO ionic
d. Cl_2 covalent
e. HF covalent
f. C_2H_6 covalent

3.2 An **element** is a pure substance that cannot be broken down into simpler substances by a chemical reaction, and in molecular art it is composed of spheres of the same color.
A **compound** is a pure substance formed by combining two or more elements together, and in molecular art it is composed of spheres of different colors.
A **molecule** is composed of atoms that are covalently bonded together.

a. **A** = CH_4, compound
B = N_2, element
C = NO, compound
D = HCl, compound
E = HCN, compound
F = O_3, element

b. All of the species are molecules composed of nonmetals joined by covalent bonds.

3.3

a. CO_2 compound, molecule
b. H_2O compound, molecule
c. NaF compound
d. $MgBr_2$ compound
e. F_2 element, molecule
f. CaO compound

3.4 Vitamin C ($C_6H_8O_6$) is likely to contain covalent bonds because it consists of the nonmetals C, H, and O.

3.5 The element contains seven protons, identifying it as nitrogen. Since there are seven protons in the nucleus and 10 electrons in the electron cloud, the ion has a –3 charge. Answer: N^{3-}.

3.6 The number of protons equals the atomic number. The charge is determined by comparing the number of protons and electrons. If the number of electrons is greater than the number of protons, the charge is negative. If the number of protons is greater than the number of electrons, the charge is positive.

a. 19 protons and 18 electrons = K^+
b. 35 protons and 36 electrons = Br^-
c. 23 protons and 21 electrons = V^{2+}

3.7 Use the identity of the element to determine the number of protons. The charge tells how many more or fewer electrons there are compared to the number of protons. A positive charge means more protons than electrons, while a negative charge means more electrons than protons.

a. Ni^{2+} = 28 protons, 26 electrons
b. Se^{2-} = 34 protons, 36 electrons
c. Fe^{3+} = 26 protons, 23 electrons

3.8 Locate the element in the periodic table. A metal in group 1A, 2A, or 3A forms a cation equal in charge to the group number. A nonmetal in group 5A, 6A, or 7A forms an anion whose charge equals 8 – (the group number).

a. magnesium (group 2A): +2 | b. iodine (group 7A): –1 | c. selenium (group 6A): –2 | d. rubidium (group 1A): +1

3.9

a. Ne b. Xe c. Kr d. Kr

3.10

a. Au^{+} = 79 protons, 78 electrons
b. Au^{3+} = 79 protons, 76 electrons
c. Sn^{2+} = 50 protons, 48 electrons
d. Sn^{4+} = 50 protons, 46 electrons

3.11

a. Mn = 25 protons, 25 electrons
b. Mn^{2+} = 25 protons, 23 electrons
c. $1s^22s^22p^63s^23p^64s^23d^5$ The two $4s^2$ valence electrons would be lost to form Mn^{2+}.

3.12 Ionic compounds are composed of a cation (derived from a metal) and an anion (derived from a nonmetal).

a. lithium (metal) and bromine (nonmetal): yes
b. chlorine and oxygen (two nonmetals): no
c. calcium and magnesium (two metals): no
d. barium (metal) and chlorine (nonmetal): yes

3.13
- Identify the cation and the anion, and use the periodic table to determine the charges.
- When ions of equal charge combine, one of each ion is needed. When ions of unequal charge combine, use the ionic charges to determine the relative number of each ion.
- Write the formula with the cation first and then the anion, omitting charges, and using subscripts to indicate the number of each ion.

a. sodium (+1) and bromine (–1) = NaBr
b. barium (+2) and oxygen (–2) = BaO
c. magnesium (+2) and iodine (–1) = 2 I^{-} anions are needed = MgI_2
d. lithium (+1) and oxygen (–2) = 2 Li^{+} cations are needed = Li_2O

3.14 a. In Na_2S, there are twice as many Na^{+} cations (smaller, darker spheres) as there are S^{2-} anions (larger, lighter spheres).
b. In $MgCl_2$, there are twice as many Cl^{-} anions (larger, lighter spheres) as there are Mg^{2+} cations (smaller, darker spheres).

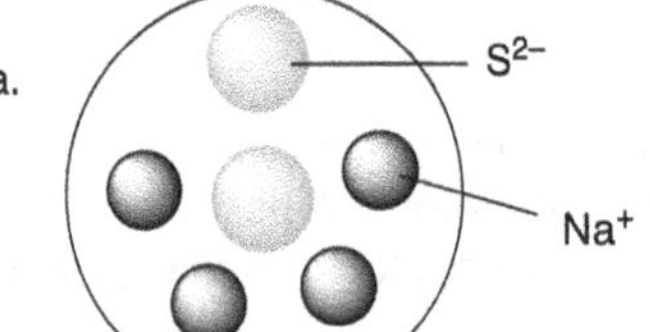

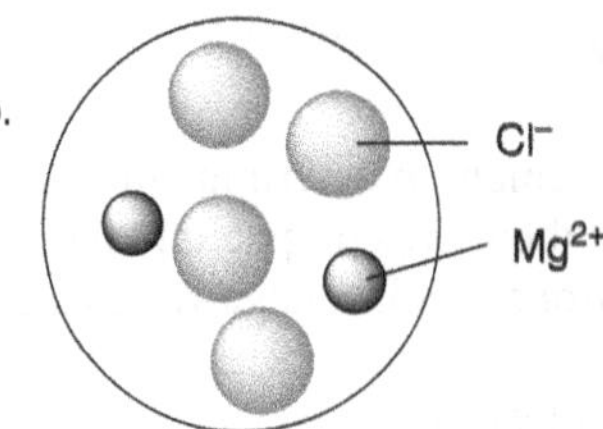

3.15 Zinc forms Zn^{2+} and oxygen forms O^{2-}; thus, zinc oxide = ZnO.

3.16 When a metal forms more than one cation, the cations are named by one of two methods.

Method [1]: Follow the name of the cation by a Roman numeral in parentheses to indicate its charge.

Method [2]: Use the suffix *-ous* for the cation with the lesser charge, and the suffix *-ic* for the cation with the higher charge. These suffixes are often added to the Latin names of the elements.

Anions are named by replacing the ending of the element name by the suffix *-ide*.

a. S^{2-} = sulfide
b. Cu^{+} = copper(I), cuprous
c. Cs^{+} = cesium
d. Al^{3+} = aluminum
e. Sn^{4+} = tin(IV), stannic

3.17

a. stannous = Sn^{2+}
b. iodide = I^{-}
c. manganese ion = Mn^{2+}
d. lead(II) = Pb^{2+}
e. selenide = Se^{2-}

3.18 Name the cation and then the anion.

a. NaF = sodium fluoride
b. MgO = magnesium oxide
c. $SrBr_2$ = strontium bromide
d. Li_2O = lithium oxide
e. TiO_2 = titanium oxide
f. $AlCl_3$ = aluminum chloride

3.19 First determine if the cation has fixed or variable charge. To name an ionic compound that contains a cation that always has the same charge, name the cation and then the anion (using the suffix *-ide*). When the metal has variable charge, use the overall anion charge to determine the charge on the cation. Then name the cation (using a Roman numeral or the suffix *-ous* or *-ic*), followed by the anion.

a. $CrCl_3$
Chromium has a variable charge, but here it must have a +3 charge to balance the three chloride ions.
chromium(III) chloride, chromic chloride

b. PbS
Lead has a variable charge, but here it must have a +2 charge to balance the sulfide ion.
lead(II) sulfide

c. SnF_4
Tin has a variable charge, but here it must have a +4 charge to balance the four fluoride ions.
tin(IV) fluoride, stannic fluoride

d. PbO_2
Lead has a variable charge, but here it must have a +4 charge to balance the two oxide ions.
lead(IV) oxide

e. $FeBr_2$
Iron has a variable charge, but here it must have a +2 charge to balance the two bromide ions.
iron(II) bromide, ferrous bromide

f. $AuCl_3$
Gold has a variable charge, but here it must have a +3 charge to balance the three chloride ions.
gold(III) chloride

3.20

a. Cu_2O = copper(I) oxide, cuprous oxide
b. CuO = copper(II) oxide, cupric oxide
c. CuCl = copper(I) chloride, cuprous chloride
d. $CuCl_2$ = copper(II) chloride, cupric chloride

3.21 Fe_2O_3 = iron(III) oxide, ferric oxide

3.22 Identify the cation and the anion and determine their charges. Balance the charges. Write the formula with the cation first, and use subscripts to show the number of each ion needed to have zero overall charge.

a. calcium bromide
Calcium is the cation (+2).
Bromide is the anion (–1).
$CaBr_2$

b. copper(I) iodide
Copper(I) is the cation (+1).
Iodide is the anion (–1).
CuI

c. ferric bromide
Iron(III) is the cation (+3).
Bromide is the anion (–1).
$FeBr_3$

d. magnesium sulfide
Magnesium is the cation (+2).
Sulfide is the anion (–2).
MgS

e. chromium(II) chloride
Chromium(II) is the cation (+2).
Chloride is the anion (–1).
$CrCl_2$

f. sodium oxide
Sodium is the cation (+1).
Oxide is the anion (–2).
Na_2O

3.23 Ionic compounds have high melting points and high boiling points. They usually dissolve in water. Their solutions conduct electricity and they form crystalline solids.

3.24 Write the formula formed from polyatomic ions with the cation first and then the anion, omitting charges. Use parentheses around polyatomic ions when more than one appears in the formula, and use subscripts to indicate the number of each ion.

a. magnesium (+2)
$MgSO_4$

b. sodium (+1)
Na_2SO_4

c. nickel (+2)
$NiSO_4$

d. aluminum (+3)
$Al_2(SO_4)_3$

e. lithium (+1)
Li_2SO_4

3.25 Use Table 3.5 to determine the charge on the polyatomic ions.

a. sodium (+1) and bicarbonate (–1): $NaHCO_3$

b. potassium (+1) and nitrate (–1): KNO_3

c. ammonium (+1) and sulfate (–2): $(NH_4)_2SO_4$

d. magnesium (+2) and phosphate (–3): $Mg_3(PO_4)_2$

e. calcium (+2) and bisulfate (–1): $Ca(HSO_4)_2$

f. barium (+2) and hydroxide (–1): $Ba(OH)_2$

3.26

a. $OH^- = KOH$
b. $NO_2^- = KNO_2$
c. $SO_4^{2-} = K_2SO_4$
d. $HSO_3^- = KHSO_3$
e. $PO_4^{3-} = K_3PO_4$
f. $CN^- = KCN$

3.27 First determine if the cation has fixed or variable charge. To name an ionic compound that contains a cation that always has the same charge, name the cation and then the anion. When the metal has variable charge, use the overall anion charge to determine the charge on the cation. Then name the cation (using a Roman numeral or the suffix *-ous* or *-ic*), followed by the anion.

a. Na_2CO_3 = sodium carbonate

b. $Ca(OH)_2$ = calcium hydroxide

c. $Mg(NO_3)_2$ = magnesium nitrate

d. $Mn(CH_3CO_2)_2$ = manganese acetate

e. $Fe(HSO_3)_3$ = iron(III) hydrogen sulfite, ferric bisulfite

f. $Mg_3(PO_4)_2$ = magnesium phosphate

3.28 Hydroxyapatite = $Ca_{10}(PO_4)_6(OH)_2$

Each Ca has a +2 charge; 10 Ca^{2+} = +20
Each PO_4 has a –3 charge; 6 PO_4^{3-} = –18
Each OH has a –1 charge; 2 OH^- = –2
Total negative charge of –20 balances a total positive charge of +20.

Solutions to Odd-Numbered End-of-Chapter Problems

3.29 Use the criteria in Problem 3.1.

a. CO_2 = covalent
b. H_2SO_4 = covalent
c. KF = ionic
d. CH_5N = covalent

3.31

a. potassium (metal) and oxygen (nonmetal) = ionic
b. sulfur and carbon (two nonmetals) = covalent
c. two bromine atoms (two nonmetals) = covalent
d. carbon and oxygen (two nonmetals) = covalent

3.33 The element contains four protons, identifying it as beryllium. Since there are four protons in the nucleus and two electrons in the electron cloud, the ion has a +2 charge. Answer: Be^{2+}.

3.35

a. 22 protons and 20 electrons = Ti^{2+}
b. 16 protons and 18 electrons = S^{2-}
c. 13 protons and 10 electrons = Al^{3+}
d. 17 protons and 18 electrons = Cl^-
e. 20 protons and 18 electrons = Ca^{2+}

3.37

a. a period 2 element that forms a +2 cation = Be
b. an ion from group 7A with 18 electrons = Cl^-
c. a cation from group 1A with 36 electrons = Rb^+

3.39

a. sodium ion = Na^+
b. selenide = Se^{2-}
c. manganese ion = Mn^{2+}
d. gold(III) = Au^{3+}
e. stannic = Sn^{4+}
f. mercurous = Hg_2^{2+}

3.41 The noble gas with the same number of electrons has the same electronic configuration as each ion.

a. O^{2-} = Ne
b. Mg^{2+} = Ne
c. Al^{3+} = Ne
d. S^{2-} = Ar
e. F^- = Ne
f. Be^{2+} = He

3.43

a. lithium = lose one electron (He)
b. iodine = gain one electron (Xe)
c. sulfur = gain two electrons (Ar)
d. strontium = lose two electrons (Kr)

3.45

	Number of Valence Electrons	Group Number	Number of Electrons Gained or Lost	Charge	Example
a. X·	1	1A	Lose 1	1+	Li
b. :Ä·	7	7A	Gain 1	1–	Cl

3.47 a. Q has two valence electrons, so it is in group 2A (2).
b. Q loses two electrons to form Q^{2+}.
c. Two F's are needed to give the compound zero overall charge: QF_2.
d. Since O forms an anion with a –2 charge, the formula is QO.

3.49 a. OH^-, hydroxide b. NH_4^+, ammonium

3.51 a. sulfate, SO_4^{2-} b. nitrite, NO_2^- c. sulfide, S^{2-}

3.53

a. sulfate = SO_4^{2-}
b. ammonium = NH_4^+
c. hydrogen carbonate = HCO_3^-
d. cyanide = CN^-

3.55

a. OH^- = 9 protons, 10 electrons
b. H_3O^+ = 11 protons, 10 electrons
c. PO_4^{3-} = 47 protons, 50 electrons

3.57 a. ClO_2^-, 33 protons, 34 electrons
b. OCN^-, 21 protons, 22 electrons

3.59

a. calcium (Ca^{2+}) and sulfur (S^{2-}) = CaS

b. aluminum (Al^{3+}) and bromine (Br^-) = $AlBr_3$

c. lithium (Li^+) and iodine (I^-) = LiI

d. nickel (Ni^{2+}) and chlorine (Cl^-) = $NiCl_2$

e. sodium (Na^+) and selenium (Se^{2-}) = Na_2Se

3.61

a. lithium (Li^+) and nitrite (NO_2^-) = $LiNO_2$

b. calcium (Ca^{2+}) and acetate (CH_3COO^-) = $Ca(CH_3COO)_2$

c. sodium (Na^+) and bisulfite (HSO_3^-) = $NaHSO_3$

d. manganese (Mn^{2+}) and phosphate (PO_4^{3-}) = $Mn_3(PO_4)_2$

e. magnesium (Mg^{2+}) and hydrogen sulfite (HSO_3^-) = $Mg(HSO_3)_2$

3.63

	Y^-	Y^{2-}	Y^{3-}
X^+	**XY**	**X_2Y**	**X_3Y**
X^{2+}	**XY_2**	**XY**	**X_3Y_2**
X^{3+}	**XY_3**	**X_2Y_3**	**XY**

3.65 a. Sulfide is –2, so the charge on each M is +1.
b. Each Br has a –1 charge, so the charge on M is +3.
c. Each nitrate has a –1 charge, so the charge on M is +2.
d. Each sulfate has a –2 charge, so the charge on M is +4.

3.67

	Br^-	OH^-	HCO_3^-	SO_3^{2-}	PO_4^{3-}
Na^+	$NaBr$	$NaOH$	$NaHCO_3$	Na_2SO_3	Na_3PO_4
Co^{2+}	$CoBr_2$	$Co(OH)_2$	$Co(HCO_3)_2$	$CoSO_3$	$Co_3(PO_4)_2$
Al^{3+}	$AlBr_3$	$Al(OH)_3$	$Al(HCO_3)_3$	$Al_2(SO_3)_3$	$AlPO_4$

3.69

a. $KHSO_4$ b. $Ba(HSO_4)_2$ c. $Al(HSO_4)_3$ d. $Zn(HSO_4)_2$

3.71

a. $Ba(CN)_2$ b. $Ba_3(PO_4)_2$ c. $BaHPO_4$ d. $Ba(H_2PO_4)_2$

3.73

a. Na_2O = sodium oxide
b. BaS = barium sulfide
c. PbS_2 = lead(IV) sulfide
d. $AgCl$ = silver chloride
e. $CoBr_2$ = cobalt(II) bromide
f. $RbBr$ = rubidium bromide
g. $PbBr_2$ = lead(II) bromide

3.75

a. $FeCl_2$ = iron(II) chloride, ferrous chloride
b. $FeBr_3$ = iron(III) bromide, ferric bromide
c. FeS = iron(II) sulfide, ferrous sulfide
d. Fe_2S_3 = iron(III) sulfide, ferric sulfide

3.77

a. sodium sulfide and sodium sulfate = Na_2S and Na_2SO_4
b. magnesium oxide and magnesium hydroxide = MgO and $Mg(OH)_2$
c. magnesium sulfate and magnesium bisulfate = $MgSO_4$ and $Mg(HSO_4)_2$

3.79

a. NH_4Cl = ammonium chloride
b. $PbSO_4$ = lead(II) sulfate
c. $Cu(NO_3)_2$ = copper(II) nitrate, cupric nitrate
d. $Ca(HCO_3)_2$ = calcium bicarbonate, calcium hydrogen carbonate
e. $Fe(NO_3)_2$ = iron(II) nitrate, ferrous nitrate

3.81

a. magnesium carbonate = $MgCO_3$
b. nickel sulfate = $NiSO_4$
c. copper(II) hydroxide = $Cu(OH)_2$
d. potassium hydrogen phosphate = K_2HPO_4
e. gold(III) nitrate = $Au(NO_3)_3$
f. lithium phosphate = Li_3PO_4
g. aluminum bicarbonate = $Al(HCO_3)_3$
h. chromous cyanide = $Cr(CN)_2$

3.83

a. $OH^- = Pb(OH)_4$
lead(IV) hydroxide
b. $SO_4^{2-} = Pb(SO_4)_2$
lead(IV) sulfate
c. $HCO_3^- = Pb(HCO_3)_4$
lead(IV) bicarbonate
d. $NO_3^- = Pb(NO_3)_4$
lead(IV) nitrate
e. $PO_4^{3-} = Pb_3(PO_4)_4$
lead(IV) phosphate
f. $CH_3CO_2^- = Pb(CH_3CO_2)_4$
lead(IV) acetate

3.85 The false statements are corrected.

a. True: Ionic compounds have high melting points.
b. False: Ionic compounds are solids at room temperature.
c. False: Most ionic compounds are soluble in water.
d. False: Ionic solids exist as crystalline lattices with the ions arranged to maximize the electrostatic interactions of anions and cations.

3.87 NaCl has a higher melting point than either CH_4 or H_2SO_4 because NaCl is an ionic compound, whereas CH_4 and H_2SO_4 are covalent. Ionic solids have higher melting points.

3.89 a. A neutral zinc atom has 30 protons and 30 electrons.
b. The Zn^{2+} cation has 30 protons and 28 electrons.
c. The electronic configuration of zinc: $1s^22s^22p^63s^23p^64s^23d^{10}$
The $4s^2$ electrons are lost to form Zn^{2+}.

3.91

Cation	a. Number of Protons	b. Number of Electrons	c. Noble Gas	d. Role
Na^+	11	10	Ne	Major cation in extracellular fluids and blood; maintains blood volume and blood pressure
K^+	19	18	Ar	Major intracellular cation
Ca^{2+}	20	18	Ar	Major cation in solid tissues like bone and teeth; required for normal muscle contraction and nerve function
Mg^{2+}	12	10	Ne	Required for normal muscle contraction and nerve function

3.93 silver (Ag^+) and nitrate (NO_3^-) = $AgNO_3$

3.95 $CaSO_3$ = calcium sulfite

3.97 ammonium (NH_4^+) and nitrate (NO_3^-) = NH_4NO_3

3.99

a. magnesium oxide (MgO) and potassium iodide (KI)
b. $CaHPO_4$ = calcium hydrogen phosphate
c. $FePO_4$ = iron(III) phosphate, ferric phosphate
d. sodium selenite = Na_2SeO_3
e. The name chromium chloride is ambiguous. Without a designation as chromium(II) or chromium(III) it's impossible to know the ratio of chromium cations to chloride anions.

Chapter 4 Covalent Compounds

Chapter Review

[1] What are the characteristic bonding features of covalent compounds? (4.1)

- Covalent bonds result from the sharing of electrons between two atoms, forming molecules. Atoms share electrons to attain the electronic configuration of the noble gas nearest them in the periodic table. For many main group elements this results in an octet of electrons.
- Covalent bonds are formed when two nonmetals combine or when a metalloid bonds to a nonmetal. Covalent bonds are preferred with elements that would have to gain or lose too many electrons to form an ion.
- Except for hydrogen, the common elements—C, N, O, and halogens—follow one rule: the number of bonds plus the number of lone pairs equals four.

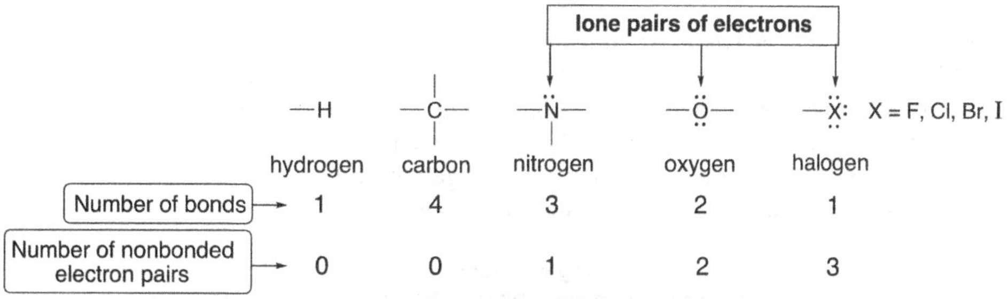

lone pairs of electrons

X = F, Cl, Br, I

	hydrogen	carbon	nitrogen	oxygen	halogen
Number of bonds	1	4	3	2	1
Number of nonbonded electron pairs	0	0	1	2	3

[2] What are Lewis structures and how are they drawn? (4.2)

- Lewis structures are electron-dot representations of molecules. Two-electron bonds are drawn with a solid line and nonbonded electrons are drawn with dots (:).
- Lewis structures contain only valence electrons. Each H gets two electrons and main group elements generally get eight.

[3] What are resonance structures? (4.4)

- Resonance structures are two or more Lewis structures having the same arrangement of atoms but a different arrangement of electrons.

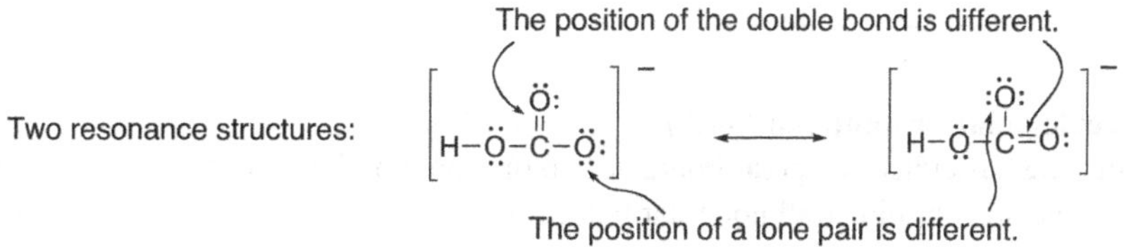

The position of the double bond is different.

Two resonance structures:

The position of a lone pair is different.

- The hybrid is a composite of all resonance structures that spreads out electron pairs in multiple bonds and lone pairs.

[4] How are covalent compounds with two elements named? (4.5)

- Name the first nonmetal by its element name and the second using the suffix *-ide*. Add prefixes to show the number of atoms of each element.

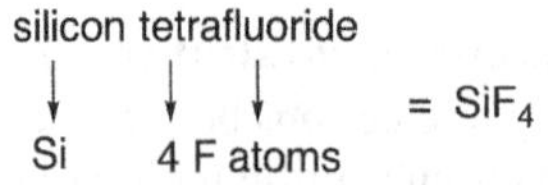

silicon tetrafluoride → Si, 4 F atoms = SiF_4

[5] How is the molecular shape around an atom determined? (4.6)

- To determine the shape around an atom, count groups—atoms and lone pairs—and keep the groups as far away from each other as possible.
- Two groups = linear, 180° bond angle; three groups = trigonal planar, 120° bond angle; four groups = tetrahedral, 109.5° bond angle.
- See Table 4.2, Common Molecular Shapes Around Atoms.

[6] How does electronegativity determine bond polarity? (4.7)

- Electronegativity is a measure of an atom's attraction for electrons in a bond.
- When two atoms have the same electronegativity value, or the difference is less than 0.5 units, the electrons are equally shared and the bond is nonpolar.
- When two atoms have very different electronegativity values—a difference of 0.5–1.9 units—the electrons are unequally shared and the bond is polar.

Increasing electronegativity (→ across; ↑ up)

1A	2A	3A	4A	5A	6A	7A	8A
H 2.1							
Li 1.0	Be 1.5	B 2.0	C 2.5	N 3.0	O 3.5	F 4.0	
Na 0.9	Mg 1.2	Al 1.5	Si 1.8	P 2.1	S 2.5	Cl 3.0	
K 0.8	Ca 1.0	Ga 1.6	Ge 1.8	As 2.0	Se 2.4	Br 2.8	
Rb 0.8	Sr 1.0	In 1.7	Sn 1.8	Sb 1.9	Te 2.1	I 2.5	

[7] When is a molecule polar or nonpolar? (4.8)

- A polar molecule has either one polar bond, or two or more bond dipoles that do not cancel.
- A nonpolar molecule has either all nonpolar bonds, or two or more bond dipoles that cancel.

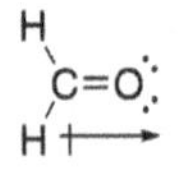

$H_2C{=}O$	BF_3	NH_3	CH_2Cl_2
one polar bond	three polar bonds All dipoles cancel. NO net dipole	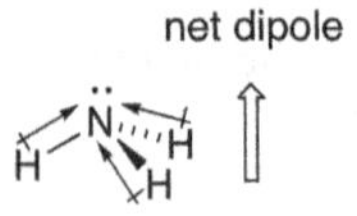 net dipole three polar bonds All dipoles reinforce.	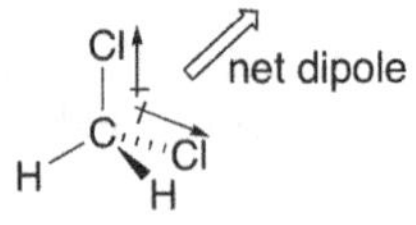 net dipole two polar bonds Two dipoles reinforce.
polar molecule	**nonpolar** molecule	**polar** molecule	**polar** molecule

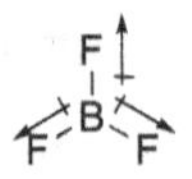

Problem Solving

[1] Introduction to Covalent Bonding (4.1)

Example 4.1 Without referring to Figure 4.1 in the text, how many covalent bonds are predicted for each atom?

a. C b. O

Analysis
Atoms with one, two, or three valence electrons form one, two, or three bonds, respectively. Atoms with four or more valence electrons form enough bonds to give an octet; that is, the predicted number of bonds = 8 – (the number of valence electrons).

Solution
a. C has four valence electrons. Thus, it is expected to form 8 – 4 = 4 bonds.
b. O has six valence electrons. Thus, it is expected to form 8 – 6 = 2 bonds.

---

[2] Lewis Structures (4.2)

Example 4.2 Draw a Lewis structure for CH_2F_2.

Analysis and Solution
[1] Arrange the atoms.

```
  H
H C F
  F
```

- Place C in the center and 2 H's and 2 F's on the periphery.
- In this arrangement, C is surrounded by four atoms, its usual number.

[2] Count the valence electrons.

1 C	x	4 e^-	=	4 e^-
2 H	x	1 e^-	=	2 e^-
2 F	x	7 e^-	=	14 e^-
				20 e^- total

[3] Add the bonds and lone pairs.

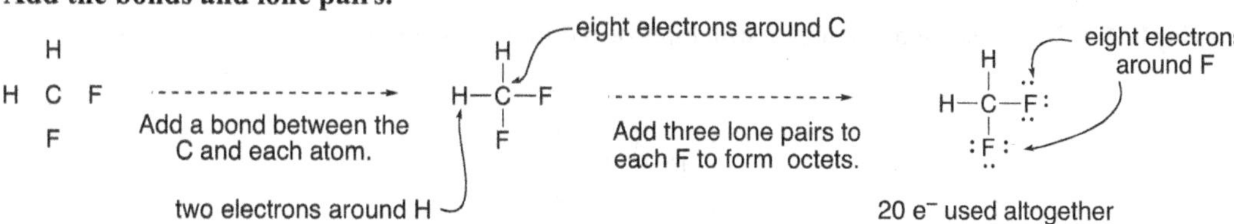

First, add four single bonds—namely, two C–H bonds and two C–F bonds. This uses eight valence electrons, and gives carbon an octet (four two-electron bonds) and each hydrogen two electrons. Next, give each F an octet by adding three lone pairs to each F atom. This uses all 20 valence electrons. To check if a Lewis structure is valid, we must answer YES to three questions.

- Have all the electrons been used?
- Is each H surrounded by two electrons?
- Is every other main group element surrounded by eight electrons?

Since the answer to all three questions is YES, we have drawn a valid Lewis structure for CH_2F_2.

Example 4.3 Draw a Lewis structure for molecular formula $C_2H_2Cl_2$ in which each carbon is bonded to one hydrogen and one chlorine.

Analysis and Solution
Follow steps [1]–[3] to draw a Lewis structure. Place five bonds between the atoms and add three lone pairs to each Cl. When the two remaining electrons are placed as a lone pair on a carbon, one C still has no octet.

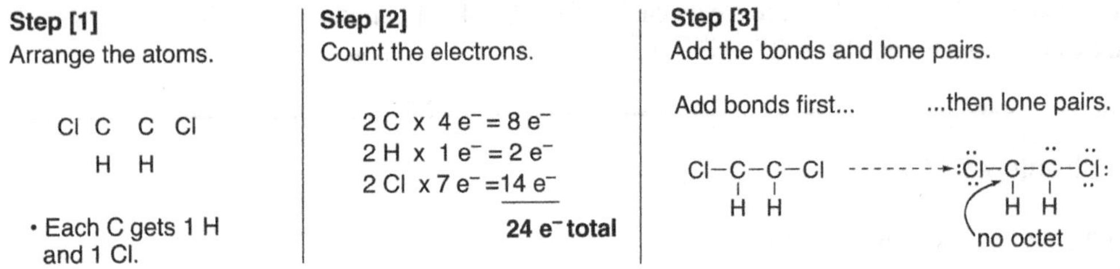

Step [1]
Arrange the atoms.

Cl C C Cl
H H

• Each C gets 1 H and 1 Cl.

Step [2]
Count the electrons.

$2\ C \times 4\ e^- = 8\ e^-$
$2\ H \times 1\ e^- = 2\ e^-$
$2\ Cl \times 7\ e^- = 14\ e^-$
$24\ e^-$ total

Step [3]
Add the bonds and lone pairs.

Add bonds first... ...then lone pairs.

Cl–C–C–Cl → :Cl–C–C–Cl:
H H

no octet

Step [4] To give both C's an octet, change *one* lone pair into *one* bonding pair of electrons between the two C atoms, forming a *double* bond.

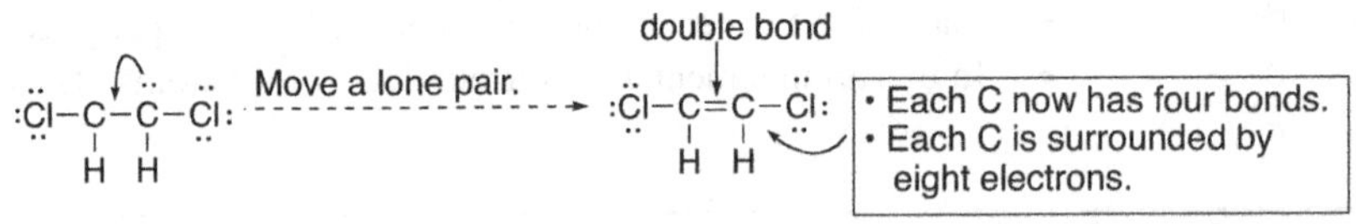

Move a lone pair.

double bond

• Each C now has four bonds.
• Each C is surrounded by eight electrons.

This uses all 24 electrons, each C and Cl have an octet, and each H has two electrons. The Lewis structure is valid.

---

[3] Resonance (4.4)

Example 4.4 Draw a second resonance structure for the following anion:

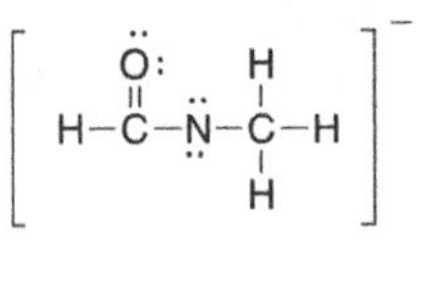

Analysis
Concentrate on the multiple bonds and lone pairs. To convert one resonance structure to another, use a lone pair to form a double bond and convert a bonding electron pair in a double bond to a lone pair.

Solution

Move an electron pair out on O.

new lone pair

A

B

Move an electron pair to form a double bond.

new double bond

The new resonance structure **B** differs from **A** in the location of a double bond and a lone pair.

---

[4] Naming Covalent Compounds (4.5)

Example 4.5 Give the formula for each compound.

a. sulfur dioxide b. nitrogen trifluoride

Analysis

- Write the symbols for the elements in the order given in the name.
- Use the prefixes to write the subscripts.

Solution

---

[5] Molecular Shape (4.6)

Example 4.6 Using the given Lewis structure, determine the shape around the indicated atom in each compound.

a. $H-CH_2-C{\equiv}N:$ b. $H-CH(Cl)-Cl$

Analysis

To predict shape around an atom, we need a valid Lewis structure, which is given in this problem. Then count groups around the atom to determine molecular shape using the information in Table 4.2.

Solution

a. The indicated C in CH_3CN is surrounded by two atoms and no lone pairs—that is, two groups. An atom surrounded by two groups is linear with a 180° bond angle.

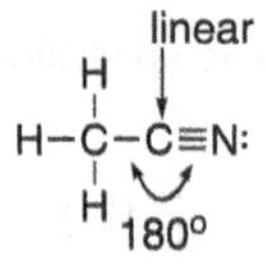

linear

H–C–C≡N:

180°

b. The C atom in CH_2Cl_2 is surrounded by four atoms—that is, four groups. An atom surrounded by four groups is tetrahedral, with 109.5° bond angles.

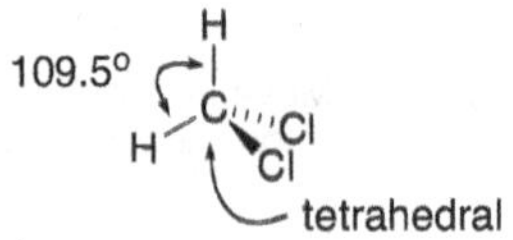

109.5°

tetrahedral

---

[6] Electronegativity and Bond Polarity (4.7)

Example 4.7 Use electronegativity values to classify each bond as nonpolar, polar, or ionic.

a. KCl b. Br_2 c. H_2O

Analysis

Calculate the electronegativity difference between the two atoms and use the following rules: less than 0.5 (nonpolar); 0.5–1.9 (polar covalent); greater than 1.9 (ionic).

Solution

	Electronegativity difference	**Bond type**
a. KCl	3.0 (Cl) – 0.8 (K) = 2.2	ionic
b. Br_2	2.8 (Br) – 2.8 (Br) = 0	nonpolar
c. H_2O	3.5 (O) – 2.1 (H) = 1.4	polar covalent

---

[7] Polarity of Molecules (4.8)

Example 4.8 Using the given Lewis structure, determine whether dimethyl ether is a polar or nonpolar molecule.

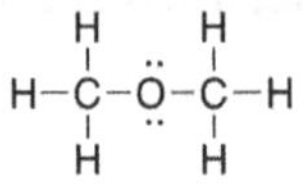

H–C–Ö–C–H (with H above and below each C)

dimethyl ether

Analysis

To determine the overall polarity of a molecule, identify the polar bonds, determine the shape around individual atoms, and then decide if the individual bond dipoles cancel or reinforce.

Solution

Ignore the C–H bonds since they are all nonpolar. Each C–O bond is polar because the electronegativity difference between O (3.5) and C (2.5) is 1.0. Since the O atom of dimethyl ether has two atoms and two lone pairs around it, dimethyl ether is a bent molecule around the O atom. The two dipoles reinforce (both point *up*), so **dimethyl ether has a net dipole;** that is, **dimethyl ether is a polar molecule.**

The two individual dipoles reinforce.

δ⁻ O, CH3 δ⁺, CH3 δ⁺ — net dipole

Do NOT draw dimethyl ether as:

$CH_3-\ddot{O}-CH_3$

The net dipole bisects the C–O–C bond angle.
The bent representation shows that the dipoles reinforce.

Note: **We must know the geometry to determine if two dipoles cancel or reinforce.** For example, do ***not*** draw dimethyl ether as a linear molecule around the O atom, because you might think that the two dipoles cancel, when in reality, they reinforce.

Self-Test

[1] Fill in the blank with one of the terms listed below.

Covalent bond (4.1)	Hybrid (4.4)	Polar (4.7)
Dipole (4.7)	Lewis structures (4.1)	Resonance structures (4.4)
Double-headed arrow (4.4)	Molecular formula (4.2)	Valence shell electron pair repulsion theory (4.6)
Electronegativity (4.7)	Nonpolar (4.7)	

1. The electron-dot structures for molecules are called _______.
2. _______ are two or more Lewis structures having the same arrangement of atoms but a different arrangement of electrons.
3. A composite of two or more resonance forms is called a _______.
4. The most stable arrangement keeps groups of electrons as far away from each other as possible; this is the basis for the _______.
5. _______ is a measure of an atom's attraction for electrons in a bond.
6. A _______ results from the sharing of electrons between two atoms.
7. Bonding between atoms of different electronegativity results in the unequal sharing of electrons, which creates a _______ bond.
8. A _______ shows that two Lewis structures are resonance structures.
9. When electrons are equally shared, the bond is said to be _______.
10. A _______ shows the number and identity of all the atoms in a compound.
11. A polar bond is said to have a _______—that is, a partial separation of charge.

[2] Fill in the blank with one of the terms listed below.

a. six electrons b. four electrons c. less electronegative d. more electronegative

12. A double bond contains _______.
13. A triple bond contains _______.
14. The symbol δ^+ is given to the _______ atom in a polar bond.
15. The symbol δ^- is given to the _______ atom in a polar bond.

[3] Fill in the blanks with the terms listed below.

a. trigonal planar
b. 109.5°
c. 120°
d. linear
e. tetrahedral
f. 180°

16. An atom surrounded by two groups is _______ and has a bond angle of _______.
17. An atom surrounded by three groups is _______ and has bond angles of _______.
18. An atom surrounded by four groups is _______ and has bond angles of _______.

[4] Match the symbol with its name.

a. solid line b. dashed line c. wedge

19. A _______ is used for bonds in the plane.
20. A _______ is used for a bond in front of the plane.
21. A _______ is used for a bond behind the plane.

[5] Fill in the table with one of the types of bonds listed below.

a. ionic b. nonpolar c. polar covalent

22. _______	Electrons are equally shared.
23. _______	Electrons are unequally shared, pulled towards the more electronegative element.
24. _______	Electrons are transferred from the less electronegative element to the more electronegative element.

Answers to Self-Test

1. Lewis structures
2. Resonance structures
3. hybrid
4. valence shell electron pair repulsion theory
5. Electronegativity
6. covalent bond
7. polar
8. double-headed arrow
9. nonpolar
10. molecular formula
11. dipole
12. b
13. a
14. c
15. d
16. d, f
17. a, c
18. e, b
19. a
20. c
21. b
22. b
23. c
24. a

Solutions to In-Chapter Problems

4.1 H is surrounded by two electrons, giving it the noble gas configuration of He. Cl is surrounded by eight electrons, giving it the noble gas configuration of Ar.

$$\text{H}\cdot + \cdot\ddot{\underset{..}{\text{Cl}}}: \longrightarrow \text{H}:\ddot{\underset{..}{\text{Cl}}}:$$

4.2 This arrangement gives each Cl an octet and the electronic configuration of Ar.

$$:\ddot{\underset{..}{\text{Cl}}}\cdot + \cdot\ddot{\underset{..}{\text{Cl}}}: \longrightarrow :\ddot{\underset{..}{\text{Cl}}}:\ddot{\underset{..}{\text{Cl}}}:$$

4.3 Atoms with one, two, or three valence electrons form one, two, or three bonds, respectively. Atoms with four or more valence electrons form enough bonds to give an octet.

a. F forms one bond. ($8 - 7$ valence $e^- = 1$ bond)
b. Si forms four bonds. ($8 - 4$ valence $e^- = 4$ bonds)
c. Br forms one bond. ($8 - 7$ valence $e^- = 1$ bond)
d. O forms two bonds. ($8 - 6$ valence $e^- = 2$ bonds)
e. P forms three bonds. ($8 - 5$ valence $e^- = 3$ bonds)
f. S forms two bonds. ($8 - 6$ valence $e^- = 2$ bonds)

4.4

a. H–C(H)(H)–C̈l: b. H–N̈(H)–Ö–H c. H–C(H)(H)–Ö–H d. :B̈r–C(H)(H)–B̈r:

4.5 Ionic bonding is observed in CaO since Ca is a metal and readily transfers electrons to the nonmetal oxygen. Covalent bonding is observed in CO_2 since carbon is a nonmetal and does not readily transfer electrons.

4.6

a.

Step [1] Arrange the atoms.	Step [2] Count the electrons.	Step [3] Add the bonds and lone pairs.
H Br	1 Br x 7 e^- = 7 e^- 1 H x 1 e^- = 1 e^- **8 e^- total**	Add bonds first... H–Br (only 2 e^- used) ⟶ ...then lone pairs. H–B̈r: (8 e^- used)

b.

Step [1] Arrange the atoms.	Step [2] Count the electrons.	Step [3] Add the bonds and lone pairs.
H C F with H above and H below C • four atoms around C	1 C x 4 e^- = 4 e^- 1 F x 7 e^- = 7 e^- 3 H x 1 e^- = 3 e^- **14 e^- total**	Add bonds first... H–C(H)(H)–F (no octet; only 8 e^- used) ⟶ ...then lone pairs. H–C(H)(H)–F̈: (14 e^- used)

c. H–Ö–Ö–H d. H–N̈(H)–N̈(H)–H e. H–C(H)(H)–C(H)(H)–H f. H–C(H)(:C̈l:)–C̈l:

4.7 Use the same steps as in Answer 4.6 to draw the Lewis structure.

H–C(H)(H)–Ö–C(H)(H)–H

4.8 After placing all electrons in bonds and lone pairs, use a lone pair to form a multiple bond if an atom does not have an octet. Follow the stepwise procedure in Example 4.3.

a. H–C≡N: b. H–C(H)=Ö

4.9 A valid Lewis structure for formic acid contains a carbon–oxygen double bond. The C atom has four bonds and each O atom has two bonds and two lone pairs.

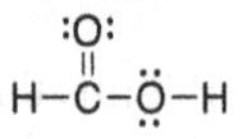

:O:
||
H–C–Ö–H

4.10

a. molecular formula: $C_2H_5NO_2$

b. Lewis structure: H–N–C–C–O–H (with H and H on the first carbon, H on N, :O: double-bonded to the second carbon)

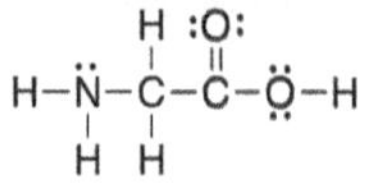

4.11 Boron is surrounded by only six electrons, so the Lewis structure for BBr_3 does not follow the octet rule.

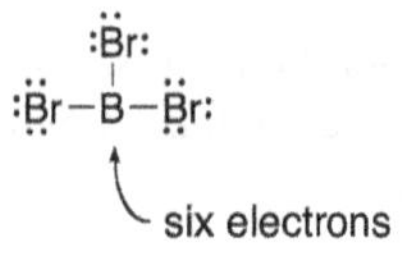

:Br–B–Br: (with a third :Br: bonded to B)

six electrons

4.12

a. H–O–P–C–N–C–C–O–H (with :O: double-bonded to P, H–O: bonded to P, H and H on the first C, H on N, H and H on the second C, :O: double-bonded to the third C)

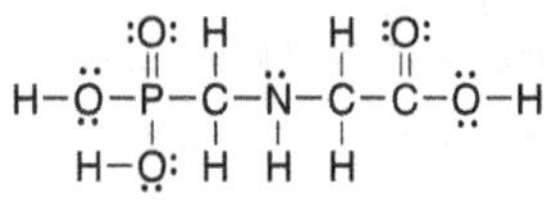

Each O atom needs two lone pairs and the N atom needs one.

b. There are 10 electrons around phosphorus.

c. Phosphorus and hydrogen do not follow the octet rule.

4.13 Resonance structures are two or more Lewis structures having the same arrangement of atoms but a different arrangement of electrons. Draw the second resonance structure as in Sample Problem 4.5 in the text, or Example 4.4. The two resonance structures in each part differ in the location of a double bond and a lone pair.

a. [H–C(H)(H)–C(=O)–O:]$^-$ ⟷ [H–C(H)(H)–C(–O:)=O]$^-$

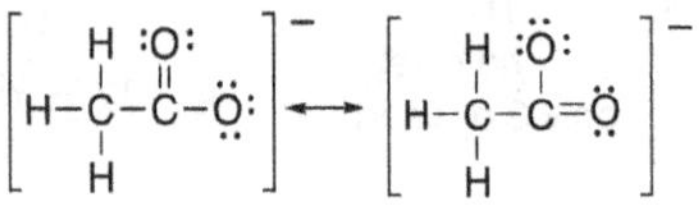

b. [H–C(=O)–N–H]$^-$ ⟷ [H–C(–O:)=N–H]$^-$

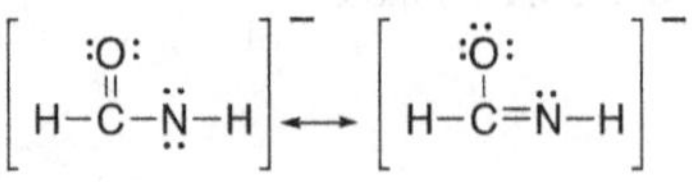

4.14

a. [:O–N=O]$^-$ ⟷ [O=N–O:]$^-$

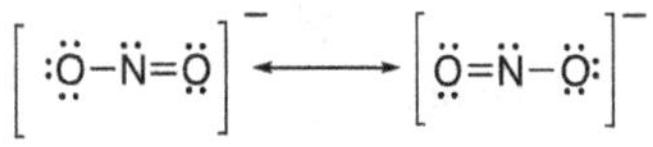

b. [O=C(–O:)–H]$^-$ ⟷ [:O–C(=O)–H]$^-$

4.15

Three resonance structures for nitrous oxide:

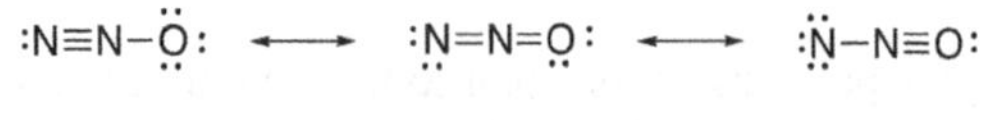

:N≡N–O: ⟷ :N=N=O: ⟷ :N–N≡O:

4.16 Name the compounds using the following two-step method:

Step [1] Name the first nonmetal by its element name and the second using the suffix *-ide.*

Step [2] Add prefixes to show the number of atoms of each element.

a. CS_2 = carbon disulfide
b. SO_2 = sulfur dioxide
c. PCl_5 = phosphorus pentachloride
d. BF_3 = boron trifluoride

4.17

a. silicon dioxide = SiO_2
b. phosphorus trichloride = PCl_3
c. sulfur trioxide = SO_3
d. dinitrogen trioxide = N_2O_3

4.18

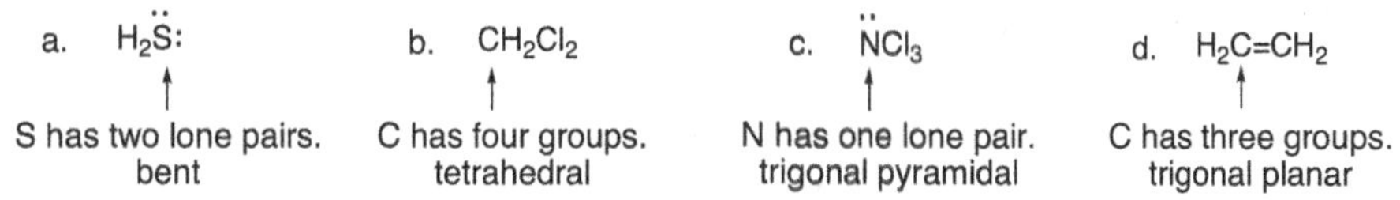

a. $H_2\ddot{S}$: — S has two lone pairs. bent
b. CH_2Cl_2 — C has four groups. tetrahedral
c. $\ddot{N}Cl_3$ — N has one lone pair. trigonal pyramidal
d. $H_2C{=}CH_2$ — C has three groups. trigonal planar

4.19

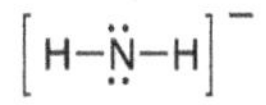

$[H{-}\ddot{\underset{..}{N}}{-}H]^-$ The ion has a bent geometry (b), because the N is surrounded by four groups.

4.20 Dihydroxyacetone has (a) two tetrahedral carbons and (b) one trigonal planar carbon.

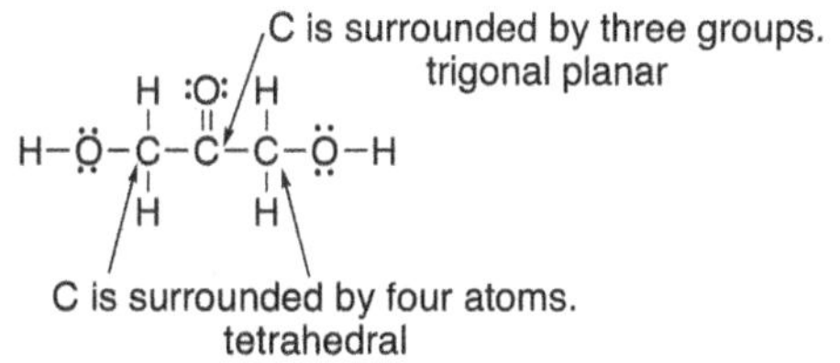

$HO{-}CH_2{-}C({=}O){-}CH_2{-}OH$

C is surrounded by three groups. trigonal planar

C is surrounded by four atoms. tetrahedral

4.21 Electronegativity *increases* left to right across a row of the periodic table as the nuclear charge increases (excluding the noble gases).
Electronegativity *decreases* down a column of the periodic table as the atomic radius increases, pushing the valence electrons farther from the nucleus.

a. Na < Li < H
b. Be < C < O
c. I < Cl < F
d. B < N < O

4.22 Calculate the electronegativity difference between the two atoms and use the following rules: less than 0.5 (nonpolar); 0.5–1.9 (polar covalent); greater than 1.9 (ionic).

a. HF = (4.0 – 2.1) = 1.9 = polar covalent
b. MgO = (3.5 – 1.2) = 2.3 = ionic
c. F_2 = (4.0 – 4.0) = 0 = nonpolar
d. ClF = (4.0 – 3.0) = 1.0 = polar covalent
e. H_2O = (3.5 – 2.1) = 1.4 = polar covalent
f. NH_3 = (3.0 – 2.1) = 0.9 = polar covalent

4.23 The head of the bond dipole arrow points towards the more electronegative atom.

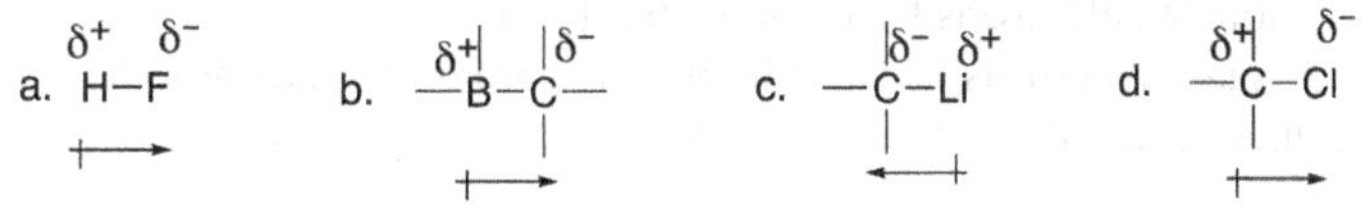

a. $\overset{\delta+}{H}{-}\overset{\delta-}{F}$ (arrow points toward F)
b. $\overset{\delta+}{B}{-}\overset{\delta-}{C}$ (arrow points toward C)
c. $\overset{\delta-}{C}{-}\overset{\delta+}{Li}$ (arrow points toward C)
d. $\overset{\delta+}{C}{-}\overset{\delta-}{Cl}$ (arrow points toward Cl)

4.24 To determine the overall polarity of a molecule, identify the polar bonds, determine the shape around individual atoms, and then decide if the individual bond dipoles cancel or reinforce.

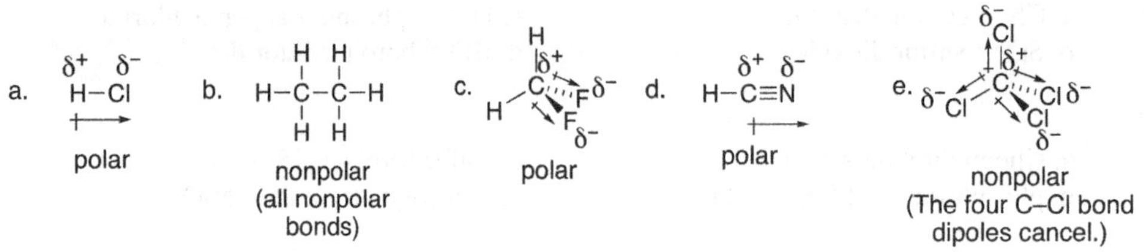

a. δ+ δ− H–Cl (dipole arrow) polar

b. H–C–C–H (ethane) nonpolar (all nonpolar bonds)

c. polar

d. δ+ δ− H–C≡N polar

e. nonpolar (The four C–Cl bond dipoles cancel.)

4.25 All C–H and C–C bonds are nonpolar.

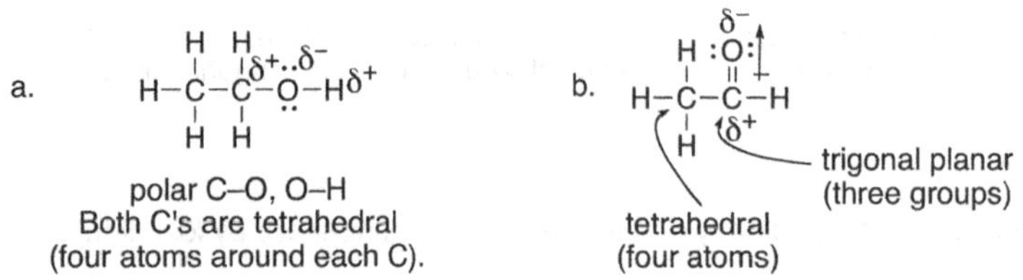

a. H–C–C–O–H with δ+ on C, δ− on O, δ+ on H
polar C–O, O–H
Both C's are tetrahedral (four atoms around each C).

b. H–C–C(=O)–H with δ− on O, δ+ on C
tetrahedral (four atoms)
trigonal planar (three groups)

4.26 a. Lewis structure of zingerone:

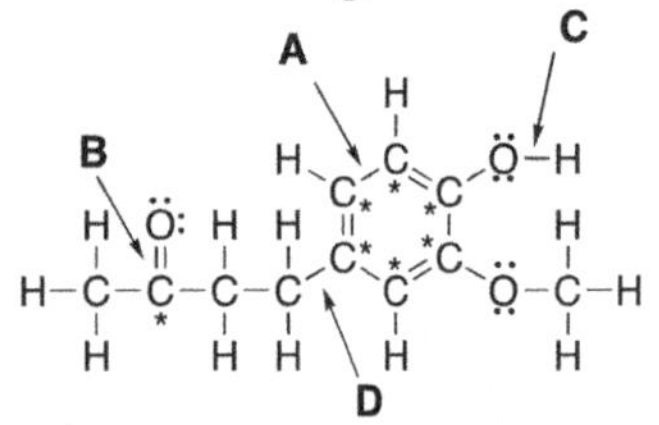

(Lewis structure with bonds labeled A, B, C, D; trigonal planar carbons marked *)

Each O atom has two lone pairs, so there is a total of six lone pairs (containing 12 electrons) in zingerone.

b. There are seven trigonal planar C's, labeled with (*).

c. Bonds **A** and **D** are between two carbon atoms so the bonds are nonpolar. Since the electronegativity of O (3.5) is much greater than the electronegativity of C (2.5) and H (2.1), the C–O bond (**B**) and O–H bond (**C**) are both polar.

Solutions to Odd-Numbered End-of-Chapter Problems

4.27 In covalent bonding, atoms share electrons to attain the electronic configuration of the noble gas closest to them in the periodic table. In ionic bonding, one atom donates electrons to the other atom.

a. LiCl: ionic; the metal Li donates electrons to the nonmetal chlorine.
HCl: covalent; H and Cl share electrons because both atoms are nonmetals and the electronegativity difference is not large enough for electron transfer to occur.

b. KBr: ionic; the metal K donates electrons to the nonmetal bromine.
HBr: covalent; H and Br share electrons because both atoms are nonmetals and the electronegativity difference is not large enough for electron transfer to occur.

4.29 Atoms with one, two, or three valence electrons form one, two, or three bonds, respectively. Atoms with four or more valence electrons form enough bonds to give an octet.

a. C = 4 bonds, 0 lone pairs
b. Se = 2 bonds, 2 lone pairs
c. I = 1 bond, 3 lone pairs
d. P = 3 bonds, 1 lone pair

4.31 Each O and S atom must get two lone pairs and each N atom must get one lone pair.

a. 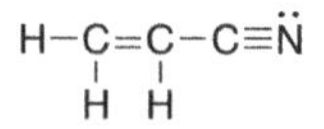

b. 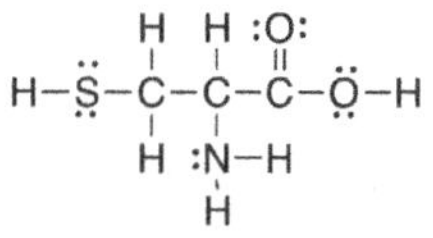

4.33 Use the common element colors (black = C, red = O, gray = H), and give each O two lone pairs.

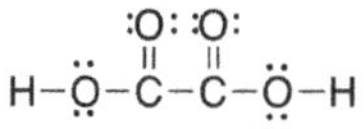

oxalic acid

4.35 Follow the steps in Examples 4.2 and 4.3 to draw the Lewis structures.

a. 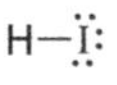

b. 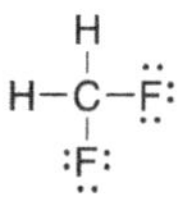

c. 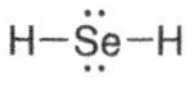

d. 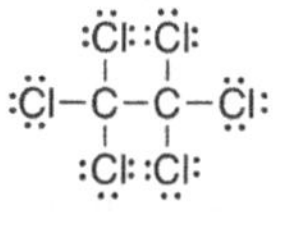

4.37 Follow the steps in Examples 4.2 and 4.3.

a. 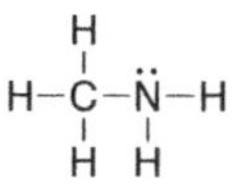

b. 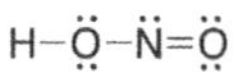

c. 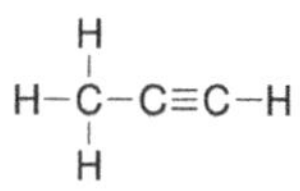

4.39 Follow the steps in Example 4.3.

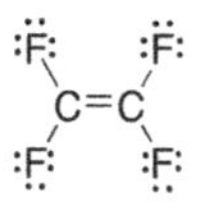

4.41

a. 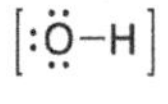

b. 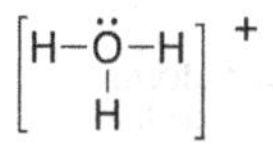

4.43

a. 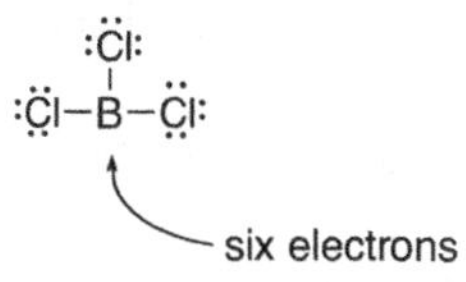

b. 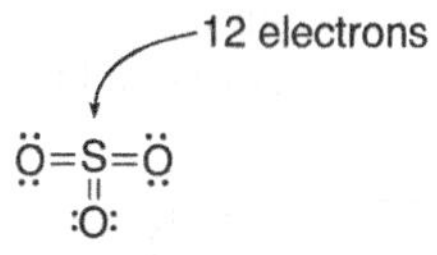

4.45 A resonance structure is one representation of electron arrangement with a given placement of atoms. A resonance hybrid is a composite of two or more resonance structures.

4.47 Use Example 4.4. In this ion, the location of a double bond and a lone pair is different in the two resonance structures.

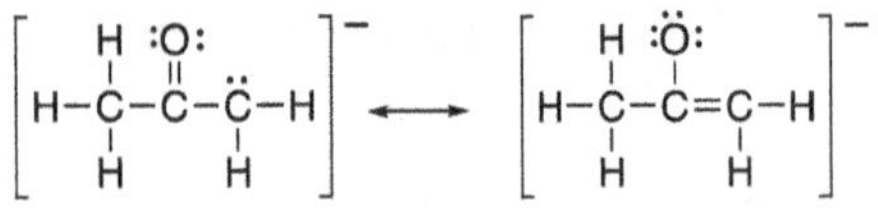

4.49 Resonance structures must have the same arrangement of atoms.

a. 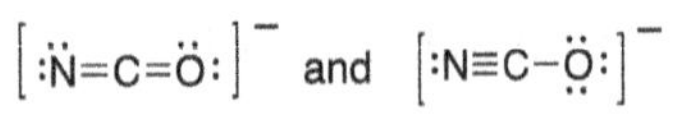 resonance structures

b. 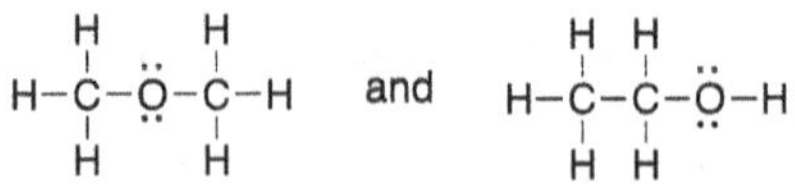 not resonance structures
The arrangement of atoms is different.

4.51 Three resonance structures for the carbonate anion:

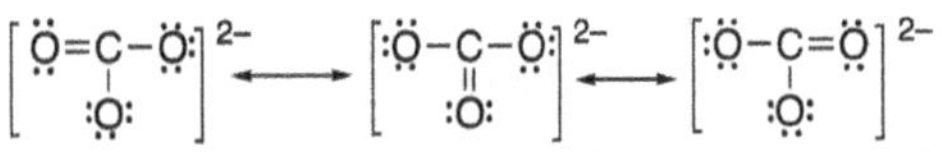

4.53 Name each compound as in Answer 4.16.

a. PBr_3 = phosphorus tribromide
b. SO_3 = sulfur trioxide
c. NCl_3 = nitrogen trichloride
d. P_2S_5 = diphosphorus pentasulfide

4.55 Work backwards to write the formula that corresponds to each name.

a. selenium dioxide = SeO_2
b. carbon tetrachloride = CCl_4
c. dinitrogen pentoxide = N_2O_5

4.57 Follow the steps in Example 4.6.

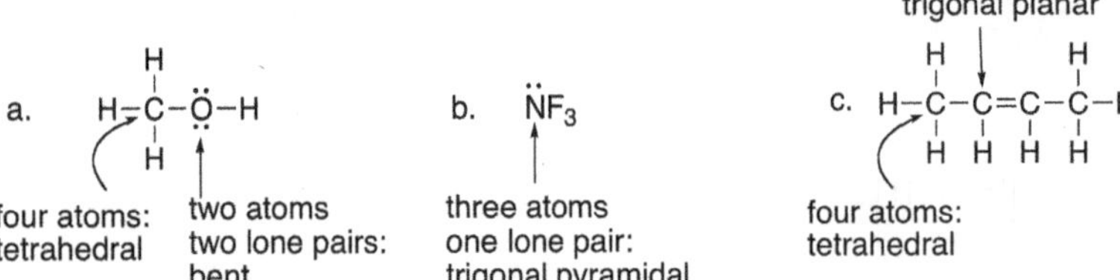

4.59 First, draw a Lewis structure; then count groups around the central atom to determine shape. With two groups, an atom is linear. With two atoms and two lone pairs, an atom has a bent shape.

a. $BeCl_2$ — two groups, linear (**A**)

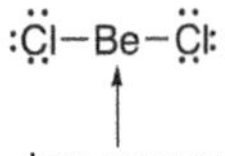

b. OF_2 — two atoms and two lone pairs, bent (**B**)

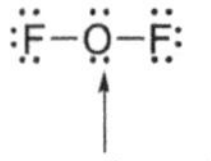

c. SCl_2 — two atoms and two lone pairs, bent (**B**)

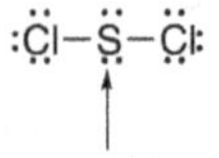

4.61 Follow the steps in Example 4.6.

a. H–N(H)–Ö–H: trigonal pyramidal, bent

b. H–C(H)(H)–C(=O)–H: tetrahedral, trigonal planar

c. $[H–C(H)(H)–N(H)(H)–H]^+$: tetrahedral

4.63

a. [1] A linear geometry means two groups, so the dark red atom has no lone pairs. [2] $BeCl_2$
b. [1] A bent geometry generally means four groups, so the dark red atom has two lone pairs. [2] H_2O

4.65 BCl_3 is trigonal planar because it has three Cl's bonded to B but no lone pairs. NCl_3 is trigonal pyramidal because it has three Cl's around N as well as a lone pair.

4.67

a. $H_3C–F$: tetrahedral: both 109.5°

b. H–C≡C–CH₂–O–H: linear: 180°; bent: 109.5°; tetrahedral: 109.5°

c. H–C(H)=C=O: trigonal planar: 120°; linear: 180°; trigonal planar: 120°

4.69 CCl_4 is tetrahedral because carbon has four groups around it. Each C in C_2Cl_4 has trigonal planar geometry because each carbon has three groups around it.

CCl_4: tetrahedral

$Cl_2C=CCl_2$: trigonal planar

4.71 Electronegativity *increases* left to right across a row of the periodic table as the nuclear charge increases (excluding the noble gases).
Electronegativity *decreases* down a column of the periodic table as the atomic radius increases, pushing the valence electrons farther from the nucleus.

a. Se < S < O   b. Na < P < Cl   c. S < Cl < F   d. P < N < O

4.73 In covalent bonding, atoms share electrons to attain the electronic configuration of the noble gas closest to them in the periodic table. In ionic bonding, one atom donates electrons to the other atom. When the electronegativity difference between the two atoms in a bond is greater than 1.9, the bond is ionic.

a. hydrogen and bromine = polar covalent (2.8 – 2.1 = 0.7)
b. nitrogen and carbon = polar covalent (3.0 – 2.5 = 0.5)
c. sodium and sulfur = polar covalent (2.5 – 0.9 = 1.6)
d. lithium and oxygen = ionic (3.5 – 1.0 = 2.5)

4.75 Calculate the electronegativity difference between the atoms and carbon and use the following rules: less than 0.5 (nonpolar); 0.5–1.9 (polar covalent); greater than 1.9 (ionic).

a. C = (2.5 – 2.5) = 0 = nonpolar
b. O = (3.5 – 2.5) = 1.0 = polar
c. Li = (2.5 – 1.0) = 1.5 = polar
d. Cl = (3.0 – 2.5) = 0.5 = polar
e. H = (2.5 – 2.1) = 0.4 = nonpolar

4.77 Calculate the electronegativity difference between the atoms to determine which bond is more polar.

a. C–O = (3.5 – 2.5) = 1.0 = more polar
C–N = (3.5 – 3.0) = 0.5
b. C–F = (4.0 – 2.5) = 1.5 = more polar
C–Cl = (3.0 – 2.5) = 0.5
c. Si–C = (2.5 – 1.8) = 0.7 = more polar
P–H = (2.1 – 2.1) = 0

4.79 The δ^+ is placed next to the *less* electronegative atom, whereas the δ^- is placed next to the *more* electronegative atom.

$\overset{\delta+}{C}-\overset{\delta-}{O}$ $\overset{\delta+}{C}-\overset{\delta-}{N}$ $\overset{\delta+}{C}-\overset{\delta-}{F}$ $\overset{\delta+}{C}-\overset{\delta-}{Cl}$ $\overset{\delta+}{Si}-\overset{\delta-}{C}$ P–H no dipole

4.81

a. $Cl_2C{=}CCl_2$
polar bonds
nonpolar compound
Bond dipoles cancel.

b. $H_3C{-}C({=}O)Cl$
polar bonds
polar compound
Bond dipoles do not cancel.

4.83 $CHCl_3$ has a net dipole, making it polar. CCl_4 has four polar bonds but no net dipole because the four bond dipoles cancel one another.

$CHCl_3$ — net dipole
polar bonds
polar compound

CCl_4
polar bonds
nonpolar compound

4.85

A a. $ClHC{=}CH_2$
b. one polar C–Cl bond
c. polar molecule
(Only one polar bond dipole makes a polar molecule.)

B a. :Cl–C≡C–Cl:
b. two polar C–Cl bonds
c. nonpolar molecule
(Two bond dipoles cancel.)

4.87

a. 20 valence electrons
[1 O atom (6 valence electrons) + 2 Cl atoms (7 valence electrons each)]

b, c:

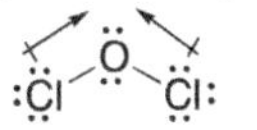

d. Cl_2O has a bent shape because O is surrounded by two atoms and two lone pairs.

e. The compound is polar because the two bond dipoles do not cancel.

4.89

a.–c.

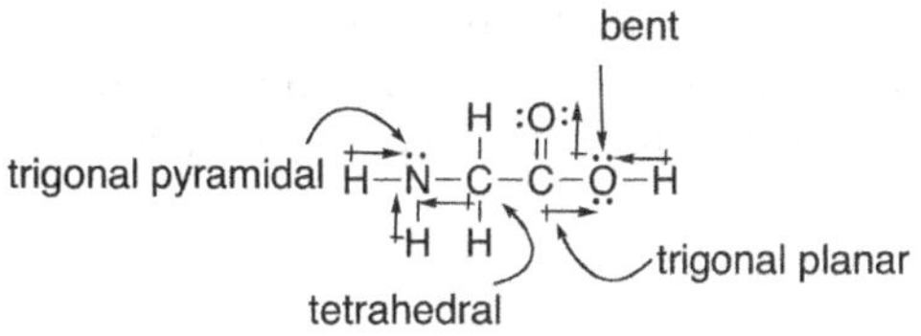

trigonal pyramidal, bent, tetrahedral, trigonal planar

2 C x 4 e^-	=	8 e^-
5 H x 1 e^-	=	5 e^-
1 N x 5 e^-	=	5 e^-
2 O x 6 e^-	=	12 e^-
		30 valence e^-

d. Glycine is a polar molecule because the many bond dipoles do not cancel.

4.91 a. Lewis structure:

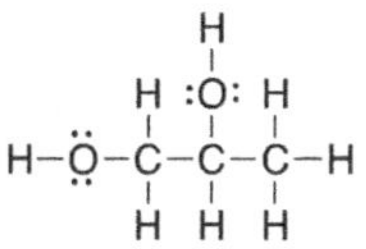

b. Both C–O and O–H bonds are polar.

c. All C's are surrounded by four atoms, making them tetrahedral. Both O's are surrounded by two atoms and two lone pairs, giving them a bent shape.

4.93 First, give the O atom two lone pairs and each N atom one lone pair. Then add double bonds to the C's that do not have four bonds.

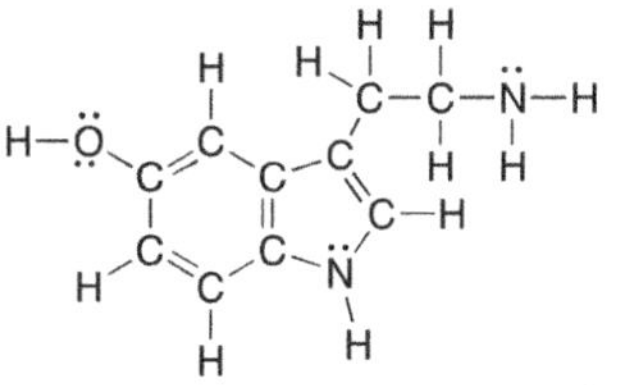

or

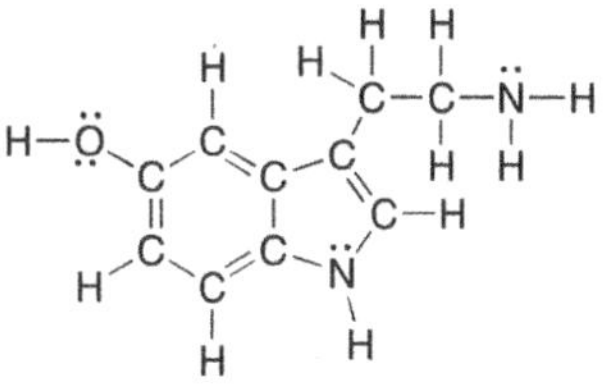

4.95

a, c.

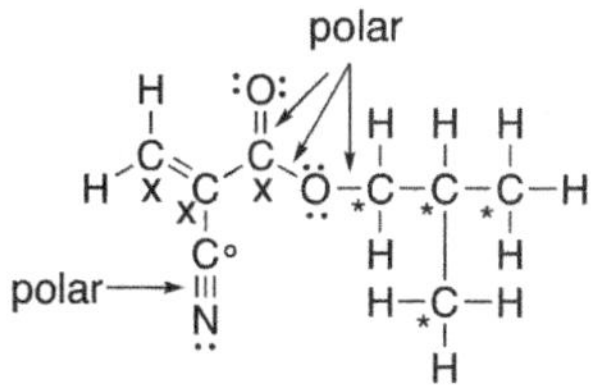

polar

polar

b. C's with x, trigonal planar
C with °, linear
C with *, tetrahedral

4.97 a. The predicted shape around each carbon is tetrahedral because each C is surrounded by four atoms.
b. The interior angle of a triangle with three equal sides is 60°.
c. The 60° bond angle deviates greatly from the theoretical tetrahedral bond angle (109.5°), making the molecule unstable.

4.99

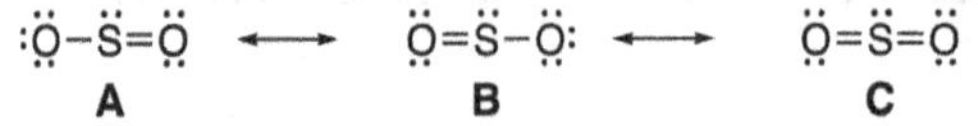

:Ö–S̈=Ö ⟷ Ö=S̈–Ö: ⟷ Ö=S̈=Ö

A **B** **C**

Structures **A** and **B** follow the octet rule. Since the S atom in **C** has 10 electrons, this structure violates the octet rule. Because each structure has an S atom surrounded by two atoms and a lone pair, the three groups much be ~120° from each other. For example, **A** could be drawn as:

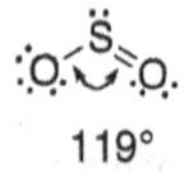

119°

Chapter 5 Chemical Reactions

Chapter Review

[1] What do the terms in a chemical equation mean and how is an equation balanced? (5.1, 5.2)

- A chemical equation contains the reactants on the left side of an arrow and the products on the right. The coefficients tell how many molecules or moles of a substance react or are formed.
- A chemical equation is balanced by placing coefficients in front of chemical formulas one at a time, beginning with the most complex formula, so that the number of atoms of each element is the same on both sides. When no number precedes a formula, the coefficient is assumed to be "1."

Chemical equation $CH_4 + 2\,O_2 \longrightarrow CO_2 + 2\,H_2O$

coefficient (the 2 in front of O_2); coefficient (the 2 in front of H_2O)

reactants: $CH_4 + 2\,O_2$ — products: $CO_2 + 2\,H_2O$

[2] List four common types of reactions. (5.3)

- In a combination reaction, two or more reactants join to form one product.
- In a decomposition reaction, one reactant is converted to two or more products.
- In a single replacement reaction, an element replaces another element in a compound to form a new element and a new compound.
- In a double replacement reaction, two compounds exchange atoms or ions to form two new compounds.

[3] What are oxidation and reduction reactions? (5.4)

- Oxidation–reduction or redox reactions are electron transfer reactions.
- **Oxidation** results in the loss of electrons. Metals and anions tend to undergo oxidation. In some reactions, oxidation results in the gain of O atoms or the loss of H atoms.
- **Reduction** results in the gain of electrons. Nonmetals and cations tend to undergo reduction. In some reactions, reduction results in the loss of O atoms or the gain of H atoms.

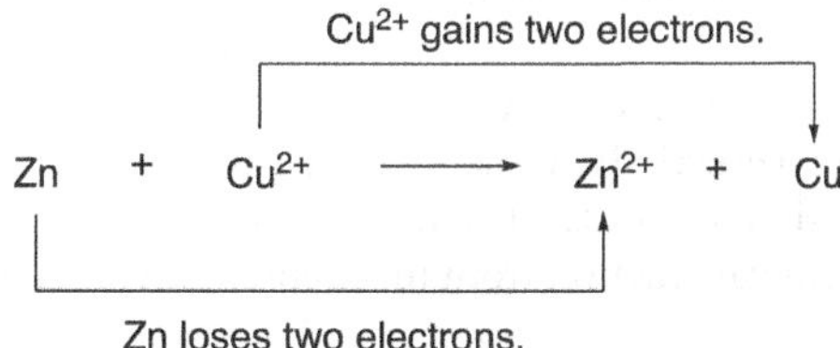
Cu2+ gains two electrons.
Zn + Cu2+ → Zn2+ + Cu
Zn loses two electrons.

- Zn loses two electrons to form Zn^{2+}, so Zn is oxidized.
- Cu^{2+} gains two electrons to form Cu metal, so Cu^{2+} is reduced.

[4] Define the terms mole and Avogadro's number. (5.5)

- A **mole** is a quantity that contains 6.02×10^{23} items—usually, atoms, molecules, or ions.
- **Avogadro's number** is the number of particles in a mole—6.02×10^{23}.
- The number of atoms or molecules in a given number of moles is calculated using Avogadro's number.

Two possible conversion factors: $\frac{1 \text{ mol}}{6.02 \times 10^{23} \text{ atoms}}$ or $\frac{6.02 \times 10^{23} \text{ atoms}}{1 \text{ mol}}$

[5] How are formula weight and molar mass calculated? (5.6)

- The **formula weight** is the sum of the atomic weights of all the atoms in a compound, reported in atomic mass units.
- The **molar mass** is the mass of one mole of a substance, reported in g/mol. The molar mass is numerically equal to the formula weight but the units are different (g/mol not amu).

[6] How are the mass of a substance and its number of moles related? (5.6)

- The molar mass is used as a conversion factor to determine how many grams are contained in a given number of moles of a substance. Similarly, the molar mass is used to determine how many moles of a substance are contained in a given number of grams.

[7] How can a balanced equation and molar mass be used to calculate the number of moles and mass of a reaction product? (5.7, 5.8)

- The coefficients in a balanced chemical equation tell us the number of moles of each reactant that combine and the number of moles of each product formed. Coefficients are used to form mole ratios that serve as conversion factors relating the number of moles of reactants and products.
- When the mass of a substance in a reaction must be calculated, first its number of moles is determined using mole ratios, and then the molar mass is used to convert moles to grams.

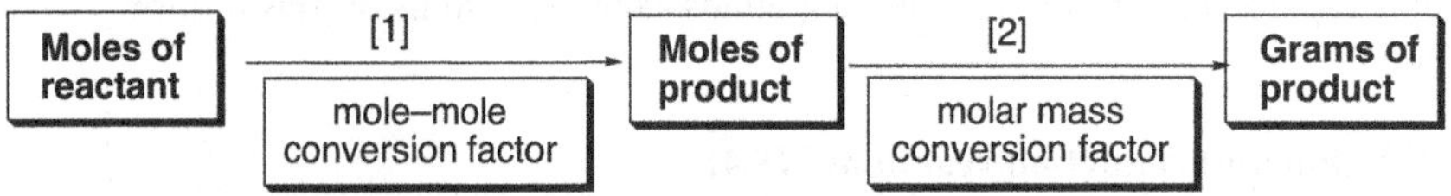

[8] What is percent yield? (5.9)

- **Percent yield** is determined by using the following equation:

$$\text{Percent yield} = \frac{\text{actual yield (g)}}{\text{theoretical yield (g)}} \times 100\%$$

- The **actual yield** is the amount of product formed in a reaction, determined by weighing a product on a balance. The **theoretical yield** is a quantity calculated from a balanced chemical equation, using mole ratios and molar mass. The theoretical yield is the maximum amount of product that can form in a chemical reaction from the amount of reactants used.

[9] What is the limiting reactant in a reaction? (5.10)

- The limiting reactant is the reactant that is completely used up in a reaction. The number of moles of limiting reactant determines the number of moles of product formed using mole ratios in the balanced equation.

Problem Solving

[1] Introduction to Chemical Reactions (5.1)

Example 5.1 Label the reactants and products, and indicate how many atoms of each type of element are present on each side of the equation.

$$3\ H_2(g) \quad + \quad N_2(g) \longrightarrow 2\ NH_3(g)$$

Analysis
Reactants are on the left side of the arrow and products are on the right side in a chemical equation. When a formula contains a subscript, multiply its coefficient by the subscript to give the total number of atoms of a given type.

Solution
In this equation, the reactants are H_2 and N_2 while the product is NH_3. To determine the number of atoms when a formula has both a coefficient and a subscript, multiply the coefficient by the subscript.

$3\ H_2$	=	6 H's	Multiply the coefficient 3 by the subscript 2.
$1\ N_2$	=	2 N's	There are 2 N's.
$2\ NH_3$	=	2 N's + 6 H's	Multiply the coefficient 2 by each subscript: 2×1 N = 2 N's; 2×3 H's = 6 H's.

Add up the atoms on each side to determine the total number for each type of element.

$$3\ H_2(g) \quad + \quad N_2(g) \longrightarrow 2\ NH_3(g)$$

Atoms in the reactants:
• 6 H's 2 N's

Atoms in the product:
• 6 H's 2 N's

[2] Balancing Chemical Reactions (5.2)

Example 5.2 Write a balanced equation for the reaction of pentane (C_5H_{12}) with oxygen (O_2) to form carbon dioxide (CO_2) and water (H_2O).

Analysis
Balance an equation with coefficients, one element at a time, beginning with the most complex formula and starting with an element that appears in only one formula on both sides of the equation. Continue placing coefficients until the number of atoms of each element is equal on both sides of the equation.

Solution
[1] Write the equation with correct formulas.

$$C_5H_{12} \quad + \quad O_2 \longrightarrow CO_2 \quad + \quad H_2O$$

- None of the elements is balanced in this equation. As an example, there are 5 C's on the left side but only 1 C on the right side.

[2] Balance the equation with coefficients one element at a time.

- Begin with C_5H_{12} because its formula is most complex. Balance the 5 C's of pentane by placing the coefficient 5 before CO_2. Balance the 12 H's by placing the coefficient 6 before H_2O.

Place a 5 to balance C's.

$$C_5H_{12} + O_2 \longrightarrow 5\ CO_2 + 6\ H_2O$$

Place a 6 to balance H's.

- The right side of the equation now has 16 O's. Place a coefficient of 8 in front of O_2 to balance the O atoms.

$$C_5H_{12} + 8\ O_2 \longrightarrow 5\ CO_2 + 6\ H_2O$$

[3] Check.

- The equation is balanced because the number of atoms of each element is the same on both sides.

Answer: $C_5H_{12} + 8\ O_2 \longrightarrow 5\ CO_2 + 6\ H_2O$

Atoms in the reactants:	Atoms in the products:
• 5 C's	• 5 C's
• 12 H's	• 12 H's (6 x 2)
• 16 O's	• 16 O's (5 x 2 O's) + (6 x 1 O)

[3] Oxidation and Reduction (5.4)

Example 5.3 Identify the species that is oxidized and the species that is reduced in the following reaction. Write out half reactions to show how many electrons are gained or lost by each species.

$$2\ Na + Cl_2 \longrightarrow 2\ Na^+ + 2\ Cl^-$$

Analysis

Metals and anions tend to lose electrons and thus undergo oxidation. Nonmetals and cations tend to gain electrons and thus undergo reduction.

Solution

The metal Na is oxidized to Na^+, losing one electron for each Na atom. Cl_2 gains two electrons, and is thus reduced to 2 Cl^-.

$$2\ Na \longrightarrow 2\ Na^+ + 2\ e^- \qquad Cl_2 + 2\ e^- \longrightarrow 2\ Cl^-$$

Na is oxidized. Cl_2 is reduced.

We need the same number of electrons lost and gained. Since each Cl_2 gains two electrons, 2 Na atoms are needed in order to lose two electrons.

[4] The Mole and Avogadro's Number (5.5)

Example 5.4 How many molecules are contained in 8.00 moles of water (H_2O)?

Analysis and Solution

[1] Identify the original quantity and the desired quantity.

8.00 mol of H_2O	? number of molecules of H_2O
original quantity	desired quantity

[2] Write out the conversion factors.

- Choose the conversion factor that places the unwanted unit, mol, in the denominator so that the units cancel.

$$\frac{1 \text{ mol}}{6.02 \times 10^{23} \text{ molecules}} \quad \text{or} \quad \boxed{\frac{6.02 \times 10^{23} \text{ molecules}}{1 \text{ mol}}}$$

Choose this conversion factor to cancel mol.

[3] Set up and solve the problem.

- Multiply the original quantity by the conversion factor to obtain the desired quantity.
- Multiplication first gives an answer that is not written in scientific notation since the coefficient (48.2) is greater than 10. Moving the decimal one place to the *left* and *increasing* the exponent by one gives the answer written in proper form.

$$8.00 \text{ mol} \times \frac{6.02 \times 10^{23} \text{ molecules}}{1 \text{ mol}} = 48.2 \times 10^{23} \text{ molecules}$$

Moles cancel. Convert to a number between 1 and 10.

$$= 4.82 \times 10^{24} \text{ molecules of } H_2O$$

Answer

[5] Molar Mass (5.6)

Example 5.5 What is the molar mass of $C_6H_8O_6$?

Analysis

Determine the number of atoms of each element from the subscripts in the chemical formula, multiply the number of atoms of each element by the atomic weight, and add up the results.

Solution

6 C atoms	×	12.01 amu	=	72.06 amu
8 H atoms	×	1.008 amu	=	8.064 amu
6 O atoms	×	16.00 amu	=	96.00 amu
Formula weight of $C_6H_8O_6$:				176.124 amu rounded to 176.12

Answer: Since the formula weight of $C_6H_8O_6$ is 176.12 amu, the molar mass is 176.12 g/mol.

---

[6] Mole Calculations in Chemical Equations (5.7)

Example 5.6 Using the balanced equation, how many moles of H_2 are produced from 1.5 mol of HCl?

$$2\ Al + 6\ HCl \longrightarrow 2\ AlCl_3 + 3\ H_2$$

Analysis and Solution

[1] Identify the original quantity and the desired quantity.

1.5 mol of HCl — original quantity

? mol of H_2 — desired quantity

[2] Write out the conversion factors.

- Use the coefficients in the balanced equation to write mole–mole conversion factors for the two compounds, HCl and H_2. Choose the conversion factor that places the unwanted unit, moles of HCl, in the denominator so that the units cancel.

$$\frac{6\ \text{mol HCl}}{3\ \text{mol } H_2} \quad \text{or} \quad \boxed{\frac{3\ \text{mol } H_2}{6\ \text{mol HCl}}}$$

Choose this conversion factor to cancel mol HCl.

[3] Set up and solve the problem.

- Multiply the original quantity by the conversion factor to obtain the desired quantity.

$$1.5\ \text{mol HCl} \times \frac{3\ \text{mol } H_2}{6\ \text{mol HCl}} = 0.75\ \text{mol } H_2$$

Answer: 0.75 mol H_2

Moles HCl cancel.

---

[7] Mass Calculations in Chemical Equations (5.8)

Example 5.7 How many grams of Fe_2O_3 (molar mass 159.7 g/mol) are produced from 3.0 mol of Fe?

$$4\ Fe + 3\ O_2 \longrightarrow 2\ Fe_2O_3$$

Analysis and Solution

[1] Convert the number of moles of reactant to the number of moles of product using a mole–mole conversion factor.

- Use the coefficients in the balanced chemical equation to write mole–mole conversion factors for the two compounds—four moles of Fe form two moles of Fe_2O_3.
- Multiply the number of moles of reactant (Fe) by the conversion factor to give the number of moles of product (Fe_2O_3).

Moles of reactant		**mole–mole conversion factor**		**Moles of product**
3.0 mol Fe	x	$\frac{2 \text{ mol } Fe_2O_3}{4 \text{ mol Fe}}$	=	1.5 mol Fe_2O_3

Moles Fe cancel.

[2] Convert the number of moles of product to the number of grams of product using the product's molar mass.

- Use the molar mass of the product (Fe_2O_3, molar mass 159.7 g/mol) to write a conversion factor.
- Multiply the number of moles of product (from step [1]) by the conversion factor to give the number of grams of product.

Moles of product		**molar mass conversion factor**		**Grams of product**
1.5 mol Fe_2O_3	x	$\frac{159.7 \text{ g } Fe_2O_3}{1 \text{ mol } Fe_2O_3}$	=	240 g Fe_2O_3 **Answer**

Moles cancel.

[8] Percent Yield (5.9)

Example 5.8 Consider the reaction of hydrogen and nitrogen to form ammonia. If the theoretical yield of ammonia is 10. g in a reaction, what is the percent yield of ammonia if only 4.5 g of ammonia are actually formed?

Analysis

Use the formula, percent yield = (actual yield/theoretical yield) × 100% to calculate the percent yield.

Solution

$$\text{Percent yield} = \frac{\text{actual yield (g)}}{\text{theoretical yield (g)}} \times 100\%$$

$$= \frac{4.5 \text{ g}}{10. \text{ g}} \times 100\% = 45\%$$

Answer

Self-Test

[1] Fill in the blank with one of the terms listed below.

Actual yield (5.9)	Law of conservation of matter (5.1)	Redox reaction (5.4)
Balanced chemical equation (5.2)	Molar mass (5.6)	Reduction (5.4)
Chemical equation (5.1)	Mole (5.5)	Theoretical yield (5.9)
Formula weight (5.6)		

1. The ______ is the mass of one mole of any substance, reported in g/mol.
2. The ______ is the amount of product formed in a reaction.
3. A ______ involves the transfer of electrons from one element to another.
4. The ______ is the sum of the atomic weights of all the atoms in a compound, reported in atomic mass units (amu).
5. ______ results in the loss of oxygen atoms or the gain of hydrogen atoms.
6. A ______ is a quantity that contains 6.02×10^{23} items—usually atoms, molecules, or ions.
7. A ______ tells us the number of *moles* of each reactant that combine and the number of *moles* of each product formed.
8. A ______ is the expression used to illustrate what substances constitute the starting materials in a reaction and what products are formed.
9. The ______ states that atoms cannot be created or destroyed in a chemical reaction.
10. The ______ is the amount of product expected from a given amount of reactant based on the coefficients in the balanced chemical equation.

[2] Label the reactants and products in the equation below.

$$2\ NH_3 \longrightarrow 2\ NH_2^- + 2\ H^+$$

11. (arrow to $2\ NH_3$)

12. (arrow to $2\ NH_2^- + 2\ H^+$)

[3] How many carbons are contained in each representation?

13. CH_3CH_2OH
14. $2\ C_2H_7N$
15. $C_6H_8O_6$

[4] Match each compound to its molar mass.

a. 151.9 g/mol b. 162.3 g/mol c. 58.4 g/mol

16. NaCl
17. $FeSO_4$
18. $C_{10}H_{14}N_2$

[5] Label each reaction as oxidation or reduction.

19. $Mg \longrightarrow Mg^{2+} + 2\,e^-$
20. $Cl_2 + 2\,e^- \longrightarrow 2\,Cl^-$
21. $O_2 + 4\,e^- \longrightarrow 2\,O^{2-}$
22. $2\,O^{2-} \longrightarrow O_2 + 4\,e^-$

[6] Fill in the coefficients to balance the chemical reaction.

a. 3 b. 4 c. 5

$C_3H_8 + \_\_O_2 \longrightarrow \_\_CO_2 + \_\_H_2O$

(23. = coefficient of O_2; 24. = coefficient of CO_2; 25. = coefficient of H_2O)

Answers to Self-Test

1. molar mass	6. mole	11. reactant	16. c	21. reduction
2. actual yield	7. balanced chemical equation	12. products	17. a	22. oxidation
3. redox reaction	8. chemical equation	13. two	18. b	23. c
4. formula weight	9. law of conservation of matter	14. four	19. oxidation	24. a
5. Reduction	10. theoretical yield	15. six	20. reduction	25. b

Solutions to In-Chapter Problems

5.1 The process is a chemical reaction because the reactants contain two gray spheres joined (indicating H_2) and two red spheres joined (indicating O_2), while the product (H_2O) contains a red sphere joined to two gray spheres (indicating O–H bonds).

5.2 The process is a physical change (freezing), because the particles in the reactants are the same as the particles in the products.

5.3 Chemical equations are written with the **reactants on the left** and the **products on the right** separated by a **reaction arrow.**

	reactants		products	
a.	$2\,H_2O_2(aq)$	$\longrightarrow$	$2\,H_2O(l) + O_2(g)$	(4 H, 4 O)
b.	$2\,C_8H_{18} + 25\,O_2$	$\longrightarrow$	$16\,CO_2 + 18\,H_2O$	(16 C, 50 O, 36 H)
c.	$2\,Na_3PO_4(aq) + 3\,MgCl_2(aq)$	$\longrightarrow$	$Mg_3(PO_4)_2(s) + 6\,NaCl(aq)$	(3 Mg, 2 P, 8 O, 6 Na, 6 Cl)

5.4 To determine the number of each type of atom when a formula has both a coefficient and a subscript, multiply the coefficient by the subscript.

For $3\,Al_2(SO_4)_3$: Al = 6 (3 × 2), S = 9 (3 × 3), O = 36 (3 × 3 × 4)

5.5 Balance the equation with coefficients one element at a time to have the same number of atoms on each side of the equation. Follow the steps in Example 5.2.

[1] Place a 2 to balance O's.

a. $2\ H_2 + O_2 \longrightarrow 2\ H_2O$

[2] Place a 2 to balance H's.

c. $CH_4 + 2\ Cl_2 \longrightarrow CH_2Cl_2 + 2\ HCl$

b. $2\ NO + O_2 \longrightarrow 2\ NO_2$

5.6 Write the balanced chemical equation for carbon monoxide and oxygen reacting to form carbon dioxide. The smallest set of whole numbers must be used.

$2\ CO + O_2 \longrightarrow 2\ CO_2$

5.7 Follow the steps in Example 5.2 to write the balanced chemical equation.

$2\ C_2H_6 + 7\ O_2 \longrightarrow 4\ CO_2 + 6\ H_2O$

5.8 Balance the equations as in Example 5.2.

a. $2\ Al + 3\ H_2SO_4 \longrightarrow Al_2(SO_4)_3 + 3\ H_2$

b. $3\ Na_2SO_3 + 2\ H_3PO_4 \longrightarrow 3\ H_2SO_3 + 2\ Na_3PO_4$

5.9 Use the common element colors to identify the elements; in this example, a blue sphere represents nitrogen and a green sphere represents chlorine. The molecular art shows that two NCl_3 molecules are converted to N_2 and three Cl_2 molecules. The reaction is a decomposition because one reactant is converted to two products.

$2\ NCl_3 \longrightarrow N_2 + 3\ Cl_2$

5.10 In a combination reaction, two or more reactants join to form one product, whereas in a decomposition reaction, one reactant is converted to two or more products.
a. combination: Cu combines with S to form CuS.
b. decomposition: $CuCO_3$ decomposes to CuO and CO_2.
c. combination: H_2 combines with Cl_2 to form HCl.
d. decomposition: $C_6H_{12}O_6$ decomposes to C_2H_6O and CO_2.

5.11 In a single replacement reaction, an element replaces another element in a compound to form a new element and a new compound. In a double replacement reaction, two compounds exchange atoms or ions to form two new compounds.

a. single replacement: The element Mg replaces Ni in $Ni(NO_3)_2$.

$\underset{\text{compound}}{Ni(NO_3)_2} + \underset{\text{element}}{Mg} \longrightarrow \underset{\text{element}}{Ni} + \underset{\text{compound}}{Mg(NO_3)_2}$

b. double replacement: K^+ and Sn^{2+} exchange to form two new compounds.

$$2\,KI + Sn(NO_3)_2 \longrightarrow SnI_2 + 2\,KNO_3$$

exchange

c. double replacement: K^+ and H exchange to form two new compounds.

$$KCN + HCl \longrightarrow KCl + HCN$$

exchange

d. single replacement: The element Na replaces Al in $Al_2(SO_4)_3$.

$$6\,Na + Al_2(SO_4)_3 \longrightarrow 3\,Na_2SO_4 + 2\,Al$$

element compound compound element

5.12 Use the definitions in Answers 5.10 and 5.11 to classify each reaction.
a. decomposition: HgO decomposes to Hg and O_2.
b. single replacement: The element Mg replaces Zn in $ZnCl_2$.
c. combination: $CH_2{=}CH_2$ combines with HBr to form a single product, CH_3CH_2Br.
d. double replacement: K^+ and H exchange to form two new compounds.

5.13

a. $C + O_2 \longrightarrow \boxed{CO_2}$

b. $MgCO_3 \longrightarrow MgO + \boxed{CO_2}$

c. $2\,Ag + \boxed{Li_2CO_3} \longrightarrow 2\,Li + Ag_2CO_3$

d. $Na_2CO_3 + BaBr_2 \longrightarrow \boxed{2\,NaBr} + \boxed{BaCO_3}$

5.14 A compound that gains electrons is reduced.
A compound that loses electrons is oxidized.

a. (oxidized) (reduced)

$$Zn(s) + 2\,H^+(aq) \longrightarrow Zn^{2+}(aq) + H_2(g)$$

$$Zn \longrightarrow Zn^{2+} + 2\,e^-$$

$$2\,H^+ + 2\,e^- \longrightarrow H_2$$

b. (reduced) (oxidized)

$$Fe^{3+}(aq) + Al(s) \longrightarrow Al^{3+}(aq) + Fe(s)$$

$$Al \longrightarrow Al^{3+} + 3\,e^-$$

$$Fe^{3+} + 3\,e^- \longrightarrow Fe$$

c. (oxidized) (reduced)

$$2\,I^- + Br_2 \longrightarrow I_2 + 2\,Br^-$$

$$2\,I^- \longrightarrow I_2 + 2\,e^-$$

$$Br_2 + 2\,e^- \longrightarrow 2\,Br^-$$

d. $$2\,AgBr \longrightarrow 2\,Ag + Br_2$$

$2\,Br^-$ (oxidized) $2\,Ag^+$ (reduced)

$$2\,Br^- \longrightarrow Br_2 + 2\,e^-$$

$$2\,Ag^+ + 2\,e^- \longrightarrow 2\,Ag$$

5.15 a. Zn is oxidized, and Hg^{2+} is reduced.

$$Zn \longrightarrow Zn^{2+} + 2\,e^- \qquad\qquad Hg^{2+} + 2\,e^- \longrightarrow Hg$$

Zn loses electrons and is oxidized. $\qquad$ Hg^{2+} gains electrons and is reduced.

b. H_2 is oxidized since it gains an O atom and $C_2H_4O_2$ is reduced since it gains hydrogen.

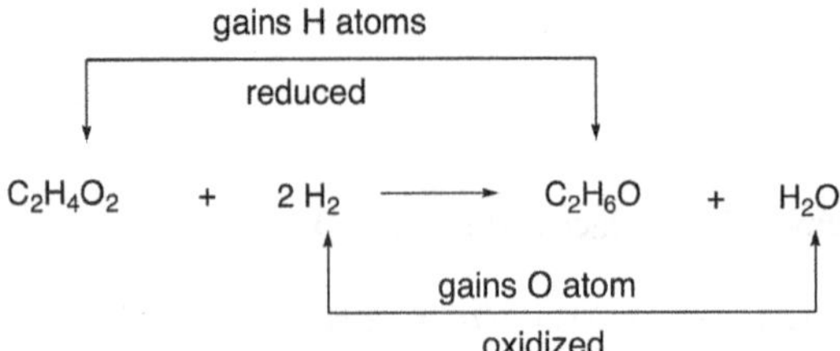

5.16 Zn is the reducing agent and Hg^{2+} is the oxidizing agent.

reduced

$$Zn + Hg^{2+} \longrightarrow Zn^{2+} + Hg$$

oxidized

5.17 One mole, abbreviated as **mol,** always contains an Avogadro's number of particles (6.02×10^{23}).

a, b, c, d: 6.02×10^{23}

5.18 Multiply the number of moles by Avogadro's number to determine the number of atoms. Avogadro's number is the conversion factor that relates moles to molecules, as in Example 5.4.

a. 2.00 mol $\times$ 6.02×10^{23} atoms/mol = 1.20×10^{24} atoms
b. 6.00 mol $\times$ 6.02×10^{23} atoms/mol = 3.61×10^{24} atoms
c. 0.500 mol $\times$ 6.02×10^{23} atoms/mol = 3.01×10^{23} atoms
d. 25.0 mol $\times$ 6.02×10^{23} atoms/mol = 1.51×10^{25} atoms

5.19 Multiply the number of moles by Avogadro's number to determine the number of molecules, as in Example 5.4.

a. 2.5 mol $\times$ 6.02×10^{23} molecules/mol = 1.5×10^{24} molecules
b. 0.25 mol $\times$ 6.02×10^{23} molecules/mol = 1.5×10^{23} molecules
c. 0.40 mol $\times$ 6.02×10^{23} molecules/mol = 2.4×10^{23} molecules
d. 55.3 mol $\times$ 6.02×10^{23} molecules/mol = 3.33×10^{25} molecules

5.20 Use Avogadro's number as a conversion factor to relate molecules to moles.

a. $$6.02 \times 10^{25}\ \text{molecules} \times \frac{1\ \text{mol}}{6.02 \times 10^{23}\ \text{molecules}} = 100.\ \text{mol}$$

b. 3.01×10^{22} molecules × $\dfrac{1 \text{ mol}}{6.02 \times 10^{23} \text{ molecules}}$ = 0.0500 mol

c. 9.0×10^{24} molecules × $\dfrac{1 \text{ mol}}{6.02 \times 10^{23} \text{ molecules}}$ = 15 mol

5.21 To calculate the formula weight, multiply the number of atoms of each element by the atomic weight and add the results.

a.

1 Ca atom	×	40.08 amu	=	40.08 amu
1 C atom	×	12.01 amu	=	12.01 amu
3 O atoms	×	16.00 amu	=	48.00 amu
Formula weight of $CaCO_3$				100.09 amu

b.

1 K atom	×	39.10 amu	=	39.10 amu
1 I atom	×	126.9 amu	=	126.9 amu
Formula weight of KI				166.00 amu rounded to 166.0 amu

5.22 Translate the ball-and-stick model to a chemical structure and answer the questions.

```
  H H H H H H H H H
  | | | | | | | | |
H-C-C-C=C-C-C-C=C-C=O
  | |       | |
  H H       H H
```

a. chemical formula $C_9H_{14}O$

b. Molecular weight

9 C atoms	×	12.01 amu	=	108.1 amu
14 H atoms	×	1.008 amu	=	14.11 amu
1 O atom	×	16.00 amu	=	16.00 amu
				138.21 amu rounded to 138.2 amu

5.23

$C_{20}H_{24}O_{10}$

20 C atoms	×	12.01 amu	=	240.2 amu
24 H atoms	×	1.008 amu	=	24.19 amu
10 O atoms	×	16.00 amu	=	160.0 amu
Molecular weight of ginkgolide B:				424.39 amu = 424.4 g/mol

5.24 Convert the moles to grams using the molar mass as a conversion factor.

a. 0.500 mol of NaCl × 58.44 g/mol = 29.2 g
b. 2.00 mol of KI × 166.0 g/mol = 332 g
c. 3.60 mol of C_2H_4 × 28.05 g/mol = 101 g
d. 0.820 mol of CH_4O × 32.04 g/mol = 26.3 g

5.25 Convert grams to moles using the molar mass as a conversion factor.

a. 100. g NaCl × $\dfrac{1 \text{ mol}}{58.44 \text{ g}}$ = 1.71 mol

b. $25.5\ \cancel{g}\ CH_4 \times \frac{1\ mol}{16.04\ \cancel{g}} = 1.59\ mol$

c. $0.250\ \cancel{g}\ C_9H_8O_4 \times \frac{1\ mol}{180.2\ \cancel{g}} = 1.39 \times 10^{-3}\ mol$

d. $25.0\ \cancel{g}\ H_2O \times \frac{1\ mol}{18.02\ \cancel{g}} = 1.39\ mol$

5.26 Use conversion factors to determine the number of molecules in 1.00 g; two 500.-mg tablets = 1.00 g.

Answer:

$1.00\ \cancel{g}\ \text{penicillin} \times \frac{6.02 \times 10^{23}\ \text{molecules}}{334.4\ \cancel{g}\ \text{penicillin}} = 1.80 \times 10^{21}$ molecules of penicillin

Grams cancel.

5.27 Use mole–mole conversion factors as in Example 5.6 and the equation below to solve the problems.

$$N_2(g) + O_2(g) \xrightarrow{\Delta} 2\ NO(g)$$

a. $(3.3\ mol\ N_2) \times (2\ mol\ NO/1\ mol\ N_2) = 6.6\ mol\ NO$
b. $(0.50\ mol\ O_2) \times (2\ mol\ NO/1\ mol\ O_2) = 1.0\ mol\ NO$
c. $(1.2\ mol\ N_2) \times (1\ mol\ O_2/1\ mol\ N_2) = 1.2\ mol\ O_2$

5.28 Use mole–mole conversion factors as in Example 5.6 and the equation below to solve the problems.

$$2\ C_2H_6(g) + 5\ O_2(g) \xrightarrow{\Delta} 4\ CO(g) + 6\ H_2O(g)$$

a. $(3.0\ mol\ C_2H_6) \times (5\ mol\ O_2/2\ mol\ C_2H_6) = 7.5\ mol\ O_2$
b. $(0.50\ mol\ C_2H_6) \times (6\ mol\ H_2O/2\ mol\ C_2H_6) = 1.5\ mol\ H_2O$
c. $(3.0\ mol\ CO) \times (2\ mol\ C_2H_6/4\ mol\ CO) = 1.5\ mol\ C_2H_6$

5.29 [1] Convert the number of moles of reactant to the number of moles of product using a mole–mole conversion factor.
[2] Convert the number of moles of product to the number of grams of product using the product's molar mass.

$$C_6H_{12}O_6(aq) \longrightarrow 2\ C_2H_6O(aq) + 2\ CO_2(g)$$

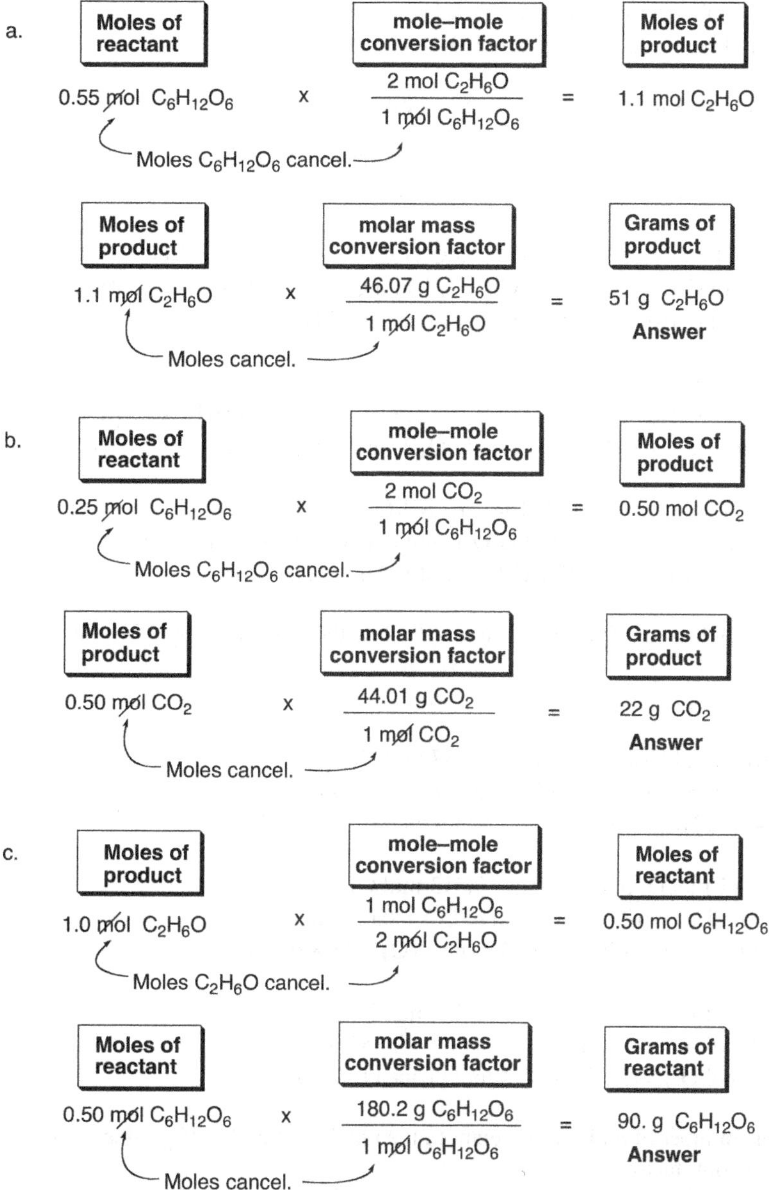

5.30 Use the steps outlined in Answer 5.29 to answer the questions.

$$C_2H_6O(l) \; + \; 3\,O_2(g) \longrightarrow 2\,CO_2(g) \; + \; 3\,H_2O(g)$$

a. $(0.50 \text{ mol } C_2H_6O) \times (2 \text{ mol } CO_2/1 \text{ mol } C_2H_6O) = 1.0 \text{ mol } CO_2$
$(1.0 \text{ mol } CO_2) \times (44.01 \text{ g } CO_2/1 \text{ mol } CO_2) = 44 \text{ g } CO_2$

b. $(2.4 \text{ mol } C_2H_6O) \times (3 \text{ mol } H_2O/1 \text{ mol } C_2H_6O) = 7.2 \text{ mol } H_2O$
$(7.2 \text{ mol } H_2O) \times (18.02 \text{ g } H_2O/1 \text{ mol } H_2O) = 130 \text{ g } H_2O$

c. (0.25 mol C_2H_6O) × (3 mol O_2/1 mol C_2H_6O) = 0.75 mol O_2
(0.75 mol O_2) × (32.00 g O_2/1 mol O_2) = 24 g O_2

5.31 Use conversion factors to solve the problems. Follow the steps in Sample Problem 5.17.

$C_7H_6O_3(s)$ (salicylic acid) + $C_2H_4O_2(l)$ (acetic acid) $\longrightarrow$ $C_9H_8O_4(s)$ (aspirin) + $H_2O(l)$

a. (55.5 g $C_7H_6O_3$) × (1 mol $C_7H_6O_3$/138.1 g $C_7H_6O_3$) = 0.402 mol $C_7H_6O_3$
(0.402 mol $C_7H_6O_3$) × (1 mol $C_9H_8O_4$/1 mol $C_7H_6O_3$) = 0.402 mol $C_9H_8O_4$
(0.402 mol $C_9H_8O_4$) × (180.2 g $C_9H_8O_4$/1 mol $C_9H_8O_4$) = 72.4 g $C_9H_8O_4$

b. (55.5 g $C_7H_6O_3$) × (1 mol $C_7H_6O_3$/138.1 g $C_7H_6O_3$) = 0.402 mol $C_7H_6O_3$
(0.402 mol $C_7H_6O_3$) × (1 mol $C_2H_4O_2$/1 mol $C_7H_6O_3$) = 0.402 mol $C_2H_4O_2$
(0.402 mol $C_2H_4O_2$) × (60.05 g $C_2H_4O_2$/1 mol $C_2H_4O_2$) = 24.1 g $C_2H_4O_2$

c. (55.5 g $C_7H_6O_3$) × (1 mol $C_7H_6O_3$/138.1 g $C_7H_6O_3$) = 0.402 mol $C_7H_6O_3$
(0.402 mol $C_7H_6O_3$) × (1 mol H_2O/1 mol $C_7H_6O_3$) = 0.402 mol H_2O
(0.402 mol H_2O) × (18.02 g H_2O/1 mol H_2O) = 7.24 g H_2O

5.32 Use conversion factors to solve the problems. Follow the steps in Sample Problem 5.17.

$N_2 + O_2 \rightarrow 2\ NO$

a. (10.0 g N_2) × (1 mol N_2/28.02 g N_2) = 0.357 mol N_2
(0.357 mol N_2) × (2 mol NO/1 mol N_2) = 0.714 mol NO
(0.714 mol NO) × (30.01 g NO/1 mol NO) = 21.4 g NO

b. (10.0 g O_2) × (1 mol O_2/32.00 g O_2) = 0.313 mol O_2
(0.313 mol O_2) × (2 mol NO/1 mol O_2) = 0.626 mol NO
(0.626 mol NO) × (30.01 g NO/1 mol NO) = 18.8 g NO

c. (10.0 g N_2) × (1 mol N_2/28.02 g N_2) = 0.357 mol N_2
(0.357 mol N_2) × (1 mol O_2/1 mol N_2) = 0.357 mol O_2
(0.357 mol O_2) × (32.00 g O_2/1 mol O_2) = 11.4 g O_2

5.33 [1] Convert the number of moles of reactant to the number of moles of product using a mole–mole conversion factor.
[2] Convert the number of moles of product to the number of grams of product—the theoretical yield—using the product's molar mass.

$C(s) + O_2(g) \longrightarrow CO_2(g)$

Moles of reactant | **mole–mole conversion factor** | **Moles of product**

$$3.50 \text{ mol C} \times \frac{1 \text{ mol } CO_2}{1 \text{ mol C}} = 3.50 \text{ mol } CO_2$$

Moles of product | **molar mass conversion factor** | **Grams of product**

$$3.50 \text{ mol } CO_2 \times \frac{44.01 \text{ g } CO_2}{1 \text{ mol } CO_2} = 154 \text{ g } CO_2$$

Theoretical yield
Answer part (a)

$$\text{Percent yield} = \frac{\text{actual yield (g)}}{\text{theoretical yield (g)}} \times 100\%$$

$$= \frac{53.5 \text{ g}}{154 \text{ g}} \times 100\% = 34.7\%$$

Answer part (b)

5.34 Use the steps in Sample Problem 5.20 to answer the questions.

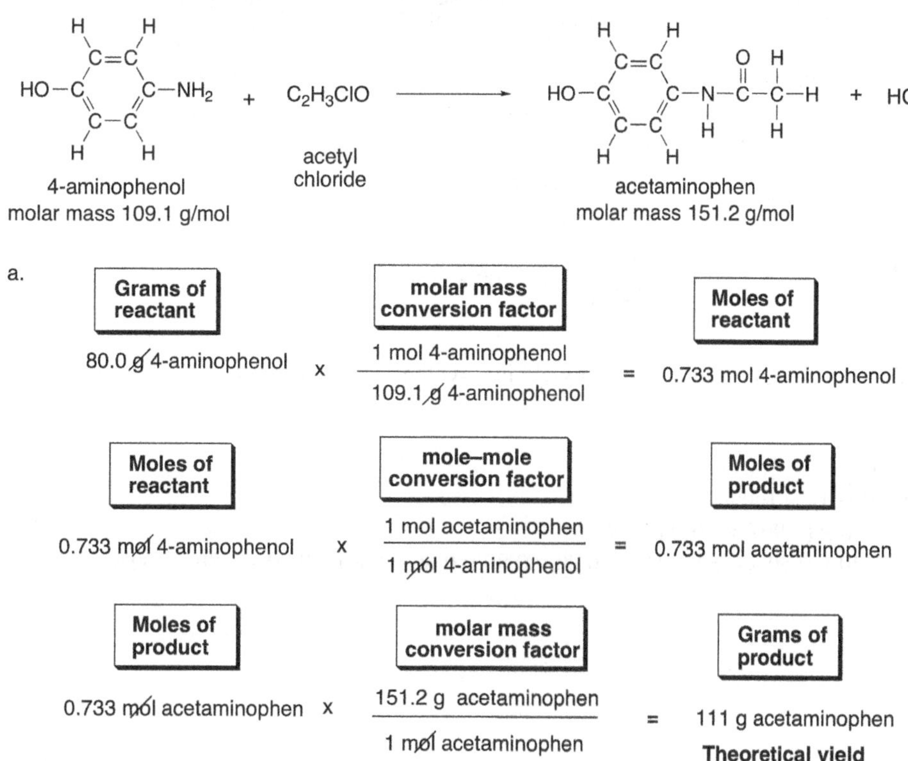

b.

$$\text{Percent yield} = \frac{\text{actual yield (g)}}{\text{theoretical yield (g)}} \times 100\%$$

$$= \frac{65.5 \text{ g}}{111 \text{ g}} \times 100\% = 59.0\%$$

Answer

5.35 Use conversion factors to solve the problem. Follow the steps in Answer 5.34.

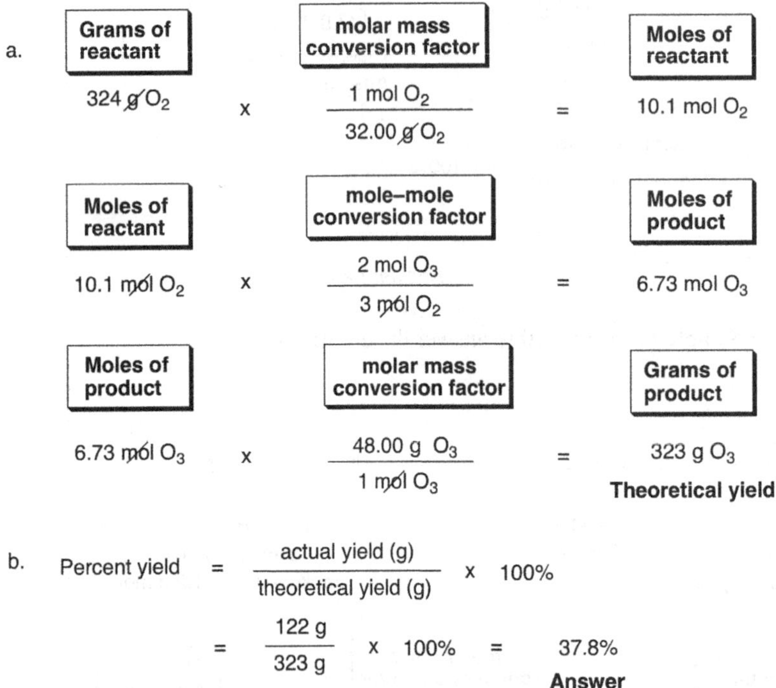

5.36 To determine the overall percent yield in a synthesis that has more than one step, multiply the percent yield for each step.

a. $(0.90)^{10} \times 100\% = 35\%$
b. $(0.80)^{10} \times 100\% = 11\%$
c. $0.50 \times (0.90)^{9} \times 100\% = 19\%$
d. $0.20 \times 0.50 \times 0.50 \times 0.80 \times 0.80 \times 0.80 \times 0.80 \times 0.80 \times 0.80 \times 0.80 \times 100\% = 1.0\%$

5.37 Use the steps in Sample Problem 5.21 to answer the questions.

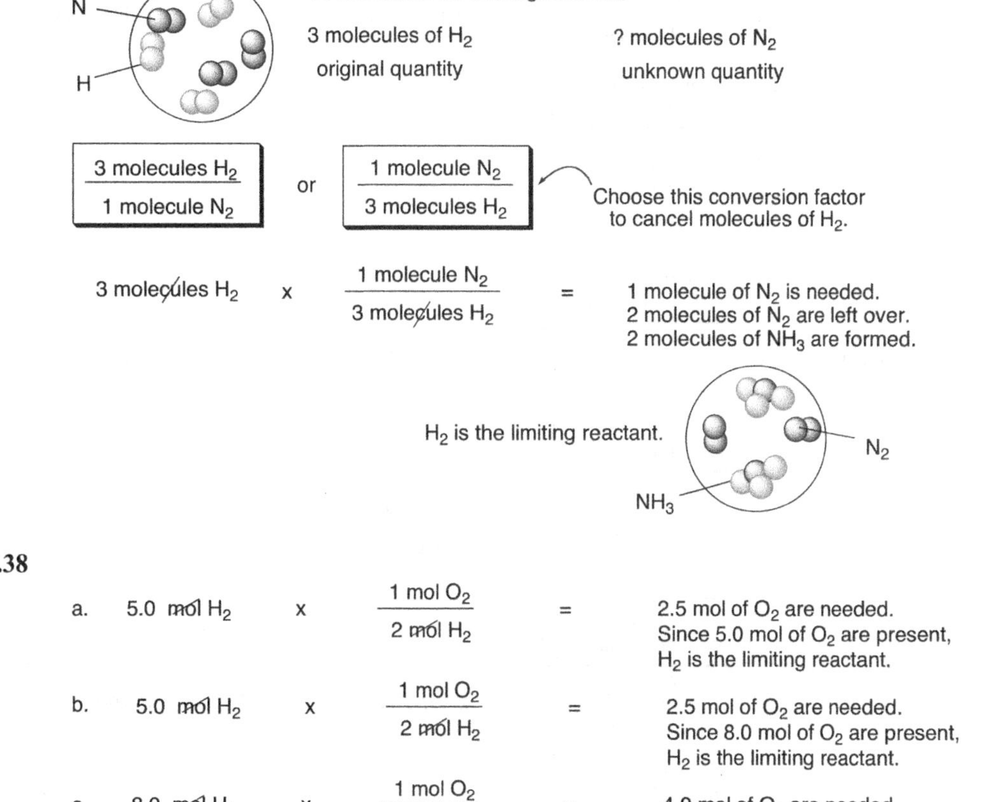

N
H
To determine the limiting reactant:
3 molecules of H2
original quantity
? molecules of N2
unknown quantity
3 molecules H2 / 1 molecule N2
or
1 molecule N2 / 3 molecules H2
Choose this conversion factor
to cancel molecules of H2.
3 molecules H2 x 1 molecule N2 / 3 molecules H2 =
1 molecule of N2 is needed.
2 molecules of N2 are left over.
2 molecules of NH3 are formed.
H2 is the limiting reactant.
N2
NH3

5.38

a. $5.0 \text{ mol } H_2 \times \dfrac{1 \text{ mol } O_2}{2 \text{ mol } H_2} =$ 2.5 mol of O_2 are needed. Since 5.0 mol of O_2 are present, H_2 is the limiting reactant.

b. $5.0 \text{ mol } H_2 \times \dfrac{1 \text{ mol } O_2}{2 \text{ mol } H_2} =$ 2.5 mol of O_2 are needed. Since 8.0 mol of O_2 are present, H_2 is the limiting reactant.

c. $8.0 \text{ mol } H_2 \times \dfrac{1 \text{ mol } O_2}{2 \text{ mol } H_2} =$ 4.0 mol of O_2 are needed. Since only 2.0 mol of O_2 are present, O_2 is the limiting reactant.

d. $2.0 \text{ mol } H_2 \times \dfrac{1 \text{ mol } O_2}{2 \text{ mol } H_2} =$ 1.0 mol of O_2 is needed. Since 5.0 mol of O_2 are present, H_2 is the limiting reactant.

5.39 Calculate the number of moles of product formed as in Sample Problem 5.22.

a. mole–mole conversion factor

$1.5 \text{ mol } H_2 \times \dfrac{1 \text{ mol } N_2}{3 \text{ mol } H_2} = 0.50 \text{ mol } N_2$ needed $\quad H_2$ is the limiting reactant.

mole–mole conversion factor

$1.5 \text{ mol } H_2 \times \dfrac{2 \text{ mol } NH_3}{3 \text{ mol } H_2} = 1.0 \text{ mol } NH_3$

Answer

b. $1.0\ \text{mol H}_2 \times \dfrac{1\ \text{mol N}_2}{3\ \text{mol H}_2}$ (mole–mole conversion factor) $= 0.33\ \text{mol N}_2$ needed H_2 is the limiting reactant.

$1.0\ \text{mol H}_2 \times \dfrac{2\ \text{mol NH}_3}{3\ \text{mol H}_2}$ (mole–mole conversion factor) $= 0.67\ \text{mol NH}_3$
Answer

c. $2.0\ \text{mol H}_2 \times \dfrac{1\ \text{mol N}_2}{3\ \text{mol H}_2}$ (mole–mole conversion factor) $= 0.67\ \text{mol N}_2$ needed H_2 is the limiting reactant.

$2.0\ \text{mol H}_2 \times \dfrac{2\ \text{mol NH}_3}{3\ \text{mol H}_2}$ (mole–mole conversion factor) $= 1.3\ \text{mol NH}_3$
Answer

d. $7.5\ \text{mol H}_2 \times \dfrac{1\ \text{mol N}_2}{3\ \text{mol H}_2}$ (mole–mole conversion factor) $= 2.5\ \text{mol N}_2$ needed N_2 is the limiting reactant.

$2.0\ \text{mol N}_2 \times \dfrac{2\ \text{mol NH}_3}{1\ \text{mol N}_2}$ (mole–mole conversion factor) $= 4.0\ \text{mol NH}_3$
Answer

5.40 Convert the number of grams of each reactant to the number of moles using molar masses. Since the mole ratio of O_2 to N_2 is 1:1, the limiting reactant has fewer moles.

a. **Grams of reactant** × **molar mass conversion factor** = **Moles of reactant**

$12.5\ \text{g N}_2 \times \dfrac{1\ \text{mol N}_2}{28.02\ \text{g N}_2} = 0.446\ \text{mol N}_2$ N_2 is the limiting reactant.

Grams of reactant × **molar mass conversion factor** = **Moles of reactant**

$15.0\ \text{g O}_2 \times \dfrac{1\ \text{mol O}_2}{32.00\ \text{g O}_2} = 0.469\ \text{mol O}_2$

b.

Grams of reactant		molar mass conversion factor		Moles of reactant	
$14.0\ \text{g}\ N_2$	x	$\frac{1\ \text{mol}\ N_2}{28.02\ \text{g}\ N_2}$	=	$0.500\ \text{mol}\ N_2$	
$13.0\ \text{g}\ O_2$	x	$\frac{1\ \text{mol}\ O_2}{32.00\ \text{g}\ O_2}$	=	$0.406\ \text{mol}\ O_2$	O_2 is the limiting reactant.

5.41 Calculate the number of moles of product formed based on the limiting reactant. Then convert moles to grams using molar mass.

a. $0.446\ \text{mol}\ N_2 \times \frac{2\ \text{mol NO}}{1\ \text{mol}\ N_2} = 0.892\ \text{mol NO}$

$0.892\ \text{mol NO} \times \frac{30.01\ \text{g}}{1\ \text{mol NO}} = 26.8\ \text{g NO}$

b. $0.406\ \text{mol}\ O_2 \times \frac{2\ \text{mol NO}}{1\ \text{mol}\ O_2} = 0.812\ \text{mol NO}$

$0.812\ \text{mol NO} \times \frac{30.01\ \text{g}}{1\ \text{mol NO}} = 24.4\ \text{g NO}$

5.42

Grams of reactant		molar mass conversion factor		Moles of reactant	
$5.00\ \text{g}\ H_2$	x	$\frac{1\ \text{mol}\ H_2}{2.016\ \text{g}\ H_2}$	=	$2.48\ \text{mol}\ H_2$	
$10.0\ \text{g}\ O_2$	x	$\frac{1\ \text{mol}\ O_2}{32.00\ \text{g}\ O_2}$	=	$0.313\ \text{mol}\ O_2$	O_2 is the limiting reactant.

$0.313\ \text{mol}\ O_2 \times \frac{2\ \text{mol}\ H_2O}{1\ \text{mol}\ O_2} = 0.626\ \text{mol}\ H_2O$

$0.626\ \text{mol}\ H_2O \times \frac{18.02\ \text{g}}{1\ \text{mol}\ H_2O} = 11.3\ \text{g}\ H_2O$

Solutions to Odd-Numbered End-of-Chapter Problems

5.43 The process is a chemical reaction because the spheres in the reactants are joined differently than the spheres in the products.

$$2\,CO + 2\,O_3 \longrightarrow 2\,CO_2 + 2\,O_2 \quad \text{(not balanced)}$$

5.45 Add up the number of atoms on each side of the equation and then label the equations as balanced or not balanced.

a. $2\,HCl(aq) + Ca(s) \longrightarrow CaCl_2(aq) + H_2(g)$

2 H, 2 Cl, 1 Ca: both sides, therefore **balanced**

b. $TiCl_4 + 2\,H_2O \longrightarrow TiO_2 + HCl$

1 Ti, 4 Cl, 4 H, 2 O — 1 Ti, 1 Cl, 1 H, 2 O

NOT balanced

c. $Al(OH)_3 + H_3PO_4 \longrightarrow AlPO_4 + 3\,H_2O$

1 Al, 1 P, 7 O, 6 H: both sides, therefore **balanced**

5.47 Write the balanced equation using the colors of the spheres to identify the atoms (gray = hydrogen and green = chlorine).

$$H_2 + Cl_2 \longrightarrow 2\,HCl$$

5.49 Balance the equation with coefficients one element at a time to have the same number of atoms on each side of the equation. Follow the steps in Example 5.2.

a. $Ni(s) + 2\,HCl(aq) \longrightarrow NiCl_2(aq) + H_2(g)$

b. $CH_4(g) + 4\,Cl_2(g) \longrightarrow CCl_4(g) + 4\,HCl(g)$

c. $2\,KClO_3 \longrightarrow 2\,KCl + 3\,O_2$

d. $Al_2O_3 + 6\,HCl \longrightarrow 2\,AlCl_3 + 3\,H_2O$

e. $4\,Al(OH)_3 + 6\,H_2SO_4 \longrightarrow 2\,Al_2(SO_4)_3 + 12\,H_2O$

5.51 Follow the steps in Example 5.2 and balance the equations.

a. $2\,C_6H_6 + 15\,O_2 \longrightarrow 12\,CO_2 + 6\,H_2O$

b. $C_7H_8 + 9\,O_2 \longrightarrow 7\,CO_2 + 4\,H_2O$

c. $2\,C_8H_{18} + 25\,O_2 \longrightarrow 16\,CO_2 + 18\,H_2O$

5.53 Use the balanced equation and the law of conservation of matter to determine the molecules of product. Each side must have the same number of O and C atoms.

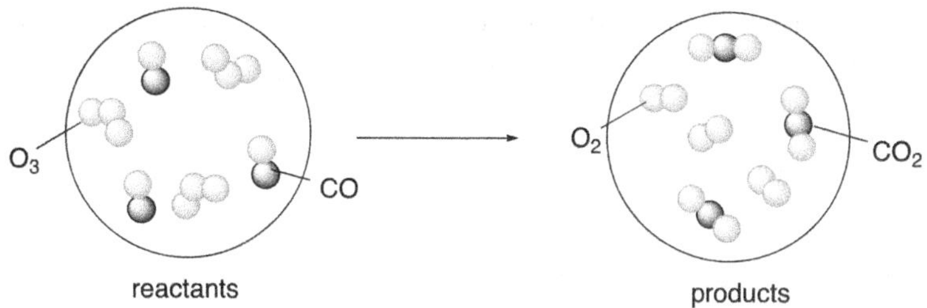

5.55 a. Since black spheres are C atoms and red spheres are O atoms, the reactants are CO and O_2 and the product is CO_2.
b.
$$2\,CO + O_2 \longrightarrow 2\,CO_2$$
c. The reaction is a combination because two reactants combine to form one product.

5.57 a. The reactant is C_2H_6O (black = C, gray = H, red = O) and the products are $H_2C{=}CH_2$ and H_2O.
b.
$$C_2H_6O \longrightarrow H_2C{=}CH_2 + H_2O$$
c. The reaction is a decomposition because one reactant decomposes to form two products.

5.59 The reaction is a double replacement because the two compounds exchange "parts" (gray spheres replace green spheres) to form two new compounds.

5.61 Use the definitions from Answers 5.10 and 5.11 to classify the reactions.
a. decomposition: BrCl decomposes to Br_2 and Cl_2.
b. single replacement: The element F_2 replaces Br in NaBr.
c. combination: Fe and O_2 combine to form Fe_2O_3.
d. double replacement: H and K^+ exchange.

5.63 Use the definitions in Answers 5.10 and 5.11 to fill in the needed reactant or product.

a. $2\,Na + \boxed{Cl_2} \longrightarrow 2\,NaCl$
b. $2\,NI_3 \longrightarrow \boxed{N_2} + \boxed{3\,I_2}$
c. $Cl_2 + 2\,KI \longrightarrow \boxed{2\,KCl} + \boxed{I_2}$
d. $KOH + HI \longrightarrow \boxed{KI} + \boxed{H_2O}$

5.65 The species that is oxidized loses one or more electrons. The species that is reduced gains one or more electrons.

a.	Fe oxidized	Cu^{2+} reduced	$Fe \longrightarrow Fe^{2+} + 2\,e^-$	$Cu^{2+} + 2\,e^- \longrightarrow Cu$
b.	Cl_2 reduced	$2\,I^-$ oxidized	$2\,I^- \longrightarrow I_2 + 2\,e^-$	$Cl_2 + 2\,e^- \longrightarrow 2\,Cl^-$
c.	2 Na oxidized	Cl_2 reduced	$2\,Na \longrightarrow 2\,Na^+ + 2\,e^-$	$Cl_2 + 2\,e^- \longrightarrow 2\,Cl^-$

5.67 The oxidizing agent gains electrons (it is reduced). The reducing agent loses electrons (it is oxidized).

$$Zn + Ag_2O \longrightarrow ZnO + 2\,Ag$$

Zn	Ag^+
oxidized	reduced
reducing agent	oxidizing agent

5.69 Acetylene is reduced because it gains hydrogen atoms.

5.71 To calculate the formula weight, multiply the number of atoms of each element by the atomic weight and add the results. The formula weight in amu is equal to the molar mass in g/mol.

a.

1 Na atom	×	22.99 amu	=	22.99 amu
1 N atom	×	14.01 amu	=	14.01 amu
2 O atoms	×	16.00 amu	=	32.00 amu
Formula weight of $NaNO_2$				69.00 amu = 69.00 g/mol

b.

2 Al atom	×	26.98 amu	=	53.96 amu
3 S atoms	×	32.07 amu	=	96.21 amu
12 O atoms	×	16.00 amu	=	192.0 amu
Formula weight of $Al_2(SO_4)_3$				342.17 amu rounded to 342.2 amu = 342.2 g/mol

c.

6 C atom	×	12.01 amu	=	72.06 amu
8 H atoms	×	1.008 amu	=	8.064 amu
6 O atoms	×	16.00 amu	=	96.00 amu
Formula weight of $C_6H_8O_6$				176.124 amu rounded to 176.12 amu = 176.12 g/mol

5.73 Determine the molecular formula of L-dopa. Then calculate the formula weight and molar mass as in Answer 5.71.

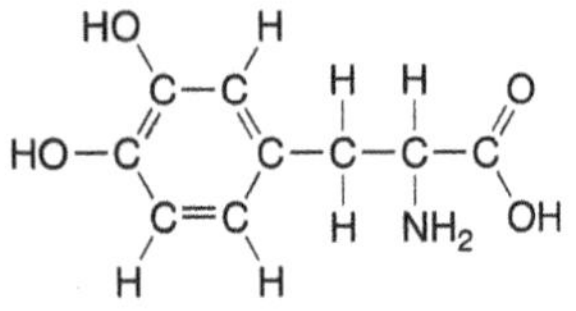

```
    HO      H
      \    /
       C—C      H   H    O
      //   \\   |   |   //
 HO—C       C—C—C—C
      \\   /    |   |   \
       C=C      H  NH2   OH
      /    \
     H      H
        L-dopa
```

a. molecular formula = $C_9H_{11}NO_4$
b. formula weight = 197.2 amu
c. molar mass = 197.2 g/mol

5.75 Convert all of the units to moles, and then compare the atomic weight or formula weight to determine the quantity with the larger mass.

a. 1 mol of Fe atoms (55.85 g) < 1 mol of Sn atoms (118.7 g)
b. 1 mol of C atoms (12.01 g) < 6.02×10^{23} N atoms = 1 mol N atoms (14.01 g)
c. 1 mol of N atoms (14.01 g) < 1 mol of N_2 molecules = 2 mol N atoms (28.02 g)
d. 1 mol of CO_2 molecules (44.01 g) > 3.01×10^{23} N_2O molecules = 0.500 mol N_2O (22.01 g)

5.77

```
    CH3O      OCH3
        \    /
         C-C
       //    \\
    H-C        C-OCH3
    H   \     /
    |    C=C
H-N-C-C/     \
  |  |  |\H    H
  H  H  H
```

a. $C_{11}H_{17}NO_3$
b. molar mass = 211.3 g/mol
c. 7.50 g × (1 mol/211.3 g) = 0.0355 mol

5.79 Calculate the molar mass of each compound as in Answer 5.71, and then multiply by 5.00 mol.

a. HCl = 182 g
b. Na_2SO_4 = 710. g
c. C_2H_2 = 130. g
d. $Al(OH)_3$ = 390. g

5.81 Convert the grams to moles using the molar mass as a conversion factor.

a. $0.500\text{ g} \times \dfrac{1\text{ mol}}{342.3\text{ g}} = 1.46 \times 10^{-3}\text{ mol}$

b. $5.00\text{ g} \times \dfrac{1\text{ mol}}{342.3\text{ g}} = 0.0146\text{ mol}$

c. $25.0\text{ g} \times \dfrac{1\text{ mol}}{342.3\text{ g}} = 0.0730\text{ mol}$

d. $0.0250\text{ g} \times \dfrac{1\text{ mol}}{342.3\text{ g}} = 7.30 \times 10^{-5}\text{ mol}$

5.83 Multiply the number of moles by Avogadro's number to determine the number of molecules, as in Example 5.4.

a. (2.00 mol) × (6.02×10^{23} molecules/mol) = 1.20×10^{24} molecules
b. (0.250 mol) × (6.02×10^{23} molecules/mol) = 1.51×10^{23} molecules
c. (26.5 mol) × (6.02×10^{23} molecules/mol) = 1.60×10^{25} molecules
d. (222 mol) × (6.02×10^{23} molecules/mol) = 1.34×10^{26} molecules
e. (5.00×10^{5} mol) × (6.02×10^{23} molecules/mol) = 3.01×10^{29} molecules

5.85 Use the molar mass as a conversion factor to convert the moles to grams. Use Avogadro's number to convert the number of molecules to moles.

a. $3.60\text{ mol} \times \dfrac{90.08\text{ g}}{1\text{ mol}} = 324\text{ g}$

b. $0.580\text{ mol} \times \dfrac{90.08\text{ g}}{1\text{ mol}} = 52.2\text{ g}$

c. $7.3 \times 10^{24}\text{ molecules} \times \dfrac{1\text{ mol}}{6.02 \times 10^{23}\text{ molecules}} \times \dfrac{90.08\text{ g}}{1\text{ mol}} = 1.1 \times 10^{3}\text{ g}$

d. $6.56 \times 10^{22}\text{ molecules} \times \dfrac{1\text{ mol}}{6.02 \times 10^{23}\text{ molecules}} \times \dfrac{90.08\text{ g}}{1\text{ mol}} = 9.82\text{ g}$

5.87

$$2\ \text{H–C≡C–H} + 5\ O_2 \longrightarrow 4\ CO_2 + 2\ H_2O$$

acetylene

a. 12.5 moles of O_2 are needed to react completely with 5.00 mol of C_2H_2.
(5.00 mol C_2H_2) × (5 mol O_2/2 mol C_2H_2) = 12.5 mol O_2

b. 12 moles of CO_2 are formed from 6.0 mol of C_2H_2.
(6.0 mol C_2H_2) × (4 mol CO_2/2 mol C_2H_2) = 12 mol CO_2

c. 0.50 moles of H_2O are formed from 0.50 mol of C_2H_2.
(0.50 mol C_2H_2) × (2 mol H_2O/2 mol C_2H_2) = 0.50 mol H_2O

d. 0.40 moles of C_2H_2 are needed to form 0.80 mol of CO_2.
(0.80 mol CO_2) × (2 mol C_2H_2/4 mol CO_2) = 0.40 mol C_2H_2

5.89 Use conversion factors as in Example 5.7 to solve the problems.

a. 220 g of CO_2 are formed from 2.5 mol of C_2H_2.
b. 44 g of CO_2 are formed from 0.50 mol of C_2H_2.
c. 4.5 g of H_2O are formed from 0.25 mol of C_2H_2.
d. 240 g of O_2 are needed to react with 3.0 mol of C_2H_2.

5.91 Use the equation to determine the percent yield.

$$\text{Percent yield} = \frac{\text{actual yield (g)}}{\text{theoretical yield (g)}} \times 100\%$$

$$= \frac{9.0\ \text{g}}{12.0\ \text{g}} \times 100\% = 75\%$$

5.93 Use the following equations to determine the percent yield.

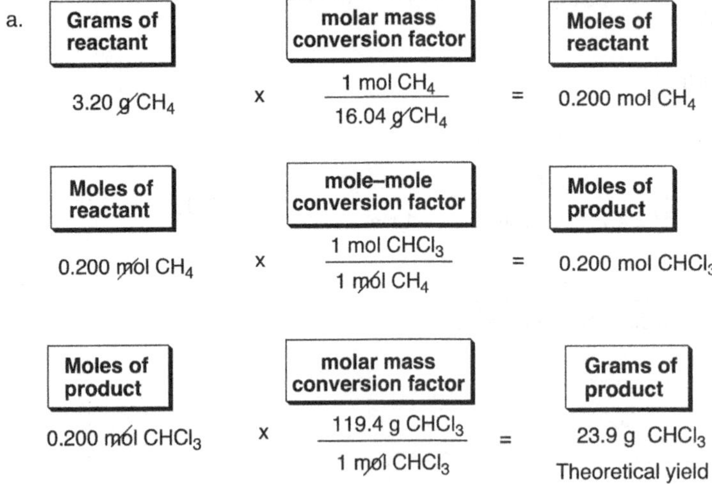

b. Percent yield $= \dfrac{\text{actual yield (g)}}{\text{theoretical yield (g)}} \times 100\%$

$= \dfrac{15.0 \text{ g } CHCl_3}{23.9 \text{ g } CHCl_3} \times 100\% = 62.8\%$

5.95

a. 4 molecules **A** $\times \dfrac{1 \text{ molecule } \mathbf{B}}{1 \text{ molecule } \mathbf{A}}$ = 4 molecules of **B** are needed.
Molecule **A** is in excess.
Molecule **B** is the limiting reactant.

b. 4 molecules **A** $\times \dfrac{1 \text{ molecule } \mathbf{B}}{2 \text{ molecules } \mathbf{A}}$ = 2 molecules of **B** are needed.
Molecule **B** is in excess.
Molecule **A** is the limiting reactant.

c. 4 molecules **A** $\times \dfrac{2 \text{ molecules } \mathbf{B}}{1 \text{ molecule } \mathbf{A}}$ = 8 molecules of **B** are needed.
Molecule **A** is in excess.
Molecule **B** is the limiting reactant.

5.97

a. 1.0 mol NO $\times \dfrac{1 \text{ mol } O_2}{2 \text{ mol NO}}$ = 0.50 mol of O_2 is needed.
O_2 is in excess.
NO is the limiting reactant.

b. 2.0 mol NO $\times \dfrac{1 \text{ mol } O_2}{2 \text{ mol NO}}$ = 1.0 mol of O_2 is needed.
NO is in excess.
O_2 is the limiting reactant.

c. 10.0 g NO $\times \dfrac{1 \text{ mol NO}}{30.01 \text{ g NO}}$ = 0.333 mol NO

10.0 g O_2 $\times \dfrac{1 \text{ mol } O_2}{32.00 \text{ g } O_2}$ = 0.313 mol O_2

0.333 mol NO $\times \dfrac{1 \text{ mol } O_2}{2 \text{ mol NO}}$ = 0.167 mol of O_2 is needed.
O_2 is in excess.
NO is the limiting reactant.

d. 28.0 g NO $\times \dfrac{1 \text{ mol NO}}{30.01 \text{ g NO}}$ = 0.933 mol NO

16.0 g O_2 $\times \dfrac{1 \text{ mol } O_2}{32.00 \text{ g } O_2}$ = 0.500 mol O_2

0.933 mol NO $\times \dfrac{1 \text{ mol } O_2}{2 \text{ mol NO}}$ = 0.467 mol of O_2 is needed.
O_2 is in excess.
NO is the limiting reactant.

5.99 Use the limiting reactant from Problem 5.97 to determine the amount of product formed. The conversion of moles of limiting reagent to grams of product is combined in a single step.

a. $1.0 \text{ mol NO} \times \dfrac{2 \text{ mol NO}_2}{2 \text{ mol NO}} \times \dfrac{46.01 \text{ g NO}_2}{1 \text{ mol NO}_2} = 46 \text{ g NO}_2$

b. $0.50 \text{ mol O}_2 \times \dfrac{2 \text{ mol NO}_2}{1 \text{ mol O}_2} \times \dfrac{46.01 \text{ g NO}_2}{1 \text{ mol NO}_2} = 46 \text{ g NO}_2$

c. $0.333 \text{ mol NO} \times \dfrac{2 \text{ mol NO}_2}{2 \text{ mol NO}} \times \dfrac{46.01 \text{ g NO}_2}{1 \text{ mol NO}_2} = 15.3 \text{ g NO}_2$

d. $0.933 \text{ mol NO} \times \dfrac{2 \text{ mol NO}_2}{2 \text{ mol NO}} \times \dfrac{46.01 \text{ g NO}_2}{1 \text{ mol NO}_2} = 42.9 \text{ g NO}_2$

5.101

a. $8.00 \text{ g C}_2\text{H}_4 \times \dfrac{1 \text{ mol C}_2\text{H}_4}{28.05 \text{ g C}_2\text{H}_4} = 0.285 \text{ mol C}_2\text{H}_4$

$12.0 \text{ g HCl} \times \dfrac{1 \text{ mol HCl}}{36.46 \text{ g HCl}} = 0.329 \text{ mol HCl}$

b. Since the mole ratio in the balanced equation is 1:1, the reactant with the smaller number of moles is the limiting reactant: C_2H_4.

c. $0.285 \text{ mol C}_2\text{H}_4 \times \dfrac{1 \text{ mol C}_2\text{H}_5\text{Cl}}{1 \text{ mol C}_2\text{H}_4} = 0.285 \text{ mol C}_2\text{H}_5\text{Cl}$

d. $0.285 \text{ mol C}_2\text{H}_5\text{Cl} \times \dfrac{64.51 \text{ g C}_2\text{H}_5\text{Cl}}{1 \text{ mol C}_2\text{H}_5\text{Cl}} = 18.4 \text{ g C}_2\text{H}_5\text{Cl}$

e. $\dfrac{10.6 \text{ g C}_2\text{H}_5\text{Cl}}{18.4 \text{ g C}_2\text{H}_5\text{Cl}} \times 100\% = 57.6\ \%$ percent yield

5.103 Refer to prior solutions to answer each part.

$$\underset{\text{sucrose}}{C_{12}H_{22}O_{11}(s)} + H_2O(l) \longrightarrow \underset{\text{ethanol}}{C_2H_6O(l)} + CO_2(g)$$

a. Calculate the molar mass as in Answer 5.71; the molar mass of sucrose = 342.3 g/mol.
b. Follow the steps in Example 5.2.

$$C_{12}H_{22}O_{11}(s) + H_2O(l) \longrightarrow 4\ C_2H_6O(l) + 4\ CO_2(g)$$

c. 8 mol of ethanol are formed from 2 mol of sucrose.
d. 10 mol of water are needed to react with 10 mol of sucrose.
e. 101 g of ethanol are formed from 0.550 mol of sucrose.
f. 18.4 g of ethanol are formed from 34.2 g of sucrose.
g. 9.21 g ethanol
h. 13.6%

5.105

a. $500 \text{ tablets} \times \frac{200.\text{ mg ibuprofen}}{1 \text{ tablet}} \times \frac{1 \text{ g}}{1000 \text{ mg}} \times \frac{1 \text{ mol ibuprofen}}{206.3 \text{ g ibuprofen}} = 0.485 \text{ mol ibuprofen}$

b. $0.485 \text{ mol ibuprofen} \times \frac{6.02 \times 10^{23} \text{ molecules}}{1 \text{ mol}} = 2.92 \times 10^{23} \text{ molecules}$

5.107

a. $20 \text{ cig} \times \frac{1.93 \text{ mg}}{1 \text{ cig}} \times \frac{1 \text{ g}}{1000 \text{ mg}} \times \frac{1 \text{ mol nicotine}}{162.3 \text{ g nicotine}} = 2.38 \times 10^{-4} \text{ mol nicotine}$

b. $2.38 \times 10^{-4} \text{ mol nicotine} \times \frac{6.02 \times 10^{23} \text{ molecules}}{1 \text{ mol}} = 1.43 \times 10^{20} \text{ molecules}$

5.109

$2400 \text{ mg} \times \frac{1 \text{ g}}{1000 \text{ mg}} \times \frac{1 \text{ mol}}{22.99 \text{ g}} \times \frac{6.02 \times 10^{23} \text{ ions}}{1 \text{ mol}} = 6.3 \times 10^{22} \text{ ions}$

5.111

$2\ C_6H_5Cl + C_2HCl_3O \longrightarrow C_{14}H_9Cl_5 + H_2O$

chlorobenzene (112.6 g/mol) — DDT

a. Calculate the molar mass as in Answer 5.71; the molar mass of DDT = 354.5 g/mol.
b. 18 g of DDT would be formed from 0.10 mol of chlorobenzene.
c. 17.8 g is the theoretical yield of DDT in grams from 11.3 g of chlorobenzene.
d. 84.3%

5.113 Use conversion factors to answer the questions about dioxin.

a. $70. \text{ kg} \times \frac{3.0 \times 10^{-2} \text{ mg}}{1 \text{ kg}} \times \frac{1 \text{ g}}{1000 \text{ mg}} = 2.1 \times 10^{-3} \text{ g dioxin}$

b. $2.1 \times 10^{-3} \text{ g dioxin} \times \frac{1 \text{ mol}}{322.0 \text{ g}} \times \frac{6.02 \times 10^{23} \text{ molecules}}{1 \text{ mol}} = 3.9 \times 10^{18} \text{ molecules}$

Chapter 6 Energy Changes, Reaction Rates, and Equilibrium

Chapter Review

[1] What is energy and what units are used to measure energy? (6.1)

- Energy is the capacity to do work. **Kinetic energy** is the energy of motion, whereas **potential energy** is stored energy.
- Energy is measured in calories (cal) or joules (J), where 1 cal = 4.184 J.
- One nutritional Calorie (Cal) = 1 kcal = 1,000 cal.

1 kcal = 1,000 cal
1 kJ = 1,000 J
1 kcal = 4.184 kJ

[2] Define bond dissociation energy and describe its relationship to bond strength. (6.2)

- The **bond dissociation energy** is the energy needed to break a covalent bond by equally dividing the electrons between the two atoms in the bond.
- The higher the bond dissociation energy, the stronger the bond.

[3] What is the heat of reaction and what is the difference between an endothermic and an exothermic reaction? (6.2)

- The **heat of reaction,** also called the **enthalpy** and symbolized by ΔH, is the energy absorbed or released in a reaction.
- In an endothermic reaction, energy is absorbed, ΔH is positive (+), and the products are higher in energy than the reactants. The bonds in the reactants are stronger than the bonds in the products.
- In an exothermic reaction, energy is released, ΔH is negative (–), and the reactants are higher in energy than the products. The bonds in the products are stronger than the bonds in the reactants.

Bond breaking is **endothermic.** Energy must be added.

$$\text{H–H} \longrightarrow \text{H}\cdot + \cdot\text{H} \qquad \Delta H = +104 \text{ kcal/mol}$$

$$\text{H}\cdot + \cdot\text{H} \longrightarrow \text{H–H} \qquad \Delta H = -104 \text{ kcal/mol}$$

Bond making is **exothermic.** Energy is released.

[4] What are the important features of an energy diagram? (6.3)

- An energy diagram illustrates the energy changes that occur during the course of a reaction. Energy is plotted on the vertical axis and reaction coordinate is plotted on the horizontal axis. The transition state is located at the top of the energy barrier that separates the reactants and products.
- The **energy of activation** is the energy difference between the reactants and the transition state. The higher the energy of activation, the slower the reaction.
- The difference in energy between the reactants and products is the ΔH.

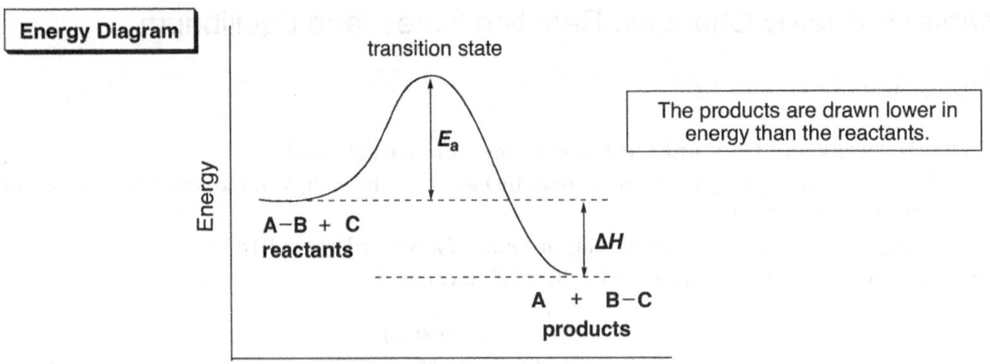

[5] How do temperature, concentration, and catalysts affect the rate of a reaction? (6.4)

- Increasing the temperature and concentration increases the reaction rate.
- A catalyst speeds up the rate of a reaction without affecting the energies of the reactants and products. Enzymes are biological catalysts that increase the rate of reactions in living organisms. Catalytic converters use a catalyst to convert automobile engine exhaust to environmentally cleaner products.

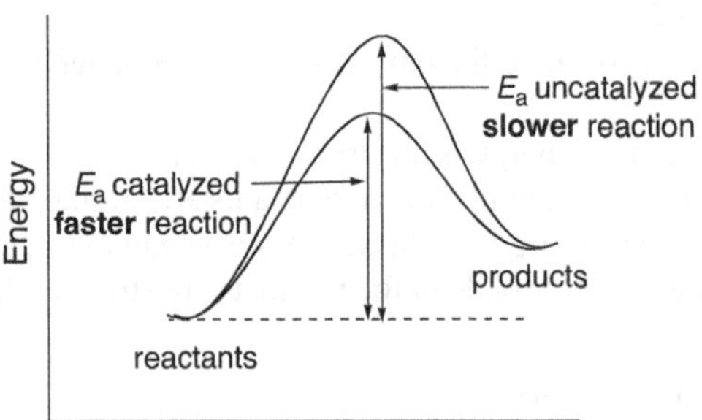

[6] What are the basic features of equilibrium? (6.5)

- At equilibrium, the rates of the forward and reverse reactions in a reversible reaction are equal and the net concentrations of all substances do not change.
- The equilibrium constant for a reaction $a\text{A} + b\text{B} \rightleftharpoons c\text{C} + d\text{D}$ is written as:

concentration of each product (mol/L)

$$\textbf{Equilibrium constant} = K = \frac{[\text{products}]}{[\text{reactants}]}$$

concentration of each reactant (mol/L)

$$K = \frac{[\text{C}]^c[\text{D}]^d}{[\text{A}]^a[\text{B}]^b}$$

- The magnitude of K tells the relative amount of reactants and products. When $K > 1$, the products are favored; when $K < 1$, the reactants are favored; when $K \approx 1$, both reactants and products are present at equilibrium.

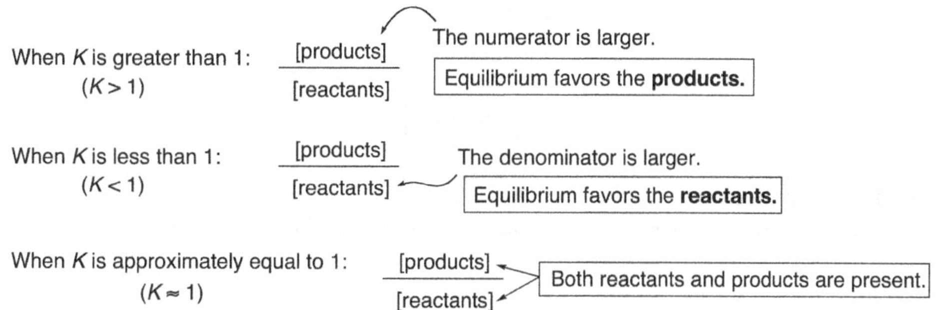

When K is greater than 1: ($K > 1$)
[products] / [reactants]
The numerator is larger.
Equilibrium favors the **products.**
When K is less than 1: ($K < 1$)
[products] / [reactants]
The denominator is larger.
Equilibrium favors the **reactants.**
When K is approximately equal to 1: ($K \approx 1$)
[products] / [reactants]
Both reactants and products are present.

[7] How does Le Châtelier's principle predict what happens when equilibrium is disturbed? (6.6)

- Le Châtelier's principle states that a system at equilibrium reacts in such a way as to counteract any disturbance to the equilibrium. How changes in concentration, temperature, and pressure affect equilibrium are summarized in Table 6.5. One example is given below:

Adding more product...

$$2\,CO(g) \quad + \quad O_2(g) \rightleftharpoons 2\,CO_2(g)$$

...drives the reaction to the left.

- Catalysts increase the rate at which equilibrium is reached, but do not alter the amount of any substance involved in the reaction.

[8] How can the principles that describe equilibrium and reaction rates be used to understand the regulation of body temperature? (6.7)

- Increasing temperature increases the rates of the reactions in the body.
- When temperature is increased, the body dissipates excess heat by dilating blood vessels and sweating. When temperature is decreased, blood vessels constrict to conserve heat, and the body shivers to generate more heat.

Problem Solving

[1] Energy (6.1)

Example 6.1 A reaction releases 573 kJ of energy. How many kilocalories does this correspond to?

Analysis and Solution

[1] Identify the original quantity and the desired quantity.

573 kJ	? kcal
original quantity	desired quantity

[2] Write out the conversion factors.

- Choose the conversion factor that places the unwanted unit, kilojoules, in the denominator so that the units cancel.

kJ–kcal conversion factors

$$\frac{4.184 \text{ kJ}}{1 \text{ kcal}} \quad \text{or} \quad \boxed{\frac{1 \text{ kcal}}{4.184 \text{ kJ}}}$$

Choose this conversion factor to cancel kJ.

[3] Set up and solve the problem.

- Multiply the original quantity by the conversion factor to obtain the desired quantity.

$$573 \cancel{\text{kJ}} \times \frac{1 \text{ kcal}}{4.184 \cancel{\text{kJ}}} = 137 \text{ kcal}$$

Kilojoules cancel.

Answer

Example 6.2 If a serving of crackers contains 2 g of protein, 5 g of fat, and 22 g of carbohydrates, estimate its number of Calories.

Analysis

Use the caloric value (Cal/g) of each class of molecule to form a conversion factor to convert the number of grams to Calories and add up the results.

Solution

[1] Identify the original quantity and the desired quantity.

2 g protein 5 g fat 22 g carbohydrates	? Cal
original quantities	desired quantity

[2] Write out the conversion factors.

- Write out conversion factors that relate the number of grams to the number of Calories for each substance. Each conversion factor must place the unwanted unit, grams, in the denominator so that the units cancel.

Cal–g conversion factor for protein

$$\frac{4 \text{ Cal}}{1 \text{ g protein}}$$

Cal–g conversion factor for carbohydrates

$$\frac{4 \text{ Cal}}{1 \text{ g carbohydrate}}$$

Cal–g conversion factor for fat

$$\frac{9 \text{ Cal}}{1 \text{ g fat}}$$

[3] Set up and solve the problem.

- Multiply the original quantities by the conversion factors for protein, fat, and carbohydrates and add up the results to obtain the desired quantity.

		Calories due to protein		Calories due to fat		Calories due to carbohydrate
Total Calories	=	2 g protein x $\frac{4\ \text{Cal}}{1\ \text{g protein}}$	+	5 g fat x $\frac{9\ \text{Cal}}{1\ \text{g fat}}$	+	22 g carbohydrate x $\frac{4\ \text{Cal}}{1\ \text{g carbohydrate}}$
		Grams cancel.		Grams cancel.		Grams cancel.
	=	8 Cal	+	45 Cal	+	88 Cal
Total Calories	=	141 Cal, rounded to 100 Cal **Answer**				

---

[2] Energy Changes in Reactions (6.2)

Example 6.3 Write the equation for the formation of HBr from H and Br atoms. Classify the reaction as endothermic or exothermic, and give the ΔH using the values in Table 6.2.

Analysis
Bond formation is exothermic, so ΔH is (–). The energy released in forming a bond is (–) the bond dissociation energy.

Solution

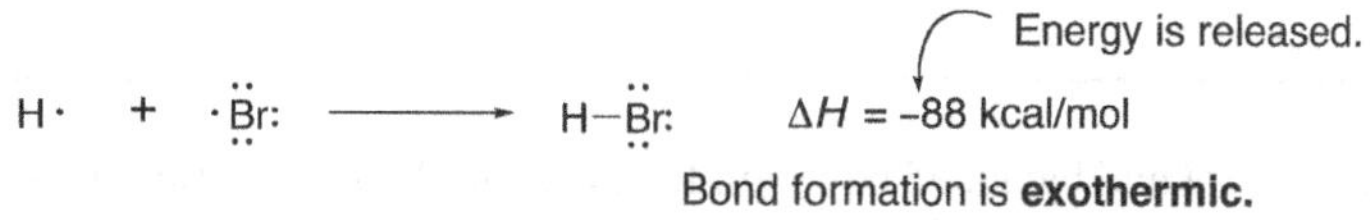

H· + ·Br: ⟶ H–Br: $\Delta H = -88$ kcal/mol

Energy is released.

Bond formation is **exothermic.**

Example 6.4 Which of the indicated carbon–halogen bonds is expected to have the lower bond dissociation energy? Which bond is weaker?

CH_3–Br: CH_3–I:

Analysis
The higher the bond dissociation energy, the stronger the bond. In comparing bonds to atoms in the same group of the periodic table, bond dissociation energies and bond strength decrease down a column.

Solution
Since I is below Br in the same group of the periodic table, the C–I bond is predicted to have the lower bond dissociation energy, thus making it weaker. The actual values for the bond dissociation energies are given and illustrate that the prediction is indeed true.

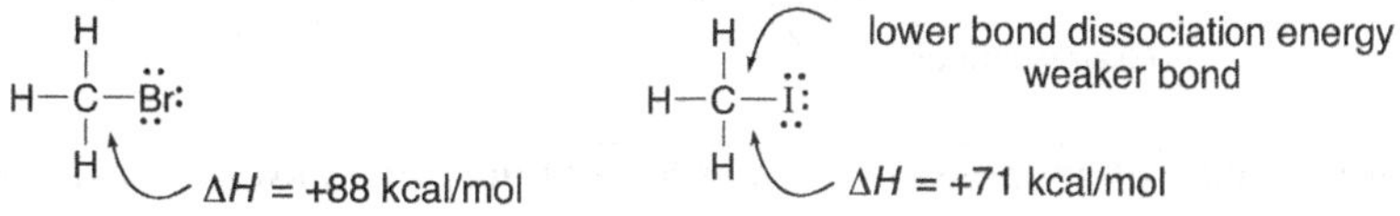

CH_3–Br: $\Delta H = +88$ kcal/mol

CH_3–I: $\Delta H = +71$ kcal/mol — lower bond dissociation energy, weaker bond

---

Example 6.5 Ammonia (NH_3) decomposes to hydrogen and nitrogen and 22.0 kcal/mol of energy is absorbed. How many kilocalories of energy are absorbed when 0.250 mol of ammonia decomposes?

$$2\ NH_3(g) \longrightarrow 3\ H_2(g) + N_2(g) \qquad \Delta H = +22.0\ \text{kcal/mol}$$

Analysis

Use the given value of ΔH to set up a conversion factor that relates the kcal of energy absorbed to the number of moles of ammonia. The value for ΔH means that the given amount of energy is absorbed for the molar quantities shown by the coefficients in the balanced chemical equation.

Solution

The given ΔH value is the amount of energy absorbed when 2 mol of ammonia decompose. Set up a conversion factor that relates kilocalories to moles of ammonia, with moles in the denominator to cancel this unwanted unit.

kcal–mol conversion factor

$$0.250\ \text{mol}\ NH_3 \times \frac{22.0\ \text{kcal}}{2\ \text{mol}\ NH_3} = 2.75\ \text{kcal of energy absorbed}$$

Moles cancel.

Answer

Example 6.6 Using the ΔH and balanced equation for the degradation of ammonia given in Example 6.5, how many kilocalories of energy are absorbed when 10.0 g of ammonia decomposes?

Analysis

To relate the number of grams of ammonia to the number of kilocalories of energy absorbed in the reaction, two operations are needed: [1] Convert the number of grams to the number of moles using the molar mass. [2] Convert the number of moles to the number of kilocalories using ΔH (kcal/mol) and the coefficients of the balanced chemical equation.

Solution

[1] Convert the number of grams of ammonia to the number of moles of ammonia.

- Use the molar mass of the reactant (NH_3, molar mass 17.03 g/mol) to write a conversion factor. Multiply the number of grams of ammonia by the conversion factor to give the number of moles of ammonia.

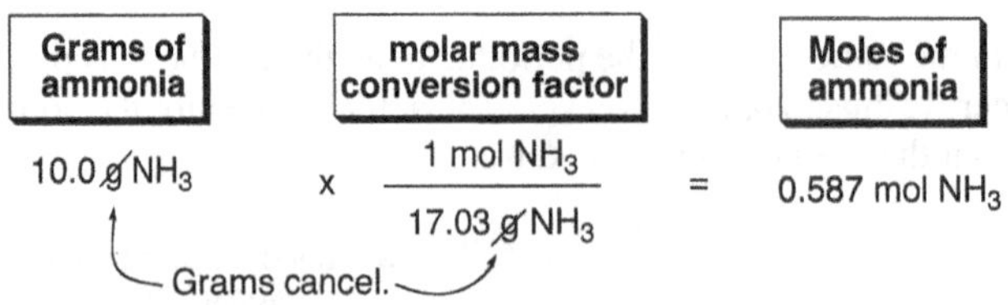

[2] Convert the number of moles of ammonia to the number of kilocalories using a kcal–mole conversion factor.

- Use the ΔH and the number of moles of ammonia in the balanced chemical equation to write a kcal–mole conversion factor—namely, two moles of ammonia (NH_3) absorb 22.0 kcal of energy. Multiply the number of moles of ammonia by the conversion factor to give the number of kilocalories of energy absorbed. This process was illustrated in Example 6.5.

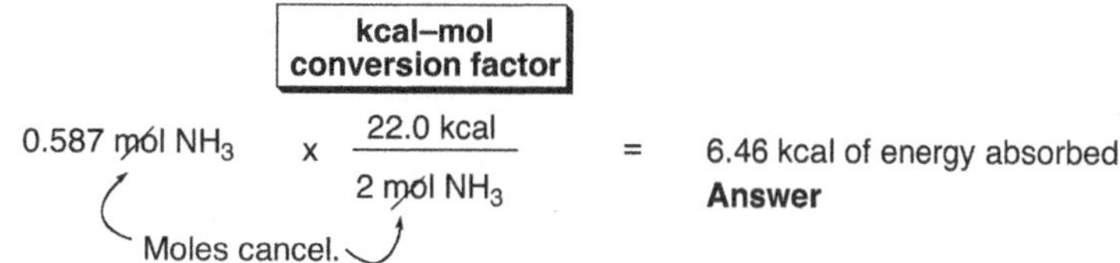

[3] Energy Diagrams (6.3)

Example 6.7 Draw an energy diagram for a reaction with a high energy of activation and a ΔH of +10 kcal/mol. Label the axes, reactants, products, transition state, E_a, and ΔH.

Analysis

A high energy of activation means a high energy barrier (a large hill) that separates reactants and products. When ΔH is (+), the products are higher in energy than the reactants.

Solution

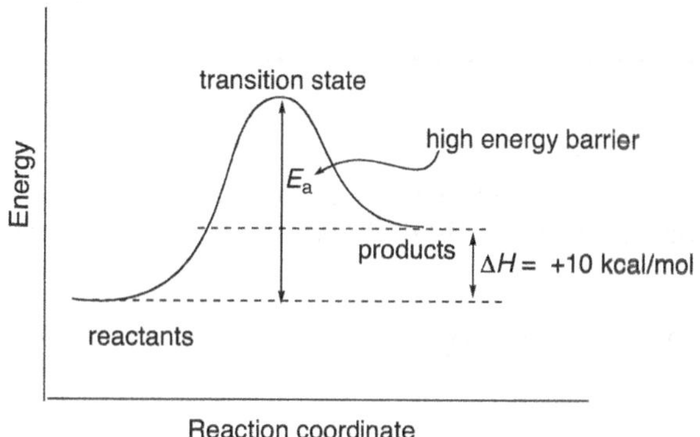

[4] Equilibrium (6.5)

Example 6.8 Write the expression for the equilibrium constant for the following balanced equation.

$$CO(g) + H_2O(g) \rightleftharpoons CO_2(g) + H_2(g)$$

Analysis

To write an expression for the equilibrium constant, multiply the concentration of the products together and divide this number by the product of the concentrations of the reactants. Each concentration term is raised to a power equal to its coefficient in the balanced chemical equation.

Solution
The concentration terms of the two products are placed in the numerator, multiplied together. The denominator contains concentration terms for the two reactants, multiplied together.

$$\text{Equilibrium constant} = K = \frac{[CO_2][H_2]}{[CO][H_2O]}$$

Example 6.9 Calculate K for the reaction of A_2 and B_2 to form AB, with the given balanced equation and the following equilibrium concentrations: $[A_2] = 0.95$ M; $[B_2] = 0.78$ M; $[AB] = 0.27$ M.

$$A_2 + B_2 \rightleftharpoons 2\,AB$$

Analysis
Write an expression for K using the balanced equation and substitute the equilibrium concentrations of all substances in the expression.

Solution

$$K = \frac{[AB]^2}{[A_2][B_2]} = \frac{(0.27)^2}{(0.95)(0.78)} = \frac{0.27 \times 0.27}{0.95 \times 0.78}$$

$$= \frac{0.0729}{0.741} = 0.0984 \text{ rounded to } 0.098$$

Answer

---

[5] Le Châtelier's Principle (6.6)

Example 6.10 In which direction is the equilibrium shifted with each of the following concentration changes for the given reaction: (a) increase $[H_2O]$; (b) increase $[CO_2]$; (c) decrease $[CO]$; (d) decrease $[H_2]$.

$$CO(g) + H_2O(g) \rightleftharpoons CO_2(g) + H_2(g)$$

Analysis
Use Le Châtelier's principle to predict the effect of a change in concentration on equilibrium. Adding more reactant or removing product drives the equilibrium to the right. Adding more product or removing reactant drives the equilibrium to the left.

Solution
a. Increasing $[H_2O]$, a reactant, drives the equilibrium to the right, thus forming more products.
b. Increasing $[CO_2]$, a product, drives the equilibrium to the left, thus forming more reactants.
c. Decreasing $[CO]$, a reactant, drives the equilibrium to the left, thus forming more reactants.
d. Decreasing $[H_2]$, a product, drives the equilibrium to the right, thus forming more products.

Self-Test

[1] Fill in the blank with one of the terms listed below.

Bond dissociation energy (6.2)
Catalyst (6.4)
Endothermic (6.2)
Energy (6.1)
Energy of activation (6.3)
Exothermic (6.2)
Heat of reaction (6.2)
Kinetic energy (6.1)
Le Châtelier's principle (6.6)
Reaction rate (6.3)
Reversible reaction (6.5)
Transition state (6.3)

1. The ______ is the energy difference between the reactants and the transition state. The higher the energy difference, the slower the reaction.
2. A ______ increases the rate at which equilibrium is reached, but does not alter the amount of any substance involved in the reaction.
3. ______ is the capacity to do work.
4. The ______, also called the enthalpy and symbolized by ΔH, is the energy absorbed or released in a reaction.
5. The ______ is located at the top of the energy barrier that separates the reactants and products.
6. ______ states that a system at equilibrium reacts in such a way as to counteract any disturbance to the equilibrium.
7. A ______ can occur in either direction, from reactants to products, or from products to reactants.
8. ______ is the energy of motion, whereas potential energy is stored energy.
9. Increasing the temperature and concentration increases ______.
10. When energy is released, a reaction is said to be ______ and ΔH is negative (–).
11. The ______ is the energy needed to break a covalent bond by equally dividing the electrons between the two atoms in the bond.
12. When energy is absorbed, a reaction is said to be ______ and ΔH is positive (+).

[2] Fill in the blank with one of the terms listed below.

a. $K > 1$ b. $K < 1$ c. $K \approx 1$

13. When ______, the concentration of the reactants is larger than the concentration of the products. We say equilibrium lies to the *left* and favors the *reactants.*
14. When______, the concentration of the products is larger than the concentration of the reactants. We say equilibrium lies to the *right* and favors the *products.*
15. When ______, anywhere in the range of 0.01–100, both reactants and products are present at equilibrium.

[3] Pick one of the terms—endothermic or exothermic—for each statement.

a. endothermic b. exothermic

16. Heat is released.
17. The products are higher in energy than the reactants.
18. The bonds formed in the products are *stronger* than the bonds broken in the reactants.
19. ΔH is positive.
20. Heat is absorbed.

[4] For each term, choose the energy diagram (or diagrams) that best match.

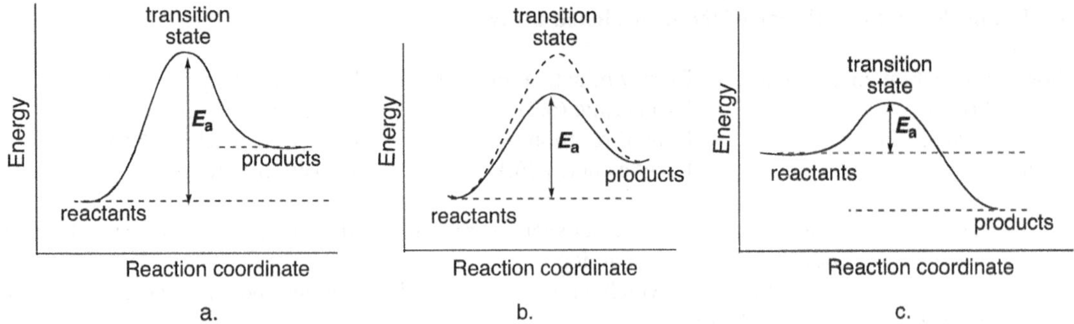

21. endothermic
22. exothermic
23. catalyzed reaction

Answers to Self-Test

1. energy of activation
2. catalyst
3. Energy
4. heat of reaction
5. transition state
6. Le Châtelier's principle
7. reversible reaction
8. Kinetic energy
9. reaction rate
10. exothermic
11. bond dissociation energy
12. endothermic
13. b
14. a
15. c
16. b
17. a
18. b
19. a
20. a
21. a, b
22. c
23. b

Solutions to In-Chapter Problems

6.1 Use conversion factors to solve each problem.

a. (42 J) × (1 cal/4.184 J) = 10. cal

b. (55.6 kcal) × (1000 cal/1 kcal) = 55,600 cal

c. (326 kcal) × (4.184 kJ/1 kcal) = 1,360 kJ

d. (25.6 kcal) × (4.184 kJ/1 kcal) × (1000 J/1 kJ) = 107,000 J

6.2 Use conversion factors to convert kcal to kJ and J.

(11.5 kcal) × (4.184 kJ/1 kcal) = 48.1 kJ
(48.1 kJ) × (1000 J/1 kJ) = 48,100 J

6.3 Use conversion factors to determine the number of Calories in 14 g of olive oil.

(14 g fat) × (9 Cal/1 g fat) = 126 Cal, or 100 Cal when rounded to one significant figure

6.4 Calculate the number of Calories as in Example 6.2.

Total Calories = (26 g carb) × (4 Cal/1 g carb) + (6 g protein) × (4 Cal/1 g protein)

Total Calories = 104 Cal + 24 Cal

Total Calories = 128 Cal = rounded to 100 Cal

6.5 Use Table 6.2 to determine the bond dissociation energy for each reaction. Forming a bond is an **exothermic reaction** and ΔH is a *negative* number. Breaking a bond is an **endothermic reaction** and ΔH is a *positive* number.

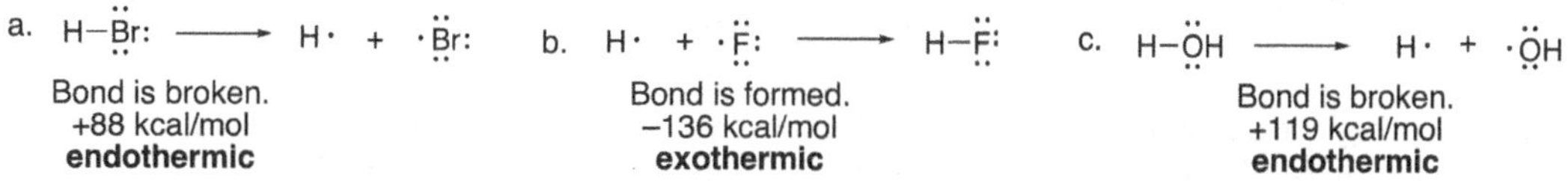

a. H–Br: ⟶ H· + ·Br:
Bond is broken.
+88 kcal/mol
endothermic

b. H· + ·F: ⟶ H–F:
Bond is formed.
−136 kcal/mol
exothermic

c. H–OH ⟶ H· + ·OH
Bond is broken.
+119 kcal/mol
endothermic

6.6 **The higher the bond dissociation energy, the stronger the bond.** In comparing bonds to atoms in the same group of the periodic table, bond dissociation energies and bond strength decrease down a column.

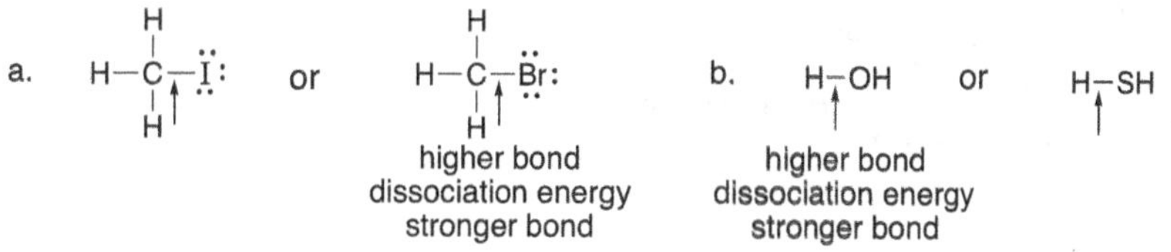

a. CH_3–I or CH_3–Br
CH_3–Br: higher bond dissociation energy, stronger bond

b. H–OH or H–SH
H–OH: higher bond dissociation energy, stronger bond

6.7 Use Table 6.3 to answer the questions for a reaction with $\Delta H = +22.0$ kcal/mol.

a. Heat is absorbed.
b. The bonds in the reactants are stronger.
c. The reactants are lower in energy.
d. The reaction is endothermic.

6.8 Use a conversion factor to solve the problem.

$C_3H_8(g) + 5\ O_2(g) \longrightarrow 3\ CO_2(g) + 4\ H_2O(l) \quad \Delta H = -531$ kcal/mol

propane

kcal–mol conversion factor

$$1.00\ \text{mol}\ O_2 \times \frac{531\ \text{kcal}}{5\ \text{mol}\ O_2} = 106\ \text{kcal of energy released}$$

Moles cancel.

Answer

6.9 Use conversion factors to solve the problems.

$$\underset{\text{glucose}}{C_6H_{12}O_6(s)} \longrightarrow \underset{\text{ethanol}}{2\ C_2H_6O(l)} + 2\ CO_2(g) \qquad \Delta H = -16\ \text{kcal/mol}$$

a. $6.0\ \cancel{\text{mol}}\ C_6H_{12}O_6 \times \dfrac{16\ \text{kcal}}{1\ \cancel{\text{mol}}\ C_6H_{12}O_6} = 96\ \text{kcal of energy released}$ **Answer**

b. $1.00\ \cancel{\text{mol}}\ C_2H_6O \times \dfrac{16\ \text{kcal}}{2\ \cancel{\text{mol}}\ C_2H_6O} = 8.0\ \text{kcal of energy released}$ **Answer**

c. $20.0\ \cancel{\text{g}}\ C_6H_{12}O_6 \times \dfrac{1\ \cancel{\text{mol}}\ C_6H_{12}O_6}{180.2\ \cancel{\text{g}}\ C_6H_{12}O_6} \times \dfrac{16\ \text{kcal}}{1\ \cancel{\text{mol}}\ C_6H_{12}O_6} = 1.8\ \text{kcal of energy released}$ **Answer**

6.10 A high energy of activation means a high energy barrier (a large hill) that separates reactants and products. When ΔH is positive, the products are higher in energy than the reactants.

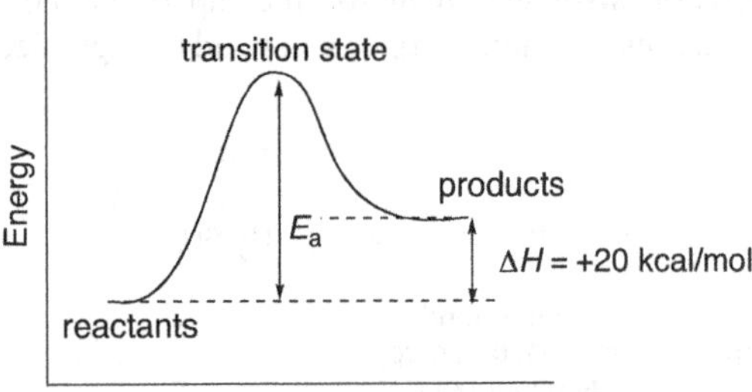

6.11 Increasing the concentration of the reactants increases the number of collisions, so the reaction rate increases. Increasing the temperature increases the reaction rate.

a. Increasing the concentration of O_3 increases the rate of the reaction.
b. Decreasing the concentration of NO decreases the rate of the reaction.
c. Increasing the temperature increases the rate of the reaction.
d. Decreasing the temperature decreases the rate of the reaction.

6.12

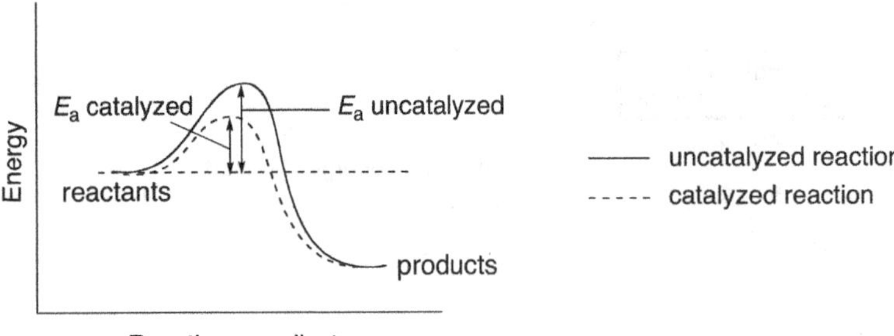

6.13 The *forward* reaction proceeds from left to right as drawn.
The *reverse* reaction proceeds from right to left as drawn.

a. $2\ SO_2(g) + O_2(g) \longrightarrow 2\ SO_3(g)$ forward reaction

$2\ SO_3(g) \longrightarrow 2\ SO_2(g) + O_2(g)$ reverse reaction

b. $N_2(g) + O_2(g) \longrightarrow 2\ NO(g)$ forward reaction

$2\ NO(g) \longrightarrow N_2(g) + O_2(g)$ reverse reaction

c. $C_2H_4O_2 + CH_4O \longrightarrow C_3H_6O_2 + H_2O$ forward reaction

$C_3H_6O_2 + H_2O \longrightarrow C_2H_4O_2 + CH_4O$ reverse reaction

6.14 To write an expression for the equilibrium constant multiply the concentration of the products together and divide this number by the product of the concentrations of the reactants. Each concentration term is raised to a power equal to its coefficient in the balanced chemical equation.

a. $PCl_3(g) + Cl_2(g) \rightleftharpoons PCl_5(g) \quad K = \dfrac{[PCl_5]}{[PCl_3][Cl_2]}$

b. $2\ SO_2(g) + O_2(g) \rightleftharpoons 2\ SO_3(g) \quad K = \dfrac{[SO_3]^2}{[SO_2]^2[O_2]}$

c. $H_2(g) + Br_2(g) \rightleftharpoons 2\ HBr(g) \quad K = \dfrac{[HBr]^2}{[Br_2][H_2]}$

d. $CH_4(g) + 3\ Cl_2(g) \rightleftharpoons CHCl_3(g) + 3\ HCl(g) \quad K = \dfrac{[HCl]^3[CHCl_3]}{[CH_4][Cl_2]^3}$

6.15 Since $K = 1$, there are equal amounts of **A** and **B** at equilibrium. **A** is represented with grey spheres, whereas **B** is the black spheres.

For $K = 1$: $\quad K = \dfrac{[B]}{[A]} = \dfrac{1}{1}$ equal amounts of **A** and **B**

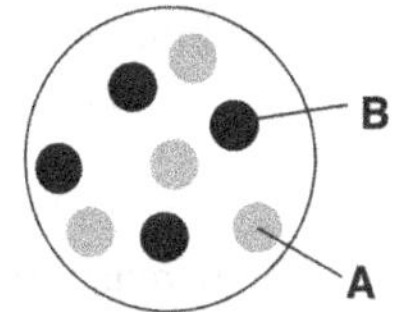

6.16 When $K > 1$, the equilibrium favors the products.
When $K < 1$, the equilibrium favors the reactants.
When $K \approx 1$, both the reactants and products are present at equilibrium.

a. 5.0×10^{-4}, $K < 1$, reactants favored
b. 4.4×10^{5}, $K > 1$, products favored
c. 350, $K > 1$, products favored
d. 0.35, $K \approx 1$, reactants and products present

6.17 Write the expression for the equilibrium constant as in Answer 6.14.

a. $K = \dfrac{[H_2S]^2}{[H_2]^2[S_2]}$

b. $K > 1$, products favored
c. ΔH would be negative because $K > 1$.
d. The products are lower in energy because ΔH is negative.
e. One cannot predict the rate of reaction without knowing the energy of activation.

6.18 First write an expression for K using the balanced equation. Then substitute the equilibrium concentrations of all substances in the expression to calculate K.

a. $K = \dfrac{[CO_2][H_2]}{[CO][H_2O]} = \dfrac{(0.0164)(0.0164)}{(0.0236)(0.00240)} = 4.75$ **Answer**

b. $K = \dfrac{[N_2O_4]}{[NO_2]^2} = \dfrac{(1.26)}{(0.0760)^2} = \dfrac{1.26}{0.0760 \times 0.0760} = 218$ **Answer**

6.19 Substitute the number of molecules of reactants and products in the expression for the equilibrium constant to see if the system is at equilibrium. If the value is smaller than K, the reaction proceeds to the right. If the value is larger than K, the reaction proceeds to the left.

a. $K = \dfrac{[\mathbf{C}][\mathbf{D}]}{[\mathbf{A}][\mathbf{B}]} = \dfrac{(4\mathbf{C})(5\mathbf{D})}{(1\mathbf{A})(2\mathbf{B})} = \dfrac{20}{2} = 10$

The system is not at equilibrium.

b. Since the value is greater than K, there is more product than would be present at equilibrium, so the reaction proceeds to the left to form more reactants.

6.20 Use Le Châtelier's principle to predict the effect of a change in concentration on equilibrium. Adding more reactant or removing product drives the equilibrium to the right. Adding more product or removing reactant drives the equilibrium to the left.

a. increase $[H_2]$ (reactant) = right
b. increase [HCl] (product) = left
c. decrease $[Cl_2]$ (reactant) = left
d. decrease [HCl] (product) = right

6.21 When temperature is increased, the reaction that removes heat is favored. When temperature is decreased, the reaction that adds heat is favored.

a. The equilibrium shifts to the **right** when the temperature is increased.
b. The equilibrium shifts to the **left** when the temperature is decreased.

6.22 When temperature is increased, the reaction that removes heat is favored. When temperature is decreased, the reaction that adds heat is favored.

a. The equilibrium shifts to the **left** when the temperature is increased.
b. The equilibrium shifts to the **right** when the temperature is decreased.

6.23 When pressure is increased, the equilibrium shifts in the direction that decreases the number of moles. When pressure is decreased, the equilibrium shifts in the direction that increases the number of moles.

a. The equilibrium shifts to the **right** when the pressure is increased.
b. The equilibrium shifts to the **left** when the pressure is decreased.

6.24 a. False. The reaction is exothermic, so increasing the temperature shifts the equilibrium to the left.
b. True.
c. False. Increasing the pressure drives the equilibrium to the left, the side with fewer moles.
d. True.

Solutions to Odd-Numbered End-of-Chapter Problems

6.25 Potential energy is stored energy, whereas kinetic energy is the energy of motion. A stationary object on a hill has potential energy, but as it moves down the hill this potential energy is converted to kinetic energy.

6.27 Use conversion factors to solve the problems.

a. (563 Cal/h) × (1000 cal/1 Cal) = 563,000 cal/h
b. 563 Cal/h = 563 kcal/h
c. (563 Cal/h) × (1000 cal/1 Cal) × (4.184 J/1 cal) = 2.36×10^6 J/h
d. (563 Cal/h) × (1000 cal/1 Cal) × (4.184 J/1 cal) × (1 kJ/1000 J) = 2,360 kJ/h

6.29 Use conversion factors to solve the problems.

a. (50 cal) × (1 kcal/1000 cal) = 0.05 kcal
b. (56 cal) × (4.184 J/1 cal) × (1 kJ/1000 J) = 0.23 kJ
c. (0.96 kJ) × (1 cal/4.184 J) × (1000 J/1 kJ) = 230 cal
d. (4,230 kJ) × (1 cal/4.184 J) × (1000 J/1 kJ) = 1.01×10^6 cal

6.31 Calculate the number of Calories as in Example 6.2.

Total Calories =	(6 g fat) × (9 Cal/1 g fat)	+ (7 g carb) × (4 Cal/1 g carb)		+ (16 g protein) × (4 Cal/1 g protein)	
Total Calories =	144 Cal	+	28 Cal	+	64 Cal
Total Calories =	236 Cal, rounded to 200 Cal				

6.33 Use a conversion factor to solve the problem.

(120 Cal) × (1 g carbohydrate/4 Cal) = 30 g carbohydrates

6.35 Calculate the number of Calories as in Example 6.2.

Total Calories =	(8 g fat) × (9 Cal/1 g fat)	+ (11 g carb) × (4 Cal/1 g carb)	+ (8 g protein) × (4 Cal/1 g protein)
Total Calories =	72 Cal	+ 44 Cal	+ 32 Cal
Total Calories =	**148 Cal, rounded to 100 Cal**		

6.37 Use the common element colors on the inside back cover to identify the atoms. In comparing bonds to atoms in the same group of the periodic table, bond dissociation energies and bond strength decrease down a column.

a. Cl_2 has a stronger bond than Br_2, because Br is below Cl in the periodic table.
b. Cl_2 has a stronger bond than I_2, because I is below Cl in the periodic table.
c. HF has a stronger bond than HBr, because Br is below F in the periodic table.

6.39 Use Table 6.3 to answer the questions.

a. ΔH is a negative value = exothermic.
b. The energy of the reactants is lower than the energy of the products = endothermic.
c. Energy is absorbed in the reaction = endothermic.
d. The bonds in the products are stronger than the bonds in the reactants = exothermic.

6.41 Use conversion factors to solve the problems.

$$C(s) + O_2(g) \longrightarrow CO_2(g) \quad \Delta H = -94 \text{ kcal/mol}$$

a. $2.5 \text{ mol C} \times \frac{94 \text{ kcal}}{1 \text{ mol C}} =$ 240 kcal of energy released **Answer**

b. $3.0 \text{ mol } O_2 \times \frac{94 \text{ kcal}}{1 \text{ mol } O_2} =$ 280 kcal of energy released **Answer**

c. $25.0 \text{ g C} \times \frac{1 \text{ mol C}}{12.01 \text{ g C}} \times \frac{94 \text{ kcal}}{1 \text{ mol C}} =$ 2.0×10^2 kcal of energy released **Answer**

6.43 Use conversion factors to solve the problems.

$$C_6H_{12}O_6(aq) + 6\,O_2(g) \rightleftharpoons 6\,CO_2(g) + 6\,H_2O(l) \quad \Delta H = -678 \text{ kcal/mol}$$

glucose
(molar mass 180.2 g/mol)

a. stronger bonds in products
b. $4.00 \text{ mol } C_6H_{12}O_6 \times \frac{678 \text{ kcal}}{1 \text{ mol } C_6H_{12}O_6} =$ 2,710 kcal **Answer**

c. $3.00 \text{ mol } O_2 \times \frac{678 \text{ kcal}}{6 \text{ mol } O_2} = 339 \text{ kcal}$ **Answer**

d. $10.0 \text{ g } C_6H_{12}O_6 \times \frac{1 \text{ mol } C_6H_{12}O_6}{180.2 \text{ g } C_6H_{12}O_6} \times \frac{678 \text{ kcal}}{1 \text{ mol } C_6H_{12}O_6} = 37.6 \text{ kcal}$ **Answer**

6.45 Label the points on the energy diagram.

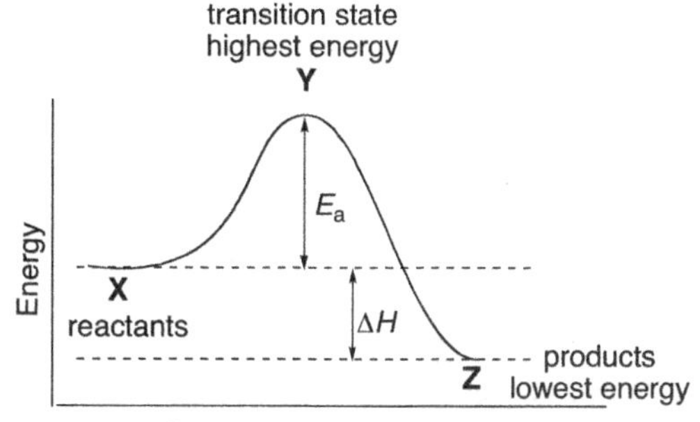

a. X
b. Z
c. Y
d. X, Y
e. X, Z
f. Y
g. Z

6.47 Draw an energy diagram to fit each description.

a.

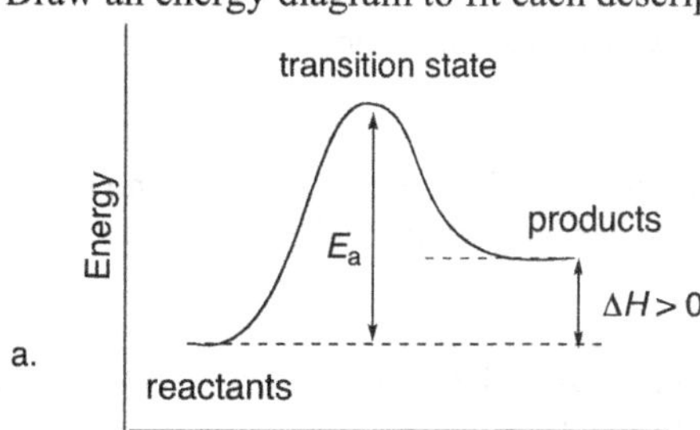

b.

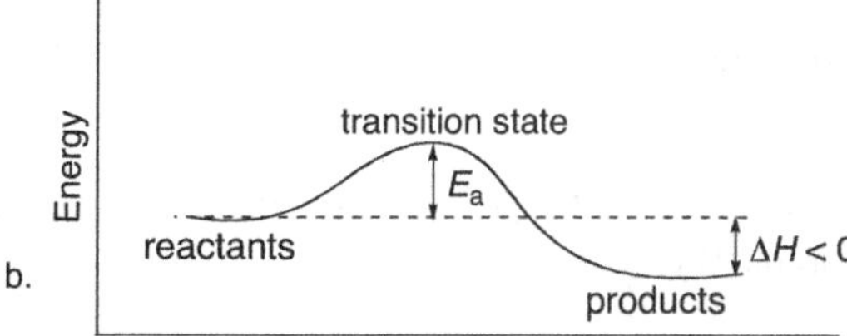

c.

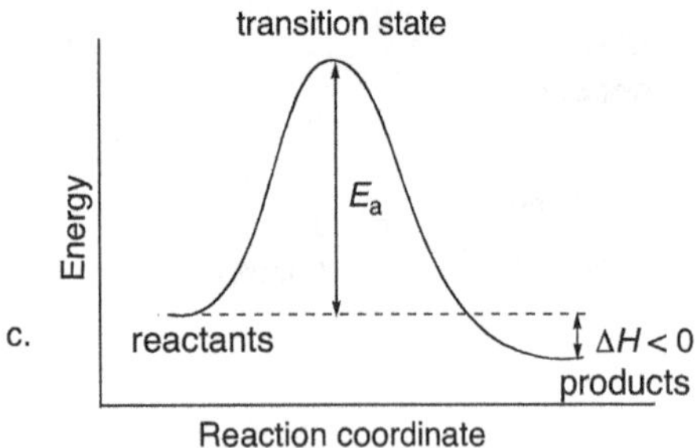

6.49 Draw an energy diagram that fits the description.

The reaction is exothermic.

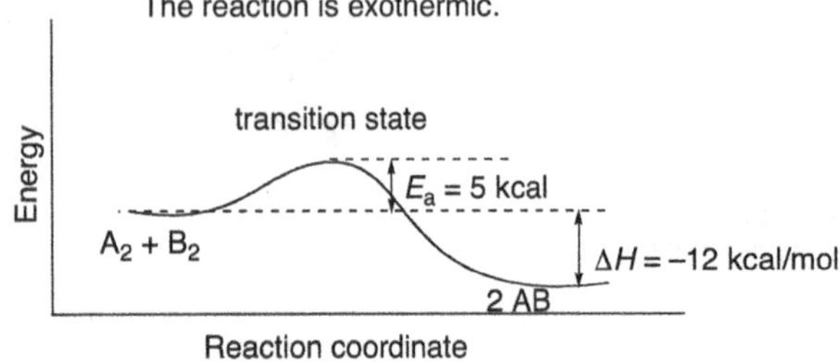

6.51 Collision orientation affects the rate of reaction because reacting molecules must have the proper orientation for new bonds to form.

6.53 Increasing temperature increases the number of collisions, and thereby increases the reaction rate. Since the average kinetic energy of the colliding molecules is larger at higher temperatures, more collisions are effective at causing reaction.

6.55 a. The reaction with $E_a = 1$ kcal will proceed faster because the energy of activation is lower.
b. K doesn't affect the reaction rate, so we cannot predict which reaction is faster.
c. One cannot predict which reaction will proceed faster from the value of ΔH.

6.57 Energy of activation (b) and temperature (c) affect the rate of reaction. K (a) doesn't affect the reaction rate.

6.59 A catalyst increases the reaction rate (a) and lowers the E_a (c). It has no effect on ΔH (b), K (d), or the relative energies of the reactants and products (e).

6.61 Use the procedure in Answer 6.19.

a. $$K = \frac{[\mathbf{C}][\mathbf{D}]}{[\mathbf{A}][\mathbf{B}]} = \frac{(1\mathbf{C})(2\mathbf{D})}{(5\mathbf{A})(6\mathbf{B})} = \frac{2}{30} = \frac{1}{15}$$

The system is not at equilibrium.

b. Since the value is smaller than the equilibrium constant ($K = 4$), there are more reactants than would be present at equilibrium, and the reaction proceeds to the right to form more products.

6.63 When $K > 1$, the equilibrium favors the products.
When $K < 1$, the equilibrium favors the reactants.
When $K \approx 1$, both the reactants and products are present at equilibrium.
When ΔH is positive, the reactants are favored.
When ΔH is negative, the products are favored.

a. $K = 5.2 \times 10^3$, so $K > 1$ and the equilibrium favors the products.
b. $\Delta H = -27$ kcal/mol, so the products are favored.
c. $K = 0.002$, so $K < 1$ and the equilibrium favors the reactants.
d. $\Delta H = +2$ kcal/mol, so the reactants are favored.

6.65 $K > 1$ is associated with a negative value of ΔH. A $K < 1$ is associated with a positive value of ΔH.

6.67 **A** is represented with black spheres, and **B** is represented with grey spheres.

a. 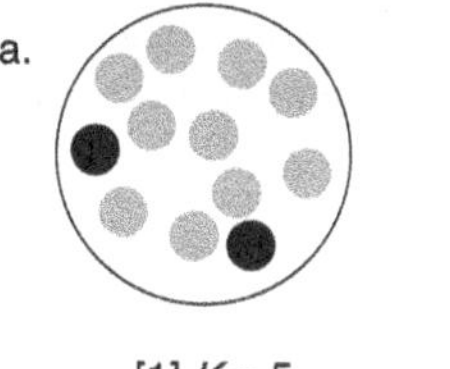
[1] $K = 5$

b. 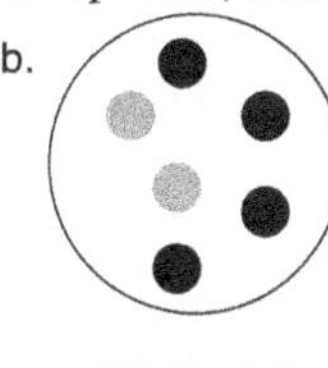
[3] $K = 0.5$

c. 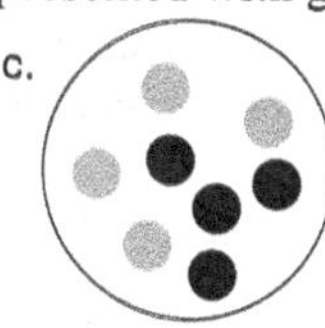
[2] $K = 1$

6.69 Use the number of molecules of reactants and products to determine K.

$$K = \frac{[AX]}{[A][X]} = \frac{4}{(2)(2)} = 1$$

$$K = \frac{[AY]}{[A][Y]} = \frac{4}{(1)(1)} = 4$$
largest equilibrium constant

$$K = \frac{[AZ]}{[A][Z]} = \frac{1}{(3)(3)} = \frac{1}{9}$$
smallest equilibrium constant

6.71 Write the expression for the equilibrium constant as in Answer 6.14.

a. $K = \dfrac{[NO_2]^2}{[NO]^2[O_2]}$

b. $K = \dfrac{[HBr]^2[CH_2Br_2]}{[CH_4][Br_2]^2}$

6.73 Work backwards from the expression for the equilibrium constant to write the chemical equation.

a. $K = \dfrac{[A_2]}{[A]^2}$ $\quad 2\,A \rightleftharpoons A_2$

b. $K = \dfrac{[AB_3]^2}{[A_2][B_2]^3}$ $\quad A_2 + 3\,B_2 \rightleftharpoons 2\,AB_3$

6.75

a. $K = \dfrac{[Br_2][H_2]}{[HBr]^2}$

b. The reactants are favored at equilibrium because $K < 1$.
c. ΔH is predicted to be positive because $K < 1$.
d. The reactants are lower in energy because the reactants are favored at equilibrium.
e. You can't predict the reaction rate from the value of K.

6.77

a. $K = \dfrac{[CO_2][H_2]}{[CO][H_2O]}$ b. $\dfrac{(0.15)(0.30)}{(0.090)(0.12)} = K = 4.2$

6.79 The concentration of reactant **A** has increased, so the equilibrium is driven to the right to form more product.

6.81 Use Le Châtelier's principle to predict the effect of a change in concentration on equilibrium. Adding more reactant or removing product drives the equilibrium to the right. Adding more product or removing reactant drives the equilibrium to the left.

a. When O_2 is increased, it drives the equilibrium to the right: increases NO and decreases N_2.
b. When NO is increased, it drives the equilibrium to the left: increases N_2 and O_2.

6.83 In an endothermic reaction, increasing the temperature drives the reaction to the right to form more product. Diagram [1] has more product, so it corresponds to the higher temperature, 200 °C.

6.85 Use Le Châtelier's principle to predict the effect of each change.

a. decrease $[O_3]$, shift to right
b. decrease $[O_2]$, shift to left
c. increase $[O_3]$, shift to left
d. decrease temperature, shift to left
e. add a catalyst, no change
f. increase pressure, shift to right

6.87 Use Le Châtelier's principle to predict the effect of each change.

a. increase $[C_2H_4]$, shift to right
b. decrease $[Cl_2]$, shift to left
c. decrease $[C_2H_4Cl_2]$, shift to right
d. decrease pressure, shift to left
e. increase temperature, shift to left
f. decrease temperature, shift to right

6.89

a. $K = \dfrac{[C_2H_6]}{[H_2][C_2H_4]}$

b. The reactants are higher in energy when ΔH is negative.
c. $K > 1$ because ΔH is negative.
d. (20.0 g ethylene) × (1 mol ethylene/28.05 g) × (28 kcal/1 mol ethylene) = 20. kcal
e. If ethylene (reactant) concentration is increased, the reaction rate will increase.
f. 1,2,4: favor shift to right; 3: favors shift to left; 5: no change, but reduces the rate of reaction.

6.91 Lactase is an enzyme that converts lactose, a naturally occurring sugar in dairy products, into the two simple sugars, glucose and galactose.

6.93 Use conversion factors to solve the problem.

$$2000 \text{ mL} \times \frac{5 \text{ g glucose}}{100 \text{ mL}} \times \frac{4 \text{ Calories}}{1 \text{ g glucose}} = 400 \text{ Calories}$$

6.95

Total Calories =	(29 g fat) × (9 Cal/1 g fat)	+ (34 g carb) × (4 Cal/1 g carb)	+ (32 g protein) × (4 Cal/1 g protein)
Total Calories =	261 Cal	+ 136 Cal	+ 128 Cal
Total Calories =	525 Calories		

(525 Cal) × (1 h/280 Cal) = 1.9 h rounded to 2 h

6.97

$$1.5\ \cancel{h} \times \frac{290\ \cancel{Cal}}{1\ \cancel{h}} = 440\ \text{Cal expended}$$

$$440\ \cancel{Cal} \times \frac{1\ \text{cookie}}{50.\ \cancel{Cal}} = 8.8\ \text{cookies}$$

6.99 Use conversion factors to solve the problem.

$$1\ \cancel{gal}\ \text{gas} \times \frac{4\ \cancel{qt}}{1\ \cancel{gal}} \times \frac{946\ \cancel{mL}}{1\ \cancel{qt}} \times \frac{0.700\ \cancel{g}}{1\ \cancel{mL}} \times \frac{1\ \cancel{mol}\ C_8H_{18}}{114.2\ \cancel{g}\ C_8H_{18}} \times \frac{1303\ \text{kcal}}{1\ \cancel{mol}} = 30{,}200\ \text{kcal}$$

Chapter 7 Gases, Liquids, and Solids

Chapter Review

[1] What is pressure and what units are used to measure it? (7.2)

- Pressure is the force per unit area. The pressure of a gas is the force exerted when gas particles strike a surface. Pressure is measured by a barometer and recorded in atmospheres (atm), millimeters of mercury (mm Hg), or pounds per square inch (psi).
- 1 atm = 760 mm Hg = 14.7 psi.

$$\textbf{Pressure} = \frac{\textbf{Force}}{\textbf{Area}} = \frac{F}{A}$$

[2] What are gas laws and how are they used to describe the relationship between the pressure, volume, and temperature of a gas? (7.3)

- Because gas particles are far apart and behave independently, a set of gas laws describes the behavior of all gases regardless of their identity. Three gas laws—**Boyle's law, Charles's law,** and **Gay–Lussac's law**—describe the relationship between the pressure, volume, and temperature of a gas. These gas laws are summarized in "Key Equations—The Gas Laws" at the end of the Chapter Review.
- For a constant amount of gas, the following relationships exist.
  [1] The pressure and volume of a gas are inversely related, so increasing the pressure decreases the volume at constant temperature.
  [2] The volume of a gas is proportional to its Kelvin temperature, so increasing the temperature increases the volume at constant pressure.
  [3] The pressure of a gas is proportional to its Kelvin temperature, so increasing the temperature increases the pressure at constant volume.

cover
P = 1 atm
V = 2 L
Decrease volume.
cover
P = 2 atm
V = 1 L
Pressure increases.

[3] Describe the relationship between the volume and number of moles of a gas. (7.4)

- **Avogadro's law** states that when temperature and pressure are held constant, the volume of a gas is proportional to its number of moles.
- One mole of any gas has the same volume, the standard molar volume of 22.4 L, at 1 atm and 273 K (STP).

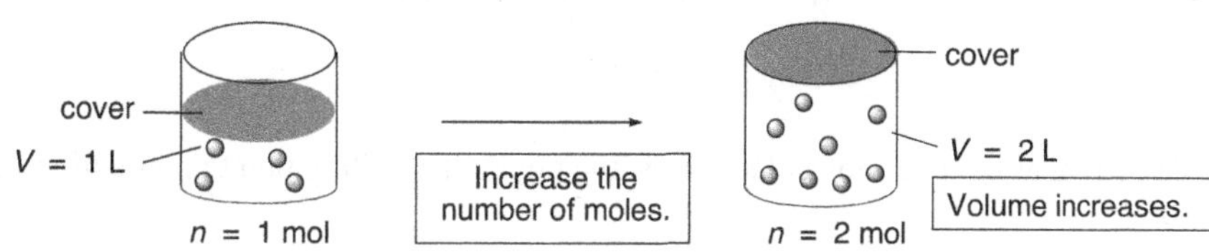

[4] What is the ideal gas law? (7.5)

- The **ideal gas law** is an equation that relates the pressure (*P*), volume (*V*), temperature (*T*), and number of moles (*n*) of a gas; $PV = nRT$, where *R* is the universal gas constant. The ideal gas law can be used to calculate any one of the four variables, as long as the other three variables are known.

$$PV = nRT$$

Ideal gas law

For atm: $R = 0.0821 \frac{\text{L} \cdot \text{atm}}{\text{mol} \cdot \text{K}}$

For mm Hg: $R = 62.4 \frac{\text{L} \cdot \text{mm Hg}}{\text{mol} \cdot \text{K}}$

[5] What is Dalton's law and how is it used to relate partial pressures and the total pressure of a gas mixture? (7.6)

- **Dalton's law** states that the total pressure of a gas mixture is the sum of the partial pressures of its component gases. The partial pressure is the pressure exerted by each component of a mixture.

$$P_{total} = P_A + P_B + P_C$$

total pressure = partial pressures of **A, B,** and **C**

[6] What types of intermolecular forces exist and how do they determine a compound's boiling point and melting point? (7.7)

- Intermolecular forces are the forces of attraction between molecules. Three types of intermolecular forces exist in covalent compounds.
  - **London dispersion forces** are due to momentary changes in electron density in a molecule.
  - **Dipole–dipole interactions** are due to permanent dipoles.

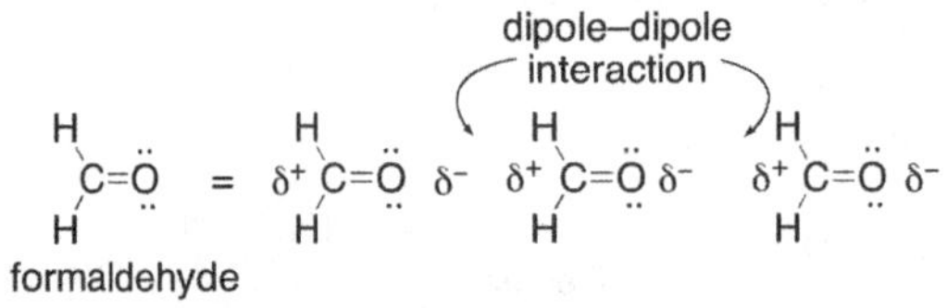

dipole–dipole interaction

$H_2C{=}O$ = δ^+ $H_2C{=}O$ δ^- δ^+ $H_2C{=}O$ δ^- δ^+ $H_2C{=}O$ δ^-

formaldehyde

  - **Hydrogen bonding,** the strongest intermolecular force, results when a H atom bonded to an O, N, or F is attracted to an O, N, or F atom in another molecule.

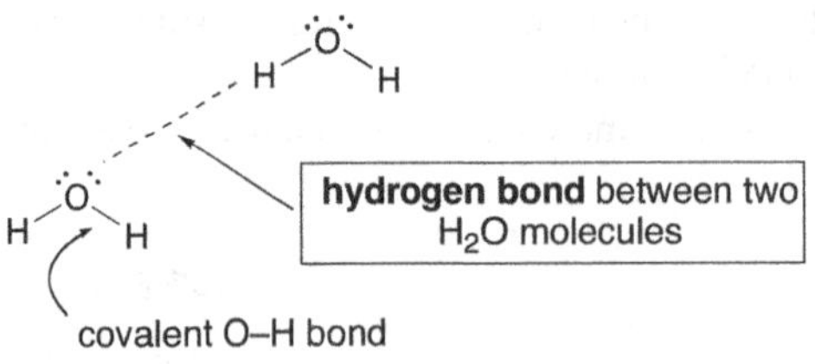

hydrogen bond between two H_2O molecules

covalent O–H bond

  - The stronger the intermolecular forces, the higher the boiling point and melting point of a compound.

[7] Describe three features of the liquid state—vapor pressure, viscosity, and surface tension. (7.8)

- **Vapor pressure** is the pressure exerted by gas molecules in equilibrium with the liquid phase. Vapor pressure increases with increasing temperature. The higher the vapor pressure at a given temperature, the lower the boiling point of a compound.

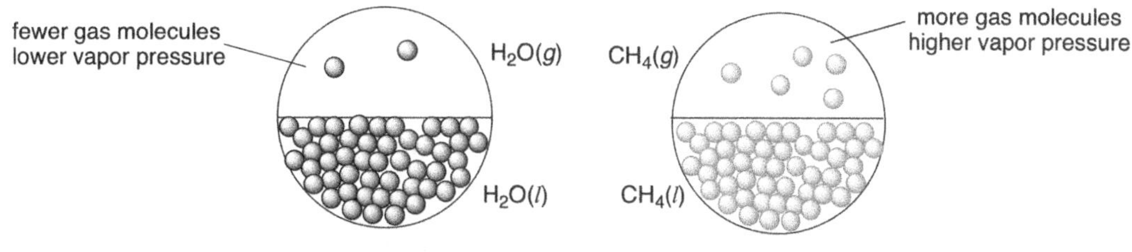

Stronger intermolecular forces keep the molecules as a liquid.

Weaker intermolecular forces allow more molecules to escape to the gas phase.

- **Viscosity** measures a liquid's resistance to flow. More viscous compounds tend to have stronger intermolecular forces or they have high molecular weights.
- **Surface tension** measures a liquid's resistance to spreading out. The stronger the intermolecular forces, the higher the surface tension.

[8] Describe the features of different types of solids. (7.9)

- Solids can be amorphous or crystalline. An **amorphous solid** has no regular arrangement of particles. A **crystalline solid** has a regular arrangement of particles in a repeating pattern. There are four types of crystalline solids: **Ionic solids** are composed of ions. **Molecular solids** are composed of individual molecules. **Network solids** are composed of vast arrays of covalently bonded atoms in a regular three-dimensional arrangement. **Metallic solids** are composed of metal cations with a cloud of delocalized electrons.

[9] What is specific heat? (7.10)

- Specific heat is the amount of energy needed to raise the temperature of 1 g of a substance by 1 °C.
- Specific heat is used as a conversion factor to calculate how much heat a known mass of a substance absorbs or how much its temperature changes.

[10] Describe the energy changes that accompany changes of state. (7.11)

- A phase change converts one state to another. Energy is absorbed when a more organized state is converted to a less organized state. Thus, energy is absorbed when a solid melts to form a liquid, or when a liquid vaporizes to form a gas.
- Energy is released when a less organized state is converted to a more organized state. Thus, energy is released when a gas condenses to form a liquid, or a liquid freezes to form a solid.
- The **heat of fusion** is the energy needed to melt 1 g of a substance, while the **heat of vaporization** is the energy needed to vaporize 1 g of a substance.

[11] What changes are depicted on heating and cooling curves? (7.12)

- A heating curve shows how the temperature of a substance changes as heat is added. Diagonal lines show the temperature increase of a single phase. Horizontal lines correspond to phase changes—solid to liquid or liquid to gas.
- A cooling curve shows how the temperature of a substance changes as heat is removed. Diagonal lines show the temperature decrease of a single phase. Horizontal lines correspond to phase changes —gas to liquid or liquid to solid.

Key Equations—The Gas Laws

Name	Equation	Variables Related	Constant Terms
Boyle's law	$P_1V_1 = P_2V_2$	P, V	T, n
Charles's law	$\frac{V_1}{T_1} = \frac{V_2}{T_2}$	V, T	P, n
Gay–Lussac's law	$\frac{P_1}{T_1} = \frac{P_2}{T_2}$	P, T	V, n
Combined gas law	$\frac{P_1V_1}{T_1} = \frac{P_2V_2}{T_2}$	P, V, T	n
Avogadro's law	$\frac{V_1}{n_1} = \frac{V_2}{n_2}$	V, n	P, T
Ideal gas law	$PV = nRT$	P, V, T, n	R

Problem Solving

[1] Gases and Pressure (7.2)

Example 7.1 Convert 200. psi to:
a. atmospheres b. mm Hg

Analysis
To solve each part, set up conversion factors that relate the two units under consideration. Use conversion factors that place the unwanted unit, psi, in the denominator to cancel.

In part (a), the conversion factor must relate psi and atm:

psi–atm conversion factor

$$\frac{1 \text{ atm}}{14.7 \text{ psi}}$$

unwanted unit (psi)

In part (b), the conversion factor must relate psi and mm Hg:

psi–mm Hg conversion factor

$$\frac{760. \text{ mm Hg}}{14.7 \text{ psi}}$$

unwanted unit (psi)

Solution
a. Convert the original unit (200. psi) to the desired unit (atm) using the conversion factor:

$$200. \text{ psi} \times \frac{1 \text{ atm}}{14.7 \text{ psi}} = 13.6 \text{ atm}$$

Psi cancels. **Answer**

b. Convert the original unit (200. psi) to the desired unit (mm Hg) using the conversion factor:

$$200. \text{ psi} \times \frac{760. \text{ mm Hg}}{14.7 \text{ psi}} = 10{,}300 \text{ mm Hg}$$

Psi cancels. **Answer**

[2] Gas Laws (7.3)

Example 7.2 A tank of compressed air contains 20. L of gas at 204 atm pressure. What volume does this gas occupy at 1.0 atm?

Analysis

Boyle's law can be used to solve this problem because an initial pressure and volume (P_1 and V_1) and a final pressure (P_2) are known, and a final volume (V_2) must be determined.

Solution

[1] Identify the known quantities and the desired quantity.

$P_1 = 204$ atm $\quad P_2 = 1.0$ atm

$V_1 = 20.$ L $\quad\quad V_2 = ?$

known quantities $\quad\quad$ desired quantity

[2] Write the equation and rearrange it to isolate the desired quantity, V_2, on one side.

$$P_1V_1 = P_2V_2$$

Solve for V_2 by dividing both sides by P_2.

$$\frac{P_1V_1}{P_2} = V_2$$

[3] Solve the problem.

- Substitute the three known quantities in the equation and solve for V_2.

$$V_2 = \frac{P_1V_1}{P_2} = \frac{(204\ \text{atm})(20.\ \text{L})}{1.0\ \text{atm}} = 4{,}080 \text{ rounded to } 4{,}100\ \text{L}$$

Atm cancels.

Answer

- Thus, the volume increased as the pressure decreased at constant temperature.

Example 7.3 A balloon that contains 0.75 L of air at 28 °C is cooled to –98 °C. What volume does the balloon now occupy?

Analysis

Since this question deals with volume and temperature, Charles's law is used to determine a final volume because three quantities are known—the initial volume and temperature (V_1 and T_1), and the final temperature (T_2).

Solution

[1] Identify the known quantities and the desired quantity.

$V_1 = 0.75$ L

$T_1 = 28$ °C $\quad T_2 = -98$ °C $\quad V_2 = ?$

known quantities $\quad\quad$ desired quantity

- Both temperatures must be converted to Kelvin temperatures using the equation **$T_K = T_C + 273$.**
- $T_1 = 28\ °C + 273 = 301\ K$
- $T_2 = -98\ °C + 273 = 175\ K$

---

[2] Write the equation and rearrange it to isolate the desired quantity, V_2, on one side.

- Use Charles's law.

$$\frac{V_1}{T_1} = \frac{V_2}{T_2}$$ Solve for V_2 by multiplying both sides by T_2.

$$\frac{V_1 T_2}{T_1} = V_2$$

---

[3] Solve the problem.

- Substitute the three known quantities in the equation and solve for V_2.

$$V_2 = \frac{V_1 T_2}{T_1} = \frac{(0.75\ L)(175\ K)}{301\ K} = 0.44\ L$$

Answer

Kelvins cancel.

- Since temperature has decreased, the volume of gas must decrease as well at constant pressure.

Example 7.4 A balloon contains 2.0 L of helium at 25 °C and 760 mm Hg. What is the volume of the balloon when it ascends to an altitude where the temperature is –40. °C and the pressure is 540 mm Hg?

Analysis

Since this question deals with pressure, volume, and temperature, the combined gas law is used to determine a final volume (V_2) because five quantities are known—the initial pressure, volume, and temperature (P_1, V_1, and T_1), and the final pressure and temperature (P_2 and T_2).

Solution

[1] Identify the known quantities and the desired quantity.

$P_1 = 760$ mm Hg $P_2 = 540$ mm Hg
$T_1 = 25$ °C $T_2 = -40.$ °C
$V_1 = 2.0$ L $V_2 = ?$

known quantities desired quantity

- Both temperatures must be converted to Kelvin temperatures.
- $T_1 = T_K = T_C + 273 = 25\ °C + 273 = 298\ K$
- $T_2 = T_K = T_C + 273 = -40.\ °C + 273 = 233\ K$

---

[2] Write the equation and rearrange it to isolate the desired quantity, V_2, on one side.

- Use the combined gas law.

$$\frac{P_1V_1}{T_1} = \frac{P_2V_2}{T_2} \quad \text{Solve for } V_2 \text{ by multiplying both sides by } \frac{T_2}{P_2}.$$

$$\frac{P_1V_1T_2}{T_1P_2} = V_2$$

[3] Solve the problem.

- Substitute the five known quantities in the equation and solve for V_2.

$$V_2 = \frac{P_1V_1T_2}{T_1P_2} = \frac{(760 \text{ mm Hg})(2.0 \text{ L})(233 \text{ K})}{(298 \text{ K})(540 \text{ mm Hg})} = 2.2 \text{ L}$$

Answer

Kelvins and mm Hg cancel.

[3] Avogadro's Law (7.4)

Example 7.5 The lungs of an average dog hold 0.15 mol of air in a volume of 3.6 L. How many moles of air do the lungs of an average cat hold if the volume is 1.2 L? Assume the temperature and pressure are constant.

Analysis

This question deals with volume and number of moles, so Avogadro's law is used to determine a final number of moles when three quantities are known—the initial volume and number of moles (V_1 and n_1), and the final volume (V_2).

Solution

[1] Identify the known quantities and the desired quantity.

known quantities		desired quantity
$V_1 = 3.6$ L	$V_2 = 1.2$ L	
$n_1 = 0.15$ mol		$n_2 = ?$

[2] Write the equation and rearrange it to isolate the desired quantity, n_2, on one side.

- Use Avogadro's law. To solve for n_2, we must invert the numerator and denominator on *both* sides of the equation, and then multiply by V_2.

$$\boxed{\frac{V_1}{n_1} = \frac{V_2}{n_2}} \xrightarrow{\text{Switch } V \text{ and } n \text{ on both sides.}} \frac{n_1}{V_1} = \frac{n_2}{V_2} \quad \text{Solve for } n_2 \text{ by multiplying both sides by } V_2.$$

$$\frac{n_1V_2}{V_1} = n_2$$

[3] Solve the problem.

- Substitute the three known quantities in the equation and solve for n_2.

$$n_2 = \frac{n_1 V_2}{V_1} = \frac{(0.15 \text{ mol})(1.2 \cancel{L})}{(3.6 \cancel{L})} = 0.050 \text{ mol}$$

Answer

Liters cancel.

---

[4] The Ideal Gas Law (7.5)

Example 7.6 What volume does 15.0 g of O_2 occupy at 1.00 atm and 32 °C?

Analysis
Use the ideal gas law to calculate *V*, since *P* and *T* are known and *n* can be determined by using the molar mass of O_2 (32.00 g/mol).

Solution
[1] Identify the known quantities and the desired quantity.

P = 1.00 atm
T = 32 °C   15.0 g O_2 V = ? L
known quantities   desired quantity

---

[2] Convert all values to proper units and choose the value of *R* that contains these units.

- Convert T_C to T_K: $T_K = T_C + 273 = 32\ °C + 273 = 305$ K.
- Use the value of *R* with atm since the pressure is given in atm; that is, $R = 0.0821$ L•atm/(mol•K).
- Convert the number of grams of O_2 to the number of moles of O_2 using the molar mass (32.00 g/mol).

molar mass conversion factor

$$15.0 \cancel{g}\ O_2 \times \frac{1 \text{ mol } O_2}{32.00 \cancel{g}\ O_2} = 0.469 \text{ mol } O_2$$

Grams cancel.

---

[3] Write the equation and rearrange it to isolate the desired quantity, *V*, on one side.

- Use the ideal gas law and solve for *V* by dividing both sides by *P*.

$$PV = nRT$$ Solve for *V* by dividing both sides by *P*.

$$V = \frac{nRT}{P}$$

---

[4] Solve the problem.

- Substitute the three known quantities in the equation and solve for *V*.

$$V = \frac{nRT}{P} = \frac{(0.469\ \cancel{mol})\left(0.0821\ \frac{L \cdot \cancel{atm}}{\cancel{mol} \cdot \cancel{K}}\right)(305\ \cancel{K})}{1.00\ \cancel{atm}} = 11.7\ L$$

Answer

---

[5] Dalton's Law and Partial Pressures (7.6)

Example 7.7 A sample of air contains four gases with the following partial pressures: N_2 (530. mm Hg), O_2 (129 mm Hg), CO_2 (48 mm Hg), and H_2O (52 mm Hg). What is the total pressure of the sample?

Analysis
Using Dalton's law, the total pressure is the sum of the partial pressures.

Solution
Adding up the four partial pressures gives the total:

530. + 129 + 48 + 52 = 759 mm Hg (total pressure)

Example 7.8 A mixture of gas contains 31% O_2, 68% N_2, and 1% CO_2 by volume. What is the partial pressure of each gas at sea level, where the total pressure is 760 mm Hg?

Analysis
Convert each percent to a decimal by moving the decimal point two places to the left. Multiply each decimal by the total pressure to obtain the partial pressure for each component.

Solution

			Partial pressure
Fraction O_2: 31% = 0.31	0.31 × 760 mm Hg	=	240 mm Hg (O_2)
Fraction N_2: 68% = 0.68	0.68 × 760 mm Hg	=	520 mm Hg (N_2)
Fraction CO_2: 1% = 0.01	0.01 × 760 mm Hg	=	8 mm Hg (CO_2)
			768 rounded to 770 mm Hg

---

[6] Intermolecular Forces, Boiling Point, and Melting Point (7.7)

Example 7.9 What types of intermolecular forces are present in each compound?
a. HBr b. CH_4 c. H_2O

Analysis
- London dispersion forces are present in all covalent compounds.
- Dipole–dipole interactions are present only in polar compounds with a permanent dipole.
- Hydrogen bonding occurs only in compounds that contain an O–H, N–H, or H–F bond.

Solution

a. $\overset{\delta+}{H}$–$\overset{\delta-}{Br}$ (polar bond)

- HBr has London forces like all covalent compounds.
- HBr has a polar bond so it exhibits dipole–dipole interactions.
- HBr has no H atom on an O, N, or F, so it has no intermolecular hydrogen bonding.

b. H–C(H)(H)–H

nonpolar molecule

- CH_4 is a nonpolar molecule because it has only nonpolar C–H bonds. Thus, it exhibits only London forces.

c. $\overset{\delta+}{H}$–$\overset{\delta-}{O}$–$\overset{\delta+}{H}$ ⇧ net dipole

- H_2O has London forces like all covalent compounds.
- H_2O has a net dipole from its two polar bonds (Section 4.8), so it exhibits dipole–dipole interactions.
- H_2O has a H atom bonded to O, so it exhibits intermolecular hydrogen bonding.

---

[7] The Liquid State (7.8)

Example 7.10 Using the given boiling points, predict which compound has the higher vapor pressure at a given temperature: ethanol (C_2H_6O, bp 78 °C) or 1-propanol (C_3H_8O, bp 97 °C).

Analysis
A higher boiling point means a compound has a lower vapor pressure and stronger forces.

Solution
Since ethanol has a lower boiling point than 1-propanol, ethanol has a higher vapor pressure and 1-propanol has a lower vapor pressure at a given temperature.

Self-Test

[1] Fill in the blank with one of the terms listed below.

Amorphous solid (7.9)	Freezing (7.11)	Surface tension (7.8)
Boiling point (7.7)	London dispersion forces (7.7)	Universal gas constant (7.5)
Condensation (7.8, 7.11)	Melting (7.11)	Vapor pressure (7.8)
Crystalline solid (7.9)	Pressure (7.2)	Vaporization (7.11)
Dipole–dipole interactions (7.7)	Standard molar volume (7.4)	Viscosity (7.8)

1. Converting a liquid to a gas is called ______.
2. ______ is the pressure exerted by gas molecules in equilibrium with the liquid phase.
3. ______ are very weak interactions due to the momentary changes in electron density in a molecule.
4. The ______ of a compound is the temperature at which a liquid is converted to the gas phase.

5. Converting a solid to a liquid is called ______.
6. The product of pressure and volume divided by the product of moles and Kelvin temperature is a constant, called the ______ and symbolized by *R*.
7. ______ is a measure of a fluid's resistance to flow freely.
8. At STP, one mole of any gas has the same volume, 22.4 L, called the ______.
9. ______ converts a gas to a liquid.
10. A ______ has a regular arrangement of particles—atoms, molecules, or ions—with a repeating structure.
11. An ______ has no regular arrangement of its closely packed particles.
12. ______ are the attractive forces between the permanent dipoles of two polar molecules.
13. ______ is the force exerted per unit area.
14. ______ converts a liquid to a solid.
15. ______ is a measure of the resistance of a liquid to spread out.

[2] Fill in each blank with one of the terms listed below.

a. pressure b. volume c. temperature

16. Boyle's law relates ______ and ______.
17. Charles's law relates ______ and ______.
18. Gay–Lussac's law relates ______ and ______.

[3] Pick the appropriate type of solid for each compound.

a. ionic solid
b. molecular solid
c. network solid
d. metallic solid

19. graphite
20. silver
21. NaBr
22. H_2O

[4] Fill in each blank with one of the terms listed below.

a. vaporization b. sublimation c. deposition

23. Condensation is the opposite of ______.
24. When a gas is converted directly to a solid, this is called ______.
25. When a solid forms a gas without passing through a liquid phase, this is called ______.

Answers to Self-Test

1. vaporization
2. Vapor pressure
3. London dispersion forces
4. boiling point
5. melting
6. universal gas constant
7. Viscosity
8. standard molar volume
9. Condensation
10. crystalline solid
11. amorphous solid
12. Dipole–dipole interactions
13. Pressure
14. Freezing
15. Surface tension
16. a and b
17. b and c
18. a and c
19. c
20. d
21. a
22. b
23. a
24. c
25. b

Solutions to In-Chapter Problems

7.1 Use Table 7.1 to compare the features of different states of methanol.

	a. Density	b. Intermolecular Spacing	c. Intermolecular Attraction
Gas	Lowest	Greatest	Lowest
Liquid	Higher	Smaller	Higher
Solid	Highest	Smallest	Highest

7.2 Use conversion factors to solve the problems.

a. $3.0\ \cancel{\text{atm}} \times \dfrac{760.\ \text{mm Hg}}{1\ \cancel{\text{atm}}} = 2{,}300\ \text{mm Hg}$ **Answer**

Atm cancels.

c. $424\ \cancel{\text{mm Hg}} \times \dfrac{1\ \text{atm}}{760.\ \cancel{\text{mm Hg}}} = 0.558\ \text{atm}$ **Answer**

mm Hg cancels.

b. $720\ \cancel{\text{mm Hg}} \times \dfrac{14.7\ \text{psi}}{760.\ \cancel{\text{mm Hg}}} = 14\ \text{psi}$ **Answer**

mm Hg cancels.

7.3 Use Boyle's law to solve the problems as in Example 7.2.

a. $V_2 = \dfrac{P_1V_1}{P_2} = \dfrac{(4.0\ \cancel{\text{atm}})(2.0\ \text{L})}{5.0\ \cancel{\text{atm}}} = 1.6\ \text{L}$ **Answer**

Atm cancels.

b. $V_2 = \dfrac{P_1V_1}{P_2} = \dfrac{(4.0\ \cancel{\text{atm}})(2.0\ \text{L})}{2.5\ \cancel{\text{atm}}} = 3.2\ \text{L}$ **Answer**

Atm cancels.

c. $V_2 = \dfrac{P_1V_1}{P_2} = \dfrac{(4.0\ \cancel{\text{atm}})(2.0\ \text{L})}{10.0\ \cancel{\text{atm}}} = 0.80\ \text{L}$ **Answer**

Atm cancels.

d. $4.0\ \cancel{\text{atm}} \times \dfrac{760.\ \text{mm Hg}}{1\ \cancel{\text{atm}}} = 3.0 \times 10^3\ \text{mm Hg}$ $\quad V_2 = \dfrac{P_1V_1}{P_2} = \dfrac{(3.0 \times 10^3\ \cancel{\text{mm Hg}})(2.0\ \text{L})}{380\ \cancel{\text{mm Hg}}} = 16\ \text{L}$ **Answer**

mm Hg cancels.

7.4 Use Boyle's law to solve the problems as in Example 7.2.

a. $P_2 = \dfrac{P_1V_1}{V_2} = \dfrac{(0.50\ \text{atm})(15.0\ \cancel{\text{mL}})}{30.0\ \cancel{\text{mL}}} = 0.25\ \text{atm}$ **Answer**

mL cancels.

b. $P_2 = \dfrac{P_1V_1}{V_2} = \dfrac{(0.50\ \text{atm})(15.0\ \cancel{\text{mL}})}{5.0\ \cancel{\text{mL}}} = 1.5\ \text{atm}$ **Answer**

mL cancels.

c. $P_2 = \frac{P_1V_1}{V_2} = \frac{(0.50 \text{ atm})(15.0 \text{ mL})}{100. \text{ mL}} = 0.075 \text{ atm}$ **Answer**

mL cancels.

d. $1.0 \text{ L} = 1.0 \times 10^3 \text{ mL}$ $P_2 = \frac{P_1V_1}{V_2} = \frac{(0.50 \text{ atm})(15.0 \text{ mL})}{1.0 \times 10^3 \text{ mL}} = 0.0075 \text{ atm}$ **Answer**

mL cancels.

7.5 Use Charles's law to solve the problem as in Example 7.3. Remember to convert temperature to T_K.

$T_K = T_C + 273$
$T_1 = 37 \text{ °C} + 273 = 310. \text{ K}$
$T_2 = 0.0 \text{ °C} + 273 = 273 \text{ K}$

$V_2 = \frac{V_1T_2}{T_1} = \frac{(0.50 \text{ L})(273 \text{ K})}{310. \text{ K}} = 0.44 \text{ L}$ **Answer**

Kelvins cancel.

7.6 Use Charles's law to solve the problems as in Example 7.3.

a. $V_2 = \frac{V_1T_2}{T_1} = \frac{(25.0 \text{ L})(450 \text{ K})}{45 \text{ K}} = 250 \text{ L}$ **Answer**

Kelvins cancel.

b. $T_K = T_C + 273$
$T_1 = 400. \text{ °C} + 273 = 673 \text{ K}$
$T_2 = 50. \text{ °C} + 273 = 323 \text{ K}$

$V_2 = \frac{V_1T_2}{T_1} = \frac{(50.0 \text{ mL})(323 \text{ K})}{673 \text{ K}} = 24 \text{ mL}$ **Answer**

Kelvins cancel.

7.7 Use Gay–Lussac's law to solve the problem.

$$\frac{P_1}{T_1} = \frac{P_2}{T_2}$$

$$\frac{P_2T_1}{P_1} = T_2$$

$T_K = T_C + 273$
$T_1 = 100. \text{ °C} + 273 = 373 \text{ K}$

$T_2 = \frac{P_2T_1}{P_1} = \frac{(1.05 \text{ atm})(373 \text{ K})}{1.00 \text{ atm}} = 392 \text{ K or } 119 \text{ °C}$ **Answer**

Atm cancels.

7.8 Use Gay–Lussac's law to solve the problem as in Sample Problem 7.4. $T_1 = 25 \text{ °C} + 273 = 298 \text{ K}$.

a. $P_2 = \frac{P_1T_2}{T_1} = \frac{(1.00 \text{ atm})(310. \text{ K})}{298 \text{ K}} = 1.04 \text{ atm}$ **Answer**

Kelvins cancel.

b. $P_2 = \frac{P_1T_2}{T_1} = \frac{(1.00\text{ atm})(150.\text{ K})}{298\text{ K}} = 0.503\text{ atm}$ **Answer**

Kelvins cancel.

c. $P_2 = \frac{P_1T_2}{T_1} = \frac{(1.00\text{ atm})(323\text{ K})}{298\text{ K}} = 1.08\text{ atm}$ **Answer**

Kelvins cancel.

d. $P_2 = \frac{P_1T_2}{T_1} = \frac{(1.00\text{ atm})(473\text{ K})}{298\text{ K}} = 1.59\text{ atm}$ **Answer**

Kelvins cancel.

7.9 Use the combined gas law to solve the problem as in Example 7.4.

$$\frac{P_1V_1}{T_1} = \frac{P_2V_2}{T_2}$$

$$\frac{P_1V_1T_2}{T_1V_2} = P_2$$

$$P_2 = \frac{P_1V_1T_2}{T_1V_2} = \frac{(750\text{ mm Hg})(1.0\text{ L})(233\text{ K})}{(298\text{ K})(2.0\text{ L})} = 290\text{ mm Hg}$$ **Answer**

7.10 Use Avogadro's law to solve the problems as in Example 7.5.

$$\frac{V_1}{n_1} = \frac{V_2}{n_2} \qquad \frac{n_2V_1}{n_1} = V_2$$ Solve for V_2 by multiplying both sides by n_2.

a. $V_2 = \frac{n_2V_1}{n_1} = \frac{(2.5\text{ mol})(3.5\text{ L})}{(5.0\text{ mol})} = 1.8\text{ L}$ **Answer**

Mol cancel.

b. $V_2 = \frac{n_2V_1}{n_1} = \frac{(3.65\text{ mol})(3.5\text{ L})}{(5.0\text{ mol})} = 2.6\text{ L}$ **Answer**

Mol cancel.

c. $V_2 = \frac{n_2V_1}{n_1} = \frac{(21.5\text{ mol})(3.5\text{ L})}{(5.0\text{ mol})} = 15\text{ L}$ **Answer**

Mol cancel.

7.11 Convert moles to volume at STP as in Sample Problem 7.7.

a. $4.5\text{ mol } O_2 \times \frac{22.4\text{ L}}{1\text{ mol}} = 1.0 \times 10^2\text{ L}$ **Answer**

b. $0.35 \text{ mol } O_2 \times \dfrac{22.4 \text{ L}}{1 \text{ mol}} = 7.8 \text{ L}$ **Answer**

c. $18.0 \text{ g } O_2 \times \dfrac{1 \text{ mol}}{32.00 \text{ g}} \times \dfrac{22.4 \text{ L}}{1 \text{ mol}} = 12.6 \text{ L}$ **Answer**

7.12 Use conversion factors to solve the problems.

a. $1.5 \text{ L} \times \dfrac{1 \text{ mol}}{22.4 \text{ L}} = 0.067 \text{ mol}$ **Answer**

b. $8.5 \text{ L} \times \dfrac{1 \text{ mol}}{22.4 \text{ L}} = 0.38 \text{ mol}$ **Answer**

c. $25 \text{ mL} \times \dfrac{1 \text{ L}}{1000 \text{ mL}} \times \dfrac{1 \text{ mol}}{22.4 \text{ L}} = 1.1 \times 10^{-3} \text{ mol}$ **Answer**

7.13 Use the ideal gas law to solve the problem as in Example 7.6.

P = 175 atm
T = 20. °C V = 5.0 L n = ? mol
known quantities desired quantity

Convert T_C to T_K. $T_K = T_C + 273 = 20.\ °C + 273 = 293 \text{ K}$

Use the value of R with atm since the pressure is given in atm; that is, $R = 0.0821 \text{ L·atm/(mol·K)}$.

$PV = nRT$ Solve for n by dividing both sides by RT.

$$n = \frac{PV}{RT} = \frac{(5.0 \text{ L})(175 \text{ atm})}{\left(0.0821 \dfrac{\text{L} \cdot \text{atm}}{\text{mol} \cdot \text{K}}\right)(293 \text{ K})} = 36 \text{ mol}$$ **Answer**

7.14 Use the ideal gas law to solve the problems as in Example 7.6 and Answer 7.13.

a. $PV = nRT$ Solve for P by dividing both sides by V. $R = 0.0821 \text{ L·atm/(mol·K)}$
$T_K = T_C + 273 = 25\ °C + 273 = 298 \text{ K}$

$$P = \frac{nRT}{V} = \frac{0.45 \text{ mol}\left(0.0821 \dfrac{\text{L} \cdot \text{atm}}{\text{mol} \cdot \text{K}}\right)(298 \text{ K})}{(10.0 \text{ L})} = 1.1 \text{ atm}$$ **Answer**

b. $PV = nRT$ Solve for P by dividing both sides by V.

$R = 0.0821$ L•atm/(mol•K)

$T_K = T_C + 273 = 20.\ °C + 273 = 293$ K

$$P = \frac{nRT}{V}$$

molar mass conversion factor

$$10.0\ \cancel{g}\ N_2 \times \frac{1\ mol\ N_2}{28.02\ \cancel{g}\ N_2} = 0.357\ mol\ N_2$$

Grams cancel.

$$P = \frac{nRT}{V} = \frac{0.357\ \cancel{mol}\left(0.0821\ \frac{\cancel{L} \cdot atm}{\cancel{mol} \cdot \cancel{K}}\right)(293\ \cancel{K})}{(5.0\ \cancel{L})} = 1.7\ atm$$

Answer

7.15 Use Dalton's law to solve the problem as in Example 7.7.

$P_{total} = P_{O_2} + P_{CO_2}$
4.0 atm = 2.5 atm $+ P_{CO_2}$
CO_2: 1.5 atm O_2: 2.5 atm

7.16 a. The percent of N_2 and O_2 is determined by how many molecules of each are present compared to the total number of molecules (eight).

N_2: $\frac{5\ N_2\ molecules}{8\ total\ molecules} = 0.625 = 62.5\%$ O_2: $\frac{3\ O_2\ molecules}{8\ total\ molecules} = 0.375 = 37.5\%$

b. Use the fractions from part (a) and the total pressure (3,500 psi) to determine the partial pressures.

N_2: 0.625 x 3500 psi = 2,200 psi O_2: 0.375 x 3500 psi = 1,300 psi

7.17 Since the atmosphere contains 21% O_2 regardless of location, use the partial pressure of O_2 (P_{O_2}) to determine the total pressure (P_{total}). Convert 21% to a decimal by moving the decimal point two places to the left (0.21).

$$P_{O_2} = 0.21 \times P_{total}$$

$$\frac{P_{O_2}}{0.21} = P_{total}$$

$$\frac{78.5\ mm\ Hg}{0.21} = 370\ mm\ Hg$$

atmospheric pressure on Kilimanjaro

7.18 London dispersion forces are very weak interactions due to the momentary changes in electron density in a molecule. All molecules exhibit these forces (a–d).

7.19 With dipole–dipole interactions, molecules align so that partial positive and partial negative charges are close together.

$\delta^+ \delta^-$ $\delta^+ \delta^-$
H–Cl H–Cl
$+\!\!\longrightarrow$ $+\!\!\longrightarrow$

7.20 Use Table 7.3 to decide which types of forces are present in each molecule. A "+" means the force is present, whereas a "–" means the force is absent.

	London Dispersion	Dipole–Dipole	Hydrogen Bonding
a. Cl_2	+	no net dipole	–
b. HCN	+	+	–
c. HF	+	+	+ (the only molecule listed with an O–H, N–H, or H–F bond)
d. CH_3Cl	+	+	–
e. H_2	+	no net dipole	–

7.21 London dispersion forces are present in all covalent compounds and atoms.
Dipole–dipole interactions are present only in polar compounds with a permanent dipole.
Hydrogen bonding occurs only in compounds that contain an O–H, N–H, or H–F bond.

a. H_2O has hydrogen bonding, which is stronger than the London forces in CO_2.
b. HBr has dipole–dipole interactions, which are stronger than the London forces in CO_2.
c. H_2O has hydrogen bonding, which is stronger than the dipole–dipole interactions in HBr.
d. Ne atoms have stronger London dispersion forces because they are larger than He atoms.

7.22 The *stronger* the intermolecular forces, the *higher* the boiling point and melting point.

a. C_2H_6 has stronger London dispersion forces, so it has a higher boiling point and melting point than CH_4.
b. CH_3OH can hydrogen bond, so it has a higher boiling point and melting point than C_2H_6.
c. Br is larger than Cl, so HBr has a higher boiling point and melting point than HCl.
d. CH_3Br has dipole–dipole interactions, so it has a higher boiling point and melting point than C_2H_6, which has only London dispersion forces.

7.23 Water has stronger intermolecular forces because it can hydrogen bond, whereas CO_2 has only London dispersion forces. As a result, water is a liquid at room temperature, whereas CO_2 is a gas.

7.24 The stronger the intermolecular forces, the lower the vapor pressure at a given temperature.

a. CH_4 has a higher vapor pressure because it has only London dispersion forces, whereas NH_3 can form hydrogen bonds.
b. CH_4 has a higher vapor pressure because it has only London dispersion forces and is a smaller molecule than C_2H_6.
c. C_2H_6 has a higher vapor pressure because it has only London dispersion forces, whereas CH_3OH can form hydrogen bonds.

7.25 The *stronger* the intermolecular forces, the stronger surface molecules are pulled down toward the interior of a liquid and the *higher* the **surface tension.**

The surface tension of gasoline should be lower than that of water, because gasoline cannot hydrogen bond, whereas water can, giving it higher surface tension.

7.26 An **ionic solid** is composed of oppositely charged ions.
A **molecular solid** is composed of individual molecules arranged regularly.
A **network solid** is composed of a vast number of atoms covalently bonded together, forming sheets or three-dimensional arrays.
A **metallic solid,** such as copper or silver, can be treated as a lattice of metal cations surrounded by a cloud of electrons that move freely.

a. $CaCl_2$ = ionic
b. Fe = metallic
c. sugar ($C_{12}H_{22}O_{11}$) = molecular
d. $NH_3(s)$ = molecular

7.27
- The larger the specific heat, the less the temperature of a substance will change when it absorbs heat energy.
- The larger the specific heat, the more heat that must be added to increase the temperature of a substance a given number of degrees.

a. Sand has the higher temperature because it has the lower specific heat.
b. Ethanol requires the larger amount of heat because it has the larger specific heat.

7.28 Water has a high specific heat, so it can absorb or release a great deal of energy with only a small temperature change.

7.29 Use specific heat as a conversion factor to calculate the amount of heat absorbed using the known mass and temperature change.

$150.\ °C - 19\ °C = 131\ °C = \Delta T$ $\quad$ cal = mass x ΔT x SH

$$= 28.0\ \cancel{g} \times 131\ \cancel{°C} \times \frac{0.107\ \text{cal}}{1\ \cancel{g} \cdot 1\ \cancel{°C}} = 392\ \text{cal}$$

(heat absorbed in cal)

$$= 392\ \cancel{\text{cal}} \times \frac{4.184\ \text{J}}{1\ \cancel{\text{cal}}} = 1{,}640\ \text{J}$$

(heat absorbed in J)

7.30

$55\ °C - 12\ °C = 43\ °C = \Delta T$ $\quad$ cal = mass x ΔT x SH

$$= 200.\ \cancel{g} \times 43\ \cancel{°C} \times \frac{1.00\ \text{cal}}{1\ \cancel{g} \cdot 1\ \cancel{°C}} = 8{,}600\ \text{cal}$$

$$= 8.6\ \text{kcal}$$

7.31 Use specific heat as a conversion factor to determine the temperature change.

Cu: $\Delta T = \dfrac{\text{cal}}{\text{g} \cdot SH}$

$= \dfrac{20.\ \text{cal}}{10.0\ \text{g} \times 0.0900\ \text{cal/(g} \cdot {}^\circ\text{C)}}$

$= 22\ {}^\circ\text{C}$

$T_2 = 15\ {}^\circ\text{C} + 22\ {}^\circ\text{C} = 37\ {}^\circ\text{C}$

Hg: $\Delta T = \dfrac{\text{cal}}{\text{g} \cdot SH}$

$= \dfrac{20.\ \text{cal}}{10.\ \text{g} \times 0.0335\ \text{cal/(g} \cdot {}^\circ\text{C)}}$

$= 60.\ {}^\circ\text{C}$

$T_2 = 15\ {}^\circ\text{C} + 60.\ {}^\circ\text{C} = 75\ {}^\circ\text{C}$

7.32 Use specific heat as a conversion factor to determine the temperature change.

$\Delta T = \dfrac{\text{cal}}{\text{g} \cdot SH}$

$= \dfrac{950.\ \text{cal}}{120.\ \text{g} \times 0.583\ \text{cal/(g} \cdot {}^\circ\text{C)}}$

$= 13.6\ {}^\circ\text{C}$

$T_2 = 20.\ {}^\circ\text{C} + 13.6\ {}^\circ\text{C} = 34\ {}^\circ\text{C}$

7.33 Use the heat of fusion as a conversion factor to determine the amount of energy, as in Sample Problem 7.17.

a. $50.0\ \text{g} \times \dfrac{79.7\ \text{cal}}{1\ \text{g}} = 3{,}985\ \text{cal}$ rounded to 3,990 cal **Answer**

b. $35.0\ \text{g} \times \dfrac{79.7\ \text{cal}}{1\ \text{g}} = 2{,}790\ \text{cal}$ **Answer**

c. $35.0\ \text{g} \times \dfrac{79.7\ \text{cal}}{1\ \text{g}} \times \dfrac{1\ \text{kcal}}{1000\ \text{cal}} = 2.79\ \text{kcal}$ **Answer**

d. $1.00\ \text{mol} \times \dfrac{18.02\ \text{g}}{1\ \text{mol}} \times \dfrac{79.7\ \text{cal}}{1\ \text{g}} = 1{,}440\ \text{cal}$ **Answer**

7.34 Use the heat of vaporization to convert grams to an energy unit, calories, as in Sample Problem 7.18.

a. $42\ \text{g} \times \dfrac{540\ \text{cal}}{1\ \text{g}} = 23{,}000\ \text{cal}$ **Answer**

b. $42\ \text{g} \times \dfrac{540\ \text{cal}}{1\ \text{g}} = 23{,}000\ \text{cal}$ **Answer**

c. $1.00\ \text{mol} \times \dfrac{18.02\ \text{g}}{1\ \text{mol}} \times \dfrac{540\ \text{cal}}{1\ \text{g}} \times \dfrac{1\ \text{kcal}}{1000\ \text{cal}} = 9.7\ \text{kcal}$ **Answer**

d. $3.5\ \text{mol} \times \dfrac{18.02\ \text{g}}{1\ \text{mol}} \times \dfrac{540\ \text{cal}}{1\ \text{g}} \times \dfrac{1\ \text{kcal}}{1000\ \text{cal}} = 34\ \text{kcal}$ **Answer**

7.35 The first graphic represents a gas (randomly placed spheres that are far apart) and the second graphic represents a liquid (closely packed but randomly arranged spheres). The molecular art, therefore, represents condensation, which is exothermic.

7.36 a. Since heat is added, the graph is a heating curve.
b. The melting point of the substance is at the plateau **B→C,** which is 40 °C.
c. The boiling point of the substance is at the plateau **D→E,** which is 77 °C.
d. Solid and liquid are present at the plateau **B→C.**
e. Only liquid is present along the **C→D** diagonal.

7.37 Follow the steps in Sample Problem 7.20. Two calculations are needed—the energy released when the water is cooled and the energy released when the water freezes to form ice.
[1] Calculate the heat released on cooling water (mass = 50.0 g) from 25 °C to 0.0 °C (25 °C – 0.0 °C = 25 °C = ΔT) using the specific heat.

$$\text{heat} = \text{mass} \times \Delta T \times \text{specific heat}$$
$$= 50.0\ \cancel{g} \times 25\ \cancel{°C} \times \frac{1.00\ \text{cal}}{1\ \cancel{g} \cdot 1\ \cancel{°C}}$$
$$= 1{,}250\ \text{cal rounded to } 1{,}300\ \text{cal}$$

[2] Calculate the heat released when 50.0 g of water freezes to ice at 0.0 °C using the heat of fusion.

$$\text{cal} = 50.0\ \cancel{g} \times \frac{79.7\ \text{cal}}{1\ \cancel{g}}$$
$$= 3{,}985\ \text{cal rounded to } 3{,}990\ \text{cal}$$

[3] Add the answer in [1] to the answer in [2].

Total energy = 1,300 + 3,990 = 5,290 cal rounded to 5,300 cal

7.38 Follow the steps in Sample Problem 7.20. Three calculations are needed—the energy needed to melt ice to water, the energy needed to heat liquid water to its boiling point, and the energy needed to vaporize liquid water to steam.
[1] Calculate the heat needed to melt ice to water (mass = 25.0 g) using the heat of fusion.

$$\text{cal} = 25.0\ \cancel{g} \times \frac{79.7\ \text{cal}}{1\ \cancel{g}}$$
$$= 1{,}993\ \text{cal rounded to } 1{,}990\ \text{cal}$$

[2] Calculate the heat needed to heat liquid water (25.0 g) from 0.0 °C to 100. °C (100. °C – 0.0 °C = 100. °C = ΔT) using the specific heat.

$$\text{heat} = \text{mass} \times \Delta T \times \text{specific heat}$$
$$= 25.0\ \cancel{g} \times 100.\ \cancel{°C} \times \frac{1.00\ \text{cal}}{1\ \cancel{g} \cdot 1\ \cancel{°C}}$$
$$= 2{,}500\ \text{cal} = 2.50 \times 10^3\ \text{cal}$$

[3] Calculate the heat needed to vaporize water to steam (25.0 g) using the heat of vaporization.

$$\text{cal} = 25.0\ \cancel{g} \times \frac{540\ \text{cal}}{1\ \cancel{g}}$$

$$= 13{,}500\ \text{cal rounded to } 14{,}000\ \text{cal}$$

[4] Add up the answers in steps [1]–[3].

$$\text{Total energy} = 1{,}990 + (2.50 \times 10^3) + 14{,}000$$

$$= 18{,}490\ \text{cal rounded to } 18{,}000\ \text{cal}$$

Solutions to Odd-Numbered End-of-Chapter Problems

7.39 a. From the pressure gauge: 2,600 psi.
b. Use a psi–mm Hg conversion factor.

$$2600\ \cancel{\text{psi}} \times \frac{760.\ \text{mm Hg}}{14.7\ \cancel{\text{psi}}} = 130{,}000\ \text{mm Hg} = 1.3 \times 10^5\ \text{mm Hg}$$

7.41 Use conversion factors to solve the problems.

a. $2.8\ \cancel{\text{atm}} \times \frac{14.7\ \text{psi}}{1\ \cancel{\text{atm}}} = 41\ \text{psi}$ **Answer**

b. $520\ \cancel{\text{mm}}\ \text{Hg} \times \frac{1\ \text{atm}}{760.\ \cancel{\text{mm}}\ \text{Hg}} = 0.68\ \text{atm}$ **Answer**

c. $20.0\ \cancel{\text{atm}} \times \frac{760.\ \text{torr}}{1\ \cancel{\text{atm}}} = 15{,}200\ \text{torr}$ **Answer**

d. $100.\ \cancel{\text{mm}}\ \text{Hg} \times \frac{101{,}325\ \text{Pa}}{760.\ \cancel{\text{mm}}\ \text{Hg}} = 13{,}300\ \text{Pa}$ **Answer**

7.43 a. Largest number of particles = highest pressure = [3].
b. Smallest number of particles = lowest pressure = [2].

7.45 a. The volume decreases from **X** to (a) with the same number of gas particles; the pressure increases.
b. The volume remains the same, but the number of particles decreases; the pressure decreases.
c. The volume remains the same, but the number of particles increases; the pressure increases.

7.47 Use the gas laws to draw the balloons.

given balloon

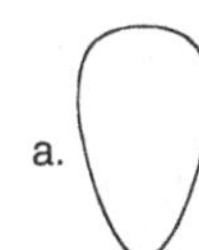

increased temperature
Volume increases.

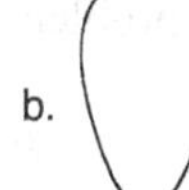

decreased pressure
Volume increases.

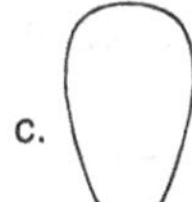

decreased pressure
Volume increases.

7.49 a. The volume increases as the outside atmospheric pressure decreases when the balloon floats to a higher altitude.
b. The volume decreases at the lower temperature when the balloon is placed in liquid nitrogen.
c. The volume decreases as the external pressure increases in the hyperbaric chamber.
d. The volume increases as the temperature increases in the microwave.

7.51 Use Boyle's law to fill in the table as in Example 7.2.

	P_1	V_1	P_2	V_2
a.	2.0 atm	3.0 L	8.0 atm	**0.75 L**
b.	55 mm Hg	0.35 L	18 mm Hg	**1.1 L**
c.	705 mm Hg	215 mL	**99.7 mm Hg**	1.52 L

7.53 Use Boyle's law to solve the problem as in Example 7.2.

$$V_2 = \frac{P_1V_1}{P_2} = \frac{(3.5\ \cancel{\text{atm}})(10.\ \text{mL})}{1.0\ \cancel{\text{atm}}} = 35\ \text{mL}$$

Answer

7.55 Use Charles's law to fill in the table as in Example 7.3.

	V_1	T_1	V_2	T_2
a.	5.0 L	310 K	**4.0 L**	250 K
b.	150 mL	45 K	**1.1 L**	45 °C
c.	60.0 L	0.0 °C	180 L	**820 K**

7.57 Use Charles's law to solve the problem as in Example 7.3.

$T_K = T_C + 273$
$T_1 = 25\ °C + 273 = 298\ K$
$T_2 = -78\ °C + 273 = 195\ K$

$$V_2 = \frac{V_1T_2}{T_1} = \frac{(2.2\ \text{L})(195\ \cancel{\text{K}})}{298\ \cancel{\text{K}}} = 1.4\ \text{L}$$

Answer

7.59 Use Gay–Lussac's law to fill in the table as in Sample Problem 7.4.

	P_1	T_1	P_2	T_2
a.	3.25 atm	298 K	**4.34 atm**	398 K
b.	550 mm Hg	273 K	**350 mm Hg**	–100. °C
c.	0.50 atm	250 °C	955 mm Hg	**1,300 K**

7.61 Use Gay–Lussac's law to solve the problem as in Sample Problem 7.4.

$T_K = T_C + 273$
$T_1 = 100.\ °C + 273 = 373\ K$
$T_2 = 150.\ °C + 273 = 423\ K$

$$P_2 = \frac{P_1T_2}{T_1} = \frac{(1.0\ \text{atm})(423\ \cancel{\text{K}})}{373\ \cancel{\text{K}}} = 1.1\ \text{atm}$$

Answer

7.63 Use the combined gas law to solve the problems as in Example 7.4.

	P_1	V_1	T_1	P_2	V_2	T_2
a.	0.90 atm	4.0 L	265 K	**1.4 atm**	3.0 L	310 K
b.	1.2 atm	75 L	5.0 °C	700. mm Hg	**110 L**	50 °C
c.	200. mm Hg	125 mL	298 K	100. mm Hg	0.62 L	**740 K**

7.65 Convert volume to moles at STP as in Sample Problem 7.7.

a. $5.0\,\cancel{L}\text{ He} \times \dfrac{1\text{ mol}}{22.4\,\cancel{L}} = 0.22\text{ mol}$ **Answer**

b. $11.2\,\cancel{L}\text{ He} \times \dfrac{1\text{ mol}}{22.4\,\cancel{L}} = 0.500\text{ mol}$ **Answer**

c. $50.0\,\cancel{mL}\text{ He} \times \dfrac{1\,\cancel{L}}{1000\,\cancel{mL}} \times \dfrac{1\text{ mol}}{22.4\,\cancel{L}} = 0.002\ 23\text{ mol}$ **Answer**

7.67 Convert moles and grams to volume at STP as in Sample Problem 7.7.

a. $4.2\,\cancel{mol}\text{ Ar} \times \dfrac{22.4\text{ L}}{1\,\cancel{mol}} = 94\text{ L}$ **Answer**

b. $3.5\,\cancel{g}\,CO_2 \times \dfrac{1\,\cancel{mol}}{44.01\,\cancel{g}} \times \dfrac{22.4\text{ L}}{1\,\cancel{mol}} = 1.8\text{ L}$ **Answer**

c. $2.1\,\cancel{g}\,N_2 \times \dfrac{1\,\cancel{mol}}{28.02\,\cancel{g}} \times \dfrac{22.4\text{ L}}{1\,\cancel{mol}} = 1.7\text{ L}$ **Answer**

7.69 Use the ideal gas law to solve the problem as in Example 7.6.

Convert T_C to T_K. $T_K = T_C + 273 = 37\text{ °C} + 273 = 310.\text{ K}$

Use the value of R with mm Hg since the pressure is given in mm Hg; that is, $R = 62.4\text{ L•mm Hg/(mol•K)}$.

$$PV = nRT$$ Solve for n by dividing both sides by RT.

$$n = \frac{PV}{RT} = \frac{(747\,\cancel{mm\,Hg})(0.45\,\cancel{L})}{\left(62.4\,\dfrac{\cancel{L}\cdot\cancel{mm\,Hg}}{\text{mol}\cdot\cancel{K}}\right)(310.\,\cancel{K})} = 0.017\text{ mol}$$ **Answer**

7.71 Use the ideal gas law to solve the problem as in Example 7.6.

Convert T_C to T_K. $T_K = T_C + 273 = 37\ °C + 273 = 310.\ K$

Use the value of R with mm Hg since the pressure is given in mm Hg; that is, $R = 62.4$ L•mm Hg/(mol•K).

$PV = nRT$ Solve for n by dividing both sides by RT.

$$n = \frac{PV}{RT} = \frac{(740\ \text{mm Hg})(5.0\ \text{L})}{\left(62.4\ \frac{\text{L} \cdot \text{mm Hg}}{\text{mol} \cdot \text{K}}\right)(310.\ \text{K})} = 0.19\ \text{mol}$$

Answer

$$0.19\ \text{mol} \times \frac{6.02 \times 10^{23}\ \text{molecules}}{1\ \text{mol}} = 1.1 \times 10^{23}\ \text{molecules}$$

Answer

7.73 Use the ideal gas law to solve the problem as in Example 7.6.

Use the value of R with mm Hg since the pressure is given in mm Hg; that is, $R = 62.4$ L•mm Hg/(mol•K).

$PV = nRT$ Solve for n by dividing both sides by RT.

$$n = \frac{PV}{RT} = \frac{(500.\ \text{mm Hg})(2.0\ \text{L})}{\left(62.4\ \frac{\text{L} \cdot \text{mm Hg}}{\text{mol} \cdot \text{K}}\right)(273\ \text{K})} = 0.059\ \text{mol} \times \frac{32.00\ \text{g}}{1\ \text{mol}} = 1.9\ \text{g}\ O_2$$

O_2 has more moles and more mass.

$$n = \frac{PV}{RT} = \frac{(650\ \text{mm Hg})(1.5\ \text{L})}{\left(62.4\ \frac{\text{L} \cdot \text{mm Hg}}{\text{mol} \cdot \text{K}}\right)(298\ \text{K})} = 0.052\ \text{mol} \times \frac{28.02\ \text{g}}{1\ \text{mol}} = 1.5\ \text{g}\ N_2$$

7.75 Use Dalton's law.

A: $\frac{3\ \text{red spheres}}{9\ \text{spheres total}} \times 630\ \text{mm Hg} = 210\ \text{mm Hg}$

B: $\frac{6\ \text{blue spheres}}{9\ \text{spheres total}} \times 630\ \text{mm Hg} = 420\ \text{mm Hg}$

7.77 Use Dalton's law to solve the problem as in Example 7.8.

Oxygen: 0.21×460 mm Hg = 97 mm Hg for O_2
Nitrogen: 0.78×460 mm Hg = 360 mm Hg for N_2

7.79 If the overall pressure is three times as great, the partial pressure is three times higher.

593 mm Hg × 3.00 = 1,780 mm Hg

7.81 Water is a liquid at room temperature because it is capable of hydrogen bonding and these strong intermolecular attractive forces give it a higher boiling point than H_2S, which can't hydrogen bond.

7.83 Use Table 7.3 to decide which types of forces are present in each molecule.

a. H–C(H)(H)–C(=O)–C(H)(H)–H

London dispersion forces
dipole–dipole interactions

b. H–C(H)(Br)–Cl

London dispersion forces
dipole–dipole interactions

7.85 Hydrogen bonding occurs only in compounds that contain an O–H, N–H, or H–F bond. Only compound (d) can hydrogen bond.

7.87 No, $H_2C{=}O$ has no H on the O atom, so it cannot form hydrogen bonds with another molecule of formaldehyde.

7.89 Undecane has the higher melting point. Although both pentane and undecane have only London dispersion forces, undecane is larger in size with greater surface area, so the London forces are stronger.

7.91 Vapor pressure and boiling point are inversely related: lower vapor pressure corresponds to a higher boiling point.

Increasing boiling point: butane < acetaldehyde < Freon-113

7.93 Glycerol is more viscous than water because it has three OH groups and many opportunities for hydrogen bonding. Acetone cannot hydrogen bond, so its intermolecular forces are weaker and it has low viscosity.

7.95 Use the definitions from Answer 7.26 to classify each solid.

a. KI: ionic
b. CO_2: molecular
c. bronze, an alloy of Cu and Sn: metallic
d. diamond: network
e. the plastic polyethylene: amorphous

7.97 The larger the specific heat, the less the temperature will change when a substance absorbs a particular amount of energy. Since the temperature of **Y** is less than the temperature of **X,** the specific heat of **Y** is higher than the specific heat of **X.**

7.99 Use specific heat as a conversion factor to calculate the energy change.

a. $\text{cal} = \text{g} \times \Delta T \times SH$

$= 50.\ \cancel{\text{g}} \times 35\ \cancel{^\circ\text{C}} \times \dfrac{1.00\ \text{cal}}{1\ \cancel{\text{g}} \cdot 1\ \cancel{^\circ\text{C}}}$

$= 1.8 \times 10^3\ \text{cal}$

$= 7.5 \times 10^3\ \text{J}$

b. $\text{cal} = \text{g} \times \Delta T \times SH$

$= 250\ \cancel{\text{g}} \times 75\ \cancel{^\circ\text{C}} \times \dfrac{0.214\ \text{cal}}{1\ \cancel{\text{g}} \cdot 1\ \cancel{^\circ\text{C}}}$

$= 4.0 \times 10^3\ \text{cal}$

$= 1.7 \times 10^4\ \text{J}$

7.101

a. H_2O $\Delta T = \dfrac{\text{cal}}{\text{g} \cdot SH} = \dfrac{200.\ \text{cal}}{100.\ \text{g} \cdot 1.00\ \text{cal}/(\text{g} \cdot {}^\circ\text{C})} = 2.00\ {}^\circ\text{C}$ $T_2 = 16.0\ {}^\circ\text{C} + 2.00\ {}^\circ\text{C} = 18.0\ {}^\circ\text{C}$

b. H_2O $\Delta T = \dfrac{350.\ \text{J}}{50.0\ \text{g} \cdot 4.18\ \text{J}/(\text{g} \cdot {}^\circ\text{C})} = 1.67\ {}^\circ\text{C}$ $T_2 = 16.0\ {}^\circ\text{C} + 1.67\ {}^\circ\text{C} = 17.7\ {}^\circ\text{C}$

7.103 The first graphic represents a liquid (closely packed but randomly arranged spheres) and the second graphic represents a gas (randomly placed spheres that are far apart). The molecular art, therefore, represents vaporization, so energy is absorbed.

7.105 a. Melting 100 g of ice is endothermic: energy is absorbed.
b. Freezing 25 g of water is exothermic: energy is released.
c. Condensing 20 g of steam is exothermic: energy is released.
d. Vaporizing 30 g of water is endothermic: energy is absorbed.

7.107

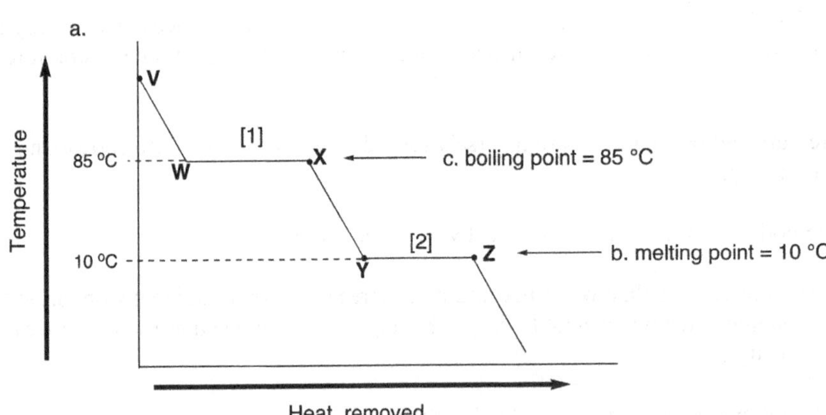

7.109

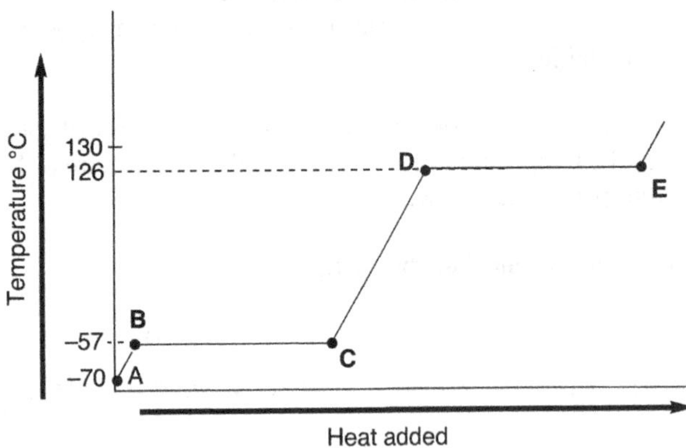

7.111 Use Sample Problem 7.20 and Answer 7.37 to solve the problem.

a. [1] cal $= 45\ \text{g} \times \dfrac{79.7\ \text{cal}}{1\ \text{g}}$

$= 3{,}600$ cal

[2] heat = mass x ΔT x SH

$= 45\ \text{g} \times 55\ °\text{C} \times \dfrac{1.00\ \text{cal}}{1\ \text{g} \cdot 1\ °\text{C}}$

$= 2{,}500$ cal

[3] Total energy $= 3{,}600 + 2{,}500 = 6{,}100$ cal

b. 6,100 cal

c. [1] cal $= 35\ \text{g} \times \dfrac{540\ \text{cal}}{1\ \text{g}} = 19{,}000$ cal

[2] heat = mass x ΔT x SH

$= 35\ \text{g} \times 100.\ °\text{C} \times \dfrac{1.00\ \text{cal}}{1\ \text{g} \cdot 1\ °\text{C}} = 3{,}500$ cal

[3] cal $= 35\ \text{g} \times \dfrac{79.7\ \text{cal}}{1\ \text{g}} = 2{,}800$ cal

[4] Total energy $= 19{,}000 + 3{,}500 + 2{,}800 = 25{,}000$ cal

$= 25$ kcal

7.113 The gases inside the bag had a volume of 250 mL at 760 mm Hg and take up a greater volume at the reduced pressure of 650 mm Hg.

$$P_1V_1 = P_2V_2 \qquad V_2 = \frac{P_1V_1}{P_2} = \frac{(760\ \text{mm Hg})(250\ \text{mL})}{650\ \text{mm Hg}} = 290\ \text{mL}$$

7.115 Water is one of the few substances that expands as it enters the solid phase. This causes the bottle to crack because the water occupies a larger volume when it freezes.

7.117 a. As a person breathes faster, he eliminates more CO_2 from the lungs; therefore, the measured value of CO_2 is lower than the normal value of 40 mm Hg.

b. Many people with advanced lung disease have lost lung tissue over time and therefore cannot exchange adequate amounts of oxygen through the lungs and into the blood, leading to a lower-than-normal partial pressure of oxygen. In addition, they often breathe more slowly and with lower volumes than normal, so they cannot eliminate enough CO_2 and therefore the partial pressure of CO_2 climbs.

7.119 Use the ideal gas law to solve the problem.

Use the value of R with atm since the pressure is given in atm; that is, $R = 0.0821$ L•atm/(mol•K).

$PV = nRT$ Solve for n by dividing both sides by RT.

$$n = \frac{PV}{RT} = \frac{(2\ \text{atm})(11.2\ \text{L})}{\left(0.0821\ \dfrac{\text{L} \cdot \text{atm}}{\text{mol} \cdot \text{K}}\right)(273\ \text{K})} = 1\ \text{mol}$$

Molar mass 4.0 g/mol = Helium

Chapter 8 Solutions

Chapter Review

[1] What are the characteristics of solutions, colloids, and suspensions? (8.1)

- A solution is a homogeneous mixture that contains small dissolved particles. Any phase of matter can form solutions. When two substances form a solution, the substance present in the lesser amount is the solute and the substance present in the larger amount is the solvent.
- A colloid is a homogeneous mixture with larger particles (1 nm–1 μm) often having an opaque appearance.
- A suspension is a heterogeneous mixture that contains large particles (> 1 μm) suspended in a liquid.

[2] What is the difference between an electrolyte and a nonelectrolyte? (8.2)

- An electrolyte dissolves in water to form ions, so it conducts an electric current. A strong electrolyte dissociates completely to form ions, whereas a weak electrolyte contains mostly uncharged molecules in water, together with a small number of ions.
- A nonelectrolyte dissolves in water to form uncharged molecules that do not conduct an electric current.

[3] What determines whether a substance is soluble in water or a nonpolar solvent? (8.3)

- One rule summarizes solubility: "Like dissolves like."
- Most ionic compounds are soluble in water. If the attractive forces between the ions and water are stronger than the attraction between the ions in the crystal, an ionic compound dissolves in water.
- Small polar compounds that can hydrogen bond are soluble in water.
- Nonpolar compounds are soluble in nonpolar solvents. Compounds with many nonpolar C–C and C–H bonds are soluble in nonpolar solvents.

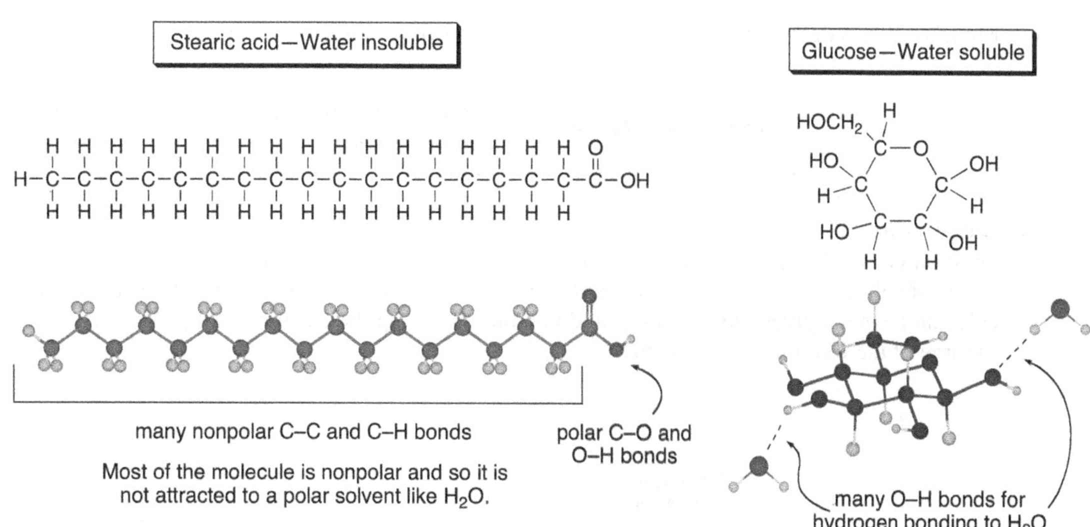

[4] What effect do temperature and pressure have on solubility? (8.4)

- The solubility of a solid in a liquid solvent generally increases with increasing temperature. The solubility of gases decreases with increasing temperature.

- Increasing pressure increases the solubility of a gas in a solvent. Pressure changes do not affect the solubility of liquids and solids.

[5] How is the concentration of a solution expressed? (8.5, 8.6)

- Concentration is a measure of how much solute is dissolved in a given amount of solution. It can be measured using mass, volume, or moles.
- Weight/volume percent (w/v)% concentration is the number of grams of solute dissolved in 100 mL of solution:

$$\textbf{Weight/volume percent concentration} \quad (w/v)\% = \frac{\text{mass of solute (g)}}{\text{volume of solution (mL)}} \times 100\%$$

- Volume/volume percent (v/v)% concentration is the number of milliliters of solute dissolved in 100 mL of solution:

$$\textbf{Volume/volume percent concentration} \quad (v/v)\% = \frac{\text{volume of solute (mL)}}{\text{volume of solution (mL)}} \times 100\%$$

- Parts per million (ppm) is the number of parts of solute in 1,000,000 parts of solution, where the units for both the solute and the solution are the same.

$$\textbf{Parts per million} \quad \text{ppm} = \frac{\text{mass of solute (g)}}{\text{mass of solution (g)}} \times 10^6$$

or

$$\text{ppm} = \frac{\text{volume of solute (mL)}}{\text{volume of solution (mL)}} \times 10^6$$

- Molarity (M) is the number of moles of solute per liter of solution.

$$\textbf{Molarity} = \textbf{M} = \frac{\textbf{moles of solute (mol)}}{\textbf{liter of solution (L)}}$$

[6] How are dilutions performed? (8.7)

- Dilution is the addition of solvent to decrease the concentration of a solute. Since the number of moles of solute is constant in carrying out a dilution, a new molarity (M_2) or volume (V_2) can be calculated from a given molarity (M_1) and volume (V_1) using the equation $M_1V_1 = M_2V_2$, as long as three of the four quantities are known.

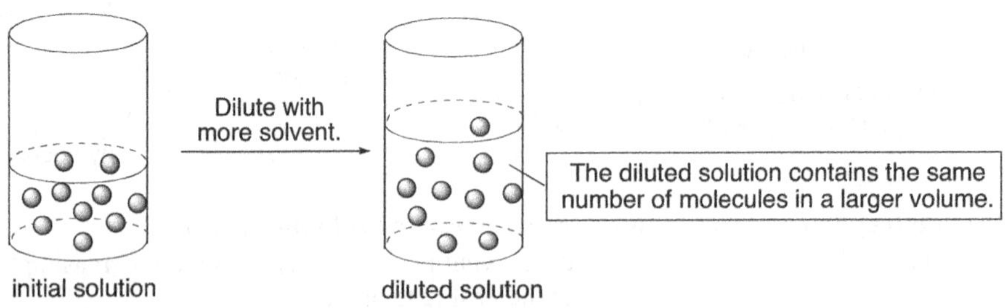
Dilute with more solvent.
The diluted solution contains the same number of molecules in a larger volume.
initial solution
diluted solution

[7] How do dissolved particles affect the boiling point and melting point of a solution? (8.8)

- A nonvolatile solute lowers the vapor pressure above a solution, thus increasing its boiling point.
- A nonvolatile solute makes it harder for solvent molecules to form a crystalline solid, thus decreasing its melting point.

[8] What is osmosis? (8.9)

- Osmosis is the selective diffusion of solvent, usually water, across a semipermeable membrane. Solvent always moves from the less concentrated solution to the more concentrated solution, until the osmotic pressure prevents additional flow of solvent.
- Since living cells contain and are surrounded by biological solutions separated by a semipermeable membrane, the osmotic pressure must be the same on both sides of the membrane. Dialysis is similar to osmosis in that it involves the selective passage of several substances—water, small molecules, and ions—across a dialyzing membrane.

Problem Solving

[1] Solubility–General Features (8.3)

Example 8.1 Predict the water solubility of each compound:

a. NaCl b. ethanol (CH_3CH_2OH) c. C_8H_{18}

Analysis

Use the general solubility rule: "Like dissolves like." Generally, ionic and polar compounds are soluble in water. Nonpolar compounds are soluble in nonpolar solvents, but insoluble in water.

Solution

a. NaCl is an ionic compound, so it dissolves in water, a polar solvent.
b. CH_3CH_2OH is a small polar molecule that contains an OH group. As a result, it can hydrogen bond to water, making it soluble.
c. C_8H_{18} has only nonpolar C–C and C–H bonds, making it a nonpolar molecule that is therefore water insoluble.

---

[2] Concentration Units–Percent Concentration (8.5)

Example 8.2 A solution contains 0.80 g of NaCl dissolved in a total volume of 25 mL. What is the weight/volume percent concentration of NaCl?

Analysis

Use the formula (w/v)% = (grams of solute)/(mL of solution) × 100%.

Solution

$$(w/v)\% = \frac{0.80 \text{ g NaCl}}{25 \text{ mL solution}} \times 100\% = 3.2\% \text{ (w/v) NaCl}$$

Answer

Example 8.3 A 1-L bottle of mouthwash contains 80 mL of ethanol. What is the volume/volume percent concentration of ethanol?

Analysis

Use the formula (v/v)% = (mL of solute)/(mL of solution) × 100%. Recall that 1 L = 1,000 mL.

Solution

$$(\text{v/v})\% = \frac{80 \text{ mL ethanol}}{1000 \text{ mL mouthwash}} \times 100\% = 8\% \text{ (v/v) ethanol}$$

Answer

Example 8.4 A solution contains 0.47% (w/v) of the antibiotic Abx in water. How many grams of antibiotic (Abx) are contained in 500. mL of this solution?

Analysis and Solution

[1] Identify the known quantities and the desired quantity.

0.47% (w/v) Abx solution 500. mL	? g Abx
known quantities	desired quantity

[2] Write out the conversion factors.

- Set up conversion factors that relate grams of antibiotic to the volume of the solution using the weight/volume percent concentration. Choose the conversion factor so that the unwanted unit, mL solution, cancels.

$$\frac{100 \text{ mL solution}}{0.47 \text{ g Abx}} \quad \text{or} \quad \boxed{\frac{0.47 \text{ g Abx}}{100 \text{ mL solution}}}$$

Choose this conversion factor to cancel mL.

[3] Solve the problem.

- Multiply the original quantity by the conversion factor to obtain the desired quantity.

$$500. \text{ mL} \times \frac{0.47 \text{ g Abx}}{100 \text{ mL solution}} = 2.4 \text{ g antibiotic}$$

Answer

Example 8.5 What volume of a 1.0% (w/v) solution of lidocaine (anesthetic) contains 20. mg?

Analysis and Solution

[1] Identify the known quantities and the desired quantity.

1.0% (w/v) lidocaine solution 20. mg	? mL lidocaine
known quantities	desired quantity

[2] Write out the conversion factors.

- Use the weight/volume percent concentration to set up conversion factors that relate grams of lidocaine to mL of solution. Since percent concentration is expressed in grams, a mg–g conversion factor is needed as well. Choose the conversion factors that place the unwanted units, mg and g, in the denominator to cancel.

mg–g conversion factors

$$\frac{1000 \text{ mg}}{1 \text{ g}} \quad \text{or} \quad \boxed{\frac{1 \text{ g}}{1000 \text{ mg}}}$$

g–mL solution conversion factors

$$\frac{1.0 \text{ g lidocaine}}{100 \text{ mL solution}} \quad \text{or} \quad \boxed{\frac{100 \text{ mL solution}}{1.0 \text{ g lidocaine}}}$$

Choose the conversion factors with the unwanted units—mg and g—in the denominator.

[3] Solve the problem.

- Multiply the original quantity by the conversion factors to obtain the desired quantity.

$$20.\ \cancel{\text{mg}} \text{ lidocaine} \times \frac{1\ \cancel{\text{g}}}{1000\ \cancel{\text{mg}}} \times \frac{100 \text{ mL solution}}{1.0\ \cancel{\text{g}} \text{ lidocaine}} = 2.0 \text{ mL solution}$$

Answer

Example 8.6 What is the concentration of **X** in parts per million in a solution that contains 45 mg of **X** in 5,000 g of solution?

Analysis

Use the formula, ppm = (g of solute)/(g of solution) × 10^6.

Solution

[1] Convert milligrams of X to grams of X so that both the solute and solution have the same unit.

$$45\ \cancel{\text{mg}}\ \mathbf{X} \times \frac{1 \text{ g}}{1000\ \cancel{\text{mg}}} = 0.045 \text{ g } \mathbf{X}$$

[2] Use the formula to calculate parts per million.

$$\frac{0.045 \text{ g } \mathbf{X}}{5{,}000 \text{ g solution}} \times 10^6 = 9 \text{ ppm } \mathbf{X}$$

Answer

[3] Concentration Units–Molarity (8.6)

Example 8.7 What is the molarity of one ampoule of sodium bicarbonate solution prepared from 4.2 g of sodium bicarbonate ($NaHCO_3$) in 50. mL of solution?

Analysis and Solution

[1] Identify the known quantities and the desired quantity.

known quantities	desired quantity
4.2 g $NaHCO_3$ 50. mL solution	? M (mol/L)

[2] Convert the number of grams of sodium bicarbonate to the number of moles using the molar mass (84.01 g/mol).

$$4.2\ \cancel{g}\ NaHCO_3 \times \frac{1\ mol}{84.01\ \cancel{g}} = 0.050\ mol\ NaHCO_3$$

Grams cancel.

- Since the volume of the solution is given in mL, it must be converted to L.

$$50.\ \cancel{mL}\ solution \times \frac{1\ L}{1000\ \cancel{mL}} = 0.050\ L\ solution$$

Milliliters cancel.

[3] Divide the number of moles of solute by the number of liters of solution to obtain the molarity.

$$\underset{\text{molarity}}{M} = \frac{\text{moles of solute (mol)}}{V\text{(L)}} = \frac{0.050\ mol\ NaHCO_3}{0.050\ L\ solution} = 1.0\ M$$

Answer

Example 8.8 How many grams of epinephrine are contained in 5.0 mL of a 0.050 M solution?

Analysis

Use the molarity to convert the volume of the solution to moles of solute. Then use the molar mass to convert moles to grams.

Solution

[1] Identify the known quantities and the desired quantity.

known quantities	desired quantity
0.050 M 5.0 mL solution	? g epinephrine

[2] Determine the number of moles of epinephrine using the molarity.

volume × molarity × mL–L conversion factor

$$5.0\ \cancel{mL}\ solution \times \frac{0.050\ mol\ epinephrine}{1\ \cancel{L}} \times \frac{1\ \cancel{L}}{1000\ \cancel{mL}} = 0.000\ 25\ mol\ epinephrine$$

[3] Convert the number of moles of epinephrine to grams using the molar mass (183.2 g/mol).

molar mass conversion factor

$$0.000\ 25\ \cancel{mol}\ \text{epinephrine} \times \frac{183.2\ \text{g epinephrine}}{1\ \cancel{mol}\ \text{epinephrine}} = 0.046\ \text{g epinephrine}$$

Answer

---

[4] Dilution (8.7)

Example 8.9 What is the concentration of a solution formed by diluting 4.0 mL of a 1.2 M glucose solution to 60.0 mL?

Analysis

Since we know an initial molarity and volume (M_1 and V_1) and a final volume (V_2), we can calculate a final molarity (M_2) using the equation, $M_1V_1 = M_2V_2$.

Solution

[1] Identify the known quantities and the desired quantity.

$M_1 = 1.2$ M
$V_1 = 4.0$ mL
$V_2 = 60.0$ mL

known quantities

$M_2 = ?$

desired quantity

---

[2] Write the equation and rearrange it to isolate the desired quantity, M_2, on one side.

$$M_1V_1 = M_2V_2$$

Solve for M_2 by dividing both sides by V_2.

$$\frac{M_1V_1}{V_2} = M_2$$

---

[3] Solve the problem.

- Substitute the three known quantities in the equation and solve for M_2.

$$M_2 = \frac{M_1V_1}{V_2} = \frac{(1.2\ \text{M})(4.0\ \cancel{mL})}{(60.0\ \cancel{mL})} = 0.080\ \text{M glucose solution}$$

Answer

---

[5] Colligative Properties (8.8)

Example 8.10 What is the boiling point of a solution that contains 0.25 mol of NaCl in 1.00 kg of water?

Analysis

Determine the number of "particles" contained in the solute. Use 0.51 °C/mol as a conversion factor to relate temperature change to the number of moles of solute particles.

Solution

Each NaCl provides two "particles," Na^+ and Cl^-.

$$\text{temperature increase} = \frac{0.51\ ^\circ C}{\cancel{mol}\ \text{particles}} \times 0.25\ \cancel{mol}\ NaCl \times \frac{2\ \cancel{mol}\ \text{particles}}{1\ \cancel{mol}\ NaCl} = 0.26\ ^\circ C$$

The boiling point of the solution is 100.0 °C + 0.26 °C = 100.26 °C, rounded to 100.3 °C.

Self-Test

[1] Fill in the blank with one of the terms listed below.

Colligative properties (8.8)	Heterogeneous mixture (8.1)	Nonvolatile (8.8)
Colloid (8.1)	Homogeneous mixture (8.1)	Solubility (8.3)
Dilution (8.7)	Molarity (8.6)	Solution (8.1)
Electrolyte (8.2)	Nonelectrolyte (8.2)	Volatile (8.8)

1. A ______ is a homogeneous mixture that contains small particles.
2. ______ is the amount of solute that dissolves in a given amount of solvent.
3. ______ is the number of moles of solute per liter of solution.
4. A ______ does not have a uniform composition throughout the sample.
5. A ______ is a homogeneous mixture with larger particles, often having an opaque appearance.
6. ______ is the addition of solvent to decrease the concentration of solute.
7. A ______ solute does not readily escape into the vapor phase, and thus it has a negligible vapor pressure at a given temperature.
8. A substance that conducts an electric current in water is called an ______.
9. ______ are properties of a solution that depend on the concentration of the solute but not its identity.
10. A substance that does not conduct an electric current in water is called a ______.
11. A ______ has a uniform composition throughout the sample.
12. A ______ solute readily escapes into the vapor phase.

[2] Fill in the blank with one of the terms listed below.

a. exothermic   b. endothermic   c. raises   d. lowers

13. When solvation releases more energy than that required to separate particles, the overall process is ______.
14. When the separation of particles requires more energy than is released during solvation, the process is ______ .
15. One mole of any nonvolatile solute ______ the freezing point of one kilogram of water the same amount, 1.86 °C.
16. One mole of any nonvolatile solute ______ the boiling point of one kilogram of water the same amount, 0.51 °C.

[3] Match the abbreviation with the definition below.

a. (v/v)%   b. (w/v)%

17. The number of grams of solute dissolved in 100 mL of solution
18. The number of milliliters of solute dissolved in 100 mL of solution

[4] Decide if the compounds are water soluble or water insoluble.

a. water soluble   b. water insoluble

19. $CH_3CH_2CH_2CH_2CH_3$
20. NaBr
21. NH_3
22. C_6H_{12}

[5] Fill in the blank with one of the terms below.

a. lower   b. higher

23. A hypertonic solution has a ______ osmotic pressure than body fluids.
24. A hypotonic solution has a ______ osmotic pressure than body fluids.
25. A fewer number of dissolved particles means a ______ osmotic pressure.

Answers to Self-Test

1. solution	6. Dilution	11. homogeneous mixture	16. c	21. a
2. Solubility	7. nonvolatile	12. volatile	17. b	22. b
3. Molarity	8. electrolyte	13. a	18. a	23. b
4. heterogeneous mixture	9. Colligative properties	14. b	19. b	24. a
5. colloid	10. nonelectrolyte	15. d	20. a	25. a

Solutions to In-Chapter Problems

8.1 A **heterogeneous mixture** does not have a uniform composition throughout a sample.
A **solution** is a homogeneous mixture that contains small particles. Liquid solutions are transparent.
A **colloid** is a homogeneous mixture with larger particles, often having an opaque appearance.

a. Cherry Garcia ice cream: heterogeneous mixture
b. mayonnaise: colloid
c, d, e. seltzer water, nail polish remover, and brass: solutions

8.2 A suspension is a heterogeneous mixture that contains large particles suspended in a liquid.

a. liquid soap—colloid (homogeneous, opaque)
b. Gatorade—solution (homogeneous, transparent)
c. Kaopectate—suspension (needs to be shaken)
d. shaving cream—colloid (homogeneous, opaque)
e. apple juice—solution (homogeneous, transparent)

8.3 A strong electrolyte dissociates completely to form ions, whereas a weak electrolyte dissolves in water to yield mostly uncharged molecules, with a few ions. A nonelectrolyte forms a solution with only uncharged molecules.

a. weak electrolyte—Molecules and ions are present.
b. nonelectrolyte—Only molecules are present.
c. strong electrolyte—Only ions are present.

8.4 A substance that conducts an electric current in water is called an **electrolyte.** If a solution contains *ions*, it will conduct electricity.
A substance that does not conduct an electric current in water is called a **nonelectrolyte.** If a solution contains *neutral molecules*, it will not conduct electricity.

a. KCl in H_2O: electrolyte
b. sucrose ($C_{12}H_{22}O_{11}$) in H_2O: nonelectrolyte
c. KI in H_2O: electrolyte

8.5 a. 1 mol Na^+ = 1 Eq Na^+
b. 1 mol Mg^{2+} = 2 Eq Mg^{2+}
c. 0.5 mol K^+ = 0.5 Eq K^+
d. 0.5 mol PO_4^{3-} = 1.5 Eq PO_4^{3-}

8.6 Use a conversion factor that relates milliequivalents to volume. A mL–L conversion factor is needed as well.

$$\frac{40.\ \text{mEq}}{1\ \cancel{\text{L}}} \times \frac{1\ \cancel{\text{L}}}{1000\ \cancel{\text{mL}}} \times 550\ \cancel{\text{mL}} = 22\ \text{mEq K}^+$$

8.7 There must be a balance between the total positive charge and the total negative charge.

Per liter: Total (+) charge = 15 mEq (Na^+) + 10. mEq K^+ + 4 mEq (Ca^{2+})
= 29 mEq
Total (–) charge = 29 mEq of Cl^- per liter

8.8 Use conversion factors that relate milliequivalents to liters and milliliters to liters.

$$\frac{1\ \cancel{\text{L}}}{125\ \cancel{\text{mEq}}} \times \frac{1000\ \text{mL}}{1\ \cancel{\text{L}}} \times 25\ \cancel{\text{mEq}} = 2.0 \times 10^2\ \text{mL}$$

8.9 Use the general solubility rule: "Like dissolves like." Generally, ionic and polar compounds are soluble in water. Nonpolar compounds are soluble in nonpolar solvents.

a. $NaNO_3$
ionic compound
water soluble

b. CH_4
nonpolar compound
NOT water soluble

c. HO–CH₂–CH₂–OH (structure: HO–C(H)(H)–C(H)(H)–OH)
polar compound
water soluble

d. KBr
ionic compound
water soluble

e. NH_2OH
polar compound
water soluble

8.10 a. Benzene (C_6H_6) and hexane (C_6H_{14}): both nonpolar compounds—would form a solution.
b. Na_2SO_4 and H_2O: one ionic compound and one polar compound—would form a solution.
c. NaCl and hexane (C_6H_{14}): one ionic and one nonpolar compound—cannot form a solution.
d. H_2O and CCl_4: one polar and one nonpolar compound—cannot form a solution.

8.11 Identify the cation and anion and use the solubility rules to predict if the ionic compound is water soluble.

a. Li_2CO_3: cation Li^+ with anion CO_3^{2-}—water soluble.
b. $MgCO_3$: cation Mg^{2+} with anion CO_3^{2-}—water insoluble.
c. KBr: cation K^+ with anion Br^-—water soluble.
d. $PbSO_4$: cation Pb^{2+} with anion SO_4^{2-}—water insoluble.
e. $CaCl_2$: cation Ca^{2+} with anion Cl^-—water soluble.
f. $MgCl_2$: cation Mg^{2+} with anion Cl^-—water soluble.

8.12 Magnesium salts are not water soluble unless the anion is a halide, nitrate, acetate, or sulfate. The interionic forces between Mg^{2+} and OH^- must be stronger than the forces of attraction between the ions and water. Therefore, milk of magnesia is a heterogeneous mixture, rather than a solution.

8.13 A soft drink becomes "flat" when CO_2 escapes from the solution. CO_2 comes out of solution faster at room temperature than at the cooler temperature of the refrigerator because the solubility of gases in liquids decreases as the temperature increases.

8.14 For most ionic and molecular **solids,** solubility generally *increases* as temperature increases.
The solubility of **gases** *decreases* with increasing temperature.
Pressure changes do not affect the solubility of liquids and solids.
The *higher* the pressure, the *higher* the solubility of a **gas** in a solvent.

	$Na_2CO_3(s)$	$N_2(g)$
a. increasing temperature	increased solubility	decreased solubility
b. decreasing temperature	decreased solubility	increased solubility
c. increasing pressure	no change	increased solubility
d. decreasing pressure	no change	decreased solubility

8.15 Use the formula in Example 8.2 to solve the problems.

$$(w/v)\% = \frac{4.3 \text{ g ethanol}}{30. \text{ mL solution}} \times 100\% = 14\% \text{ (w/v) ethanol}$$
Answer

$$(w/v)\% = \frac{0.021 \text{ g antiseptic}}{30. \text{ mL solution}} \times 100\% = 0.070\% \text{ (w/v) antiseptic}$$
Answer

8.16 Use the formula in Example 8.2 to solve the problem. One liter equals 1,000 mL.

$$(w/v)\% = \frac{35 \text{ g NaCl}}{1000 \text{ mL sea water}} \times 100\% = 3.5\%$$
Answer

8.17 Use the formula in Example 8.3 to solve the problem.

$$(v/v)\% = \frac{21 \text{ mL ethanol}}{250 \text{ mL mouthwash}} \times 100\% = 8.4\% \text{ (v/v) ethanol}$$

Answer

8.18 Use the stepwise analysis in Example 8.5 to solve the problem.

0.50% (w/v) vitamin C
1,000. mg vitamin C
known quantities

? mL solution
desired quantity

mg–g conversion factors

$$\frac{1000 \text{ mg}}{1 \text{ g}} \text{ or } \boxed{\frac{1 \text{ g}}{1000 \text{ mg}}}$$

g–mL solution conversion factors

$$\frac{0.50 \text{ g vitamin C}}{100 \text{ mL solution}} \text{ or } \boxed{\frac{100 \text{ mL solution}}{0.50 \text{ g vitamin C}}}$$

Choose the conversion factors with the unwanted units—mg and g—in the denominator.

$$1000. \text{ mg vitamin C} \times \frac{1 \text{ g}}{1000 \text{ mg}} \times \frac{100 \text{ mL solution}}{0.50 \text{ g vitamin C}} = 2.0 \times 10^2 \text{ mL solution}$$

Answer

8.19 Use the volume/volume percent concentration to determine the number of mL of ethanol.

$$8.0\text{-oz bottle} \times \frac{29.6 \text{ mL}}{1 \text{ oz}} \times \frac{62 \text{ mL ethanol}}{100 \text{ mL solution}} = 150 \text{ mL ethanol}$$

8.20 a. Use the formula in Example 8.2. Use a conversion factor to convert mg to g.

$$(w/v)\% = \frac{100 \text{ mg}}{5 \text{ mL}} \times \frac{1 \text{ g}}{1000 \text{ mg}} \times 100\% = 2\% \text{ (w/v) ibuprofen}$$

b.

$$\frac{100 \text{ mg}}{5 \text{ mL}} \times \frac{5 \text{ mL}}{1 \text{ tsp}} \times \frac{2.5 \text{ tsp}}{1 \text{ dose}} = 250 \text{ mg}$$

(rounded to 300 mg, one significant figure)

c. The medicine must be shaken because it is a suspension, and the medication will settle out on standing.

8.21 Use the formula, ppm = (g of solute)/(g of solution) × 10^6 to calculate parts per million as in Example 8.6.

a. $$0.042 \text{ mg DDT} \times \frac{1 \text{ g}}{1000 \text{ mg}} = 0.000\,042 \text{ g DDT} \qquad \frac{0.000\,042 \text{ g DDT}}{1{,}400 \text{ g plankton}} \times 10^6 = 0.030 \text{ ppm DDT}$$

Answer

b. $$1.0 \text{ kg tissue} \times \frac{1000 \text{ g}}{1.0 \text{ kg}} = 1{,}000 \text{ g tissue} \qquad \frac{5 \times 10^{-4} \text{ g DDT}}{1{,}000 \text{ g tissue}} \times 10^6 = 0.5 \text{ ppm DDT}$$

Answer

c. $2.0\ \text{mg DDT} \times \dfrac{1\ \text{g}}{1000\ \text{mg}} = 0.0020\ \text{g DDT}$ $\dfrac{0.0020\ \text{g DDT}}{1{,}000\ \text{g tissue}} \times 10^6 = 2.0\ \text{ppm DDT}$ **Answer**

d. $225\ \mu\text{g DDT} \times \dfrac{1\ \text{g}}{1{,}000{,}000\ \mu\text{g}} = 0.000\ 225\ \text{g DDT}$ $\dfrac{0.000\ 225\ \text{g DDT}}{1.0 \times 10^3\ \text{g breast milk}} \times 10^6 = 0.23\ \text{ppm DDT}$ **Answer**

8.22 Calculate the molarity of each aqueous solution as in Example 8.7.

a. $\underset{\text{molarity}}{M} = \dfrac{\text{moles of solute (mol)}}{V\text{(L)}} = \dfrac{1.0\ \text{mol NaCl}}{0.50\ \text{L solution}} = 2.0\ \text{M}$ **Answer**

b. $250\ \text{mL} \times \dfrac{1\ \text{L}}{1000\ \text{mL}} = 0.25\ \text{L}$ $\underset{\text{molarity}}{M} = \dfrac{2.0\ \text{mol NaCl}}{0.25\ \text{L solution}} = 8.0\ \text{M}$ **Answer**

c. $5.0\ \text{mL} \times \dfrac{1\ \text{L}}{1000\ \text{mL}} = 0.0050\ \text{L}$ $\underset{\text{molarity}}{M} = \dfrac{0.050\ \text{mol NaCl}}{0.0050\ \text{L solution}} = 10.\ \text{M}$ **Answer**

d. $12.0\ \text{g NaCl} \times \dfrac{1\ \text{mol NaCl}}{58.44\ \text{g NaCl}} = 0.205\ \text{mol NaCl}$ $\underset{\text{molarity}}{M} = \dfrac{0.205\ \text{mol NaCl}}{2.0\ \text{L solution}} = 0.10\ \text{M}$ **Answer**

e. $24.4\ \text{g NaCl} \times \dfrac{1\ \text{mol NaCl}}{58.44\ \text{g NaCl}} = 0.418\ \text{mol NaCl}$

$350\ \text{mL} \times \dfrac{1\ \text{L}}{1000\ \text{mL}} = 0.35\ \text{L}$

$\underset{\text{molarity}}{M} = \dfrac{0.418\ \text{mol NaCl}}{0.35\ \text{L solution}} = 1.2\ \text{M}$ **Answer**

f. $60.0\ \text{g NaCl} \times \dfrac{1\ \text{mol NaCl}}{58.44\ \text{g NaCl}} = 1.03\ \text{mol NaCl}$

$750\ \text{mL} \times \dfrac{1\ \text{L}}{1000\ \text{mL}} = 0.75\ \text{L}$

$\underset{\text{molarity}}{M} = \dfrac{1.03\ \text{mol NaCl}}{0.75\ \text{L solution}} = 1.4\ \text{M}$ **Answer**

8.23 Convert the number of grams of acetic acid to moles using the molar mass. Convert mL to L to determine molarity.

$5.00\ \text{g acetic acid} \times \dfrac{1\ \text{mol}}{60.05\ \text{g}} = 0.0833\ \text{mol acetic acid}$

$100.\ \text{mL} \times \dfrac{1\ \text{L}}{1000\ \text{mL}} = 0.100\ \text{L}$

$M = \dfrac{\text{mol}}{\text{L}} = \dfrac{0.833\ \text{mol}}{0.100\ \text{L}} = 0.833\ \text{M acetic acid}$

8.24 Use the molarity as a conversion factor to determine the number of milliliters as in Sample Problem 8.11.

a. 1.5 M solution, 0.15 mol glucose (known quantities); ? V (L) solution (desired quantity)

$$V(\text{L}) = \frac{\text{moles of solute (mol)}}{\text{M}} = \frac{0.15\ \text{mol glucose}}{1.5\ \text{mol/L}} = 0.10\ \text{L solution}$$

$$0.10\ \text{L solution} \times \frac{1000\ \text{mL}}{1\ \text{L}} = 1.0 \times 10^2\ \text{mL glucose solution}$$

Answer

b. $$V(\text{L}) = \frac{\text{moles of solute (mol)}}{\text{M}} = \frac{0.020\ \text{mol glucose}}{1.5\ \text{mol/L}} = 0.013\ \text{L solution} \times \frac{1000\ \text{mL}}{1\ \text{L}} = 13\ \text{mL glucose solution}$$

Answer

c. $$V(\text{L}) = \frac{\text{moles of solute (mol)}}{\text{M}} = \frac{0.0030\ \text{mol glucose}}{1.5\ \text{mol/L}} = 0.0020\ \text{L solution} \times \frac{1000\ \text{mL}}{1\ \text{L}} = 2.0\ \text{mL glucose solution}$$

Answer

d. $$V(\text{L}) = \frac{\text{moles of solute (mol)}}{\text{M}} = \frac{3.0\ \text{mol glucose}}{1.5\ \text{mol/L}} = 2.0\ \text{L solution} \times \frac{1000\ \text{mL}}{1\ \text{L}} = 2.0 \times 10^3\ \text{mL glucose solution}$$

Answer

8.25 Use the molarity as a conversion factor to determine the number of moles.

a. mol = V (L) • M = (2.0 L)(2.0 mol/L) = 4.0 mol NaCl

b. mol = V (L) • M = (2.5 L)(0.25 mol/L) = 0.63 mol NaCl

c. mol = V (L) • M = (0.025 L)(2.0 mol/L) = 0.050 mol NaCl

d. mol = V (L) • M = (0.25 L)(0.25 mol/L) = 0.063 mol NaCl

8.26 Use the molarity to convert the volume of the solution to moles of solute. Then use the molar mass to convert moles to grams as in Example 8.8.

volume | molarity | molar mass conversion factor

a. $$0.10\ \text{L solution} \times \frac{1.25\ \text{mol NaCl}}{1\ \text{L}} = 0.125\ \text{mol NaCl} \times \frac{58.44\ \text{g NaCl}}{1\ \text{mol NaCl}} = 7.3\ \text{g NaCl}$$

Answer

b. $2.0\ \cancel{L}\ \text{solution} \times \frac{1.25\ \text{mol NaCl}}{1\ \cancel{L}} = 2.5\ \cancel{\text{mol}}\ \text{NaCl} \times \frac{58.44\ \text{g NaCl}}{1\ \cancel{\text{mol}}\ \text{NaCl}} = 150\ \text{g NaCl}$ **Answer**

c. $0.55\ \cancel{L}\ \text{solution} \times \frac{1.25\ \text{mol NaCl}}{1\ \cancel{L}} = 0.69\ \cancel{\text{mol}}\ \text{NaCl} \times \frac{58.44\ \text{g NaCl}}{1\ \cancel{\text{mol}}\ \text{NaCl}} = 40.\ \text{g NaCl}$ **Answer**

d. $50.\ \cancel{\text{mL}}\ \text{solution} \times \frac{1\ \cancel{L}}{1000\ \cancel{\text{mL}}} \times \frac{1.25\ \cancel{\text{mol}}\ \text{NaCl}}{1\ \cancel{L}} \times \frac{58.44\ \text{g NaCl}}{1\ \cancel{\text{mol}}\ \text{NaCl}} = 3.7\ \text{g NaCl}$ **Answer**

8.27 Use the molar mass to convert grams to moles. Then use the molarity to convert moles of solute to volume of solution. Sample Problem 8.12 is similar, but the order of steps is different.

molar mass conversion factor | molarity conversion factor

a. $0.500\ \cancel{g}\ \text{sucrose} \times \frac{1\ \cancel{\text{mol}}\ \text{sucrose}}{342.3\ \cancel{g}\ \text{sucrose}} \times \frac{1\ \cancel{L}}{0.25\ \cancel{\text{mol}}\ \text{sucrose}} \times \frac{1000\ \text{mL}}{1\ \cancel{L}} = 5.8\ \text{mL solution}$ **Answer**

b. $2.0\ \cancel{g}\ \text{sucrose} \times \frac{1\ \cancel{\text{mol}}\ \text{sucrose}}{342.3\ \cancel{g}\ \text{sucrose}} \times \frac{1\ \cancel{L}}{0.25\ \cancel{\text{mol}}\ \text{sucrose}} \times \frac{1000\ \text{mL}}{1\ \cancel{L}} = 23\ \text{mL solution}$ **Answer**

c. $1.25\ \cancel{g}\ \text{sucrose} \times \frac{1\ \cancel{\text{mol}}\ \text{sucrose}}{342.3\ \cancel{g}\ \text{sucrose}} \times \frac{1\ \cancel{L}}{0.25\ \cancel{\text{mol}}\ \text{sucrose}} \times \frac{1000\ \text{mL}}{1\ \cancel{L}} = 15\ \text{mL solution}$ **Answer**

d. $50.0\ \cancel{\text{mg}}\ \text{sucrose} \times \frac{1\ \cancel{g}}{1000\ \cancel{\text{mg}}} \times \frac{1\ \cancel{\text{mol}}\ \text{sucrose}}{342.3\ \cancel{g}\ \text{sucrose}} \times \frac{1\ \cancel{L}}{0.25\ \cancel{\text{mol}}\ \text{sucrose}} \times \frac{1000\ \text{mL}}{1\ \cancel{L}} = 0.58\ \text{mL solution}$ **Answer**

8.28 Calculate a new molarity (M_2) using the equation, $M_1V_1 = M_2V_2$.

$$M_2 = \frac{M_1V_1}{V_2} = \frac{(3.8\ \text{M})(25.0\ \cancel{\text{mL}})}{(275\ \cancel{\text{mL}})} = 0.35\ \text{M glucose solution}$$ **Answer**

8.29 Calculate the volume needed (V_1) using the equation, $M_1V_1 = M_2V_2$.

a. $V_1 = \frac{M_2V_2}{M_1} = \frac{(2.5\ \cancel{M})(525\ \text{mL})}{6.0\ \cancel{M}} = 220\ \text{mL}$ **Answer**

b. $V_1 = \frac{M_2V_2}{M_1} = \frac{(4.0\ \cancel{M})(750\ \text{mL})}{6.0\ \cancel{M}} = 5.0 \times 10^2\ \text{mL}$ **Answer**

c. $V_1 = \frac{M_2V_2}{M_1} = \frac{(0.10\ M)(450\ mL)}{6.0\ M} = 7.5\ mL$ **Answer**

d. $V_1 = \frac{M_2V_2}{M_1} = \frac{(3.5\ M)(25\ mL)}{6.0\ M} = 15\ mL$ **Answer**

8.30 Calculate the new concentration of ketamine, and determine the volume needed to supply 75 mg.

$$C_2 = \frac{C_1V_1}{V_2} = \frac{(100.\ mg/mL)(2.0\ mL)}{10.0\ mL} = 20.\ mg/mL$$

$$75\ mg \times \frac{1\ mL}{20.\ mg} = 3.8\ mL$$ **Answer**

8.31 Determine the number of "particles" contained in the solute. Use 0.51 °C/mol as a conversion factor to relate the temperature change to the number of moles of solute particles, as in Example 8.10.

a. 2.0 mol of sucrose molecules

$$\frac{0.51\ ^{\circ}C}{mol\ particles} \times 2.0\ mol = 1.0\ ^{\circ}C$$ Boiling point = 100.0 °C + 1.0 °C = 101.0 °C

b. 2.0 mol of KNO_3

$$\frac{0.51\ ^{\circ}C}{mol\ particles} \times 2.0\ mol\ KNO_3 \times \frac{2\ mol\ particles}{mol\ KNO_3} = 2.0\ ^{\circ}C$$

Boiling point = 100.0 °C + 2.0 °C = 102.0 °C

c. 2.0 mol of $CaCl_2$

$$\frac{0.51\ ^{\circ}C}{mol\ particles} \times 2.0\ mol\ CaCl_2 \times \frac{3\ mol\ particles}{mol\ CaCl_2} = 3.1\ ^{\circ}C$$

Boiling point = 100.0 °C + 3.1 °C = 103.1 °C

d. 20.0 g of NaCl

$$20.0\ g\ NaCl \times \frac{1\ mol\ NaCl}{58.44\ g\ NaCl} = 0.342\ mol\ NaCl$$

$$\frac{0.51\ ^{\circ}C}{mol\ particles} \times 0.342\ mol\ NaCl \times \frac{2\ mol\ particles}{mol\ NaCl} = 0.35\ ^{\circ}C$$

Boiling point = 100.0 °C + 0.35 °C = 100.35 °C rounded to 100.4 °C

8.32 Determine the number of "particles" contained in the solute. Use 1.86 °C/mol as a conversion factor to relate temperature change to the number of moles of solute particles, as in Sample Problem 8.16.

a. 2.0 mol of sucrose molecules

$$\frac{1.86\ ^\circ\text{C}}{\text{mol particles}} \times 2.0\ \text{mol} = 3.7\ ^\circ\text{C} \qquad \text{Melting point} = -3.7\ ^\circ\text{C}$$

b. 2.0 mol of KNO_3

$$\frac{1.86\ ^\circ\text{C}}{\text{mol particles}} \times 2.0\ \text{mol KNO}_3 \times \frac{2\ \text{mol particles}}{\text{mol KNO}_3} = 7.4\ ^\circ\text{C} \qquad \text{Melting point} = -7.4\ ^\circ\text{C}$$

c. 2.0 mol of $CaCl_2$

$$\frac{1.86\ ^\circ\text{C}}{\text{mol particles}} \times 2.0\ \text{mol CaCl}_2 \times \frac{3\ \text{mol particles}}{\text{mol CaCl}_2} = 11\ ^\circ\text{C} \qquad \text{Melting point} = -11\ ^\circ\text{C}$$

d. 20.0 g of NaCl

$$20.0\ \text{g NaCl} \times \frac{1\ \text{mol NaCl}}{58.44\ \text{g NaCl}} = 0.342\ \text{mol NaCl}$$

$$\frac{1.86\ ^\circ\text{C}}{\text{mol particles}} \times 0.342\ \text{mol NaCl} \times \frac{2\ \text{mol particles}}{\text{mol NaCl}} = 1.27\ ^\circ\text{C} \qquad \text{Melting point} = -1.27\ ^\circ\text{C}$$

8.33 Determine the number of moles of ethylene glycol. Use 1.86 °C/mol as a conversion factor to relate temperature change to the number of moles of solute particles, as in Sample Problem 8.16.

$$250\ \text{g ethylene glycol} \times \frac{1\ \text{mol}}{62.07\ \text{g}} = 4.0\ \text{mol ethylene glycol}$$

$$\text{temperature decrease} = \frac{1.86\ ^\circ\text{C}}{\text{mol particles}} \times 4.0\ \text{mol C}_2\text{H}_6\text{O}_2 = -7.4\ ^\circ\text{C}$$

8.34 The solvent (water) flows from the less concentrated solution to the more concentrated solution.

a. A 5.0% sugar solution has greater osmotic pressure than a 1.0% sugar solution.
b. 4.0 M NaCl has greater osmotic pressure than 3.0 M NaCl.
c. 0.75 M NaCl has greater osmotic pressure than 1.0 M glucose solution because NaCl has two particles for each NaCl, whereas glucose has only one.

8.35 a. Water flows across the membrane.
b. More water initially flows from the 1.0 M side to the more concentrated 1.5 M side. When equilibrium is reached there is equal flow of water in both directions.
c. The height of the 1.5 M side will be higher and the height of the 1.0 M NaCl will be lower.

8.36 A *hypotonic* solution has a lower osmotic pressure than body fluids.
A *hypertonic* solution has a higher osmotic pressure than body fluids.

a. A 3% (w/v) glucose solution is hypotonic, so it will cause hemolysis.
b. A 0.15 M KCl solution is isotonic, so no change results.
c. A 0.15 M Na_2CO_3 solution is hypertonic, so it will cause crenation.

Solutions to Odd-Numbered End-of-Chapter Problems

8.37 Use the particle sizes in Table 8.1 to answer the questions. Convert meters to nanometers (nm) or micrometers (μm) for comparison.

a. $1.5 \times 10^{-7}\ \cancel{m} \times \dfrac{1\ nm}{10^{-9}\ \cancel{m}} = 150\ nm$ colloid

b. $2.0 \times 10^{-5}\ \cancel{m} \times \dfrac{1\ \mu m}{10^{-6}\ \cancel{m}} = 20\ \mu m$ suspension ($> 1\ \mu m$)

c. $1.2 \times 10^{-10}\ \cancel{m} \times \dfrac{1\ nm}{10^{-9}\ \cancel{m}} = 0.12\ nm$ solution ($< 1\ nm$)

8.39 **A** shows KI dissolved in water, because the K^+ and I^- ions are separated when KI is dissolved. The solution contains ions, so it conducts an electric current. **B** shows KI as a molecule, without being separated into ions, a situation that does not occur.

8.41 Use the definitions from Answers 8.1 and 8.2 to classify each substance as a solution, a colloid, or a suspension.

a. bronze (alloy of Sn and Cu): solution
b. diet soda: solution
c. orange juice with pulp: suspension
d. household ammonia: solution
e. gasoline : solution
f. fog: colloid

8.43 Use the definitions from Answer 8.3.

a. weak electrolyte
b. strong electrolyte

8.45 2.5 mol K^+ = 2.5 Eq K^+
2.5 mol Cl^- = 2.5 Eq Cl^-
2.5 mol Ca^{2+} = 5.0 Eq Ca^{2+}
2.5 mol SO_4^{2-} = 5.0 Eq SO_4^{2-}

8.47 Use the concentration of Na^+ in mEq/L to determine the number of milliequivalents in a given volume.

a. $\dfrac{132\ mEq}{1\ \cancel{L}} \times 1.0\ \cancel{L} = 130\ mEq\ Na^+$

b. $\dfrac{132\ mEq}{1\ \cancel{L}} \times 2.0\ \cancel{L} = 260\ mEq\ Na^+$

c. $\dfrac{132\ mEq}{1\ \cancel{L}} \times \dfrac{1.0\ \cancel{L}}{1000\ \cancel{mL}} \times 440\ \cancel{mL} = 58\ mEq\ Na^+$

8.49 The general solubility rule is "like dissolves like."
a. polar solute and polar solvent: **A,** soluble
b. nonpolar solute and nonpolar solvent: **A,** soluble
c. nonpolar solute and polar solvent: **B,** insoluble
d. polar solute and nonpolar solvent: **B,** insoluble

8.51 A solution that has less than the maximum number of grams of solute is said to be **unsaturated.**
A solution that has the maximum number of grams of solute that can dissolve is said to be **saturated.**
A solution that has more than the maximum number of grams of solute is said to be **supersaturated.**

a. adding 30 g to 100 mL of H_2O at 20 °C: unsaturated
b. adding 65 g to 100 mL of H_2O at 50 °C: saturated
c. adding 20 g to 50 mL of H_2O at 20 °C: saturated
d. adding 42 g to 100 mL of H_2O at 50 °C and slowly cooling to 20 °C to give a clear solution with no precipitate: supersaturated

8.53 Use the general solubility rule: "Like dissolves like." Generally, ionic and polar compounds are soluble in water. Nonpolar compounds are soluble in nonpolar solvents.

a. LiCl	b. C_7H_8	c. Na_3PO_4
ionic water soluble	nonpolar NOT water soluble	ionic water soluble

8.55 Methanol will hydrogen bond to water.

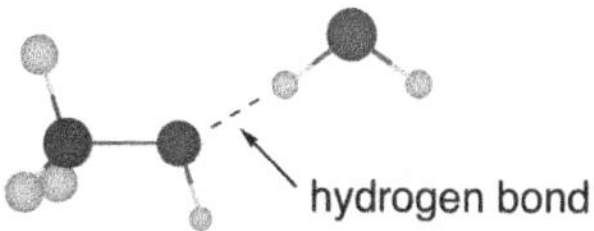

8.57 Water-soluble compounds are ionic or are small polar molecules that can hydrogen bond with the water solvent, but nonpolar compounds, such as oil, are soluble in nonpolar solvents.

8.59 Cholesterol is not water soluble because it is a large nonpolar molecule with a single OH group.

8.61 The solubility of gases decreases with increasing temperature. The H_2O molecules are omitted in each representation.

a.

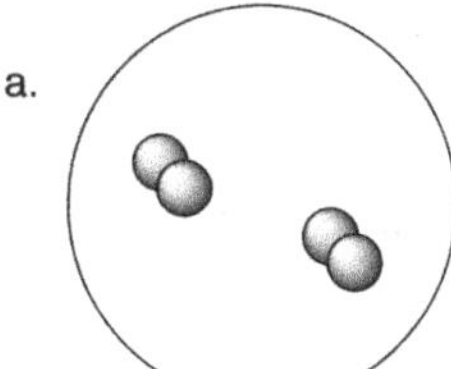

Less O_2 is dissolved at 50 °C.

b.

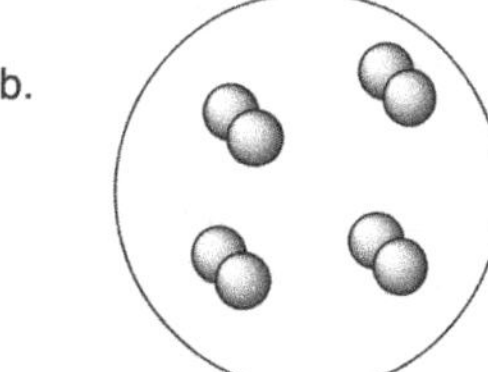

More O_2 is dissolved at 10 °C.

8.63 For most ionic and molecular **solids,** solubility generally *increases* as temperature increases. The solubility of **gases** *decreases* as temperature increases. Pressure changes do not affect the solubility of liquids and solids. The *higher* the pressure, the *higher* the solubility of a **gas** in a solvent.

	NaCl(*s*)
a. increasing temperature	increased solubility
b. decreasing temperature	decreased solubility
c. increasing pressure	no change
d. decreasing pressure	no change

8.65 Identify the cation and anion and use the solubility rules to predict if the ionic compound is water soluble.

a. K_2SO_4: cation K^+ with anion SO_4^{2-}—water soluble.
b. $MgSO_4$: cation Mg^{2+} with anion SO_4^{2-}—water soluble.
c. $ZnCO_3$: cation Zn^{2+} with anion CO_3^{2-}—not water soluble.
d. KI: cation K^+ with anion I^-—water soluble.
e. $Fe(NO_3)_3$: cation Fe^{3+} with anion NO_3^-—water soluble.
f. $PbCl_2$: cation Pb^{2+} with anion Cl^-—not water soluble.
g. CsCl : cation Cs^+ with anion Cl^-—water soluble.
h. $Ni(HCO_3)_2$: cation Ni^{2+} with anion HCO_3^-—not water soluble.

8.67 a. Convert mg to g (160 mg = 0.160 g) and use the formula in Example 8.2.

$$(w/v)\% = \frac{0.160\text{ g}}{5.0\text{ mL}} \times 100\% = 3.2\%$$

b.

$$\frac{160\text{ mg}}{5.0\text{ mL}} \times \frac{29.6\text{ mL}}{1\text{ fl oz}} \times 4.0\text{ fl oz} = 3{,}800\text{ mg}$$

$$3800\text{ mg} \times \frac{1\text{ g}}{1000\text{ mg}} = 3.8\text{ g}$$

8.69 Use the formula in Example 8.2 to solve the problems.

a. $$(w/v)\% = \frac{10.0\text{ g LiCl}}{750\text{ mL solution}} \times 100\% = 1.3\%\ (w/v)\text{ LiCl}$$
Answer

b. $$(w/v)\% = \frac{25\text{ g NaNO}_3}{150\text{ mL solution}} \times 100\% = 17\%\ (w/v)\text{ NaNO}_3$$
Answer

c. $$(w/v)\% = \frac{40.0\text{ g NaOH}}{500.\text{ mL solution}} \times 100\% = 8.00\%\ (w/v)\text{ NaOH}$$
Answer

8.71 Use the formula in Example 8.3 to solve the problem.

$$\text{(v/v)\%} = \frac{\text{25 mL ethyl acetate}}{\text{150 mL solution}} \times 100\% = \text{17\% (v/v) ethyl acetate}$$

Answer

8.73 Calculate the molarity of each aqueous solution as in Example 8.7.

a. $$\underset{\text{molarity}}{M} = \frac{\text{moles of solute (mol)}}{V\text{(L)}} = \frac{\text{3.5 mol KCl}}{\text{1.50 L solution}} = \text{2.3 M}$$

Answer

b. $$\text{855 mL solution} \times \frac{\text{1 L}}{\text{1000 mL}} = \text{0.855 L solution}$$

$$\underset{\text{molarity}}{M} = \frac{\text{moles of solute (mol)}}{V\text{(L)}} = \frac{\text{0.44 mol NaNO}_3}{\text{0.855 L solution}} = \text{0.51 M}$$

Answer

c. $$\text{25.0 g NaCl} \times \frac{\text{1 mol NaCl}}{\text{58.44 g}} = \text{0.428 mol NaCl}$$

$$\text{650 mL solution} \times \frac{\text{1 L}}{\text{1000 mL}} = \text{0.65 L solution}$$

$$\underset{\text{molarity}}{M} = \frac{\text{moles of solute (mol)}}{V\text{(L)}} = \frac{\text{0.428 mol NaCl}}{\text{0.65 L solution}} = \text{0.66 M}$$

Answer

d. $$\text{10.0 g NaHCO}_3 \times \frac{\text{1 mol NaHCO}_3}{\text{84.01 g}} = \text{0.119 mol NaHCO}_3$$

$$\underset{\text{molarity}}{M} = \frac{\text{moles of solute (mol)}}{V\text{(L)}} = \frac{\text{0.119 mol NaHCO}_3}{\text{3.3 L solution}} = \text{0.036 M}$$

Answer

8.75

a. Add 12 g of acetic acid to the flask and then water to bring the volume to 250 mL.

$$\text{(w/v)\%} = \frac{x\text{ g acetic acid}}{\text{250 mL solution}} \times 100\% = \text{4.8\% (w/v) acetic acid} \qquad x = \text{12 g acetic acid}$$

Answer

b. Add 55 mL of ethyl acetate to the flask and then water to bring the volume to 250 mL.

$$\text{(v/v)\%} = \frac{x\text{ mL ethyl acetate}}{\text{250 mL solution}} \times 100\% = \text{22\% (v/v) ethyl acetate} \qquad x = \text{55 mL ethyl acetate}$$

Answer

c. Add 37 g of NaCl to the flask and then water to bring the volume to 250 mL.

$$\underset{\text{molarity}}{M} = \frac{\text{moles of solute (mol)}}{V\,(L)}$$

moles of solute (mol) = (M)(V) = (2.5 M) (0.25 L solution) = 0.63 mol NaCl

$$0.63\ \text{mol NaCl} \times \frac{58.44\ g}{1\ \text{mol NaCl}} = 37\ g\ NaCl$$

Answer

8.77 Calculate the moles of solute in each solution.

a. moles of solute (mol) = (M)(V) = (0.25 M) (0.15 L solution) = 0.038 mol $NaNO_3$
Answer

b. moles of solute (mol) = (M)(V) = (2.0 M) (0.045 L solution) = 0.090 mol HNO_3
Answer

c. moles of solute (mol) = (M)(V) = (1.5 M) (2.5 L solution) = 3.8 mol HCl
Answer

8.79 Calculate the number of grams of each solute.

a. $0.038\ \text{mol}\ NaNO_3 \times \frac{85.00\ g\ NaNO_3}{1\ \text{mol}\ NaNO_3} = 3.2\ g\ NaNO_3$
Answer

b. $0.090\ \text{mol}\ HNO_3 \times \frac{63.02\ g\ HNO_3}{1\ \text{mol}\ HNO_3} = 5.7\ g\ HNO_3$
Answer

c. $3.8\ \text{mol HCl} \times \frac{36.46\ g\ HCl}{1\ \text{mol HCl}} = 140\ g\ HCl$
Answer

8.81 Calculate the number of milliliters of ethanol in the bottle of wine.

$$(v/v)\% = \frac{x\ \text{mL ethanol}}{750\ \text{mL solution}} \times 100\% = 11.0\%\ (v/v)\ \text{ethanol}$$

x = 83 mL ethanol
Answer

8.83 Use the formula, ppm = (g of solute)/(g of solution) × 10^6, to calculate parts per million as in Example 8.6.

a. $80\ \mu g\ CHCl_3 \times \frac{1\ g}{1{,}000{,}000\ \mu g} = 0.000\ 08\ g\ CHCl_3$ $\quad \frac{0.000\ 08\ g\ CHCl_3}{1{,}000\ g} \times 10^6 = 0.08\ ppm\ CHCl_3$
Answer

b. $700\ \mu g\ \text{glyphosate} \times \frac{1\ g}{1{,}000{,}000\ \mu g} = 0.0007\ g\ \text{glyphosate}$

$$\frac{0.0007\ g\ \text{glyphosate}}{1{,}000\ g} \times 10^6 = 0.7\ \text{ppm glyphosate}$$

Answer

8.85 When solution **X** is diluted, the volume will increase, but the amount of solute will stay the same (**B**).

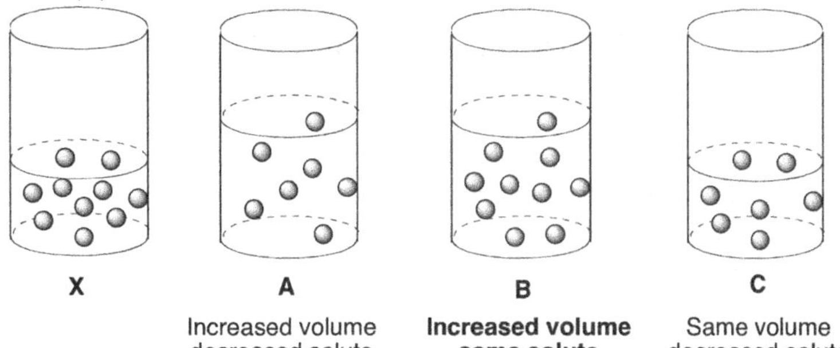

8.87

a.

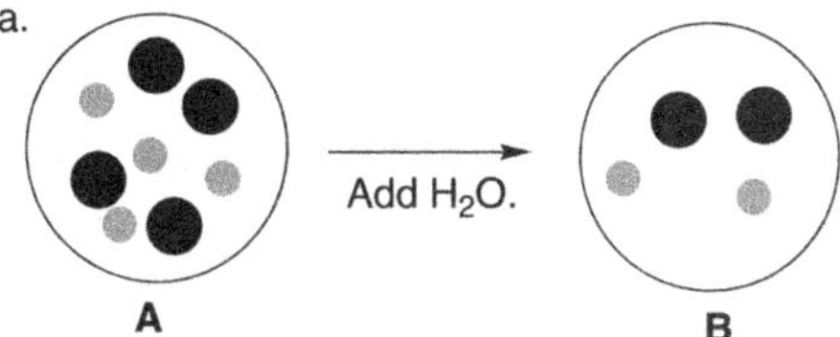

B has half as much NaCl, and is therefore half the molarity: 0.05 M.

b. $M_1V_1 = M_2V_2 = (0.1\ \text{M})(50.0\ \text{mL}) = (0.05\ \text{M})(x\ \text{mL})$

$x = 100\ \text{mL}$ Since the original volume was 50 mL, **50 mL** of water is added.

8.89 Use the equation, $(C_1)(V_1) = (C_2)(V_2)$, to calculate the new concentration after dilution.

a. $C_2 = \dfrac{C_1V_1}{V_2} = \dfrac{(30.0\%)(100.\ \cancel{\text{mL}})}{200.\ \cancel{\text{mL}}} = 15.0\%\ \text{(w/v)}$

b. $C_2 = \dfrac{C_1V_1}{V_2} = \dfrac{(30.0\%)(100.\ \cancel{\text{mL}})}{500.\ \cancel{\text{mL}}} = 6.00\%\ \text{(w/v)}$

c. $C_2 = \dfrac{C_1V_1}{V_2} = \dfrac{(30.0\%)(250\ \cancel{\text{mL}})}{1500\ \cancel{\text{mL}}} = 5.0\%\ \text{(w/v)}$

d. $C_2 = \dfrac{C_1V_1}{V_2} = \dfrac{(30.0\%)(350\ \cancel{\text{mL}})}{750\ \cancel{\text{mL}}} = 14\%\ \text{(w/v)}$

8.91 Use the equation, $(M_1)(V_1) = (M_2)(V_2)$, to calculate the volume needed to prepare each solution.

a. $V_1 = \dfrac{M_2V_2}{M_1} = \dfrac{(1.0\ \cancel{\text{M}})(25\ \text{mL})}{2.5\ \cancel{\text{M}}} = 10.\ \text{mL}$

b. $V_1 = \dfrac{M_2V_2}{M_1} = \dfrac{(0.75\ \cancel{\text{M}})(1500\ \text{mL})}{2.5\ \cancel{\text{M}}} = 450\ \text{mL}$

c. $V_1 = \dfrac{M_2V_2}{M_1} = \dfrac{(0.25\ \cancel{M})(15\ \text{mL})}{2.5\ \cancel{M}} = 1.5\ \text{mL}$

d. $V_1 = \dfrac{M_2V_2}{M_1} = \dfrac{(0.025\ \cancel{M})(250\ \text{mL})}{2.5\ \cancel{M}} = 2.5\ \text{mL}$

8.93 Determine the number of "particles" contained in the solute. Use 0.51 °C/mol as a conversion factor to relate temperature change to the number of moles of solute particles, as in Example 8.10.

a. 3.0 mol of fructose molecules

$$\frac{0.51\ ^\circ\text{C}}{\cancel{\text{mol}}\ \text{particles}} \times 3.0\ \cancel{\text{mol}} = 1.5\ ^\circ\text{C} \qquad 100.0\ ^\circ\text{C} + 1.5\ ^\circ\text{C} = 101.5\ ^\circ\text{C}$$

b. 1.2 mol of KI

$$\frac{0.51\ ^\circ\text{C}}{\cancel{\text{mol}}\ \text{particles}} \times 1.2\ \cancel{\text{mol}}\ \text{KI} \times \frac{2\ \cancel{\text{mol}}\ \text{particles}}{\cancel{\text{mol}}\ \text{KI}} = 1.2\ ^\circ\text{C} \qquad 100.0\ ^\circ\text{C} + 1.2\ ^\circ\text{C} = 101.2\ ^\circ\text{C}$$

c. 1.5 mol Na_3PO_4

$$\frac{0.51\ ^\circ\text{C}}{\cancel{\text{mol}}\ \text{particles}} \times 1.5\ \cancel{\text{mol}}\ \text{Na}_3\text{PO}_4 \times \frac{4\ \cancel{\text{mol}}\ \text{particles}}{\cancel{\text{mol}}\ \text{Na}_3\text{PO}_4} = 3.1\ ^\circ\text{C} \qquad 100.0\ ^\circ\text{C} + 3.1\ ^\circ\text{C} = 103.1\ ^\circ\text{C}$$

8.95 Determine the number of "particles" contained in the solute. Use 1.86 °C/mol as a conversion factor to relate temperature change to the number of moles of solute particles, as in Sample Problem 8.16.

$$\frac{150\ \cancel{\text{g}}\ \text{ethylene glycol}}{(\text{C}_2\text{H}_6\text{O}_2)} \times \frac{1\ \text{mol ethylene glycol}}{62.07\ \cancel{\text{g}}\ \text{ethylene glycol}} = 2.4\ \text{mol ethylene glycol}$$

$$\frac{1.86\ ^\circ\text{C}}{\cancel{\text{mol}}\ \text{particles}} \times 2.4\ \cancel{\text{mol}} = 4.5\ ^\circ\text{C} \qquad \text{Melting point} = -4.5\ ^\circ\text{C}$$

8.97
a. The NaCl solution has a higher boiling point because it contains two particles per mole, whereas the glucose solution contains one per mole.
b. The glucose solution has the higher melting point because it contains one particle per mole, whereas the NaCl solution contains two particles per mole.
c. The NaCl solution has the higher osmotic pressure because it contains two particles per mole, whereas the glucose solution contains one per mole.
d. The glucose solution has a higher vapor pressure at a given temperature because it contains one particle per mole, whereas the NaCl solution contains two particles per mole.

8.99 Determine the number of "particles" contained in the solute. If necessary, use 1.86 °C/mol as a conversion factor to relate the temperature change to the number of moles of solute particles.

a. A 0.10 M glucose solution has a higher melting point than 0.10 M NaOH even though the solutions have the same molar concentration, because the NaOH solution contains twice as many particles. Therefore the NaOH solution will have its melting point reduced by twice as much as the glucose solution.

b. A 0.20 M NaCl solution has a higher melting point than a 0.15 M $CaCl_2$ solution.

$$\frac{1.86\ ^{\circ}\text{C}}{\cancel{\text{mol}}\ \text{particles}} \times 0.20\ \cancel{\text{mol}}\ \text{NaCl} \times \frac{2\ \cancel{\text{mol}}\ \text{particles}}{\cancel{\text{mol}}\ \text{NaCl}} = 0.74\ ^{\circ}\text{C}$$

Melting point = –0.74 °C
higher melting point

$$\frac{1.86\ ^{\circ}\text{C}}{\cancel{\text{mol}}\ \text{particles}} \times 0.15\ \cancel{\text{mol}}\ \text{CaCl}_2 \times \frac{3\ \cancel{\text{mol}}\ \text{particles}}{\cancel{\text{mol}}\ \text{CaCl}_2} = 0.84\ ^{\circ}\text{C}$$

Melting point = –0.84 °C

c. A 0.10 M Na_2SO_4 solution has a higher melting point than 0.10 M Na_3PO_4 even though the solutions have the same molar concentration, because the Na_2SO_4 solution contains fewer particles (three vs. four for Na_3PO_4).

d. A 0.10 M glucose solution has a higher melting point than a 0.20 M glucose solution because it has a lower molar concentration, which depresses the melting point less than the more concentrated solution.

8.101 a. **A > B.** Since solution **A** contains a dissolved solute, water will flow from compartment **B** to compartment **A.**

b. **B > A.** Since solution **B** is more concentrated, water will flow from compartment **A** to compartment **B.**

c. **No change** will occur because both solutions have an equal number of dissolved particles.

d. **A > B.** Solution **A** contains more dissolved particles than solution **B.** Therefore, water will flow from compartment **B** to compartment **A.**

e. **No change** will occur because both solutions have an equal number of dissolved particles (NaCl has two particles per mole, making it equivalent to the glucose solution).

8.103 At warmer temperatures, CO_2 is less soluble in water and more is in the gas phase and escapes as the can is opened and the pressure is reduced.

8.105 Calculate the weight/volume percent concentration of glucose, and the molarity.

$$(\text{w/v})\% = \frac{0.09\ \text{g glucose}}{100\ \text{mL blood}} \times 100\% = 0.09\%\ (\text{w/v})\ \text{glucose}$$

Answer

$$0.09\ \cancel{\text{g}}\ \text{glucose} \times \frac{1\ \text{mol glucose}}{180.2\ \cancel{\text{g}}\ \text{glucose}} = 0.0005\ \text{mol glucose}$$

$$\frac{0.0005\ \text{mol glucose}}{0.1\ \text{L blood}} = 0.005\ \text{M}$$

Answer

8.107 a. 280 mL of mannitol solution would have to be given.

$$(\text{w/v})\% = \frac{70.\ \text{g mannitol}}{x\ \text{mL solution}} \times 100\% = 25\%\ (\text{w/v})\ \text{mannitol}$$

x = 280 mL
Answer

b. The hypertonic mannitol solution draws water out of swollen brain cells and thus reduces the pressure on the brain.

8.109 When a cucumber is placed in a concentrated salt solution, water moves out of the cells of the cucumber to the hypertonic salt solution, so the cucumber shrinks and loses its crispness.

8.111 Convert ounces to milliliters, and then calculate the weight/volume percent concentration.

$$8.0\ \not{oz} \times \frac{29.6\ \text{mL}}{1\ \not{oz}} = 240\ \text{mL}$$

$$(\text{w/v})\% = \frac{15\ \text{g complex carbohydrates}}{240\ \text{mL solution}} \times 100\% = 6.3\%\ (\text{w/v})\ \text{complex carbohydrates}$$

Answer

8.113 Use the molarity to determine the number of moles of HCl, and then convert this to grams.

$$\text{mol} = \text{M} \times V = 0.10\ \text{M HCl} \times 2.0\ \text{L} = 0.20\ \text{mol HCl}$$

$$0.20\ \not{\text{mol}}\ \text{HCl} \times \frac{36.46\ \text{g HCl}}{1\ \not{\text{mol}}\ \text{HCl}} = 7.3\ \text{g HCl}$$

Answer

8.115

$$5.0\ \text{L} = 5.0 \times 10^3\ \text{mL blood}$$

$$(\text{w/v})\% = \frac{x\ \text{g ethanol}}{5.0 \times 10^3\ \text{mL blood}} \times 100\% = 0.08\%\ (\text{w/v})\ \text{ethanol} \qquad x = 4\ \text{g ethanol}$$

$$4\ \not{g}\ \text{ethanol} \times \frac{1000\ \text{mg}}{1\ \not{g}} = 4{,}000\ \text{mg ethanol}$$

Answer

8.117

a. $$\frac{x\ \text{g acetaminophen}}{1\ \text{mL}} \times 10^6 = 15\ \text{ppm acetaminophen} \qquad x = 0.000\ 015\ \text{g acetaminophen}$$

$$0.000\ 015\ \not{g}\ \text{acetaminophen} \times \frac{10^6\ \mu\text{g}}{1\ \not{g}} = 15\ \mu\text{g acetaminophen, which is in the therapeutic range}$$

b. $$\frac{0.000\ 015\ \not{g}\ \text{acetaminophen}}{1\ \not{\text{mL}}\ \text{blood}} \times 5000\ \not{\text{mL}}\ \text{blood} \times \frac{1\ \text{mol}}{151.2\ \not{g}} = 0.000\ 50\ \text{mol acetaminophen}$$

Answer

Chapter 9 Acids and Bases

Chapter Review

[1] Describe the principal features of acids and bases. (9.1)

- A Brønsted–Lowry acid is a proton donor, often symbolized by HA. A Brønsted–Lowry acid must contain one or more hydrogen atoms.

This proton is donated to H_2O.

$$HCl(g) + H_2O(l) \longrightarrow H_3O^+(aq) + Cl^-(aq)$$

Brønsted–Lowry acid

- A Brønsted–Lowry base is a proton acceptor, often symbolized by B:. To form a bond to a proton, a Brønsted–Lowry base must contain a lone pair of electrons.

This electron pair forms a new bond to a H from H_2O.

$$NH_3 + H_2O(l) \longrightarrow [NH_4]^+ + OH^-(aq)$$

Brønsted–Lowry base

[2] What are the principal features of an acid–base reaction? (9.2)

- In a Brønsted–Lowry acid–base reaction, a proton is transferred from the acid (HA) to the base (B:). In this reaction, the acid loses a proton to form its conjugate base and the base gains a proton to form its conjugate acid.

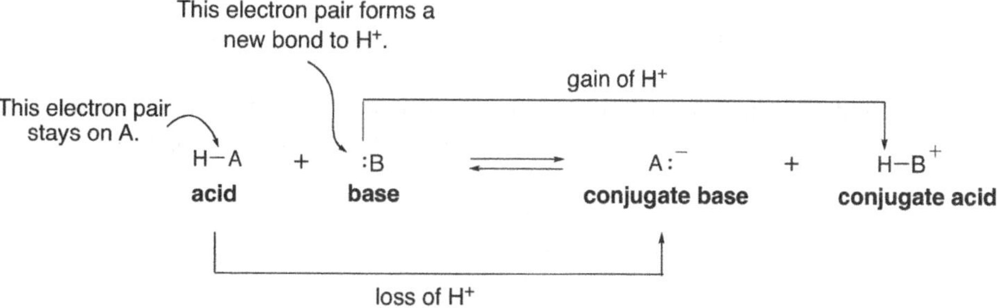

[3] How is acid strength related to the direction of equilibrium in an acid–base reaction? (9.3)

- A strong acid readily donates a proton, and when dissolved in water, 100% of the acid dissociates into ions.

• Use a single reaction arrow.
• The product is greatly favored at equilibrium.

$$\underset{\text{strong acid}}{HCl(g)} + H_2O(l) \longrightarrow H_3O^+(aq) + \underset{\text{conjugate base}}{Cl^-(aq)}$$

$$\underset{\text{strong acid}}{H_2SO_4(l)} + H_2O(l) \longrightarrow H_3O^+(aq) + \underset{\text{conjugate base}}{HSO_4^-(aq)}$$

- A strong base readily accepts a proton. When a strong base like NaOH is dissolved in water, 100% of the base dissociates into ions.

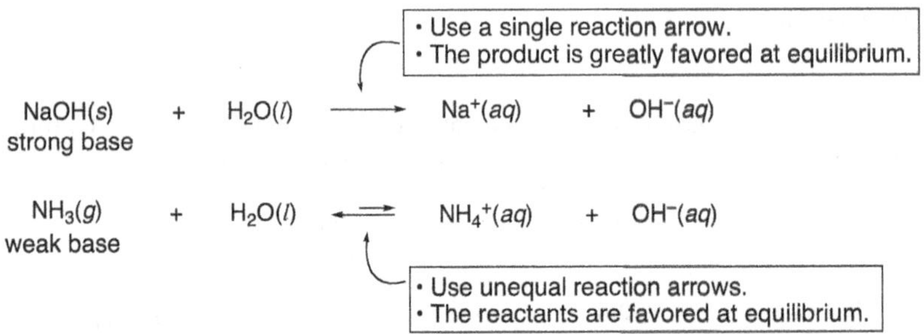
• Use a single reaction arrow.
• The product is greatly favored at equilibrium.
NaOH(s) + H2O(l) → Na+(aq) + OH−(aq)
strong base
NH3(g) + H2O(l) ⇌ NH4+(aq) + OH−(aq)
weak base
• Use unequal reaction arrows.
• The reactants are favored at equilibrium.

- An inverse relationship exists between acid and base strength. **A strong acid forms a weak conjugate base, whereas a weak acid forms a strong conjugate base.**
- **In an acid–base reaction, the stronger acid reacts with the stronger base to form the weaker acid and the weaker base.**

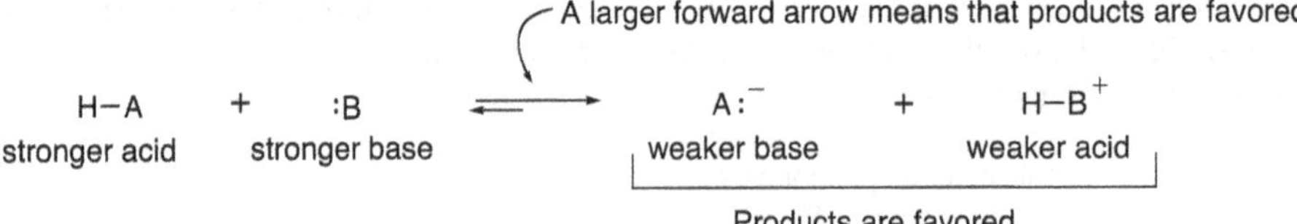
A larger forward arrow means that products are favored.
H−A + :B ⇌ A:− + H−B+
stronger acid
stronger base
weaker base
weaker acid
Products are favored.

[4] What is the acid dissociation constant and how is it related to acid strength? (9.4)

- For a general acid HA, the acid dissociation constant K_a is defined by the following equation:

$$K_a = \frac{[H_3O^+][A:^-]}{[HA]}$$

- **The *stronger* the acid, the *larger* the K_a.** Equilibrium in an acid–base reaction favors the formation of the acid with the smaller K_a value.

[5] What is the ion–product of water and how is it used to calculate hydronium or hydroxide ion concentration? (9.5)

- The ion–product of water, K_w, is a constant for all aqueous solutions; $K_w = [H_3O^+][OH^-] = 1.0 \times 10^{-14}$ at 25 °C. If either $[H_3O^+]$ or $[OH^-]$ is known, then the other value can be calculated from K_w.

$$K_w = [H_3O^+][OH^-]$$
$$K_w = (1.0 \times 10^{-7}) \times (1.0 \times 10^{-7})$$
$$K_w = 1.0 \times 10^{-14}$$

[6] What is pH? (9.6)

- The pH of a solution measures the concentration of H_3O^+.

$$\boxed{pH = -\log [H_3O^+]}$$

- A pH = 7 means $[H_3O^+] = [OH^-]$ and the solution is neutral.
- A pH < 7 means $[H_3O^+] > [OH^-]$ and the solution is acidic.
- A pH > 7 means $[H_3O^+] < [OH^-]$ and the solution is basic.

[7] Draw the products of some common acid–base reactions. (9.7)

- In a Brønsted–Lowry acid–base reaction with hydroxide bases (MOH), the acid HA donates a proton to OH^- to form H_2O. The anion from the acid HA combines with the cation M^+ of the base to form the salt MA. This reaction is called a **neutralization** reaction.

$$\underset{\text{acid}}{HA(aq)} + \underset{\text{base}}{MOH(aq)} \longrightarrow \underset{\text{water}}{H\text{–}OH(l)} + \underset{\text{salt}}{MA(aq)}$$

- In acid–base reactions with bicarbonate (HCO_3^-) or carbonate (CO_3^{2-}) bases, carbonic acid (H_2CO_3) is formed, which decomposes to form H_2O and CO_2.

$$H^+(aq) + \underset{\text{bicarbonate}}{HCO_3^-(aq)} \longrightarrow [\,H_2CO_3(aq)\,] \longrightarrow H_2O(l) + \underset{\text{bubbles of } CO_2}{CO_2(g)}$$

[8] What happens to the pH of an aqueous solution when a salt is dissolved? (9.8)

- A salt can form an acidic, basic, or neutral solution depending on whether its cation and anion are derived from strong or weak acids and bases.

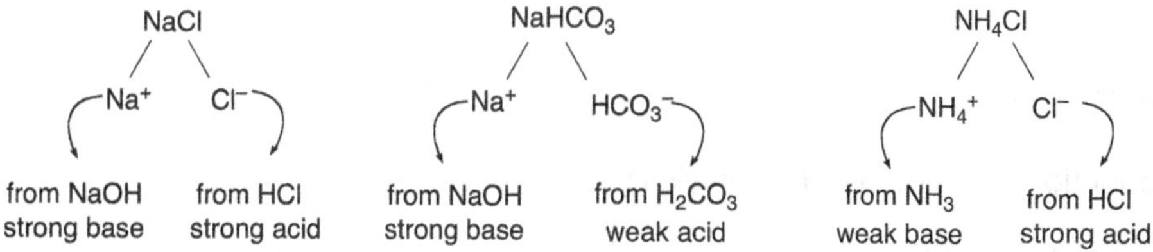

- A salt derived from a strong acid and strong base forms a neutral solution with pH = 7. When one ion of a salt is derived from a weak acid or base, the ion derived from the stronger acid or base determines whether the solution is acidic or basic.

For $NaHCO_3$: $HCO_3^-(aq) + H_2O(l) \rightleftharpoons H_2CO_3(aq) + \boxed{OH^-(aq)}$

Hydroxide makes the solution basic, so the pH > 7.

For NH_4Cl: $NH_4^+(aq) + H_2O(l) \rightleftharpoons NH_3(aq) + \boxed{H_3O^+(aq)}$

H_3O^+ makes the solution acidic, so the pH < 7.

[9] How is a titration used to determine the concentration of an acid or base? (9.9)

- A titration is a procedure that uses a base of known volume and molarity to react with a known volume of acid of unknown molarity. The volume and molarity of the base are used to calculate the number of moles of base that react, and from this value, the molarity of the acid can be determined. A titration can also be used to determine the concentration of a base by using an acid of known molarity.

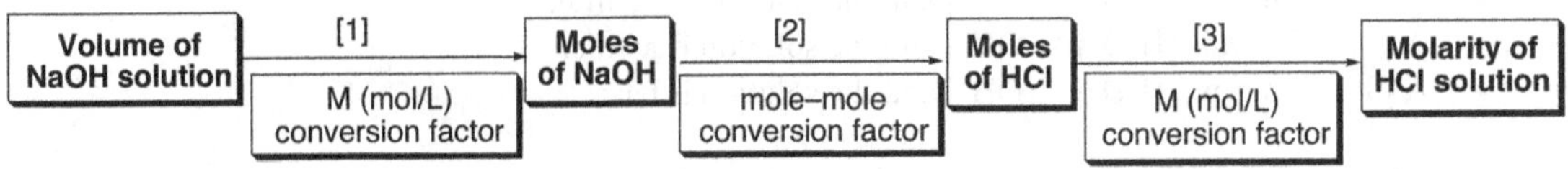

[10] What is a buffer? (9.10)

- A buffer is a solution whose pH changes very little when acid or base is added. Most buffers are composed of approximately equal amounts of a weak acid and the salt of its conjugate base.

[11] What is the principal buffer present in the blood? (9.11)

- The principal buffer in the blood is carbonic acid/bicarbonate. Since carbonic acid (H_2CO_3) is in equilibrium with dissolved CO_2, the amount of CO_2 in the blood affects its pH, which is normally maintained in the range of 7.35–7.45. When the CO_2 concentration in the blood is higher than normal, the acid–base equilibrium shifts to form more H_3O^+ and the pH decreases. When the CO_2 concentration in the blood is lower than normal, the acid–base equilibrium shifts to consume $[H_3O^+]$, so $[H_3O^+]$ decreases, and the pH increases.

$$CO_2(g) + H_2O(l) \rightleftharpoons \underset{\text{carbonic acid}}{H_2CO_3(aq)} \overset{H_2O}{\rightleftharpoons} H_3O^+(aq) + \underset{\text{bicarbonate}}{HCO_3^-(aq)}$$

(carbonic acid and bicarbonate: principal buffer in the blood)

Problem Solving

[1] Introduction to Acids and Bases (9.1)

Example 9.1 Which of the following species can be Brønsted–Lowry acids?

a. HBr b. H_2SO_4 c. I_2

Analysis

A Brønsted–Lowry acid must contain a hydrogen atom, but it may be neutral or contain a net positive or negative charge.

Solution

a. HBr is a Brønsted–Lowry acid because it contains a H.
b. H_2SO_4 is a Brønsted–Lowry acid because it contains a H.
c. I_2 is not a Brønsted–Lowry acid because it does not contain a H.

Example 9.2 Which of the following species can be Brønsted–Lowry bases?

a. KOH b. Br^- c. C_2H_6

Analysis

A Brønsted–Lowry base must contain a lone pair of electrons, but it may be neutral or have a net negative charge.

Solution

a. KOH is a base because it contains hydroxide, OH^-, which has three lone pairs on its O atom.
b. Br^- is a base because it has four lone pairs.
c. C_2H_6 is not a base because it has no lone pairs.

Example 9.3 Classify each reactant as a Brønsted–Lowry acid or base.

a. $Cl^-(aq) + HSO_4^-(aq) \rightleftharpoons HCl(aq) + SO_4^{2-}(aq)$

b. $HPO_4^{2-}(aq) + OH^-(aq) \rightleftharpoons PO_4^{3-}(aq) + H_2O(l)$

Analysis

In each equation, the Brønsted–Lowry acid is the species that loses a proton and the Brønsted–Lowry base is the species that gains a proton.

Solution

a. HSO_4^- is the acid because it loses a proton (H^+) to form SO_4^{2-}, and Cl^- is the base because it gains a proton to form HCl.

loss of H^+

$Cl^-(aq) + HSO_4^-(aq) \rightleftharpoons HCl(aq) + SO_4^{2-}(aq)$

base acid

gain of H^+

b. HPO_4^{2-} is the acid because it loses a proton (H^+) to form PO_4^{3-}, and OH^- is the base because it gains a proton to form H_2O.

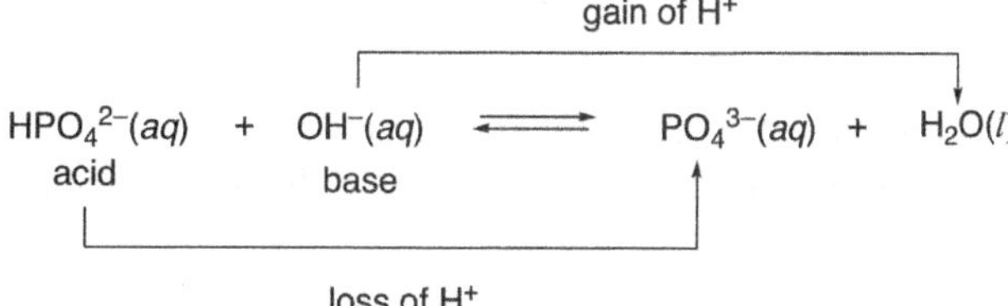

[2] The Reaction of a Brønsted–Lowry Acid with a Brønsted–Lowry Base (9.2)

Example 9.4 Draw the conjugate acid of each base:

a. Br^- b. HPO_4^{2-}

Analysis

Conjugate acid–base pairs differ by the presence of a proton. To draw a conjugate acid from a base, *add* a proton, H^+. Then add +1 to the charge of the base to give the charge on the conjugate acid.

Solution

a. $Br^- + H^+$ gives HBr as the conjugate acid. HBr has no charge because a proton with a +1 charge is added to an anion with a –1 charge.
b. $HPO_4^{2-} + H^+$ gives $H_2PO_4^-$ as the conjugate acid. $H_2PO_4^-$ has a –1 charge because a proton with a +1 charge is added to an anion with a –2 charge.

Example 9.5 Draw the conjugate base of each acid:

a. H_3O^+ b. H_2Se

Analysis

Conjugate acid–base pairs differ by the presence of a proton. To draw a conjugate base from an acid, *remove* a proton, H^+. Then, add –1 to the charge of the acid to give the charge on the conjugate base.

Solution

a. Remove H^+ from H_3O^+ to form H_2O, the conjugate base. H_2O has no charge because –1 is added to a cation that has a +1 charge to begin with.
b. Remove H^+ from H_2Se to form HSe^-, the conjugate base. HSe^- has a –1 charge because –1 is added to a molecule that had no charge to begin with.

---

[3] Acid and Base Strength (9.3)

Example 9.6 Using Table 9.2:

a. Is H_2SO_4 or NH_4^+ the stronger acid?
b. Draw the conjugate base of each acid and predict which base is stronger.

Analysis

Use Table 9.2 to determine the stronger acid. The stronger the acid, the weaker the conjugate base.

Solution

a. H_2SO_4 is located above NH_4^+ in Table 9.2, making it the stronger acid.
b. To draw each conjugate base, remove a proton (H^+). Since NH_4^+ is the weaker acid, NH_3 is the stronger conjugate base.

H_2SO_4 stronger acid	lose H^+ →	HSO_4^-	weaker base
NH_4^+ weaker acid	lose H^+ →	NH_3	stronger base

---

[4] Equilibrium and Acid Dissociation Constants (9.4)

Example 9.7 Which acid is stronger: HF or H_3PO_4?

Analysis
Use Table 9.3 to find the K_a for each acid. The acid with the larger K_a is the stronger acid.

Solution

HF	H_3PO_4
$K_a = 7.2 \times 10^{-4}$	$K_a = 7.5 \times 10^{-3}$
	larger K_a
	stronger acid

---

[5] The Dissociation of Water (9.5)

Example 9.8 Calculate the value of $[H_3O^+]$ and $[OH^-]$ in a 0.001 M KOH solution.

Analysis
KOH is a strong base that completely dissociates to form K^+ and OH^-, so the concentration of KOH gives the concentration of OH^- ions. The value of $[OH^-]$ can then be used to calculate $[H_3O^+]$ from the expression for K_w.

Solution
The value of $[OH^-]$ in a 0.001 M KOH solution is 0.001 M = 1×10^{-3} M.

$$[H_3O^+] = \frac{K_w}{[OH^-]} = \frac{1 \times 10^{-14}}{1 \times 10^{-3}} = 1 \times 10^{-11}\ M$$

concentration of OH^- (points to 1×10^{-3}); 1×10^{-11} M = concentration of H_3O^+

---

[6] Common Acid–Base Reactions (9.7)

Example 9.9 Write a balanced equation for the reaction of HNO_3 with LiOH.

Analysis
The acid and base react to form a salt and water.

Solution
HNO_3 is the acid and LiOH is the base. H^+ from the acid reacts with OH^- from the base to give H_2O. A salt ($LiNO_3$) is also formed from the cation of the base (Li^+) and the anion of the acid (NO_3^-).

$$\underset{\text{acid}}{HNO_3(aq)} + \underset{\text{base}}{LiOH(aq)} \longrightarrow \underset{\text{salt}}{LiNO_3(aq)} + H_2O(l)$$

---

[7] The Acidity and Basicity of Salt Solutions (9.8)

Example 9.10 Determine whether each salt forms an acidic, basic, or neutral solution when dissolved in water.

a. NaBr b. KCH_3COO

Analysis
Determine what type of acid and base (strong or weak) are used to form the salt. When the ions in the salt come from a strong acid and strong base, the solution is neutral. When the ions come from acids and bases of different strength, the ion derived from the stronger reactant determines the acidity.

Solution

a. NaBr → Na^+ and Br^-

Na^+ from NaOH, **strong base**; Br^- from HBr, **strong acid**

neutral solution

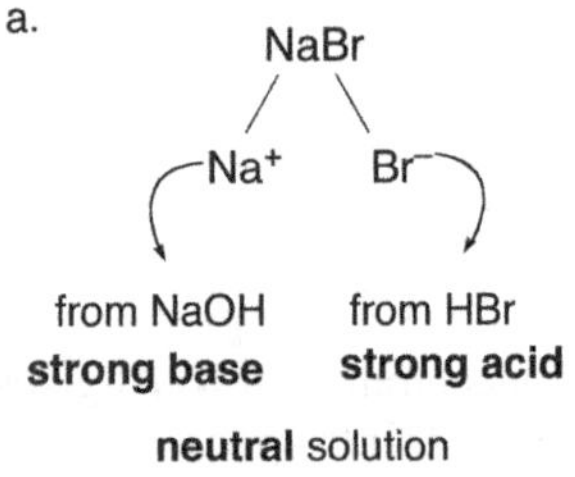

b. KCH_3COO → K^+ and CH_3COO^-

K^+ from KOH, **strong base**; CH_3COO^- from CH_3COOH, weak acid

basic solution

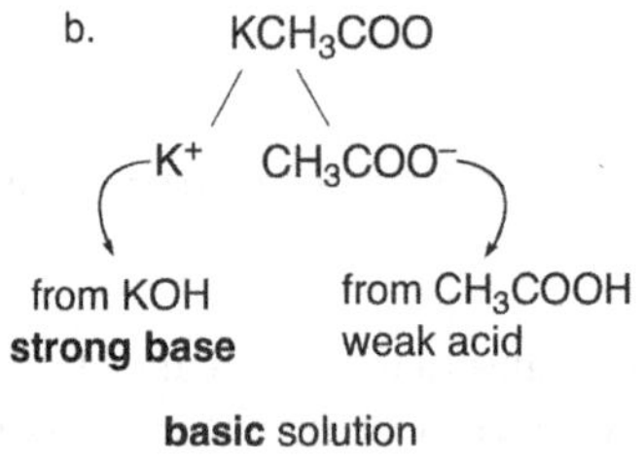

[8] Titration (9.9)

Example 9.11 What is the molarity of an HCl solution if 17.2 mL of 0.15 M NaOH are needed to neutralize 5.00 mL of the sample?

$$HCl(aq) + NaOH(aq) \longrightarrow NaCl(aq) + H_2O(l)$$

Analysis and Solution
[1] Determine the number of moles of base used to neutralize the acid.

- Use the molarity (M) and volume (V) of the base to calculate the number of moles (mol = MV).

$$17.2\ \cancel{mL}\ NaOH \times \frac{1\ \cancel{L}}{1000\ \cancel{mL}} \times \frac{0.15\ mol\ NaOH}{1\ \cancel{L}} = 0.0026\ mol\ NaOH$$

---

[2] Determine the number of moles of acid that react from the balanced chemical equation.

- Since each HCl molecule contains one proton, *one* mole of the acid HCl reacts with *one* mole of the base NaOH in the neutralization reaction. The coefficients in the balanced equation form a mole ratio to calculate the number of moles of acid that react.

$$0.0026 \text{ mol NaOH} \times \frac{1 \text{ mol HCl}}{1 \text{ mol NaOH}} = 0.0026 \text{ mol HCl}$$

[3] Determine the molarity of the acid from the number of moles and known volume.

$$\underset{\text{molarity}}{\text{M}} = \frac{\text{mol}}{\text{L}} = \frac{0.0026 \text{ mol HCl}}{5.00 \text{ mL solution}} \times \frac{1000 \text{ mL}}{1 \text{ L}} = \underset{\textbf{Answer}}{0.52 \text{ M HCl}}$$

[9] Buffers (9.10)

Example 9.12 What is the pH of a buffer that contains 0.15 M NaH_2PO_4 and 0.15 M Na_2HPO_4?

Analysis

Use the K_a of the weak acid of the buffer from Table 9.6 and the expression $[H_3O^+] = K_a([HA]/[A:^-])$ to calculate $[H_3O^+]$. Calculate the pH using the expression $pH = -\log [H_3O^+]$.

Solution

[1] Substitute the given concentrations of NaH_2PO_4 and Na_2HPO_4 for [HA] and [A:⁻], respectively. K_a for NaH_2PO_4 is 6.2×10^{-8}.

$$[H_3O^+] = K_a \times \frac{[NaH_2PO_4]}{[Na_2HPO_4]} = (6.2 \times 10^{-8}) \times \frac{[0.15 \text{ M}]}{[0.15 \text{ M}]}$$

$$[H_3O^+] = 6.2 \times 10^{-8} \text{ M}$$

[2] Use an electronic calculator to convert the H_3O^+ concentration to pH.

$$pH = -\log [H_3O^+] = -\log(6.2 \times 10^{-8})$$

$$pH = 7.21$$

Self-Test

[1] Fill in the blank with one of the terms listed below.

Acid (9.1)	Brønsted–Lowry base (9.1)	Monoprotic acid (9.1)
Amphoteric (9.2)	Buffer (9.10)	Net ionic equation (9.7)
Base (9.1)	Conjugate acid (9.2)	Proton transfer reaction (9.2)
Brønsted–Lowry acid (9.1)	Conjugate base (9.2)	Triprotic acid (9.1)

1. A ______ contains only the species involved in a reaction.
2. A ______ is a proton acceptor.
3. According to the Arrhenius definition, an ______ contains a hydrogen atom and dissolves in water to form a hydrogen ion, H^+.

4. A ______ is a solution whose pH changes very little when acid or base is added.
5. A compound that contains both a hydrogen atom and a lone pair of electrons can be either an acid or a base, depending on the particular reaction. Such a compound is said to be ______.
6. The product formed by loss of a proton from an acid is called its ______.
7. A Brønsted–Lowry acid–base reaction is a ______ because it always results in the transfer of a proton from an acid to a base.
8. A ______ contains *three* acidic protons.
9. The product formed by gain of a proton by a base is called its ______.
10. According to the Arrhenius definition, a ______ contains hydroxide and dissolves in water to form OH^-.
11. A ______ is a proton donor.
12. A ______ contains *one* acidic proton.

[2] Fill in the blank with one of the terms listed below.

a. strong acid
b. weak acid
c. strong base
d. weak base

13. A ______ donates a proton, but when dissolved in water, only a small fraction of it dissociates into ions.
14. A ______ accepts a proton, but when dissolved in water, only a small fraction of it forms ions.
15. A ______ readily accepts a proton, forming a *weak* conjugate acid.
16. A ______ readily donates a proton, forming a *weak* conjugate base.

[3] Fill in the blank with one of the terms listed below.

a. an acidic solution
b. a basic solution
c. a neutral solution

17. A salt derived from a *strong base* and a weak acid forms ______.
18. A salt derived from a weak base and a *strong acid* forms ______.
19. A solution with a pH = 7 is ______.
20. A solution with a pH > 7 is ______.
21. A solution with a pH < 7 is ______.

[4] Decide if each compound is a Brønsted–Lowry acid, a Brønsted–Lowry base, or both.

a. Brønsted–Lowry acid
b. Brønsted–Lowry base

22. $Ca(OH)_2$

23. H_2SO_4

24. $H_2\ddot{O}:$

25. H—C(H)(H)—C(=Ö:)—:Ö—H

Answers to Self-Test

1. net ionic equation	6. conjugate base	11. Brønsted–Lowry acid	16. a	21. a
2. Brønsted–Lowry base	7. proton transfer reaction	12. monoprotic acid	17. b	22. b
3. acid	8. triprotic acid	13. b	18. a	23. a
4. buffer	9. conjugate acid	14. d	19. c	24. a, b
5. amphoteric	10. base	15. c	20. b	25. a, b

Solutions to In-Chapter Problems

9.1 a. F^- is fluoride, so HF is **hydrofluoric acid.**
b. NO_3^- is nitrate, so HNO_3 is **nitric acid.**
c. CN^- is cyanide, so HCN is **hydrocyanic acid.**

9.2 The name of the anion (ClO_2^-) ends in *-ite,* so the acid ($HClO_2$) is chlorous acid.

9.3 A Brønsted–Lowry acid must contain a hydrogen atom, but it may be neutral or contain a net positive or negative charge. Use Example 9.1 to help determine which of the compounds are Brønsted–Lowry acids.

a. HI: contains a H atom, a Brønsted–Lowry acid
b. SO_4^{2-}: no H atom
c. $H_2PO_4^-$: contains a H atom, a Brønsted–Lowry acid
d. Cl^-: no H atom

9.4 A Brønsted–Lowry base must contain a lone pair of electrons, but it may be neutral or have a net negative charge. Use Example 9.2 to help determine which of the compounds are Brønsted–Lowry bases.

a. $Al(OH)_3$: lone pairs on OH, a Brønsted–Lowry base
b. Br^-: lone pairs on Br, a Brønsted–Lowry base
c. NH_4^+: no lone pair of electrons
d. CN^-: lone pairs on C and N, a Brønsted–Lowry base

9.5 In each equation, the Brønsted–Lowry acid is the species that loses a proton and the Brønsted–Lowry base is the species that gains a proton. Use Example 9.3 to help determine which reactant is an acid and which is a base.

a. $\underset{\text{acid}}{HCl(g)} + \underset{\text{base}}{NH_3(g)} \longrightarrow Cl^-(aq) + NH_4^+(aq)$

b. $\underset{\text{acid}}{CH_3COOH(l)} + \underset{\text{base}}{H_2O(l)} \longrightarrow CH_3COO^-(aq) + H_3O^+(aq)$

c. $\underset{\text{base}}{OH^-(aq)} + \underset{\text{acid}}{HSO_4^-(aq)} \longrightarrow H_2O(l) + SO_4^{2-}(aq)$

9.6 Use Example 9.4 to draw the conjugate acid of each species. Conjugate acid–base pairs differ by the presence of a proton. To draw a conjugate acid from a base, *add* a proton, H^+. Then add +1 to the charge of the base to give the charge on the conjugate acid.

a. H_2O: add one H^+ to make H_3O^+.
b. I^-: add one H^+ to make HI.
c. HCO_3^-: add one H^+ to make H_2CO_3.

9.7 Use Example 9.5 to draw the conjugate base of each species. Conjugate acid–base pairs differ by the presence of a proton. To draw a conjugate base from an acid, *remove* a proton, H^+. Then, add –1 to the charge of the acid to give the charge on the conjugate base.

a. H_2S: remove one H^+ to make HS^-.
b. HCN: remove one H^+ to make CN^-.
c. HSO_4^-: remove one H^+ to make SO_4^{2-}.

9.8 Use the common element colors on the inside back cover to draw the acid; then remove H^+.

a. HBr (acid), Br^- (conjugate base)

b. H–O–C(=O)–C(=O)–O–H (acid) H–O–C(=O)–C(=O)–O^- (conjugate base)

9.9 Use Sample Problem 9.6 to label the acid and base in each reaction. The Brønsted–Lowry acid loses a proton to form its conjugate base. The Brønsted–Lowry base gains a proton to form its conjugate acid.

a. $H_2O(l)$ (base) + $HI(g)$ (acid) $\rightleftharpoons$ $I^-(aq)$ (conjugate base) + $H_3O^+(aq)$ (conjugate acid)

b. $CH_3COOH(l)$ (acid) + $NH_3(g)$ (base) $\rightleftharpoons$ $CH_3COO^-(aq)$ (conjugate base) + $NH_4^+(aq)$ (conjugate acid)

c. $Br^-(aq)$ (base) + $HNO_3(aq)$ (acid) $\rightleftharpoons$ $HBr(aq)$ (conjugate acid) + $NO_3^-(aq)$ (conjugate base)

9.10 Draw the conjugate acid and base of ammonia as in Examples 9.4 and 9.5.

a. NH_3: add one H^+ to make NH_4^+.

b. NH_3: remove one H^+ to make NH_2^-.

9.11 a. H_2O is the acid because it loses a proton to form OH^-. This mean HSO_4^- must be the base, which gains a proton.

gain of H^+

HSO_4^- (base) + H_2O (acid) $\rightleftharpoons$ H_2SO_4 + OH^-

loss of H^+

b. H_2O is the base because it gains a proton to form H_3O^+. This means HSO_4^- is the acid, which loses a proton.

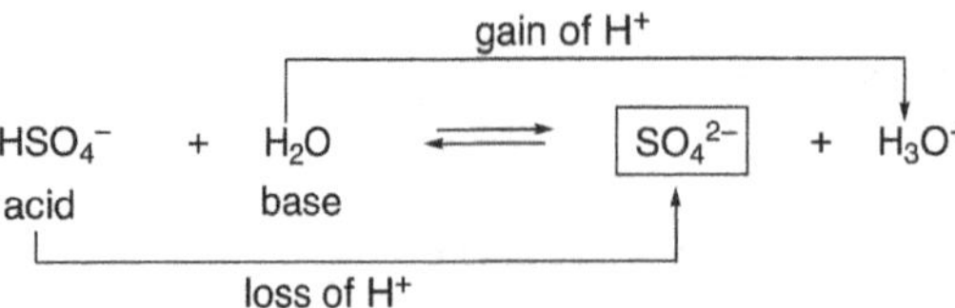

9.12 The stronger the acid HA, the more readily it dissociates to form its conjugate base. In molecular art, the strongest acid has the most A^- and H_3O^+ ions, and the fewest molecules of undissociated HA.

D (smallest amount of A^- and H_3O^+) < **F** < **E** (largest amount of A^- and H_3O^+)

9.13 Use Table 9.2 to determine which acid is stronger. The stronger the acid, the weaker the conjugate base.

a. H_2SO_4 is the stronger acid; H_3PO_4 has the stronger conjugate base.
b. HCl is the stronger acid; HF has the stronger conjugate base.
c. H_2CO_3 is the stronger acid; NH_4^+ has the stronger conjugate base.
d. HF is the stronger acid; HCN has the stronger conjugate base.

9.14 To determine if the reactants or products are favored at equilibrium:

- Identify the acid in the reactants and the conjugate acid in the products.
- Determine the relative strength of the acid and the conjugate acid.
- Equilibrium favors the formation of the weaker acid.

a. $HF(g) + OH^-(aq) \rightleftharpoons F^-(aq) + H_2O(l)$

HF(*g*): acid
$H_2O(l)$: conjugate acid, weaker acid

Products are favored.

b. $NH_4^+(aq) + Cl^-(aq) \rightleftharpoons NH_3(g) + HCl(aq)$

$NH_4^+(aq)$: acid, weaker acid
HCl(*aq*): conjugate acid

Reactants are favored.

c. $HCO_3^-(aq) + H_3O^+(aq) \rightleftharpoons H_2CO_3(aq) + H_2O(l)$

$H_3O^+(aq)$: acid
$H_2CO_3(aq)$: conjugate acid, weaker acid

Products are favored.

9.15 Compare the acid and conjugate acid to determine if the reactants or products are favored.

a. $C_3H_6O_3(aq) + H_2O(l) \rightleftharpoons C_3H_5O_3^-(aq) + H_3O^+(aq)$

$C_3H_6O_3(aq)$: acid, weaker acid
$H_3O^+(aq)$: conjugate acid

Reactants are favored.

b. $C_3H_6O_3(aq) + HCO_3^-(aq) \rightleftharpoons C_3H_5O_3^-(aq) + H_2CO_3(aq)$

$C_3H_6O_3(aq)$: acid
$H_2CO_3(aq)$: conjugate acid, weaker acid

Products are favored.

9.16 Use Table 9.3 to find the K_a for each acid as in Example 9.7. The acid with the larger K_a is the stronger acid.

a. increasing acid strength: $HPO_4^{2-} < H_2PO_4^- < H_3PO_4$
b. increasing acid strength: $HCN < CH_3COOH < HF$

9.17 To determine the direction of equilibrium, identify the acid in the reactants and the conjugate acid in the products as in Sample Problem 9.10. Then compare their K_a values. **Equilibrium favors the formation of the acid with the smaller K_a value.**

reactants favored

$$HCO_3^-(aq) + NH_3(aq) \rightleftharpoons CO_3^{2-}(aq) + NH_4^+(aq)$$

acid — $K_a = 5.6 \times 10^{-11}$ — smaller K_a — **weaker acid**

conjugate acid — $K_a = 5.6 \times 10^{-10}$ — larger K_a — **stronger acid**

9.18 Compare the acids by comparing their K_a values from Table 9.3.
a. H_2CO_3 has the larger K_a.
b. H_2CO_3 is stronger.
c. HCN has the stronger conjugate base.
d. H_2CO_3 has the weaker conjugate base.
e. When H_2CO_3 is dissolved in water, the equilibrium lies further to the right.

9.19 Use the equation $[OH^-] = K_w/[H_3O^+]$ to calculate the hydroxide ion concentration as in Sample Problem 9.11. When $[OH^-] > [H_3O^+]$, the solution is basic. When $[OH^-] < [H_3O^+]$, the solution is acidic.

a. $[OH^-] = \dfrac{K_w}{[H_3O^+]} = \dfrac{1.0 \times 10^{-14}}{10^{-3}} = 10^{-11}\ M$ acidic

b. $[OH^-] = \dfrac{K_w}{[H_3O^+]} = \dfrac{1.0 \times 10^{-14}}{10^{-11}} = 10^{-3}\ M$ basic

c. $[OH^-] = \dfrac{K_w}{[H_3O^+]} = \dfrac{1.0 \times 10^{-14}}{2.8 \times 10^{-10}} = 3.6 \times 10^{-5}\ M$ basic

d. $[OH^-] = \dfrac{K_w}{[H_3O^+]} = \dfrac{1.0 \times 10^{-14}}{5.6 \times 10^{-4}} = 1.8 \times 10^{-11}\ M$ acidic

9.20 Use the equation $[H_3O^+] = K_w/[OH^-]$ to calculate the hydronium ion concentration as in Sample Problem 9.11.

a. $[H_3O^+] = \dfrac{K_w}{[OH^-]} = \dfrac{1.0 \times 10^{-14}}{10^{-6}} = 10^{-8}\ M$ basic

b. $[H_3O^+] = \dfrac{K_w}{[OH^-]} = \dfrac{1.0 \times 10^{-14}}{10^{-9}} = 10^{-5}\ M$ acidic

c. $[H_3O^+] = \dfrac{K_w}{[OH^-]} = \dfrac{1.0 \times 10^{-14}}{5.2 \times 10^{-11}} = 1.9 \times 10^{-4}$ M
acidic

d. $[H_3O^+] = \dfrac{K_w}{[OH^-]} = \dfrac{1.0 \times 10^{-14}}{7.3 \times 10^{-4}} = 1.4 \times 10^{-11}$ M
basic

9.21 Since NaOH is a strong base that completely dissociates to form Na^+ and OH^-, the concentration of NaOH gives the concentration of OH^- ions. The $[OH^-]$ can then be used to calculate $[H_3O^+]$ from the expression for K_w. Similarly, since HCl is a strong acid and completely dissociates, the $[H_3O^+]$ can then be used to calculate $[OH^-]$ from the expression for K_w. See Example 9.8.

a. $[H_3O^+] = \dfrac{K_w}{[OH^-]} = \dfrac{1 \times 10^{-14}}{10^{-3}} = 10^{-11}$ M
concentration of H_3O^+
(10^{-3}: concentration of OH^-)

b. $[OH^-] = \dfrac{K_w}{[H_3O^+]} = \dfrac{1 \times 10^{-14}}{10^{-3}} = 10^{-11}$ M
concentration of OH^-
(10^{-3}: concentration of H_3O^+)

c. $[OH^-] = \dfrac{K_w}{[H_3O^+]} = \dfrac{1 \times 10^{-14}}{1.5} = 6.7 \times 10^{-15}$ M
concentration of OH^-
(1.5: concentration of H_3O^+)

d. $[H_3O^+] = \dfrac{K_w}{[OH^-]} = \dfrac{1 \times 10^{-14}}{3.0 \times 10^{-1}} = 3.3 \times 10^{-14}$ M
concentration of H_3O^+
(3.0×10^{-1}: concentration of OH^-)

9.22 When the coefficient of a number written in scientific notation is one, the pH equals the value of x in 10^{-x}.

a. 1×10^{-6} M: pH = 6
b. 1×10^{-12} M: pH = 12
c. 0.000 01 M = 10^{-5}: pH = 5
d. 0.000 000 000 01 M = 10^{-11}: pH = 11

9.23 The pH equals the value of x in 10^{-x}, and this gives the H_3O^+ concentration. An *acidic* solution has a pH < 7. A *basic* solution has a pH > 7. A *neutral* solution has a pH = 7.

a. pH = 13, $[H_3O^+] = 1 \times 10^{-13}$ M: basic
b. pH = 7, $[H_3O^+] = 1 \times 10^{-7}$ M: neutral
c. pH = 3, $[H_3O^+] = 1 \times 10^{-3}$ M: acidic

9.24 Use a calculator to determine the antilogarithm of (– pH); $[H_3O^+]$ = antilog(–pH).

a. $[H_3O^+]$ = antilog(–pH) = antilog(–10.2)

$[H_3O^+]$ = 6×10^{-11} M

b. $[H_3O^+]$ = antilog(–pH) = antilog(–7.8)

$[H_3O^+]$ = 2×10^{-8} M

c. $[H_3O^+]$ = antilog(–pH) = antilog(–4.3)

$[H_3O^+]$ = 5×10^{-5} M

9.25 Use a calculator to determine the logarithm of a number that contains a coefficient other than one in scientific notation; pH = –log $[H_3O^+]$.

a. pH = –log $[H_3O^+]$ = –log(1.8×10^{-6})

= –(–5.74) = 5.74

b. pH = –log $[H_3O^+]$ = –log(9.21×10^{-12})

= –(–11.036) = 11.036

c. pH = –log $[H_3O^+]$ = –log(8.8×10^{-5})

= –(–4.06) = 4.06

d. pH = –log $[H_3O^+]$ = –log(7.62×10^{-11})

= –(–10.118) = 10.118

9.26 Use a calculator to determine the antilogarithm of (–pH).

$[H_3O^+]$ = antilog (–pH)

= antilog (–3.30)

$[H_3O^+]$ = 5.0×10^{-4} M

9.27 Write the balanced equations as in Section 9.7A.

a. $HNO_3(aq)$ + $NaOH(aq)$ ⟶ $H_2O(l)$ + $NaNO_3(aq)$

water salt

b. $H_2SO_4(aq)$ + $KOH(aq)$ ⟶ $H_2O(l)$ + $K_2SO_4(aq)$

water salt

$H_2SO_4(aq)$ + 2 $KOH(aq)$ ⟶ 2 $H_2O(l)$ + $K_2SO_4(aq)$

Place a 2 to balance K. Place a 2 to balance H and O.

9.28 A net ionic equation contains only the species involved in a reaction.

$H^+(aq)$ + $OH^-(aq)$ ⟶ $H_2O(l)$ for Problem 9.27a, b

9.29 The acid and base react to form a salt and carbonic acid (H_2CO_3), which decomposes to CO_2 and H_2O as in Sample Problem 9.17.

$H_2SO_4(aq)$ + $CaCO_3(s)$ ⟶ $CaSO_4(aq)$ + $H_2O(l)$ + $CO_2(g)$

from H_2CO_3

9.30 The acid and base react to form a salt and carbonic acid (H_2CO_3), which decomposes to CO_2 and H_2O as in Sample Problem 9.17.

a. $HNO_3(aq)$ + $NaHCO_3(aq)$ ⟶ $NaNO_3(aq)$ + $H_2O(l)$ + $CO_2(g)$
(from H_2CO_3)

b. 2 $HNO_3(aq)$ + $MgCO_3(s)$ ⟶ $Mg(NO_3)_2(aq)$ + $H_2O(l)$ + $CO_2(g)$
(from H_2CO_3)

9.31 Follow Example 9.10 to determine the acidity when a salt is dissolved in water. Determine what types of acid and base (strong or weak) are used to form the salt. When the ions in the salt come from a strong acid and strong base, the solution is neutral. When the ions come from acids and bases of different strength, the ion derived from the stronger reactant determines the acidity.

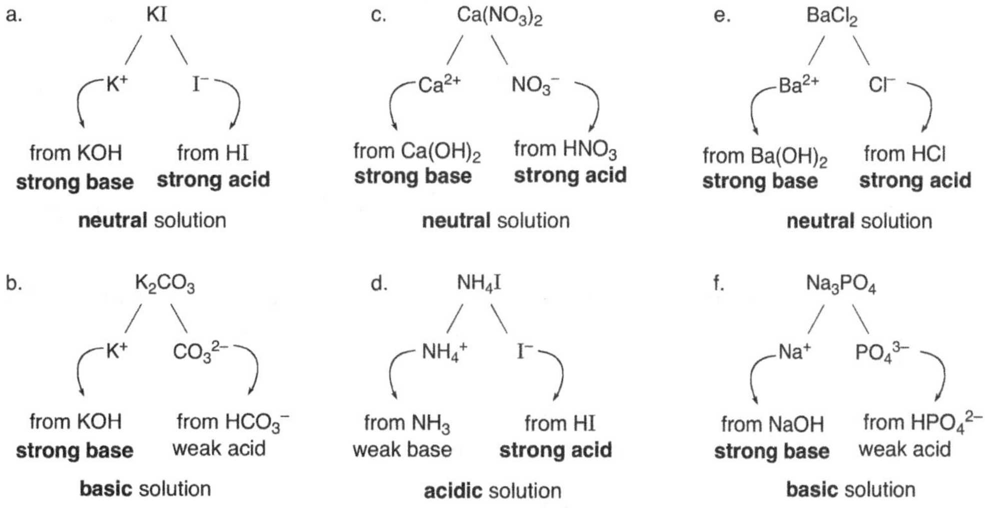

a. KI
- K^+ from KOH — **strong base**
- I^- from HI — **strong acid**

neutral solution

c. $Ca(NO_3)_2$
- Ca^{2+} from $Ca(OH)_2$ — **strong base**
- NO_3^- from HNO_3 — **strong acid**

neutral solution

e. $BaCl_2$
- Ba^{2+} from $Ba(OH)_2$ — **strong base**
- Cl^- from HCl — **strong acid**

neutral solution

b. K_2CO_3
- K^+ from KOH — **strong base**
- CO_3^{2-} from HCO_3^- — weak acid

basic solution

d. NH_4I
- NH_4^+ from NH_3 — weak base
- I^- from HI — **strong acid**

acidic solution

f. Na_3PO_4
- Na^+ from NaOH — **strong base**
- PO_4^{3-} from HPO_4^{2-} — weak acid

basic solution

9.32 A basic solution has a pH > 7.

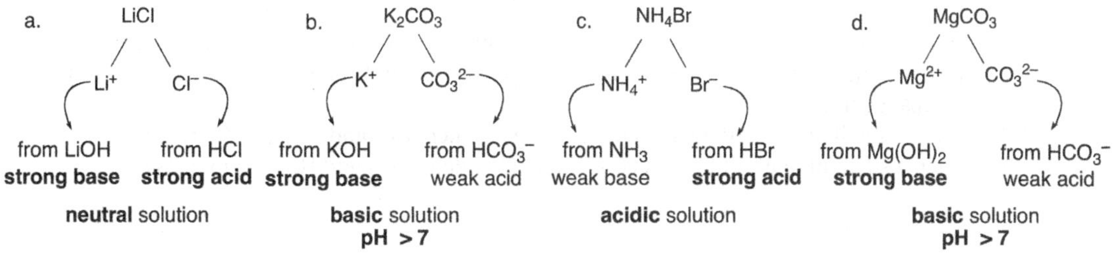

a. LiCl
- Li^+ from LiOH — **strong base**
- Cl^- from HCl — **strong acid**

neutral solution

b. K_2CO_3
- K^+ from KOH — **strong base**
- CO_3^{2-} from HCO_3^- — weak acid

basic solution
pH > 7

c. NH_4Br
- NH_4^+ from NH_3 — weak base
- Br^- from HBr — **strong acid**

acidic solution

d. $MgCO_3$
- Mg^{2+} from $Mg(OH)_2$ — **strong base**
- CO_3^{2-} from HCO_3^- — weak acid

basic solution
pH > 7

9.33 Calculate the molarity of the solution using a titration as in Example 9.11.

$$25.5\ \cancel{mL}\ NaOH \times \frac{1\ \cancel{L}}{1000\ \cancel{mL}} \times \frac{0.24\ mol\ NaOH}{1\ \cancel{L}} = 0.0061\ mol\ NaOH$$

$$0.0061\ \cancel{mol}\ NaOH \times \frac{1\ mol\ HCl}{1\ \cancel{mol}\ NaOH} = 0.0061\ mol\ HCl$$

$$\underset{\text{molarity}}{M} = \frac{mol}{L} = \frac{0.0061\ mol\ HCl}{15\ \cancel{mL}\ solution} \times \frac{1000\ \cancel{mL}}{1\ L} = 0.41\ M\ HCl$$

Answer

9.34 Calculate the number of milliliters of NaOH solution needed.

$$5.0\ \cancel{mL}\ H_2SO_4 \times \frac{1\ \cancel{L}}{1000\ \cancel{mL}} \times \frac{6.0\ mol\ H_2SO_4}{1\ \cancel{L}} = 0.030\ mol\ H_2SO_4$$

$$0.030\ \cancel{mol}\ H_2SO_4 \times \frac{2\ mol\ NaOH}{1\ \cancel{mol}\ H_2SO_4} = 0.060\ mol\ NaOH$$

$$0.060\ \cancel{mol}\ NaOH \times \frac{1\ \cancel{L}}{2.0\ \cancel{mol}\ NaOH} \times \frac{1000\ mL}{1\ \cancel{L}} = 30.\ mL\ NaOH\ solution$$

9.35 **A *buffer* is a solution whose pH changes very little when acid or base is added.** Most buffers are solutions composed of approximately equal amounts of a weak acid and the salt of its conjugate base.

a. A solution containing HBr and NaBr is not a buffer, because it contains a strong acid, HBr.
b. A solution containing HF and KF is a buffer because HF is a weak acid and F^- is its conjugate base.
c. A solution containing CH_3COOH alone is not a buffer because it contains a weak acid only.

9.36

a. HCO_3^- and CO_3^{2-} are both needed because the H_3O^+ concentration depends on two terms—K_a, which is a constant, and the ratio of the concentrations of the weak acid and its conjugate base. If these concentrations do not change much, the concentration of H_3O^+ and therefore the pH do not change much.
b. When a small amount of acid is added, $[HCO_3^-]$ increases and $[CO_3^{2-}]$ decreases.
c. When a small amount of base is added, $[HCO_3^-]$ decreases and $[CO_3^{2-}]$ increases.

9.37 Calculate the pH of the buffer as in Example 9.12. The pH of a buffer is the same value whenever equal concentrations of the weak acid and conjugate base are present.

a. $$[H_3O^+] = K_a \times \frac{[H_2PO_4^-]}{[HPO_4^{2-}]} = (6.2 \times 10^{-8}) \times \frac{[0.10\ M]}{[0.10\ M]}$$

$$[H_3O^+] = 6.2 \times 10^{-8}\ M$$

$$pH = -\log[H_3O^+] = -\log(6.2 \times 10^{-8})$$

$$pH = 7.21$$

b. $$[H_3O^+] = K_a \times \frac{[H_2PO_4^-]}{[HPO_4^{2-}]} = (6.2 \times 10^{-8}) \times \frac{[1.0\ M]}{[1.0\ M]}$$

$$[H_3O^+] = 6.2 \times 10^{-8}\ M$$

$$pH = -\log[H_3O^+] = -\log(6.2 \times 10^{-8})$$

$$pH = 7.21$$

c. $$[H_3O^+] = K_a \times \frac{[H_2PO_4^-]}{[HPO_4^{2-}]} = (6.2 \times 10^{-8}) \times \frac{[0.50\ M]}{[0.50\ M]}$$

$$[H_3O^+] = 6.2 \times 10^{-8}\ M$$

$$pH = -\log[H_3O^+] = -\log(6.2 \times 10^{-8})$$

$$pH = 7.21$$

9.38 Calculate the pH of the buffer as in Example 9.12.

$$[H_3O^+] = K_a \times \frac{[CH_3COOH]}{[CH_3COO^-]} = (1.8 \times 10^{-5}) \times \frac{[0.20\ M]}{[0.15\ M]}$$

$$[H_3O^+] = 2.4 \times 10^{-5}\ M$$

$$pH = -\log[H_3O^+] = -\log(2.4 \times 10^{-5})$$

$$pH = 4.62$$

Solutions to Odd-Numbered End-of-Chapter Problems

9.39 A Brønsted–Lowry acid must contain a hydrogen atom, but it may be neutral or contain a net positive or negative charge. Use Example 9.1 to help determine which of the compounds are Brønsted–Lowry acids.

a. HBr: contains a H atom, a Brønsted–Lowry acid
b. Br_2: no H atom
c. $AlCl_3$: no H atom
d. HCOOH: contains a H atom, a Brønsted–Lowry acid
e. NO_2^-: no H atom
f. HNO_2: contains a H atom, a Brønsted–Lowry acid

9.41 A Brønsted–Lowry base must contain a lone pair of electrons, but it may be neutral or have a net negative charge. Use Example 9.2 to help determine which of the compounds are Brønsted–Lowry bases.

a. OH^-: lone pairs on O, a Brønsted–Lowry base
b. Ca^{2+}: no lone pair of electrons
c. C_2H_6: no lone pair of electrons
d. PO_4^{3-}: lone pairs on O, a Brønsted–Lowry base
e. OCl^-: lone pairs on O and Cl, a Brønsted–Lowry base
f. $MgCO_3$: lone pairs on O, a Brønsted–Lowry base

9.43 Draw the conjugate acid of each species as in Example 9.4.

a. HS^-: add one H^+ to make H_2S.
b. CO_3^{2-}: add one H^+ to make HCO_3^-.
c. NO_2^-: add one H^+ to make HNO_2.
d. $CH_3\ddot{N}H_2$ (H–C(H)(H)–N̈(H)–H) add one H+ to make $[CH_3NH_3]^+$ (H–C(H)(H)–N(H)(H)–H)

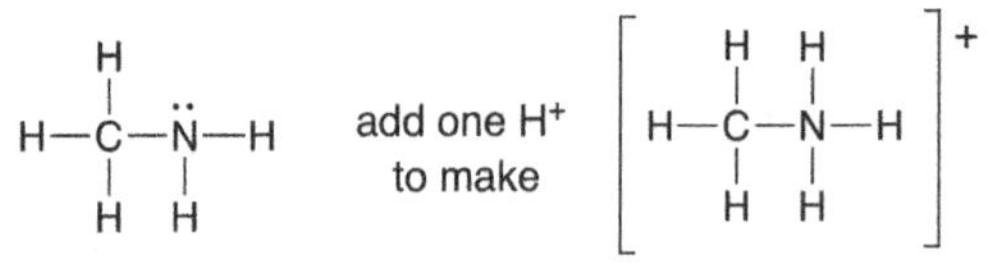

9.45 Draw the conjugate base of each species as in Example 9.5.

a. HNO_2: remove one H^+ to make NO_2^-.
b. NH_4^+: remove one H^+ to make NH_3.
c. H_2O_2: remove one H^+ to make HO_2^-.

9.47 Draw the acid and then remove H^+.

a. H–O–N=O (HNO_2) —(– H+)→ $^-$O–N=O conjugate base (NO_2^-)

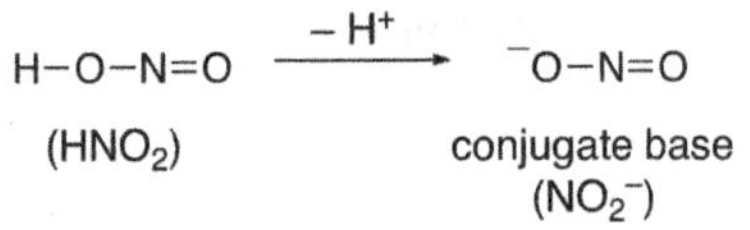

b. $^-$O–C(=O)–O–H (HCO_3^-) —(– H+)→ $^-$O–C(=O)–O$^-$ conjugate base (CO_3^{2-})

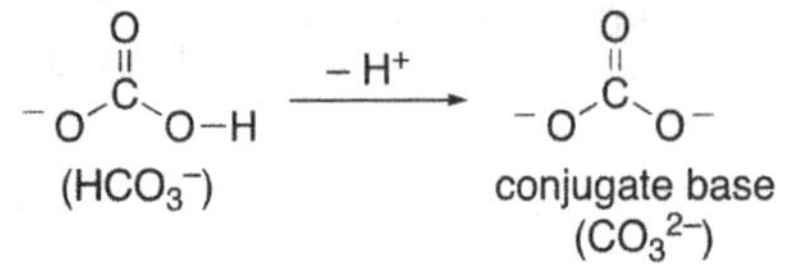

9.49 The acid loses a proton (a gray sphere) to form its conjugate base, while the base gains a proton to form its conjugate acid.

a. acid + base → conjugate base + conjugate acid

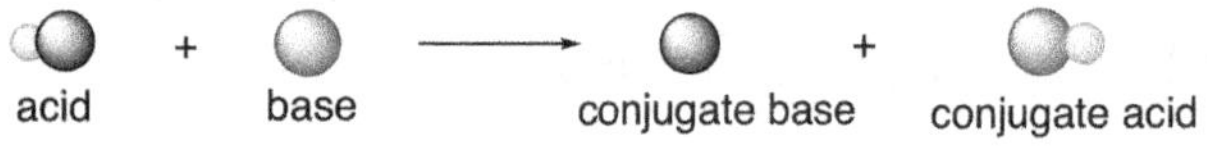

b. base + acid → conjugate acid + conjugate base

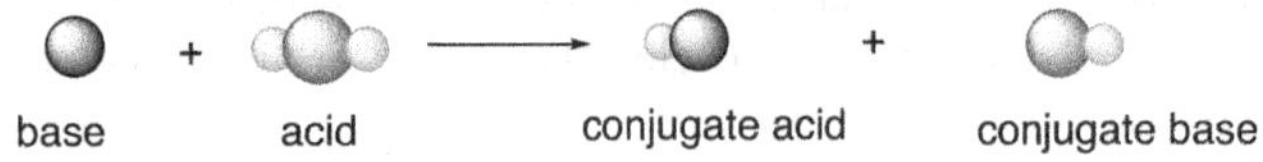

9.51 Label the conjugate acid–base pairs in each reaction as in Answer 9.9.

a. $HI(g)$ acid — $I^-(aq)$ conjugate base
$NH_3(g)$ base — $NH_4^+(aq)$ conjugate acid

b. $HCOOH(l)$ acid — $HCOO^-(aq)$ conjugate base
$H_2O(l)$ base — $H_3O^+(aq)$ conjugate acid

c. $HSO_4^-(aq)$ base — $H_2SO_4(aq)$ conjugate acid
$H_2O(l)$ acid — $OH^-(aq)$ conjugate base

9.53 Draw the conjugate acid and base of HCO_3^-.

a. conjugate acid: H_2CO_3 b. conjugate base: CO_3^{2-}

9.55

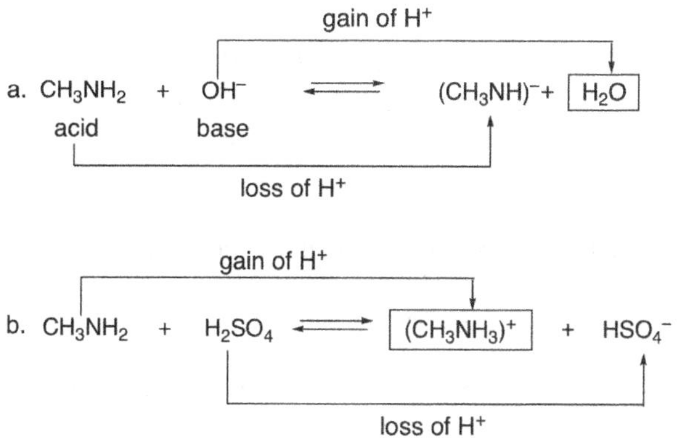

gain of H+

a. CH_3NH_2 + OH^- ⇌ $(CH_3NH)^-$ + H_2O
acid base

loss of H+

gain of H+

b. CH_3NH_2 + H_2SO_4 ⇌ $(CH_3NH_3)^+$ + HSO_4^-

loss of H+

9.57 Draw the acid–base reaction.

$$HNO_3(aq) + H_2O(l) \longrightarrow H_3O^+(aq) + NO_3^-(aq)$$

9.59 **A** represents HCl (a strong acid) because it shows a fully dissociated acid. **B** represents HF (a weak acid), because it is only partially dissociated.

9.61 a. **B** represents a strong acid because HZ is completely dissociated, forming H_3O^+ and Z^-.
b. **A** represents a weak acid because most of the acid HZ remains, and few ions (H_3O^+ and Z^-) are formed.

9.63 Use Tables 9.2 and 9.3 to determine the stronger acid as in Example 9.7. The acid with the larger K_a is the stronger acid. The weaker acid has the stronger conjugate base.

a. CH_3COOH is a stronger acid than H_2O. H_2O has the stronger conjugate base.
b. H_3PO_4 is a stronger acid than HCO_3^-. HCO_3^- has the stronger conjugate base.
c. H_2SO_4 is a stronger acid than HSO_4^-. HSO_4^- has the stronger conjugate base.

9.65 a. An acid that dissociates to a greater extent in water is a stronger acid: **A is the stronger acid.**
b. An acid with a smaller K_a is weaker (**A** is weaker): **B is the stronger acid.**
c. An acid with a stronger conjugate base is a weaker acid (**A** is weaker): **B is the stronger acid.**

9.67

Acid	Formula	Conjugate Base	Formula
a. strong	H_2SO_4	weak	HSO_4^-
b. weak	HOBr	strong	OBr^-
c. strong	HNO_3	weak	NO_3^-
d. weak	HCO_2H	strong	HCO_2^-

9.69 The acid with the larger K_a is the stronger acid. The stronger acid has a weaker conjugate base.

a. HSO_4^- — $K_a = 1.2 \times 10^{-2}$, larger K_a, **stronger acid**; SO_4^{2-} — conjugate base

$H_2PO_4^-$ — $K_a = 6.2 \times 10^{-8}$; HPO_4^{2-} — conjugate base, **stronger base**

b. CH_3COOH — $K_a = 1.8 \times 10^{-5}$, larger K_a, **stronger acid**; CH_3COO^- — conjugate base

CH_3CH_2COOH — $K_a = 1.3 \times 10^{-5}$; $CH_3CH_2COO^-$ — conjugate base, **stronger base**

9.71 The stronger acid has more dissociated ions. The stronger acid has the larger K_a.

a. **B** has more dissociated ions (A^- and H_3O^+) and is therefore the stronger acid.
b. **B** has the larger K_a since it is the stronger acid.
c. The weaker acid has the stronger conjugate base. **A** has the stronger conjugate base.

9.73 To determine if the reactants or products are favored at equilibrium:

- Identify the acid in the reactants and the conjugate acid in the products.
- Determine the relative strength of the acid and the conjugate acid.
- Equilibrium favors the formation of the weaker acid.

a. $H_3PO_4(aq) + CN^-(aq) \rightleftharpoons H_2PO_4^-(aq) + HCN(aq)$

H_3PO_4: acid; HCN: conjugate acid, weaker acid

Products are favored.

b. $Br^-(aq) + HSO_4^-(aq) \rightleftharpoons SO_4^{2-}(aq) + HBr(aq)$

HSO_4^-: acid, weaker acid; HBr: conjugate acid

Reactants are favored.

c. $CH_3COO^-(aq) + H_2CO_3(aq) \rightleftharpoons CH_3COOH(aq) + HCO_3^-(aq)$

H_2CO_3: acid, weaker acid; CH_3COOH: conjugate acid

Reactants are favored.

9.75 Use the equation $[OH^-] = K_w/[H_3O^+]$ to calculate the hydroxide ion concentration as in Answer 9.19. When $[OH^-] > [H_3O^+]$, the solution is basic. When $[OH^-] < [H_3O^+]$, the solution is acidic.

a. $[OH^-] = \dfrac{K_w}{[H_3O^+]} = \dfrac{1.0 \times 10^{-14}}{10^{-8}} = 10^{-6}$ M
basic

b. $[OH^-] = \dfrac{K_w}{[H_3O^+]} = \dfrac{1.0 \times 10^{-14}}{10^{-10}} = 10^{-4}$ M
basic

c. $[OH^-] = \dfrac{K_w}{[H_3O^+]} = \dfrac{1.0 \times 10^{-14}}{3.0 \times 10^{-4}} = 3.3 \times 10^{-11}$ M
acidic

d. $[OH^-] = \dfrac{K_w}{[H_3O^+]} = \dfrac{1.0 \times 10^{-14}}{2.5 \times 10^{-11}} = 4.0 \times 10^{-4}$ M
basic

9.77 Use the equation $[H_3O^+] = K_w/[OH^-]$ to calculate the hydronium ion concentration as in Sample Problem 9.11.

a. $[H_3O^+] = \dfrac{K_w}{[OH^-]} = \dfrac{1.0 \times 10^{-14}}{10^{-2}} = 10^{-12}$ M
basic

b. $[H_3O^+] = \dfrac{K_w}{[OH^-]} = \dfrac{1.0 \times 10^{-14}}{4.0 \times 10^{-8}} = 2.5 \times 10^{-7}$ M
acidic

c. $[H_3O^+] = \dfrac{K_w}{[OH^-]} = \dfrac{1.0 \times 10^{-14}}{6.2 \times 10^{-7}} = 1.6 \times 10^{-8}$ M
basic

d. $[H_3O^+] = \dfrac{K_w}{[OH^-]} = \dfrac{1.0 \times 10^{-14}}{8.5 \times 10^{-13}} = 1.2 \times 10^{-2}$ M
acidic

9.79 Use a calculator to determine the logarithm of a number that contains a coefficient other than one in scientific notation; pH = $-\log [H_3O^+]$ as in Answer 9.25.

a. pH $= -\log [H_3O^+] = -\log(10^{-12})$
$= -(-12) = 12$

b. pH $= -\log [H_3O^+] = -\log(2.5 \times 10^{-7})$
$= -(-6.60) = 6.60$

c. pH $= -\log [H_3O^+] = -\log(1.6 \times 10^{-8})$
$= -(-7.80) = 7.80$

d. pH $= -\log [H_3O^+] = -\log(1.2 \times 10^{-2})$
$= -(-1.92) = 1.92$

9.81

$[H_3O^+]$	$[OH^-]$	pH	Classification
5.3×10^{-3}	1.9×10^{-12}	2.28	acidic
5.0×10^{-7}	2.0×10^{-8}	6.30	acidic
4×10^{-5}	2.5×10^{-10}	4.4	acidic
1.5×10^{-5}	6.8×10^{-10}	4.82	acidic

9.83 Use a calculator to determine the antilogarithm of (–pH); $[H_3O^+]$ = antilog(–pH).

a. $[H_3O^+]$ = antilog(–pH) = antilog(–12)

$[H_3O^+] = 1 \times 10^{-12}$ M

b. $[H_3O^+]$ = antilog(–pH) = antilog(–1)

$[H_3O^+] = 1 \times 10^{-1}$ M

c. $[H_3O^+]$ = antilog(–pH) = antilog(–1.80)

$[H_3O^+] = 1.6 \times 10^{-2}$ M

d. $[H_3O^+]$ = antilog(–pH) = antilog(–8.90)

$[H_3O^+] = 1.3 \times 10^{-9}$ M

9.85 Use the equations in Answers 9.83 and 9.75 to calculate the concentrations of H_3O^+ and OH^- in the sample.

$[H_3O^+]$ = antilog(–pH) = antilog(–5.90)

$[H_3O^+] = 1.3 \times 10^{-6}$ M

$$[OH^-] = \frac{K_w}{[H_3O^+]} = \frac{1.0 \times 10^{-14}}{1.3 \times 10^{-6}} = 7.7 \times 10^{-9}\text{ M}$$

9.87

$[H_3O^+]$ = antilog(–pH) = antilog(–8.5)

$[H_3O^+] = 3 \times 10^{-9}$ M

$$[OH^-] = \frac{K_w}{[H_3O^+]} = \frac{1.0 \times 10^{-14}}{3 \times 10^{-9}} = 0.3 \times 10^{-5} = 3 \times 10^{-6}\text{ M}$$

9.89 Use a calculator to determine the logarithm of a number that contains a coefficient other than one in scientific notation; pH = –log $[H_3O^+]$.

a. pH = –log $[H_3O^+]$ = $-\log(2.5 \times 10^{-3})$

= –(–2.60) = 2.60

b. $$[H_3O^+] = \frac{K_w}{[OH^-]} = \frac{1.0 \times 10^{-14}}{1.5 \times 10^{-2}} = 6.7 \times 10^{-13}\text{ M}$$

pH = –log $[H_3O^+]$ = $-\log(6.7 \times 10^{-13})$

= –(–12.17) = 12.17

9.91 The pH of 0.10 M HCl is lower than the pH of 0.1 M CH_3COOH solution (1.0 vs. 2.9) because HCl (a strong acid) is fully dissociated to H_3O^+ and Cl^-, whereas CH_3COOH (a weak acid) is only partially dissociated.

9.93 Write the balanced equations as in Section 9.7.

a. HBr(*aq*) + KOH(*aq*) ⟶ KBr(*aq*) + H_2O(*l*)

b. $2\ HNO_3(aq) + Ca(OH)_2(aq) \longrightarrow 2\ H_2O(l) + Ca(NO_3)_2(aq)$

c. $HCl(aq) + NaHCO_3(aq) \longrightarrow NaCl(aq) + H_2O(l) + CO_2(g)$

d. $H_2SO_4(aq) + Mg(OH)_2(aq) \longrightarrow 2\ H_2O(l) + MgSO_4(aq)$

9.95 Write the balanced equation.

$2\ HNO_3(aq) + CaCO_3(s) \longrightarrow Ca(NO_3)_2(aq) + CO_2(g) + H_2O(l)$

9.97 Follow Example 9.10 to determine the acidity when a salt is dissolved in water. Determine what types of acid and base (strong or weak) are used to form the salt. When the ions in the salt come from a strong acid and strong base, the solution is neutral. When the ions come from acids and bases of different strength, the ion derived from the stronger reactant determines the acidity.

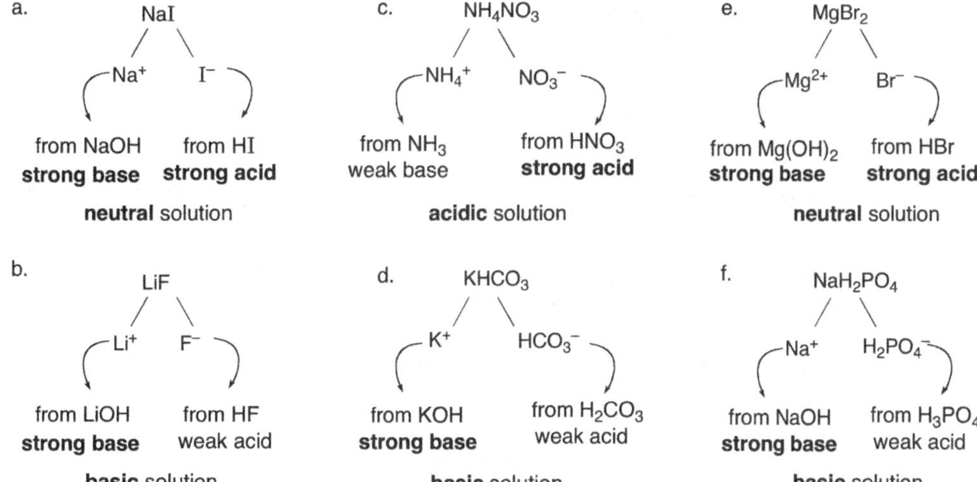

9.99 Calculate the molarity of the solution using a titration as in Example 9.11.

$$35.5\ \text{mL NaOH} \times \frac{1\ \text{L}}{1000\ \text{mL}} \times \frac{0.10\ \text{mol NaOH}}{1\ \text{L}} = 0.0036\ \text{mol NaOH}$$

$$0.0036\ \text{mol NaOH} \times \frac{1\ \text{mol HCl}}{1\ \text{mol NaOH}} = 0.0036\ \text{mol HCl}$$

$$\underset{\text{molarity}}{M} = \frac{\text{mol}}{\text{L}} = \frac{0.0036\ \text{mol HCl}}{25.0\ \text{mL solution}} \times \frac{1000\ \text{mL}}{1\ \text{L}} = 0.14\ M\ \text{HCl}$$

Answer

9.101 Calculate the molarity of the solution using a titration as in Example 9.11.

$$15.5\ \text{mL NaOH} \times \frac{1\ \text{L}}{1000\ \text{mL}} \times \frac{0.20\ \text{mol NaOH}}{1\ \text{L}} = 0.0031\ \text{mol NaOH}$$

$$0.0031\ \text{mol NaOH} \times \frac{1\ \text{mol CH}_3\text{COOH}}{1\ \text{mol NaOH}} = 0.0031\ \text{mol CH}_3\text{COOH}$$

$$\underset{\text{molarity}}{\text{M}} = \frac{\text{mol}}{\text{L}} = \frac{0.0031\ \text{mol CH}_3\text{COOH}}{25.0\ \text{mL solution}} \times \frac{1000\ \text{mL}}{1\ \text{L}} = 0.12\ \text{M CH}_3\text{COOH}$$

Answer

9.103 Calculate the number of milliliters of solution needed.

$$10.0\ \text{mL CH}_3\text{COOH} \times \frac{1\ \text{L}}{1000\ \text{mL}} \times \frac{2.5\ \text{mol CH}_3\text{COOH}}{1\ \text{L}} = 0.025\ \text{mol CH}_3\text{COOH}$$

$$0.025\ \text{mol CH}_3\text{COOH} \times \frac{1\ \text{mol NaOH}}{1\ \text{mol CH}_3\text{COOH}} = 0.025\ \text{mol NaOH}$$

$$0.025\ \text{mol NaOH} \times \frac{1\ \text{L}}{1.0\ \text{mol NaOH}} \times \frac{1000\ \text{mL}}{1\ \text{L}} = 25\ \text{mL NaOH solution}$$

9.105 **B** is a buffer because it contains equal amounts of the acid and its conjugate base. **A** is not a buffer because it contains only acid.

9.107 When a small amount of strong acid is added, the base Z^- reacts with the acid H_3O^+ to form HZ (c).

$$Z^- + H_3O^+ \longrightarrow HZ + H_2O$$

9.109
a. Both HNO_2 and NO_2^- are needed to prepare the buffer because HNO_2 is a proton donor that will react with added base and NO_2^- is a proton acceptor that will react with added acid.
b. When a small amount of acid is added to the buffer, the concentration of HNO_2 increases and the concentration of NO_2^- decreases.
c. The concentration of HNO_2 decreases and the concentration of NO_2^- increases when a small amount of base is added to the buffer.

9.111 Calculate the pH of the buffer as in Example 9.12.

a. $[H_3O^+] = K_a \times \frac{[Na_2HPO_4]}{[Na_3PO_4]} = 2.2 \times 10^{-13} \times \frac{[0.10\ M]}{[0.10\ M]}$

$[H_3O^+] = 2.2 \times 10^{-13}\ M$

$pH = -\log [H_3O^+] = -\log(2.2 \times 10^{-13})$

$pH = 12.66$

b. $[H_3O^+] = K_a \times \frac{[NaHCO_3]}{[Na_2CO_3]} = 5.6 \times 10^{-11} \times \frac{[0.22\ M]}{[0.22\ M]}$

$[H_3O^+] = 5.6 \times 10^{-11}\ M$

$pH = -\log [H_3O^+] = -\log(5.6 \times 10^{-11})$

$pH = 10.25$

9.113 Calculate the pH of the buffer as in Example 9.12.

a. $[H_3O^+] = K_a \times \frac{[CH_3COOH]}{[NaCH_3COO]} = 1.8 \times 10^{-5} \times \frac{[0.20\ M]}{[0.20\ M]}$

$[H_3O^+] = 1.8 \times 10^{-5}\ M$

$pH = -\log [H_3O^+] = -\log(1.8 \times 10^{-5})$

$pH = 4.74$

b. $[H_3O^+] = K_a \times \frac{[CH_3COOH]}{[NaCH_3COO]} = 1.8 \times 10^{-5} \times \frac{[0.20\ M]}{[0.40\ M]}$

$[H_3O^+] = 9.0 \times 10^{-6}\ M$

$pH = -\log [H_3O^+] = -\log(9.0 \times 10^{-6})$

$pH = 5.05$

c. $[H_3O^+] = K_a \times \frac{[CH_3COOH]}{[NaCH_3COO]} = 1.8 \times 10^{-5} \times \frac{[0.20\ M]}{[0.10\ M]}$

$$[H_3O^+] = 3.6 \times 10^{-5}\ M$$

$$pH = -\log [H_3O^+] = -\log(3.6 \times 10^{-5})$$

$$pH = 4.44$$

9.115 The pH of unpolluted rainwater is lower than that of pure water because CO_2 combines with rainwater to form H_2CO_3, which is acidic, lowering the pH.

9.117 Calculate the concentrations.

$$[H_3O^+] = \text{antilog}(-pH) = \text{antilog}(-7.50)$$

$$[H_3O^+] = 3.2 \times 10^{-8}\ M$$

$$[OH^-] = \frac{K_w}{[H_3O^+]} = \frac{1.0 \times 10^{-14}}{3.2 \times 10^{-8}} = 3.1 \times 10^{-7}\ M$$

9.119 By breathing into a bag, the individual breathes in air with a higher CO_2 concentration. Thus, the CO_2 concentration in the lungs and the blood increases, thereby lowering the pH.

9.121 A rise in CO_2 concentration leads to an increase in $H^+(aq)$ concentration by the following equilibria:

$$CO_2(g) + H_2O(l) \rightleftharpoons H_2CO_3(aq) \rightleftharpoons HCO_3^-(aq) + H^+(aq)$$

9.123 Write the acid–base reaction that occurs when OCl^- dissolves in water.

$$OCl^-(aq) + H_2O(l) \longrightarrow HOCl(aq) + OH^-(aq)$$

This reaction forms OH^-, which makes the pool water basic.

Chapter 10 Nuclear Chemistry

Chapter Review

[1] Describe the different types of radiation emitted by a radioactive nucleus. (10.1)

- A radioactive nucleus can emit alpha (α) particles, beta (β) particles, positrons, or gamma (γ) rays.
- An α particle is a high-energy nucleus that contains two protons and two neutrons.

alpha particle: α or ${}^{4}_{2}He$

- A β particle is a high-energy electron that has a –1 charge and negligible mass.

beta particle: β or ${}^{0}_{-1}e$

- A positron is an antiparticle of a β particle. A positron has a +1 charge and negligible mass.

Symbol: ${}^{0}_{+1}e$ or β^{+} (positron)

Formation: $${}^{1}_{1}p \longrightarrow {}^{1}_{0}n + {}^{0}_{+1}e$$
proton → neutron + positron

- A gamma ray is high-energy radiation with no mass or charge.

gamma ray: γ

[2] How are equations for nuclear reactions written? (10.2)

- In an equation for a nuclear reaction, the sum of the mass numbers (A) must be equal on both sides of the equation. The sum of the atomic numbers (Z) must be equal on both sides of the equation as well.
- In the following equation, the sum of the mass numbers on each side of the equation is 18, and the sum of the atomic numbers on each side of the equation is 9.

$${}^{18}_{9}F \longrightarrow {}^{0}_{+1}e + {}^{18}_{8}O$$

[3] What is the half-life of a radioactive isotope? (10.3)

- The half-life ($t_{1/2}$) is the time it takes for one-half of a radioactive sample to decay. Knowing the half-life and the amount of a radioactive substance, one can calculate how much of a sample remains after a period of time. For example, P-32 decays to S-32 with a half-life of 14 days.

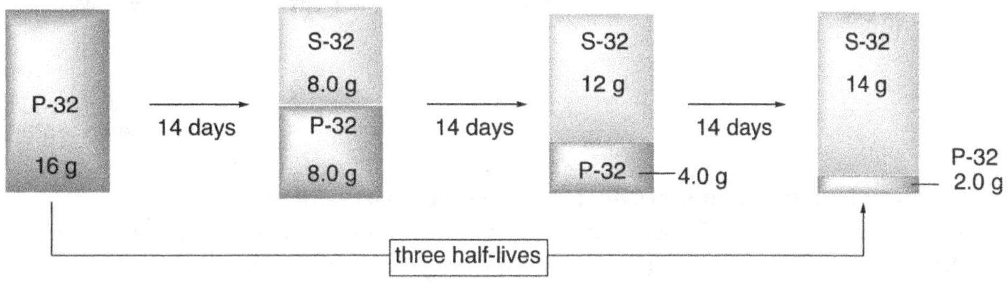

- The half-life of radioactive C-14 can be used to date archaeological artifacts.

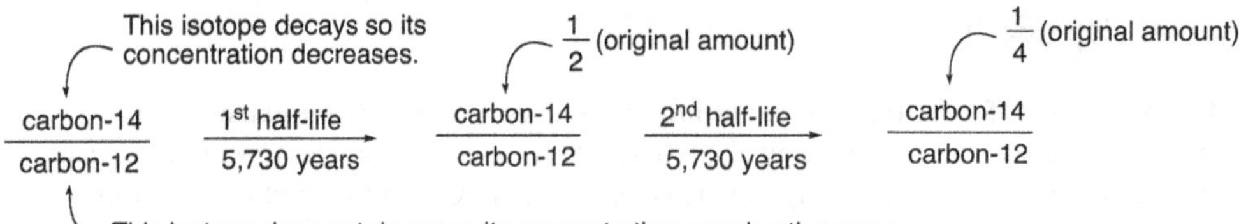
This isotope decays so its concentration decreases.
1/2 (original amount)
1/4 (original amount)
carbon-14 / carbon-12
1st half-life 5,730 years
carbon-14 / carbon-12
2nd half-life 5,730 years
carbon-14 / carbon-12
This isotope does not decay so its concentration remains the same.

[4] What units are used to measure radioactivity? (10.4)

- Radiation in a sample is measured by the number of disintegrations per second, most often using the curie (Ci); 1 Ci = 3.7×10^{10} disintegrations/s. The Becquerel (Bq) is also used; 1 Bq = 1 disintegration/s; 1 Ci = 3.7×10^{10} Bq.
- The exposure of a substance to radioactivity is measured with the rad (radiation absorbed dose) or the rem (radiation equivalent for man).

[5] Give examples of common radioisotopes used in medicine. (10.5)

- Iodine-131 is used to diagnose and treat thyroid disease.
- Technetium-99m is used to evaluate the functioning of the gall bladder and bile ducts, and in bone scans to look for the spread of cancer to other sites in the body.
- Red blood cells tagged with technetium-99m are used to find the site of a gastrointestinal bleed.
- Thallium-201 is used to diagnose coronary artery disease.
- Cobalt-60 is used as an external source of radiation for cancer treatment.
- Iodine-125 and iridium-192 are used in the internal radiation treatment of prostate cancer and breast cancer, respectively.
- Carbon-11, oxygen-15, nitrogen-13, and fluorine-18 are used in positron emission tomography.

[6] What are nuclear fission and nuclear fusion? (10.6)

- Nuclear fission is the splitting apart of a heavy nucleus into lighter nuclei and neutrons.
- Nuclear fusion is the joining together of two light nuclei to form a larger nucleus.
- Both nuclear fission and nuclear fusion release a great deal of energy. Nuclear fission is used in nuclear power plants to generate electricity. Nuclear fusion occurs in stars.
- One common nuclear reaction is the fission of uranium-235 into krypton-91 and barium-142.

$$^{235}_{92}\text{U} + {}^{1}_{0}\text{n} \longrightarrow {}^{91}_{36}\text{Kr} + {}^{142}_{56}\text{Ba} + 3\,{}^{1}_{0}\text{n}$$

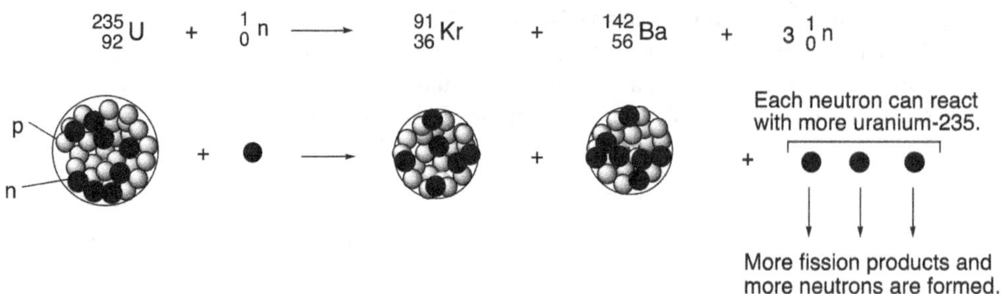
p
n
Each neutron can react with more uranium-235.
More fission products and more neutrons are formed.

[7] What medical imaging techniques do not use radioactivity? (10.7)

- X-rays and CT scans both use X-rays, a high-energy form of electromagnetic radiation.
- MRI (magnetic resonance imaging) uses low-energy radio waves to image soft tissue.

Problem Solving

[1] Introduction (10.1)

Example 10.1 Thallium-201 and iridium-192 are radioactive isotopes that can be produced using a nuclear reactor. Complete the following table for both isotopes.

	Atomic Number	Mass Number	Number of Protons	Number of Neutrons	Isotope Symbol
Thallium-201					
Iridium-192					

Analysis

- The atomic number (Z) = the number of protons.
- The mass number (A) = the number of protons + the number of neutrons.
- Isotopes are written with the mass number to the upper left of the element symbol and the atomic number to the lower left of the element symbol.

Solution

	Atomic Number	Mass Number	Number of Protons	Number of Neutrons	Isotope Symbol
Thallium-201	81	201	81	201 – 81 = 120	${}^{201}_{81}Tl$
Iridium-192	77	192	77	192 – 77 = 115	${}^{192}_{77}Ir$

---

[2] Nuclear Reactions (10.2)

Example 10.2 Write a balanced nuclear equation for the β emission of strontium-90, a radioisotope used to produce β particles.

Analysis

Balance the atomic numbers and mass numbers on both sides of a nuclear equation. With β emission, treat the β particle as an electron with zero mass in balancing mass numbers, and a –1 charge when balancing the atomic numbers.

Solution

[1] Write an incomplete equation with the original nucleus on the left and the particle emitted on the right.

- Use the identity of the element to determine the atomic number; strontium has an atomic number of 38.

$$ {}^{90}_{38}Sr \longrightarrow {}^{0}_{-1}e \quad + \quad ? $$

[2] Calculate the mass number and the atomic number of the newly formed nucleus on the right.

- Mass number: Since a β particle has no mass, the masses of the new particle and the original particle are the same, 90.

- Atomic number: Since β emission converts a neutron into a proton, the new nucleus has one more proton than the original nucleus; 38 = –1 + ?. Thus, the new nucleus has an atomic number of 39.

[3] Use the atomic number to identify the new nucleus and complete the equation.

- From the periodic table, the element with an atomic number of 39 is yttrium, Y.
- Write the mass number and the atomic number with the element symbol to complete the equation.

$$^{90}_{38}\text{Sr} \longrightarrow {}^{0}_{-1}\text{e} + {}^{90}_{39}\text{Y}$$

Example 10.3 Write a balanced nuclear equation for the positron emission of rubidium-77.

Analysis

Balance the atomic numbers and mass numbers on both sides of a nuclear equation. With positron emission, treat the positron as a particle with zero mass when balancing mass numbers, and a +1 charge when balancing the atomic numbers.

Solution

[1] Write an incomplete equation with the original nucleus on the left and the particle emitted on the right.

- Use the identity of the element to determine the atomic number; rubidium has an atomic number of 37.

$$^{77}_{37}\text{Rb} \longrightarrow {}^{0}_{+1}\text{e} + ?$$

[2] Calculate the mass number and the atomic number of the newly formed nucleus on the right.

- Mass number: Since a β^+ particle has no mass, the masses of the new particle and the original particle are the same, 77.
- Atomic number: Since β^+ emission converts a proton into a neutron, the new nucleus has one fewer proton than the original nucleus; 37 – 1 = 36. Thus, the new nucleus has an atomic number of 36.

[3] Use the atomic number to identify the new nucleus and complete the equation.

- From the periodic table, the element with an atomic number of 36 is krypton, Kr.
- Write the mass number and the atomic number with the element symbol to complete the equation.

$$^{77}_{37}\text{Rb} \longrightarrow {}^{0}_{+1}\text{e} + {}^{77}_{36}\text{Kr}$$

[3] Detecting and Measuring Radiation (10.4)

Example 10.4 A patient must be given a 7.5-mCi dose of iodine-131, which is available as a solution that contains 2.5 mCi/mL. What volume of solution must be administered?

Analysis

Use the amount of radioactivity (mCi/mL) as a conversion factor to convert the dose of radioactivity from millicuries to a volume in milliliters.

Solution

The dose of radioactivity is known in millicuries, and the amount of radioactivity per unit volume (2.5 mCi/mL) is also known. Use 2.5 mCi/mL as a millicurie–milliliter conversion factor.

mCi–mL conversion factor

$$7.5 \text{ mCi dose} \times \frac{1 \text{ mL}}{2.5 \text{ mCi}} = 3.0 \text{ mL dose}$$

Answer

Millicuries cancel.

Self-Test

[1] Fill in the blank with one of the terms listed below.

Alpha (α) particle (10.1)
Beta (β) particle (10.1)
Gamma (γ) ray (10.1)
Geiger counter (10.4)
Half-life (10.3)
Nuclear fission (10.6)
Nuclear fusion (10.6)
Positron (10.1)
Rad (10.4)
Radioactive isotope (10.1)
Radioactivity (10.1)
Radiocarbon dating (10.3)
Rem (10.4)
X-rays (10.7)

1. The ______ of a radioactive isotope is the time it takes for one-half of the sample to decay.
2. ______ is the joining together of two light nuclei to form a larger nucleus.
3. A ______ is a small portable device used for measuring radioactivity.
4. An ______ is a high-energy particle that contains two protons and two neutrons.
5. A ______ is unstable and spontaneously emits energy to form a more stable nucleus.
6. ______ is based on the fact that the ratio of radioactive carbon-14 to stable carbon-12 is a constant value in a living organism.
7. A ______ is a high-energy electron.
8. ______ are a high-energy type of electromagnetic radiation.
9. A ______ is called an antiparticle of a β particle, because their charges are different but their masses are the same.
10. A ______ is the amount of radiation absorbed by one gram of a substance. The amount of energy absorbed varies with both the nature of the substance and the type of radiation.
11. ______ is the splitting apart of a heavy nucleus into lighter nuclei and neutrons.
12. A ______ is high-energy radiation released from a radioactive nucleus.
13. ______ is any form of nuclear radiation emitted by a radioactive isotope.
14. A ______ is the amount of radiation that also factors in its energy and potential to damage tissue.

[2] Fill in the blanks in the table below.

	Atomic Number	Mass Number	Number of Protons	Number of Neutrons
$^{60}_{27}Co$	15. ______	60	17. ______	18. ______
$^{99m}_{43}Tc$	43	16. ______	43	19. ______

[3] Pick the type of radioactivity that matches each phrase. A question may have more than one correct answer.

a. alpha particle
b. beta particle
c. positron
d. gamma ray

20. Has a positive charge
21. Has a negative charge
22. Has no charge
23. Has a mass equivalent to that of helium
24. Has no mass
25. Is formed by the decay of a radioactive nucleus

Answers to Self-Test

1. half-life
2. Nuclear fusion
3. Geiger counter
4. alpha particle
5. radioactive isotope
6. Radiocarbon dating
7. beta particle
8. X-rays
9. positron
10. rad
11. Nuclear fission
12. gamma ray
13. Radioactivity
14. rem
15. 27
16. 99
17. 27
18. 33
19. 56
20. a, c
21. b
22. d
23. a
24. b, c, d
25. a, b, c, d

Solutions to In-Chapter Problems

10.1 Refer to Example 10.1 to answer the question.

- The atomic number (Z) = the number of protons.
- The mass number (A) = the number of protons + the number of neutrons.
- Isotopes are written with the mass number to the upper left of the element symbol and the atomic number to the lower left of the element symbol.

	Atomic Number	Mass Number	Number of Protons	Number of Neutrons	Isotope Symbol
Cobalt-59	27	59	27	32	$^{59}_{27}Co$
Cobalt-60	27	60	27	33	$^{60}_{27}Co$

10.2 Fill in the table as in Example 10.1, using the rules in Answer 10.1.

		Atomic Number	Mass Number	Number of Protons	Number of Neutrons
a.	$^{85}_{38}Sr$	38	85	38	47
b.	$^{67}_{31}Ga$	31	67	31	36
c.	Selenium-75	34	75	34	41

10.3 An α particle has no electrons around the nucleus and a +2 charge, whereas a helium atom has two electrons around the nucleus and is neutral.

10.4 Use Table 10.1 to identify Q in each of the following symbols.

a. ${}^{0}_{-1}Q$ β particle

b. ${}^{4}_{2}Q$ α particle

c. ${}^{0}_{+1}Q$ positron

10.5 Write a balanced nuclear equation as in Example 10.2.

[1] Write an incomplete equation with the original nucleus on the left and the particle emitted on the right. Radon-222 has an atomic number of 86.

$${}^{222}_{86}Rn \longrightarrow {}^{4}_{2}He + ?$$

[2] Calculate the mass number and the atomic number of the newly formed nucleus on the right.

- Radon-222 emits an α particle.
- Mass number: Subtract the mass of an α particle (4) to obtain the mass of the new nucleus; 222 – 4 = 218.
- Atomic number: Subtract the two protons of an α particle to obtain the atomic number of the new nucleus; 86 – 2 = 84.

[3] Use the atomic number to identify the new nucleus and complete the equation.

- From the periodic table, the element with an atomic number of 84 is polonium, Po.
- Write the mass number and the atomic number with the element symbol to complete the equation.

$${}^{222}_{86}Rn \longrightarrow {}^{4}_{2}He + {}^{218}_{84}Po$$

10.6 Work backwards to write the equation that produces radon-222.

$$? \longrightarrow {}^{4}_{2}He + {}^{222}_{86}Rn$$

Mass number = 222 + 4 = 226

Atomic number = 86 + 2 = 88

$${}^{226}_{88}Ra \longrightarrow {}^{4}_{2}He + {}^{222}_{86}Rn$$

10.7 Write the balanced nuclear equation for each isotope as in Example 10.2 or Answer 10.5.

a. ${}^{218}_{84}Po \longrightarrow {}^{4}_{2}He + {}^{214}_{82}Pb$

b. ${}^{230}_{90}Th \longrightarrow {}^{4}_{2}He + {}^{226}_{88}Ra$

c. ${}^{252}_{99}Es \longrightarrow {}^{4}_{2}He + {}^{248}_{97}Bk$

10.8 Write a balanced nuclear equation for the β emission of iodine-131 as in Example 10.2.

[1] Write an incomplete equation with the original nucleus on the left and the particle emitted on the right.

- Use the identity of the element to determine the atomic number; iodine has an atomic number of 53.

$${}^{131}_{53}I \longrightarrow {}^{0}_{-1}e + ?$$

[2] Calculate the mass number and the atomic number of the newly formed nucleus on the right.

- Mass number: Since a β particle has no mass, the masses of the new particle and the original particle are the same, 131.
- Atomic number: Since β emission converts a neutron into a proton, the new nucleus has one more proton than the original nucleus; 53 = –1 + ?. Thus, the new nucleus has an atomic number of 54.

[3] Use the atomic number to identify the new nucleus and complete the equation.

- From the periodic table, the element with an atomic number of 54 is xenon, Xe.
- Write the mass number and the atomic number with the element symbol to complete the equation.

$$^{131}_{53}\text{I} \longrightarrow \, ^{0}_{-1}\text{e} + \, ^{131}_{54}\text{Xe}$$

10.9 Write a balanced nuclear equation for the β emission of each isotope as in Example 10.2 and Answer 10.8.

a. $^{20}_{9}\text{F} \longrightarrow \, ^{0}_{-1}\text{e} + \, ^{20}_{10}\text{Ne}$

b. $^{92}_{38}\text{Sr} \longrightarrow \, ^{0}_{-1}\text{e} + \, ^{92}_{39}\text{Y}$

c. $^{55}_{24}\text{Cr} \longrightarrow \, ^{0}_{-1}\text{e} + \, ^{55}_{25}\text{Mn}$

10.10 Write a balanced nuclear equation for positron emission as in Example 10.3.

[1] Write an incomplete equation with the original nucleus on the left and the particle emitted on the right.

- Use the identity of the element to determine the atomic number; arsenic has an atomic number of 33.

$$^{74}_{33}\text{As} \longrightarrow \, ^{0}_{+1}\text{e} + \, ?$$

[2] Calculate the mass number and the atomic number of the newly formed nucleus on the right.

- Mass number: Since a β^+ particle has no mass, the masses of the new particle and the original particle are the same, 74.
- Atomic number: Since β^+ emission converts a proton into a neutron, the new nucleus has one fewer proton than the original nucleus; 33 – 1 = 32. Thus, the new nucleus has an atomic number of 32.

[3] Use the atomic number to identify the new nucleus and complete the equation.

- From the periodic table, the element with an atomic number of 32 is germanium, Ge.
- Write the mass number and the atomic number with the element symbol to complete the equation.

$$^{74}_{33}\text{As} \longrightarrow \, ^{0}_{+1}\text{e} + \, ^{74}_{32}\text{Ge}$$

b. Use the same steps as in part (a).

$$^{15}_{8}O \longrightarrow ^{0}_{+1}e + ^{15}_{7}N$$

10.11 When β and γ emission occur together, the atomic number increases by one (due to the β particle), and a γ ray is released.

$$^{192}_{77}Ir \longrightarrow ^{192}_{78}Pt + ^{0}_{-1}e + \gamma$$

Both β particles and γ rays are emitted.

10.12 When γ emission occurs alone, there is no change in the atomic number or the mass number. When β and γ emission occur together, the atomic number increases by one (due to the β particle), and a γ ray is released.

a. $^{11}_{5}B \longrightarrow ^{11}_{5}B + \gamma$

γ emission alone
no change in atomic number or mass number

b. $^{40}_{19}K \longrightarrow ^{40}_{20}Ca + ^{0}_{-1}e + \gamma$

Atomic number increases, without a change in the mass number.
Both β particles and γ rays are emitted.

10.13

a. 8 blue spheres $\times \frac{1}{2} \times \frac{1}{2}$ = 2 blue spheres

two half-lives

Two half-lives have elapsed.

b. If two half-lives have elapsed and each half-life takes 22 minutes, the total time elapsed is:
2×22 min = 44 min.

10.14 To calculate the amount of radioisotope present after the given number of half-lives, multiply the initial mass by ½ for each half-life.

a. 1.00 g (initial mass) $\times \frac{1}{2} \times \frac{1}{2}$ = 0.250 g of phosphorus-32 remains.

The mass is halved two times.

b. 1.00 g (initial mass) $\times \frac{1}{2} \times \frac{1}{2} \times \frac{1}{2} \times \frac{1}{2}$ = 0.0625 g of phosphorus-32 remains.

The mass is halved four times.

c. 1.00 g (initial mass) $\times \frac{1}{2} \times \frac{1}{2} \times \frac{1}{2} \times \frac{1}{2} \times \frac{1}{2} \times \frac{1}{2} \times \frac{1}{2} \times \frac{1}{2}$

The mass is halved eight times.

= 0.003 91 g of phosphorus-32 remains.

10.15 To calculate the amount of radioisotope present, first determine the number of half-lives that occur in the given amount of time. Then multiply the initial mass by ½ for each half-life to determine the amount present.

a. $6.0\ \text{hours} \times \dfrac{1\ \text{half-life}}{6.0\ \text{hours}} = 1.0\ \text{half-life}$ $\quad \underset{\text{initial mass}}{160.\ \text{mg}} \times \dfrac{1}{2} = 80.\ \text{mg of Tc-99m}$

b. $18.0\ \text{hours} \times \dfrac{1\ \text{half-life}}{6.0\ \text{hours}} = 3.0\ \text{half-lives}$ $\quad \underset{\text{initial mass}}{160.\ \text{mg}} \times \dfrac{1}{2} \times \dfrac{1}{2} \times \dfrac{1}{2} = 20.0\ \text{mg of Tc-99m}$

c. $24.0\ \text{hours} \times \dfrac{1\ \text{half-life}}{6.0\ \text{hours}} = 4.0\ \text{half-lives}$ $\quad \underset{\text{initial mass}}{160.\ \text{mg}} \times \dfrac{1}{2} \times \dfrac{1}{2} \times \dfrac{1}{2} \times \dfrac{1}{2} = 10.0\ \text{mg of Tc-99m}$

d. $2\ \text{days} \times \dfrac{24\ \text{hours}}{1\ \text{day}} \times \dfrac{1\ \text{half-life}}{6.0\ \text{hours}} = 8\ \text{half-lives}$

$$\underset{\text{initial mass}}{160.\ \text{mg}} \times \frac{1}{2} \times \frac{1}{2} \times \frac{1}{2} \times \frac{1}{2} \times \frac{1}{2} \times \frac{1}{2} \times \frac{1}{2} \times \frac{1}{2} = 0.6\ \text{mg of Tc-99m}$$

10.16 If an artifact has 1/8 of the amount of C-14 compared to living organisms, it has decayed by three half-lives (½ × ½ × ½).

$$3\ \text{half-lives} \times \frac{5{,}730\ \text{years}}{1\ \text{half-life}} = 17{,}200\ \text{years}$$

10.17 Use the amount of radioactivity (mCi/mL) as a conversion factor to convert the dose of radioactivity from millicuries to a volume in milliliters.

mCi–mL conversion factor

$$110\ \text{mCi dose} \times \frac{1\ \text{mL}}{25\ \text{mCi}} = 4.4\ \text{mL}$$

Answer

Millicuries cancel.

10.18 First, determine how many half-lives have occurred. Then, multiply the initial activity by ½ for each half-life.

$$24\ \text{days} \times \frac{1\ \text{half-life}}{8.0\ \text{days}} = 3.0\ \text{half-lives}$$

$$240\ \text{mCi} \times \underbrace{\frac{1}{2} \times \frac{1}{2} \times \frac{1}{2}}_{\text{three half-lives}} = 30.\ \text{mCi remaining activity}$$

Answer

10.19 Use the conversion factor given to convert mrem to rem.

a. $200 \text{ mrem} \times \frac{1 \text{ rem}}{1000 \text{ mrem}} = 0.2 \text{ rem}$ ← larger dose

b. $0.014 \text{ rem} \times \frac{1000 \text{ mrem}}{1 \text{ rem}} = 14 \text{ mrem}$

10.20 Calculate the number of half-lives that pass in nine days.

$$9 \text{ days} \times \frac{1 \text{ half-life}}{3 \text{ days}} = 3 \text{ half-lives} \qquad \frac{1}{2} \times \frac{1}{2} \times \frac{1}{2} = \frac{1}{8}$$

10.21

a. ${}^{153}_{62}\text{Sm}$: 153 – 62 = 91 neutrons; 62 protons, 62 electrons

b. $${}^{153}_{62}\text{Sm} \longrightarrow {}^{153}_{63}\text{Eu} + {}^{0}_{-1}\text{e}$$

Atomic number increases. One β particle is emitted.

c. $150 \text{ mCi} \times \frac{1}{2} \times \frac{1}{2} \times \frac{1}{2} \times \frac{1}{2} = 9.4 \text{ mCi}$

initial activity; 4 half-lives

10.22 The emission of a positron decreases the atomic number by one, but the mass number stays the same. Use Example 10.3.

$${}^{13}_{7}\text{N} \longrightarrow {}^{0}_{+1}\text{e} + {}^{13}_{6}\text{C}$$

10.23 Nuclear fission splits an atom into two lighter nuclei. Write the equation with the information given, and then balance the equation.

$${}^{235}_{92}\text{U} + {}^{1}_{0}\text{n} \longrightarrow {}^{133}_{51}\text{Sb} + ? + 3\,{}^{1}_{0}\text{n}$$

The atomic number must be 41 = niobium:
92 = 51 + X
The mass number of niobium must be 100:
235 + 1 = 133 + X + 3, X = 100

$${}^{235}_{92}\text{U} + {}^{1}_{0}\text{n} \longrightarrow {}^{133}_{51}\text{Sb} + {}^{100}_{41}\text{Nb} + 3\,{}^{1}_{0}\text{n}$$

10.24 Balance each reaction.

a. $${}^{1}_{1}\text{H} + {}^{1}_{1}\text{H} \longrightarrow {}^{2}_{1}\text{H} + {}^{0}_{+1}\text{e}$$

b. $${}^{1}_{1}\text{H} + {}^{2}_{1}\text{H} \longrightarrow {}^{3}_{2}\text{He}$$

c. $${}^{1}_{1}\text{H} + {}^{3}_{2}\text{He} \longrightarrow {}^{4}_{2}\text{He} + {}^{0}_{+1}\text{e}$$

Solutions to Odd-Numbered End-of-Chapter Problems

10.25 Refer to Example 10.1 to answer the question.
- The atomic number (Z) = the number of protons.
- The mass number (A) = the number of protons + the number of neutrons.
- Isotopes are written with the mass number to the upper left of the element symbol and the atomic number to the lower left of the element symbol.

	a. Atomic Number	b. Number of Protons	c. Number of Neutrons	d. Mass Number	Isotope Symbol
Fluorine-18	9	9	9	18	$^{18}_{9}F$
Fluorine-19	9	9	10	19	$^{19}_{9}F$

10.27 Fill in the table using Example 10.1 and the definitions in Answer 10.25.

	Atomic Number	Mass Number	Number of Protons	Number of Neutrons	Isotope Symbol
a. Chromium-51	24	51	24	27	$^{51}_{24}Cr$
b. Palladium-103	46	103	46	57	$^{103}_{46}Pd$
c. Potassium-42	19	42	19	23	$^{42}_{19}K$
d. Xenon-133	54	133	54	79	$^{133}_{54}Xe$

10.29 Use the definitions in Section 10.1B and Table 10.1 to fill in the table.

	Change in Mass	Change in Charge
a. α particle	–4	–2
b. β particle	0	+1
c. γ ray	0	0
d. positron	0	–1

10.31 Use the definitions in Section 10.1B and Table 10.1 to fill in the table.

	Mass	Charge
a. α	4	+2
b. n	1	0
c. γ	0	0
d. β	0	–1

10.33 Complete the equation. The white circles represent neutrons and the black circles represent protons.

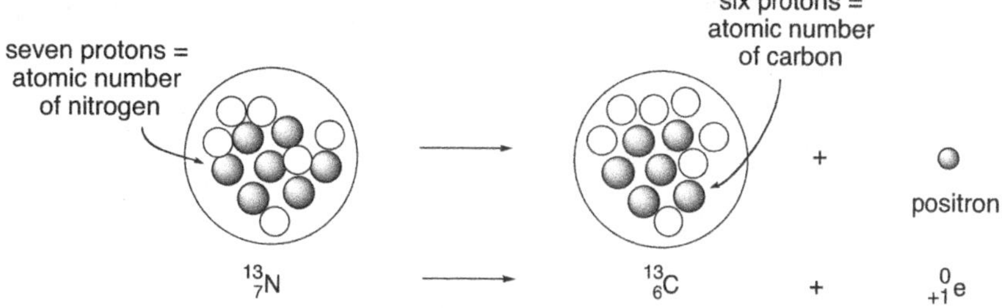

10.35 Complete the nuclear equation as in Example 10.2.

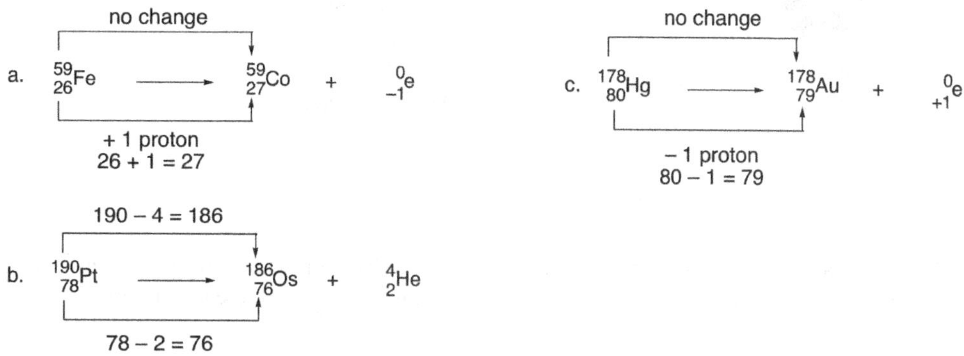

10.37 Complete each nuclear equation.

a. $${}^{90}_{39}Y \longrightarrow {}^{90}_{40}Zr + {}^{0}_{-1}e$$

+ 1 proton; ${}^{0}_{-1}e$ = β particle

b. $${}^{135}_{60}Nd \longrightarrow {}^{135}_{59}Pr + {}^{0}_{+1}e$$

Work backwards: 59 + 1 = 60; ${}^{0}_{+1}e$ = positron

c. $${}^{210}_{83}Bi \longrightarrow {}^{206}_{81}Tl + {}^{4}_{2}He$$

210 – 4 = 206; 83 – 2 = 81; ${}^{4}_{2}He$ = α particle

10.39 Write the two nuclear equations for the decay of bismuth-214.

$${}^{214}_{83}Bi \longrightarrow {}^{214}_{84}Po + {}^{0}_{-1}e$$

${}^{0}_{-1}e$ = β particle

$${}^{214}_{83}Bi \longrightarrow {}^{210}_{81}Tl + {}^{4}_{2}He$$

${}^{4}_{2}He$ = α particle

10.41 Write the chemical equation for each nuclear reaction.

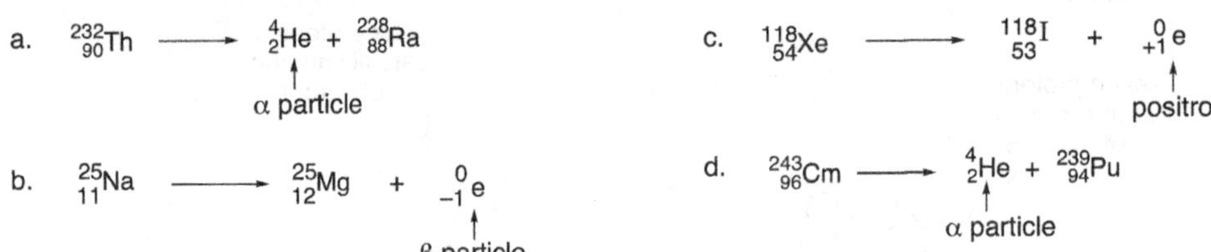

10.43

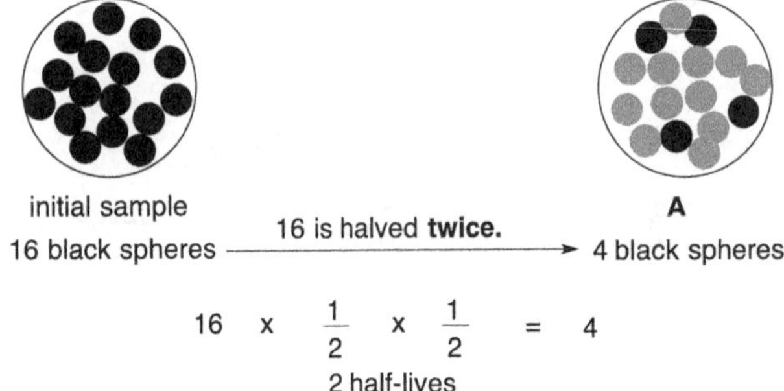

$$16 \times \frac{1}{2} \times \frac{1}{2} = 4$$

2 half-lives

10.45 After each half-life, the mass is halved. In (a) and (b) work backwards to determine the mass of **A** that decayed to the given mass of **B.** The total mass of **A** + **B** in each box must be 4.4 g.

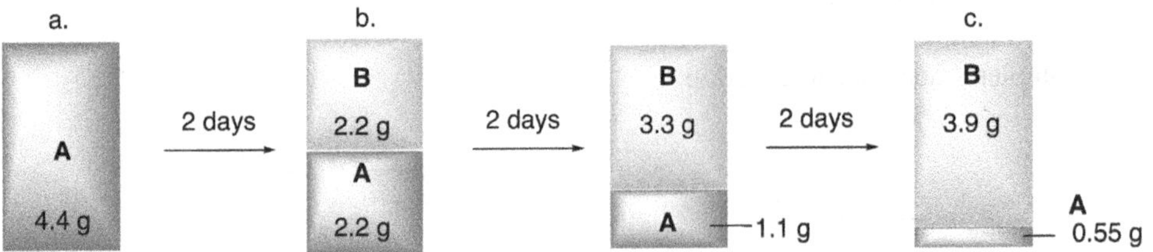

10.47 Determine the number of half-lives that have passed in 12 days to determine the length of each half-life.

$$\underset{\text{initial mass}}{2.4\text{ g}} \times \frac{1}{2} \times \frac{1}{2} \times \frac{1}{2} = 0.30\text{ g} \qquad \frac{12\text{ days}}{3\text{ half-lives}} = 4.0\text{ days in one half-life}$$

The mass is halved three times.

10.49 To calculate the amount of iodine present after the given number of half-lives, multiply the initial mass by ½ for each half-life. The total mass is always 64 mg.

a. 8.0 days = 1 half-life

$$\underset{\text{initial mass}}{64\text{ mg}} \times \frac{1}{2} = 32\text{ mg of iodine-131} \qquad 64 - 32 = 32\text{ mg of xenon-131}$$

b. 16 days = 2 half-lives

$$\underset{\text{initial mass}}{64\text{ mg}} \times \frac{1}{2} \times \frac{1}{2} = 16\text{ mg of iodine-131} \qquad 64 - 16 = 48\text{ mg of xenon-131}$$

c. 24 days = 3 half-lives

$$\underset{\text{initial mass}}{64\text{ mg}} \times \frac{1}{2} \times \frac{1}{2} \times \frac{1}{2} = 8.0\text{ mg of iodine-131} \qquad 64 - 8.0 = 56\text{ mg of xenon-131}$$

d. 32 days = 4 half-lives

$$\underset{\text{initial mass}}{64\text{ mg}} \times \frac{1}{2} \times \frac{1}{2} \times \frac{1}{2} \times \frac{1}{2} = 4.0\text{ mg of iodine-131} \qquad 64 - 4.0 = 60.\text{ mg of xenon-131}$$

10.51 In artifacts over 50,000 years old, the percentage of carbon-14 is too small to accurately measure.

10.53 Use the conversion factors in Table 10.4 to convert mCi to disintegrations/second.

$$5.0\text{ mCi} \times \frac{3.7 \times 10^{10}\text{ disintegrations/s}}{1\text{ Ci}} \times \frac{1\text{ Ci}}{1000\text{ mCi}} = 1.9 \times 10^{8}\text{ disintegrations/second}$$

10.55 Use the concentration as a conversion factor as in Example 10.4.

$$28\text{ mCi} \times \frac{1\text{ mL}}{12\text{ mCi}} = 2.3\text{ mL solution}$$

10.57 Use conversion factors to solve the problem; 1 mCi = 1,000 μCi.

$$68\text{ kg} \times \frac{190\ \mu\text{Ci}}{1\text{ kg}} \times \frac{1\text{ mCi}}{1000\ \mu\text{Ci}} \times \frac{1\text{ mL}}{5\text{ mCi}} = 2.6\text{ mL solution}$$

10.59 Calculate the amount of radioactivity present after each amount of time has elapsed.

a. $$\underset{\text{initial activity}}{20\text{ mCi}} \times \underbrace{\frac{1}{2}}_{6\text{ h = one half-life}} = 10\text{ mCi}$$

b. $$\underset{\text{initial activity}}{20\text{ mCi}} \times \underbrace{\frac{1}{2} \times \frac{1}{2}}_{12\text{ h = two half-lives}} = 5\text{ mCi}$$

c. $$\underset{\text{initial activity}}{20\text{ mCi}} \times \underbrace{\frac{1}{2} \times \frac{1}{2} \times \frac{1}{2} \times \frac{1}{2}}_{24\text{ h = four half-lives}} = 1\text{ mCi}$$

10.61 The curie measures the number of disintegrations per second, whereas the number of rads indicates the radiation absorbed by one gram of a substance.

10.63 This represents a fatal dose because 600 rem is uniformly fatal.

$$20\ \cancel{Sv} \times \frac{100\text{ rem}}{1\ \cancel{Sv}} = 2{,}000\text{ rem}$$

10.65 a. Nuclear fusion is a reaction that occurs in the sun.
b. Nuclear fission occurs when a neutron is used to bombard a nucleus.
c. In both nuclear fission and nuclear fusion a large amount of energy is released.
d. Nuclear fusion reactions require very high temperatures.

10.67 Nuclear fission splits an atom into two lighter nuclei. Balance each equation.

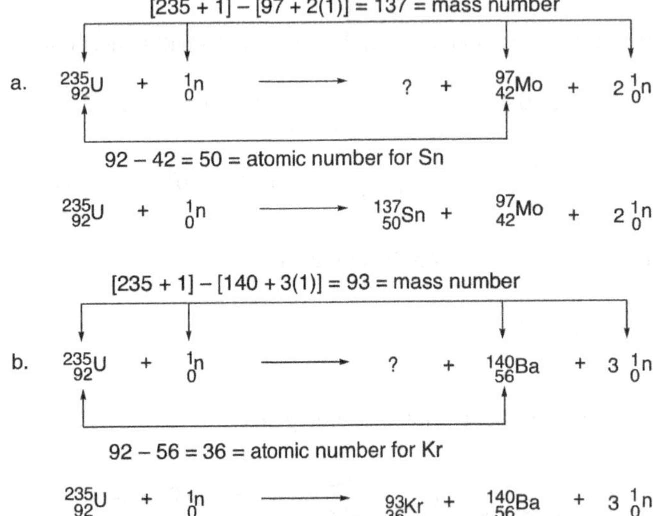

10.69

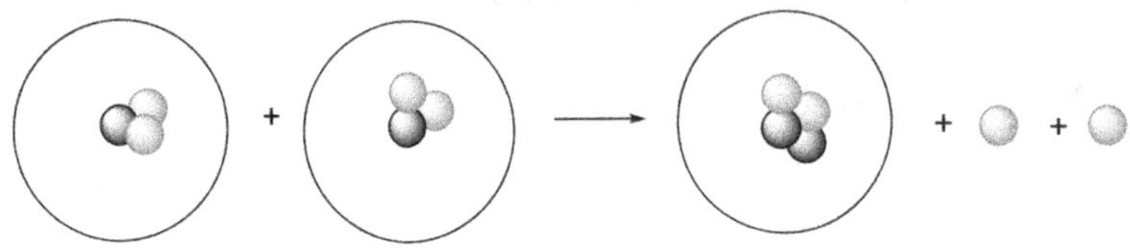

10.71 Write out the equation from the information given. Then balance the equation.

$$^{2}_{1}H + ^{2}_{1}H \longrightarrow \underset{\text{tritium}}{^{3}_{1}H} + ^{1}_{1}H$$

10.73 Two problems with generating electricity from nuclear power plants include the containment of radiation leaks and the disposal of radioactive waste.

10.75

a. $^{74}_{33}\text{As} \longrightarrow \; ^{0}_{+1}\text{e} + \; ^{74}_{32}\text{Ge}$

b. $90.\ \text{days} \times \dfrac{1\ \text{half-life}}{18\ \text{days}} = 5\ \text{half-lives}$ $\quad \underset{\text{initial mass}}{120\ \text{mg}} \times \dfrac{1}{2} \times \dfrac{1}{2} \times \dfrac{1}{2} \times \dfrac{1}{2} \times \dfrac{1}{2} = 3.8\ \text{mg of As-74}$

c. $7.5\ \text{mCi} \times \dfrac{2.0\ \text{mL}}{10.0\ \text{mCi}} = 1.5\ \text{mL}$

10.77

a. $^{89}_{38}\text{Sr}$ — 89 – 38 = 51 neutrons; 38 protons, 38 electrons

b. $^{89}_{38}\text{Sr} \longrightarrow \; ^{89}_{39}\text{Y} + \; ^{0}_{-1}\text{e}$

c. $\underset{\text{initial acitivity}}{10.0\ \mu\text{g}} \times \dfrac{1}{2} \times \dfrac{1}{2} \times \dfrac{1}{2} \times \dfrac{1}{2} = 0.625\ \mu\text{g}$

10.79

a. $^{192}_{77}\text{Ir} \longrightarrow \; ^{192}_{78}\text{Pt} + \; ^{0}_{-1}\text{e} + \gamma$

b. $150\ \text{days} \times \dfrac{1\ \text{half-life}}{74\ \text{days}} = 2.0\ \text{half-lives}$ $\quad \underset{\text{initial mass}}{120\ \text{mg}} \times \dfrac{1}{2} \times \dfrac{1}{2} = 30.\ \text{mg of Ir-192}$

c. $300\ \text{days} \times \dfrac{1\ \text{half-life}}{74\ \text{days}} = 4.0\ \text{half-lives}$ $\quad \underset{\text{initial activity}}{36\ \text{Ci}} \times \dfrac{1}{2} \times \dfrac{1}{2} \times \dfrac{1}{2} \times \dfrac{1}{2} = 2.3\ \text{Ci}$

10.81

$^{238}_{92}\text{U} + \textbf{X} = \; ^{246}_{98}\text{Cf} + 4\,^{1}_{0}\text{n}$

$$\begin{aligned} 238 + x &= 246 + 4 \\ x &= 12 = \text{mass number} \\ 92 + y &= 98 \\ y &= 6 = \text{atomic number} \end{aligned}$$

$\textbf{X} = \; ^{12}_{6}\text{C}$

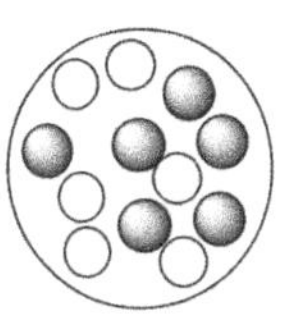

10.83 Balance the nuclear equation.

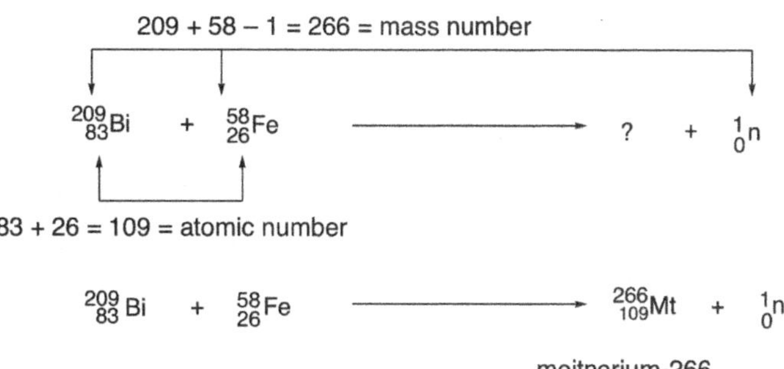

$^{209}_{83}\text{Bi} + \; ^{58}_{26}\text{Fe} \longrightarrow \; ^{266}_{109}\text{Mt} + \; ^{1}_{0}\text{n}$

meitnerium-266

10.85 a. Iodine-131 is used for the treatment and diagnosis of thyroid diseases.
b. Iridium-192 is used for the treatment of breast cancer.
c. Thallium-201 is used for the diagnosis of heart disease.

10.87 a. Iodine-125, with its longer half-life, is used for the treatment of prostate cancers with implanted radioactive seeds.
b. Iodine-131, with its shorter half-life, is used for the diagnostic and therapeutic treatment of thyroid diseases and tumors. A patient is administered radioactive iodine-131, which is then incorporated into the thyroid hormone, thyroxine. Since its half-life is short, the radioactive iodine isotope decays so that little remains after a month or so.

10.89 Radiology technicians must be shielded to avoid exposure to excessive and dangerous doses of radiation.

10.91 High doses of stable iodine-127 will prevent the absorption and uptake of the radioactive iodine-131.

10.93 Use conversion factors to solve the problem.

$$1.0 \cancel{\text{g}} \text{ coal} \times \frac{1 \cancel{\text{kg}}}{1000 \cancel{\text{g}}} \times \frac{2.2 \cancel{\text{lb}}}{1 \cancel{\text{kg}}} \times \frac{3.4 \times 10^8 \text{ kcal}}{2000. \cancel{\text{lb}}} = 370 \text{ kcal}$$

Chapter 11 Introduction to Organic Molecules and Functional Groups

Chapter Review

[1] What are the characteristic features of organic compounds? (11.2)

- Organic compounds contain carbon atoms and most contain hydrogen atoms. Carbon forms four bonds.

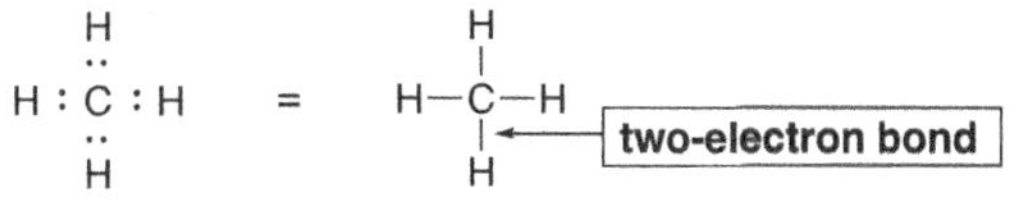

H : C : H (with H above and below) = H—C—H (with H above and below) ← two-electron bond

methane

- Carbon forms single, double, and triple bonds to itself and other atoms.

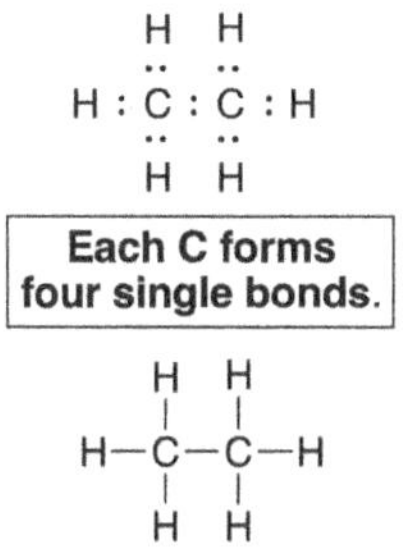

Each C forms four single bonds. — ethane

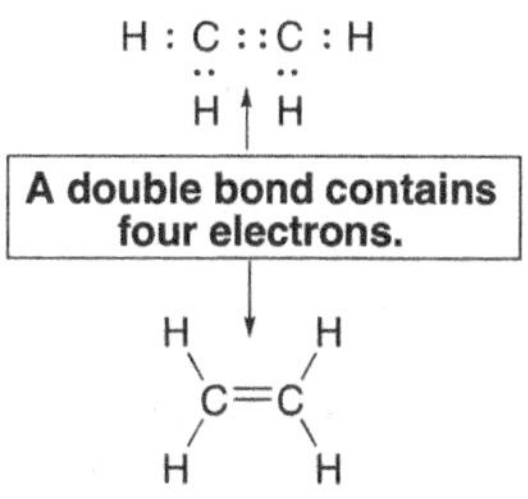

A double bond contains four electrons. — ethylene

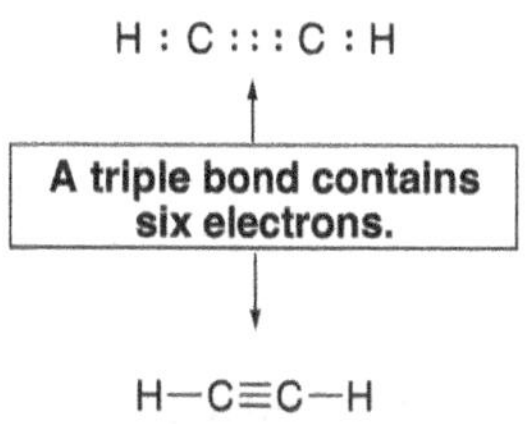

A triple bond contains six electrons. — H—C≡C—H — acetylene

- Carbon atoms can bond to form chains or rings.

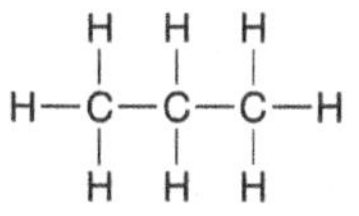

propane 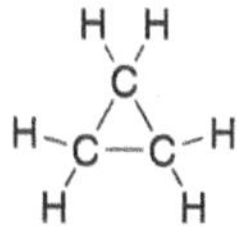 — cyclopropane

- Organic compounds often contain heteroatoms, commonly N, O, and the halogens.

[2] How can we predict the shape around an atom in an organic molecule? (11.3)

- The shape around an atom is determined by counting groups (bonded atoms and lone pairs), and then arranging the groups to keep them as far away from each other as possible.

Number of Groups	Number of Atoms	Number of Lone Pairs	Shape	Bond Angle	Example
2	2	0	linear	180°	HC≡CH
3	3	0	trigonal planar	120°	CH_2=CH_2
4	4	0	tetrahedral	109.5°	CH_3CH_3
4	3	1	trigonal pyramidal	~109.5°	CH_3NH_2 (N atom)
4	2	2	bent	~109.5°	CH_3OH (O atom)

[3] What shorthand methods are used to draw organic molecules? (11.4)

- In condensed structures, atoms are drawn in but the two-electron bonds are generally omitted. Lone pairs are omitted as well. Parentheses are used around like groups bonded together or to the same atom.

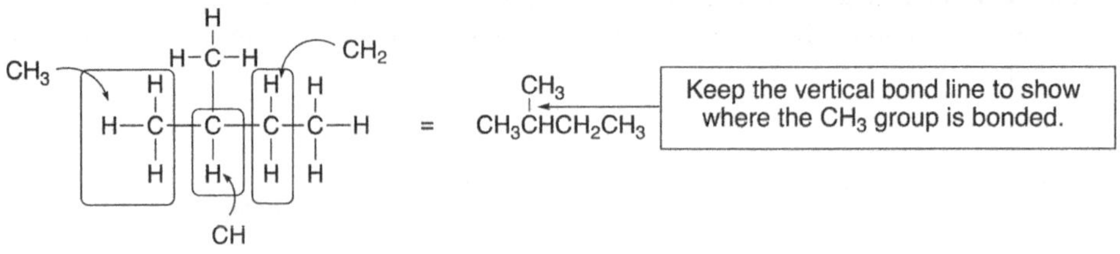

CH_3 H H–C–H CH_2 H H H H–C–C–C–C–H H H H H CH = CH_3 $CH_3CHCH_2CH_3$

Keep the vertical bond line to show where the CH_3 group is bonded.

- Three assumptions are used in drawing skeletal structures: [1] There is a carbon at the intersection of two lines or at the end of any line. [2] Each carbon has enough hydrogens to give it four bonds. [3] Heteroatoms and the hydrogens bonded to them are drawn in.

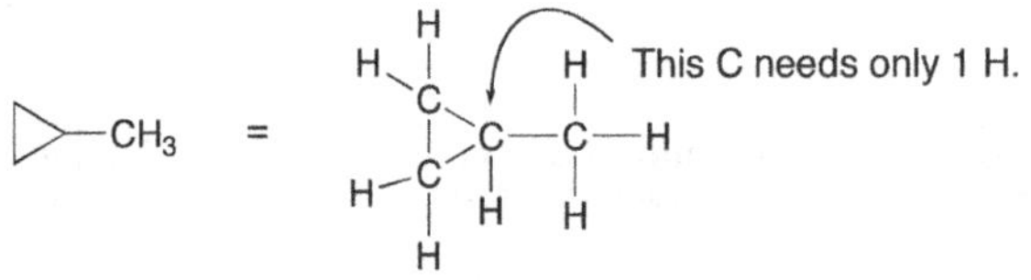

$-CH_3$ = H H C H–C H H H C–C–H H H

This C needs only 1 H.

[4] What is a functional group and why are functional groups important? (11.5)

- A functional group is an atom or a group of atoms with characteristic chemical and physical properties.
- A functional group determines all of the properties of a molecule—its shape, physical properties, and the type of reactions it undergoes.

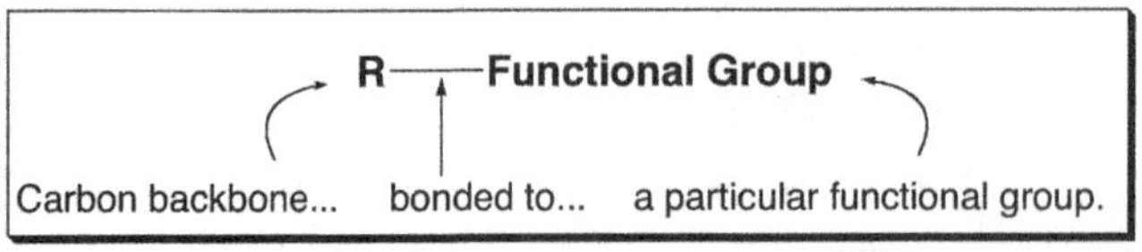

R—Functional Group

Carbon backbone... bonded to... a particular functional group.

[5] How do organic compounds differ from ionic inorganic compounds? (11.6)

- The main difference between organic compounds and ionic inorganic compounds is their bonding. Organic compounds are composed of discrete molecules with covalent bonds. Ionic inorganic compounds are composed of ions held together by the strong attraction of oppositely charged ions. Other properties that are consequences of these bonding differences are summarized in Table 11.5.

[6] How can we determine if an organic molecule is polar or nonpolar? (11.6A)

- A nonpolar molecule has either no polar bonds or two or more bond dipoles that cancel.
- A polar molecule has either one polar bond, or two or more bond dipoles that do not cancel.

H H
H—C—C—H
H H

ethane

no polar bonds
nonpolar molecule

H H δ−
δ+
H—C—C—Cl:
H H polar bond

chloroethane

one polar bond
polar molecule

Two bond dipoles cancel.

:Cl—C≡C—Cl:

dichloroacetylene

nonpolar molecule

Two bond dipoles do *not* cancel.

H_3C—O—H net dipole

methanol

polar molecule

[7] What are the solubility properties of organic compounds? (11.6B)

- Most organic molecules are soluble in organic solvents.
- Hydrocarbons and other nonpolar organic molecules are insoluble in water.
- Polar organic molecules are water soluble only if they are small (less than six carbons) and contain a nitrogen or oxygen atom that can hydrogen bond with water.

CH_3CH_2—OH

ethanol

H_2O soluble

CH_3
CH_3 $CH(CH_2)_3CH(CH_3)_2$
CH_3
HO

cholesterol

There are too many nonpolar C–C and C–H bonds for cholesterol to be soluble in water.

H_2O insoluble

[8] What are vitamins and when is a vitamin fat soluble or water soluble? (11.7)

- A vitamin is an organic compound that cannot be synthesized by the body but is needed in small amounts for cell function.
- A fat-soluble vitamin has many nonpolar C–C and C–H bonds and few polar bonds, making it insoluble in water. Vitamin A is fat soluble.
- A water-soluble vitamin has many polar bonds so it dissolves in water. Vitamin C is water soluble.

Problem Solving

[1] Characteristic Features of Organic Compounds (11.2)

Example 11.1 Draw in all H's and lone pairs in each compound.

a. Br—C—C b. C—C≡N

Analysis

Each C and heteroatom must be surrounded by eight electrons. Use the common bonding patterns in Table 11.1 to fill in the needed H's and lone pairs. C needs four bonds; Br needs one bond and three lone pairs; N needs three bonds and one lone pair.

Solution

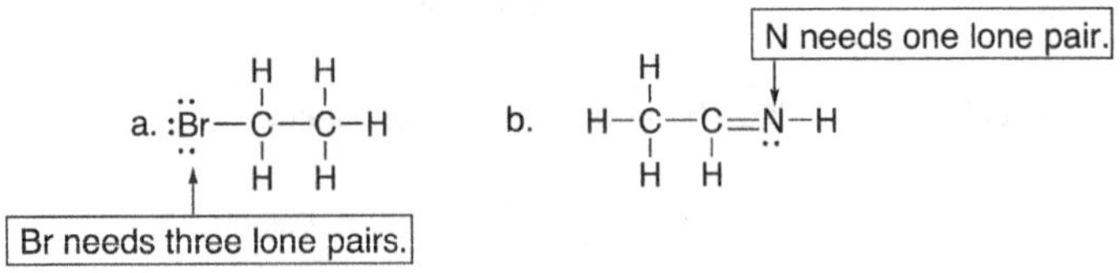

a. :Br–C–C–H (each C bearing two H's)

Br needs three lone pairs.

b. H–C–C=N–H (first C bearing two H's, second C bearing one H)

N needs one lone pair.

[2] Shapes of Organic Molecules (11.3)

Example 11.2 Determine the shape around each atom in acetonitrile.

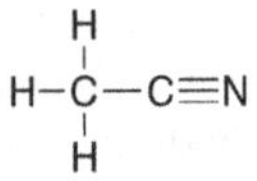

H–C(H)(H)–C≡N

acetonitrile

Analysis

First, draw in all lone pairs on the heteroatom. A heteroatom needs eight electrons around it, so add one lone pair to N. Then count groups to determine the molecular shape.

Solution

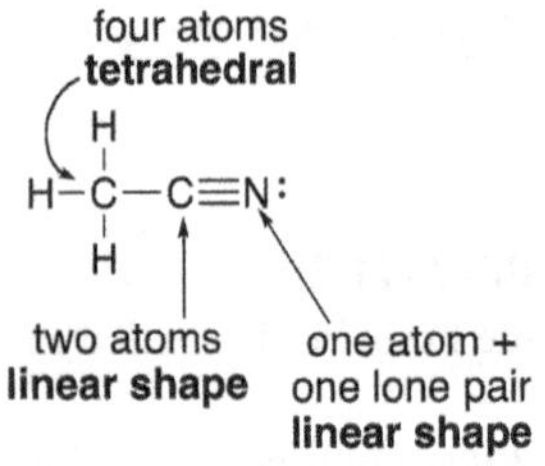

four atoms **tetrahedral**

two atoms **linear shape**

one atom + one lone pair **linear shape**

- One C is surrounded by four atoms (3 H's and 1 C), making it tetrahedral.
- One C is surrounded by two atoms (1 C and 1 N), making it linear.
- The N atom is surrounded by 1 C and one lone pair (two groups), giving it a linear shape.

Example 11.3 Determine the shape around each indicated atom in vitamin C.

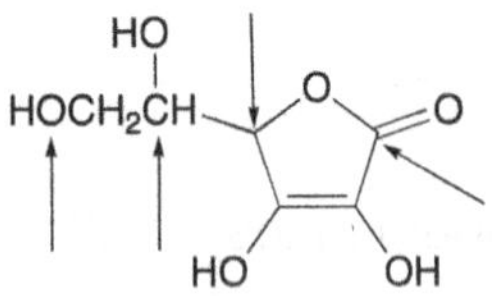

vitamin C
(ascorbic acid)

Analysis

First, add lone pairs of electrons to each heteroatom so that each is surrounded by eight electrons. Each O gets two lone pairs. Draw in the C–H bonds around the indicated atoms. Then count groups to predict shape.

Solution

Add electron pairs around O atoms:

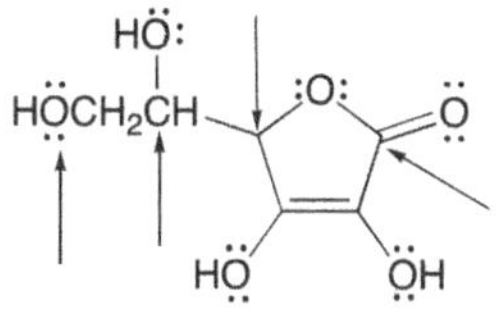

Draw in C–H bonds:

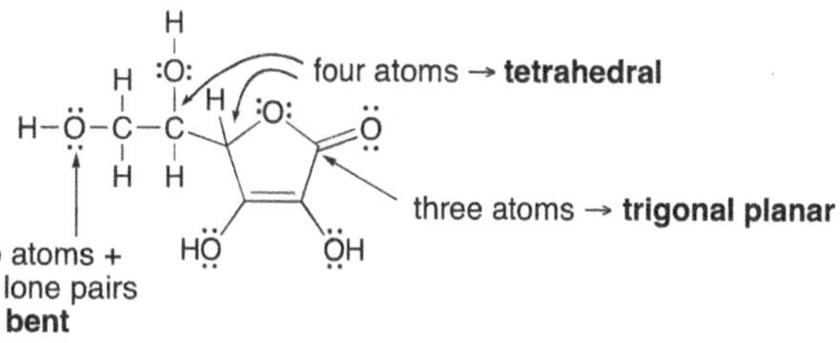

four atoms → **tetrahedral**

three atoms → **trigonal planar**

two atoms + two lone pairs **bent**

---

[3] Drawing Organic Molecules (11.4)

Example 11.4 Convert each compound into a condensed structure.

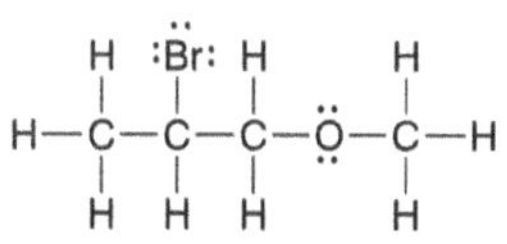

Analysis

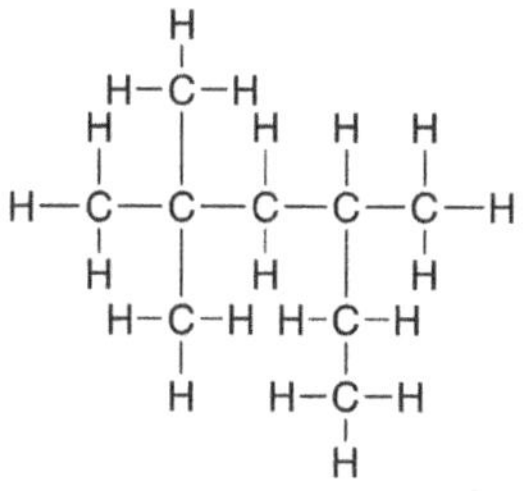

Start at the left and proceed to the right, making sure that each carbon has four bonds. Omit lone pairs on the heteroatoms O and Br. When like groups are bonded together or bonded to the same atom, use parentheses to further simplify the structure.

Solution

	Condensed structure	Further simplified
a. 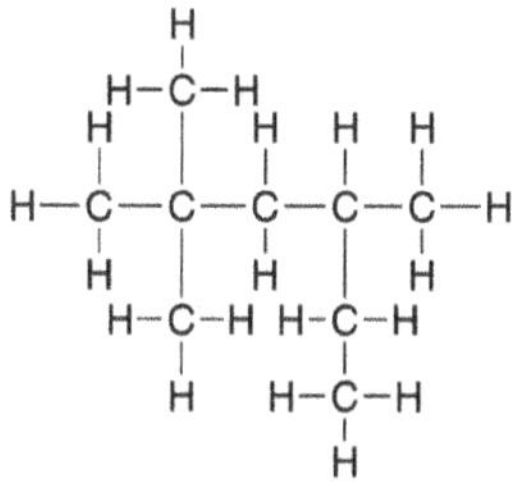	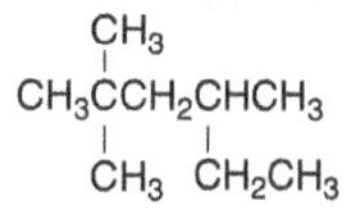 $CH_3C(CH_3)_2CH_2CH(CH_2CH_3)CH_3$ (drawn as $CH_3CCH_2CHCH_3$ with CH_3 above and below the second C and CH_2CH_3 below the CH)	$(CH_3)_3CCH_2CH(CH_2CH_3)CH_3$
b. 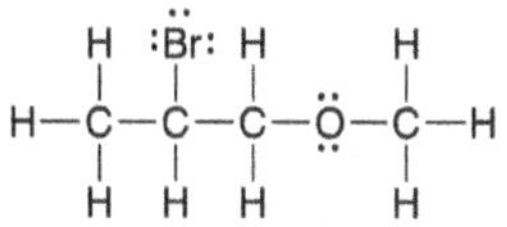 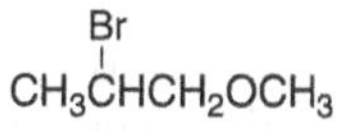	$CH_3CHCH_2OCH_3$ with Br on the second C	

Example 11.5 Convert each skeletal structure to a complete structure with all C's, H's, and lone pairs drawn in.

a. [cyclopentane ring]–CH_3

b. [six-membered ring containing O]

Analysis

To draw each complete structure, place a C atom at the corner of each polygon and add H's to give each carbon four bonds. The O atom has two bonds so it needs two lone pairs to have eight electrons around it.

Solution

a. This C needs only 1 H.

b. Each C needs 2 H's.

[4] Functional Groups (11.5)

Example 11.6 Identify the functional groups in each compound.

a. $CH_3C(=O)C(=O)CH_3$

A
butanedione
(component of butter flavor)

b. HO–C_6H_4–N(H)–C(=O)–CH_3

B
acetaminophen
(analgesic in Tylenol)

Analysis

Concentrate on the multiple bonds and heteroatoms and refer to Tables 11.2, 11.3, and 11.4.

Solution

a. **A**

b. **B**

A contains two carbonyl groups. Each carbonyl carbon is bonded to two other carbons, making it a ketone. Thus, **A** contains two ketones.

B has a carbon atom bonded to a hydroxyl group (OH), as well as a benzene ring, a six-membered ring with three double bonds. It also has an amide functional group (N bonded to a C=O group). Thus, **B** has three functional groups.

[5] Properties of Organic Compounds (11.6)

Example 11.7 Explain why ethanol (CH_3CH_2OH) is a polar molecule.

Analysis
First, locate the polar bonds in CH_3CH_2OH. Then, determine the shape to check whether bond dipoles cancel or reinforce.

Solution
The C–C and C–H bonds in CH_3CH_2OH are nonpolar, while the C–O and O–H bonds are polar. Since the O atom is surrounded by two atoms (C and H) and two lone pairs, it has a bent shape and the individual bond dipoles do *not* cancel. This makes CH_3CH_2OH a polar molecule with a net dipole.

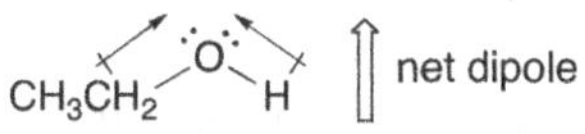

CH_3CH_2–Ö–H ⇑ net dipole

Self-Test

[1] Fill in the blank with one of the terms listed below.

Aldehyde (11.5)	Functional group (11.5)	Organic chemistry (11.1)
Carboxylic acid (11.5)	Heteroatom (11.2)	Polar (11.6)
Ester (11.5)	Hydrocarbon (11.5)	Vitamins (11.7)
Fat-soluble vitamin (11.7)	Nonpolar (11.6)	Water-soluble vitamin (11.7)

1. An ______ has a hydrogen atom bonded directly to the carbonyl carbon.
2. A ______ is an atom or a group of atoms with characteristic chemical and physical properties.
3. If the individual bond dipoles cancel in a molecule, the molecule is ______.
4. A vitamin that dissolves in water and has many polar bonds is classified as a ______.
5. Any atom that is not carbon or hydrogen is called a ______.
6. An ______ contains an OR group bonded directly to the carbonyl carbon.
7. ______ are organic compounds needed in small amounts for normal cell function.
8. A ______ is a compound that contains only the elements of carbon and hydrogen.
9. If the individual bond dipoles do not cancel, a molecule is ______.
10. ______ is the study of compounds that contain the element carbon.
11. A vitamin that dissolves in an organic solvent but is insoluble in water is classified as a ______.
12. A ______ contains an OH group bonded directly to the carbonyl carbon.

[2] Match the structure with the name of the functional group.

a. 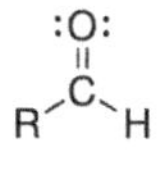 R–C(=O:)–H

b. R–C(=O:)–R

c. R–C(=O:)–N̈(H (or R))–H (or R)

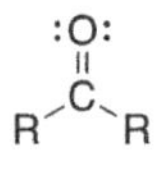

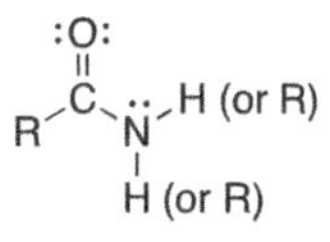

d. R–ÖH

e. R–Ö–R

13. Amide
14. Alcohol
15. Ether
16. Ketone
17. Aldehyde

[3] Fill in the blank with one of the terms listed below.

a. Alkenes
b. Alkanes
c. Alkynes
d. Aromatic hydrocarbons

18. ______ have only C–C single bonds and no functional group.
19. ______ contain a benzene ring—a six-membered ring with three double bonds.
20. ______ have a C–C triple bond as a functional group.
21. ______ have a C–C double bond as a functional group.

[4] Identify the functional group present in each compound.

a. aldehyde
b. ketone
c. amide
d. carboxylic acid
e. amine
f. ester

22. 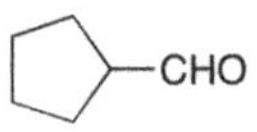 (cyclopentane)–CHO

24. 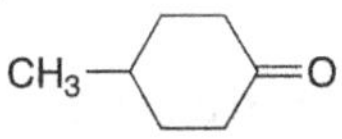 CH_3–(cyclohexane)=O

23. 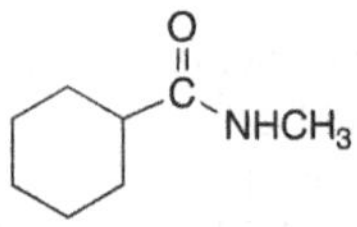 (cyclohexane)–C(=O)–NHCH$_3$

25. 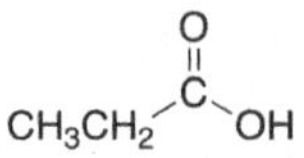 CH_3CH_2–C(=O)–OH

Answers to Self-Test

1. aldehyde	6. ester	11. fat-soluble vitamin	16. b	21. a
2. functional group	7. Vitamins	12. carboxylic acid	17. a	22. a
3. nonpolar	8. hydrocarbon	13. c	18. b	23. c
4. water-soluble vitamin	9. polar	14. d	19. d	24. b
5. heteroatom	10. Organic chemistry	15. e	20. c	25. d

Solutions to In-Chapter Problems

11.1 Organic compounds contain the element carbon.

a. C_6H_{12}—organic
b. H_2O—inorganic
c. KI—inorganic
d. $MgSO_4$—inorganic
e. CH_4O—organic
f. NaOH—inorganic

11.2 Draw in all the H's and lone pairs as in Example 11.1. Each C and heteroatom must be surrounded by eight electrons.

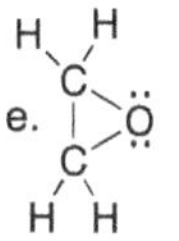

11.3 To determine the shape around each atom, draw in all the lone pairs on any heteroatom and then count groups as in Example 11.2.

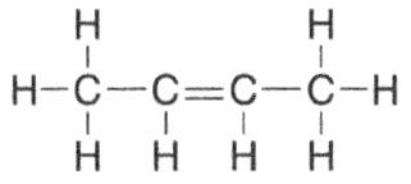

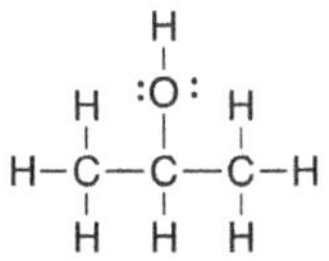

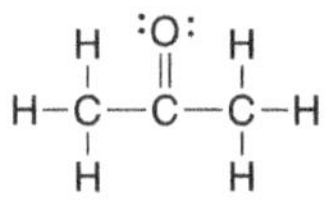

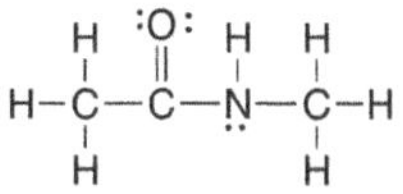

a. 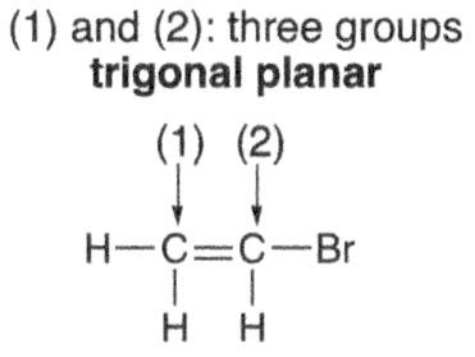 (1) and (2): three groups **trigonal planar**

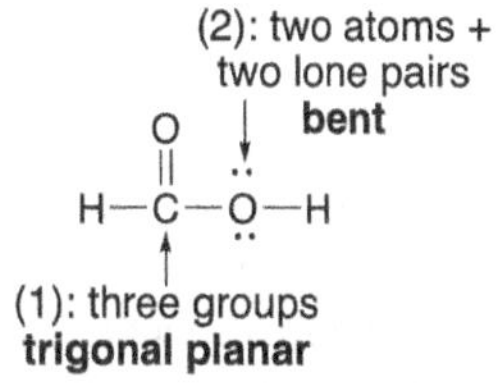

b. 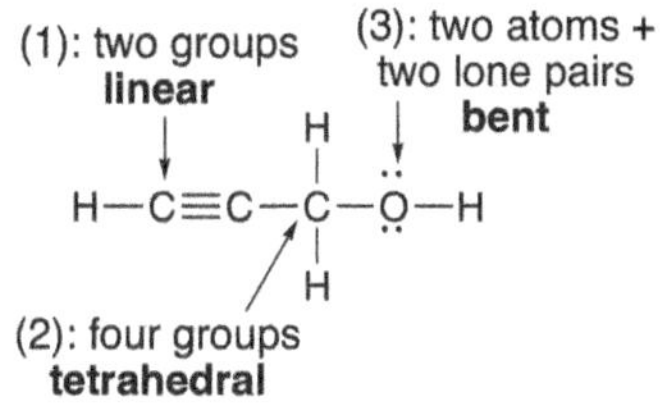 (1): two groups **linear**; (2): four groups **tetrahedral**; (3): two atoms + two lone pairs **bent**

c. (1): three groups **trigonal planar**; (2): two atoms + two lone pairs **bent**

d. 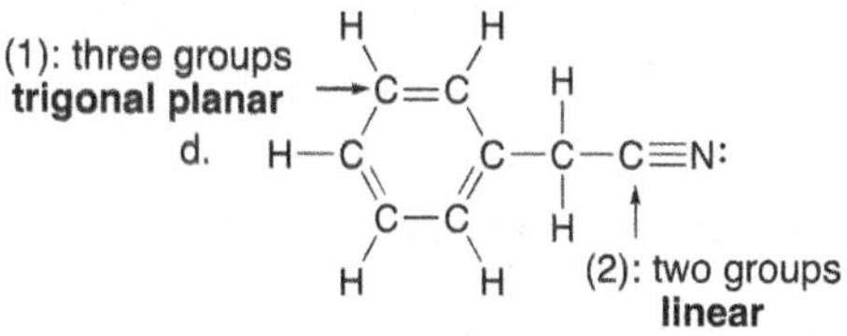 (1): three groups **trigonal planar**; (2): two groups **linear**

11.4 To determine the bond angles, first use the steps in Example 11.2, and then use the values in Table 11.1.

a. 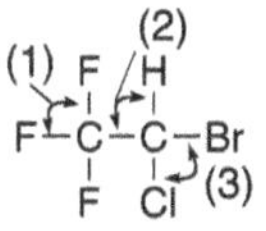 (1), (2), and (3): four groups tetrahedral **109.5°**

b. 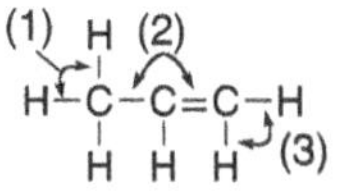 (1): four groups tetrahedral **109.5°**

(2) and (3): three groups trigonal planar **120°**

c. 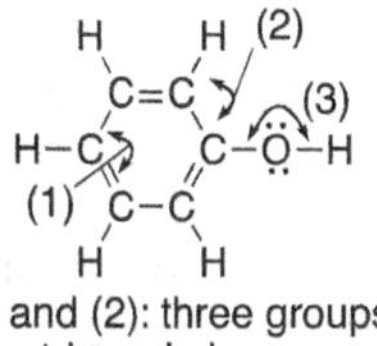 (1) and (2): three groups trigonal planar **120°**

(3): two atoms + two lone pairs bent **109.5°**

11.5 When drawing three-dimensional structures:

- A solid line is used for bonds in the plane.
- A wedge is used for a bond in front of the plane.
- A dashed line is used for a bond behind the plane.

a. 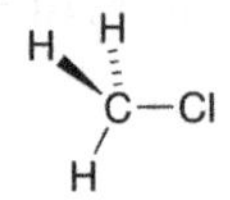 [structure: CH_3Cl three-dimensional drawing]

b. 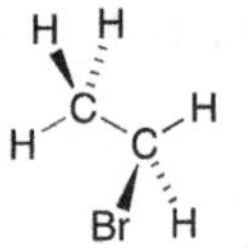 [structure: CH_2BrCH_3 three-dimensional drawing]

11.6 Count the groups around each atom to determine the molecular shape as in Example 11.3.

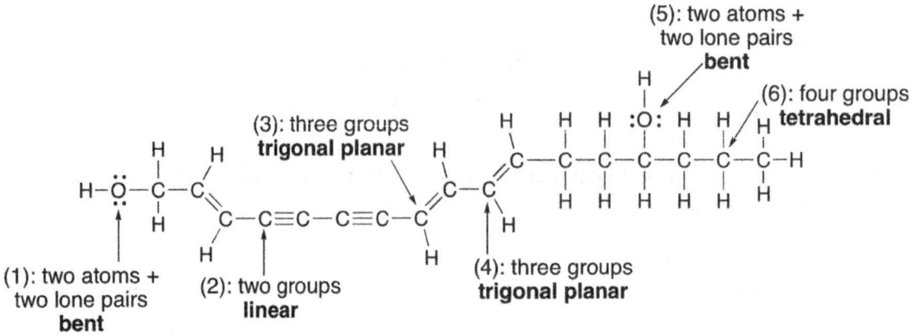

(1): two atoms + two lone pairs **bent**

(2): two groups **linear**

(3): three groups **trigonal planar**

(4): three groups **trigonal planar**

(5): two atoms + two lone pairs **bent**

(6): four groups **tetrahedral**

11.7 a. Each N atom (in blue) has one lone pair and each O atom (in red) has two lone pairs. Lidocaine contains 2 N atoms (two lone pairs total) and 1 O atom (two lone pairs), so it has four lone pairs altogether.

b. Count groups to predict shape.

atom (1): 4 atoms = 4 groups, tetrahedral
atom (2): 3 atoms + 1 lone pair = 4 groups, trigonal pyramidal
atom (3): 3 atoms + 0 lone pairs = 3 groups, trigonal planar
atom (4): 3 atoms + 0 lone pairs = 3 groups, trigonal planar

11.8 Convert each compound to a condensed formula as in Example 11.4.

a. 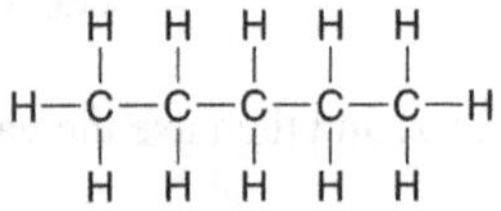 [expanded structure] = $CH_3CH_2CH_2CH_2CH_3$ = $CH_3(CH_2)_3CH_3$

b. 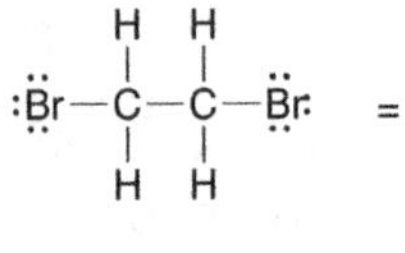 [expanded structure] = $BrCH_2CH_2Br$ = $Br(CH_2)_2Br$

c. 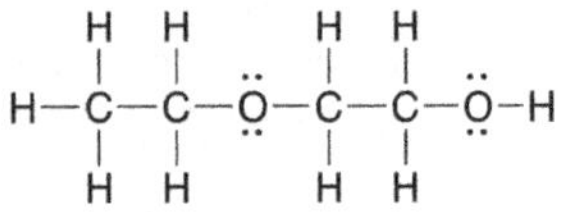 [expanded structure] = $CH_3CH_2OCH_2CH_2OH$ = $CH_3CH_2O(CH_2)_2OH$

d. 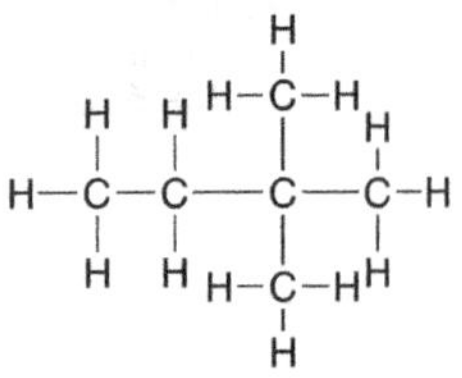 [expanded structure] = $CH_3CH_2C(CH_3)(CH_3)CH_3$ = $CH_3CH_2C(CH_3)_3$

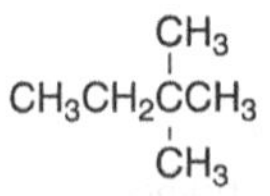

11.9 Work backwards to convert each condensed formula to a complete structure.

a. $CH_3(CH_2)_8CH_3$ = [complete structure: ten C chain, each C with H above and below, H at each end]

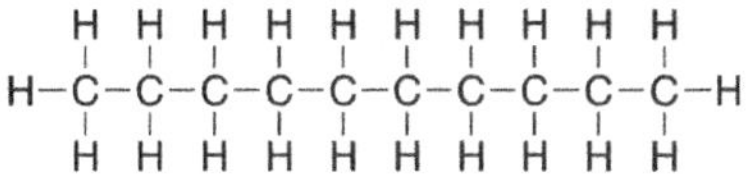

b. $CH_3(CH_2)_4OH$ = [complete structure: H–C–C–C–C–C–Ö–H]

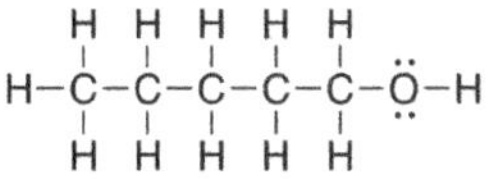

c. CH_3CCl_3 = [complete structure: H–C–C(Cl)₃ with three H on first C]

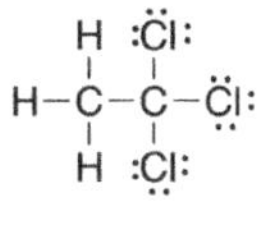

d. $CH_3(CH_2)_4CH(CH_3)_2$ = [complete structure]

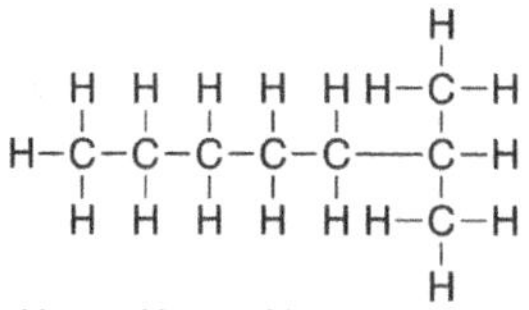

e. $(CH_3)_2CHCH_2NH_2$ = [complete structure]

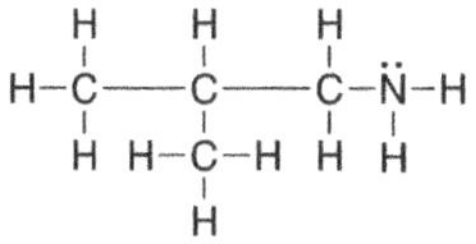

11.10 To convert each skeletal structure to a complete structure, place a C atom at the corner of each polygon and add H's to give each carbon four bonds as in Example 11.5.

a. [ring structure with all H's] b. [ring structure with Cl's] c. [ring structure with OH]

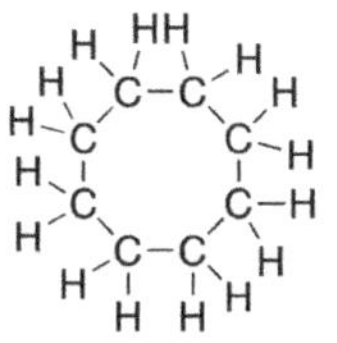

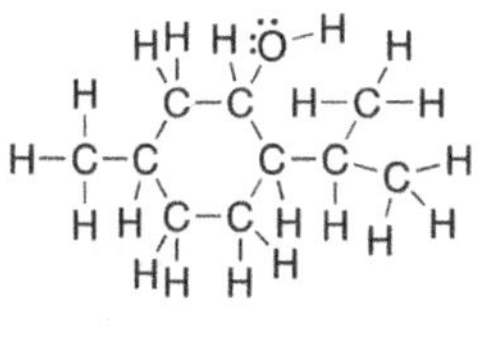

11.11 Each carbon must have four bonds. To determine the number of H's bonded to each C, count the number of bonds, and then add H's to equal four.

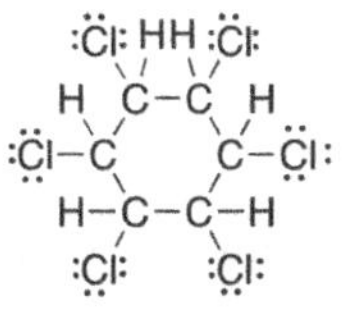

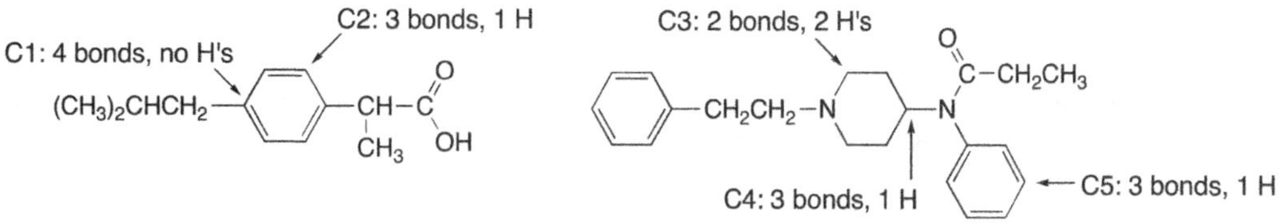

C1: 4 bonds, no H's
C2: 3 bonds, 1 H
$(CH_3)_2CHCH_2$–[benzene ring]–$CH(CH_3)$–$COOH$

C3: 2 bonds, 2 H's
C4: 3 bonds, 1 H
C5: 3 bonds, 1 H
[benzene ring]–CH_2CH_2–N[piperidine ring]–N(C(=O)–CH_2CH_3)–[benzene ring]

11.12 Identify the functional groups in each compound. Concentrate on the heteroatoms and multiple bonds.

a. $CH_3CH(OH)CH_3$ — OH hydroxyl; alcohol

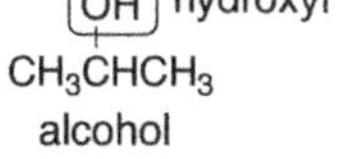

b. [benzene ring]–CH=CH₂ — aromatic ring; alkene

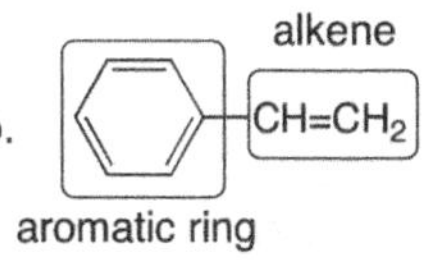

c. $H_2NCH_2CH_2CH_2CH_2NH_2$ — amine, amine

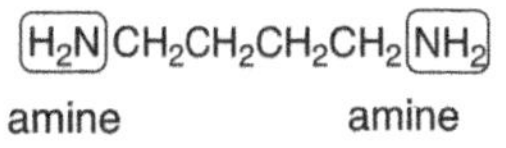

11.13 Identify the functional groups as in Example 11.6 and then draw out the complete structures.

a. aldehyde

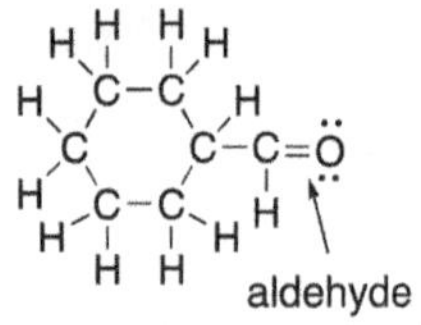

c. ketone

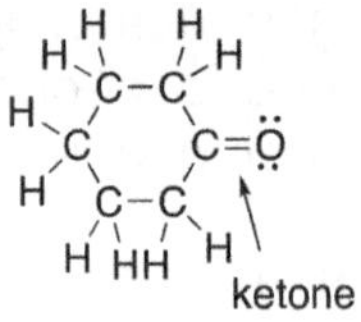

e. amide

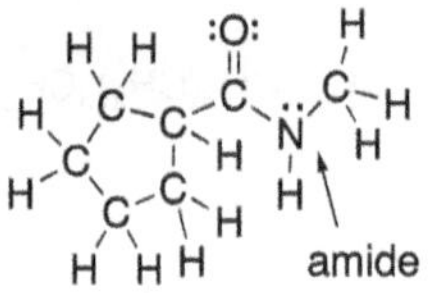

b. H–C–C–C–C–Ö–H (carboxylic acid)

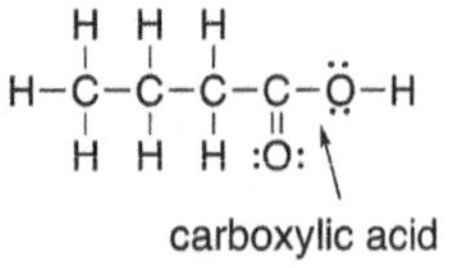

d. H–C–C–C–Ö–C–C–H (ester)

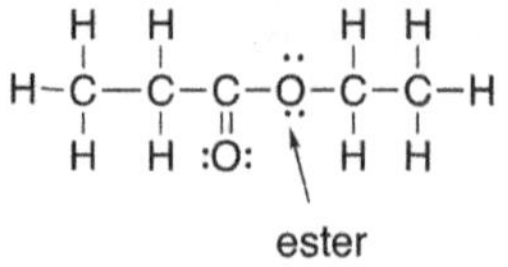

11.14 Both a carboxylic acid (RCO_2H) and an alcohol (ROH) contain OH groups, but a carboxylic acid has a C=O (a carbonyl group) bonded to the OH group, whereas an alcohol does not.

A = $CH_3-CH(CH_3)-CH_2OH$ — alcohol

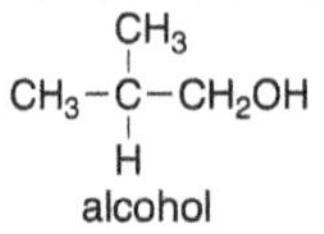

B = $CH_3-CH(CH_3)-C(=O)OH$ — carboxylic acid

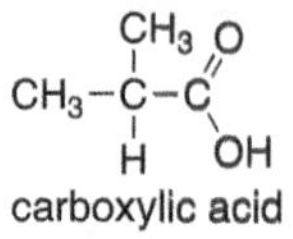

11.15 Both an amine and an amide have N atoms, but an amide has a C=O (a carbonyl group) bonded to the N, whereas an amine does not.

C = $CH_3-CH(CH_3)-CH_2NH_2$ — amine

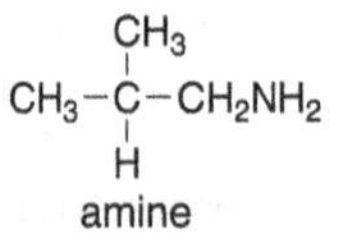

D = $CH_3CH_2CH_2C(=O)NH_2$ — amide

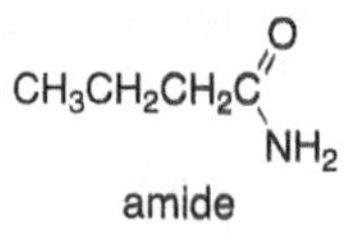

11.16 Both an ether and an ester contain OR groups, but an ester has a C=O (a carbonyl group) bonded to the OR group, whereas an ether does not.

E = $CH_3CH_2CH_2C(=O)OCH_3$ — ester

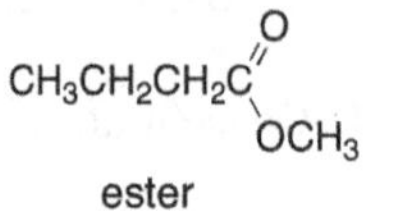

F = $CH_3-C(CH_3)_2-OCH_3$ — ether

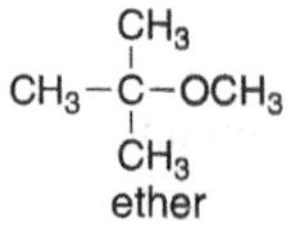

11.17 Identify the functional groups.

$H_2NC(=O)-CH_2-C_6H_4-O-CH_2CH(OH)CH_2NHCH(CH_3)_2$

amide; aromatic ring; ether; amine; hydroxyl group

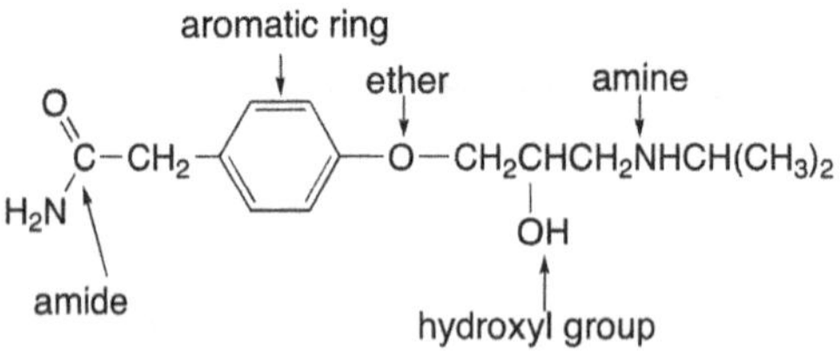

11.18 Use the common element colors shown on the inside back cover to convert the ball-and-stick model to a shorthand representation, and then identify the functional groups.

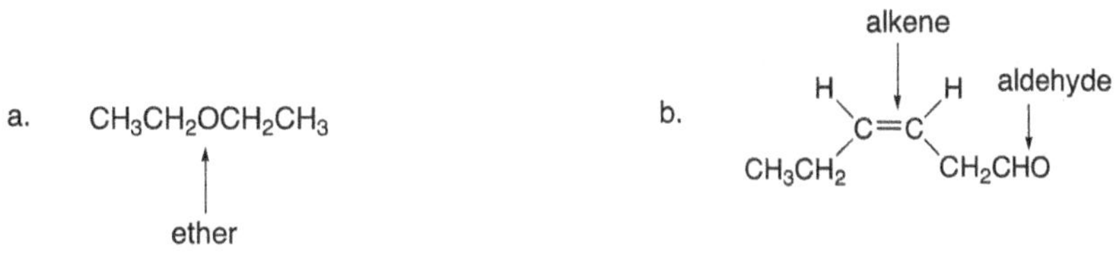

a. $CH_3CH_2OCH_2CH_3$ — ether

b. $CH_3CH_2(H)C=C(H)CH_2CHO$ — alkene; aldehyde

11.19 Draw the skeletal structure as in Sample Problem 11.9.

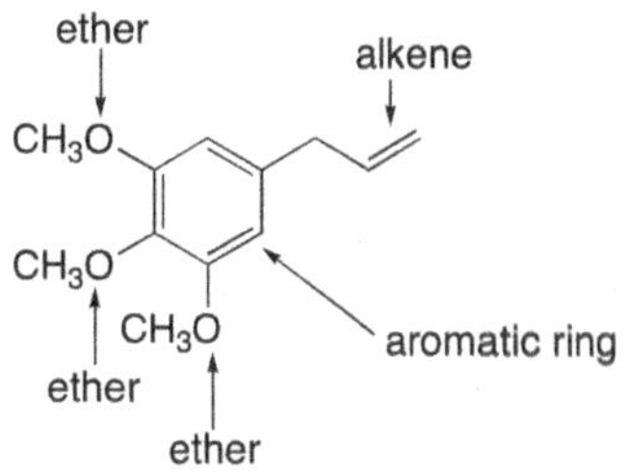

Labels: ether (CH_3O), ether (CH_3O), ether (CH_3O), alkene, aromatic ring

11.20 All of the bonds except C–C and C–H bonds are polar. The symbol δ^+ is given to the less electronegative atom, and the symbol δ^- is given to the more electronegative atom.

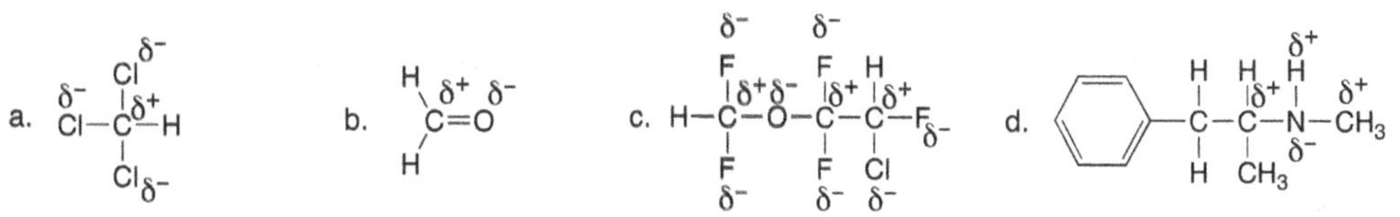

a. CCl_3H: C δ^+, each Cl δ^-

b. $H_2C=O$: C δ^+, O δ^-

c. $H-CF_2-O-CF_2-CHClF$: each C bonded to F/O/Cl δ^+; O δ^-; F δ^-; Cl δ^-

d. $C_6H_5-CH_2-CH(CH_3)-NH-CH_3$: C δ^+, N δ^-, H δ^+, CH_3 δ^+

11.21 With organic compounds that have more than one polar bond, the shape of the molecule determines the overall polarity.

- If the individual bond dipoles cancel in a molecule, the molecule is nonpolar.
- If the individual bond dipoles do not cancel, the molecule is polar.

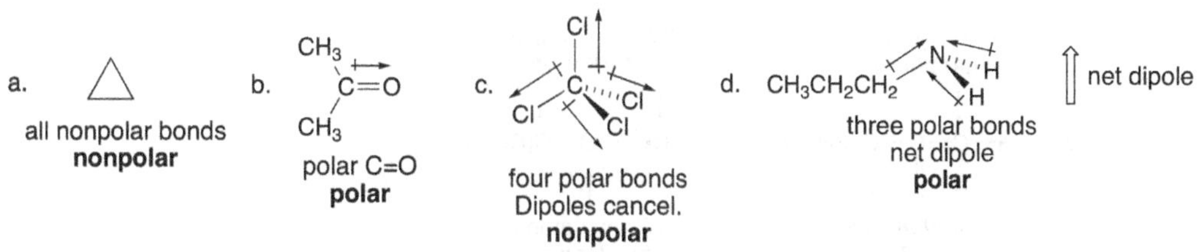

a. (cyclopropane) all nonpolar bonds **nonpolar**

b. $(CH_3)_2C=O$ polar C=O **polar**

c. CCl_4 four polar bonds Dipoles cancel. **nonpolar**

d. $CH_3CH_2CH_2NH_2$ three polar bonds net dipole **polar**

net dipole

11.22 Draw the molecule in three dimensions around the O atom to determine why dimethyl ether is polar.

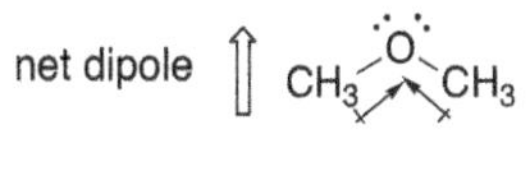

net dipole CH_3OCH_3

The two C–O bonds are polar.
The molecule is bent, so there is a net dipole.
Therefore, the molecule is polar.

11.23 Use the rule "like dissolves like" to determine if the compounds are soluble in water.
a. Octane is a hydrocarbon, and therefore nonpolar. This means it is **insoluble** in water.
b. Acetone is a small polar molecule due to the C=O. This means it is water **soluble.**
c. Stearic acid is a fatty acid with a polar functional group, but because it has a very long hydrocarbon chain, it is **insoluble** in water.

11.24 DDT is a nonpolar, fat-soluble compound that is soluble in tissues. Therefore, once it is ingested by birds, it stays in their tissues for long periods of time, making it detectable.

11.25 Niacin is soluble in water because it is a small molecule that contains many polar bonds (C–N, C=O, C–O, and O–H), making it polar overall.

11.26 Vitamin K_1 is not water soluble because it has relatively few polar bonds compared to the number of nonpolar bonds. This makes it soluble in organic solvents.

Solutions to Odd-Numbered End-of-Chapter Problems

11.27 Organic compounds contain the element carbon.

a. H_2SO_4—inorganic   b. Br_2—inorganic   c. C_5H_{12}—organic

11.29 Draw in all the H's and lone pairs as in Example 11.1. Each C and heteroatom must be surrounded by eight electrons.

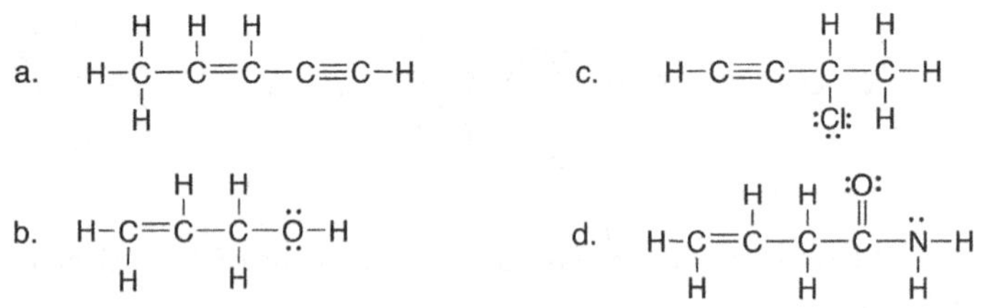

```
     H  H  H                          H  H
     |  |  |                          |  |
a.  H-C-C=C-C≡C-H          c.  H-C≡C-C-C-H
     |                                |  |
     H                               :Cl: H

      H  H                         H  H  :O:
      |  |  ..                     |  |  ||  ..
b.  H-C=C-C-O-H            d.  H-C=C-C-C-N-H
      |    |  ..                   |    |     |
      H    H                       H    H     H
```

11.31 To determine the shape around each atom, draw in all the lone pairs on any heteroatom and then count groups as in Example 11.2.

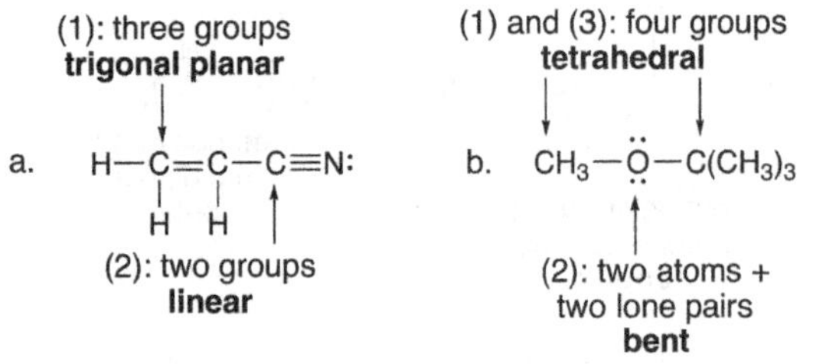

```
     (1): three groups                  (1) and (3): four groups
     trigonal planar                    tetrahedral
          ↓                              ↓          ↓
                                              ..
a.  H—C=C—C≡N:                   b.  CH3—O—C(CH3)3
      |  |   ↑                                ..
      H  H                                    ↑
     (2): two groups                    (2): two atoms +
     linear                             two lone pairs
                                        bent
```

11.33 To determine the bond angles, first use the steps in Example 11.2 and then use the values in Table 11.1.

a. 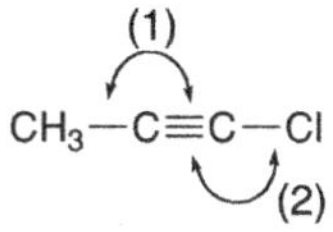 $CH_3-C\equiv C-Cl$

(1) and (2): two groups
linear
180°

b. 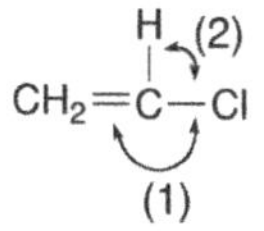 $CH_2=C(H)-Cl$

(1) and (2): three groups
trigonal planar
120°

c. 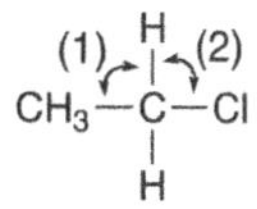 CH_3-CH_2-Cl

(1) and (2): four groups
tetrahedral
109.5°

11.35 To determine the shape around each atom, draw in all the lone pairs on any heteroatom and then count groups as in Example 11.3.

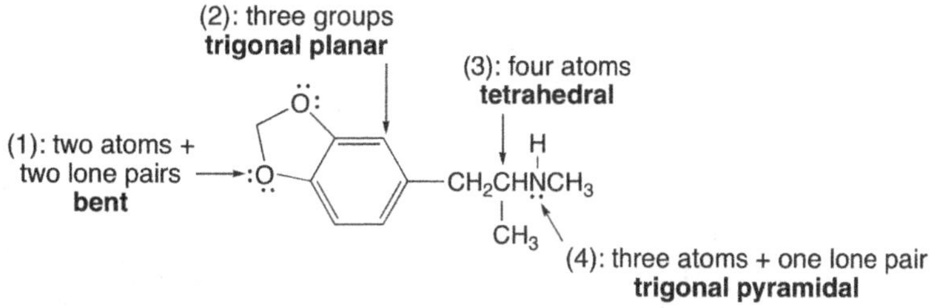

(1): two atoms + two lone pairs **bent**

(2): three groups **trigonal planar**

(3): four atoms **tetrahedral**

(4): three atoms + one lone pair **trigonal pyramidal**

11.37 Each C in benzene is surrounded by three groups, so each carbon is trigonal planar, giving 120° bond angles.

11.39 Convert each compound to a condensed structure as in Example 11.4.

a. 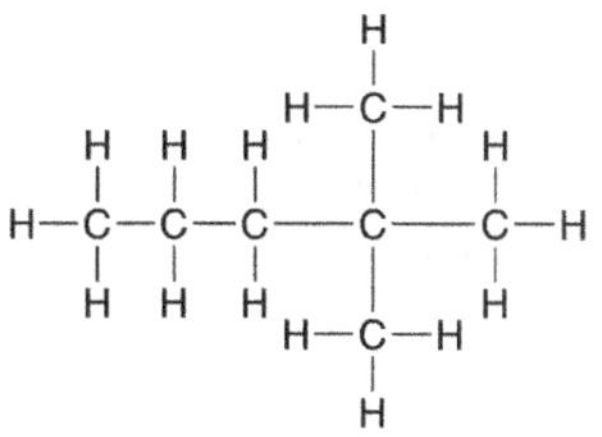 = $CH_3CH_2CH_2C(CH_3)_2CH_3$ = $CH_3(CH_2)_2C(CH_3)_3$

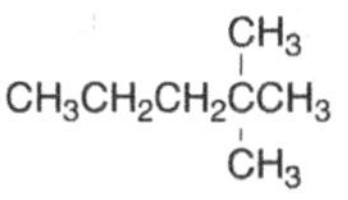

b. 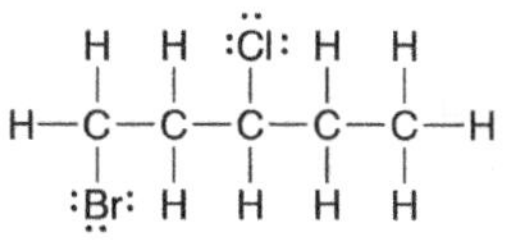 = $BrCH_2CH_2CH(Cl)CH_2CH_3$ = $Br(CH_2)_2CH(Cl)CH_2CH_3$

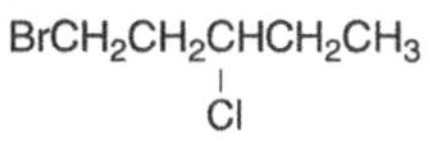

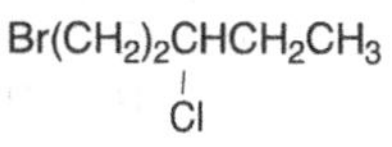

c. 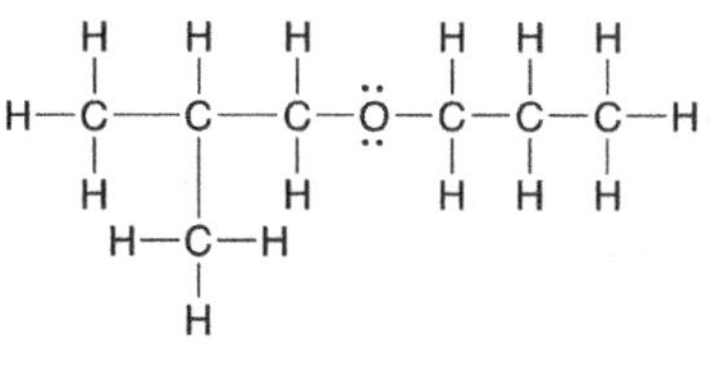 = $CH_3CH(CH_3)CH_2OCH_2CH_2CH_3$ = $(CH_3)_2CHCH_2O(CH_2)_2CH_3$

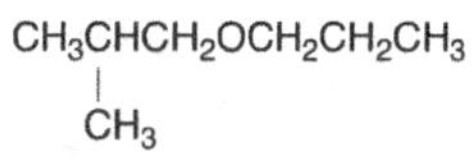

d. 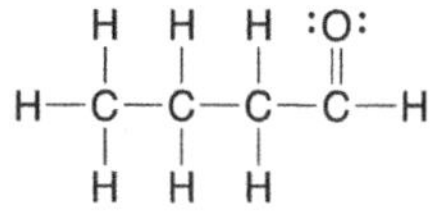 = $CH_3CH_2CH_2CHO$ = $CH_3(CH_2)_2CHO$

11.41 To convert each compound to a skeletal structure, remove all C's and H's bonded to C's. Remove the lone pairs.

a. 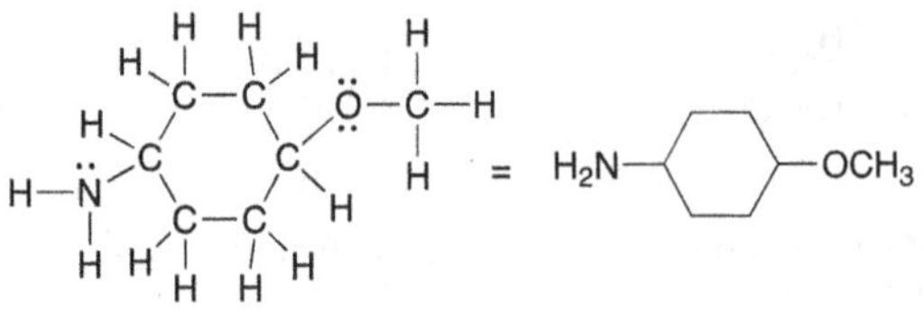 = H_2N–(cyclohexane)–OCH_3

b. 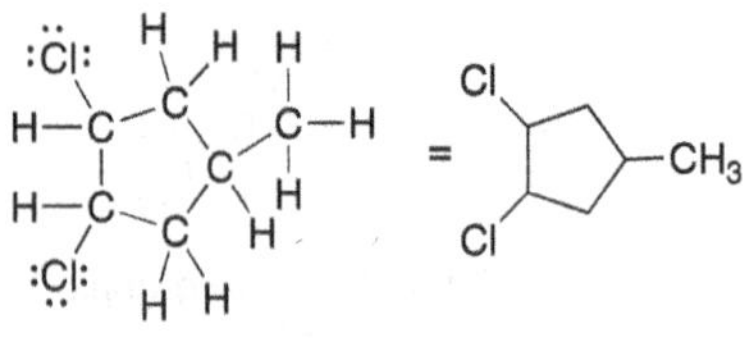 = Cl, Cl, CH_3

11.43 Work backwards to convert each shorthand structure to a complete structure.

a. $(CH_3)_2CH(CH_2)_6CH_3$ =

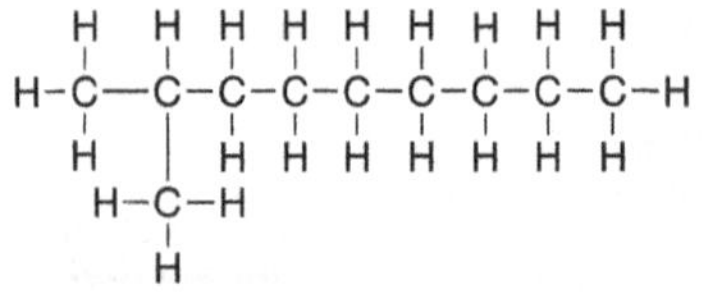

b. $(CH_3)_3COH$ =

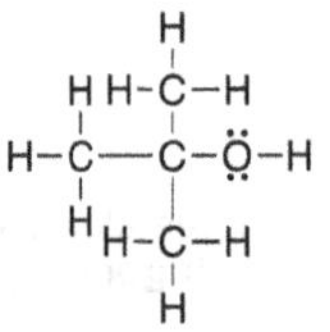

c. $CH_3CO_2(CH_2)_3CH_3$ =

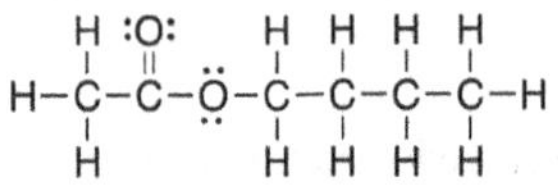

d. 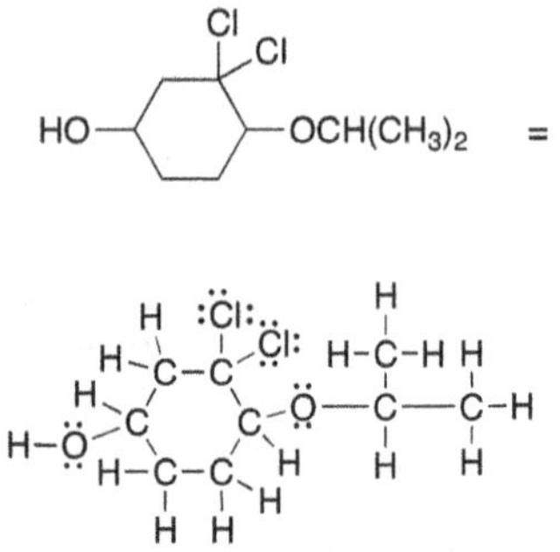 HO–(ring)(Cl)(Cl)–$OCH(CH_3)_2$ =

11.45

A a.

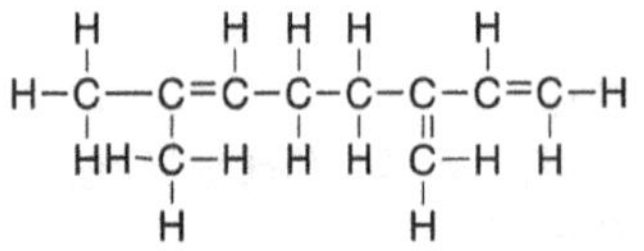

b. $(CH_3)_2C{=}CH(CH_2)_2CCH{=}CH_2$ (with $=CH_2$ on C)

or

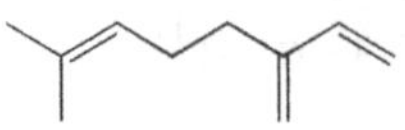

B a.

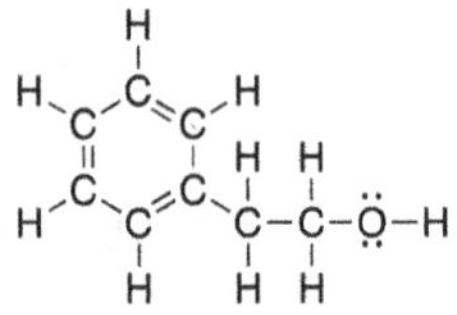

b. 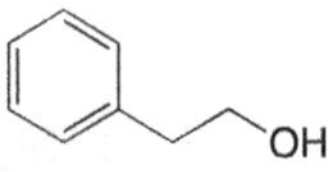 OH

11.47 Remember that C can have only four bonds.

a. $(CH_3)_3CHCH_2CH_3$ — Five bonds to C

b. $CH_3(CH_2)_4CH_3OH$ — Five bonds to C

c. $(CH_3)_2C{=}CH_3$ — Five bonds to C

d. 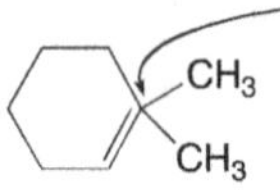 CH_3, CH_3 — Five bonds to C

e. Five bonds to C — CH_3, CH_3, CH_3

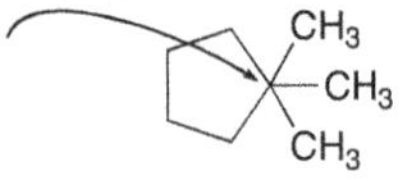

11.49 Convert the shorthand structures to complete structures by drawing in C's, H's, and lone pairs.

HO–(benzene ring)–CH_2CHCO_2H with NH_2 = H–Ö–C(ring, H's on each CH)–C–C(H)(H)–C(H)(:N–H, H)–C(=:O:)–Ö–H

11.51 Identify the functional groups in each molecule.

a. H(H)C=C(CH_3)–CH=C(H)(H) — alkenes

b. $CH_3C(=O)(CH_2)_4CH_3$ — ketone

c. $CH_3C(=O)OCH_2CH_2CH_3$ — ester

11.53 Identify the functional groups in each molecule.

a. $HC(=O)OCH_2CH(CH_3)_2$ — ester

b. $CH_2{=}CHCHCH_2–C{\equiv}C–C{\equiv}C–CH_2CH{=}CH(CH_5)_5CH_3$ with OH — alkene, hydroxyl group, two alkynes, alkene

c. CH_3O (ether), HO (hydroxyl group), aromatic ring, C(=O)H (aldehyde)

11.55 **Functional groups in dopamine:** two hydroxyl groups (OH groups with red spheres), an aromatic ring (six-membered ring with three C=C's), and one amine (NH_2 group with a blue sphere).

(HO)$_2$–(benzene ring)–$CH_2CH_2NH_2$ — hydroxyl groups, aromatic ring, amine

11.57

$(CH_3)_2C{=}CHCH_2CH_2C(CH_3){=}CHCH{=}O$ — alkene, alkene, aldehyde

11.59 Draw an organic compound that fits each set of criteria.

a. a hydrocarbon having molecular formula C_3H_4 that contains a triple bond: $HC{\equiv}CCH_3$
b. an alcohol containing three carbons: $CH_3CH_2CH_2OH$
c. an aldehyde containing three carbons: CH_3CH_2CHO
d. a ketone having molecular formula C_4H_8O: $CH_3CH_2C(=O)CH_3$

11.61 Draw the three five-carbon structures.

$CH_3CH_2CH_2CH_2CH_3$	$CH_3{-}CH_2{-}CH{=}CH{-}CH_3$	$CH_3{-}CH_2{-}C{\equiv}C{-}CH_3$
alkane	alkene	alkyne
C_5H_{12}	C_5H_{10}	C_5H_8

11.63 Solubility properties can help you differentiate unknown compounds. NaCl dissolves in water but not in dichloromethane, whereas cholesterol dissolves in dichloromethane but not in water. Therefore, when you test the solubility, the water-soluble compound is NaCl, and the water-insoluble (dichloromethane-soluble) compound is cholesterol.

11.65 Use the solubility rule, "Like dissolves like."
a. Pentane and water are not soluble, so the beaker contains two layers with the less dense liquid, pentane, on top.
b. Ethanol and pentane are both organic compounds, so the ethanol is soluble in pentane and a single layer forms.
c. 1-Decanol and pentane are both organic compounds, so the 1-decanol is soluble in pentane and a single layer forms.

11.67 Net dipoles and molecular geometry will determine whether a molecule is polar or nonpolar. When a molecule has polar bonds, you must determine whether the bond dipoles cancel or reinforce. To determine if they cancel or reinforce, you must know the shape of the molecule.

11.69 All of the bonds except C–C and C–H bonds are polar. The symbol δ^+ is given to the less electronegative atom, whereas the symbol δ^- is given to the more electronegative atom.

a. $CFCl_3$: F δ^-, each Cl δ^-, C δ^+

b. $H{-}O{-}CH_2CH(OH)CH_2{-}O{-}H$: each O δ^-, each H on O δ^+, each C bonded to O δ^+

11.71 With organic compounds that have more than one polar bond, the shape of the molecule determines the overall polarity.
- If the individual bond dipoles cancel in a molecule, the molecule is nonpolar.
- If the individual bond dipoles do not cancel, the molecule is polar.

a. 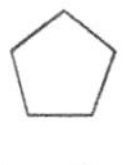 all nonpolar bonds **nonpolar**

b. 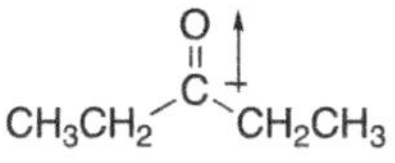 $CH_3CH_2C(=O)CH_2CH_3$ polar C=O **polar**

c. 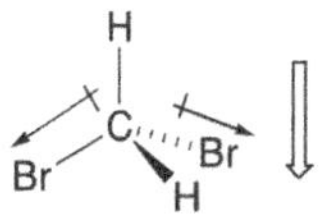 two polar bonds dipoles reinforce **polar** (net dipole)

11.73 Ethylene glycol is more water soluble than butane because ethylene glycol contains two polar hydroxyl groups capable of hydrogen bonding. Butane has no polar groups and cannot hydrogen bond, and is therefore sparingly soluble in water.

11.75 Sucrose has multiple OH groups, making it water soluble. 1-Dodecanol has a long nonpolar hydrocarbon chain and a single OH group, so it is insoluble in water.

11.77 Fat-soluble vitamins will persist in tissues and accumulate, whereas any excess of water-soluble vitamins will be excreted in the urine. Therefore, if large quantities of fat-soluble vitamins are ingested, they can build up to toxic levels because they are not excreted readily in the urine, whereas large quantities of water-soluble vitamins will be eliminated much more readily.

11.79 Vitamin E is a fat-soluble vitamin, making it soluble in organic solvents and insoluble in water.

11.81 a. C_3H_8: Yes, there are just enough H's to give each C four bonds.
b. C_3H_9: No, there are too many H's. The maximum number of H's for 3 C's is 8 H's.
c. C_3H_6: Yes, if you put the three C's in a ring, each C gets two H's and each C has four bonds.

Examples:

a. 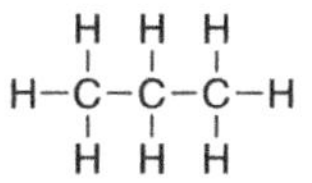 C_3H_8

b. 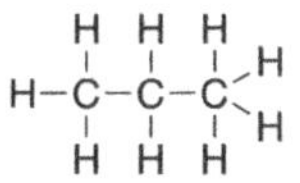 C_3H_9 One C has 5 bonds. invalid structure

c. 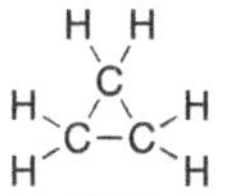 C_3H_6

11.83

a. $CH_3CH_2CH_2NHCH_3$

b. 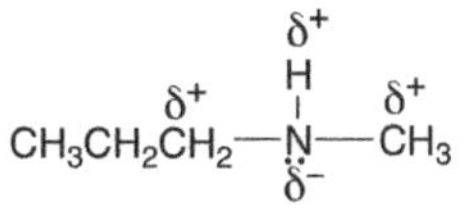 $CH_3CH_2CH_2—N(H)—CH_3$ with δ^+ on H, δ^+ on C atoms, δ^- on N

c. Four groups around N (three atoms and one lone pair); **trigonal pyramidal** around N.

d. This is a **polar molecule**. The bond dipoles don't cancel.

11.85

a. $C_9H_{11}NO_2$
b. Five lone pairs are drawn in on benzocaine.
c. There are seven trigonal planar carbons (*).
d. Benzocaine is water soluble due to the multiple polar bonds.
e. Polar bonds are drawn in bold.

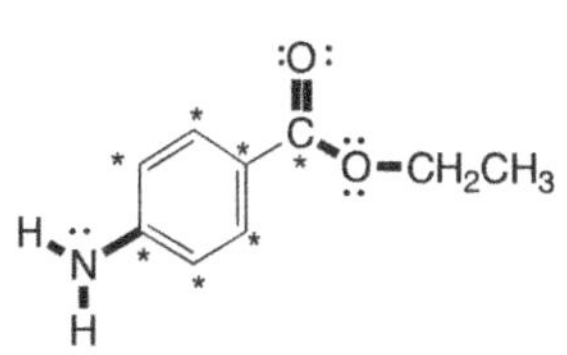

11.87 Answer each question about aldosterone.

a.

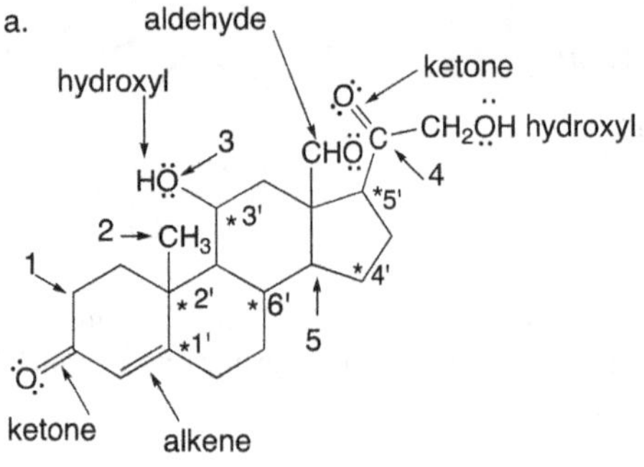

b. Each O needs two lone pairs.
c. There are 21 C's.
d. Number of H's at carbon:

1': 0 H's
2': 0 H's
3': 1 H
4': 2 H's
5': 1 H
6': 1 H

e. Shape at each carbon

1: Tetrahedral – 4 atoms
2: Tetrahedral – 4 atoms
3: Bent – 2 atoms, 2 lone pairs
4: Trigonal planar – 3 atoms
5: Tetrahedral – 4 atoms

f.

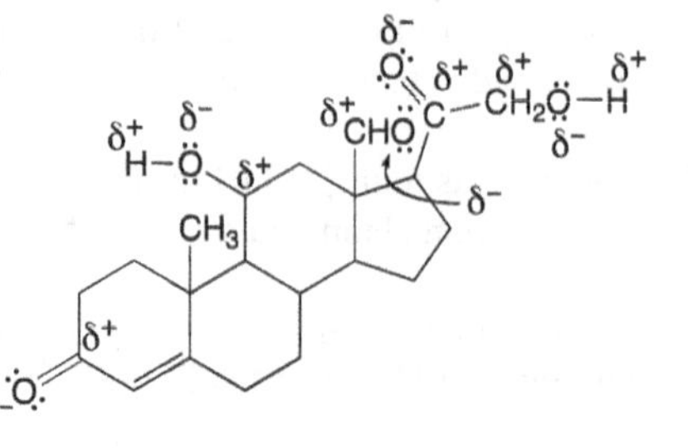

11.89 The waxy coating on cabbage leaves will prevent loss of water and keep the leaves crisp.

11.91 THC is insoluble in water. THC is fat soluble and will therefore persist in tissues for an extended period of time. Ethanol is water soluble and will be quickly excreted in the urine. As a result, THC is detectable many weeks after exposure, but ethanol is not.

Chapter 12 Alkanes

Chapter Review

[1] What are the characteristics of an alkane? (12.1, 12.2)

- Alkanes are hydrocarbons having only nonpolar C–C and C–H single bonds.
- Alkanes may be cyclic or acyclic. Acyclic alkanes (C_nH_{2n+2}) have no rings, whereas cycloalkanes (C_nH_{2n}) have one or more rings.

Acyclic alkanes

$CH_3CH_2CH_2CH_3$

Cycloalkanes

$-CH_3$

- A carbon is classified as 1°, 2°, 3°, or 4° by the number of carbons bonded to it. A 1° C is bonded to one carbon, a 2° C is bonded to two carbons, and so forth.

Classification of carbon atoms

1° C 2° C 3° C 4° C

Example

1° C 2° C 3° C 4° C

[2] What are constitutional isomers? (12.2)

- Isomers are different compounds with the same molecular formula.
- Constitutional isomers differ in the way the atoms are connected to each other. $CH_3CH_2CH_2CH_3$ and $HC(CH_3)_3$ are constitutional isomers because they both have molecular formula C_4H_{10}, but one compound has a chain of four carbons in a row and the other does not.

Constitutional isomers

[3] How are alkanes named? (12.4, 12.5)

- Alkanes are named using the IUPAC system of nomenclature. A name has three parts: the parent, the suffix, and the prefix. The parent indicates the number of carbons in the longest chain or the ring; the suffix indicates the functional group (*-ane* = alkane); and the prefix tells the number and location of substituents coming off the chain or ring.

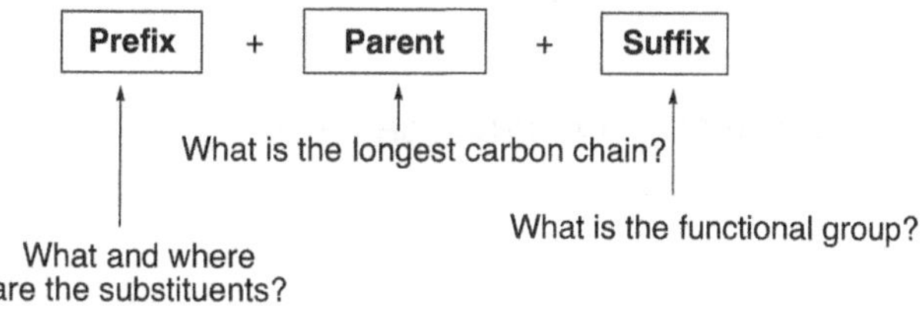
Prefix
+
Parent
+
Suffix
What is the longest carbon chain?
What is the functional group?
What and where
are the substituents?

- Alkyl groups are formed by removing one hydrogen from an alkane. Alkyl groups are named by changing the *-ane* ending of the parent alkane to the suffix *-yl.*

[4] Characterize the physical properties of alkanes. (12.7)

- Alkanes are nonpolar, so they have weak intermolecular forces, low melting points, and low boiling points.
- The melting points and boiling points of alkanes increase as the number of carbons increases due to increased surface area.

$CH_3CH_2CH_2CH_3$	$CH_3CH_2CH_2CH_2CH_3$	$CH_3CH_2CH_2CH_2CH_2CH_3$
butane	pentane	hexane
bp = –0.5 °C	bp = 36 °C	bp = 69 °C

Increasing surface area
Increasing boiling point

- Alkanes are insoluble in water.

[5] What are the products of the complete combustion and incomplete combustion of an alkane? (12.8)

- Alkanes burn in the presence of air. Combustion forms CO_2 and H_2O as products. Incomplete combustion forms CO and H_2O.

Complete combustion

$$2\ (CH_3)_3CCH_2CH(CH_3)_2 + 25\ O_2 \xrightarrow[\text{flame}]{} 16\ CO_2 + 18\ H_2O + \text{energy}$$

Incomplete combustion

$$2\ CH_4 + 3\ O_2 \xrightarrow[\text{flame}]{} 2\ CO + 4\ H_2O + \text{energy}$$

- Alkane combustion has increased the level of CO_2, a greenhouse gas, in the atmosphere.
- The incomplete combustion of alkanes forms carbon monoxide (CO), a toxin and air pollutant.

[6] What products are formed during the halogenation of an alkane? (12.9)

- Reaction of an alkane with a halogen (X_2) forms an alkyl halide (RX) and a hydrogen halide (HX). Halogenation is a substitution reaction in which a halogen atom X replaces a hydrogen atom.

Problem Solving

[1] Simple Alkanes (12.2)

Example 12.1 Are the compounds in each pair constitutional isomers or are they not isomers of each other?

a. $CH_3CH_2CH_2CH_3$ and $CH_3CH_2CH_2CH_2CH_3$

b. $CH_3CH_2CH_2CH_2CH_3$ and [cyclopentane]

Analysis

First compare molecular formulas; two compounds are isomers only if they have the same molecular formula. Then check how the atoms are connected to each other. Constitutional isomers have atoms bonded to different atoms.

Solution

a. $CH_3CH_2CH_2CH_3$ and $CH_3CH_2CH_2CH_2CH_3$

molecular formula C_4H_{10} — molecular formula C_5H_{12}

The two compounds have a different number of C's and H's, so they have different molecular formulas. Thus, they are *not* isomers of each other.

b. $CH_3CH_2CH_2CH_2CH_3$ and (cyclopentane)

molecular formula C_5H_{12} — molecular formula C_5H_{10}

The two compounds have the same number of C's but a different number of H's, so they have different molecular formulas. Thus, they are *not* isomers of each other.

Example 12.2 Draw two isomers with molecular formula C_4H_{10}.

Analysis

Since isomers are different compounds with the same molecular formula, we can draw one molecule with four carbons in a straight chain, but the other must have a one-carbon branch. Then add enough H's to give each C four bonds.

Solution

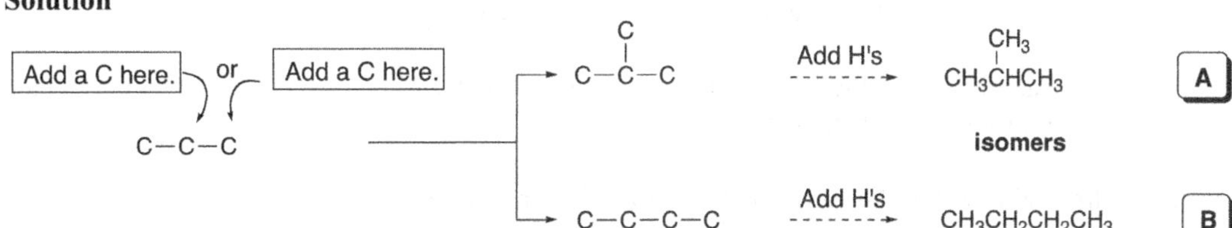

Compounds **A** and **B** are isomers because they have the same molecular formula but their atoms are connected together differently.

Example 12.3 Classify each carbon atom in the following molecule as 1°, 2°, 3°, or 4°.

$$\begin{array}{c} CH_3 \\ | \\ CH_3CH_2CHCHCH_3 \\ | \\ CH_3 \end{array}$$

Analysis

To classify a carbon atom, count the number of C's bonded to it. Draw a complete structure with all bonds and atoms to clarify the structure if necessary.

Solution

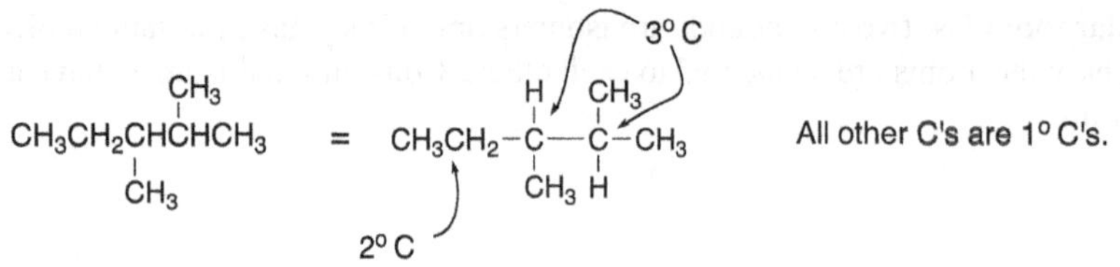

$$CH_3CH_2CH(CH_3)CH(CH_3)CH_3 = CH_3CH_2{-}\underset{CH_3}{\overset{H}{C}}{-}\underset{H}{\overset{CH_3}{C}}{-}CH_3$$

2° C, 3° C

All other C's are 1° C's.

[2] Alkane Nomenclature (12.4)

Example 12.4 Give the IUPAC name for the following compound.

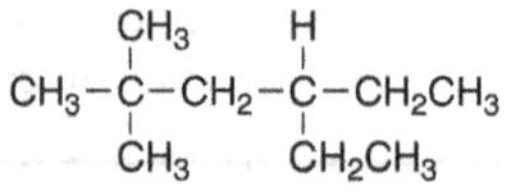

$$CH_3{-}\underset{CH_3}{\overset{CH_3}{C}}{-}CH_2{-}\underset{CH_2CH_3}{\overset{H}{C}}{-}CH_2CH_3$$

Analysis and Solution

[1] Name the parent and use the suffix *-ane* since the molecule is an alkane.

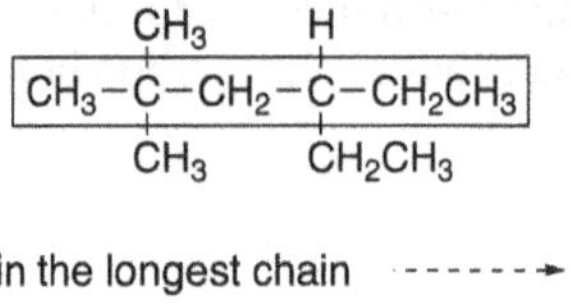

$$\boxed{CH_3{-}\underset{CH_3}{\overset{CH_3}{C}}{-}CH_2{-}\underset{CH_2CH_3}{\overset{H}{C}}{-}CH_2CH_3}$$

6 C's in the longest chain -------→ **hexane**

- Box in the atoms of the longest chain to clearly show which carbons are part of the longest chain and which carbons are substituents.

[2] Number the chain to give the first substituent the lower number.

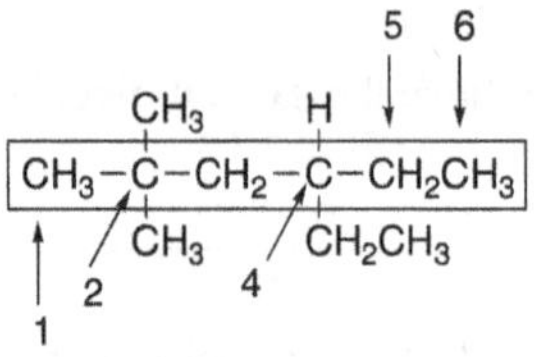

$$\boxed{CH_3{-}\underset{CH_3}{\overset{CH_3}{C}}{-}CH_2{-}\underset{CH_2CH_3}{\overset{H}{C}}{-}CH_2CH_3}$$

1, 2, 4, 5, 6

- Numbering from left to right puts the first substituent at C2.

[3] Name and number the substituents.

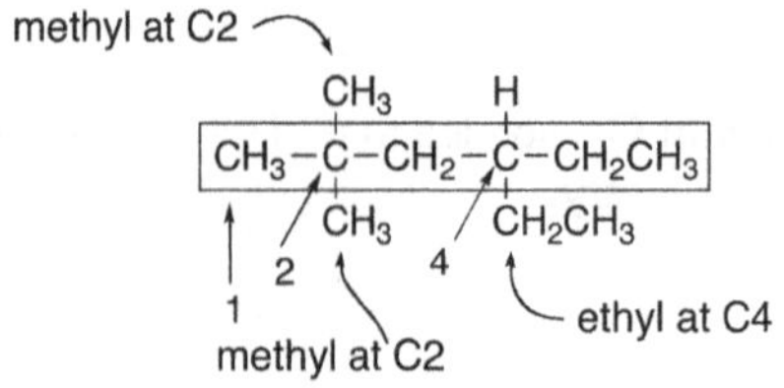

methyl at C2

$$\boxed{CH_3{-}\underset{CH_3}{\overset{CH_3}{C}}{-}CH_2{-}\underset{CH_2CH_3}{\overset{H}{C}}{-}CH_2CH_3}$$

1, 2, 4

methyl at C2

ethyl at C4

- This compound has three substituents: two methyl groups at C2 and an ethyl group at C4.

[4] Combine the parts.

- Write the name as one word and use the prefix di- before methyl since there are two methyl groups.
- Alphabetize the *e* of **e**thyl before the *m* of **m**ethyl. The prefix di- is ignored when alphabetizing.

methyl at C2

$$\mathrm{CH_3{-}\overset{CH_3}{\underset{CH_3}{C}}{-}CH_2{-}\overset{H}{\underset{CH_2CH_3}{C}}{-}CH_2CH_3}$$ ← **hexane**

1, 2, 4

methyl at C2 ethyl at C4

Answer: 4-ethyl-2,2-dimethylhexane

Example 12.5 Give the structure of the compound with the following IUPAC name: 2,2,3,4-tetramethylhexane.

Analysis

To derive a structure from a name, first look at the end of the name to find the parent name and the suffix. From the parent we know the number of C's in the longest chain, and the suffix tells us the functional group; the suffix *-ane* = an alkane. Then, number the carbon chain from either end and add the substituents. Finally, add enough H's to give each C four bonds.

Solution

2,2,3,4-Tetramethyl**hexane** has **hexane** (6 C's) as the longest chain and four methyl groups at carbons 2, 2, 3, and 4.

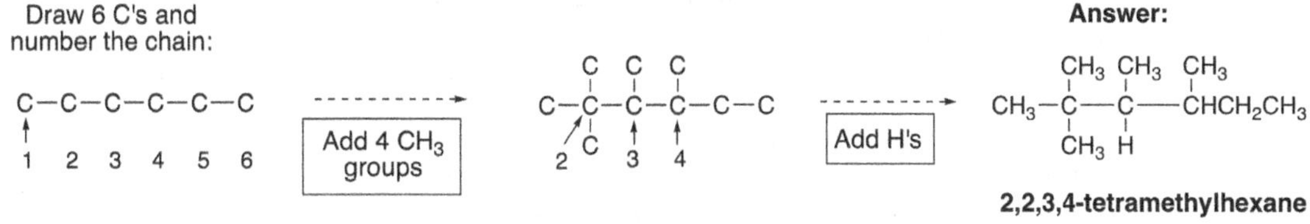

---

[3] Cycloalkanes (12.5)

Example 12.6 Give the IUPAC name for the following compound.

$CH_2CH_2CH_3$

$CH_2CH_2CH_2CH_3$

Analysis and Solution

[1] Name the ring. The ring has 6 C's so the molecule is named as a cyclohexane.

[2] Name and number the substituents.

- There are two substituents: $CH_3CH_2CH_2$– is a propyl group and $CH_3CH_2CH_2CH_2$– is a butyl group.
- Number to put the two groups at C1 and C2 (not C1 and C6).
- Place the butyl group at C1 because the *b* of **butyl** comes before the *p* of **propyl** in the alphabet.

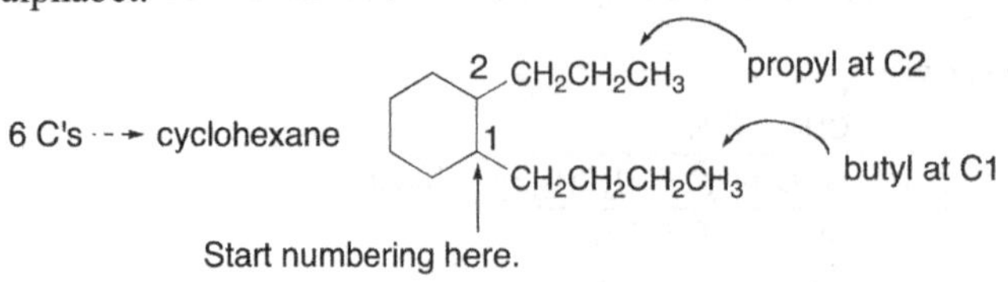

6 C's → cyclohexane

2 $CH_2CH_2CH_3$ propyl at C2

1 $CH_2CH_2CH_2CH_3$ butyl at C1

Start numbering here.

Answer: 1-butyl-2-propylcyclohexane

Self-Test

[1] Fill in the blank with one of the terms listed below.

Acyclic alkanes (12.1)	Constitutional isomers (12.2)	Oxidation (12.8)
Alkanes (12.1)	Cycloalkanes (12.1)	Pheromone (12.1)
Branched-chain alkane (12.2)	Greenhouse gas (12.8)	Reduction (12.8)
Combustion (12.8)	Isomers (12.2)	Saturated hydrocarbons (12.1)

1. A ______ is a chemical substance used for communication in a specific animal species, most commonly an insect population.
2. ______ contain carbons joined in one or more rings.
3. CO_2 is a ______ because it absorbs thermal energy that normally radiates from the earth's surface, and redirects it back to the surface.
4. ______ results in an increase in the number of C–O bonds or a decrease in the number of C–H bonds.
5. Alkanes that contain chains of carbon atoms but no rings are called ______.
6. ______ differ in the way the atoms are connected to each other.
7. A ______ is an alkane that contains one or more carbon substituents bonded to a carbon chain.
8. ______ results in a decrease in the number of C–O bonds or an increase in the number of C–H bonds.
9. Acyclic alkanes are also called ______ because they have the maximum number of hydrogen atoms per carbon.
10. ______ are hydrocarbons having only C–C and C–H single bonds.
11. ______ are two different compounds with the same molecular formula.
12. Alkanes undergo ______—that is, they burn in the presence of oxygen to form carbon dioxide (CO_2) and water.

[2] Label each carbon as 1°, 2°, 3°, or 4°.

$CH_3CH_2CH(CH_3)CH(CH_3)CH_2CH_3$ — 13. (arrow at the CH bearing a CH_3)

1,2-dimethylcyclohexane — 14., 15.

$(CH_3)_3CC(CH_3)_3$ — 16.

ethylcyclopropane — 17. (ring carbon), 18. (CH_3 of the ethyl group)

[3] Label each pair of compounds as constitutional isomers, identical molecules, or not isomers of each other.

a. constitutional isomers
b. identical molecules
c. not isomers

19. $CH_3CH_2CH_2$–$CH_2CH_2CH_3$ (drawn as CH_2–CH_2 with CH_3CH_2 and CH_2CH_3 below) and $CH_3(CH_2)_4CH_3$

20. CH_3–C(=O)–OCH_3 and CH_3CH_2–C(=O)–OH

21. $CH_3CH_2CH_2CH_3$ and methylcyclopropane (ring–CH_3)

[4] Pick the correct IUPAC name from each pair.

22. (a) 4-methylpentane or (b) 2-methylpentane
23. (a) 1,2-diethylcyclohexane or (b) 1,6-diethylcyclohexane
24. (a) 3-methylhexane or (b) 1,3-dimethylpentane
25. (a) 2-ethyl-2-methylcycloheptane or (b) 1-ethyl-1-methylcycloheptane

Answers to Self-Test

1. pheromone	6. Constitutional isomers	11. Isomers	16. 4°	21. c
2. Cycloalkanes	7. branched-chain alkane	12. combustion	17. 3°	22. b
3. greenhouse gas	8. Reduction	13. 3°	18. 1°	23. a
4. Oxidation	9. saturated hydrocarbons	14. 2°	19. b	24. a
5. acyclic alkanes	10. Alkanes	15. 3°	20. a	25. b

Solutions to In-Chapter Problems

12.1
- An acyclic alkane has the molecular formula $\mathbf{C_nH_{2n+2}}$, where n is the number of carbons it contains.
- A cycloalkane has two fewer H's than an acyclic alkane with the same number of carbons, so its general formula is $\mathbf{C_nH_{2n}}$.

a. An acyclic alkane with three carbons has eight hydrogen atoms; C_3H_8.
b. A cycloalkane with four carbons has eight hydrogen atoms; C_4H_8.
c. A cycloalkane with nine carbons has 18 hydrogen atoms; C_9H_{18}.
d. An acyclic alkane with seven carbons has 16 hydrogen atoms; C_7H_{16}.

12.2 An acyclic alkane has the molecular formula $\mathbf{C_nH_{2n+2}}$, whereas a cycloalkane has the molecular formula $\mathbf{C_nH_{2n}}$.

a. C_5H_{12}, $5 \times 2 + 2 = 12$: acyclic
b. C_4H_8, $4 \times 2 = 8$: cyclic
c. $C_{12}H_{24}$, $12 \times 2 = 24$: cyclic
d. $C_{10}H_{22}$, $10 \times 2 + 2 = 22$: acyclic

12.3 To determine if the compounds are constitutional isomers, first compare molecular formulas as in Example 12.1; two compounds are isomers only if they have the same molecular formula. Then check how the atoms are connected to each other. Constitutional isomers have atoms bonded to different atoms.

a. $CH_3CH_2CH_2CH_3$ and $CH_3CH_2CH_3$
C_4H_{10} C_3H_8
different molecular formulas
not isomers

b. $CH_3CH_2CH_2OH$ and $CH_3OCH_2CH_3$
C_3H_8O C_3H_8O
same molecular formula
different arrangement of atoms
constitutional isomers

c. 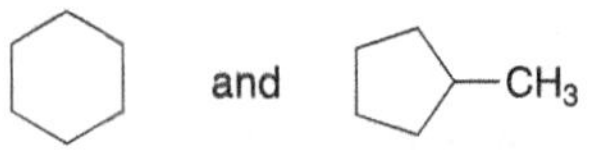

C_6H_{12} C_6H_{12}
same molecular formula
different arrangement of atoms
constitutional isomers

12.4 Draw two isomers as in Example 12.2. Since isomers are different compounds with the same molecular formula, add a one-carbon branch to two different carbons to form two different molecules. Then add enough H's to give each C four bonds.

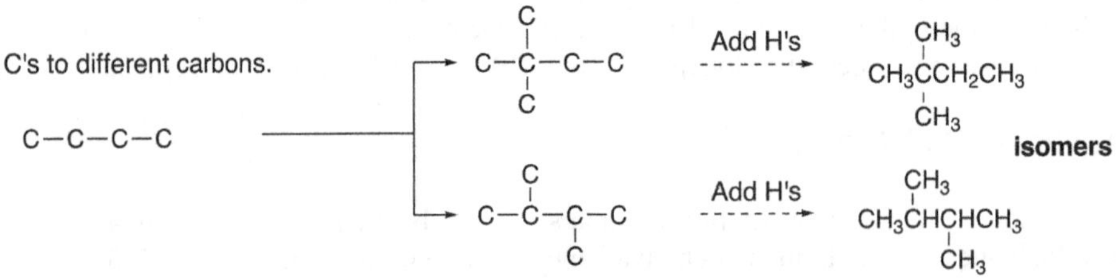

12.5 Isopentane has 4 C's in a row with a one-carbon branch.

a. 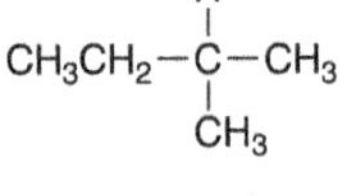
isopentane

b. 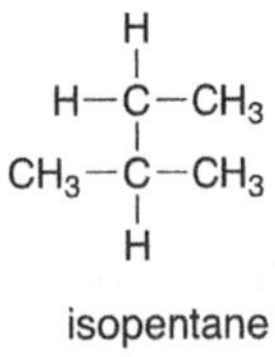
isopentane

c. $CH_3CH_2CH(CH_3)_2$
isopentane

d. 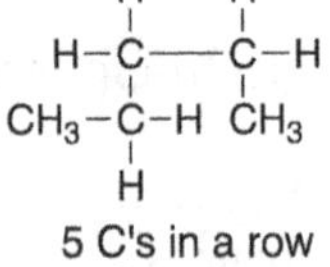
5 C's in a row
not isopentane

12.6 To classify a carbon atom, count the number of C's bonded to it as in Example 12.3. Draw a complete structure with all bonds and atoms to clarify the structure if necessary.

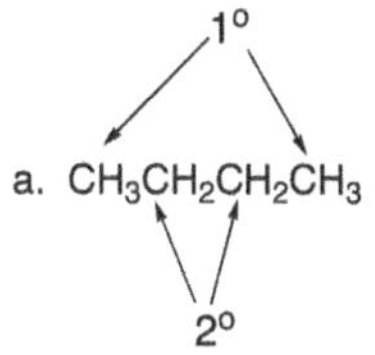

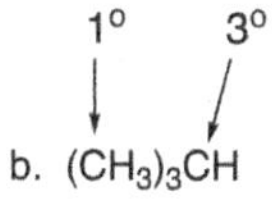

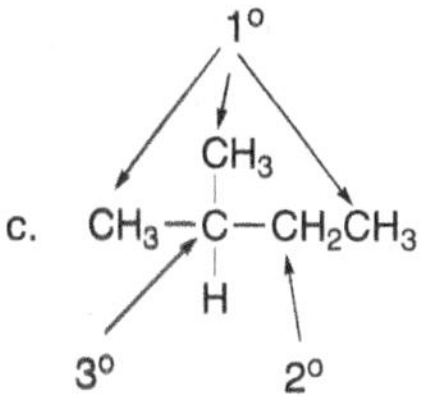

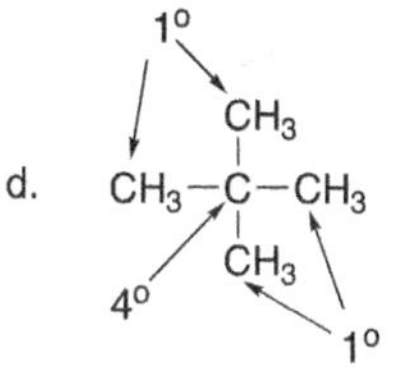

a. $CH_3CH_2CH_2CH_3$ (1°, 2°)

b. $(CH_3)_3CH$ (1°, 3°)

c. $CH_3-C(CH_3)(H)-CH_2CH_3$ (1°, 3°, 2°)

d. $CH_3-C(CH_3)_2-CH_3$ (1°, 4°)

12.7 Draw a structure to fit each description.

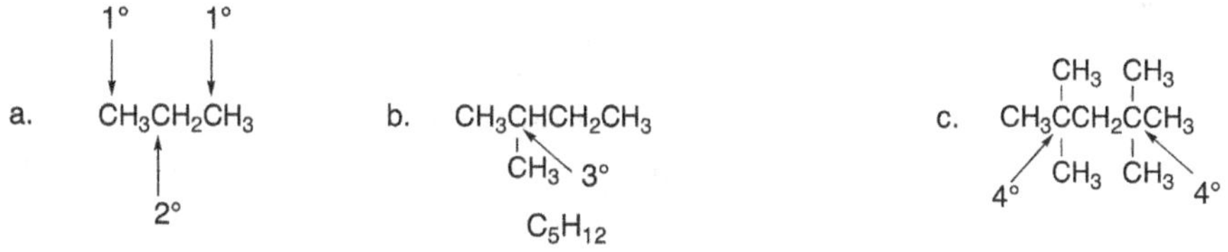

a. $CH_3CH_2CH_3$ (1°, 1°, 2°)

b. $CH_3CH(CH_3)CH_2CH_3$ (3°) C_5H_{12}

c. $CH_3C(CH_3)_2CH_2C(CH_3)_2CH_3$ (4°, 4°)

12.8 Draw a skeletal structure for each compound.

a. $CH_3CH_2CH_2CH_2CH_2CH_3$ =

b. $CH_3(CH_2)_5CH_3$ =

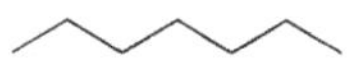

12.9 Convert each skeletal structure to a complete structure with all atoms and bond lines.

a.

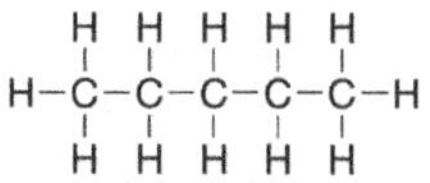

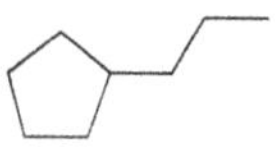

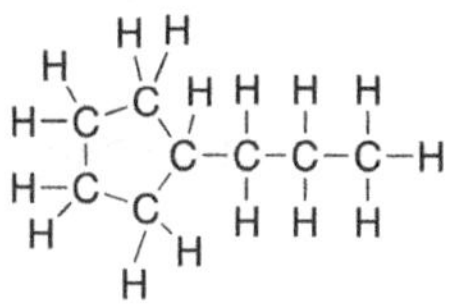

b.

c.

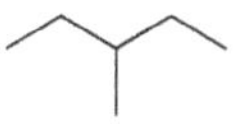

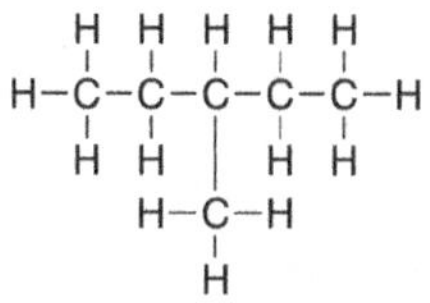

12.10 Re-drawing **A** and **B** as condensed structures shows that they both contain six carbons in a row, making them identical.

$CH_3CH_2CH_2CH_2CH_2CH_3$
A

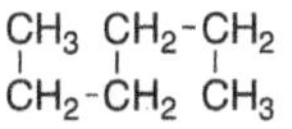

$CH_3\ CH_2-CH_2$
$CH_2-CH_2\ CH_3$
B

12.11 To give the IUPAC name for each compound, follow the steps in Example 12.4:
[1] Name the parent and use the suffix *-ane* since the molecule is an alkane.
[2] Number the chain to give the first substituent the lower number.
[3] Name and number the substituents.
[4] Combine the parts.

a. $CH_3CH_2CHCH_2CH_3$ with CH_3 substituent → numbered 1, 2, 3, 5 → $CH_3CH_2CHCH_2CH_3$ (CH_3 at 3) → **3-methylpentane**

5 C's in the longest chain
pentane

methyl at C3

b. $CH_3CH_2CH_2-C(H)(CH_3)-C(H)(CH_2CH_2CH_3)-CH_2CH_2CH_2CH_3$ → numbered 1, 2, 4, 5

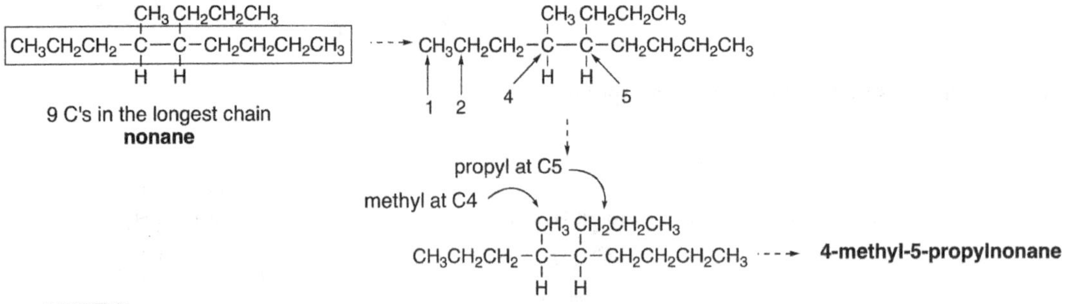

9 C's in the longest chain
nonane

propyl at C5
methyl at C4

→ **4-methyl-5-propylnonane**

c. $H-C(CH_2CH_3)(CH_3)-CH_2-CHCH_3$ with CH_3 → numbered 4, 2, 1 → **2,4-dimethylhexane**

6 C's in the longest chain
hexane

methyls at C2 and C4

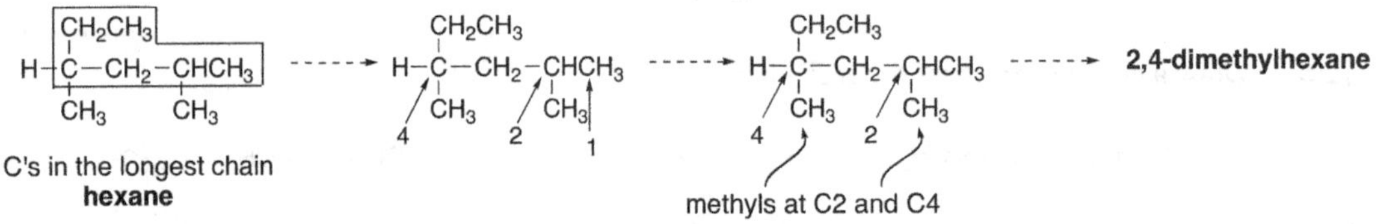

12.12 To give the IUPAC name for each compound follow the steps in Example 12.4.

a. $(CH_3)_2CHCH(CH_3)_2$ → $CH_3-C(H)(CH_3)-C(H)(CH_3)-CH_3$ → numbered 1, 2, 4 → **2,3-dimethylbutane**

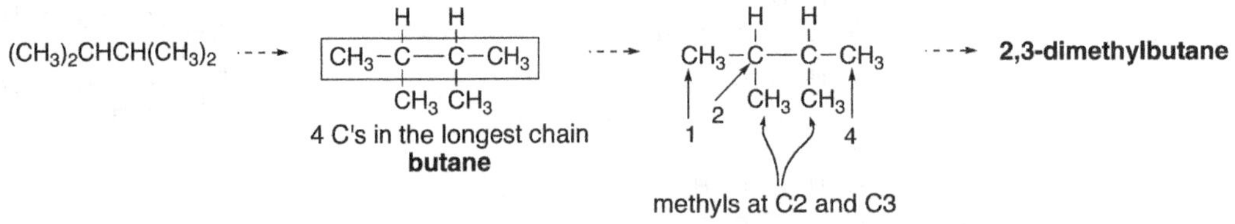

4 C's in the longest chain
butane

methyls at C2 and C3

b. $CH_3CH_2CH_2CHCH_2C(H)(CH_2CH_2CH_2CH_3)-H$ with CH_3CH_2 → numbered 1, 2, 4, 6 → **4-ethyldecane**

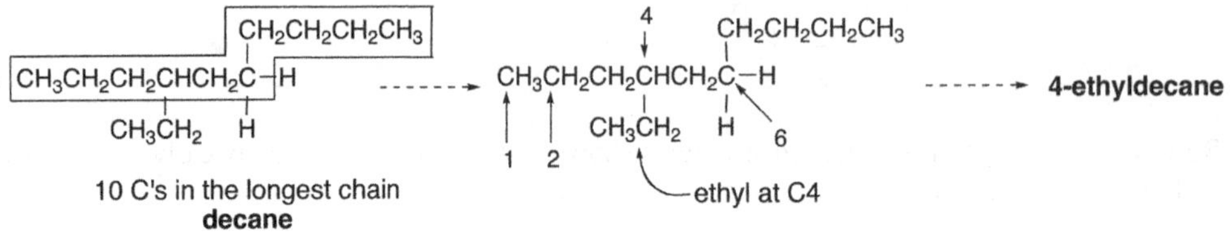

10 C's in the longest chain
decane

ethyl at C4

c. $CH_3CH_2CH_2CH_2-C(CH_3)(CH_2CH_3)-C(H)(CH_3)-CH_2CH_3$ → numbered 4, 3, 2, 1 → **4-ethyl-3,4-dimethyloctane**

methyl at C4

8 C's in the longest chain
octane

ethyl at C4
methyl at C3

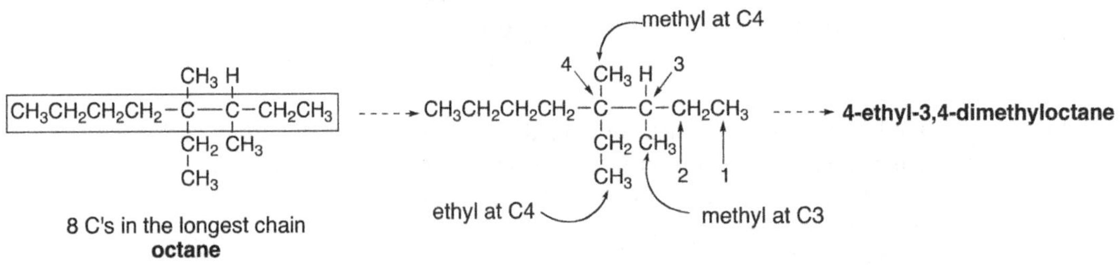

12.13 To draw the structure corresponding to each IUPAC name, follow the steps in Example 12.5.

a. 3-methyl**hexane**

Draw 6 C's and number the chain:

C–C–C–C–C–C (1 2 3 4 5 6) → [Add 1 CH_3 group] → C–C–C(C)–C–C–C (2, 3) → [Add H's] → $CH_3CH_2CHCH_2CH_2CH_3$ with CH_3 on C3

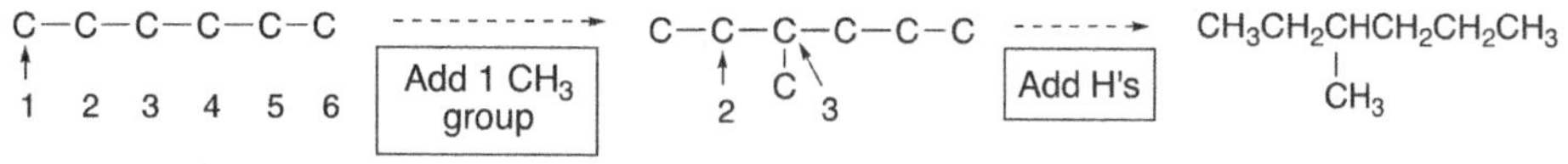

b. 3,3-dimethyl**pentane**

Draw 5 C's and number the chain:

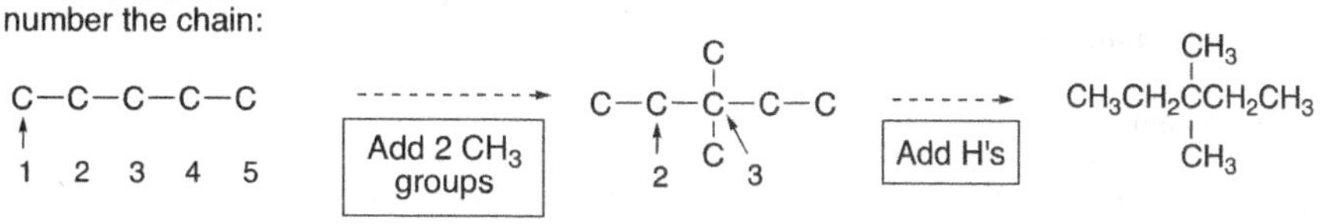

C–C–C–C–C (1 2 3 4 5) → [Add 2 CH_3 groups] → C–C–C(C)(C)–C–C (2, 3) → [Add H's] → $CH_3CH_2CCH_2CH_3$ with two CH_3 on C3

c. 3,5,5-trimethyl**octane**

Draw 8 C's and number the chain:

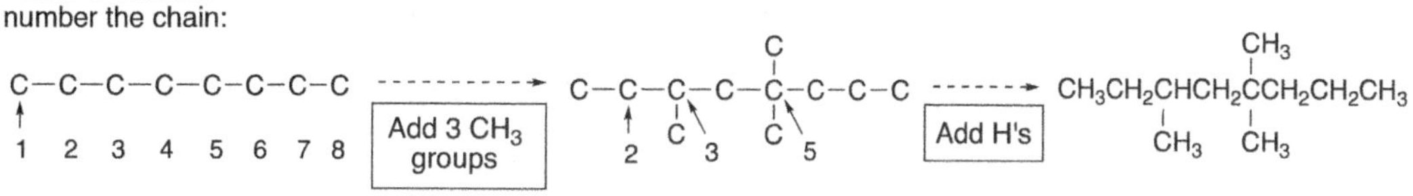

C–C–C–C–C–C–C–C (1 2 3 4 5 6 7 8) → [Add 3 CH_3 groups] → C–C–C(C)–C–C(C)(C)–C–C–C (2, 3, 5) → [Add H's] → $CH_3CH_2CHCH_2CCH_2CH_2CH_3$ with CH_3 on C3 and two CH_3 on C5

d. 3-ethyl-4-methyl**hexane**

Draw 6 C's and number the chain:

C–C–C–C–C–C (1 2 3 4 5 6) → [Add 2 alkyl groups] → C–C–C(C–C)–C(C)–C–C (2, 3, 4) → [Add H's] → $CH_3CH_2CHCHCH_2CH_3$ with CH_2CH_3 on C3 and CH_3 on C4

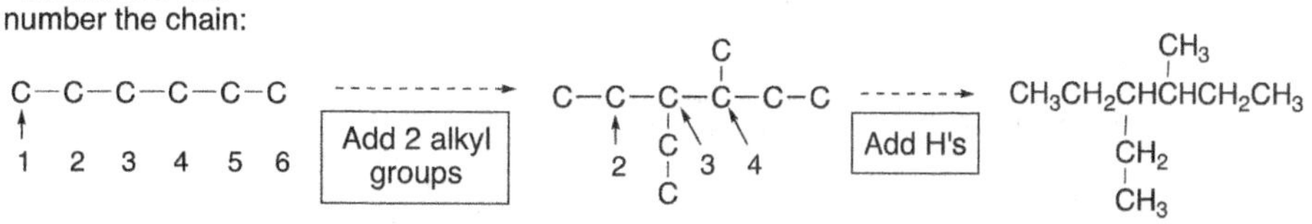

12.14 To draw the structure corresponding to each IUPAC name, follow the steps in Example 12.5.

a. 2,2-dimethyl**butane**

Draw 4 C's and number the chain:

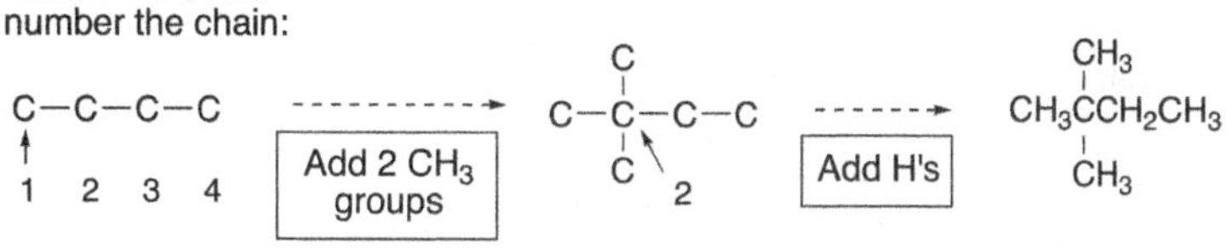

C–C–C–C (1 2 3 4) → [Add 2 CH_3 groups] → C–C(C)(C)–C–C (2) → [Add H's] → $CH_3CCH_2CH_3$ with two CH_3 on C2

b. 6-butyl-3-methyl**decane**

Draw 10 C's and number the chain:

C–C–C–C–C–C–C–C–C–C
1 2 3 4 5 6 7 8 9 10

Add 2 alkyl groups

C–C–C–C–C–C–C–C–C–C with C on carbon 3 and C–C–C–C on carbon 6

Add H's

$CH_3CH_2CHCH_2CH_2CHCH_2CH_2CH_2CH_3$ with CH_3 on C3 and $CH_2CH_2CH_2CH_3$ on C6

c. 4,4,5,5-tetramethyl**nonane**

Draw 9 C's and number the chain:

C–C–C–C–C–C–C–C–C
1 2 3 4 5 6 7 8 9

Add 4 CH_3 groups

C–C–C–C–C–C–C–C–C with two C's on carbon 4 and two C's on carbon 5

Add H's

$CH_3CH_2CH_2C—CCH_2CH_2CH_2CH_3$ with two CH_3 groups on each central C

d. 3-ethyl-5-propyl**nonane**

Draw 9 C's and number the chain:

C–C–C–C–C–C–C–C–C
1 2 3 4 5 6 7 8 9

Add 2 alkyl groups

C–C–C–C–C–C–C–C–C with C–C on carbon 3 and C–C–C on carbon 5

Add H's

$CH_3CH_2CHCH_2CHCH_2CH_2CH_2CH_3$ with CH_2CH_3 on C3 and $CH_2CH_2CH_3$ on C5

12.15 Label the functional groups.

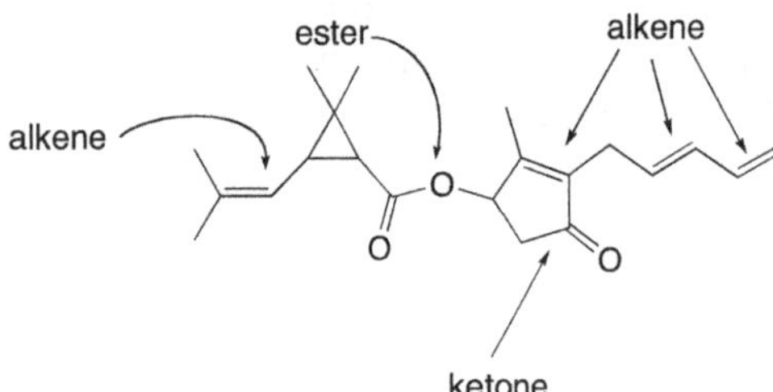

12.16 Vitamin D_3 has many nonpolar C–C and C–H bonds, which makes it water insoluble but fat soluble.

12.17 To give the IUPAC name for each compound, follow the steps in Example 12.6:
[1] Name the ring.
[2] Name and number the substituents.

a.

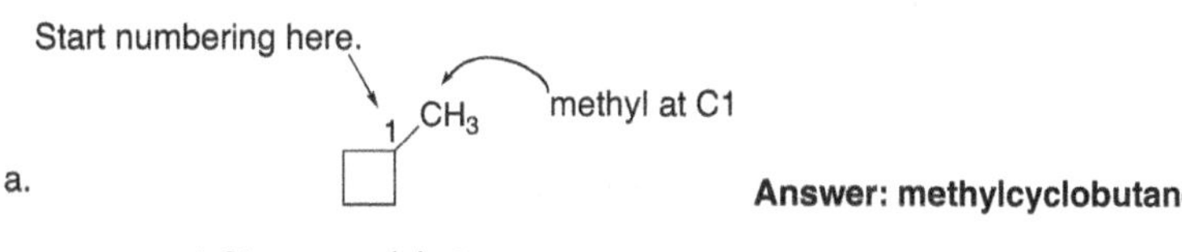

Answer: methylcyclobutane

4 C's → cyclobutane

b.

Start numbering here.
1
CH_3
CH_3
two methyls at C1

Answer: 1,1-dimethylcyclohexane

6 C's → cyclohexane

c.

Start numbering here.
CH_3CH_2 1 3 $CH_2CH_2CH_3$
propyl at C3

Answer: 1-ethyl-3-propylcyclopentane

5 C's → cyclopentane
ethyl at C1

d.

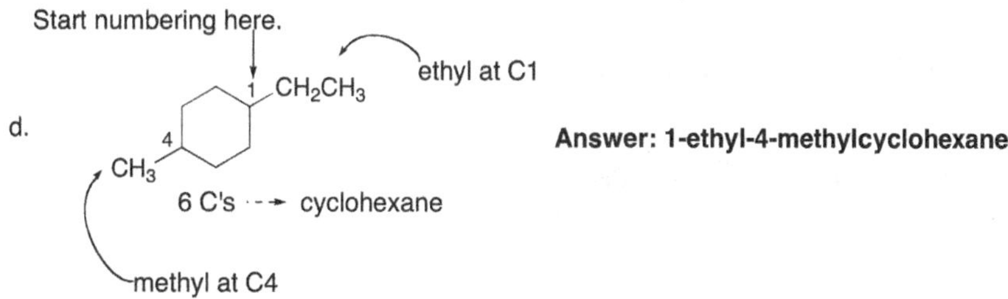

Answer: 1-ethyl-4-methylcyclohexane

12.18 Give the structure corresponding to each IUPAC name.

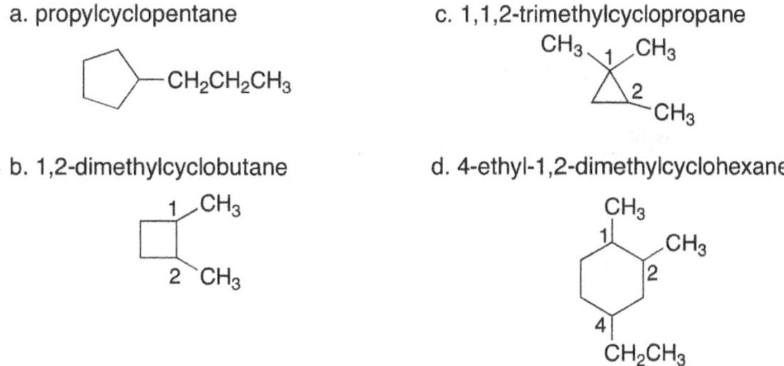

12.19 The melting points and boiling points of alkanes increase as the number of carbons increases.

a. Highest boiling point, most carbons: decane
b. Lowest boiling point, fewest carbons: pentane
c. Highest melting point, most carbons: decane
d. Lowest melting point, fewest carbons: pentane

12.20 Vaseline is a complex mixture of hydrocarbons, and is nonpolar, so it is insoluble in a polar solvent like water. In a weakly polar solvent like dichloromethane, it is soluble.

12.21 Elastol is a high molecular weight alkane, making it nonpolar. By the solubility rule, "like dissolves like," nonpolar oil has similar intermolecular forces to nonpolar polyisobutylene, so it is soluble in Elastol.

12.22 Write a balanced equation for each reaction. Combustion reactions release CO_2 and H_2O.

a. $CH_3CH_2CH_3 + 5\,O_2 \xrightarrow{\text{flame}} 3\,CO_2 + 4\,H_2O$

b. $2\,CH_3CH_2CH_2CH_3 + 13\,O_2 \xrightarrow{\text{flame}} 8\,CO_2 + 10\,H_2O$

12.23 Write a balanced equation for the reaction. Incomplete combustion reactions release CO and H_2O.

$2\,CH_3CH_3 + 5\,O_2 \longrightarrow 4\,CO + 6\,H_2O$

12.24 Halogenation replaces a hydrogen atom by a halogen.

a. $CH_3-C(CH_3)_2-CH_3 + Br_2 \xrightarrow{\text{light}} CH_3-C(CH_3)_2-CH_2Br + HBr$

b. cyclopropane $+ Cl_2 \xrightarrow{\text{heat}}$ cyclopropyl–Cl $+ HCl$

12.25 Butane contains two different types of C's (1° and 2°), so two different alkyl halides can form.

$CH_3CH_2CH_2CH_2\text{–}\boxed{H} + Cl_2 \xrightarrow{\text{light}} CH_3CH_2CH_2CH_2\text{–}\boxed{Cl} + HCl$

or

$CH_3CH_2CH(\boxed{H})CH_3 + Cl_2 \xrightarrow{\text{light}} CH_3CH_2CH(\boxed{Cl})CH_3 + HCl$

12.26 Work backwards and replace the halogen atom of the alkyl halide by a hydrogen to identify the alkane needed to make it.

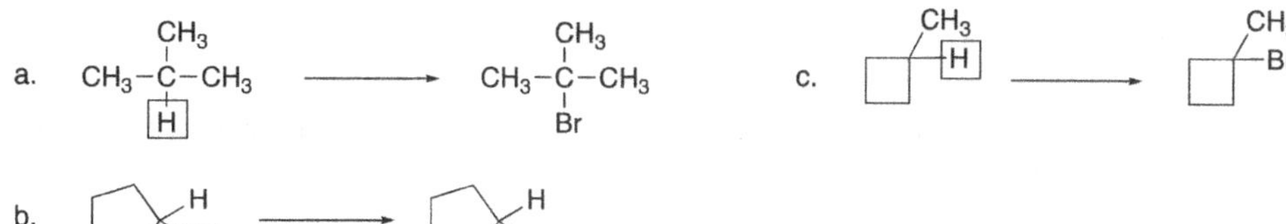

Solutions to Odd-Numbered End-of-Chapter Problems

12.27 Use the formula C_nH_{2n+2} to determine the number of hydrogen atoms.

$(31 \times 2) + 2 = 64$ H's

12.29 Use the formula C_nH_{2n+2} to determine the number of hydrogen atoms.

A C_{26} alkane has 54 H's $[(26 \times 2) + 2 = 54]$.
A C_{27} alkane has 56 H's $[(27 \times 2) + 2 = 56]$.
A C_{28} alkane has 58 H's $[(28 \times 2) + 2 = 58]$.
A C_{29} alkane has 60 H's $[(29 \times 2) + 2 = 60]$.
A C_{30} alkane has 62 H's $[(30 \times 2) + 2 = 62]$.

12.31 To classify a carbon atom, count the number of C's bonded to it as in Example 12.3. Draw a complete structure with all bonds and atoms to clarify the structure if necessary.

a. $CH_3(CH_2)_3CH_3$ — 1° (both terminal CH_3), 2° (CH_2)

b. $CH_3CH_2CHCHCH_3$ with CH_3 substituents — 1°, 2°, 3°, 1°

c. $(CH_3)_3CC(CH_3)_3$ — 1°, 1°, 4°

d. 1°, 3°, 4°, CH_3, CH_3, CH_3 — 1°, 2°

12.33 To determine if the compounds are isomers, first compare molecular formulas as in Example 12.1; two compounds are isomers only if they have the same molecular formula. Then check how the atoms are connected to each other. Constitutional isomers have atoms bonded to different atoms.

a. 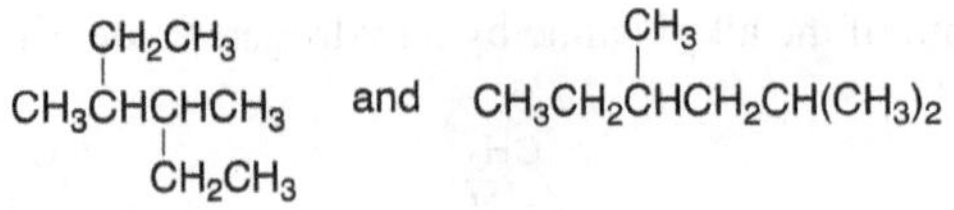 $CH_3CH(CH_2CH_3)CH(CH_2CH_3)CH_3$ and $CH_3CH_2CH(CH_3)CH_2CH(CH_3)_2$

C_8H_{18} C_8H_{18}

same molecular formula
constitutional isomers

b. 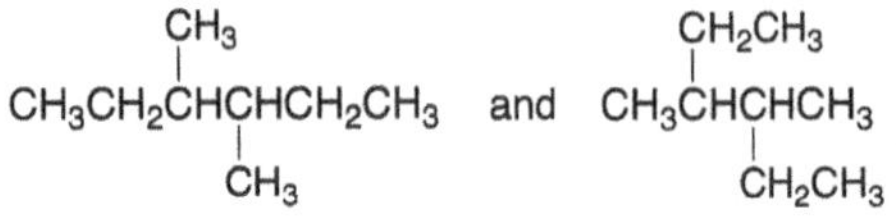 $CH_3CH_2CH(CH_3)CH(CH_3)CH_2CH_3$ and $CH_3CH(CH_2CH_3)CH(CH_3)CH_2CH_3$

C_8H_{18} C_8H_{18}

same molecular formula
identical connectivity
identical

c. 1,1-dimethylcyclopropane (ring with CH_3, CH_3) and cyclopropane–CH_2CH_3

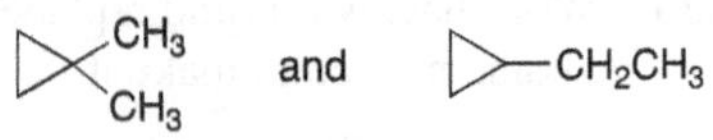

C_5H_{10} C_5H_{10}

same molecular formula
constitutional isomers

d. 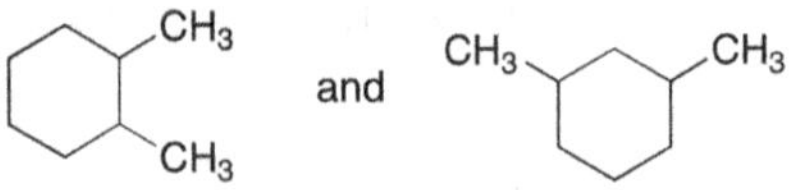 cyclohexane with CH_3 and CH_3 on adjacent carbons and cyclohexane with CH_3 and CH_3 (1,3)

C_8H_{16} C_8H_{16}

same molecular formula
constitutional isomers

12.35 **A, B,** and **C** all have molecular formula C_8H_{18}. **A** and **B** both have a six-carbon long chain and an ethyl group bonded to C3 (3-ethylhexane). **C** has a seven-carbon long chain with a methyl group bonded to C3 (3-methylheptane).

a. **A** and **B:** identical
b. **A** and **C:** constitutional isomers
c. **B** and **C:** constitutional isomers

12.37 To determine if the molecules are constitutional isomers or identical, determine if the atoms are connected to each other in the same way.

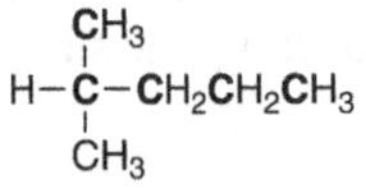

$H{-}C(CH_3)(CH_3){-}CH_2CH_2CH_3$ 5 C's in a row
CH_3 bonded to second C

a. 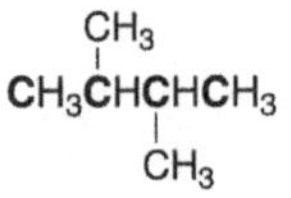 $CH_3CH(CH_3)CH(CH_3)CH_3$
4 C's in a row
constitutional isomers

b. 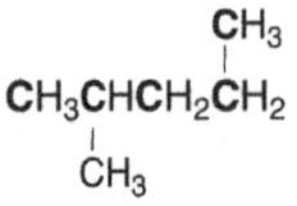 $CH_3CH(CH_3)CH_2CH_2CH_3$
5 C's in a row
identical

c. 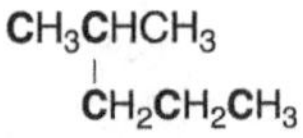 $CH_3CH(CH_3)CH_2CH_2CH_3$
5 C's in a row
identical

d. 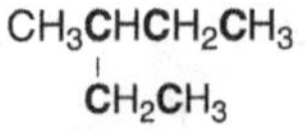 $CH_3CH(CH_2CH_3)CH_2CH_3$
5 C's in a row
CH_3 bonded to middle C
constitutional isomers

e.  (skeletal structure)
5 C's in a row
identical

12.39 Draw structures that fit each description.

a. 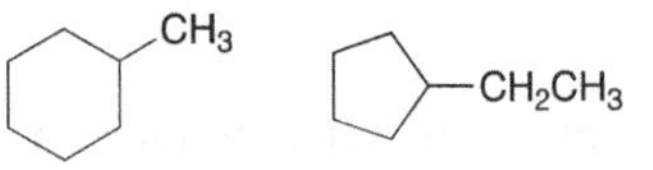 methylcyclohexane (CH_3) and ethylcyclopentane ($-CH_2CH_3$)

cycloalkanes
constitutional isomers
C_7H_{14}

b. $CH_3CH_2CH_2CH_2CH_2OH$ $CH_3OCH_2CH_2CH_2CH_3$

alcohol ether

constitutional isomers
$C_5H_{12}O$

c. $CH_3CH_2CH_2Cl$ $CH_3CH(Cl)CH_3$

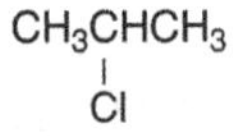

constitutional isomers
C_3H_7Cl

12.41 Draw all of the structures that satisfy the criteria.

$CH_3CH(CH_3)CH_2CH_2CH_2CH_2CH_3$ $CH_3CH_2CH(CH_3)CH_2CH_2CH_2CH_3$ $CH_3CH_2CH_2CH(CH_3)CH_2CH_2CH_3$

molecular formula C_8H_{18}
7 C's in the longest chain
one CH_3 group bonded to each chain

12.43 Draw all of the structures that satisfy the criteria.

molecular formula C_3H_6O:

cyclopropyl–OH	$CH_3C(=O)CH_3$	epoxide–CH_3
alcohol	ketone	cyclic ether

12.45 Give the IUPAC name for each molecule.

$CH_3CH_2CH_2CH_2CH_2CH_3$	$CH_3CH_2CH_2-CH(CH_3)-CH_3$	$CH_3CH_2-CH(CH_3)-CH_2CH_2$	$CH_3CH_2-C(CH_3)_2-CH_3$	$CH_3-CH(CH_3)-CH(CH_3)-CH_3$
hexane	2-methylpentane	3-methylpentane	2,2-dimethylbutane	2,3-dimethylbutane

12.47 Give the IUPAC name for each molecule.

a.

1 4
$CH_3CH_2CH_2-CH(CH_3)-CH_2CH_2CH_3$

7 C's = heptane
4-methylheptane

b.

cyclopentane ring with CH_2CH_3

5 C's in a ring = cyclopentane
ethylcyclopentane

12.49 To give the IUPAC name for each compound, follow the steps in Example 12.4.

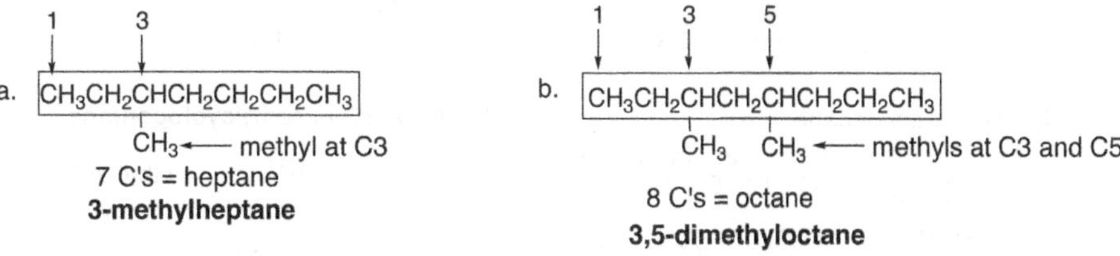

c. $CH_3CH_2CH_2C(CH_2CH_3)_3$

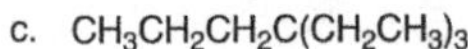

CH_2CH_3 ← two ethyls at C3

$CH_3CH_2CH_2C{-}CH_2CH_3$

CH_2CH_3

6 C's = hexane

3,3-diethylhexane

d. methyls at C2 and C6

$CH_3CHCH_2CHCH_2CH_3$ with CH_2CHCH_3 (bearing CH_3) and CH_2CH_3 ← ethyl at C4

8 C's = octane

4-ethyl-2,6-dimethyloctane

e. $(CH_3CH_2)_2CHCH_2CH_2CH_2CH(CH_3)_2$

ethyl at C6 → CH_3CH_2

$CH_3CH_2{-}CCH_2CH_2CH_2CCH_3$ (H, H; CH_3 ← methyl at C2)

8 C's = octane

6-ethyl-2-methyloctane

f. $CH_3CH_2{-}C(CH_3)_2{-}CH_2CH_2CH_2CH_2CH_2{-}C(CH_3)_2{-}CH_3$

← two methyls at C2, two methyls at C8

10 C's = decane

2,2,8,8-tetramethyldecane

12.51 To give the IUPAC name for each compound, follow the steps in Example 12.6.

a. 8 C's = **cyclooctane**

b. $-CH_3$ ← methyl

3 C's = cyclopropane

methylcyclopropane

c. 4 C's = cyclobutane

1,1,2,2-tetramethylcyclobutane

d. CH_3CH_2, CH_2CH_3 at positions 1 and 3

6 C's = cyclohexane

1,3-diethylcyclohexane

12.53 Give the structure corresponding to each IUPAC name.

a. 3-ethyl**hexane**

$CH_3CH_2CHCH_2CH_2CH_3$ with CH_2CH_3 ← ethyl at C3

b. 3-ethyl-3-methyl**octane**

CH_3 ← methyl at C3

$CH_3CH_2CCH_2CH_2CH_2CH_2CH_3$

CH_2CH_3 ← ethyl at C3

c. 2,3,4,5-tetramethyl**decane**

four methyl groups at C2, C3, C4, and C5

$CH_3CHCHCHCHCH_2CH_2CH_2CH_2CH_3$ with four CH_3 groups

d. cyclo**nonane**

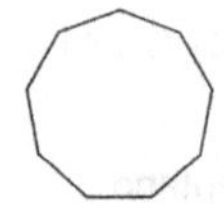

e. 1,1,3-trimethyl**cyclohexane**

CH_3 CH_3 at C1, CH_3 at C3 — three methyl groups

f. 1-ethyl-2,3-dimethyl**cyclopentane**

CH_2CH_3 ← ethyl at C1

CH_3 at C2, CH_3 at C3 ← two methyl groups

12.55 Explain why each IUPAC name is incorrect.

a. 2-methylbutane: Number to give CH_3 the lower number, 2 not 3.

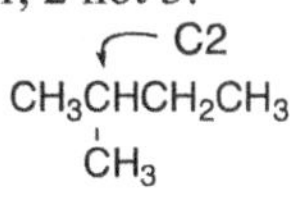

C2

$CH_3CHCH_2CH_3$

CH_3

b. methylcyclopentane: No number is assigned if there is only one substituent.

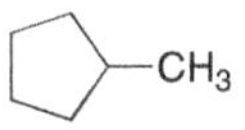

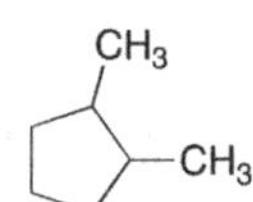

c. 2-methylpentane: The parent has five carbons, not four.

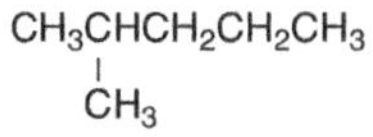

$CH_3CHCH_2CH_2CH_3$

CH_3

d. 2,5-dimethylheptane: The longest chain was not chosen.

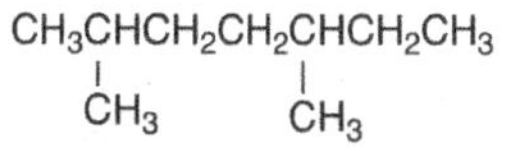

$CH_3CHCH_2CH_2CHCH_2CH_3$

CH_3 CH_3

e. 1,3-dimethylcyclohexane: Number to give the second substituent the lower number.

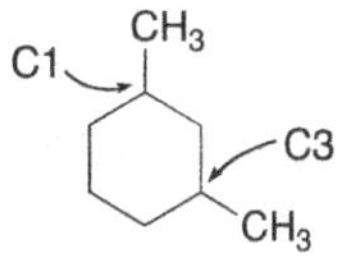

CH_3 C1 C3 CH_3

f. 1-ethyl-2-propylcyclopentane: The lower number is assigned alphabetically.

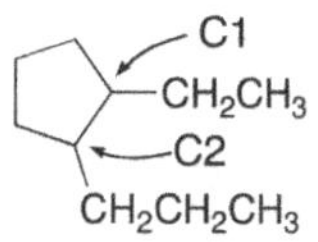

C1 CH_2CH_3 C2 $CH_2CH_2CH_3$

12.57 Draw the isomers and then give the IUPAC name.

CH_3 CH_3

1,2-dimethylcyclopentane

CH_3 CH_3

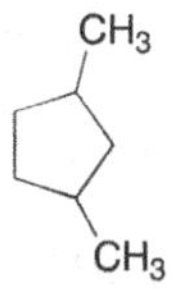

1,3-dimethylcyclopentane

CH_3 CH_3

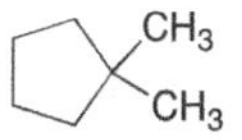

1,1-dimethylcyclopentane

12.59 Draw a skeletal structure for each compound.

a. octane

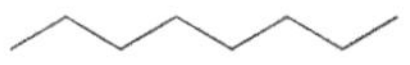

b. 1,2-dimethylcyclopentane

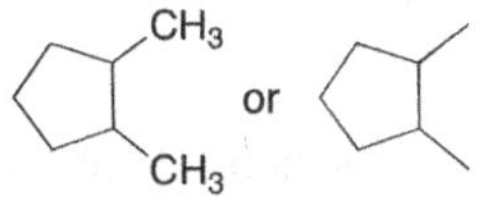

CH_3 CH_3 or

c. $CH_3CHCH_2CH_2CH_2CH_2CH_3$

CH_3

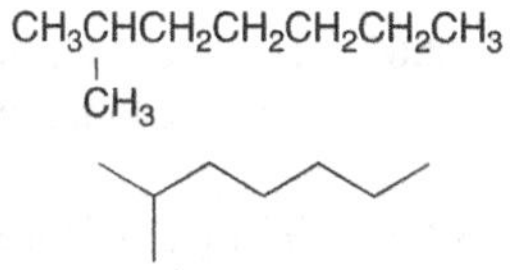

12.61 Convert each structure to a complete structure with all atoms drawn in.

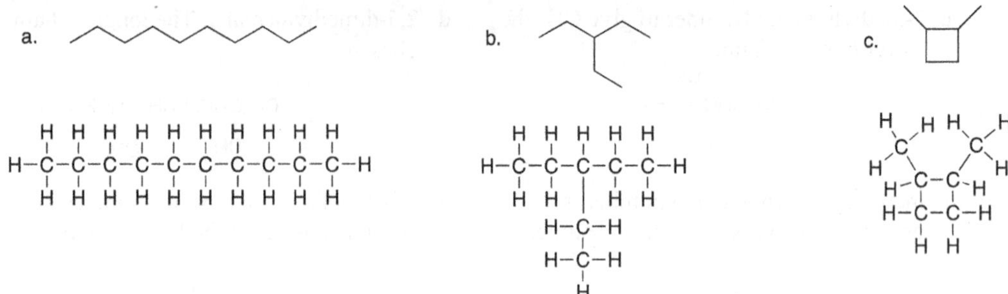

12.63 The melting points and boiling points of alkanes increase as the number of carbons increases.

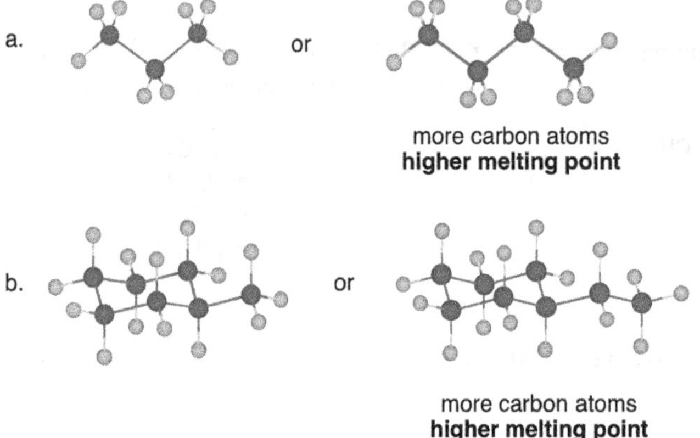

12.65 Branched alkanes have lower boiling points than linear alkanes with the same number of carbons.

a. increasing boiling point: $(CH_3)_4C$ < $(CH_3)_2CHCH_2CH_3$ < $CH_3CH_2CH_2CH_2CH_3$
most branching — no branching

b. increasing boiling point: $(CH_3)_2CHCH(CH_3)_2$ < $CH_3CH_2CH_2CH(CH_3)_2$ < $CH_3(CH_2)_4CH_3$
most branching — no branching

12.67 Hexane is a nonpolar hydrocarbon, making it soluble in organic solvents like dichloromethane, but insoluble in water.

12.69 Write a balanced equation for each reaction. Combustion reactions form CO_2 and H_2O.

a. $2\ CH_3CH_3 + 7\ O_2 \longrightarrow 4\ CO_2 + 6\ H_2O$

b. $(CH_3)_2CHCH_2CH_3 + 8\ O_2 \longrightarrow 5\ CO_2 + 6\ H_2O$

12.71 Write a balanced equation for each reaction. Incomplete combustion reactions form CO and H_2O.

a. $2\ CH_3CH_2CH_3 + 7\ O_2 \longrightarrow 6\ CO + 8\ H_2O$

b. $2\ CH_3CH_2CH_2CH_3 + 9\ O_2 \longrightarrow 8\ CO + 10\ H_2O$

12.73 Write a balanced equation for the oxidation of glucose to form CO_2 and H_2O.

$$C_6H_{12}O_6 + 6\ O_2 \longrightarrow 6\ CO_2 + 6\ H_2O$$

12.75 Replace a hydrogen atom with a bromine atom to draw the product of halogenation.

a. $CH_4 + Br_2 \xrightarrow{\text{heat}} CH_3Br + HBr$

b. cycloheptane $+ Br_2 \xrightarrow{\text{heat}}$ bromocycloheptane $+ HBr$

c. cyclobutane $+ Br_2 \xrightarrow{\text{heat}}$ bromocyclobutane $+ HBr$

12.77

a. $2\ (CH_3)_3CC(CH_3)_3 + 25\ O_2 \xrightarrow{\text{flame}} 16\ CO_2 + 18\ H_2O$

b. $CH_3\text{-}C(CH_3)_2\text{—}C(CH_3)_2\text{-}CH_3 + Cl_2 \xrightarrow{\text{light}} CH_3\text{-}C(CH_3)_2\text{—}C(CH_3)_2\text{-}CH_2Cl + HCl$

c. $CH_3\text{-}C(CH_3)_2\text{—}C(CH_3)_2\text{-}CH_3 + Br_2 \xrightarrow{\text{heat}} CH_3\text{-}C(CH_3)_2\text{—}C(CH_3)_2\text{-}CH_2Br + HBr$

12.79 Higher molecular weight alkanes in warmer weather means less evaporation. Lower molecular weight alkanes in colder weather means the gasoline won't freeze.

12.81 The mineral oil can prevent the body's absorption of important fat-soluble vitamins. The vitamins dissolve in the mineral oil, and are thus not absorbed. Instead, they are expelled with the mineral oil.

12.83 The nonpolar asphalt will be most soluble in the paint thinner (c) because "like dissolves like." The liquid alkanes of the paint thinner dissolve the high molecular weight hydrocarbons of the asphalt.

12.85 Answer each question about the compound.

a. $CH_3CH_2CH_2\text{—}C(CH_3)_2\text{—}CH(CH_2CH_3)\text{—}CH_2CH_3$

7 C's = heptane

3-ethyl-4,4-dimethylheptane

b. $CH_3CH_2CH(CH_3)\text{—}CH(CH_3)\text{—}CH(CH_2CH_3)\text{—}CH_2CH_3$

constitutional isomer

c. not water soluble

d. soluble in organic solvents

e. $C_{11}H_{24} + 17\ O_2 \longrightarrow 11\ CO_2 + 12\ H_2O$

f. (skeletal structure)

12.87

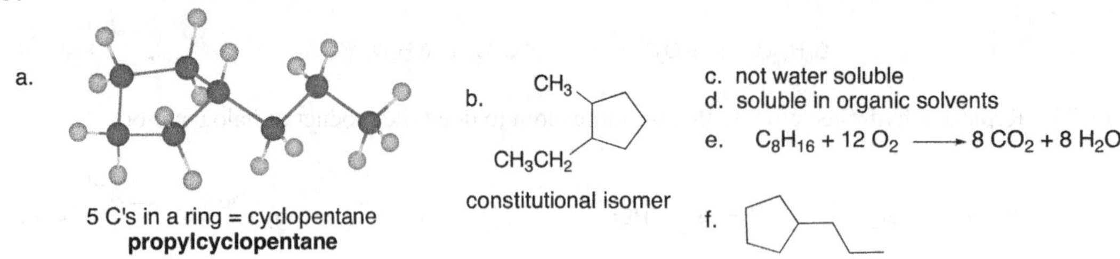

12.89 A compound with 10 carbons and two rings will have $2n - 2$ H's.
$(10 \times 2) - 2 = 18$
$C_{10}H_{18}$

12.91 Cyclopentane has a more rigid structure. The rings can get closer together because they are not floppy, resulting in an increased force of attraction and a higher boiling point.

Chapter 13 Unsaturated Hydrocarbons

Chapter Review

[1] What are the characteristics of alkenes, alkynes, and aromatic compounds?

- Alkenes are unsaturated hydrocarbons that contain a carbon–carbon double bond and have molecular formula C_nH_{2n}. Each carbon of the double bond is trigonal planar. (13.1)

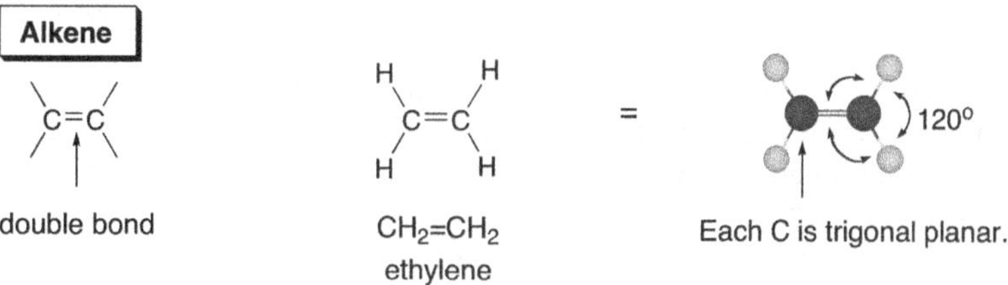

- Alkynes are unsaturated hydrocarbons that contain a carbon–carbon triple bond and have molecular formula C_nH_{2n-2}. Each carbon of the triple bond is linear. (13.1)

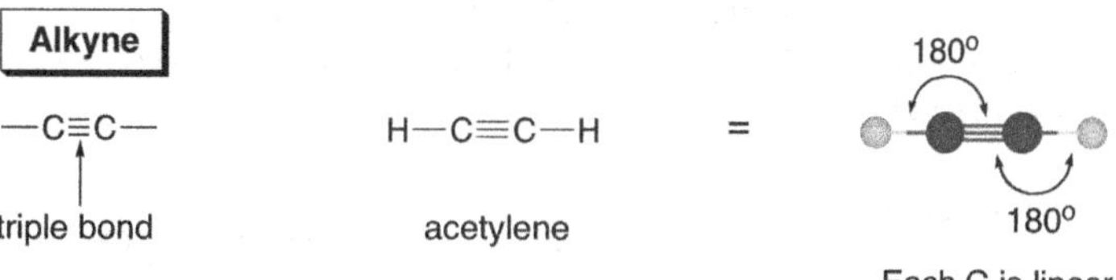

- Benzene, molecular formula C_6H_6, is the most common aromatic hydrocarbon. Benzene is a stable hybrid of two resonance structures, each containing a six-membered ring and three double bonds. Each carbon of benzene is trigonal planar. (13.9)

[2] How are alkenes, alkynes, and substituted benzenes named?

- An alkene is identified by the suffix *-ene*, and the carbon chain is numbered to give the C=C the lower number. (13.2)
- An alkyne is identified by the suffix *-yne*, and the carbon chain is numbered to give the C≡C the lower number. (13.2)
- Substituted benzenes are named by naming the substituent and adding the word *benzene*. When two substituents are bonded to the ring, the prefixes ortho, meta, and para are used to show the relative position of the two groups: 1,2-, 1,3-, or 1,4-, respectively. With three substituents on a benzene ring, number to give the lowest possible numbers. (13.10)

[3] What is the difference between constitutional isomers and stereoisomers? How are cis and trans isomers different? (13.3)

- Constitutional isomers differ in the way the atoms are bonded to each other.
- Stereoisomers differ only in the three-dimensional arrangement of the atoms.
- Cis and trans isomers are one type of stereoisomer. A cis alkene has two alkyl groups on the same side of the double bond. A trans alkene has two alkyl groups on opposite sides of the double bond.

General structure	Two possible arrangements	
$CH_3CH{=}CHCH_3$	$CH_3(H)C{=}C(H)CH_3$	$CH_3(H)C{=}C(CH_3)H$
2-butene	two CH_3 groups on the **same** side	two CH_3 groups on **opposite** sides
	cis isomer	**trans** isomer

[4] How do saturated and unsaturated fatty acids differ? (13.3B)

- Fatty acids are carboxylic acids (RCOOH) with long carbon chains. Saturated fatty acids have no double bonds in the carbon chain, whereas unsaturated fatty acids have one or more double bonds in their long carbon chains.
- Generally, double bonds in naturally occurring fatty acids are cis.
- As the number of double bonds in the fatty acid increases, the melting point decreases.

[5] What types of reactions do alkenes undergo? (13.6)

- Alkenes undergo addition reactions with reagents X–Y. One bond of the double bond and the X–Y bond break and two new single bonds (C–X and C–Y) are formed.
- Alkenes react with four different reagents—H_2 (Pd catalyst), X_2 (X = Cl or Br), HX (X = Cl or Br), and H_2O (with H_2SO_4).

General addition reaction:

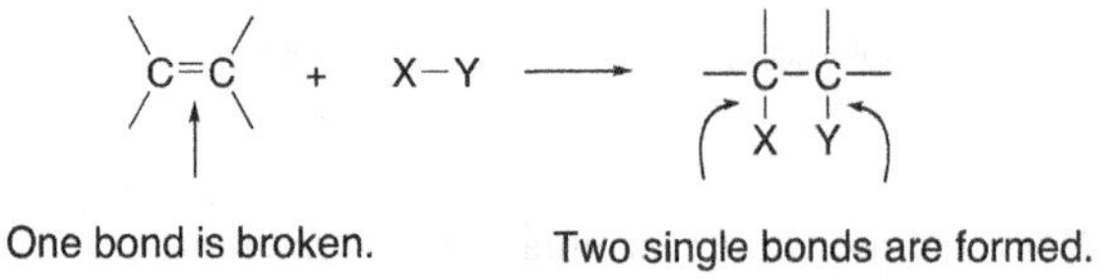

$$C{=}C + X{-}Y \longrightarrow -\overset{|}{\underset{X}{C}}-\overset{|}{\underset{Y}{C}}-$$

One bond is broken. Two single bonds are formed.

Hydrogenation (13.6A):	$CH_2{=}CH_2$	+ H_2	$\xrightarrow{Pd}$	$CH_2(H){-}CH_2(H)$	
Halogenation (13.6B):	$CH_2{=}CH_2$	+ X_2	$\longrightarrow$	$CH_2(X){-}CH_2(X)$	(X = Cl or Br)
Hydrohalogenation (13.6C):	$CH_2{=}CH_2$	+ HX	$\longrightarrow$	$CH_2(H){-}CH_2(X)$	(X = Cl or Br)
Hydration (13.6D):	$CH_2{=}CH_2$	+ H_2O	$\xrightarrow[H_2SO_4]{}$	$CH_2(H){-}CH_2(OH)$	

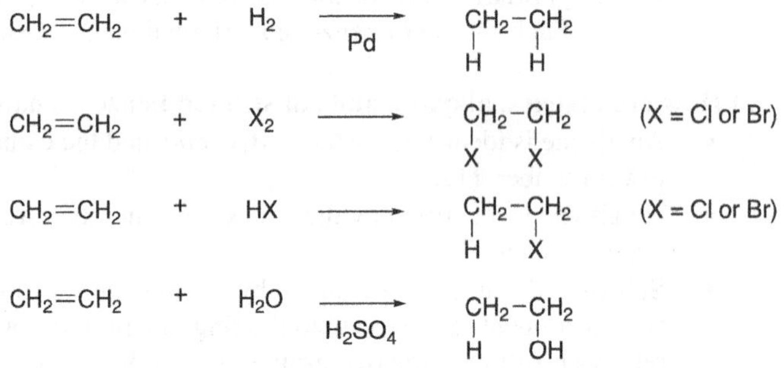

[6] What is Markovnikov's rule? (13.6)

- Markovnikov's rule explains the selectivity observed when an unsymmetrical reagent HZ (Z = Cl, Br, or OH) adds to an unsymmetrical alkene like propene ($CH_3CH{=}CH_2$). The H of HZ is added to the end of the C=C that has more H's to begin with, forming $CH_3CH(Z)CH_3$.

Example

This C has no H's so the OH bonds here.

CH_3-cyclohexene + H–OH $\xrightarrow{H_2SO_4}$ product (C with CH_3 and OH; adjacent C with H and H)

This C has 1 H so the H bonds here.

[7] What products are formed when a vegetable oil is partially hydrogenated? (13.7)

- When an unsaturated oil is partially hydrogenated, some but not all of the cis C=C's add H_2, reducing the number of double bonds and increasing the melting point.
- Some of the cis double bonds are converted to trans double bonds, thus forming trans fats, whose shape and properties closely resemble those of saturated fats.

[8] What are polymers, and how are they formed from alkene monomers? (13.8)

- Polymers are large molecules made up of repeating smaller molecules called monomers covalently bonded together. When alkenes are polymerized, one bond of the double bond breaks, and two new single bonds join the alkene monomers together in long carbon chains.

Three monomer units joined together

$$CH_2{=}CHZ + CH_2{=}CHZ + CH_2{=}CHZ \longrightarrow \text{–}CH_2CH(Z)\text{–}CH_2CH(Z)\text{–}CH_2CH(Z)\text{–}$$

Repeating unit: $\left[CH_2CH(Z)\right]_n$ — Many monomers are joined together.

shorthand structure

[9] What types of reactions does benzene undergo? (13.13)

- To keep the stable aromatic ring intact, benzene undergoes substitution, not addition, reactions. One H atom on the ring is replaced by another atom or group of atoms. Reactions include chlorination (substitution by Cl), nitration (substitution by NO_2), and sulfonation (substitution by SO_3H).

Chlorination (13.13A): $C_6H_6 + Cl_2 \xrightarrow{FeCl_3} C_6H_5Cl + HCl$

Nitration (13.13B): $C_6H_6 + HNO_3 \xrightarrow{H_2SO_4} C_6H_5NO_2 + H_2O$

Sulfonation (13.13C): $C_6H_6 + SO_3 \xrightarrow{H_2SO_4} C_6H_5SO_3H$

Problem Solving

[1] Alkenes and Alkynes (13.1)

Example 13.1 Draw a complete structure for each alkene or alkyne.

a. $(CH_3)_2C{=}CHCH_2CH_2CH_3$ b. $CH_3CH_2CH_2C{\equiv}CCH_3$

Analysis
First, draw the multiple bond in each structure. Draw an alkene so that each C of the double bond has three atoms around it. Draw an alkyne so that each C of the triple bond has two atoms around it. All other C's have four single bonds.

Solution
a. One alkene C has bonds to two CH_3 groups. The other alkene C has one bond to 1 H and one bond to a $CH_2CH_2CH_3$ group.

$(CH_3)_2C{=}CHCH_2CH_2CH_3$ → (Draw the double bond.) C=C (Add 2 CH_3 groups. Add H and $CH_2CH_2CH_3$.) → [complete structure]

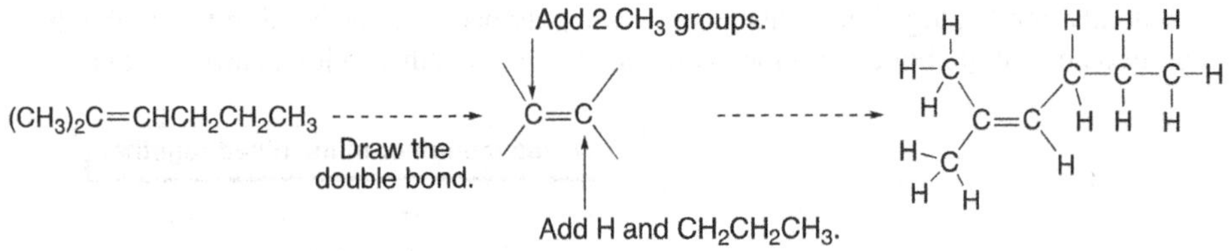

b. One alkyne C has a single bond to a $CH_2CH_2CH_3$ group. The other alkyne C has a single bond to a CH_3 group.

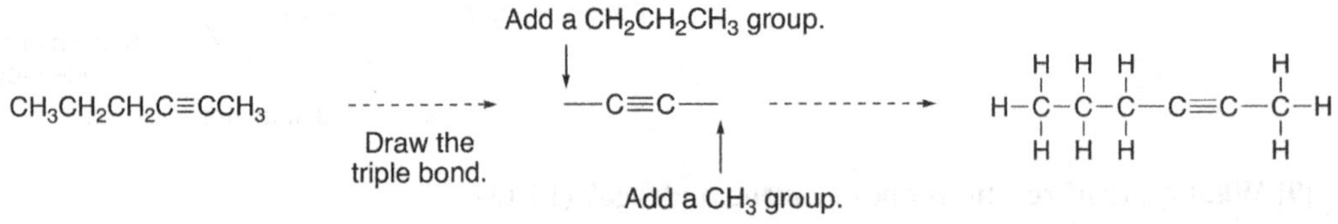

$CH_3CH_2CH_2C{\equiv}CCH_3$ → (Draw the triple bond.) —C≡C— (Add a $CH_2CH_2CH_3$ group. Add a CH_3 group.) → [complete structure]

[2] Nomenclature of Alkenes and Alkynes (13.2)

Example 13.2 Give the IUPAC name for the following compound.

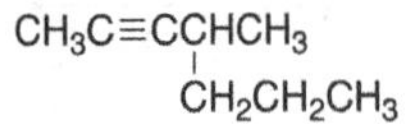

$CH_3C{\equiv}CCHCH_3$ with $CH_2CH_2CH_3$ on the CH carbon

Analysis and Solution
[1] Find the longest chain containing both carbon atoms of the multiple bond.

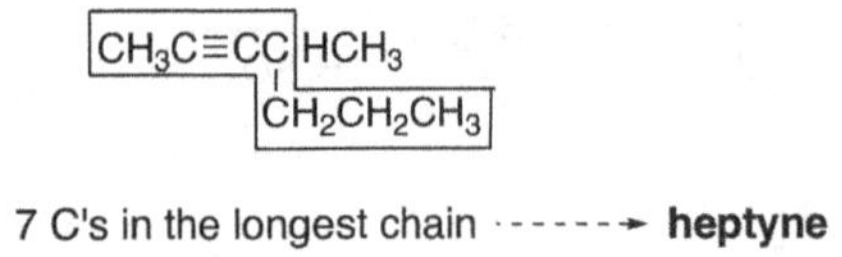

$CH_3C{\equiv}CCHCH_3$ / $CH_2CH_2CH_3$ (longest chain boxed)

7 C's in the longest chain → **heptyne**

[2] Number the chain to give the triple bond the lower number.

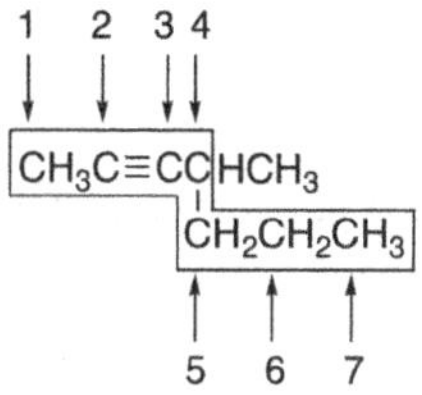

1 2 3 4

$CH_3C{\equiv}CCHCH_3$
$CH_2CH_2CH_3$

5 6 7

- Numbering from left to right is preferred since the triple bond begins at C2 (not C5). The molecule is named as a **2-heptyne.**

[3] Name and number the substituents and write the complete name.

- The alkyne has one methyl group located at C4.

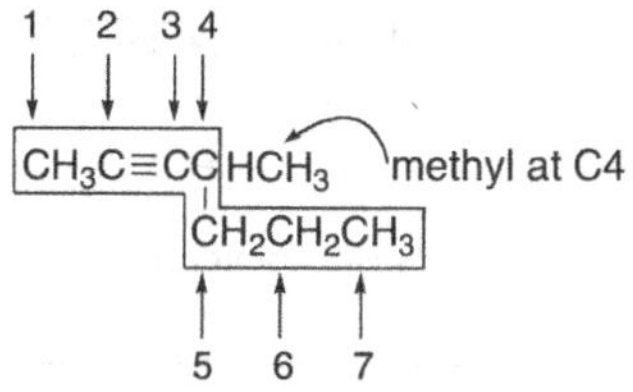

1 2 3 4

$CH_3C{\equiv}CCHCH_3$ methyl at C4
$CH_2CH_2CH_3$

5 6 7

Answer: 4-methyl-2-heptyne

Example 13.3 Draw the structure corresponding to the IUPAC name: 1,3-dimethylcyclopentene.

Analysis

First identify the parent name to find the longest carbon chain or ring, and then use the suffix to determine the functional group; the suffix *-ene* = an alkene and *-yne* = an alkyne. Then number the carbon chain or ring and place the functional group at the indicated carbon. For a cycloalkene, the double bond is located between C1 and C2. Add the substituents and enough hydrogens to give each carbon four bonds.

Solution

1,3-Dimethylcyclopentene has 5 C's in a ring (cyclopent-) and a double bond that begins at C1. Two methyl groups are bonded to C1 and C3.

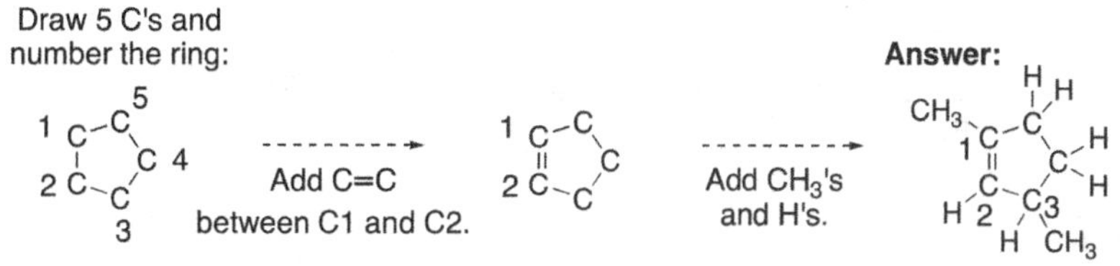

Draw 5 C's and number the ring:

Add C=C between C1 and C2.

Add CH_3's and H's.

Answer:

1,3-dimethylcyclopentene

[3] Cis–Trans Isomers (13.3)

Example 13.4 Draw *cis*- and *trans*-3-heptene.

Analysis
First, use the parent name to draw the carbon skeleton, and place the double bond at the correct carbon; 3-heptene indicates a 7 C chain with the double bond beginning at C3. Then use the definitions of cis and trans to draw the isomers.

Solution
Each C of the double bond is bonded to an alkyl group and a hydrogen. A cis isomer has the two alkyl groups bonded to the same side of the double bond. A trans isomer has the two alkyl groups bonded to the opposite sides of the double bond.

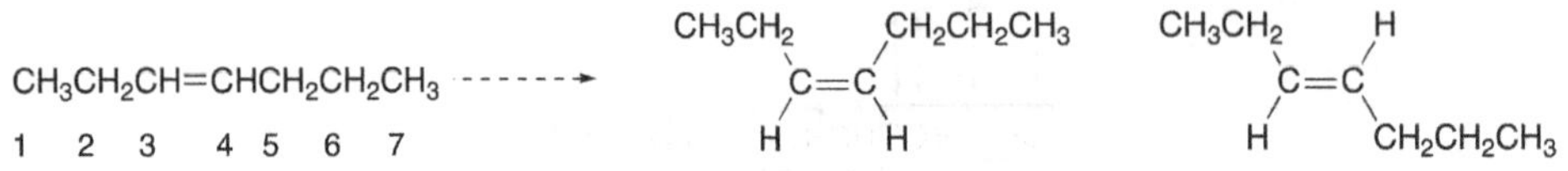

$CH_3CH_2CH{=}CHCH_2CH_2CH_3$ (1 2 3 4 5 6 7) ------→

3-heptene

CH_3CH_2 and $CH_2CH_2CH_3$ on top, H and H on bottom of C=C: *cis*-3-heptene

CH_3CH_2 and H on top, H and $CH_2CH_2CH_3$ on bottom of C=C: *trans*-3-heptene

---

[4] Reactions of Alkenes (13.6)

Example 13.5 Draw the product of the following reaction.

(cyclopentylidene)$=CH_2$ + H_2 $\xrightarrow{Pd}$

Analysis
To draw the product of a hydrogenation reaction:

- Locate the C=C and mentally break one bond in the double bond.
- Mentally break the H–H bond of the reagent.
- Add one H atom to each C of the C=C, thereby forming two new C–H single bonds.

Solution

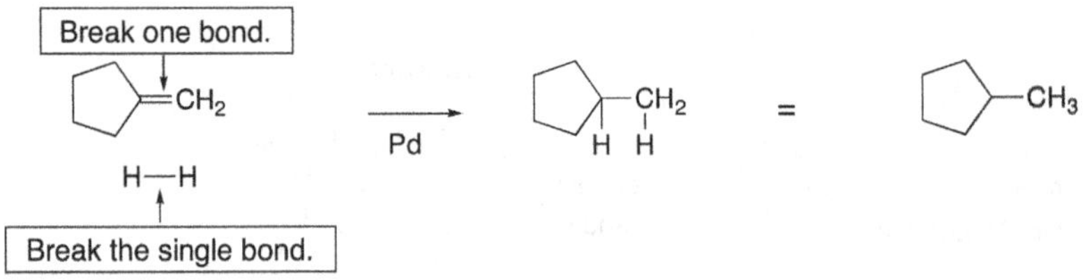

Break one bond. → (cyclopentylidene)$=CH_2$

H—H ← Break the single bond.

$\xrightarrow{Pd}$ (cyclopentyl, H)–CH_2(H) = (cyclopentyl)–CH_3

Example 13.6 Draw the product of the following reaction.

$(CH_3)_2C{=}CHCH_2CH_3$ + HBr ⟶

Analysis
Alkenes undergo addition reactions, so the elements of H and Br must be added to the double bond. Since the alkene is unsymmetrical, the H atom of HBr bonds to the carbon that has more H's to begin with.

Solution

$(CH_3)_2C{=}CHCH_2CH_3$

Draw out.

$(CH_3)_2C{=}C(H)CH_2CH_3 + H{-}Br \longrightarrow CH_3{-}C(CH_3)(Br){-}C(H)(H){-}CH_2CH_3$

This C has no H's so the Br bonds here.

This C has 1 H so the H bonds here.

[5] Polymers—The Fabric of Modern Society (13.8)

Example 13.7 What polymer is formed when $CH_2{=}CHNHCOCH_3$ is polymerized?

Analysis
Draw three or more alkene molecules and arrange the carbons of the double bonds next to each other. Break one bond of each double bond, and join the alkenes together with single bonds. With unsymmetrical alkenes, substituents are bonded to every other carbon.

Solution

Join these 2 C's. Join these 2 C's.

$CH_2{=}C(H)NHCOCH_3 \quad CH_2{=}C(H)NHCOCH_3 \quad CH_2{=}C(H)NHCOCH_3 \longrightarrow$

Break one bond that joins each C=C.

Answer:

$-CH_2{-}CH(NHCOCH_3){-}CH_2{-}CH(NHCOCH_3){-}CH_2{-}CH(NHCOCH_3)-$

[6] Nomenclature of Benzene Derivatives (13.10)

Example 13.8 Name each of the following aromatic compounds.

a. $CH_3(CH_2)_3{-}C_6H_4{-}(CH_2)_3CH_3$ b. benzene ring with CH_3, Br, I on adjacent carbons

Analysis
Name the substituents on the benzene ring. With two groups, alphabetize the substituent names and use the prefix ortho, meta, or para to indicate their location. With three substituents, alphabetize the substituent names, and number to give the lowest set of numbers.

Solution

a. $CH_3(CH_2)_3$–[benzene ring]–$(CH_2)_3CH_3$
butyl group   butyl group

- The two substituents are located 1,4- or **para** to each other.
- **Answer: *p*-dibutylbenzene**

b. [toluene ring with CH_3 at position 1, Br at 2, I at 3]
toluene

- Since a CH_3– group is bonded to the ring, name the molecule as a derivative of toluene.
- Place the CH_3 group at the "1" position, and number to give the lowest set of numbers.
- **Answer: 2-bromo-3-iodotoluene**

Self-Test

[1] Fill in the blank with one of the terms listed below.

Addition reaction (13.6)	Hydration (13.6)	Stereoisomers (13.3)
Alkenes (13.1)	Halogenation (13.6)	Substitution reaction (13.13)
Alkynes (13.1)	Hydrogenation (13.6)	Trans isomer (13.3)
Cis isomer (13.3)	Hydrohalogenation (13.6)	Unsaturated hydrocarbons (13.1)
Fats (13.3)	Oils (13.3)	

1. _______ are compounds that contain a carbon–carbon triple bond.
2. When two alkyl groups are on the *same side* of a double bond, the compound is called the _______.
3. _______ are liquids at room temperature and are formed from fatty acids having a large number of double bonds.
4. _______ are compounds that contain a carbon–carbon double bond.
5. _______ is the addition of halogen (X_2) to an alkene.
6. _______ is the addition of hydrogen (H_2) to an alkene.
7. _______ is the addition of HX (X = Cl or Br) to an alkene.
8. _______ are isomers that differ *only* in the three-dimensional arrangement of atoms.
9. _______ are solids at room temperature and are generally formed from fatty acids having few double bonds.
10. _______ are compounds that contain fewer than the maximum number of hydrogen atoms per carbon.
11. When two alkyl groups are on *opposite sides* of a double bond, the compound is called the _______.
12. A _______ is a reaction in which an atom is *replaced* by another atom or a group of atoms.
13. An _______ is a reaction in which elements are added to a compound.
14. _______ is the addition of water to an alkene.

[2] Label the substituents on each benzene ring as ortho, meta, or para.

a. ortho   b. meta   c. para

15. 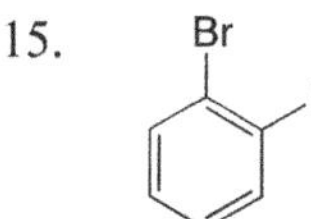

16. Cl, Br (para-substituted benzene)

17. Cl, Br (meta-substituted benzene)

[3] Label each statement as true (T) or false (F).

18. Cis alkenes have two alkyl groups on the same side of the double bond.
19. Stereoisomers have atoms bonded to different atoms.
20. Trans alkenes have two alkyl groups on the same side of the double bond.
21. Two compounds with the same molecular formula can be stereoisomers.

[4] For each reaction, pick the missing reagent needed to produce the products.

a. HNO_3 b. Br_2 c. Cl_2 d. H_2O

22. $CH_2{=}CHCH_3$ + ___?___ ⟶ $BrCH_2CH(Br)CH_3$

23. benzene + ___?___ $\xrightarrow[H_2SO_4]{}$ nitrobenzene (NO_2) + H_2O

24. benzene + ___?___ $\xrightarrow[FeCl_3]{}$ chlorobenzene (Cl) + HCl

25. $CH_3CH{=}CHCH_3$ + ___?___ $\xrightarrow[H_2SO_4]{}$ $CH_3CH_2CH(OH)CH_3$

Answers to Self-Test

1. Alkynes	6. Hydrogenation	11. trans isomer	16. c	21. T
2. cis isomer	7. Hydrohalogenation	12. substitution reaction	17. b	22. b
3. Oils	8. Stereoisomers	13. addition reaction	18. T	23. a
4. Alkenes	9. Fats	14. Hydration	19. F	24. c
5. Halogenation	10. Unsaturated hydrocarbons	15. a	20. F	25. d

Solutions to In-Chapter Problems

13.1 To draw a complete structure with all atoms, bond lines, and lone pairs for each condensed structure, first draw in the multiple bonds. Then draw in all the other C's and H's, as in Example 13.1.

a. $CH_2{=}CHCH_2OH$ = H–C(H)=C(H)–C(H)(H)–Ö–H

b. $(CH_3)_2C{=}CH(CH_2)_2CH_3$ = H–C(H)(H)–C(CH₃)=C(H)–C(H)(H)–C(H)(H)–C(H)(H)–H

c. $(CH_3)_2CHC{\equiv}CCH_2C(CH_3)_3$ = H–C–C–C≡C–C–C–C–H (complete structure with all H atoms)

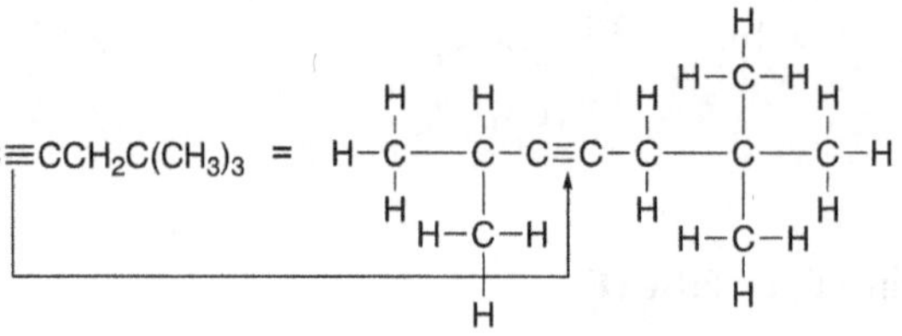

13.2 Use the general formulas [saturated hydrocarbon (C_nH_{2n+2}), alkene (C_nH_{2n}), and alkyne (C_nH_{2n-2})] to determine the molecular formula for each compound.

a. alkene = C_nH_{2n}, $4 \times 2 = 8$, C_4H_8
b. saturated hydrocarbon = C_nH_{2n+2}, $(6 \times 2) + 2 = 14$, C_6H_{14}
c. alkyne = C_nH_{2n-2}, $(7 \times 2) - 2 = 12$, C_7H_{12}
d. alkene = C_nH_{2n}, $5 \times 2 = 10$, C_5H_{10}

13.3 To draw a complete structure give each C enough H's to have four bonds.

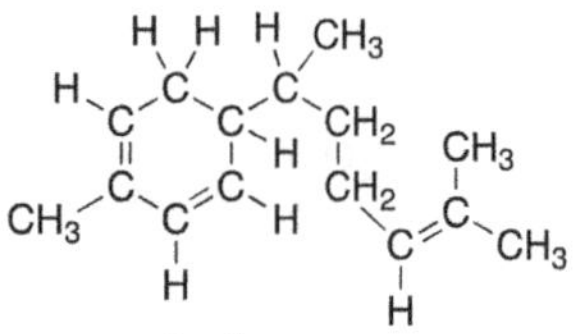

zingiberene

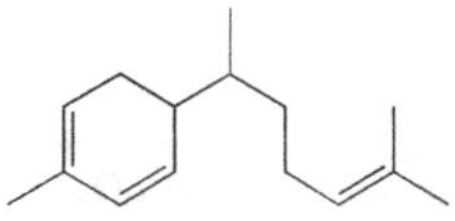

skeletal structure

13.4 Give the IUPAC name for each compound using the following steps, as in Example 13.2:
[1] Find the longest chain containing both carbon atoms of the multiple bond.
[2] Number the chain to give the double bond the lower number.
[3] Name and number the substituents and write the complete name.

a. $CH_2{=}CHCHCH_2CH_3$ (with CH_3 on C3) → 1 2 3 4 5 $CH_2{=}CHCHCH_2CH_3$ → **Answer: 3-methyl-1-pentene**

5 C's in the longest chain
pentene

double bond at C1

methyl at C3

b. $(CH_3CH_2)_2C{=}CHCH_2CH_2CH_3$ → $CH_3CH_2C{=}CHCH_2CH_2CH_3$ (with CH_2CH_3 on C3) → **Answer: 3-ethyl-3-heptene**

7 C's in the longest chain
heptene

ethyl at C3

double bond at C3

c. $CH_3CH_2CH{=}CHCH{=}CHCH_3$ → 7 6 5 4 3 2 1 $CH_3CH_2CH{=}CHCH{=}CHCH_3$ → **Answer: 2,4-heptadiene**

7 C's in the longest chain
heptadiene

double bonds at C2 and C4

d. [cyclopentene ring] CH_2CH_3 --→ [ring numbered 1, 2, 3, 4, 5] CH_2CH_3 --→ **Answer: 3-ethylcyclopentene**

5 C's in the ring
cyclopentene

ethyl at C3

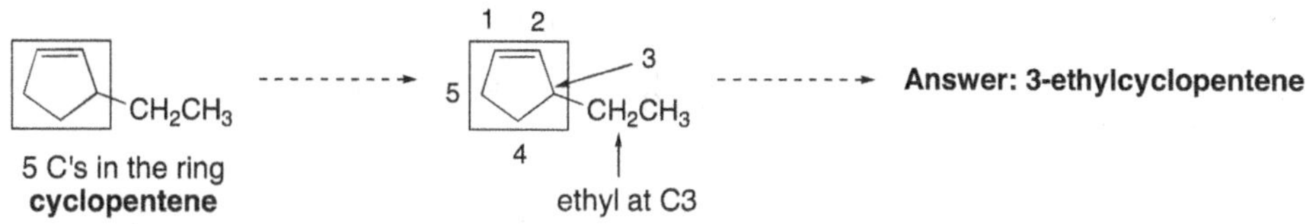

13.5 Give the IUPAC name for each compound using the following steps, as in Example 13.2:
[1] Find the longest chain containing both carbon atoms of the multiple bond.
[2] Number the chain to give the multiple bond the lower number.
[3] Name and number the substituents and write the complete name.

a. $CH_3CH_2CH_2CH_2CH_2C{\equiv}CCH(CH_3)_2$

$CH_3CH_2CH_2CH_2CH_2C{\equiv}CCHCH_3$ (with CH_3 on CH) --→ $CH_3CH_2CH_2CH_2CH_2C{\equiv}CCHCH_3$ (numbered 5 4 3 2 1) --→ **Answer: 2-methyl-3-nonyne**

9 C's in the longest chain
nonyne

triple bond at C3

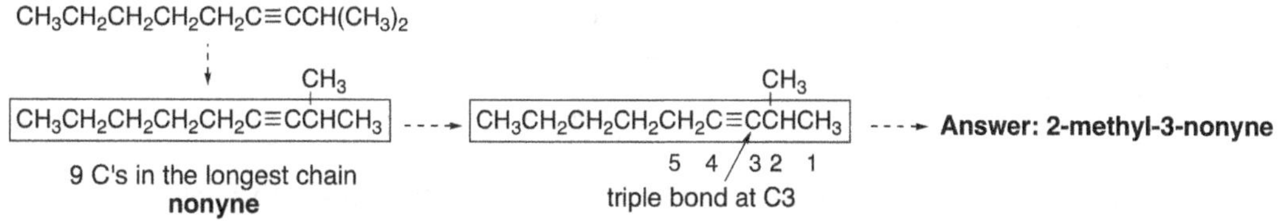

b. $CH_3CH_2-C{\equiv}C-CH_2-C(CH_2CH_3)(CH_3)-CH_3$ --→ $CH_3CH_2-C{\equiv}C-CH_2-C(CH_2CH_3)(CH_3)-CH_3$ (numbered 1 2 3 4 5 6 7 8) --→ **Answer: 6,6-dimethyl-3-octyne**

8 C's in the longest chain
octyne

triple bond at C3

2 methyl groups at C6

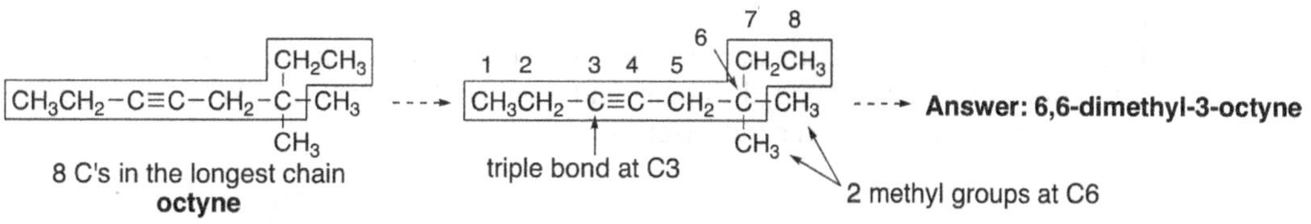

13.6 To draw the structure corresponding to each name, follow the steps in Example 13.3.

- Identify the parent name to find the longest carbon chain or ring, and then use the suffix to determine the functional group; the suffix *-ene* = alkene and *-yne* = alkyne.
- Number the carbon chain or ring and place the functional group at the indicated carbon. Add the substituents and enough hydrogens to give each carbon four bonds.

a. 4-methyl-1-**hexene** --→ C=C–C–C(CH₃)–C–C (numbered 1 2 3 4 5 6) --→ $CH_2{=}CHCH_2CH(CH_3)CH_2CH_3$

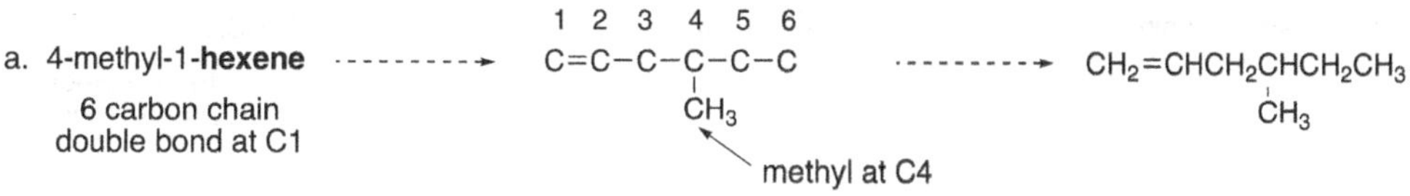

6 carbon chain
double bond at C1

methyl at C4

b. 5-ethyl-2-methyl-2-**heptene** --→ C–C(CH₃)=C–C–C(CH₂CH₃)–C–C (numbered 1 2 3 4 5 6 7) --→ $CH_3C(CH_3){=}CHCH_2CH(CH_2CH_3)CH_2CH_3$

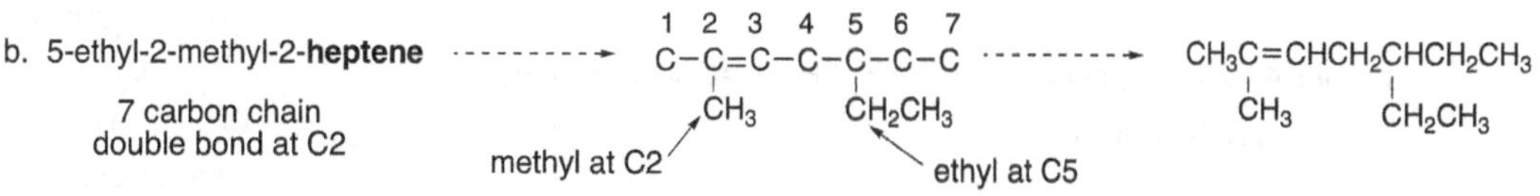

7 carbon chain
double bond at C2

methyl at C2

ethyl at C5

c. 2,5-dimethyl-3-**hexyne** --→ C–C(CH₃)–C≡C–C(CH₃)–C (numbered 1 2 3 4 5 6) --→ $CH_3CH(CH_3)C{\equiv}CCH(CH_3)CH_3$

6 carbon chain
triple bond at C3

methyl at C2

methyl at C5

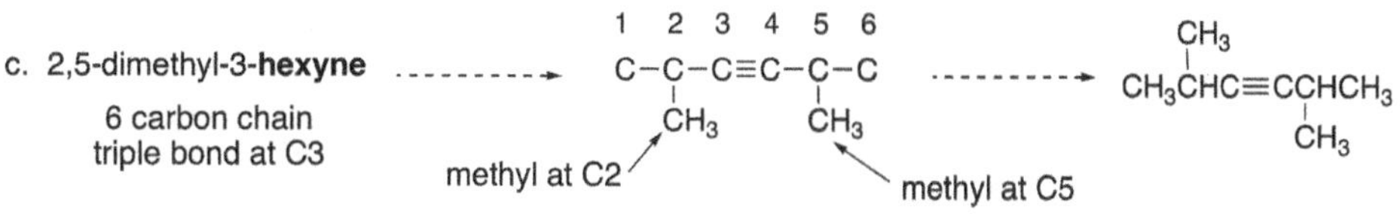

d. 1-propyl**cyclobutene** ---------→ 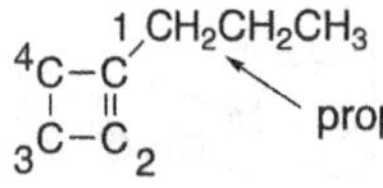 propyl at C1 ---------→ 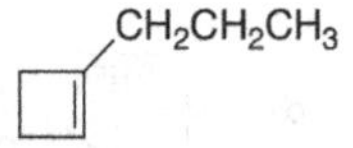

4 carbon ring
double bond at C1

e. 1,3-**cyclohexadiene** ---------→ 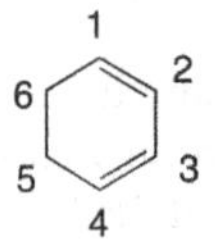

6 carbon ring
2 double bonds (C1 and C3)

f. 4-ethyl-1-**decyne** ---→ 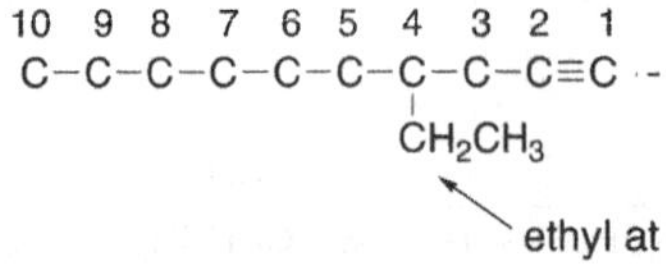 ---→ 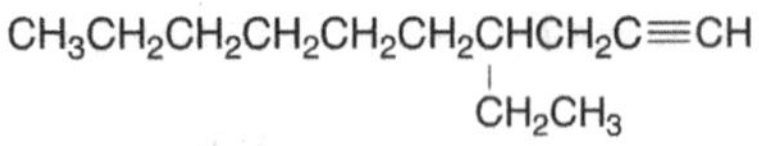

10 carbon chain
triple bond at C1

ethyl at C4

13.7 To draw the structures of the cis and trans isomers, follow the steps in Example 13.4.

- Use the parent name to draw the carbon skeleton and place the double bond at the correct carbon.
- Use the definitions of cis and trans to draw the isomers. When the two alkyl groups are on the same side of the double bond, the compound is called the cis isomer. When they are on opposite sides, it is called the trans isomer.

a. *cis*-2-**octene** -------→ $CH_3CH{=}CHCH_2CH_2CH_2CH_2CH_3$ -------→ 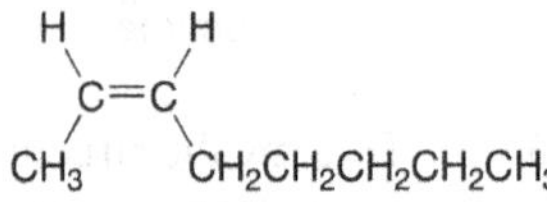

8 carbon chain — 1 2 3 4 5 6 7 8

cis isomer
both alkyl groups on the same side

b. *trans*-3-**heptene** -------→ $CH_3CH_2CH{=}CHCH_2CH_2CH_3$ -------→ 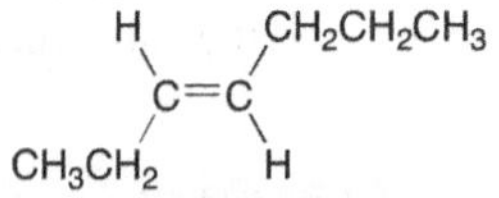

7 carbon chain — 1 2 3 4 5 6 7

trans isomer
alkyl groups on opposite sides

c. *trans*-4-methyl-2-**pentene** -------→ $CH_3CH{=}CHCH(CH_3)CH_3$ -------→ 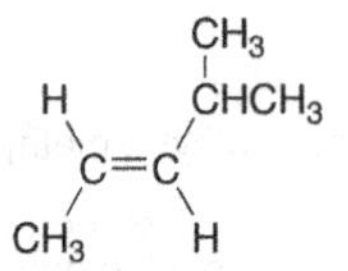

5 carbon chain — 1 2 3 4 5

trans isomer
alkyl groups on opposite sides

13.8 When the two alkyl groups are on the same side of the double bond, the compound is called the cis isomer. When they are on opposite sides, it is called the trans isomer.

cis

H H

H C=C

C=C $CH_2CH_2CH_3$

$HOCH_2(CH_2)_7CH_2$ H

trans

13.9 Stereoisomers differ only in the three-dimensional arrangement of atoms. Constitutional isomers differ in the way the atoms are bonded to each other.

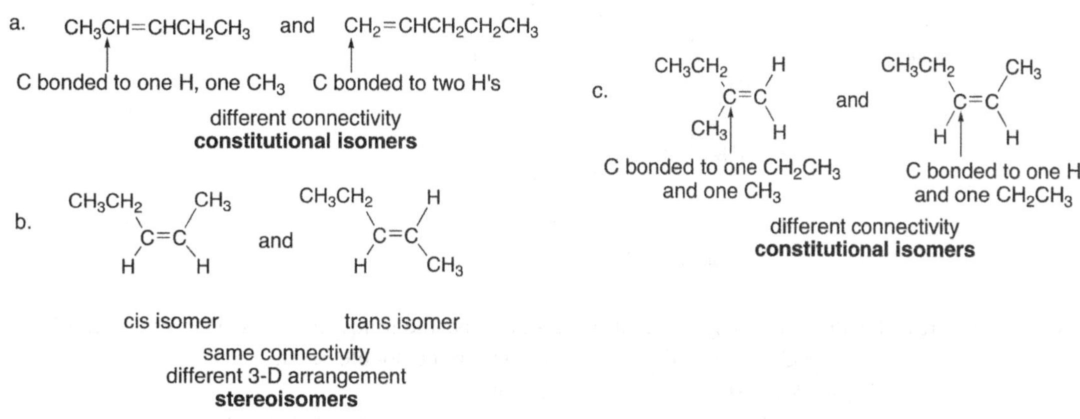

13.10

a. cis →

b. trans ← cis

13.11 Double bonds in naturally occurring fatty acids are cis.

H H H H H H H H

C=C C=C C=C C=C O

$CH_3CH_2CH_2CH_2CH_2$ CH_2 CH_2 CH_2 $CH_2CH_2CH_2COH$

arachidonic acid

13.12 Fats are solids at room temperature because of their higher melting points. They are formed from fatty acids with few double bonds. Oils are liquids at room temperature because of their lower melting points. They are also formed from fatty acids, but have more double bonds.

$CH_3(CH_2)_{14}COOH$	$CH_3(CH_2)_5CH{=}CH(CH_2)_7COOH$
palmitic acid	palmitoleic acid
no double bonds	one double bond
higher melting point	lower melting point
63 °C	1 °C

13.13

a. 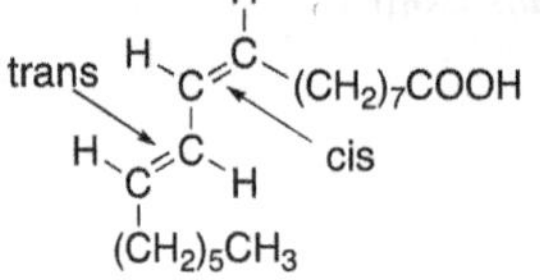 trans; cis; $(CH_2)_7COOH$; $(CH_2)_5CH_3$

b. 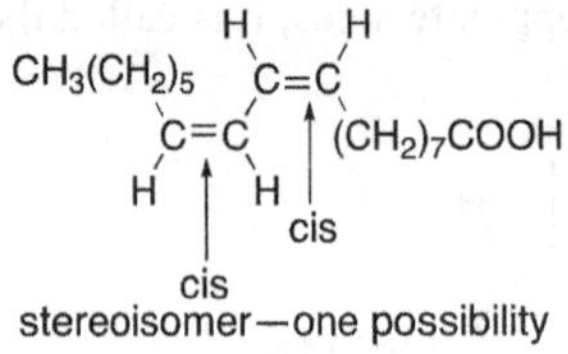 $CH_3(CH_2)_5$ C=C C=C $(CH_2)_7COOH$; cis; cis

stereoisomer—one possibility

c. This stereoisomer would have a lower melting point because both double bonds are cis.

d. **A** is a constitutional isomer of linoleic acid because the double bonds are at different places along the carbon chain.

13.14 The functional groups in tamoxifen are labeled.

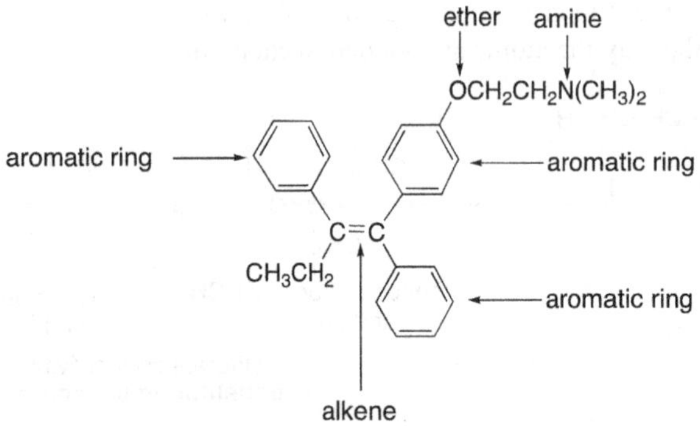

ether, amine: $OCH_2CH_2N(CH_3)_2$

aromatic ring; aromatic ring; aromatic ring

CH_3CH_2 C=C

alkene

13.15 To draw the product of a hydrogenation reaction, use the following steps, as in Example 13.5:

- Locate the C=C and mentally break one bond in the double bond.
- Mentally break the H–H bond of the reagent.
- Add one H atom to each C of the C=C, thereby forming two new C–H single bonds.

a. $CH_3CH_2CH{=}CHCH_2CH_3 \xrightarrow[Pd]{H_2} CH_3CH_2CH_2CH_2CH_2CH_3$

b. 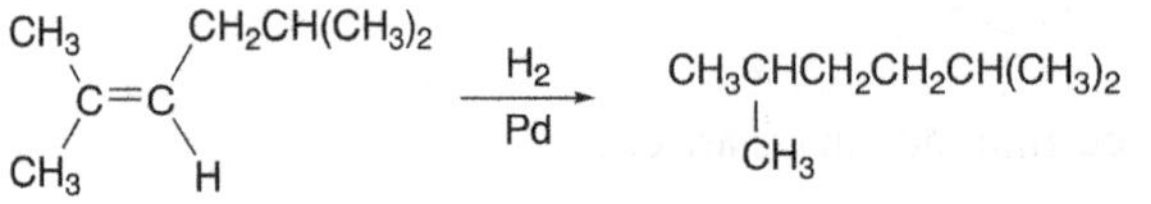 $(CH_3)_2C{=}CHCH_2CH(CH_3)_2 \xrightarrow[Pd]{H_2} CH_3CH(CH_3)CH_2CH_2CH(CH_3)_2$

c. 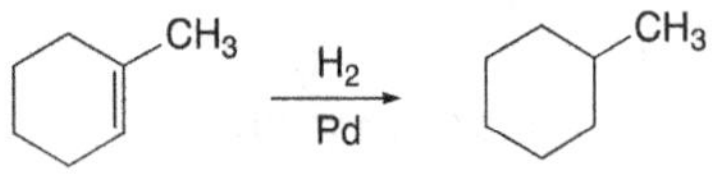 1-methylcyclohexene (CH_3) $\xrightarrow[Pd]{H_2}$ methylcyclohexane (CH_3)

13.16 To draw the product of each halogenation reaction, add a halogen to both carbons of the double bond.

a. 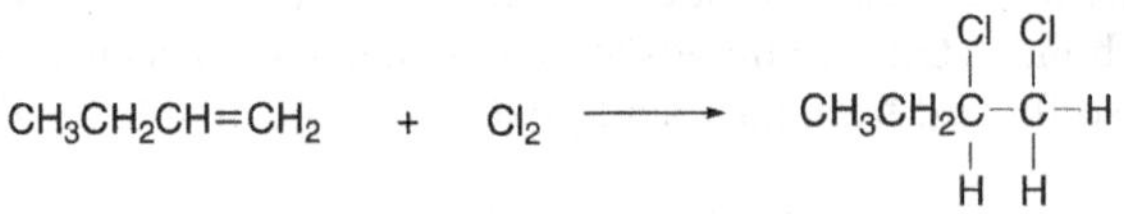 $CH_3CH_2CH{=}CH_2 + Cl_2 \longrightarrow CH_3CH_2CHClCH_2Cl$

b. (1,2-dimethylcyclohexene) + Br_2 ⟶ (1,2-dibromo-1,2-dimethylcyclohexane)

13.17 In hydrohalogenation reactions, the elements of H and Br (or H and Cl) must be added to the double bond. When the alkene is unsymmetrical, the H atom of HX bonds to the carbon that has more H's to begin with.

a. $CH_3CH{=}CHCH_3$ + HBr ⟶ $CH_3CHBr{-}CH_2CH_3$

This C does not have any H's.
Add Br here.

b. (1-methylcyclohexene) + HBr ⟶ (1-bromo-1-methylcyclohexane)

This C has more H's.
Add H here.

This C does not have any H's.
Add Cl here.

c. $(CH_3)_2C{=}CHCH_3$ + HCl ⟶ $CH_3{-}C(CH_3)(Cl){-}CH_2CH_3$

This C has more H's.
Add H here.

d. (cyclopentene) + HCl ⟶ (chlorocyclopentane)

13.18 In hydration reactions, the elements of H and OH are added to the double bond. In unsymmetrical alkenes, the H atom bonds to the less substituted carbon.

a. $CH_3CH{=}CHCH_3 \xrightarrow[H_2SO_4]{H{-}OH} CH_3CH(OH){-}CH_2CH_3$

This C has one H so the OH bonds here.

b. $CH_3CH_2CH{=}CH_2 \xrightarrow[H_2SO_4]{H{-}OH} CH_3CH_2CH(OH){-}CH_2{-}H$

This C has 2 H's so the H bonds here.

c. (1,2-dimethylcyclohexene) $\xrightarrow[H_2SO_4]{H{-}OH}$ (1,2-dimethylcyclohexan-1-ol)

13.19 Draw the products of each reaction.

a. $CH_3CH_2CH_2CH{=}CH_2 \xrightarrow[Pd]{H_2} CH_3CH_2CH_2CH_2CH_3$

H H
$CH_3CH_2CH_2C{-}C{-}H$
H H

b. $CH_3CH_2CH_2CH{=}CH_2 \xrightarrow{Cl_2}$

H H
$CH_3CH_2CH_2C{-}C{-}H$
Cl Cl

c. $CH_3CH_2CH_2CH{=}CH_2 \xrightarrow{Br_2}$

Br Br
$CH_3CH_2CH_2C{-}C{-}H$
H H

d. $CH_3CH_2CH_2CH{=}CH_2 \xrightarrow{H{-}Br}$

H H
$CH_3CH_2CH_2C{-}C{-}H$
Br H

e. $CH_3CH_2CH_2CH{=}CH_2 \xrightarrow{H{-}Cl}$

H H
$CH_3CH_2CH_2C{-}C{-}H$
Cl H

f. $CH_3CH_2CH_2CH{=}CH_2 \xrightarrow[H_2SO_4]{H{-}OH}$

HO H
$CH_3CH_2CH_2C{-}C{-}H$
H H

13.20 Draw the products of the hydrogenation reactions.

a.

H H H H H H
C=C C=C C=C
CH_3CH_2 CH_2 CH_2 $CH_2CH_2CH_2CH_2CH_2CH_2CH_2COOH$

$\downarrow$ H_2, Pd

H H H H
C=C C=C
CH_3CH_2 $CH_2CH_2CH_2CH_2$ $CH_2CH_2CH_2CH_2CH_2CH_2CH_2COOH$

+

H H H H
C=C C=C
CH_3CH_2 CH_2 $CH_2CH_2CH_2CH_2CH_2CH_2CH_2CH_2CH_2CH_2COOH$

b. $CH_3CH_2CH_2CH_2CH_2CH_2CH_2CH_2CH_2CH_2CH_2CH_2CH_2CH_2CH_2CH_2CH_2COOH$

13.21 To draw the polymers, draw three or more alkene molecules and arrange the carbons of the double bonds next to each other. Break one bond of each double bond, and join the alkenes together with single bonds. With unsymmetrical alkenes, substituents are bonded to every other carbon. Use Example 13.7 as a guide.

a. Join these 2 C's. Join these 2 C's.

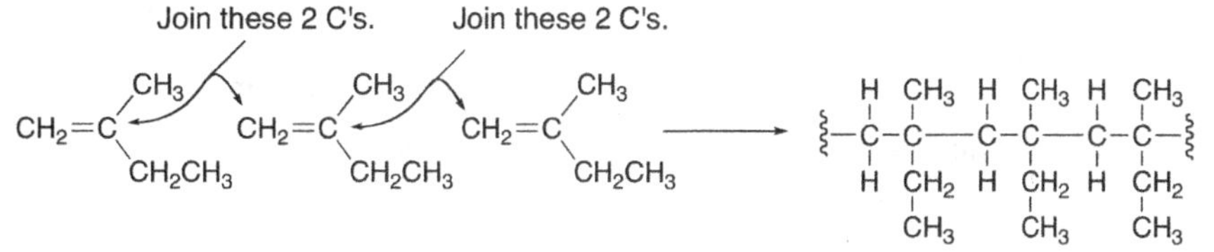

$CH_2{=}C(CH_3)CH_2CH_3$ + $CH_2{=}C(CH_3)CH_2CH_3$ + $CH_2{=}C(CH_3)CH_2CH_3$ → $-[C(H)_2-C(CH_3)(CH_2CH_3)]_3-$

b. Join these 2 C's. Join these 2 C's.

$CH_2{=}C(CH_3)CN$ + $CH_2{=}C(CH_3)CN$ + $CH_2{=}C(CH_3)CN$ → $-[C(H)_2-C(CH_3)(CN)]_3-$

c. Join these 2 C's. Join these 2 C's.

$CH_2{=}CH(C_6H_4Cl)$ + $CH_2{=}CH(C_6H_4Cl)$ + $CH_2{=}CH(C_6H_4Cl)$ → $-[C(H)_2-C(H)(C_6H_4Cl)]_3-$

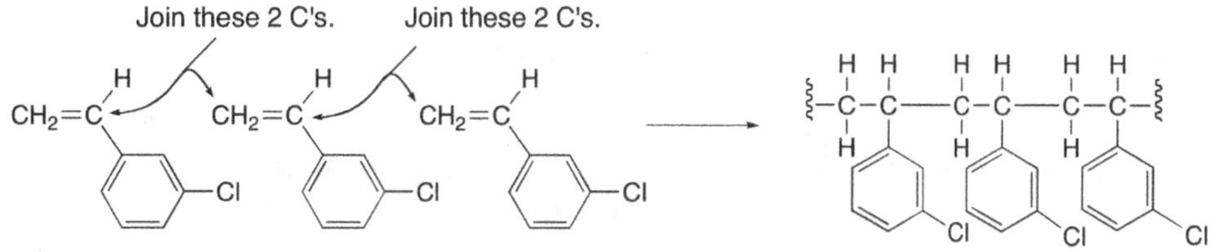

13.22 Work backwards to determine what monomer is used to form the polymer.

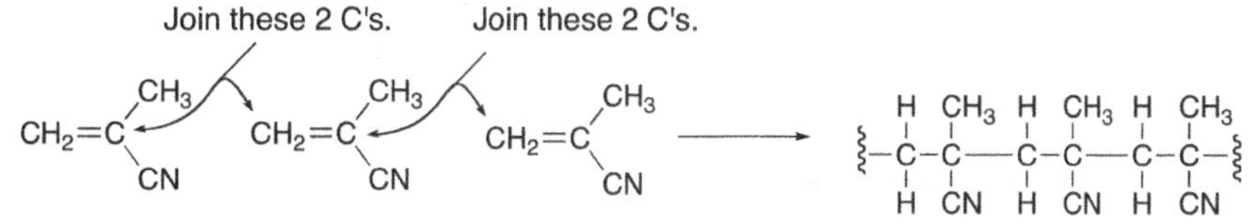

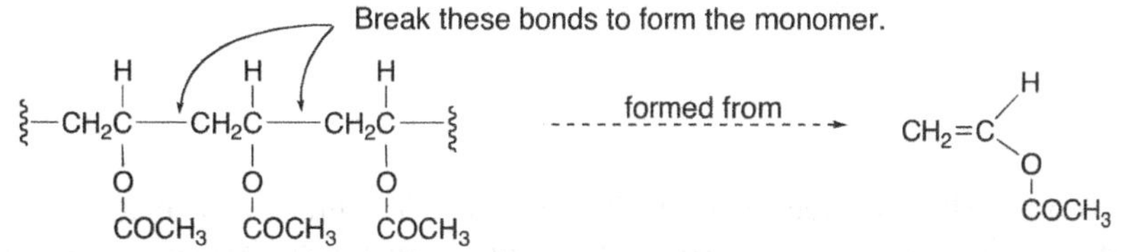

Break these bonds to form the monomer.

$-CH_2CH(OCOCH_3)-CH_2CH(OCOCH_3)-CH_2CH(OCOCH_3)-$

poly(vinyl acetate)

formed from

$CH_2{=}CH(OCOCH_3)$

13.23 Name each aromatic compound as in Example 13.8. Name the substituents on the benzene ring. With two groups, alphabetize the substituent names and use the prefix ortho, meta, or para to indicate their location. With three substituents, alphabetize the substituent names, and number to give the lowest set of numbers.

a. $CH_2CH_2CH_3$ ← propyl

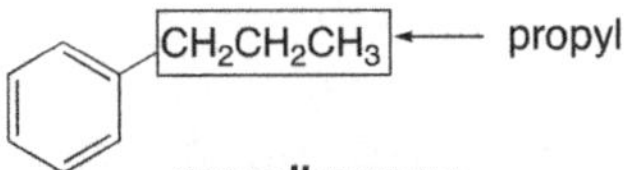

propylbenzene

c. OH ← OH on benzene ring = phenol

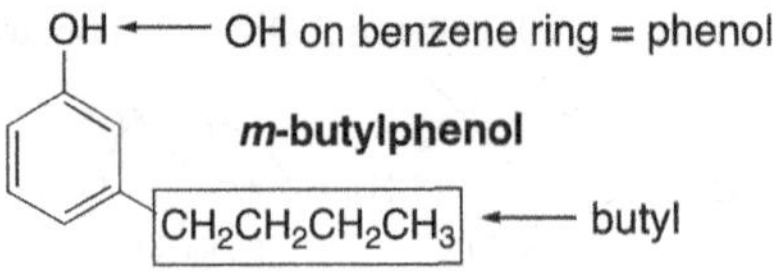

***m*-butylphenol**

$CH_2CH_2CH_2CH_3$ ← butyl

b. CH_2CH_3 ← ethyl

I ← iodo

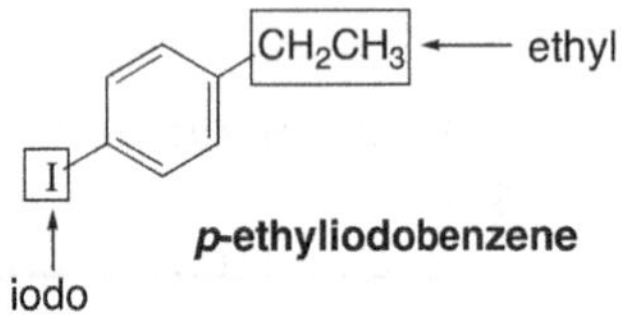

***p*-ethyliodobenzene**

d. CH_3 ← CH_3 on benzene ring = toluene

Br ← bromo

Cl ← chloro

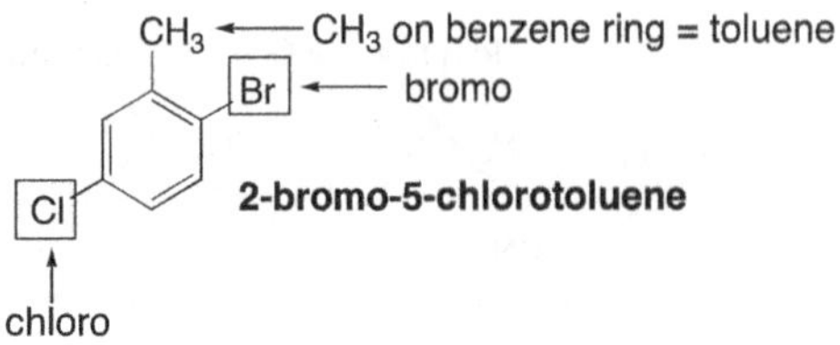

2-bromo-5-chlorotoluene

13.24 Draw the structure corresponding to each name.

a. pentylbenzene — $CH_2CH_2CH_2CH_2CH_3$

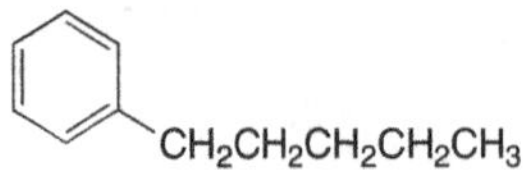

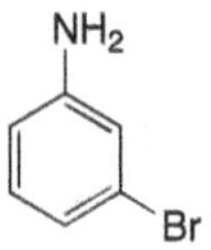

b. *o*-dichlorobenzene — Cl, Cl

c. *m*-bromoaniline — NH_2, Br

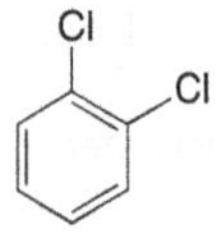

d. 4-chloro-1,2-diethylbenzene — CH_2CH_3, CH_2CH_3, Cl

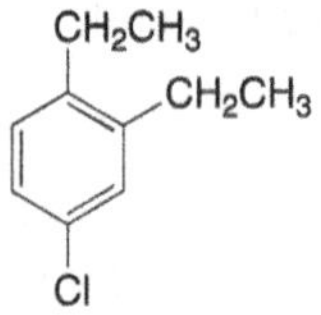

13.25 Commercially available sunscreens contain a benzene ring. Therefore, compound (a) might be found in a sunscreen since it contains two aromatic rings. Compound (b) does not contain any aromatic rings.

13.26 Phenols are antioxidants because the OH group on the benzene ring prevents unwanted oxidation reactions from occurring. Of the compounds listed, only curcumin (b) contains a phenol group (OH on a benzene ring), making it an antioxidant.

13.27 Draw the products of each substitution reaction.

- Chlorination replaces one of the H's on the benzene ring with Cl.
- Nitration replaces one of the H's on the benzene ring with NO_2.
- Sulfonation replaces one of the H's on the benzene ring with SO_3H.

a. Cl_2, $FeCl_3$ → Cl, Cl, Cl

c. SO_3, H_2SO_4 → Cl, Cl, SO_3H

b. HNO_3, H_2SO_4 → Cl, Cl, NO_2

13.28 Draw the products of the substitution reaction. The Cl can replace any of the H's on the benzene ring, giving three different products.

CH_3 — Cl_2, $FeCl_3$ → CH_3, Cl + CH_3, Cl + CH_3, Cl

o-chlorotoluene *m*-chlorotoluene *p*-chlorotoluene

Solutions to Odd-Numbered End-of-Chapter Problems

13.29

a. molecular formula: $C_{10}H_{12}O$
b. aromatic ring, alkene, ether
c. trans
d. Tetrahedral C's are indicated. All other C's are trigonal planar.

aromatic ring, ether, trans alkene, CH_3 ← tetrahedral, CH_3 ↑ tetrahedral, anethole

13.31 Use the general formulas [saturated hydrocarbon (C_nH_{2n+2}), alkene (C_nH_{2n}), and alkyne (C_nH_{2n-2})] to determine the molecular formula for each compound with 10 C's.

a. $(10 \times 2) + 2 = 22$: molecular formula $C_{10}H_{22}$
b. $10 \times 2 = 20$: molecular formula $C_{10}H_{20}$
c. $(10 \times 2) - 2 = 18$: molecular formula $C_{10}H_{18}$

13.33 Draw three alkynes with molecular formula C_5H_8.

$HC{\equiv}CCH_2CH_2CH_3$ $CH_3C{\equiv}CCH_2CH_3$ $HC{\equiv}C{-}C(CH_3)_2{-}H$

13.35 Label each carbon as tetrahedral, trigonal planar, or linear by counting groups.

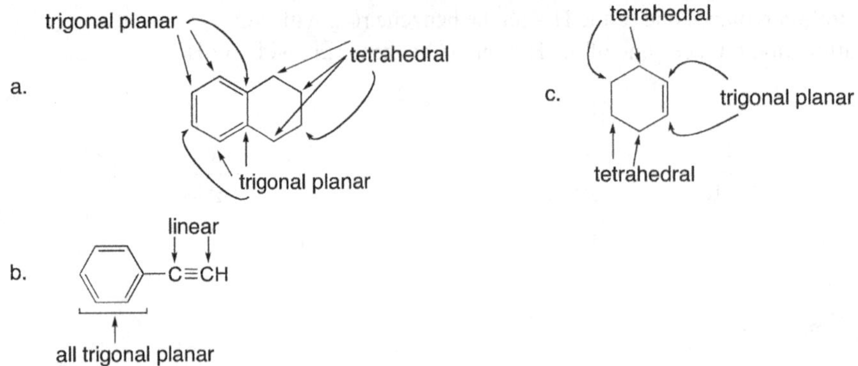

13.37 Give the IUPAC name for each compound.

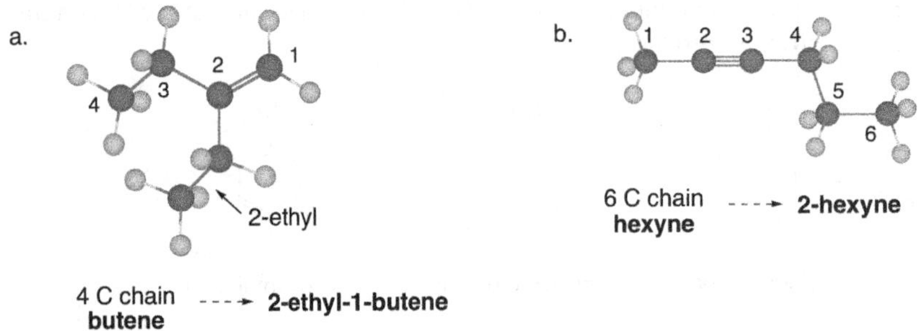

13.39 Give the IUPAC name for each compound using the following steps, as in Example 13.2:
[1] Find the longest chain containing both carbon atoms of the multiple bond.
[2] Number the chain to give the multiple bond the lower number.
[3] Name and number the substituents and write the complete name.

a. $CH_2{=}CHCH_2CH_2C(CH_3)_3$

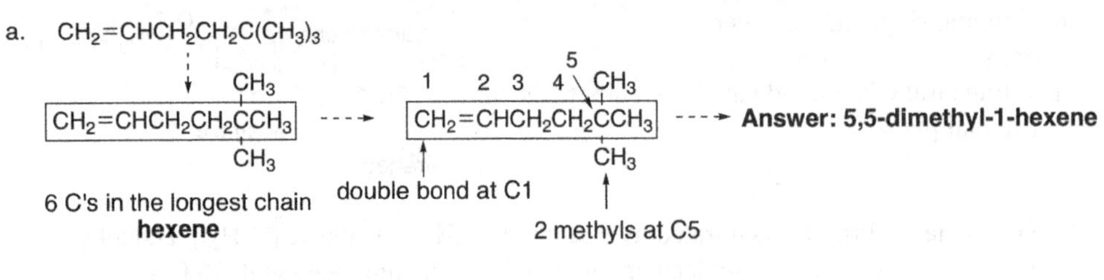

b. $(CH_3CH_2)_2C{=}CHCHCH_2CHCH_3$ (with CH_3 on C5 and C7)

CH_3CH_2 CH_3 CH_3

$CH_3CH_2C{=}CHCHCH_2CHCH_3$

8 C's in the longest chain
octene

ethyl at C3 2 methyls at C5 and C7

CH_3CH_2 CH_3 CH_3

$CH_3CH_2C{=}CHCHCH_2CHCH_3$

1 2 3 4 5 6 7 8

double bond at C3

Answer:
3-ethyl-5,7-dimethyl-3-octene

c. $CH_2{=}C(CH_2CH_3)CH_2CH_2CH_2CH_2CH_3$ ----→ numbered 1 2 (C1=C2), 3 4 5 6 7 along $CH_2CH_2CH_2CH_2CH_3$ ----→ **Answer: 2-ethyl-1-heptene**

7 C's in the longest chain
heptene

double bond at C1

ethyl at C2

d. $CH_3C{\equiv}CCH_2C(CH_3)_3$

$CH_3C{\equiv}CCH_2C(CH_3)_2CH_3$ ----→ numbered 1 2 3 … 5 … 6 ----→ **Answer: 5,5-dimethyl-2-hexyne**

6 C's in the longest chain
hexyne

triple bond at C2

2 methyls at C5

e. $CH_3C{\equiv}C{-}CH_2{-}CH(CH_3){-}C(CH_3)(CH_2CH_3){-}CH_2CH_3$ ----→ numbered 1 2 3 4 5 6 … 8 ----→ **Answer: 6-ethyl-5,6-dimethyl-2-octyne**

8 C's in the longest chain
octyne

triple bond at C2

2 methyls at C5 and C6

ethyl at C6

f. $CH_2{=}CHCH_2{-}C(CH_3)_2{-}CH{=}CH_2$ ----→ numbered 6 5 4 3 … 1 ----→ **Answer: 3,3-dimethyl-1,5-hexadiene**

6 C's in the longest chain
hexadiene

2 methyls at C3
double bonds at C1 and C5

13.41 Give the IUPAC name for each compound using the steps in Answer 13.39 and Example 13.2.

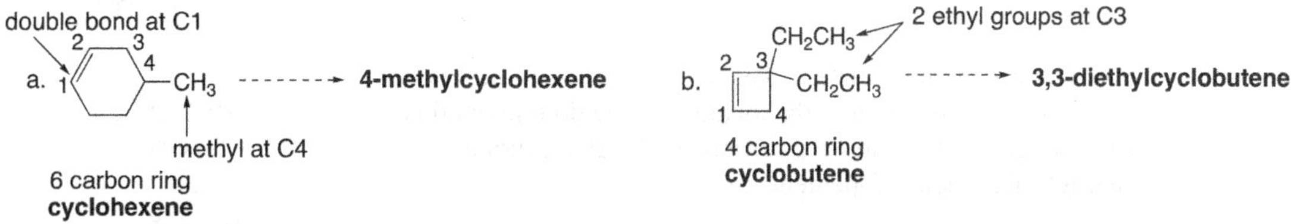

13.43 To draw the structure corresponding to each name, follow the steps in Example 13.3.

- Identify the parent name to find the longest carbon chain or ring, and then use the suffix to determine the functional group; the suffix *-ene* = alkene and *-yne* = alkyne.
- Number the carbon chain or ring and place the functional group at the indicated carbon. Add the substituents and enough hydrogens to give each carbon four bonds.

a. 3-methyl-1-**octene** ------→ 1 2 3 4 5 6 7 8: C=C–C(CH$_3$)–C–C–C–C–C ------→ $CH_2{=}CHCH(CH_3)CH_2CH_2CH_2CH_2CH_3$

8 carbon chain
double bond at C1

methyl at C3

b. 1-ethyl**cyclobutene** → C=C(CH₂CH₃) / C–C ring; ethyl at C1 → cyclobutene with CH_2CH_3
4 carbon ring
double bond at C1

c. 2-methyl-3-**hexyne** → 1 2 3 4 5 6: C–C(CH₃)–C≡C–C–C; methyl at C2 → $CH_3-CH(CH_3)-C\equiv CCH_2CH_3$
6 carbon chain
triple bond at C3

d. 3,5-diethyl-2-methyl-3-**heptene** → 1 2 3 4 5 6 7: C–C(CH₃)–C(CH₂CH₃)=C–C(CH₂CH₃)–C–C; methyl at C2; 2 ethyl groups at C3 and C5 → $CH_3-CH(CH_3)-C(CH_2CH_3)=CH-CH(CH_2CH_3)CH_2CH_3$
7 carbon chain
double bond at C3

e. 1,3-**heptadiene** → 1 2 3 4 5 6 7: C=C–C=C–C–C–C → $CH_2=CH-CH=CHCH_2CH_2CH_3$
7 carbon chain
double bonds at C1 and C3

f. *cis*-7-methyl-2-**octene** → 1 2 3 4 5 6 7 8: C–C=C–C–C–C–C(CH₃)–C; methyl at C7 → $CH_3(H)C=C(H)CH_2CH_2CH_2CH(CH_3)CH_3$ (cis)
8 carbon chain
double bond at C2

13.45 Correct each of the incorrect IUPAC names.

a. The name 5-methyl-4-hexene places the double bond at C4 instead of C2. Assign the lower number to the alkene: 2-methyl-2-hexene.

$CH_3C(CH_3)=CHCH_2CH_2CH_3$
2-methyl-2-hexene

b. The name 1-methylbutene makes the last carbon in the chain a substituent. In addition, the location of the double bond is not specified. There are five carbons in the chain (not a methyl substituent): 2-pentene.

$CH_3CH=CHCH_2CH_3$
2-pentene

c. The name 2,3-dimethylcyclohexene starts numbering substituents at C2 instead of C1. Number to put the C=C between C1 and C2, and then give the first substituent the lower number: 1,6-dimethylcyclohexene.

(cyclohexene ring numbered 1–6, CH_3 at C1 and C6)
1,6-dimethylcyclohexene

d. The name 3-butyl-1-butyne does not name the longest chain. Name the seven carbon chain: 3-methyl-1-heptyne.

$HC\equiv CCH(CH_3)CH_2CH_2CH_2CH_3$ (1, 2, 3 numbered)
3-methyl-1-heptyne

13.47 When the two alkyl groups are on the same side of the double bond, the compound is called the cis isomer. When they are on opposite sides it is called the trans isomer.

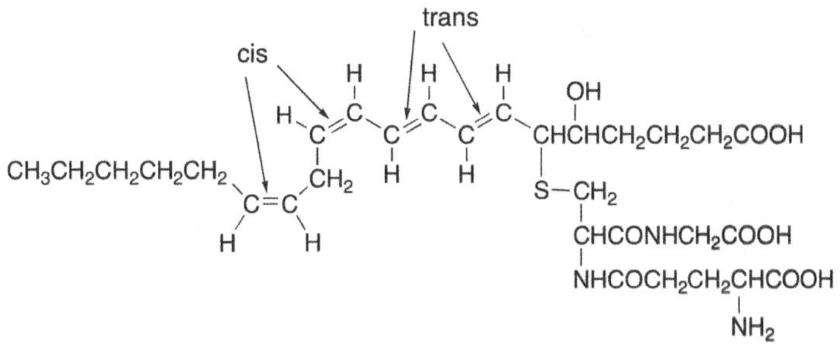

cis; trans; trans

$CH_3CH_2CH_2CH_2CH_2$–CH=CH–CH_2–CH=CH–CH=CH–CH=CH–CH(OH)$CH_2CH_2CH_2COOH$; S–CH_2–$CHCONHCH_2COOH$; $NHCOCH_2CH_2CHCOOH$; NH_2

13.49 Give the IUPAC name for the alkene. Use the definition in Answer 13.47 to determine if it is the cis or trans isomer.

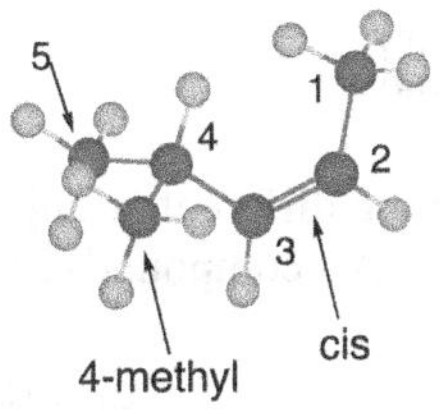

5, 4, 1, 2, 3; 4-methyl; cis

cis-4-methyl-2-pentene

13.51 Draw the cis and trans isomers for each compound, as in Example 13.4.

a. H, H; C=C; CH_3, $CH_2CH_2CH_2CH_2CH_2CH_3$

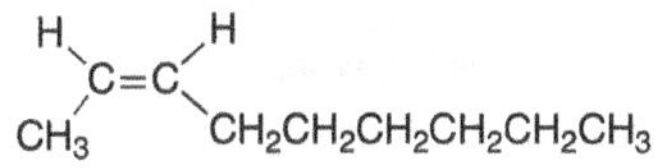

***cis*-2-nonene**

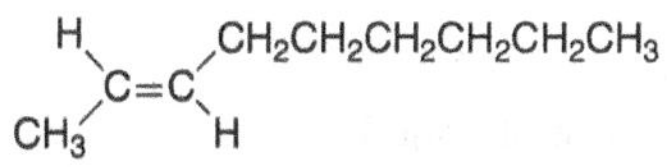

H, $CH_2CH_2CH_2CH_2CH_2CH_3$; C=C; CH_3, H

***trans*-2-nonene**

b. H, H; C=C; CH_3CH(CH_3), $CH_2CH_2CH_3$

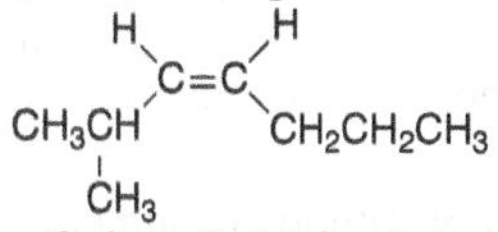

***cis*-2-methyl-3-heptene**

H, $CH_2CH_2CH_3$; C=C; CH_3CH(CH_3), H

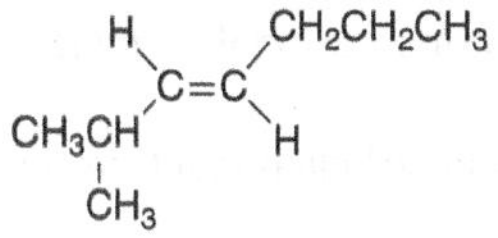

***trans*-2-methyl-3-heptene**

13.53

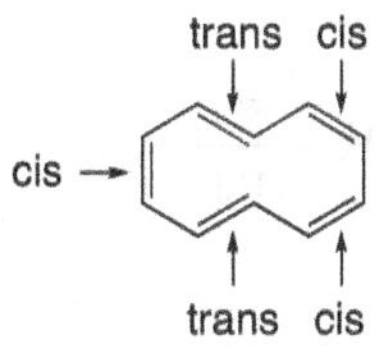

trans cis

cis →

trans cis

13.55 Determine if the molecules are constitutional isomers, stereoisomers, or identical.

a. 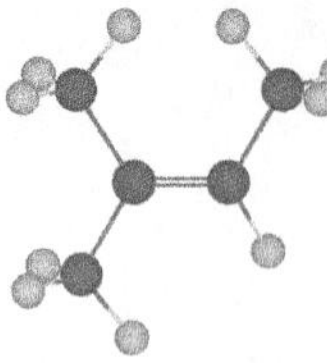 and

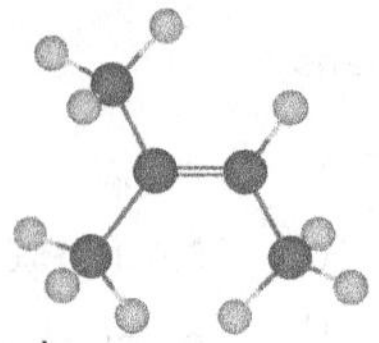

same molecular formula
same connectivity
identical

b. 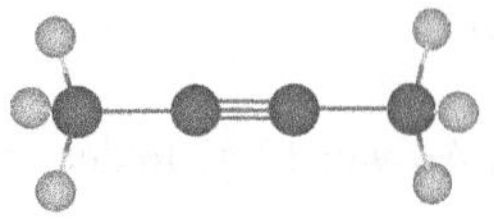 and

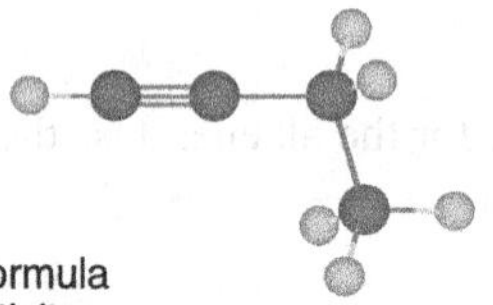

same molecular formula
different connectivity
constitutional isomers

13.57 Isomers have the same molecular formula but a different arrangement of atoms. Stereoisomers differ only in the 3-D arrangement of groups. All compounds have molecular formula C_6H_{12}.

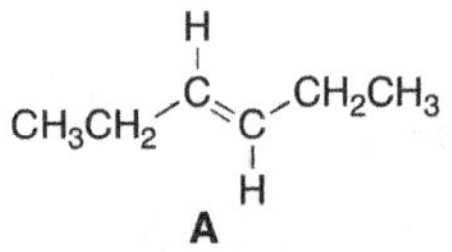

A *trans*-3-hexene — $CH_3CH_2(H)C{=}C(H)CH_2CH_3$

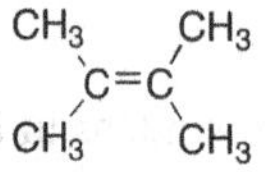

B 2,3-dimethyl-2-butene — $(CH_3)_2C{=}C(CH_3)_2$

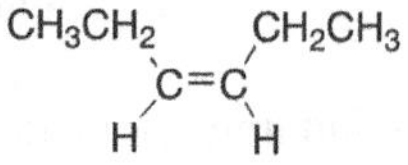

C *cis*-3-hexene — $CH_3CH_2(H)C{=}C(H)CH_2CH_3$

a. **A** and **B** are constitutional isomers.
b. **A** and **C** are stereoisomers.
c. **B** and **C** are constitutional isomers.

13.59 Draw the products of each reaction by adding H_2 to the double bond.

a. $CH_2{=}CHCH_2CH_2CH_2CH_3 \xrightarrow[Pd]{H_2} CH_3CH_2CH_2CH_2CH_2CH_3$

b. $(CH_3)_2C{=}CHCH_2CH_2CH_3 \xrightarrow[Pd]{H_2} (CH_3)_2CHCH_2CH_2CH_2CH_3$

c. 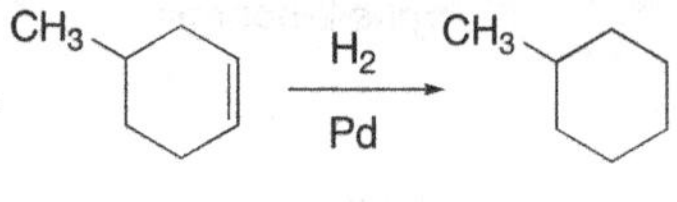 3-methylcyclohexene (CH$_3$) $\xrightarrow[Pd]{H_2}$ methylcyclohexane (CH$_3$)

d. 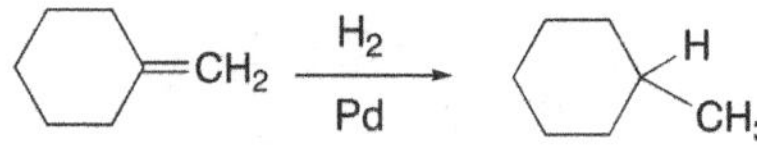 methylenecyclohexane (=CH$_2$) $\xrightarrow[Pd]{H_2}$ methylcyclohexane (H, CH$_3$)

13.61 Draw the products of each reaction by adding HCl to the double bond.

a. 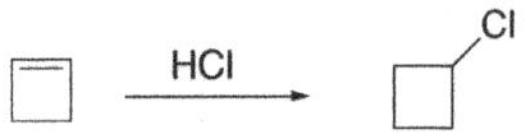 cyclobutene $\xrightarrow{HCl}$ chlorocyclobutane (Cl)

b. $(CH_3)_2C{=}C(CH_3)_2 \xrightarrow{HCl} (CH_3)_2C(Cl){-}C(H)(CH_3)_2$

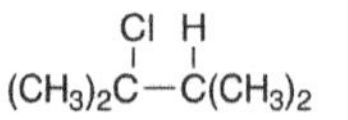

c. $CH_2{=}CHCH_2CH(CH_3)_2 \xrightarrow{HCl} CH_3CH(Cl)CH_2CH(CH_3)_2$

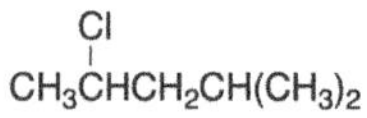

d. 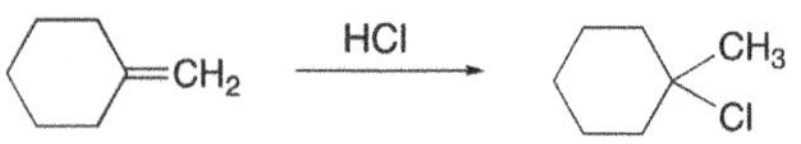 methylenecyclohexane (=CH$_2$) $\xrightarrow{HCl}$ 1-chloro-1-methylcyclohexane (CH$_3$, Cl)

13.63 Draw the products of each reaction by adding the specified reagent to the double bond.

a. 1-ethylcyclohexene $\xrightarrow[Pd]{H_2}$ product with H, CH_2CH_3, H, H

b. 1-ethylcyclohexene $\xrightarrow{Cl_2}$ product with Cl, CH_2CH_3, Cl, H

c. 1-ethylcyclohexene $\xrightarrow{Br_2}$ product with Br, CH_2CH_3, Br, H

d. 1-ethylcyclohexene $\xrightarrow{HCl}$ product with Cl, CH_2CH_3, H, H

e. 1-ethylcyclohexene $\xrightarrow{HBr}$ product with Br, CH_2CH_3, H, H

f. 1-ethylcyclohexene $\xrightarrow[H_2SO_4]{H_2O}$ product with OH, CH_2CH_3, H, H

13.65 Work backwards to determine what alkene is needed as a starting material to prepare each of the alkyl halides or dihalides.

a. $CH_2{=}CH_2 \xrightarrow{HBr} CH_3CH_2Br$

b. cyclohexene $\xrightarrow{HCl}$ chlorocyclohexane (Cl)

c. 1-methylcyclopentene (CH_3) $\xrightarrow{Cl_2}$ dichloro product (Cl, Cl, CH_3)

d. $CH_2{=}CHCH_2CH(CH_3)_2 \xrightarrow{Br_2} BrCH_2CHCH_2CH(CH_3)_2$ (Br on C2)

13.67 Work backwards to determine what reagent is needed to convert 2-methylpropene to each product.

a. $(CH_3)_2C{=}CH_2 \xrightarrow{HCl} (CH_3)_3CCl$

b. $(CH_3)_2C{=}CH_2 \xrightarrow[Pd]{H_2} (CH_3)_3CH$

c. $(CH_3)_2C{=}CH_2 \xrightarrow[H_2SO_4]{H_2O} (CH_3)_3COH$

d. $(CH_3)_2C{=}CH_2 \xrightarrow{HBr} (CH_3)_3CBr$

e. $(CH_3)_2C{=}CH_2 \xrightarrow{Br_2} (CH_3)_2CCH_2Br$ (Br on C2)

f. $(CH_3)_2C{=}CH_2 \xrightarrow{Cl_2} (CH_3)_2CCH_2Cl$ (Cl on C2)

13.69 To draw the polymer, draw three or more alkene molecules and arrange the carbons of the double bonds next to each other. Break one bond of each double bond, and join the alkenes together with single bonds. With unsymmetrical alkenes, substituents are bonded to every other carbon. Use Example 13.7 as a guide.

Join these 2 C's. Join these 2 C's.

$CH_2{=}C(H)(COOH)$ $CH_2{=}C(H)(COOH)$ $CH_2{=}C(H)(COOH)$ $\longrightarrow$ $-CH_2C(COOH)(H)-CH_2C(COOH)(H)-CH_2C(COOH)(H)-$

13.71 Translate the 3-D model to a structure, and draw the polymers using the steps in Example 13.7.

a. Join these 2 C's. Join these 2 C's.

$CH_2{=}C(CH_2CH_3)H$ + $CH_2{=}C(CH_2CH_3)H$ + $CH_2{=}C(CH_2CH_3)H$ ⟶ $-CH_2C(CH_2CH_3)H-CH_2C(CH_2CH_3)H-CH_2C(CH_2CH_3)H-$

b. Join these 2 C's. Join these 2 C's.

$CH_2{=}C(Cl)CN$ + $CH_2{=}C(Cl)CN$ + $CH_2{=}C(Cl)CN$ ⟶ $-CH_2C(Cl)(CN)-CH_2C(Cl)(CN)-CH_2C(Cl)(CN)-$

13.73 Work backwards to determine what monomer was used to form the polymer.

Each one of these units is from the monomer:

$-CH_2-C(Br)(Cl)-CH_2-C(Br)(Cl)-CH_2-C(Br)(Cl)-$ ⟶ $CH_2{=}C(Cl)Br$

13.75 Draw two resonance structures by moving the double bonds.

Cl-benzene ⟷ Cl-benzene

13.77 Name each aromatic compound as in Example 13.8. Name the substituents on the benzene ring. With two groups, alphabetize the substituent names and use the prefix ortho, meta, or para to indicate their location. With three substituents, alphabetize the substituent names, and number to give the lowest set of numbers.

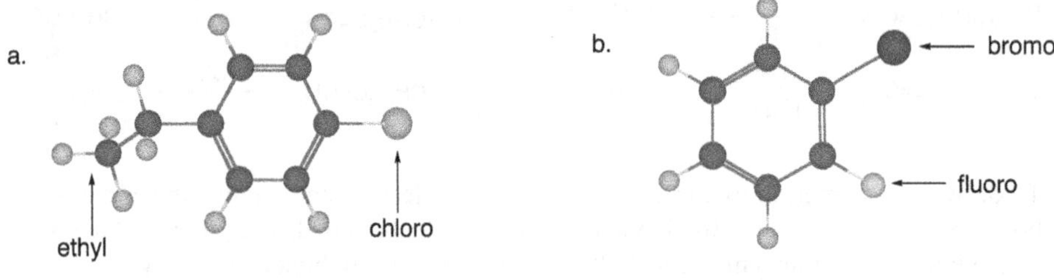

***p*-chloroethylbenzene** ***o*-bromofluorobenzene**

13.79 Name each aromatic compound as in Example 13.8 and Answer 13.77.

a. 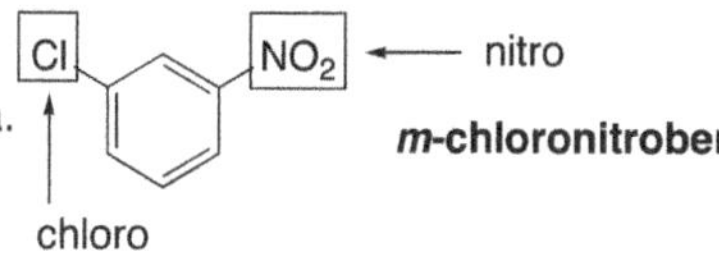 Cl ← chloro; NO_2 ← nitro — ***m*-chloronitrobenzene**

b. 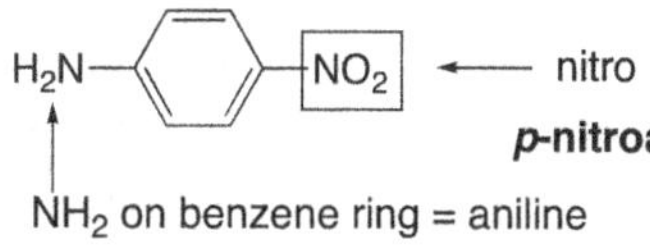 H_2N ← NH_2 on benzene ring = aniline; NO_2 ← nitro — ***p*-nitroaniline**

c. 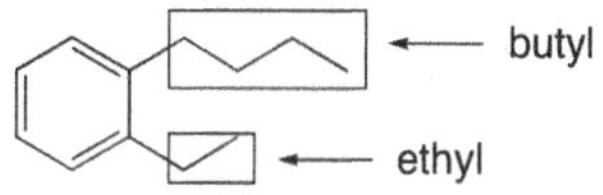 ← butyl; ← ethyl — ***o*-butylethylbenzene**

d. 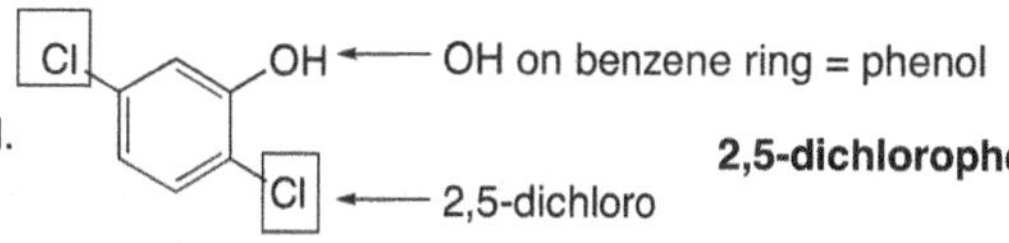 Cl; OH ← OH on benzene ring = phenol; Cl ← 2,5-dichloro — **2,5-dichlorophenol**

13.81 Draw and name the three isomers with Cl and NH_2 as substituents. Recall that a benzene ring with an NH_2 group is named aniline.

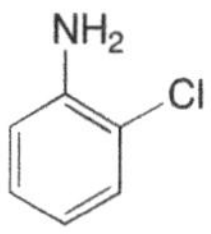

o-chloroaniline 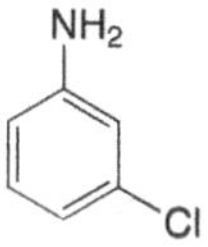 — *m*-chloroaniline  — *p*-chloroaniline

13.83 Work backwards to draw the structure from the IUPAC name.

a. 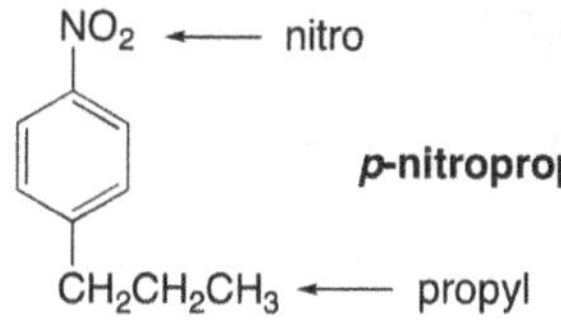 NO_2 ← nitro; $CH_2CH_2CH_3$ ← propyl — ***p*-nitropropylbenzene**

b. 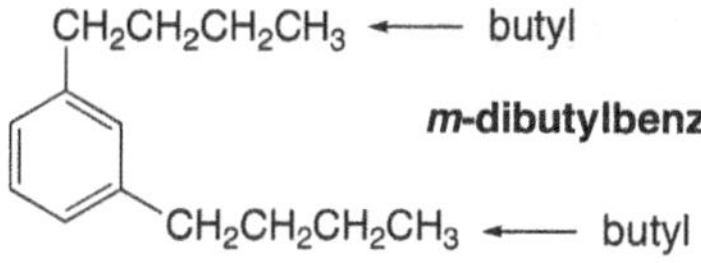 $CH_2CH_2CH_2CH_3$ ← butyl; $CH_2CH_2CH_2CH_3$ ← butyl — ***m*-dibutylbenzene**

c. 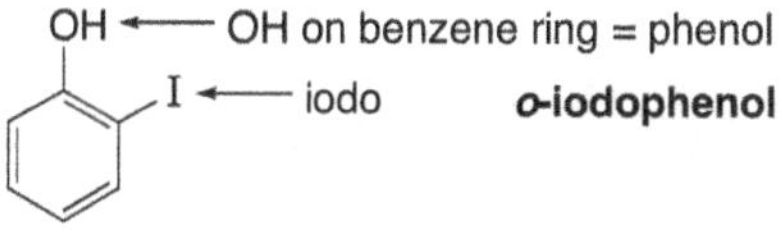 OH ← OH on benzene ring = phenol; I ← iodo — ***o*-iodophenol**

d. 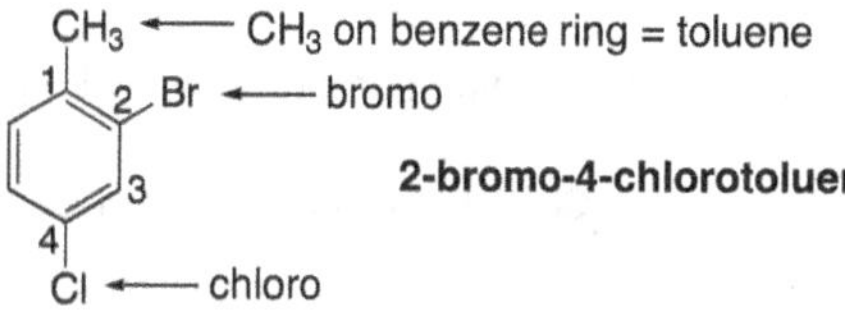 CH_3 ← CH_3 on benzene ring = toluene; Br ← bromo; Cl ← chloro — **2-bromo-4-chlorotoluene**

e. 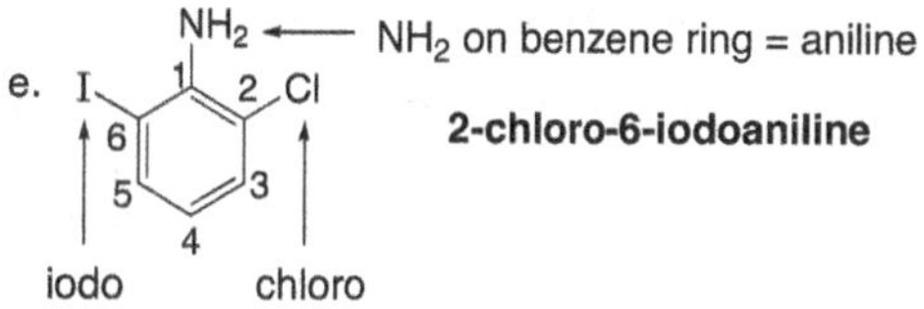 NH_2 ← NH_2 on benzene ring = aniline; I ← iodo; Cl ← chloro — **2-chloro-6-iodoaniline**

13.85 Draw the products of each reaction.

a. 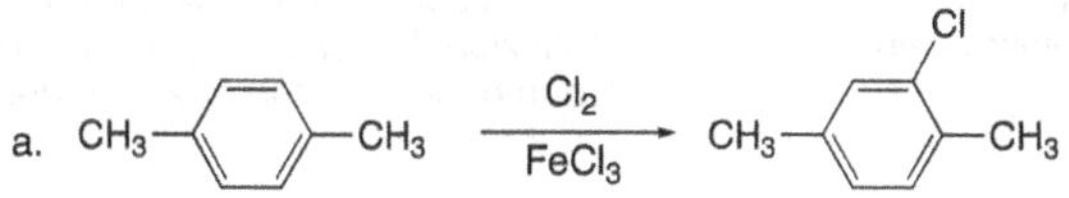 $CH_3-C_6H_4-CH_3$ $\xrightarrow[FeCl_3]{Cl_2}$ $CH_3-C_6H_3(Cl)-CH_3$

b. $CH_3-C_6H_4-CH_3$ $\xrightarrow[H_2SO_4]{HNO_3}$ $CH_3-C_6H_3(NO_2)-CH_3$

c. $CH_3-C_6H_4-CH_3$ $\xrightarrow[H_2SO_4]{SO_3}$ $CH_3-C_6H_3(SO_3H)-CH_3$

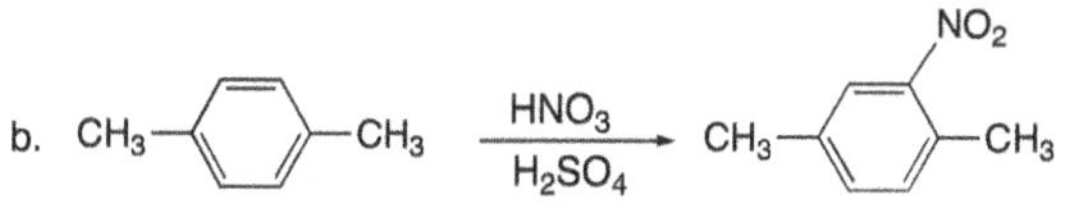

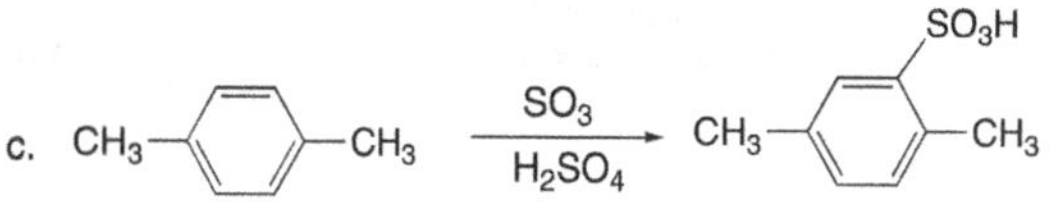

13.87 Draw the three products formed in the reaction of bromobenzene.

Bromobenzene $\xrightarrow[H_2SO_4]{HNO_3}$ o-bromonitrobenzene (Br, NO_2) + m-bromonitrobenzene (Br, NO_2) + p-bromonitrobenzene (Br, NO_2)

13.89 Vitamin E is an antioxidant because of the phenol, which has an OH bonded to the benzene ring.

13.91 Methoxychlor is more water soluble than DDT, because methoxychlor's OCH_3 groups can hydrogen bond to water. This increase in water solubility makes methoxychlor more biodegradable.

13.93

a. $CH_3CH_2CH_2CH_2CH_2CH{=}CHCH_2CH{=}CHCH_2CH_2CH_2CH_2CH_2CH_2CH_2COOH$ (both C=C with H, H cis)

$\downarrow$ H_2, Pd

$CH_3CH_2CH_2CH_2CH_2CH_2CH_2CH_2CH{=}CHCH_2CH_2CH_2CH_2CH_2CH_2CH_2COOH$ (H, H cis)

\+

$CH_3CH_2CH_2CH_2CH_2CH{=}CHCH_2CH_2CH_2CH_2CH_2CH_2CH_2CH_2CH_2CH_2COOH$ (H, H cis)

Partial hydrogenation adds hydrogen to one of the double bonds.

b. Complete hydrogenation adds hydrogen to both of the double bonds, forming:

$CH_3CH_2CH_2CH_2CH_2CH_2CH_2CH_2CH_2CH_2CH_2CH_2CH_2CH_2CH_2CH_2CH_2COOH$

c. one possibility:

$CH_3CH_2CH_2CH_2CH_2CH_2CH_2CH_2(H)C{=}C(H)CH_2CH_2CH_2CH_2CH_2CH_2CH_2COOH$

trans

13.95 Recall from Section 13.12 that many phenols are antioxidants.

a. C_6H_5–$CH_2CH_2CH_2OH$
an alcohol
not an antioxidant

b. C_6H_5–OCH_3
an ether
not an antioxidant

c. HO–C_6H_4–OCH_3
This compound could be an antioxidant because it has an OH group bonded to the aromatic ring.

13.97 When benzene is oxidized to phenol, it is converted to a more water-soluble compound that can then be excreted in the urine.

13.99

a. 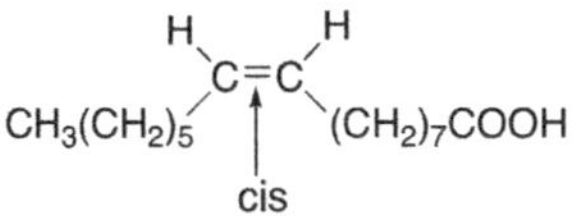 cis-structure: H and H on one side of C=C; $CH_3(CH_2)_5$ and $(CH_2)_7COOH$ on the other. cis — **palmitoleic acid**

b. 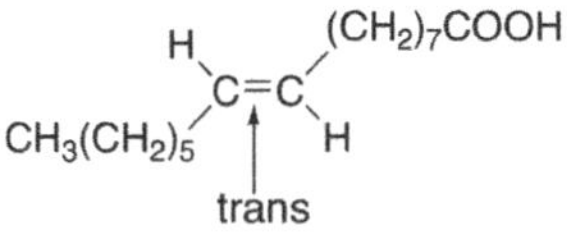 trans-structure: H / $(CH_2)_7COOH$ on top; $CH_3(CH_2)_5$ / H on bottom. trans — **stereoisomer**

c. 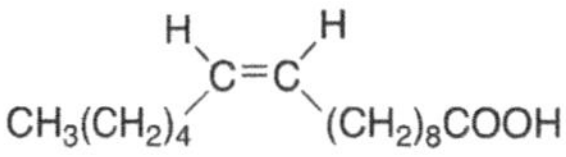 H, H on top; $CH_3(CH_2)_4$ / $(CH_2)_8COOH$ on bottom. **constitutional isomer** one possibility

13.101 All the carbons in benzene are trigonal **planar** with 120° bond angles, resulting in a flat ring. In cyclohexane, all the carbons are tetrahedral, so the ring puckers in order to have bond angles close to the tetrahedral bond angle of 109.5°.

13.103

a. $CH_2{=}CH(CH_2)_4CH_3$ 7 C chain **1-heptene**

b. $CH_2{=}CH(CH_2)_4CH_3 \xrightarrow{H_2} CH_3(CH_2)_5CH_3$

c. $CH_2{=}CH(CH_2)_4CH_3 \xrightarrow{H_2O} CH_3CH(OH)CH_2CH_2CH_2CH_2CH_3$

d. polymerization: $\sim CH_2C(R)(H){-}CH_2C(R)(H){-}CH_2C(R)(H) \sim$ R = $(CH_2)_4CH_3$

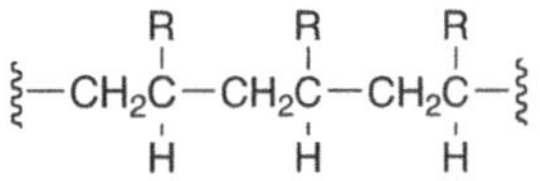

13.105

	[1] Name	[2] Functional Group	[3] Fact
a. HDPE	high-density polyethylene	none	HDPE is rigid and used in milk containers and water jugs.
b. PS	polystyrene	aromatic ring	Polystyrene is used in molded trays and packaging.
c. PAH	polycyclic aromatic hydrocarbon	aromatic ring	PAHs are produced when organic compounds are burned.

13.107 *cis*-2-Hexene and *trans*-3-hexene are constitutional isomers because the double bond is located in a different place on the carbon chain (C2 vs. C3).

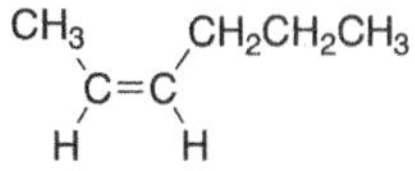

CH_3 and $CH_2CH_2CH_3$ on top, H and H on bottom of C=C — *cis*-2-hexene

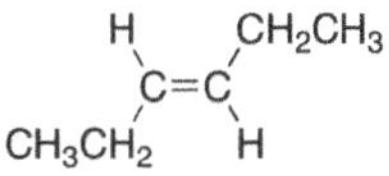

H and CH_2CH_3 on top, CH_3CH_2 and H on bottom of C=C — *trans*-3-hexene

Chapter 14 Organic Compounds That Contain Oxygen, Halogen, or Sulfur

Chapter Review

[1] What are the characteristics of alcohols, ethers, alkyl halides, and thiols?

- Alcohols contain a hydroxyl group (OH group) bonded to a tetrahedral carbon. Since the O atom is surrounded by two atoms and two lone pairs, alcohols have a bent shape around O. (14.2)
- Ethers have two alkyl groups bonded to an oxygen atom. Since the O atom is surrounded by two atoms and two lone pairs, ethers have a bent shape around O. (14.7)
- Alkyl halides contain a halogen atom (X = F, Cl, Br, or I) singly bonded to a tetrahedral carbon. (14.9)
- Thiols contain a sulfhydryl group (SH group) bonded to a tetrahedral carbon. Since the S atom is surrounded by two atoms and two lone pairs, thiols have a bent shape around S. (14.10)

[2] How are alcohols and alkyl halides classified?

- Alcohols and alkyl halides are classified by the number of C's bonded to the C with the functional group.
- RCH_2OH = 1° alcohol; R_2CHOH = 2° alcohol; R_3COH = 3° alcohol. (14.2)

Alcohol	Classification of alcohols		
—C—O—H (hydroxyl group)	CH_3—C(H)(H)—OH	CH_3—C(CH_3)(H)—OH	CH_3—C(CH_3)(CH_3)—OH
	1°	2°	3°

- RCH_2X = 1° alkyl halide; R_2CHX = 2° alkyl halide; R_3CX = 3° alkyl halide. (14.9)

Alkyl halide	Classification of alkyl halides		
—C—X:	CH_3—C(H)(H)—X:	CH_3—C(CH_3)(H)—X:	CH_3—C(CH_3)(CH_3)—X:
X = F, Cl, Br, I	1°	2°	3°

[3] What are the properties of alcohols, ethers, alkyl halides, and thiols?

- Alcohols have a bent shape and polar C–O and O–H bonds, so they have a net dipole. Their OH bond allows for intermolecular hydrogen bonding between two alcohol molecules or between an alcohol molecule and water. As a result, alcohols have the strongest intermolecular forces of the four families of molecules in this chapter. (14.2)

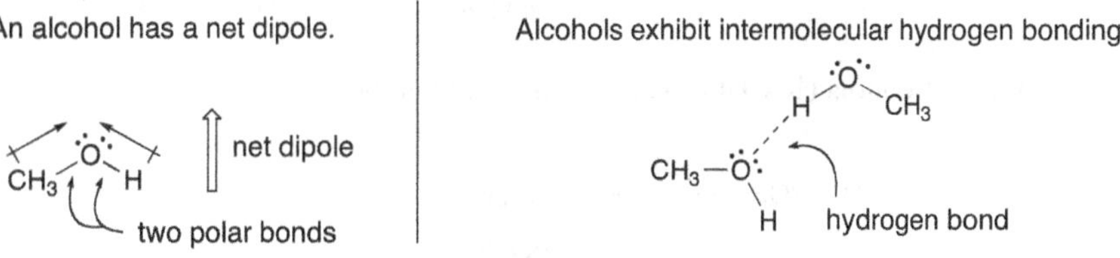

- Ethers have a bent shape and two polar C–O bonds so they have a net dipole. (14.7)

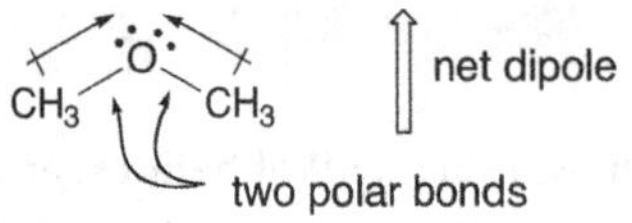

net dipole

two polar bonds

- Alkyl halides with one halogen have one polar bond and a net dipole. (14.9)

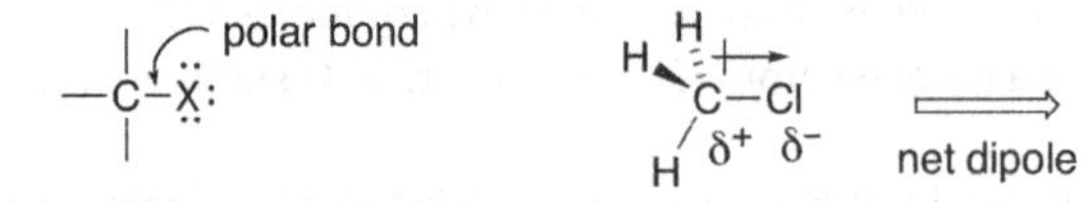

polar bond

X = F, Cl, Br, I

net dipole

- Thiols have lower boiling points than alcohols with the same number of carbons. (14.10)

[4] How are alcohols, ethers, alkyl halides, and thiols named?

- Alcohols are identified by the suffix *-ol.* (14.3)
- Ethers are named in two ways. Simple ethers are named by naming the alkyl groups bonded to the ether oxygen and adding the word *ether*. More complex ethers are named as *alkoxy alkanes*; that is, the simpler alkyl group is named as an alkoxy group (RO) bonded to an alkane. (14.7B)
- Alkyl halides are named as *halo alkanes*; that is, the halogen is named as a substituent (halo group) bonded to an alkane. (14.9B)
- Thiols are identified by the suffix *-thiol.* (14.10)

[5] What products are formed when an alcohol undergoes dehydration? (14.5A)

- Alcohols form alkenes on treatment with strong acid. The elements of H and OH are lost from two adjacent atoms and a new carbon–carbon double bond is formed.

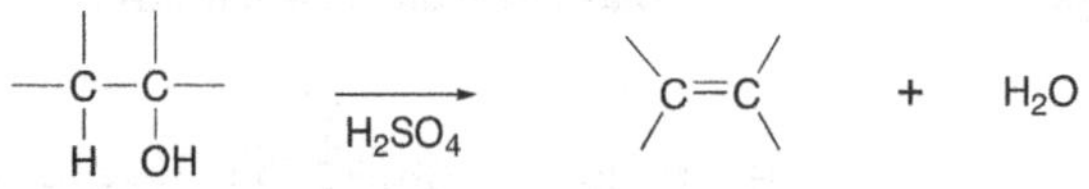

$$\text{H–C–C–OH} \xrightarrow{H_2SO_4} \text{C=C} + H_2O$$

- Dehydration follows the Zaitsev rule: When more than one alkene can be formed, the major product of elimination is the alkene that has more alkyl groups bonded to it.

[6] What products are formed when an alcohol is oxidized? (14.5B)

- Primary alcohols (RCH_2OH) are oxidized to aldehydes (RCHO), which are further oxidized to carboxylic acids (RCO_2H).

$$RCH_2OH \xrightarrow{[O]} \underset{\text{aldehyde}}{RCHO} \xrightarrow{[O]} \underset{\text{carboxylic acid}}{RCOOH}$$

- Secondary alcohols (R_2CHOH) are oxidized to ketones.

$$R_2CHOH \xrightarrow{[O]} \underset{\text{ketone}}{R_2CO}$$

- Tertiary alcohols have no C–H bond on the carbon with the OH group, so they are not oxidized.

$$R_3COH \xrightarrow{[O]} \text{No reaction}$$

[7] What product is formed when a thiol is oxidized? (14.10)

- Thiols (RSH) are oxidized to disulfides (RSSR).
- Disulfides are reduced to thiols.

$$\underset{\text{thiol}}{2\ R{-}S{-}H} \underset{[H]}{\overset{[O]}{\rightleftharpoons}} \underset{\text{disulfide}}{R{-}S{-}S{-}R}$$

Problem Solving

[1] Structure and Properties of Alcohols (14.2)

Example 14.1 Classify each alcohol as 1°, 2°, or 3°.

a. 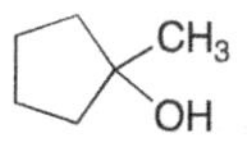 1-methylcyclopentanol (cyclopentane ring with CH_3 and OH on the same carbon)

b. $CH_3CH_2CH_2OH$

Analysis

To determine whether an alcohol is 1°, 2°, or 3°, locate the C with the OH group and count the number of C's bonded to it. A 1° alcohol has the OH group on a C bonded to one C, and so forth.

Solution

Draw out the structure or add H's to the skeletal structure to clearly see how many C's are bonded to the C bearing the OH group.

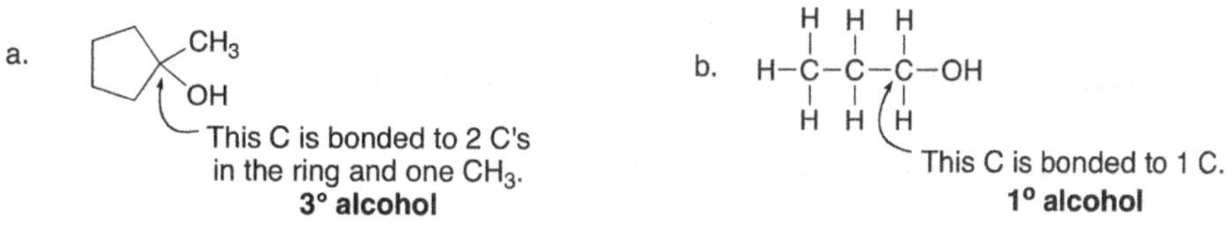

a. (cyclopentane ring with CH_3 and OH on the same carbon)
This C is bonded to 2 C's in the ring and one CH_3.
3° alcohol

b. H–C(H)(H)–C(H)(H)–C(H)(H)–OH
This C is bonded to 1 C.
1° alcohol

---

[2] Nomenclature of Alcohols (14.3)

Example 14.2 Give the IUPAC name of the following alcohol.

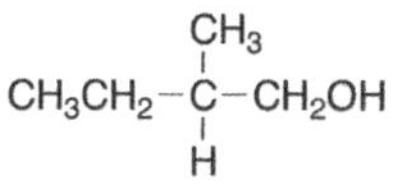

$$CH_3CH_2{-}\overset{CH_3}{\underset{H}{C}}{-}CH_2OH$$

Analysis and Solution

[1] Find the longest carbon chain that contains the carbon bonded to the OH group.

$$\begin{array}{c} CH_3 \\ | \\ \boxed{CH_3CH_2-C-CH_2}OH \\ | \\ H \end{array}$$

4 C's in the longest chain ------→ **butanol**

- Change the *-e* ending of the parent alkane to the suffix *-ol*.

[2] Number the carbon chain to give the OH group the lower number, and apply all of the other rules of nomenclature.

a. **Number** the chain.

$$\begin{array}{c} CH_3 \\ | \\ \boxed{CH_3CH_2-C-CH_2}OH \\ | \\ H \end{array}$$

4 3 2 1

1-butanol

b. **Name** and **number** the substituent.

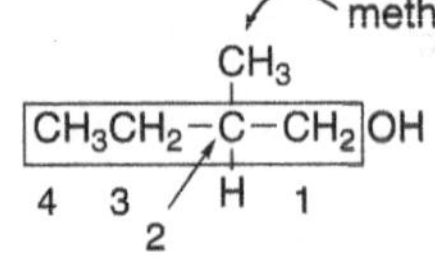

Answer: 2-methyl-1-butanol

[3] Reactions of Alcohols (14.5)

Example 14.3 Draw all possible products of dehydration of the following alcohol, and predict which one is the major product.

$$\begin{array}{c} CH_3 \\ | \\ CH_3CH_2-C-CH_2CH_3 \\ | \\ OH \end{array}$$

Analysis

To draw the products of dehydration:

- First find the carbon bonded to the OH group, and then identify all carbons with H's bonded to this carbon.
- Remove the elements of H and OH from two adjacent C's, and draw a double bond between these C's in the product.
- When two different alkenes are formed, the major product has more C's bonded to the C=C.

Solution

In this example, there are two different C's bonded to the C with the OH. Elimination of H and OH forms two different alkenes. The major product is 3-methyl-2-pentene because it has three C's bonded to the C=C, whereas 2-ethyl-1-butene has only two C's bonded to the C=C.

$$\text{CH}_3\text{CH}_2\text{–C(OH)(CH}_3\text{)–CH}_2\text{CH}_3 \xrightarrow{H_2SO_4} \text{CH}_3\text{CH}_2\text{–C(=CH}_2\text{)–CH}_2\text{CH}_3 + H_2O$$

2-ethyl-1-butene
2 C's bonded to the double bond
minor product

$$\text{CH}_3\text{–CH(H)–C(CH}_3\text{)(OH)–CH}_2\text{CH}_3 \xrightarrow{H_2SO_4} \text{CH}_3\text{–CH=C(CH}_3\text{)–CH}_2\text{CH}_3 + H_2O$$

3-methyl-2-pentene
3 C's bonded to the double bond
major product

Example 14.4 Draw the carbonyl products formed when each alcohol is oxidized with $K_2Cr_2O_7$.

a. cyclopentanol (cyclopentane ring–OH)

b. CH_3CHCH_2OH with CH_3 on the second carbon

Analysis

Classify the alcohol as 1°, 2°, or 3° by drawing in all the H atoms on the C with the OH. Then concentrate on the C with the OH group and replace H atoms by bonds to O. Keep in mind:

- RCH_2OH (1° alcohols) are oxidized to RCHO, which are then oxidized to RCOOH.
- R_2CHOH (2° alcohols) are oxidized to R_2CO.
- R_3COH (3° alcohols) are not oxidized because they have no H atom on the C with the OH.

Solution

a. Since this is a 2° alcohol with only one H atom on the C bonded to the OH group, it is oxidized to a ketone.

only 1 H atom

2° alcohol (cyclopentanol, C bearing H and OH) $\xrightarrow{[O]}$ ketone (cyclopentanone, C=O)

b. $(CH_3)_2CHCH_2OH$ is a 1° alcohol with two H atoms on the C bonded to the OH group. Thus, it is first oxidized to an aldehyde and then to a carboxylic acid.

$$\text{CH}_3\text{–CH(CH}_3\text{)–CH}_2\text{OH} \xrightarrow{[O]} \text{CH}_3\text{–CH(CH}_3\text{)–CHO} \xrightarrow{[O]} \text{CH}_3\text{–CH(CH}_3\text{)–COOH}$$

1° alcohol (2 H atoms) — aldehyde — carboxylic acid

[4] Ethers (14.7)

Example 14.5 Rank the following compounds in order of increasing boiling point.

$CH_3CH_2CH_2{-}O{-}CH_3$ (**A**) $CH_3CH_2CH_2CH_2OH$ (**B**) $CH_3CH_2CH_2CH_2CH_3$ (**C**)

Analysis
Look at the functional groups to determine the strength of the intermolecular forces—**the stronger the forces, the higher the boiling point.**

Solution
C is an alkane with only nonpolar C–C and C–H bonds, so it has the weakest intermolecular forces and therefore the lowest boiling point. **B** is an alcohol capable of intermolecular hydrogen bonding, so it has the strongest intermolecular forces and the highest boiling point. **A** is an ether, so it contains a net dipole but is incapable of intermolecular hydrogen bonding. **A,** therefore, has intermolecular forces of intermediate strength and has a boiling point between the boiling points of **B** and **C.**

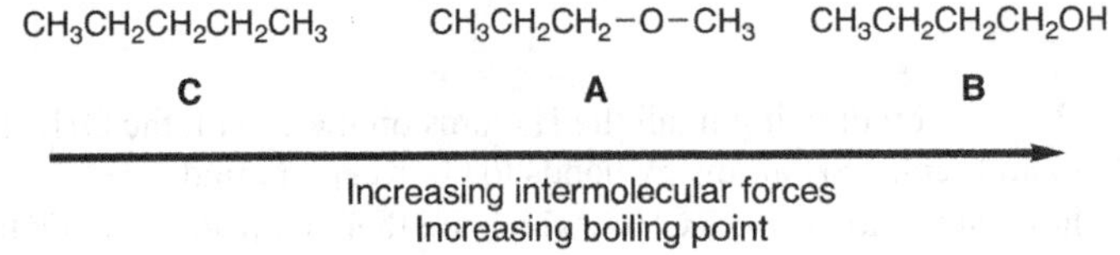

Example 14.6 Give the IUPAC name for the following ether.

$CH_3CH(CH_3)CH_2{-}O{-}CH_3$

Analysis and Solution

[1] Name the longer chain as an alkane and the shorter chain as an alkoxy group.

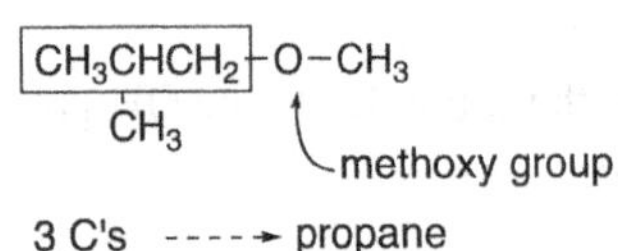

[2] Apply other nomenclature rules to complete the name.

3 2 1 $CH_3CH(CH_3)CH_2{-}O{-}CH_3$ — methyl at C2

Answer: 1-methoxy-2-methylpropane

[5] Organic Compounds That Contain Sulfur (14.10)

Example 14.7 Give the IUPAC name for the following thiol.

$CH_3CH_2CH(CH_3)CH_2CH(SH)CH_2CH_3$

Analysis and Solution

[1] Find the longest carbon chain that contains the carbon bonded to the SH group.

CH_3 (on C5), SH (on C3)

$CH_3CH_2CHCH_2CHCH_2CH_3$

7 C's in the longest chain ---→ **heptane**

- Name the alkane and add the suffix *-thiol*: **heptanethiol.**

[2] Number the carbon chain to give the SH group the lower number and apply all of the other rules of nomenclature.

a. **Number** the chain.

CH_3 SH

$CH_3CH_2CHCH_2CHCH_2CH_3$

7 6 5 4 3 2 1

3-heptanethiol

b. **Name** and **number** the substituent.

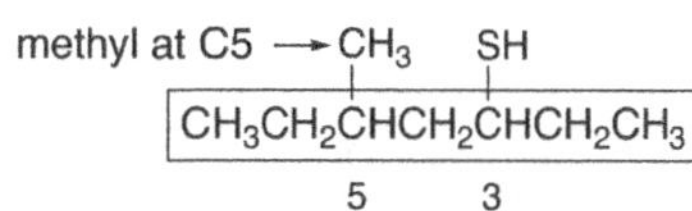

Answer: 5-methyl-3-heptanethiol

Self-Test

[1] Fill in the blank with one of the terms listed below.

Alcohols (14.1)	Disulfides (14.10)	Heterocycle (14.7)
Alkyl halides (14.1)	Elimination (14.5)	Oxidation (14.5)
Chlorofluorocarbons (14.9)	Ethers (14.1)	Sulfhydryl group (14.1)
Dehydration (14.5)	Glycols (14.3)	Thiols (14.1)

1. ______ are organic compounds that have two alkyl groups bonded to an oxygen atom.
2. ______ are compounds that contain a sulfur–sulfur bond.
3. ______ results in an increase in the number of C–O bonds or a decrease in the number of C–H bonds.
4. Loss of H_2O from a starting material is called ______.
5. ______ contain a sulfhydryl group (SH group) bonded to a tetrahedral carbon atom.
6. ______ contain a hydroxyl group (OH group) bonded to a tetrahedral carbon atom.
7. ______ are organic molecules containing a halogen atom bonded to a tetrahedral carbon atom.
8. A ring that contains a heteroatom is called a ______.
9. ______ is a reaction in which elements of the starting material are "lost" and a new multiple bond is formed.
10. Compounds with two hydroxyl groups are called diols (using the IUPAC system) or ______.
11. A ______ is the SH group.
12. ______ are simple halogen-containing compounds having the general molecular formula CF_xCl_{4-x}.

[2] Classify each alkyl halide as 1°, 2°, or 3°.

a. 1° alkyl halide b. 2° alkyl halide c. 3° alkyl halide

13. $CH_3CH_2CH_2-\overset{\displaystyle H}{\underset{\displaystyle CH_3}{C}}-Br$

14. $H-\overset{\displaystyle Br}{\underset{\displaystyle CH_2CH_2CH_3}{C}}-H$

15. $CH_3CH_2\overset{\displaystyle Br}{C}HCH_2CH_3$

16. $CH_3CH_2\overset{\displaystyle Br}{\underset{\displaystyle CH_3}{C}}CH_2CH_2CH_2CH_3$

[3] Match the reactants with the products.

17. $CH_3CH_2CH_2-\overset{\displaystyle H}{\underset{\displaystyle CH_3}{C}}-OH \xrightarrow{H_2SO_4}$

18. $CH_3CH_2CH_2-\overset{\displaystyle H}{\underset{\displaystyle CH_3}{C}}-OH \xrightarrow{[O]}$

19. $CH_3CH_2CH_2-\overset{\displaystyle H}{\underset{\displaystyle H}{C}}-OH \xrightarrow[\text{(first step)}]{[O]}$

20. $CH_3CH_2CH_2-\overset{\displaystyle O}{\overset{\|}{C}}-H \xrightarrow{[O]}$

21. $CH_3CH_2CH_2-\overset{\displaystyle H}{\underset{\displaystyle H}{C}}-OH \xrightarrow{H_2SO_4}$

22. $CH_3CH_2CH_2-\overset{\displaystyle OH}{\underset{\displaystyle H}{C}}-CH_2CH_3 \xrightarrow{[O]}$

a. $CH_3CH_2CH_2-\overset{\displaystyle O}{\overset{\|}{C}}-CH_3$

b. $CH_3CH_2CH{=}CHCH_3$

c. $CH_3CH_2CH{=}CH_2$

d. $CH_3CH_2CH_2-\overset{\displaystyle O}{\overset{\|}{C}}-CH_2CH_3$

e. $CH_3CH_2CH_2-\overset{\displaystyle O}{\overset{\|}{C}}-H$

f. $CH_3CH_2CH_2-\overset{\displaystyle O}{\overset{\|}{C}}-OH$

[4] Rank the compounds in each group in order of increasing boiling point.

23. **A** $CH_3CH_2CH_2CH_2OH$ **B** $CH_3CH_2CH_2OCH_3$ **C** $CH_3CH_2CH_2CH_2CH_3$

24. **A** $CH_3CH_2-\overset{\displaystyle OCH_2CH_3}{\underset{\displaystyle H}{C}}-CH_2CH_2CH_3$ **B** $CH_3CH_2-\overset{\displaystyle CH_2CH_2CH_3}{\underset{\displaystyle H}{C}}-CH_2CH_2CH_3$ **C** $CH_3CH_2CH_2-\overset{\displaystyle OH}{\underset{\displaystyle H}{C}}-CH_2CH_2CH_2CH_3$

25. **A** $CH_3CH_2CH_2CH_2CH(CH_3)_2$ **B** $CH_3CH_2CH_2CH(CH_3)CH_2OH$ **C** $CH_3CH_2CH(CH_3)CH_2OCH_3$

Answers to Self-Test

1. Ethers	6. Alcohols	11. sulfhydryl group	16. c	21. c
2. Disulfides	7. Alkyl halides	12. Chlorofluorocarbons	17. b	22. d
3. Oxidation	8. heterocycle	13. b	18. a	23. **C** < **B** < **A**
4. dehydration	9. Elimination	14. a	19. e	24. **B** < **A** < **C**
5. Thiols	10. glycols	15. b	20. f	25. **A** < **C** < **B**

Solutions to In-Chapter Problems

14.1 Label the –OH groups, –SH groups, halogens, and ether oxygens in each compound.

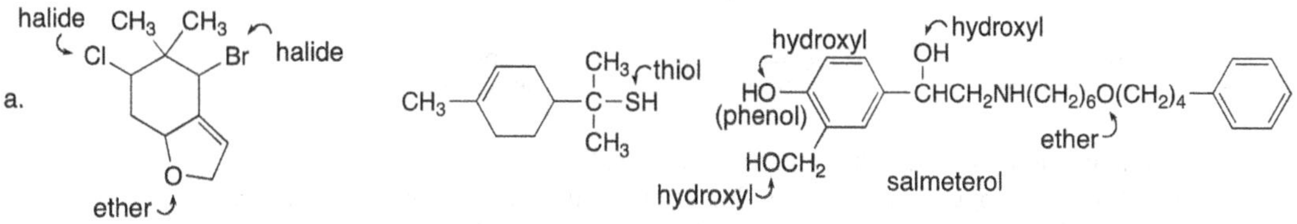

a. halide, halide, ether; thiol; hydroxyl, hydroxyl (phenol), hydroxyl, ether — salmeterol

b. The OH on the benzene ring of salmeterol is part of a phenol, so it is *not* an alcohol.

14.2 To determine whether an alcohol is 1°, 2°, or 3°, locate the C with the OH group and count the number of C's bonded to it. A 1° alcohol has the OH group on a C bonded to one C, and so forth, as in Example 14.1.

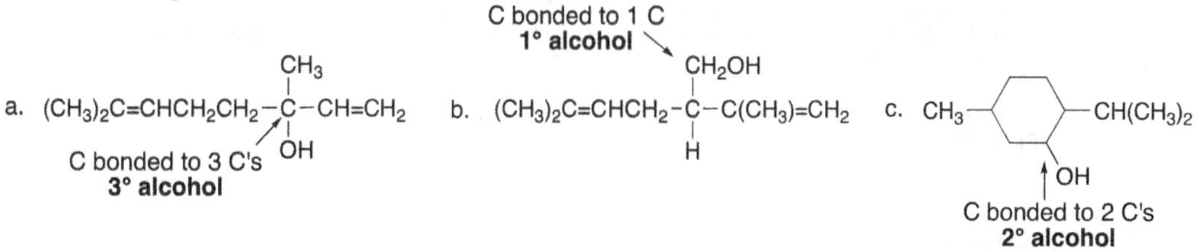

a. $(CH_3)_2C{=}CHCH_2CH_2{-}C(CH_3)(OH){-}CH{=}CH_2$ — C bonded to 3 C's, **3° alcohol**

b. $(CH_3)_2C{=}CHCH_2{-}C(H)(CH_2OH){-}C(CH_3){=}CH_2$ — C bonded to 1 C, **1° alcohol**

c. CH_3–cyclohexane–$CH(CH_3)_2$ with OH — C bonded to 2 C's, **2° alcohol**

14.3 Use the definition in Example 14.1 and Answer 14.2 to label the hydroxyl groups.

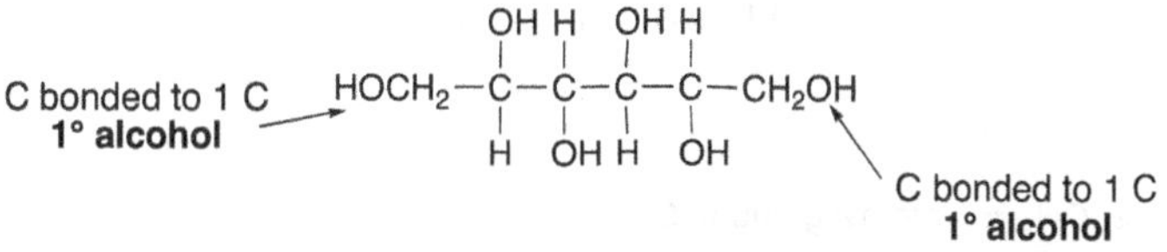

$HOCH_2{-}CH(OH){-}CH(OH){-}CH(OH){-}CH(OH){-}CH_2OH$

C bonded to 1 C, **1° alcohol** (each terminal CH_2OH)

All other hydroxyl groups are on C's bonded to 2 C's.
2° alcohols

14.4 Alcohols have stronger intermolecular forces and therefore higher boiling points than hydrocarbons of comparable size and shape.

a. 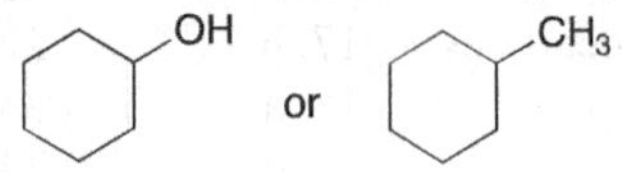 cyclohexanol (cyclohexane–OH) or methylcyclohexane (cyclohexane–CH_3)
alcohol
higher boiling point

b. $(CH_3)_3C-OH$ or $(CH_3)_4C$
alcohol
higher boiling point

14.5 Hydrocarbons are insoluble in water. Low molecular weight alcohols (< 6 C's) are water soluble, but higher molecular weight alcohols (≥ 6 C's) are insoluble in water.

a. 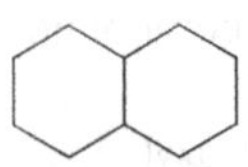 (bicyclic fused rings) hydrocarbon **insoluble**

b. $(CH_3)_3C-OH$ alcohol with 4 C's **soluble**

c. $CH_2=CHCH_2CH_3$ hydrocarbon **insoluble**

14.6 To name alcohols using the IUPAC system, follow the steps in Example 14.2:
[1] Find the longest carbon chain that contains the carbon bonded to the OH group.
[2] Number the carbon chain to give the OH group the lower number, and apply all other rules of nomenclature.

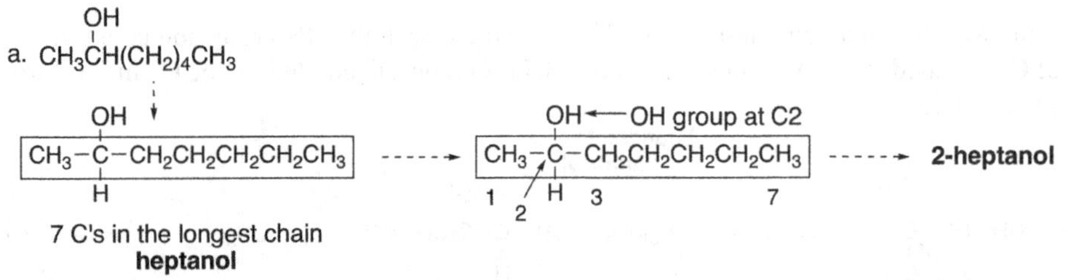

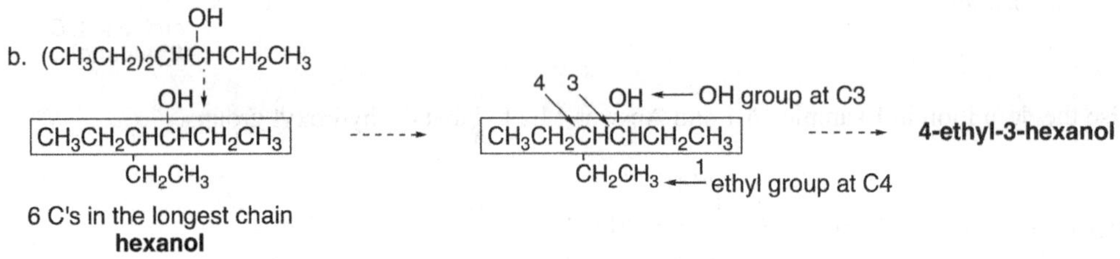

a. $CH_3CH(OH)(CH_2)_4CH_3$

$CH_3-C(OH)(H)-CH_2CH_2CH_2CH_2CH_3$
7 C's in the longest chain
heptanol

→ $CH_3-C(OH)(H)-CH_2CH_2CH_2CH_2CH_3$ (numbered 1, 2, 3 … 7); OH group at C2 → **2-heptanol**

b. $(CH_3CH_2)_2CHCH(OH)CH_2CH_3$

$CH_3CH_2CH(CH_2CH_3)CH(OH)CH_2CH_3$
6 C's in the longest chain
hexanol

→ numbered 4, 3 … 1; OH group at C3; ethyl group at C4 → **4-ethyl-3-hexanol**

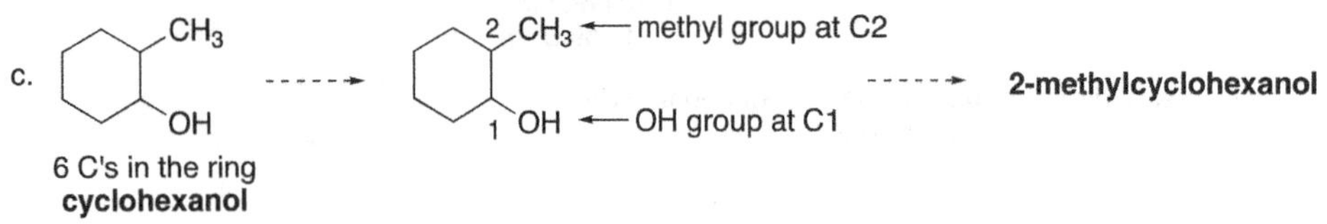

c. 2-methylcyclohexane ring with CH_3 and OH
6 C's in the ring
cyclohexanol

→ CH_3 at 2 ← methyl group at C2; OH at 1 ← OH group at C1 → **2-methylcyclohexanol**

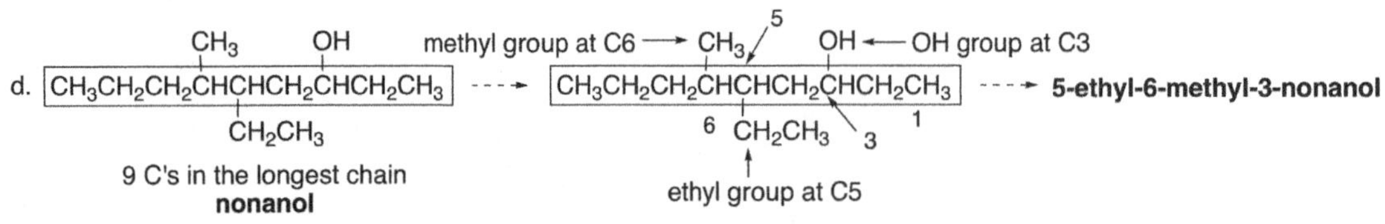

d. $CH_3CH_2CH_2CH(CH_3)CH(CH_2CH_3)CH_2CH(OH)CH_2CH_3$
9 C's in the longest chain
nonanol

→ methyl group at C6 → CH_3; OH ← OH group at C3; 6, 5, 3, 1 numbering; ethyl group at C5 → **5-ethyl-6-methyl-3-nonanol**

14.7 Work backwards to draw the structure corresponding to each name.

a. 7,7-dimethyl-4-**octanol** ------→ C–C–C–C(OH)–C–C–C(CH_3)(CH_3)–C (numbered 1 2 3 4 5 6) ------→ $CH_3CH_2CH_2CH(OH)CH_2CH_2C(CH_3)_2CH_3$

8 carbon chain
OH at C4

two methyl groups at C7

b. 5-methyl-4-propyl-3-**heptanol** --→ C–C–C(HO)–C($CH_2CH_2CH_3$)–C(CH_3)–C–C (numbered 1 2 3 4 … 7) ------→ $CH_3CH_2CH(HO)CH(CH_2CH_2CH_3)CH(CH_3)CH_2CH_3$

methyl group at C5

7 carbon chain
OH at C3

propyl group at C4

c. 2-ethyl-3-methyl**cyclohexanol** ------→ (cyclohexane ring: 1 OH, 2 CH_2CH_3, 3 CH_3)

6 carbon ring
OH at C1

ethyl group at C2

methyl group at C3

d. 1,3-**cyclohexanediol** ------→ (cyclohexane ring: 1 OH, 2, 3 OH)

6 carbon ring
OH at C1 and C3

14.8 To draw the products of the dehydration of each alcohol, follow the steps in Example 14.3.

- First find the carbon bonded to the OH group, and then identify all carbons with H's bonded to this carbon.
- Remove the elements of H and OH from two adjacent C's, and draw a double bond between these C's in the product.
- When two different alkenes are formed, the major product has more C's bonded to the C=C.

a. $CH_3-C(H)(OH)-CH_3$ $\xrightarrow{H_2SO_4}$ $CH_3-C(H)=CH_2$

These 2 C's are identical. When either one loses a H, the same product is formed.

c. (cyclopentyl)–OH $\xrightarrow{H_2SO_4}$ (cyclopentene)

Both adjacent C's are identical. When either one loses a H, the same product is formed.

b. (benzene ring)–CH_2CH_2OH $\xrightarrow{H_2SO_4}$ (benzene ring)–$CH=CH_2$

This C loses a H to form the product.

14.9 The Zaitsev rule states that when two different alkenes are formed, the major product has more C's bonded to the C=C.

a. $CH_3CH(OH)CH_2CH_2CH_3 \xrightarrow{H_2SO_4} CH_2{=}CHCH_2CH_2CH_3 + CH_3CH{=}CHCH_2CH_3$

$CH_2{=}CHCH_2CH_2CH_3$: 1 C bonded to the double bond

$CH_3CH{=}CHCH_2CH_3$: 2 C's bonded to the double bond — **major product**

b. $CH_3{-}C(CH_3)(OH){-}CH_2CH_3 \xrightarrow{H_2SO_4} CH_2{=}C(CH_3){-}CH_2CH_3 + CH_3C(CH_3){=}CHCH_3$

$CH_2{=}C(CH_3){-}CH_2CH_3$: 2 C's bonded to the double bond

$CH_3C(CH_3){=}CHCH_3$: 3 C's bonded to the double bond — **major product**

c. 1-methylcyclohexanol (CH_3, OH) $\xrightarrow{H_2SO_4}$ methylenecyclohexane ($=CH_2$) + 1-methylcyclohexene (CH_3)

methylenecyclohexane: 2 C's bonded to the double bond

1-methylcyclohexene: 3 C's bonded to the double bond — **major product**

14.10 **B** has no H on the carbon adjacent to the carbon with the OH group, so H_2O cannot be lost.

A: $C_6H_5{-}C(CH_3)(H){-}OH$ — H for dehydration (on CH_3)

B: $C_6H_5{-}CH_2OH$ — no H, no dehydration

14.11 Draw the products when each alcohol is oxidized, as in Example 14.4.

- RCH_2OH (1° alcohols) are oxidized to RCHO, which are then oxidized to RCOOH.
- R_2CHOH (2° alcohols) are oxidized to R_2CO.
- R_3COH (3° alcohols) are not oxidized because they have no H atom on the C with the OH.

a. $CH_3CH(OH)CH_2CH_2CH_3$ (2° alcohol) $\xrightarrow{[O]}$ $CH_3C(=O)CH_2CH_2CH_3$ (ketone)

b. $(CH_3)_3CCH_2CH_2OH$ (1° alcohol) $\xrightarrow{[O]}$ $(CH_3)_3CCH_2C(=O){-}OH$ (carboxylic acid)

c. cyclopentanol (2° alcohol) $\xrightarrow{[O]}$ cyclopentanone (ketone)

d. 1-methylcyclohexanol (CH_3, OH) (3° alcohol) $\xrightarrow{[O]}$ No reaction

14.12 Ethylene glycol is oxidized to a dicarboxylic acid.

$HOCH_2CH_2OH$ (ethylene glycol) $\xrightarrow{[O]}$ $HO{-}C(=O){-}C(=O){-}OH$

14.13 Draw the three constitutional isomers.

$CH_3{-}O{-}CH_2CH_2CH_3$ $\quad CH_3CH_2{-}O{-}CH_2CH_3$ $\quad (CH_3)_2CHOCH_3$

14.14 Look at the functional groups to determine the strength of the intermolecular forces, as in Example 14.5. **The stronger the forces, the higher the boiling point.**

a. $CH_3(CH_2)_6CH_3$ or $CH_3(CH_2)_5OCH_3$

hydrocarbon — polar, **higher boiling point**

b.  cyclohexane or tetrahydropyran (ring with O)

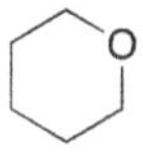

hydrocarbon — polar, **higher boiling point**

c. $CH_3(CH_2)_6OH$ or $CH_3(CH_2)_5OCH_3$

polar, can hydrogen bond, **higher boiling point** — polar, no hydrogen bonding

d. $CH_3(CH_2)_5OCH_3$ or CH_3OCH_3

polar, larger molecule, **higher boiling point** — polar

14.15 Low molecular weight ethers are water soluble. If the ether has more than five carbons, the ether is insoluble in water.

a. $CH_3CH_2-O-CH_3$

3 C ether
water soluble

b. 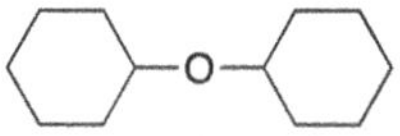 cyclohexyl–O–cyclohexyl

12 C ether
water insoluble

c. 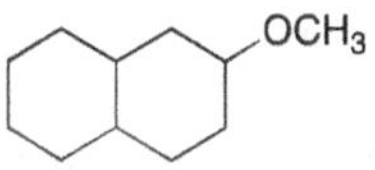 bicyclic ring with OCH_3

11 C ether
water insoluble

14.16 Name each ether as in Example 14.6.

a. CH_3-O-[$CH_2CH_2CH_2CH_3$] ----→ CH_3-O-[$CH_2CH_2CH_2CH_3$] (numbered 1 2 3 4) ----→ **1-methoxybutane** (or butyl methyl ether)

4 C's in the longer chain: butane
methoxy group

b. 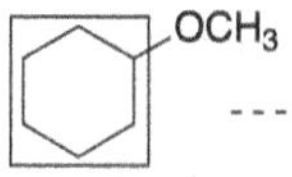 [cyclohexane ring]–OCH_3 -------→ [cyclohexane ring]–OCH_3 -------→ **methoxycyclohexane**

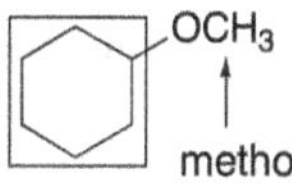

6 C's in the ring: cyclohexane
methoxy group

c. $CH_3CH_2CH_2-O-CH_2CH_2CH_3$ -------→ **dipropyl ether**

two propyl groups

14.17 Work backwards from the IUPAC name to the structure.

a. **dibutyl** ether ----------→ $CH_3CH_2CH_2CH_2-O-CH_2CH_2CH_2CH_3$

two butyl groups on the ether
butyl group — butyl group

b. **ethyl propyl** ether ---------→ $CH_3CH_2CH_2-O-CH_2CH_3$

one ethyl, one propyl group on the ether
propyl group — ethyl group

c. 1-methoxy**pentane** → $\underset{5\ \ 4\ \ 3\ \ 2\ \ 1}{\boxed{CH_3CH_2CH_2CH_2CH_2}}-O-CH_3$ (methoxy group)
5 carbon chain

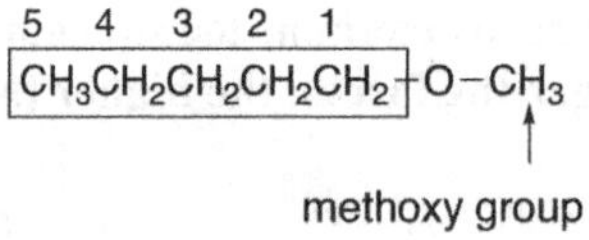

d. 3-ethoxy**hexane** → $\boxed{CH_3CH_2CHCH_2CH_2CH_3}$ with OCH_2CH_3 ← ethoxy group at C3
6 carbon chain
1 2 3 6

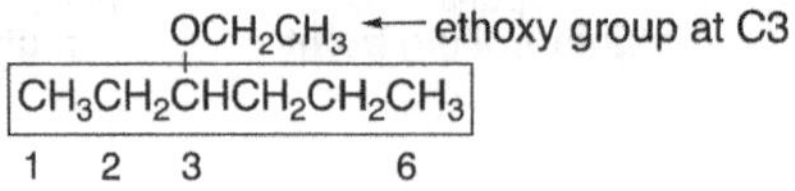

14.18

a. $:\ddot{F}-C(\ddot{F}:)(\ddot{F}:)-C(\ddot{F}:)(H)-\ddot{O}-C(H)(\ddot{F}:)-\ddot{F}:$

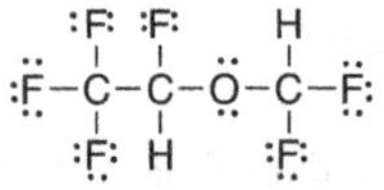

b. Desflurane is organic, and by "like dissolves like," organic compounds are soluble in organic solvents.

14.19 Draw the three-dimensional structure of halothane. Recall that solid lines represent bonds in the plane of the page, wedges represent bonds in front, and dashed lines represent bonds behind.

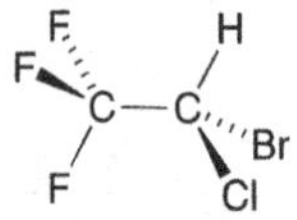

F, F, F — C — C — H, Br, Cl

14.20 Classify each alkyl halide as 1°, 2°, or 3°.

- A primary (1°) alkyl halide has a halogen on a carbon bonded to one carbon.
- A secondary (2°) alkyl halide has a halogen on a carbon bonded to two carbons.
- A tertiary (3°) alkyl halide has a halogen on a carbon bonded to three carbons.

a. $CH_3CH_2CH_2CH_2CH_2-Br$
C bonded to 1 C
1° alkyl halide

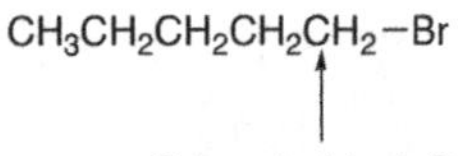

b. (cyclohexane with CH_3 and F)
C bonded to 3 C's
3° alkyl halide

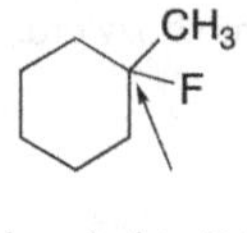

c. $CH_3-C(CH_3)(CH_3)-CHCH_3$ with Cl
C bonded to 2 C's
2° alkyl halide

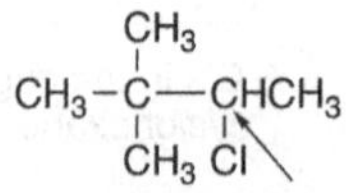

14.21 Label the alkyl halide and each alcohol as 1°, 2°, or 3°.

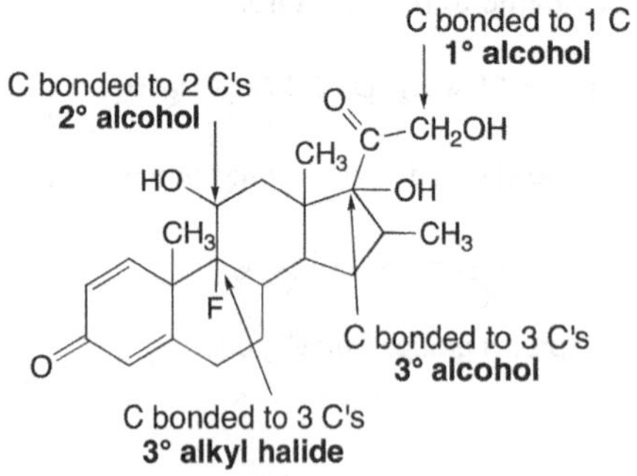

C bonded to 2 C's
2° alcohol

C bonded to 1 C
1° alcohol

C bonded to 3 C's
3° alcohol

C bonded to 3 C's
3° alkyl halide

14.22 The boiling point of alkyl halides increases with the size of the alkyl group as well as the size of the halogen.

a. $CH_3CH_2CH_2F$, $CH_3CH_2CH_2Cl$, $CH_3CH_2CH_2I$ → Increasing boiling point; Increasing size of the halogen

b. $CH_3(CH_2)_4CH_3$, $CH_3(CH_2)_5Cl$, $CH_3(CH_2)_5Br$ → Increasing boiling point

14.23 Give the IUPAC name for each compound. Always start by finding the longest chain.

a. $CH_3CH(Br)CH_2CH_2CH_2CH_3$ — 6 carbon chain **hexane** → bromo at C2 → **2-bromohexane**

b. $(CH_3)_2CHCH(Cl)CH_2CH_3$ → $CH_3CH(CH_3)CH(Cl)CH_2CH_3$ — 5 carbon chain **pentane** → methyl at C2; chloro at C3 → **3-chloro-2-methylpentane**

c. Cyclohexane ring with CH_3 and Br — 6 carbon ring **cyclohexane** → methyl at C2; bromo at C1 → **1-bromo-2-methylcyclohexane**

14.24 Work backwards from the name to the structure.

a. 3-chloro-2-methyl**hexane** — 6 carbon chain → $CH_3CH(CH_3)CH(Cl)CH_2CH_2CH_3$ (chloro at C3; methyl at C2)

b. 4-ethyl-5-iodo-2,2-dimethyl**octane** — 8 carbon chain → $CH_3C(CH_3)_2CH_2CH(CH_2CH_3)CH(I)CH_2CH_2CH_3$ (ethyl at C4; iodo at C5; two methyl groups at C2)

c. 1,1,3-tribromo**cyclohexane** — 6 carbon ring → cyclohexane ring with Br, Br (two bromo groups at C1) and Br (one bromo group at C3)

d. **propyl** chloride — 3 carbon chain → $CH_3CH_2CH_2$—Cl (one chloro group)

14.25 **Chlorofluorocarbons (CFCs)** have the general molecular structure $\mathbf{CF_xCl_{4-x}}$. **Hydrochlorofluorocarbons (HCFCs)** contain the elements of H, Cl, and F bonded to carbon, whereas **hydrofluorocarbons (HFCs)** contain the elements of H and F bonded to carbon.

a. CF_3Cl is a CFC. b. $CHFCl_2$ is an HCFC. c. CH_2F_2 is an HFC.

14.26 Name the thiols using the steps in Example 14.7.

a. $CH_3CH_2CHCH_2CH_3$ with SH on C3 → numbered 1, 2, 3; SH at C3 → **3-pentanethiol**

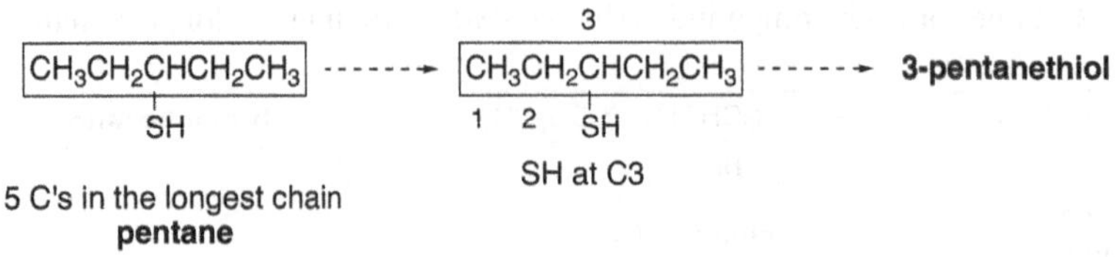

5 C's in the longest chain
pentane

b. Cyclohexane ring with SH → SH at C1 → **cyclohexanethiol**

6 C's in the ring
cyclohexane

c. $(CH_3CH_2)_2CHCH_2CH_2CH_2CH_2SH$

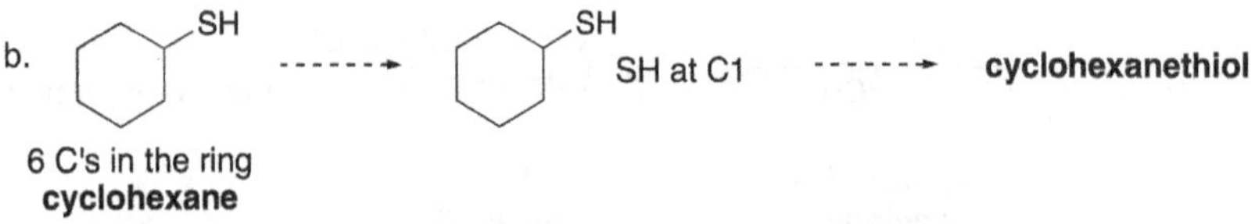

$CH_3CH_2CH(CH_2CH_3)CH_2CH_2CH_2CH_2SH$ → numbered 1, 2, ..., 5; ethyl at C5; SH at C1 → **5-ethyl-1-heptanethiol**

7 C's in the longest chain
heptane

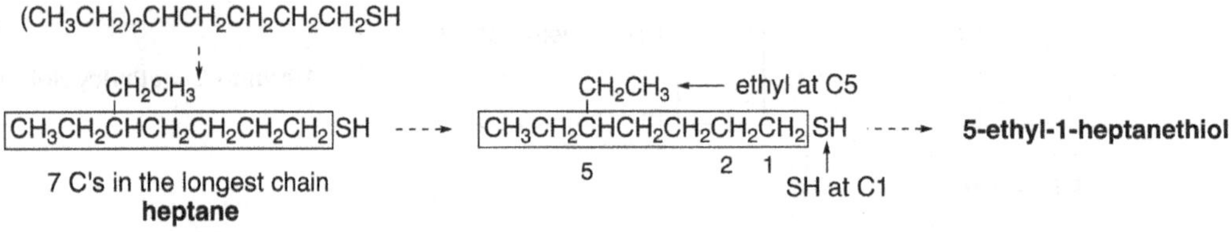

14.27 When thiols are oxidized, they form disulfides. Disulfides are converted back to thiols with a reducing agent.

a. $2\ CH_3CH_2CH_2{-}SH \xrightarrow{[O]} CH_3CH_2CH_2{-}S{-}S{-}CH_2CH_2CH_3$

b. $CH_3CH_2CH_2CH_2{-}S{-}S{-}CH_2CH_3 \xrightarrow{[H]} CH_3CH_2CH_2CH_2SH + CH_3CH_2SH$

14.28

$CH_2{=}CHCH_2{-}S{-}S{-}CH_2CH{=}CH_2 \xrightarrow{[H]} 2\ CH_2{=}CHCH_2SH$

Solutions to Odd-Numbered End-of-Chapter Problems

14.29 To determine whether an alcohol is 1°, 2°, or 3°, locate the C with the OH group and count the number of C's bonded to it. A 1° alcohol has the OH group on a C bonded to one C, and so forth, as in Example 14.1.

a. $CH_3CH_2CH_2OH$ — C bonded to 1 C — **1° alcohol**

b. $(CH_3CH_2)_3COH$ — C bonded to 3 C's — **3° alcohol**

c. 1-methylcyclohexanol with OH and CH_3 on adjacent ring carbons (2-methylcyclohexanol) — C bonded to 2 C's — **2° alcohol**

d. $CH_3CH_2CH(CH_3)CH(OH)CH_3$ — C bonded to 2 C's — **2° alcohol**

14.31 Classify each alkyl halide as 1°, 2°, or 3°.

- A primary (1°) alkyl halide has a halogen on a carbon bonded to one carbon.
- A secondary (2°) alkyl halide has a halogen on a carbon bonded to two carbons.
- A tertiary (3°) alkyl halide has a halogen on a carbon bonded to three carbons.

a. $CH_3(CH_2)_5CHClCH_3$ — C bonded to 2 C's — **2° alkyl halide**

b. $CH_3CH_2C(CH_3)(Cl)CH_2CH_3$ — C bonded to 3 C's — **3° alkyl halide**

c. iodocyclohexane — C bonded to 2 C's — **2° alkyl halide**

d. $CH_3CH_2-C(CH_3)_2-CH_2Br$ — C bonded to 1 C — **1° alkyl halide**

14.33 Draw a structure that fits each description.

a. $CH_3CH_2CH(OH)CH_2CH_2CH_3$ — 2° alcohol

b. $CH_3-O-CH_2CH_2CH_2CH_2CH_3$ — ether with a methoxy group

c. $CH_3CH_2C(CH_3)_2-Br$ — 3° alkyl halide

14.35 Draw the structure of the six constitutional isomers of molecular formula $C_5H_{12}O$ that contain an ether functional group.

$CH_3-O-CH(CH_3)CH_2CH_3$

$CH_3-O-CH_2CH_2CH_2CH_3$

$CH_3-C(CH_3)_2-O-CH_3$

$CH_3CH_2CH_2OCH_2CH_3$

$CH_3-CH(CH_3)-O-CH_2CH_3$

$(CH_3)_2CHCH_2OCH_3$

14.37

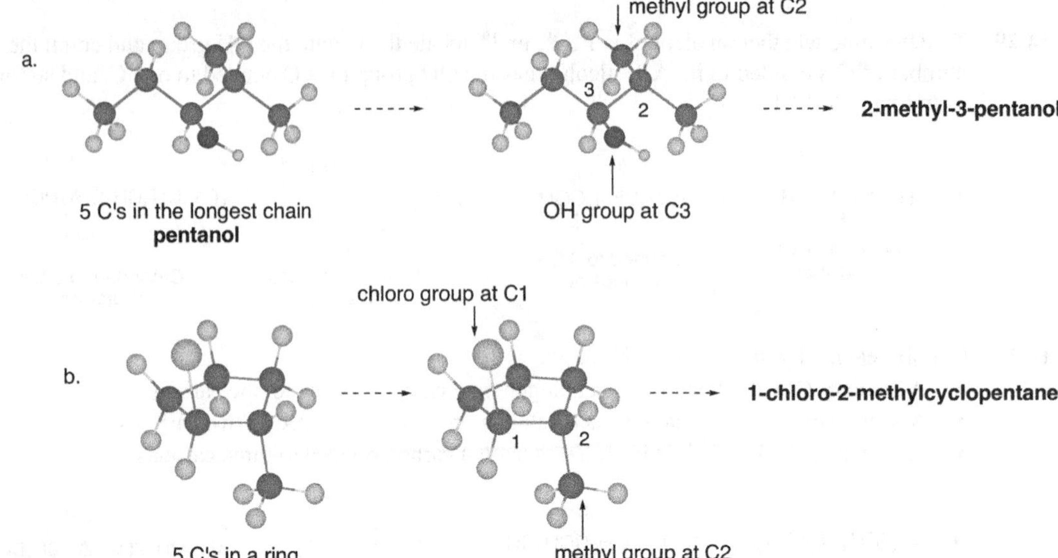
a.
methyl group at C2
3
2
2-methyl-3-pentanol
5 C's in the longest chain
pentanol
OH group at C3
chloro group at C1
b.
1
2
1-chloro-2-methylcyclopentane
5 C's in a ring
cyclopentane
methyl group at C2

14.39

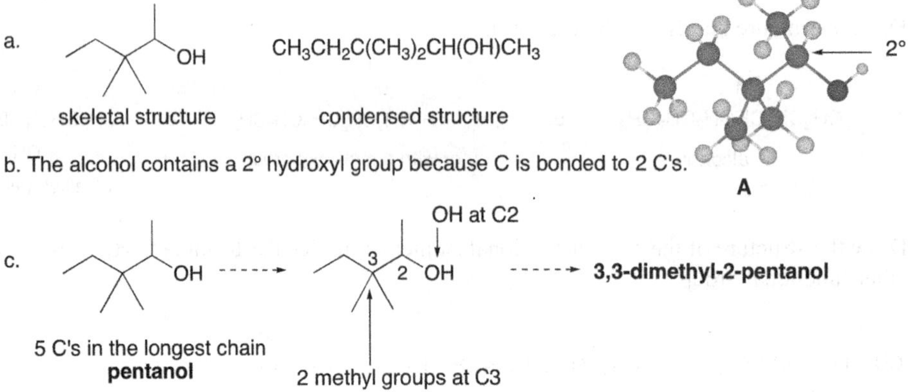
a.
OH
CH3CH2C(CH3)2CH(OH)CH3
2°
skeletal structure
condensed structure
b. The alcohol contains a 2° hydroxyl group because C is bonded to 2 C's.
A
OH at C2
c.
OH
3
2 OH
3,3-dimethyl-2-pentanol
5 C's in the longest chain
pentanol
2 methyl groups at C3

14.41 To name alcohols using the IUPAC system, follow the steps in Example 14.2:
[1] Find the longest carbon chain that contains the carbon bonded to the OH group.
[2] Number the carbon chain to give the OH group the lower number, and apply all other rules of nomenclature.

a. $CH_3CH(OH)CH_2CH_2CH_3$ ----→ $CH_3CH(OH)CH_2CH_2CH_3$ ----→ **2-pentanol**

5 C's in the longest chain **pentanol**; OH group at C2

b. $CH_3CH(OH)C(CH_3)_2CH_2CH_2CH_3$ ----→ ----→ **3,3-dimethyl-2-hexanol**

6 C's in the longest chain **hexanol**; OH group at C2; two methyl groups at C3

c. $CH_3CH(CH_3)CH_2CH(CH_2CH_3)CH_2CH_2CH_2OH$ ----→ ----→ **4-ethyl-6-methyl-1-heptanol**

7 C's in the longest chain **heptanol**; methyl at C6; ethyl at C4; OH group at C1

d. HO–cyclohexane–$CH_2CH_2CH_2CH_3$ ----→ ----→ **4-butylcyclohexanol**

6 C's in the ring **cyclohexanol**; OH group at C1; butyl group at C4

e. cyclobutane with OH and two CH_3 ----→ ----→ **2,2-dimethylcyclobutanol**

4 C's in the ring **cyclobutanol**; OH group at C1; two methyl groups at C2

f. CH_3CH_2–cyclopentane(OH)–CH_2CH_3 ----→ ----→ **2,5-diethylcyclopentanol**

5 C's in the ring **cyclopentanol**; OH group at C1; two ethyl groups at C2, C5

14.43 Work backwards to draw the structure corresponding to each name.

a. 3-**hexanol** ------→ $CH_3CH_2CH(OH)CH_2CH_2CH_3$ (OH at C3)

6 C chain with OH at C3

b. **propyl** alcohol ------→ $CH_2CH_2CH_2OH$

3 C chain with OH at C1

c. 2-methyl**cyclopropanol** ------→ cyclopropane ring with OH, CH_3

3 C ring with OH group at C1

methyl at C2 → CH_3

d. 1,2-**butanediol** ------→ $CH_2(OH)CH(OH)CH_2CH_3$

4 C chain with OH groups at C1 and C2

e. 4,4,5-trimethyl-3-**heptanol** ------→ $CH_3CH_2-C(H)(CH_3)-C(CH_3)_2-C(H)(OH)-CH_2CH_3$

7 C chain with OH at C3

3

three methyl groups at C4 and C5

f. 3,5-dimethyl-1-**heptanol** ------→ $HOCH_2CH_2C(H)(CH_3)CH_2C(H)(CH_3)CH_2CH_3$

1, 3, 5

7 C chain with OH at C1

two methyl groups at C3 and C5

14.45 Name each ether as in Example 14.6.

a. CH_3CH_2O[$CH_2CH_2CH_2CH_3$] ----→ CH_3CH_2O[$CH_2CH_2CH_2CH_3$] ----→ **1-ethoxybutane** (or butyl ethyl ether)

1 2 3 4

4 C's in the longer chain butane

ethoxy group

b. [$CH_3CH_2CH_2CH(OCH_2CH_3)CH_3$] ----→ [$CH_3CH_2CH_2CH(OCH_2CH_3)CH_3$] ----→ **2-ethoxypentane**

OCH_2CH_3 ← ethoxy group at C2

2 1

5 C's in the longer chain pentane

c. [cyclopentane]$-OCH_2CH_3$ ----→ [cyclopentane]$-OCH_2CH_3$ ----→ **ethoxycyclopentane**

5 C's in the ring cyclopentane

ethoxy group

14.47 Draw the structures and then name the isomers.

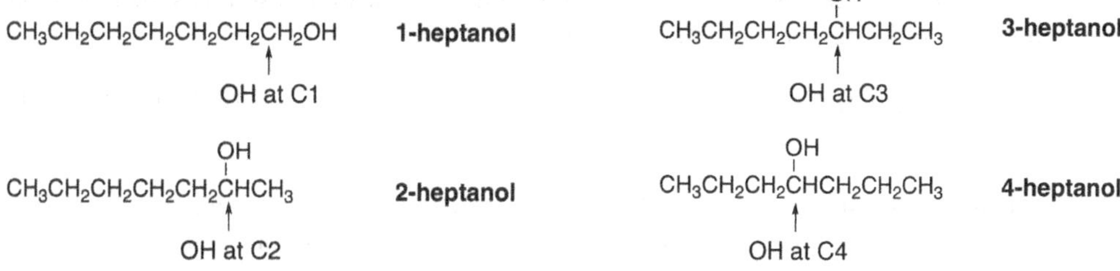

14.49 Name each compound using the IUPAC system.

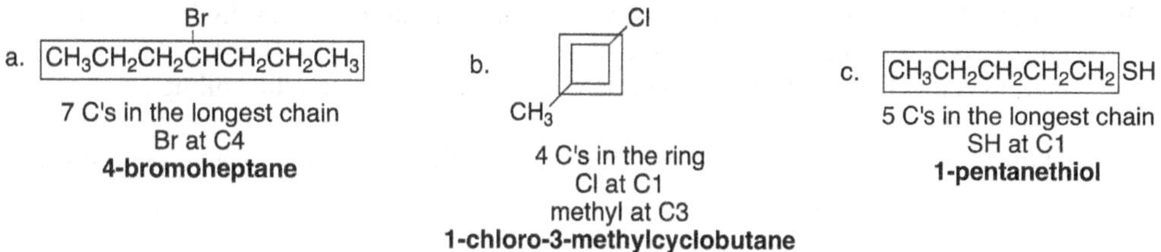

14.51 Work backwards from the name to draw each structure.

a. 2-methoxy**propane** ---→ $CH_3-C(OCH_3)(H)-CH_3$
3 C chain
methoxy at C2

b. **cyclobutyl** ethyl ether ---→ OCH_2CH_3 (on cyclobutane)
4 C ring
ethyl

c. 1-ethoxy-2-ethyl**cyclohexane** ---→ OCH_2CH_3, CH_2CH_3 (on cyclohexane)
6 C ring
ethoxy at C1
ethyl

d. **butyl** chloride---→ $CH_3CH_2CH_2CH_2Cl$
4 C chain
chlorine at C1

e. 2-methyl**cyclohexanethiol** ---→ SH, CH_3 (on cyclohexane)
6 C ring with SH group
methyl at C2

f. 1-ethyl-2-fluoro**cyclobutane** ---→ CH_2CH_3, F (on cyclobutane)
4 C ring
ethyl at C1
fluoro at C2

14.53 Alcohols have higher boiling points than hydrocarbons of comparable size and shape. The boiling points of alkyl halides increase with the size of the alkyl group and the size of the halogen.

a. $CH_3CH_2CH_2I$ (larger halogen)
b. $HOCH_2CH_2OH$ (two hydroxyl groups)
c. $CH_3CH_2CH_2OH$ (can hydrogen bond)

14.55 **C** has the highest boiling point because it has an OH group that can intermolecularly hydrogen bond. **B** is a hydrocarbon with the weakest intermolecular forces, so it has the lowest boiling point. In order of increasing boiling point:

$$\underset{\textbf{B}}{CH_3CH_2CH_2CH_2CH_3} \quad < \quad \underset{\textbf{A}}{CH_3CH_2OCH_2CH_3} \quad < \quad \underset{\textbf{C}}{CH_3CH_2CH_2CH_2OH}$$

14.57 Ethanol is a polar molecule capable of hydrogen bonding with itself and water. Dimethyl ether is polar, but cannot hydrogen bond with itself. The stronger intermolecular forces make the boiling point of ethanol higher. Both ethanol and dimethyl ether have only two carbons and can hydrogen bond to H_2O, so both are water soluble.

14.59 **B** ($CH_3CH_2CH_2CH_2OH$) has an OH group capable of intermolecular hydrogen bonding, so its boiling point is higher than the boiling point of **A** ($CH_3CH_2CH_2CH_2SH$), which has weaker intermolecular forces. Thus, the boiling point of **A** is 98 °C and the boiling point of **B** is 118 °C.

14.61 To draw the products of dehydration of each alcohol, follow the steps in Example 14.3.

a. Cyclobutanol (OH) $\xrightarrow{H_2SO_4}$ cyclobutene

This C loses a H to form the product.

b. C_6H_5–CH(OH)CH_2CH_3 $\xrightarrow{H_2SO_4}$ C_6H_5–CH=CHCH$_3$

This C loses a H to form the product.

c. $CH_3CH(OH)CH_2CH_2CH_2CH_3 \xrightarrow{H_2SO_4}$ $CH_3CH{=}CHCH_2CH_2CH_3$ + $CH_2{=}CH(CH_2)_3CH_3$

Either C can lose a H.

$CH_3CH{=}CHCH_2CH_2CH_3$: 2 C's bonded to the double bond, **major product**

$CH_2{=}CH(CH_2)_3CH_3$: 1 C bonded to the double bond

d. $CH_3CH_2C(OH)(CH_3)CH_2CH_3 \xrightarrow{H_2SO_4}$ $CH_3CH{=}C(CH_3)CH_2CH_3$ + $CH_3CH_2C({=}CH_2)CH_2CH_3$

Either C can lose a H.

$CH_3CH{=}C(CH_3)CH_2CH_3$: 3 C's bonded to the double bond, **major product**

$CH_3CH_2C({=}CH_2)CH_2CH_3$: 2 C's bonded to the double bond

14.63 Draw the products of dehydration, as in Example 14.3.

a. $CH_3CH_2CH(OH)CH_2CH_2CH_3 \xrightarrow{H_2SO_4} CH_3CH{=}CHCH_2CH_2CH_3 + CH_3CH_2CH{=}CHCH_2CH_3$

b. The carbons of both double bonds are bonded to an equal number of C's; as a result, roughly equal amounts of both isomers are formed.

14.65 Work backwards to determine what alcohol can be used to form each product.

a. $C_6H_5CH{=}CHC_6H_5 \leftarrow C_6H_5CH(OH)CH_2C_6H_5$

b. cyclooctene $\leftarrow$ cyclooctanol

14.67 Work backwards to determine what alcohols can be used to form propene. The OH group can be bonded to either the middle or the end carbon.

$CH_3CH{=}CH_2 \leftarrow CH_3CH_2CH_2OH$ or $CH_3CH(OH)CH_3$

14.69 Draw the products when each alcohol is oxidized, as in Example 14.4.

- RCH_2OH (1° alcohols) are oxidized to RCHO, which are then oxidized to RCOOH.
- R_2CHOH (2° alcohols) are oxidized to R_2CO.
- R_3COH (3° alcohols) are not oxidized because they have no H atom on the C with the OH.

a. $CH_3(CH_2)_6CH_2OH$ (1° alcohol) $\xrightarrow{[O]}$ $CH_3(CH_2)_6COOH$ (carboxylic acid)

b. $(CH_3CH_2)_2CHOH$ (2° alcohol) $\xrightarrow{[O]}$ $CH_3CH_2COCH_2CH_3$ (ketone)

c. $CH_3CH_2CH(CH_3)CH_2OH$ (1° alcohol) $\xrightarrow{[O]}$ $CH_3CH_2CH(CH_3)COOH$ (carboxylic acid)

d. $CH_3CH_2C(CH_3)_2OH$ (3° alcohol) $\xrightarrow{[O]}$ No reaction

14.71 Draw the product of oxidation.

cortisol $\xrightarrow{[O]}$ product (3° OH not oxidized; the HO on the ring is oxidized to C=O and the CH_2OH is oxidized to COH=O)

14.73 Work backwards to determine what alcohol can be used to prepare each carbonyl compound.

a. cycloheptanone (ketone) $\leftarrow$ cycloheptanol (2° alcohol)

b. C_6H_5COOH (carboxylic acid) $\leftarrow$ $C_6H_5CH_2OH$ (1° alcohol)

14.75 Draw the products of each reaction.

a. $CH_3CH_2CH_2CH(OH)CH_2CH_2CH_3 \xrightarrow{H_2SO_4} CH_3CH_2CH_2CH{=}CHCH_2CH_3$

b. $CH_3CH_2CH_2CH(OH)CH_2CH_2CH_3 \xrightarrow{K_2Cr_2O_7} CH_3CH_2CH_2C(=O)CH_2CH_2CH_3$

14.77

$CH_3CH_2CH_2CH_2CH_2CH{=}CH_2 \xleftarrow{-H_2O} CH_3CH_2CH_2CH_2CH_2CH_2CH_2OH$ (alcohol)

$CH_3CH_2CH_2CH_2CH_2CH_2CH_2OH \xrightarrow{K_2Cr_2O_7} CH_3CH_2CH_2CH_2CH_2CH_2CO_2H$

14.79 Draw the structure of the alcohol that fits the description. Since no reaction occurs with an oxidizing agent, the compound must be a 3° alcohol. Since only one alkene is formed when treated with sulfuric acid, all carbons adjacent to the C–OH must be identical.

$(CH_3)_3COH$

3° alcohol
All 3 C's adjacent to the C–OH are identical.

14.81 Draw the disulfides formed by thiol oxidation.

a. $C_6H_{11}SH \xrightarrow{[O]} C_6H_{11}S{-}SC_6H_{11}$ (cyclohexyl)

b. $CH_3(CH_2)_4SH \xrightarrow{[O]} CH_3(CH_2)_4S{-}S(CH_2)_4CH_3$

14.83 Draw the thiols formed by disulfide reduction.

a. $CH_3S{-}SCH_2CH_2CH_3 \xrightarrow{[H]} CH_3SH + HSCH_2CH_2CH_3$

b. $CH_2{=}CHCH_2S{-}SCH_2CH_2CH_3 \xrightarrow{[H]} CH_2{=}CHCH_2SH + HSCH_2CH_2CH_3$

14.85 Write a balanced equation for the combustion of diethyl ether.

$$C_4H_{10}O + 6\,O_2 \longrightarrow 4\,CO_2 + 5\,H_2O$$

diethyl ether

14.87 a. CFCs are chlorofluorocarbons with a general formula of CF_xCl_{4-x}; HCFCs have fluorine, chlorine, and hydrogen bonded to carbon; and HFCs have only hydrogen and fluorine bonded to carbon.

b. CFCs destroy the ozone layer whereas HCFCs and HFCs decompose more readily before ascending to the ozone layer.

14.89 PEG is capable of hydrogen bonding with water, so PEG is water soluble. PVC cannot hydrogen bond to water, so PVC is water insoluble, even though it has many polar bonds.

14.91 The greater the blood alcohol level, the greater the color change from red-orange to green with the breathalyzer.

$$CH_3CH_2OH + K_2Cr_2O_7 \longrightarrow CH_3COH\ (C{=}O) + Cr^{3+}$$

ethanol (orange) (green)

14.93 Draw the product of the oxidation of propylene glycol.

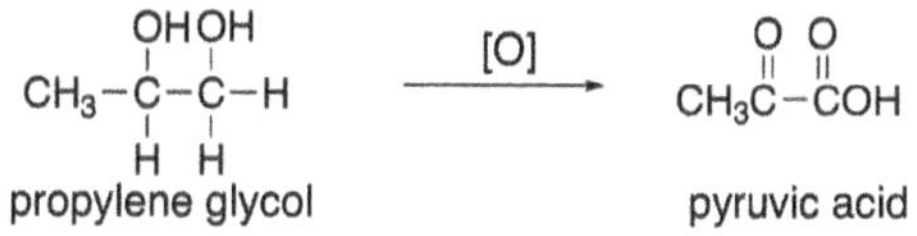

$CH_3-CH(OH)-CH(OH)-H$ (propylene glycol) $\xrightarrow{[O]}$ $CH_3C(=O)-C(=O)OH$ (pyruvic acid)

14.95 Draw the oxidation products formed from ethanol. Antabuse blocks the conversion of acetaldehyde to acetic acid, and the accumulation of acetaldehyde makes people ill.

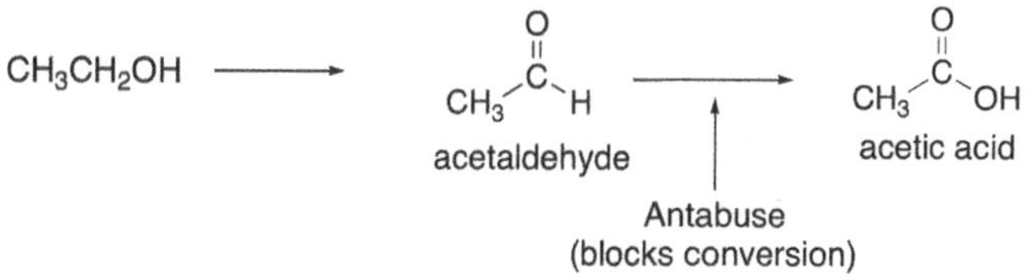

$CH_3CH_2OH \longrightarrow CH_3CHO$ (acetaldehyde) $\longrightarrow CH_3COOH$ (acetic acid)

Antabuse (blocks conversion)

14.97 Ether molecules have no H's on O's capable of hydrogen bonding to other ether molecules, but the hydrogens in water can hydrogen bond to the oxygen of the ether.

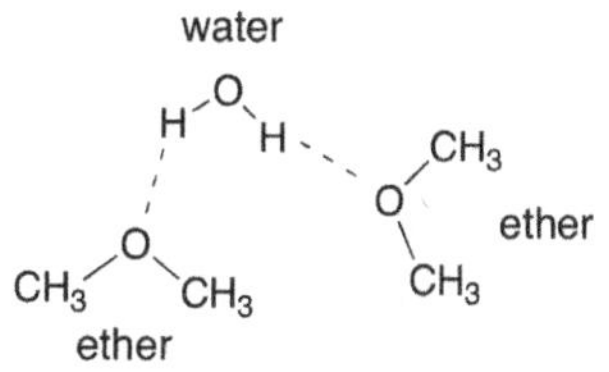

water

CH_3OCH_3 ether

CH_3OCH_3 ether

14.99 Answer each question about the alcohol.

a. and b. 2° alcohol

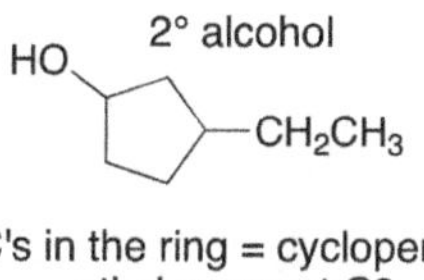

5 C's in the ring = cyclopentanol
ethyl group at C3
3-ethylcyclopentanol

c. 3-ethylcyclopentanol $\xrightarrow{H_2SO_4}$ 4-ethylcyclopentene (with CH_2CH_3) + 3-ethylcyclopentene (with CH_2CH_3)

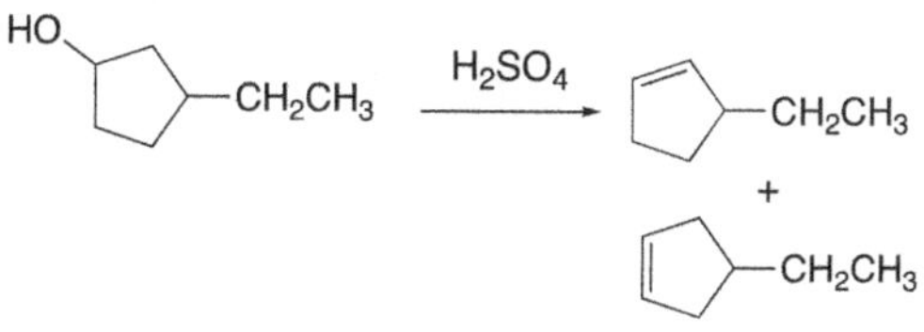

d. 3-ethylcyclopentanol $\xrightarrow{[O]}$ 3-ethylcyclopentanone (CH_2CH_3)

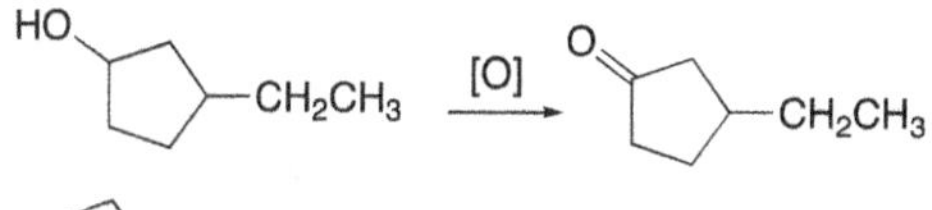

e. 2-ethylcyclopentanol (CH_2CH_3, OH) constitutional isomer

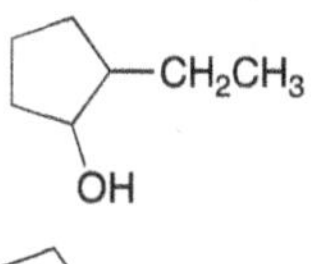

f. cyclopentyl–OCH_2CH_3 constitutional isomer

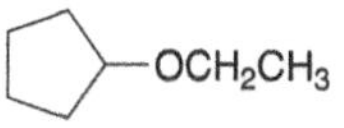

14.101

a. and b. **A** $CH_3CH_2C(CH_3)(H)CH_2OH$ $\xrightarrow{H_2SO_4}$ $CH_3CH_2C(CH_3)=CH_2$

2-methyl-1-butene

B $CH_3CH_2C(CH_3)(OH)CH_3$ $\xrightarrow{H_2SO_4}$ $CH_3CH_2C(CH_3)=CH_2$ + $CH_3CH=C(CH_3)CH_3$

2-methyl-1-butene
minor product

major product

Chapter 15 The Three-Dimensional Shape of Molecules

Chapter Review

[1] When is a molecule chiral or achiral? (15.2)

- A chiral molecule is not superimposable on its mirror image.

$CH_3CHBrCl$

mirror
not superimposable

$CH_3CHBrCl$ is a chiral molecule.

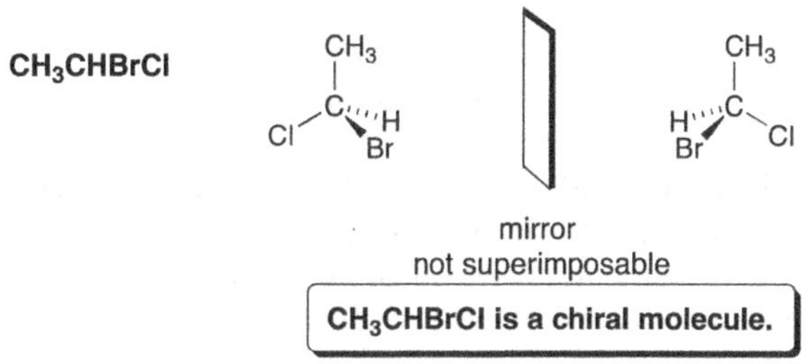

- An achiral molecule is superimposable on its mirror image.

CH_2BrCl

mirror
superimposable

CH_2BrCl is an achiral molecule.

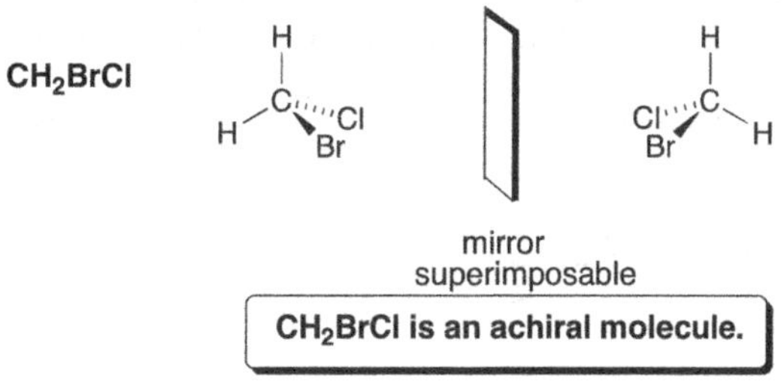

- To determine whether a molecule is chiral, draw it and its mirror image and see if all the atoms and bonds align.

[2] What is a chirality center? (15.2, 15.3)

- A chirality center is a carbon with four different groups around it.

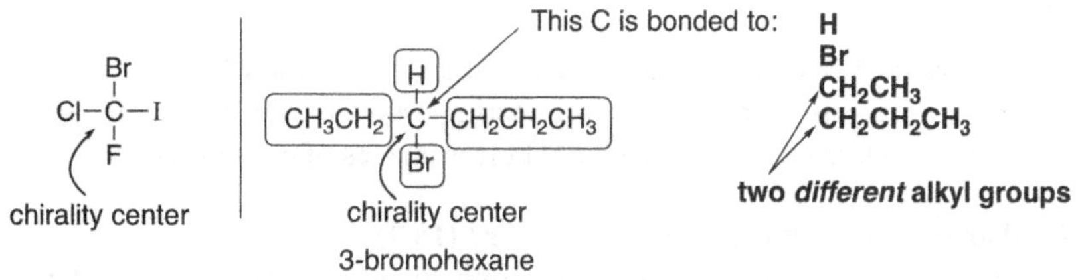

chirality center

This C is bonded to: **H**, **Br**, **CH_2CH_3**, **$CH_2CH_2CH_3$**

chirality center
3-bromohexane

two *different* alkyl groups

- A molecule with one chirality center is chiral.

[3] What are enantiomers? (15.2, 15.3)

- Enantiomers are mirror images that are not superimposable on each other. To draw two enantiomers, draw one molecule in three dimensions around the chirality center. Then draw the mirror image so that the substituents are a reflection of the groups in the first molecule.

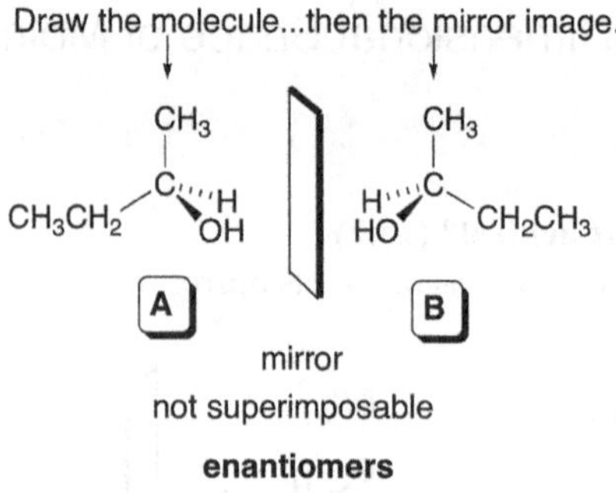

Draw the molecule...then the mirror image.

CH_3 | CH_3CH_2–C–H, OH **A** | mirror | CH_3 | H–C–CH_2CH_3, HO **B**

mirror
not superimposable
enantiomers

[4] Why do some chiral drugs have different properties from their mirror image isomers? (15.5)

- When a chiral drug must interact with a chiral receptor, only one enantiomer fits the receptor properly and evokes a specific response. Ibuprofen, naproxen, and L-dopa are examples of chiral drugs in which the two enantiomers have very different properties.

[5] What is a Fischer projection? (15.6)

- A Fischer projection is a specific way of depicting a chirality center. The chirality center is located at the intersection of a cross. The horizontal lines represent bonds that come out of the plane on wedges, and the vertical lines represent bonds that go back on dashed lines.

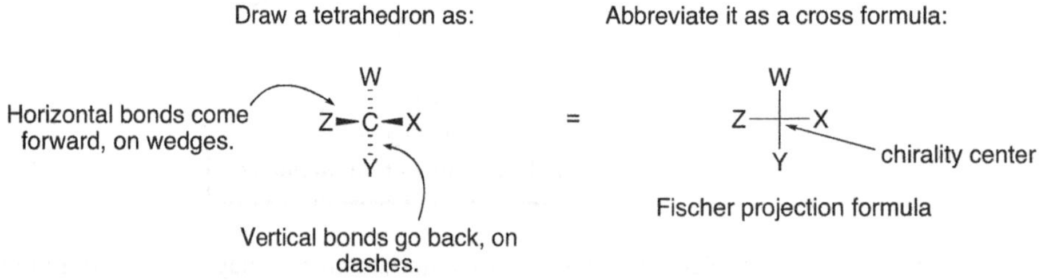

Draw a tetrahedron as:

Horizontal bonds come forward, on wedges.

W
Z–C–X
Y

Vertical bonds go back, on dashes.

=

Abbreviate it as a cross formula:

W
Z–+–X
Y

chirality center

Fischer projection formula

[6] How do the physical properties of two enantiomers compare? (15.7)

- Two enantiomers have identical physical properties except for their interaction with plane-polarized light.
- The specific rotation of two enantiomers is the same numerical value, but opposite in sign.

[7] What is the difference between an enantiomer and a diastereomer? (15.8)

- Enantiomers are stereoisomers that are nonsuperimposable mirror images of each other.
- Diastereomers are stereoisomers that are *not* mirror images of each other.

[8] How is the shape of a molecule related to its odor? (15.9)

- It is thought that the odor of a molecule is determined more by its shape than the presence of a particular functional group, so that compounds of similar shape have similar odors. For an odor to be perceived, a molecule must bind to an olfactory receptor, resulting in a nerve impulse that travels to the brain. Enantiomers may have different odors because they bind with chiral receptors and each enantiomer fits the chiral receptors in a different way.

Problem Solving

[1] Chirality Centers (15.3)

Example 15.1 Locate the chirality centers in each molecule.

a. HOOC–CH(CH_3)–CH(CH_3)–CH_2Cl (drawn with each C showing its substituents: O=C–OH; CH_3–C–H; CH_3–C–H; H–C–H; Cl)

b. cyclopentyl–C(H)(Cl)–C(Cl)(CH_3)–H

Analysis

In compounds with many C's, look at each C individually and eliminate those C's that can't be chirality centers. Thus, omit all CH_2 and CH_3 groups and all multiply bonded C's. Check all remaining C's to see if they are bonded to four different groups.

Solution

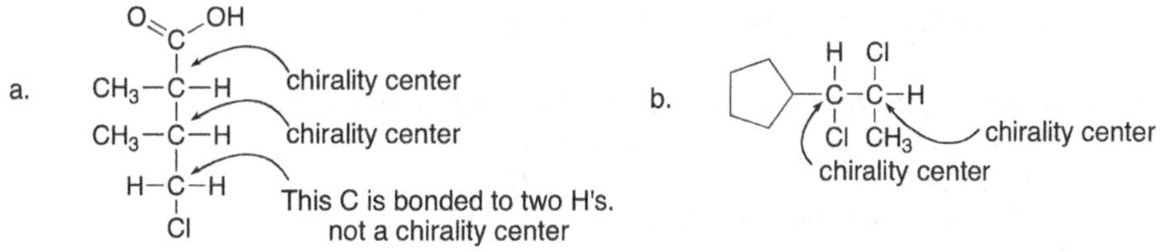

a. CH_3–C–H: chirality center
CH_3–C–H: chirality center
H–C–H: This C is bonded to two H's. not a chirality center

b. C bonded to H and Cl: chirality center
C bonded to Cl and CH_3: chirality center

---

[2] Fischer Projections (15.6)

Example 15.2 Draw both enantiomers of the following compound using Fischer projection formulas.

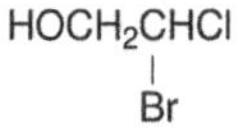

$HOCH_2CHCl$ (with Br on the CH carbon)

Analysis

- Draw the tetrahedron of one enantiomer with the horizontal bonds on wedges and the vertical bonds on dashed lines. Arrange the four groups on the chirality center—H, Br, Cl, and CH_2OH—arbitrarily in the first enantiomer.
- Draw the second enantiomer by arranging the substituents in the mirror image so they are a reflection of the groups in the first molecule.
- Replace the chirality center with a cross to draw the Fischer projections.

Solution

chirality center

$HOCH_2CHCl$ with Br on the central carbon

Draw the compound.

Cl above; H—C—Br; CH_2OH below

• Horizontal bonds come forward.
• Vertical bonds go back.

Then, draw the mirror image.

Cl above; Br—C—H; CH_2OH below

Fischer projection of one enantiomer

Fischer projection of the second enantiomer

[3] Compounds With Two or More Chirality Centers (15.8)

Example 15.3 Considering the four stereoisomers of 3,4-heptanediol (**A–D**):
a. Which compound is an enantiomer of **A?**
b. Which compound is an enantiomer of **B?**
c. What two compounds are diastereomers of **C?**

Four stereoisomers of 3,4-heptanediol:

	A	B	C	D
	CH_2CH_3	CH_2CH_3	CH_2CH_3	CH_2CH_3
C3	H—C—OH	H—C—OH	HO—C—H	HO—C—H
C4	H—C—OH	HO—C—H	HO—C—H	H—C—OH
	CH_2	CH_2	CH_2	CH_2
	CH_2	CH_2	CH_2	CH_2
	CH_3	CH_3	CH_3	CH_3

Analysis
Keep in mind the definitions:

- **Enantiomers are stereoisomers that are nonsuperimposable mirror images of each other.** Look for two compounds that when placed side-by-side have the groups on both chirality centers drawn as a *reflection* of each other. A given compound has only one possible enantiomer.
- **Diastereomers are stereoisomers that are *not* mirror images of each other.**

Solution
a. **A** and **C** are enantiomers because they are nonsuperimposable mirror images of each other.
b. **B** and **D** are enantiomers for the same reason.

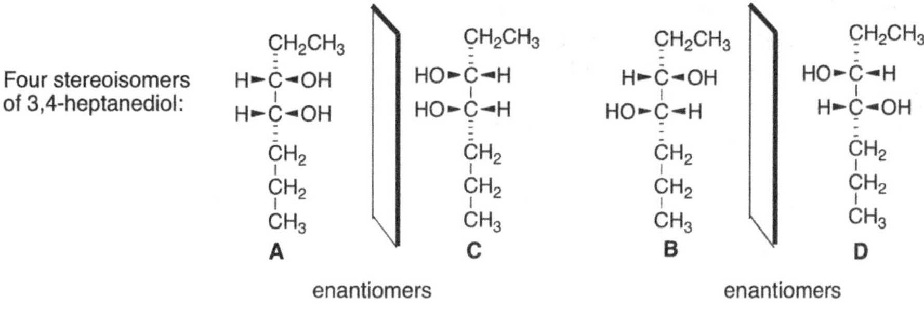

c. Any compound that is a stereoisomer of **C** but not its mirror image is a diastereomer. Thus, **B** and **D** are diastereomers of **C.**

Self-Test

[1] Fill in the blank with one of the terms listed below.

Achiral (15.2)	Diastereomers (15.8)	Stereochemistry (15.1)
Chiral (15.2)	Enantiomers (15.2)	Stereoisomers (15.1)
Chirality center (15.2)	Racemic mixture (15.5)	

1. An equal mixture of two enantiomers is called a ______.
2. A molecule (or object) that is *not* superimposable on its mirror image is said to be ______.
3. ______ are stereoisomers but they are *not* mirror images of each other.
4. ______ differ *only* in the three-dimensional arrangement of atoms.
5. A molecule (or object) that *is* superimposable on its mirror image is said to be ______.
6. ______ is the three-dimensional structure of molecules.
7. ______ are mirror images that are not superimposable.
8. A carbon atom surrounded by four different groups is a ______.

[2] Label the molecules as chiral or achiral.

9. $I{-}C(Br)(F){-}CH_3$

10. $Cl{-}C(OH)(CH_2CH_3){-}CH(H){-}CH_3$

11. $Br{-}C(CH_3)_2{-}C(Cl)(H){-}C(CH_3)_2{-}CH_3$

12. $Br{-}C(OH)(CH_3){-}C(Cl)(H){-}H$

[3] How are the compounds related? Are they enantiomers or diastereomers?

A	**B**	**C**	**D**
CO_2H	CO_2H	CO_2H	CO_2H
Cl–C–H	H–C–Cl	Cl–C–H	H–C–Cl
H–C–OH	H–C–OH	HO–C–H	HO–C–H
CH_3	CH_3	CH_3	CH_3

13. What is the relationship between **A** and **B**?
14. What is the relationship between **A** and **C**?
15. What is the relationship between **B** and **C**?
16. What is the relationship between **C** and **D**?
17. What is the relationship between **A** and **D**?

[4] How are the compounds in each pair related? Are they identical molecules or enantiomers?

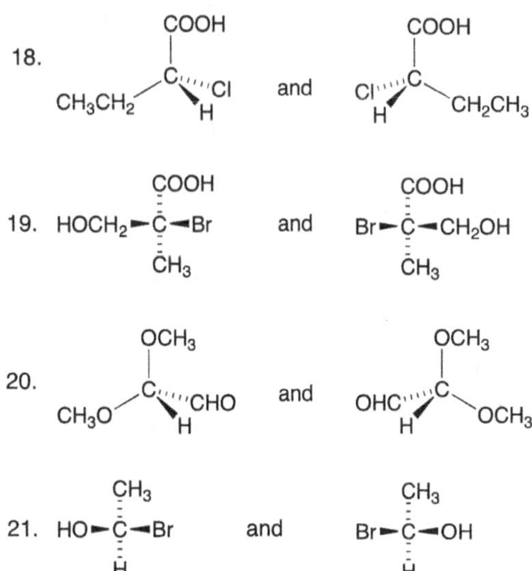

[5] Label each object as chiral or achiral.

22. wine glass 23. pencil 24. iPod 25. face

Answers to Self-Test

1. racemic mixture	6. Stereochemistry	11. chiral	16. diastereomers	21. enantiomers
2. chiral	7. Enantiomers	12. chiral	17. enantiomers	22. achiral
3. Diastereomers	8. chirality center	13. diastereomers	18. enantiomers	23. achiral
4. Stereoisomers	9. chiral	14. diastereomers	19. enantiomers	24. chiral
5. achiral	10. achiral	15. enantiomers	20. identical	25. chiral

Solutions to In-Chapter Problems

15.1 *Constitutional isomers* differ in the way the atoms are connected to one another. *Stereoisomers* differ *only* in the three-dimensional arrangement of atoms.

a. $CH_3CH_2CH(CH_3)CH(CH_3)CH_3$ and $CH_3CH(CH_3)CH_2CH(CH_3)CH_3$

different connection of atoms
constitutional isomers

b. $CH_3CH_2(H)C{=}C(H)CH_3$ (cis) and $CH_3CH_2(H)C{=}C(CH_3)H$ (trans)

stereoisomers

c. $CH_3CH{=}CHCH_2CH_3$ and $CH_3CH_2CH{=}CHCH_3$

identical

d. [oxetane] and [cyclopropyl]–OH

different connection of atoms
constitutional isomers

15.2 Draw the isomers of *trans*-2-hexene.

trans-2-hexene: $CH_3(H)C{=}C(H)CH_2CH_2CH_3$

a. $CH_3(H)C{=}C(H)CH_2CH_2CH_3$ — *cis*-2-hexene, stereoisomer

b. $CH_2{=}CHCH_2CH_2CH_2CH_3$ — different connection of atoms, constitutional isomer

c. cyclohexane — different connection of atoms, constitutional isomer

15.3 A molecule (or object) that is *not* superimposable on its mirror image is said to be *chiral.* A molecule (or object) that *is* superimposable on its mirror image is said to be *achiral.*

a. nail—achiral
b. screw—chiral
c. glove—chiral
d. pencil—achiral

15.4 A molecule that is chiral is not superimposable on its mirror image. A chirality center is a carbon with four different groups bonded to it.

15.5 To locate the chirality centers, look at each C individually and eliminate those C's that can't be chirality centers. Thus, omit all CH_2 and CH_3 groups and all multiply bonded C's. Check all remaining C's to see if they are bonded to four different groups, as in Example 15.1.

a. CH_3–C(H)(Cl)–CH_2CH_3

b. $(CH_3)_3CH$ none

c. $CH_2{=}CH$–C(H)(OH)–CH_3

d. C_6H_5–C(H)(Br)–CH_3

15.6 Label each chirality center, as in Example 15.1.

a. Br–C_6H_4–C(H)(2-pyridyl)–$CH_2CH_2N(CH_3)_2$

b. C_6H_5–C(H)($CH_2CH_2NHCH_3$)–O–C_6H_4–CF_3

15.7

$HOCH_2$–C(=O)–C(H)(OH)–CH_2OH — chirality center

15.8 Label each chirality center, as in Example 15.1.

a. $CH_3CH_2CH_2$–C(H)(OH)–CH_3

b. H_2N–C(COOH)(H)–C(CH_3)(H)–CH_2CH_3

c. CH_3CH_2–C(Br)(H)–CH_2CH_2–C(Cl)(H)–CH_3

15.9 Label the two chirality centers in vitamin K.

O

CH_3

$CH_2CH{=}CCH_2CH_2CH_2CHCH_2CH_2CH_2CHCH_2CH_2CH_2CH(CH_3)_2$

O CH_3 CH_3 CH_3

15.10 To draw both enantiomers:

[1] Draw one enantiomer by arbitrarily placing the four different groups on the bonds to the chirality center.

[2] Draw a mirror plane and arrange the substituents in the mirror image so that they are a reflection of the groups in the first molecule.

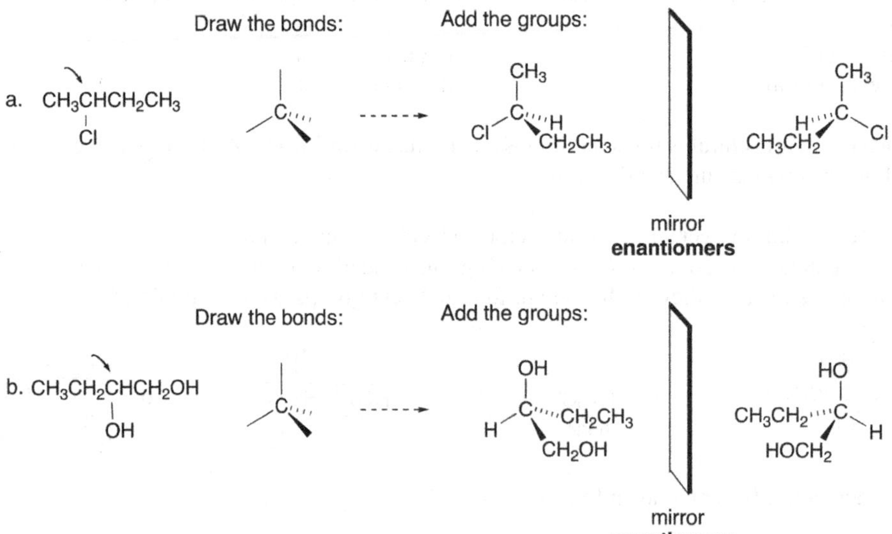

15.11 Use the directions in Answer 15.10.

CH_3

$HOCH_2CH_2{-}C{-}CH_2CH_2CH_3$

SH

chirality center

$CH_3CH_2CH_2$ | $HOCH_2CH_2$ | C | CH_3 | SH

$CH_2CH_2CH_3$ | CH_3 | C | HS | CH_2CH_2OH

enantiomers

15.12 Locate the chirality centers in each compound.

a. OH

b. CH_2CH_3

none
All C's are part of multiple bonds or are bonded to two or three H's.

c. Cl, Cl

15.13 Label the two chirality centers.

CH_3NH — Cl — Cl

15.14 Answer each question about propranolol.

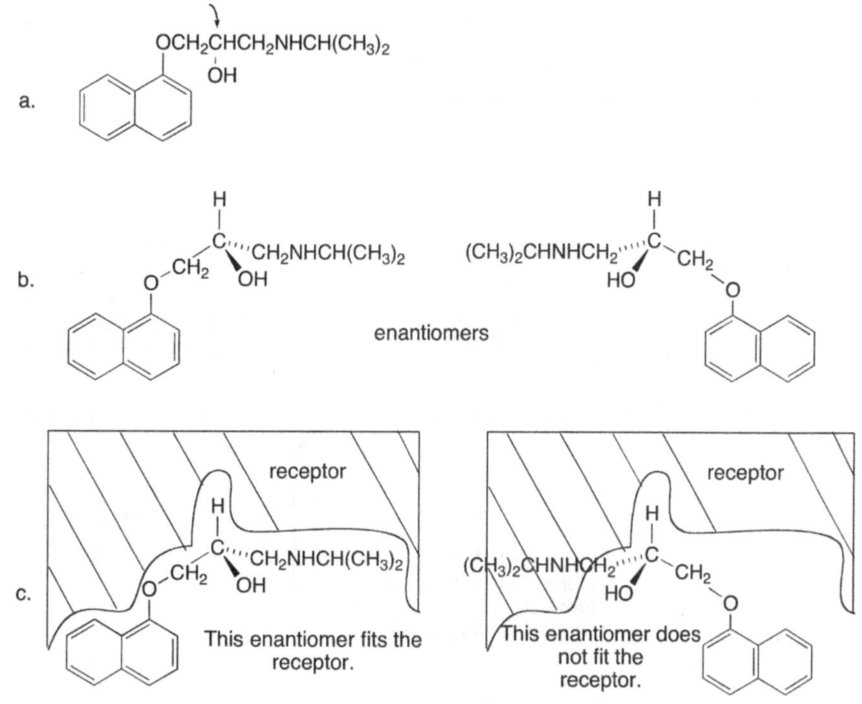

15.15 Convert each molecule to a Fisher projection formula by replacing the chirality center with a cross.

a. H–C–Br with CH_3 above and Cl below → H–(cross)–Br with CH_3 above and Cl below

b. H–C–OH with COOH above and CH_2Cl below → H–(cross)–OH with COOH above and CH_2Cl below

15.16 Work backwards to draw a structure with wedges (horizontal bonds) and dashes (vertical bonds).

a. CH_3CH_2–[cross: CH_3 top, Br right, Cl bottom] ------→ CH_3CH_2–C(wedge)–Br, with CH_3 (dashed) above and Cl (dashed) below

b. CH_3CH_2–[cross: H top, CH_3 right, NH_2 bottom] ------→ CH_3CH_2–C(wedge)–CH_3, with H (dashed) above and NH_2 (dashed) below

15.17 Convert each molecule to a Fisher projection formula as in Example 15.2.

- Draw the tetrahedron of one enantiomer with the horizontal bonds on wedges and the vertical bonds on dashed lines. Arrange the four groups on the chirality center arbitrarily in the first enantiomer.
- Draw the second enantiomer by arranging the substituents in the mirror image so they are a reflection of the groups in the first molecule.
- Replace the chirality center with a cross to draw the Fischer projections.

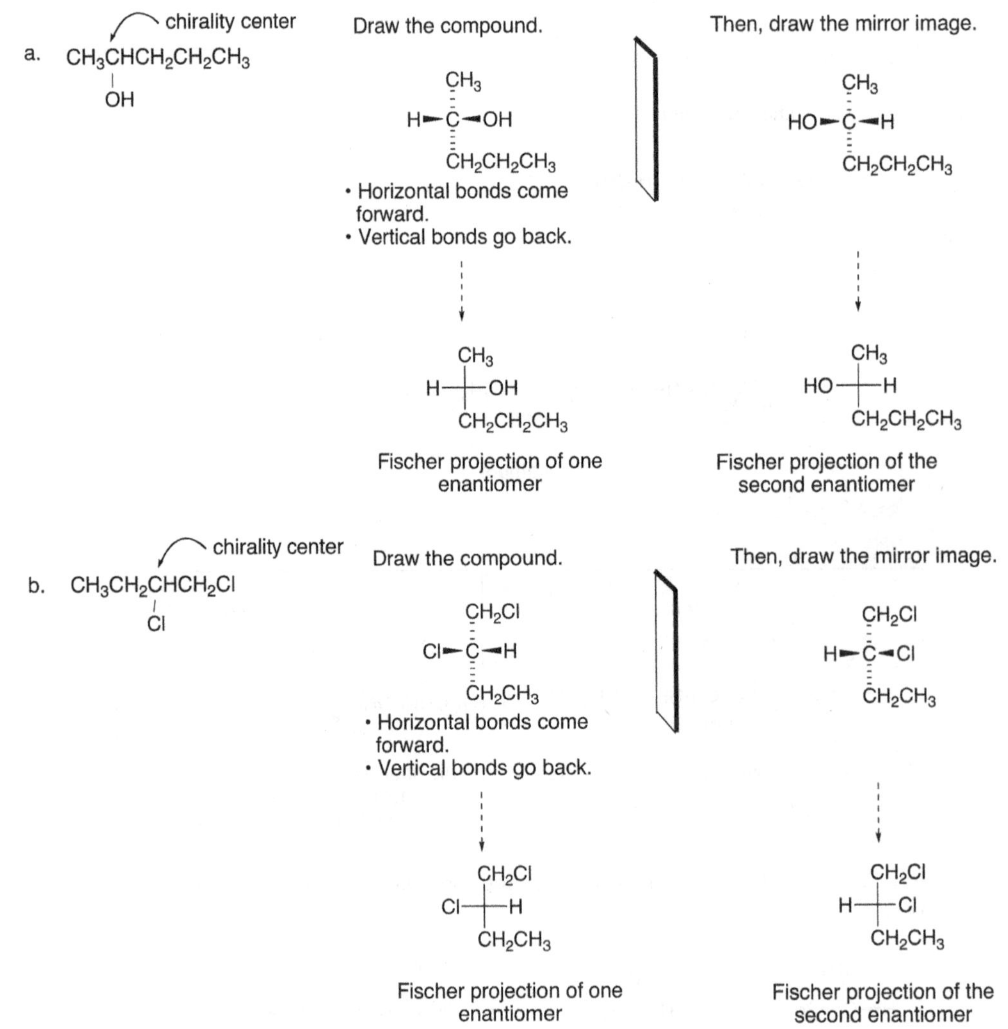

15.18 A chiral compound is optically active and an achiral compound is optically inactive.

a. $CH_3(CH_2)_6CH_3$ — achiral, optically inactive

b. COOH, C, H, CH_3, OH — chiral, optically active

c. CH_3, C, H, CH_3, OH — achiral, optically inactive

d. (cyclohexane) — achiral, optically inactive

15.19 Because two enantiomers rotate the plane of polarized light to an equal extent but in opposite directions, the rotation of one enantiomer is cancelled by the rotation of the other enantiomer and the sample is optically inactive.

15.20 Answer the questions about cholesterol as in Sample Problem 15.3.
a. The specific rotation of cholesterol is a negative value, so cholesterol is levorotatory.
b. The specific rotation of the enantiomer has the same value, but is opposite in sign—that is, +32.

15.21 Answer each question about alanine, as in Sample Problem 15.3.

a. COO^-, H, C, CH_3, $H_3\overset{+}{N}$
(*R*)-alanine

b. Two enantiomers have the same melting point, so the melting point of (*R*)-alanine is 297 °C.
c. The specific rotation of (*R*)-alanine is –8.5.

15.22 **Enantiomers are stereoisomers that are nonsuperimposable mirror images of each other.** Look for two compounds that when placed side-by-side have the groups on both chirality centers drawn as a *reflection* of each other. A given compound has only one possible enantiomer. **Diastereomers are stereoisomers that are *not* mirror images of each other.**

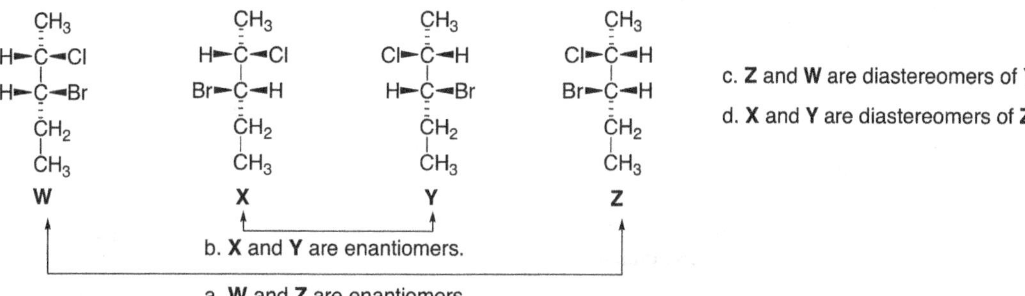
W
X
Y
Z
c. Z and W are diastereomers of Y.
d. X and Y are diastereomers of Z.
b. X and Y are enantiomers.
a. W and Z are enantiomers.

15.23 a. **A** and **B** are stereoisomers but not mirror images, so they are diastereomers.
b. **A** and **C** are mirror images that are not superimposable, so they are enantiomers.
c. **B** and **C** are stereoisomers but not mirror images, so they are diastereomers.

15.24

CH_3 — H $CH_2CH_2CH_3$ — chirality center — O

CH_3 — H $CH_2CH_2CH_3$ — O | $CH_3CH_2CH_2$ — H — CH_3 — O

enantiomers

Solutions to Odd-Numbered End-of-Chapter Problems

15.25 Use the definitions in Answer 15.3 to label each object as chiral or achiral.

a. chalk—achiral
b. shoe—chiral
c. soccer ball—achiral

15.27 Draw the mirror image for each compound.

a. CF_3, Br, C, Cl, H — molecule; CF_3, Cl, C, Br, H — mirror image

nonsuperimposable
chiral

b. H, Cl—C—H, Cl — molecule; H, H—C—Cl, Cl — mirror image

superimposable
achiral

15.29 **A** and **B** are mirror images of each other—enantiomers. **A** and **C** are superimposable and identical.

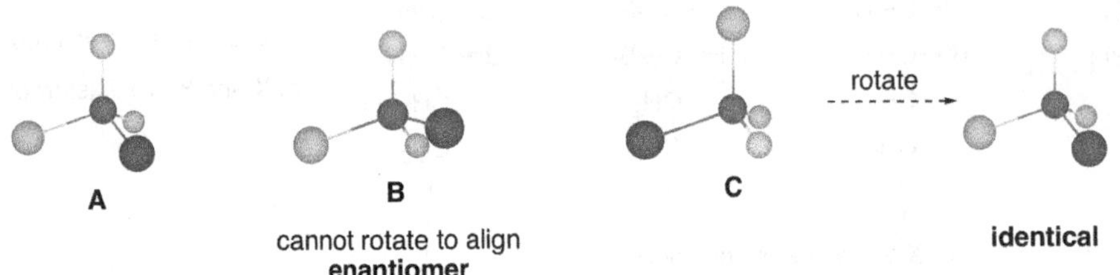

15.31 To locate the chirality centers, look at each C individually and eliminate those C's that can't be chirality centers. Thus, omit all CH_2 and CH_3 groups and all multiply bonded C's. Check all remaining C's to see if they are bonded to four different groups, as in Example 15.1.

a. CH_3CH_2CHCl with Cl on the end C

two Cl's bonded to end C
no chirality center

b. $CH_3CCH_2CHCH_3$ with CH_3 and H on C2, Cl on C4

four different
groups bonded to C

c. CH_3 on cyclopentene ring

d. CH_3 on cyclopentene ring

none

15.33

a. H OH (cyclohexane ring)

no chirality center

b. HO, H (cyclohexene ring)

chirality center

15.35 Draw a compound to fit each description.

a. $CH_3CH_2-C(CH_3)(H)-CH_2CH_2CH_3$

C_7H_{16}
one chirality center

b. $CH_3CH_2-C(OH)(H)-CH_2CH_2CH_3$

$C_6H_{14}O$
one chirality center

15.37 Locate the chirality centers in each compound.

a. (pyridine ring bonded to N-methylpyrrolidine ring; N, N, CH_3)

b. $HO_2CCH_2CH(NH_2)C(=O)NHCH(CO_2CH_3)CH_2-C_6H_5$

15.39 Locate the chirality center and draw the enantiomers.

$C_6H_5-CH_2CH(CH_3)NHCH_3$ (H on N)

methamphetamine

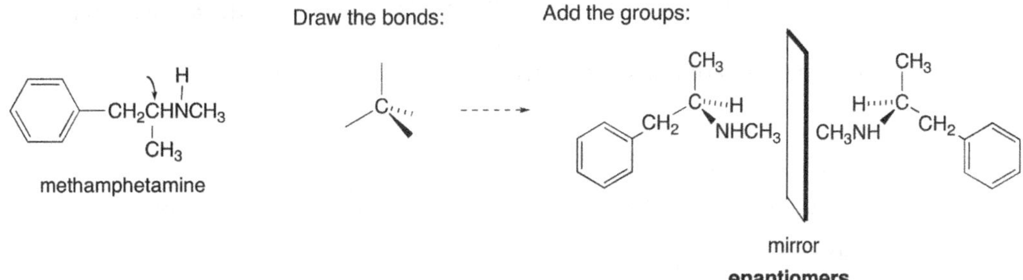

enantiomers

15.41 Enantiomers are stereoisomers that are nonsuperimposable mirror images of each other.

a. $CH_3-C(CH_3)(OH)-H$ and $H-C(CH_3)(OH)-CH_3$

The central C is bonded to two CH_3 groups.
There is no chirality center.
identical molecules

b. $CH_3-C(NH_2)(H)-CO_2H$ and $HO_2C-C(NH_2)(H)-CH_3$

The central C is bonded to four different groups.
The compounds are nonsuperimposable mirror images.
enantiomers

15.43 Enantiomers are stereoisomers that are nonsuperimposable mirror images of each other. Diastereomers are stereoisomers that are *not* mirror images of each other.

a. $CH_3CH_2CH_2OCH_3$ and $CH_3CH_2OCH_2CH_3$
different connectivity
constitutional isomers

c. 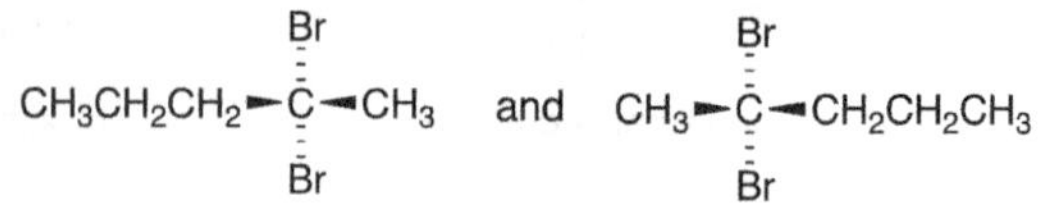 $CH_3CH_2CH_2$–C(Br)(Br)–CH_3 and CH_3–C(Br)(Br)–$CH_2CH_2CH_3$
There is no chirality center.
identical molecules

b. 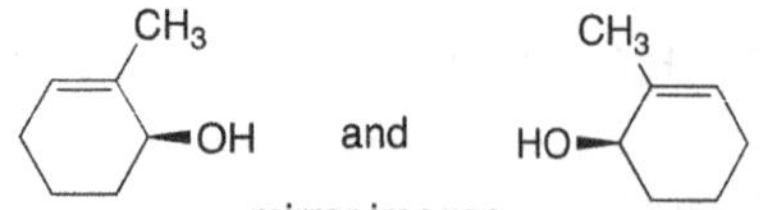 (structures) OH and HO
mirror images
enantiomers

d. 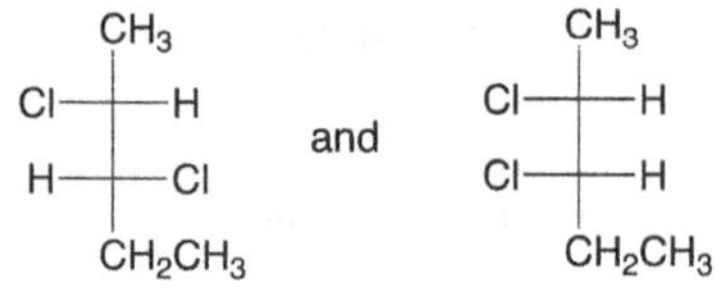 Fischer projections: CH_3 / Cl–H / H–Cl / CH_2CH_3 and CH_3 / Cl–H / Cl–H / CH_2CH_3
two chirality centers
not mirror images
diastereomers

15.45 Answer each question with a compound of molecular formula $C_5H_{10}O_2$.

a. 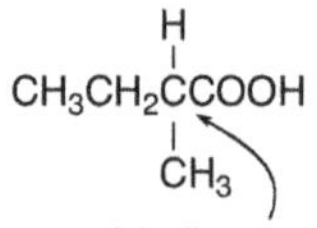 $CH_3CH_2CH(CH_3)COOH$
chirality center

b. $CH_3CH_2CH_2CH_2COOH$
no chirality center
same molecular formula
constitutional isomer

c. 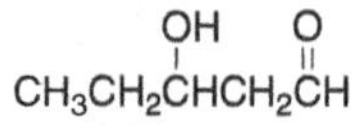 $CH_3CH_2CH(OH)CH_2CHO$
no COOH
same molecular formula
constitutional isomer

15.47 Draw the constitutional isomers and then label the chirality centers.

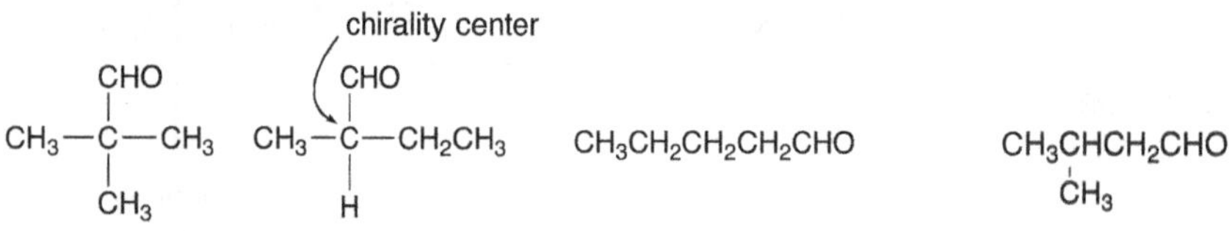

CH_3–C(CHO)(CH_3)–CH_3 CH_3–C(CHO)(H)–CH_2CH_3 (chirality center) $CH_3CH_2CH_2CH_2CHO$ $CH_3CH(CH_3)CH_2CHO$

15.49 Convert the given ball-and-stick model to a Fischer projection by replacing the chirality center with a cross.

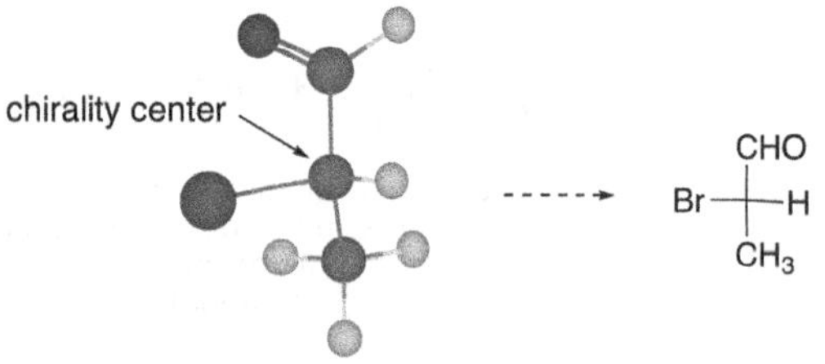

chirality center

CHO / Br–H / CH_3

15.51 Convert each molecule to a Fisher projection formula by replacing the chirality center with a cross.

a. H–C–OH (CHO top, CH_3 bottom) → Fischer projection: CHO top, H left, OH right, CH_3 bottom

b. OCH_3 top; CH_3–C–H; CH_3–C–OH; CH_2CH_3 bottom → Fischer projection: OCH_3 top; CH_3 left, H right; CH_3 left, OH right; CH_2CH_3 bottom

15.53 Work backwards to draw three-dimensional representations from the Fischer projections.

a. Fischer projection: CH_3 top, H left, F right, $CH(CH_3)_2$ bottom → H–C–F (CH_3 top, $CH(CH_3)_2$ bottom)

b. Fischer projection: COOH top; H left, NH_2 right; CH_3 left, H right; CH_2CH_3 bottom → H–C–NH_2; CH_3–C–H (COOH top, CH_2CH_3 bottom)

15.55 Enantiomers are stereoisomers that are nonsuperimposable mirror images of each other.

a. CH_3 top, Cl left, H right, CH_3 bottom and CH_3 top, H left, Cl right, CH_3 bottom

The central C is bonded to two CH_3 groups.
There is no chirality center.
identical molecules

b. COOH top, CH_3 left, OH right, H bottom and COOH top, HO left, CH_3 right, H bottom

mirror images
enantiomers

c. CO_2H top, H left, NH_2 right, CH_2OH bottom and CO_2H top, H_2N left, H right, CH_2OH bottom

mirror images
enantiomers

d. $C(CH_3)_3$ top, $(CH_3)_3C$ left, OH right, H bottom and $C(CH_3)_3$ top, HO left, $C(CH_3)_3$ right, H bottom

The central C is bonded to two $C(CH_3)_3$ groups.
There is no chirality center.
identical molecules

15.57

a. An optically active compound rotates the plane of polarized light. An optically inactive compound does not rotate the plane of polarized light.
b. One possibility: $CH_3CH_2CH_2CH_2OH$ is achiral and optically inactive.
c. One possibility:

CH_3CH_2–C(CH_3)(H)(OH)

chiral and optically active

15.59

a. 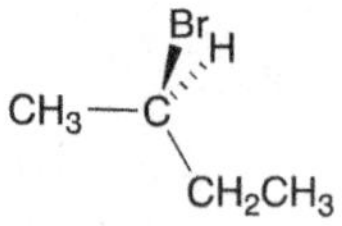 CH_3—C(Br)(H)—CH_2CH_3

chiral
optically active

b. 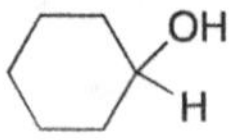 cyclohexane with OH and H

achiral
optically inactive

15.61

a. Lactose is dextrorotatory.
b. The specific rotation of two enantiomers is the same numerical value, but opposite in sign. Therefore, the specific rotation of the enantiomer of lactose is –52.
c. Two enantiomers have identical physical properties except for their interaction with plane-polarized light, so 22 g of the enantiomer of lactose dissolves in 100 mL of water.

15.63 **Enantiomers are stereoisomers that are nonsuperimposable mirror images of each other.** Look for two compounds that when placed side-by-side have the groups on both chirality centers drawn as a *reflection* of each other. A given compound has only one possible enantiomer. **Diastereomers are stereoisomers that are *not* mirror images of each other.**

a.

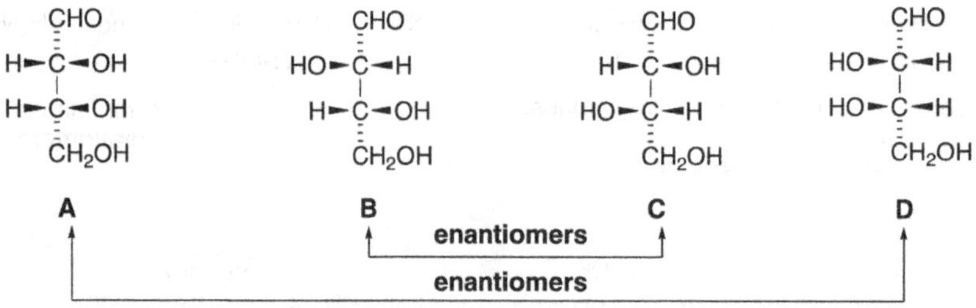

A	B	C	D
CHO	CHO	CHO	CHO
H–C–OH	HO–C–H	H–C–OH	HO–C–H
H–C–OH	H–C–OH	HO–C–H	HO–C–H
CH_2OH	CH_2OH	CH_2OH	CH_2OH

B and **C**: **enantiomers**
A and **D**: **enantiomers**

The figure above shows the enantiomers. All other relationships are diastereomers.
[1] **A and B,** diastereomers; [2] **A** and **C,** diastereomers; [3] **A** and **D,** enantiomers; [4] **B** and **C,** enantiomers; [5] **B** and **D,** diastereomers; [6] **C** and **D,** diastereomers.

b.

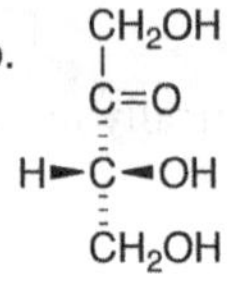

CH_2OH
C=O
H–C–OH
CH_2OH

15.65 Enantiomers are stereoisomers that are mirror images. *cis*-2-Butene and *trans*-2-butene are diastereomers because they are stereoisomers that are not mirror images of one another.

CH_3(H)C=C(H)CH_3 (cis) — CH_3(H)C=C(CH_3)H (trans)

cis
trans

not mirror images
diastereomers

15.67 Answer each question about lactic acid.

a. CH_3CHCO_2H with OH on the second carbon — chirality center

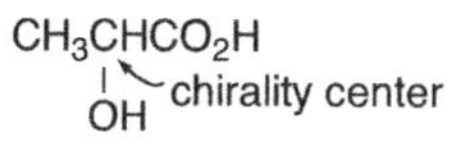

b. CH_3–C(H)(OH)–COOH   HOOC–C(H)(OH)–CH_3

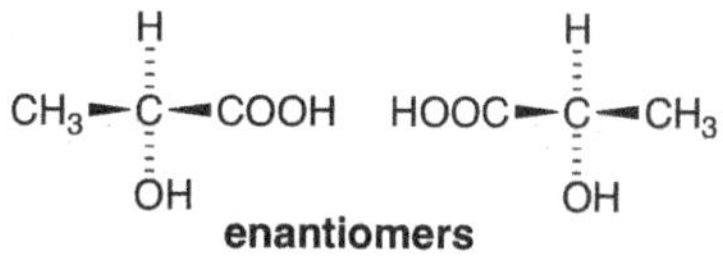

enantiomers

c. CH_3–C(H)(OH)–COOH   HOOC–C(H)(OH)–CH_3

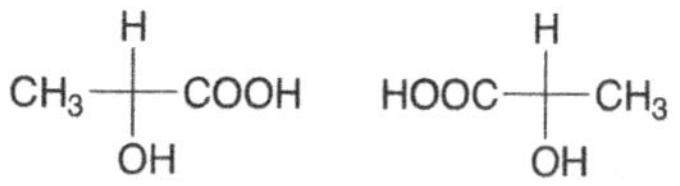

Fischer projections

15.69 Answer each question about hydroxydihydrocitronellal.

a. $CH_3C(OH)(CH_3)CH_2CH_2CH_2CH(CH_3)–CH_2CHO$ — chirality center

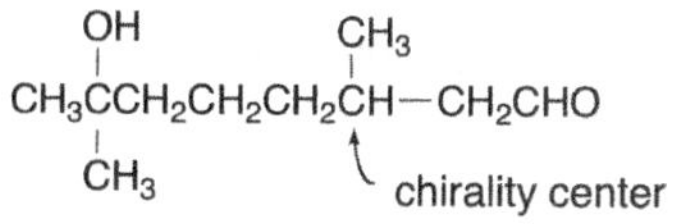

b. $CH_3C(HO)(CH_3)CH_2CH_2CH_2$–C($CH_3$)(H)–$CH_2CHO$ $OHCCH_2$–C(CH_3)(H)–$CH_2CH_2CH_2C(OH)(CH_3)CH_3$

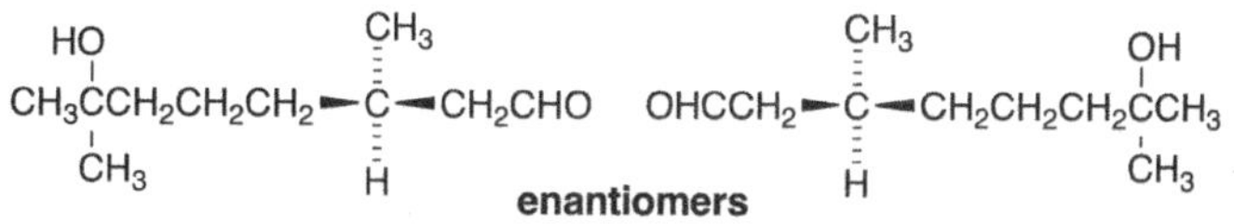

enantiomers

c. $CH_3C(HO)(CH_3)CH_2CH_2CH_2$–C($CH_3$)(H)–$CH_2CHO$ $OHCCH_2$–C(CH_3)(H)–$CH_2CH_2CH_2C(OH)(CH_3)CH_3$

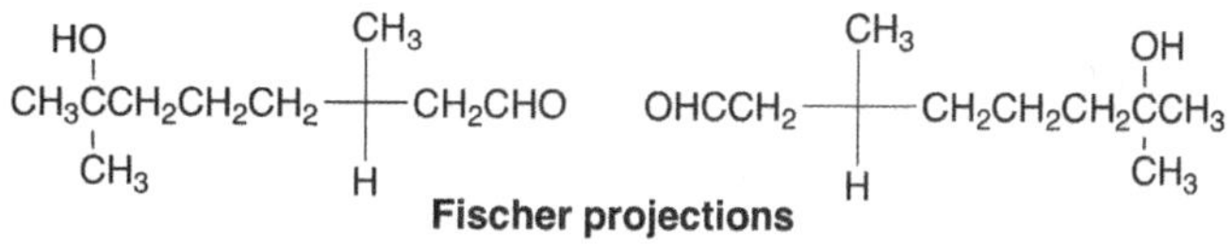

Fischer projections

15.71 Answer each question about Darvon.

a. chirality center

CH_2(phenyl)

$CH_3CH_2CO_2$–C–(phenyl)

CH_3–C–H

$CH_2N(CH_3)_2$

chirality center

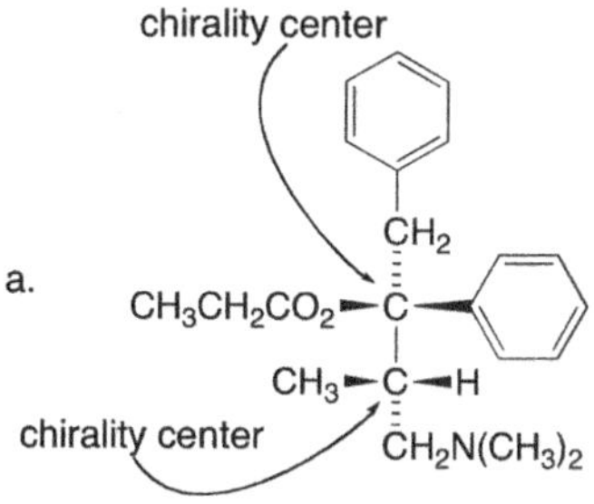

b. CH_2(phenyl)

(phenyl)–C–$O_2CCH_2CH_3$

H–C–CH_3

$CH_2N(CH_3)_2$

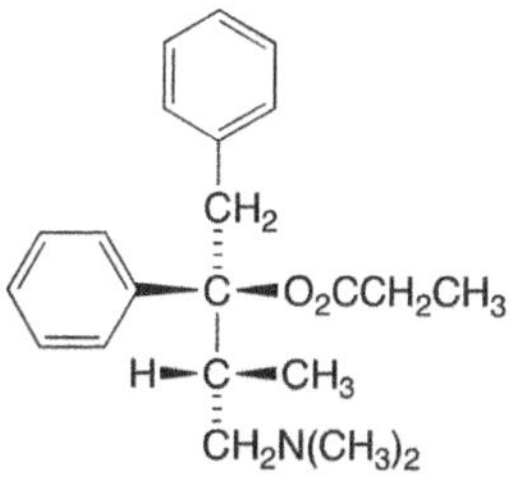

enantiomer

(Novrad is Darvon spelled backwards.)

c. 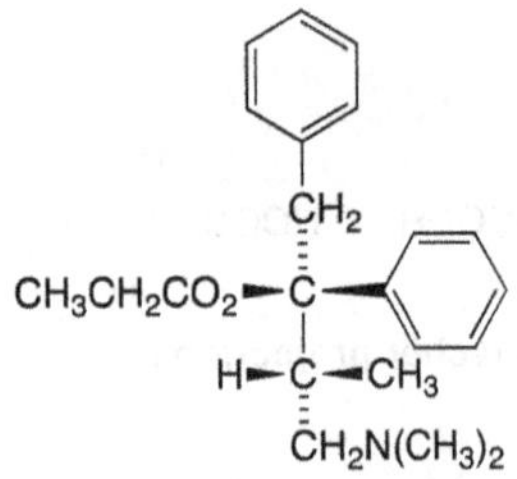 diastereomer

d. 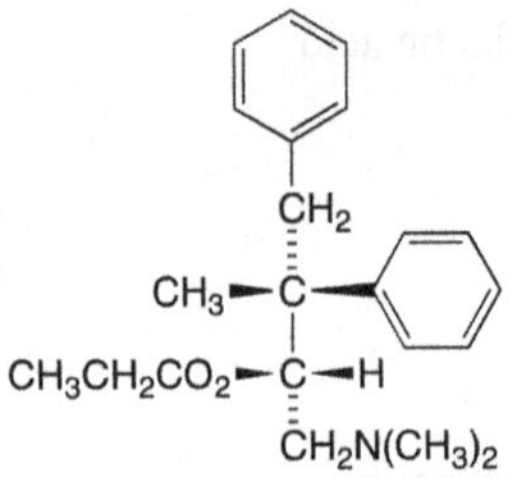 constitutional isomer

e. 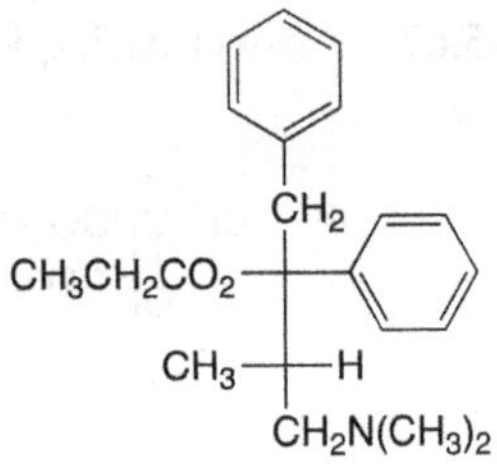 Fischer projection

15.73 Each chirality center in sucrose is labeled with a star.

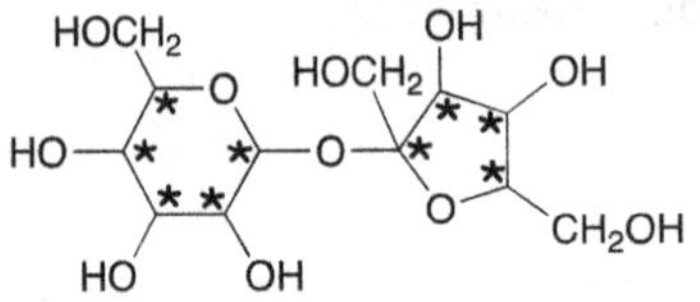

★ = chirality center

Chapter 16 Aldehydes and Ketones

Chapter Review

[1] What are the characteristics of aldehydes and ketones?

- An aldehyde has the general structure RCHO, and contains a carbonyl group (C=O) bonded to at least one hydrogen atom. (16.1)
- A ketone has the general structure RCOR', and contains a carbonyl group (C=O) bonded to two carbon atoms. (16.1)

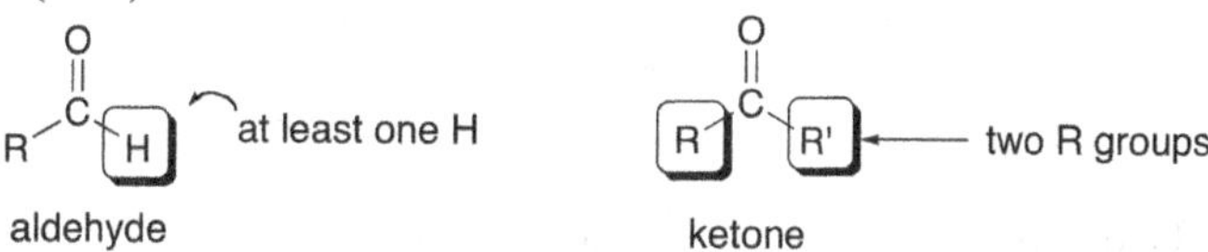

- The carbonyl carbon is trigonal planar, with bond angles of approximately 120°. (16.1)
- The carbonyl group is polar, giving an aldehyde or ketone stronger intermolecular forces than hydrocarbons. (16.3)
- Aldehydes and ketones have lower boiling points than alcohols, but higher boiling points than hydrocarbons of comparable size. (16.3)
- Aldehydes and ketones are soluble in organic solvents, but those having less than six C's are water soluble, too. (16.3)

[2] How are aldehydes and ketones named? (16.2)

- Aldehydes are identified by the suffix *-al*, and the carbon chain is numbered to put the carbonyl group at C1.
- Ketones are identified by the suffix *-one*, and the carbon chain is numbered to give the carbonyl group the lower number.

[3] Give examples of useful aldehydes and ketones. (16.4)

- Formaldehyde (CH_2=O) is an irritant in smoggy air, a preservative, and a disinfectant, and it is a starting material for synthesizing polymers.
- Acetone [$(CH_3)_2C$=O] is an industrial solvent and polymer starting material.
- Several aldehydes—cinnamaldehyde, vanillin, geranial, and citronellal—have characteristic odors and occur naturally in fruits and plants.
- Dihydroxyacetone [$(HOCH_2)_2C$=O] is the active ingredient in artificial tanning agents, and many other ketones are useful sunscreens.
- Amygdalin, a naturally occurring carbonyl derivative, forms toxic HCN on hydrolysis.

[4] What products are formed when aldehydes are oxidized? (16.5)

- Aldehydes are oxidized to carboxylic acids (RCOOH) with $K_2Cr_2O_7$ or Tollens reagent.

$$\text{RCHO} \xrightarrow[\text{or Tollens reagent}]{K_2Cr_2O_7} \underset{\text{carboxylic acid}}{\text{RCOOH}}$$

- Ketones are not oxidized because they contain no H atom on the carbonyl carbon.

[5] What products are formed when aldehydes and ketones are reduced? (16.6)

- Aldehydes are reduced to 1° alcohols with H_2 and a Pd catalyst.

$$\text{RCHO} \xrightarrow[\text{Pd}]{H_2} \underset{1^\circ\text{ alcohol}}{\text{RCH}_2\text{OH}}$$

- Ketones are reduced to 2° alcohols with H_2 and a Pd catalyst.

$$\text{R}_2\text{CO} \xrightarrow[\text{Pd}]{H_2} \underset{2^\circ\text{ alcohol}}{\text{R}_2\text{CHOH}}$$

- Biological reduction occurs with the coenzyme NADH.

$$\text{C}{=}\text{O} \xrightarrow[\text{enzyme}]{\text{NADH}} \text{H}-\text{C}-\text{OH} + \text{NAD}^+$$

[6] What reactions occur during vision? (16.7)

- When light hits the retina, the crowded C=C in 11-*cis*-retinal is isomerized to a more stable trans double bond, and a nerve impulse is generated. All-*trans*-retinal is converted back to the cis isomer by a three-step sequence—reduction, isomerization, and oxidation.

[7] What are hemiacetals and acetals, and how are they prepared? (16.8)

- Hemiacetals contain an OH group and an OR group bonded to the same carbon.
- Acetals contain two OR groups bonded to the same carbon.
- Treatment of an aldehyde or ketone with an alcohol (ROH) first forms an unstable hemiacetal, which reacts with more alcohol to form an acetal.

$$\text{RC(=O)H (or R)} \underset{}{\overset{\text{R'OH, H}_2\text{SO}_4}{\rightleftharpoons}} \underset{\text{hemiacetal}}{\text{R}-\text{C(OH)(H or R)}-\text{OR'}} \overset{\text{R'OH, H}_2\text{SO}_4}{\rightleftharpoons} \underset{\text{acetal}}{\text{R}-\text{C(OR')(H or R)}-\text{OR'}} + \text{H}_2\text{O}$$

- Cyclic hemiacetals are stable compounds present in carbohydrates and some drugs.

$$\underset{\text{5-hydroxypentanal}}{\text{HOCH}_2\text{CH}_2\text{CH}_2\text{CH}_2\text{CHO}} \xrightarrow{\text{re-draw}} \text{(ring form: CHO ... OH)} \longrightarrow \text{cyclic hemiacetal}$$

- Acetals are converted back to aldehydes and ketones by hydrolysis with water and acid.

Problem Solving

[1] Nomenclature (16.2)

Example 16.1 Give the IUPAC name for each aldehyde.

a. 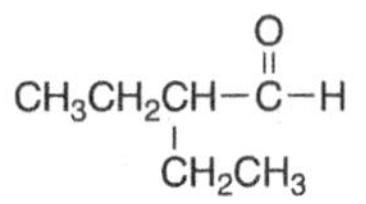 $CH_3CH_2CH(CH_2CH_3)-C(=O)-H$

b. 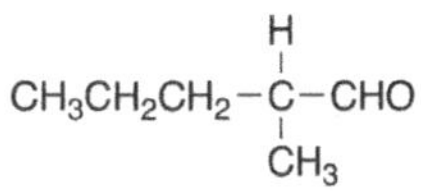 $CH_3CH_2CH_2-CH(CH_3)-CHO$

Analysis and Solution

a. [1] Find and name the longest chain containing the CHO:

$CH_3CH_2CH(CH_2CH_3)-C(=O)-H$

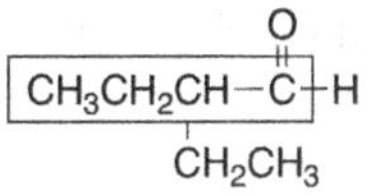

butan*e* ----→ butan*al*
(4 C's)

[2] Number and name substituents, making sure the CHO group is at C1:

$\overset{3}{C}H_3CH_2\overset{2}{C}H(CH_2CH_3)-\overset{1}{C}(=O)-H$

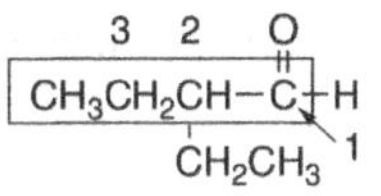

Answer: 2-ethylbutanal

b. [1] Find and name the longest chain containing the CHO:

$CH_3CH_2CH_2-CH(CH_3)-C(=O)-H$

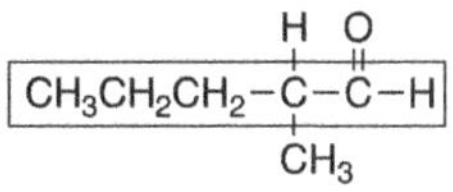

pentan*e* ----→ pentan*al*
(5 C's)

[2] Number and name substituents, making sure the CHO group is at C1:

$CH_3CH_2CH_2-\overset{2}{C}H(CH_3)-\overset{1}{C}(=O)-H$

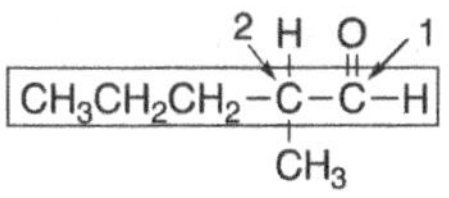

Answer: 2-methylpentanal

Example 16.2 Give the IUPAC name for each ketone.

a. 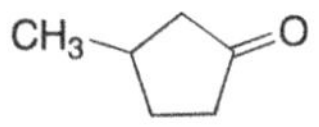 3-methyl-substituted cyclopentanone (CH_3 on ring, $=O$ on ring)

b. 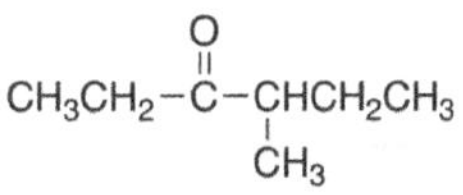 $CH_3CH_2-C(=O)-CH(CH_3)CH_2CH_3$

Analysis and Solution

a. [1] Name the ring:

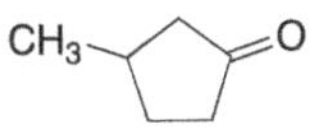

cyclopentan*e* ----→ cyclopentan*one*
(5 C's)

[2] Number and name the substituents, making sure the carbonyl carbon is at C1:

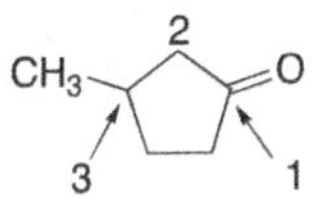

(ring numbered 1 at C=O, 2, 3 at CH_3)

Answer: 3-methylcyclopentanone

b. [1] Find and name the longest chain containing the carbonyl group:

$CH_3CH_2-\overset{O}{\overset{\|}{C}}-\underset{CH_3}{\underset{|}{C}}HCH_2CH_3$

hexane ----→ hexan*one*
(6 C's)

[2] Number and name the substituents, making sure the carbonyl carbon has the lower number:

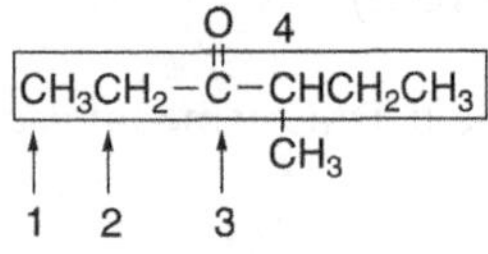

Answer: 4-methyl-3-hexanone

---

[2] Reactions of Aldehydes and Ketones (16.5)

Example 16.3 What product is formed when each carbonyl compound is treated with $K_2Cr_2O_7$?

a. $CH_3-\overset{O}{\overset{\|}{C}}-\underset{CH_3}{\underset{|}{C}}HCH_2CH_3$ b. cyclobutyl–CHO

Analysis

Compounds that contain a C–H and C–O bond on the *same* carbon are oxidized with $K_2Cr_2O_7$. Thus:

- Aldehydes (RCHO) are oxidized to RCO_2H.
- Ketones (R_2CO) are *not* oxidized with $K_2Cr_2O_7$.

Solution

The ketone in part (a) is inert to oxidation, but the aldehyde in part (b) is oxidized with $K_2Cr_2O_7$ to a carboxylic acid.

a. $CH_3-\overset{O}{\overset{\|}{C}}-\underset{CH_3}{\underset{|}{C}}HCH_2CH_3 \xrightarrow{K_2Cr_2O_7}$ No reaction

This C is bonded only to other C's.

b. cyclobutyl–CHO $\xrightarrow{K_2Cr_2O_7}$ cyclobutyl–COOH

Replace 1 C–H bond by 1 C–O bond.

Example 16.4 What product is formed when each compound is treated with Tollens reagent (Ag_2O, NH_4OH)?

a. cyclohexyl–CHO b. $CH_3-\overset{O}{\overset{\|}{C}}-CH_2CH_3$

Analysis

Only aldehydes (RCHO) react with Tollens reagent. Ketones and alcohols are inert to oxidation.

Solution

The aldehyde in part (a) is oxidized to RCO_2H, but the ketone in part (b) does not react with Tollens reagent.

a. Cyclohexyl–CHO (C–H) $\xrightarrow[NH_4OH]{Ag_2O}$ Cyclohexyl–COOH (C–OH)

Replace 1 C–H bond by 1 C–O bond.

b. $CH_3COCH_2CH_3$ $\xrightarrow[NH_4OH]{Ag_2O}$ No reaction

This C=O is bonded only to other C's.

[3] Reduction of Aldehydes and Ketones (16.6)

Example 16.5 What alcohol is formed when each aldehyde or ketone is treated with H_2 in the presence of a Pd catalyst?

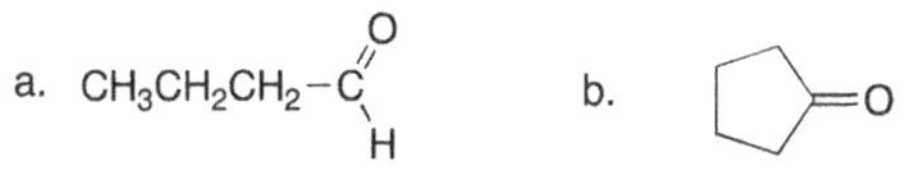

a. $CH_3CH_2CH_2-CHO$　　b. cyclopentanone (cyclopentyl=O)

Analysis

To draw the products of reduction:

- Locate the C=O and mentally break one bond in the double bond.
- Mentally break the H–H bond of the reagent.
- Add one H atom to each atom of the C=O, forming new C–H and O–H single bonds.

Solution

The aldehyde (RCHO) in part (a) forms a 1° alcohol (RCH_2OH) and the ketone in part (b) forms a 2° alcohol (R_2CHOH).

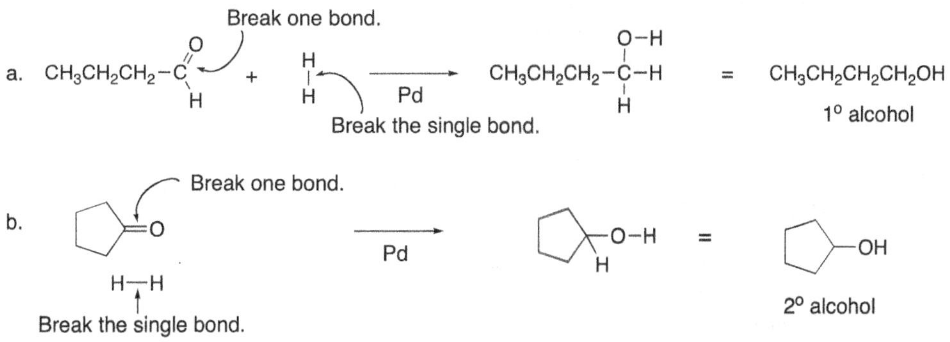

a. $CH_3CH_2CH_2-CHO$ (Break one bond.) + H–H (Break the single bond.) $\xrightarrow{Pd}$ $CH_3CH_2CH_2-CH(H)-O-H$ = $CH_3CH_2CH_2CH_2OH$

1° alcohol

b. cyclopentanone (Break one bond.) + H–H (Break the single bond.) $\xrightarrow{Pd}$ cyclopentyl(H)–O–H = cyclopentyl–OH

2° alcohol

[4] Acetals (16.8)

Example 16.6 Draw the hemiacetal and acetal formed when the following ketone is treated with methanol (CH_3OH) in the presence of H_2SO_4.

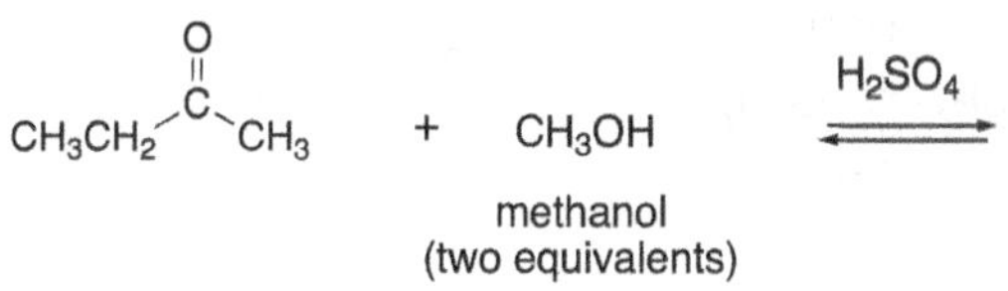

$CH_3CH_2COCH_3$ + CH_3OH $\overset{H_2SO_4}{\rightleftharpoons}$

methanol
(two equivalents)

Analysis

To form a hemiacetal and acetal from a carbonyl compound:

- Locate the C=O in the starting material.
- Break one C–O bond and add one equivalent of CH_3OH across the double bond, placing the OCH_3 group on the carbonyl carbon. This forms the hemiacetal.
- Replace the OH group of the hemiacetal by OCH_3 to form the acetal.

Solution

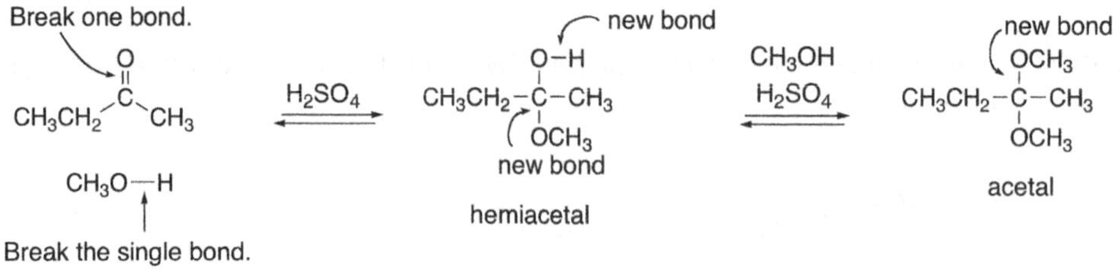

Break one bond.

$CH_3CH_2C(=O)CH_3$ + CH_3O—H $\xrightleftharpoons{H_2SO_4}$ $CH_3CH_2C(OH)(OCH_3)CH_3$ $\xrightleftharpoons[H_2SO_4]{CH_3OH}$ $CH_3CH_2C(OCH_3)_2CH_3$

Break the single bond.

new bond (O–H); new bond (OCH_3)

hemiacetal

new bond (OCH_3)

acetal

Example 16.7 Identify each compound as an ether, hemiacetal, or acetal.

a. $CH_3-O-CH_2CH_2CH_3$

b. 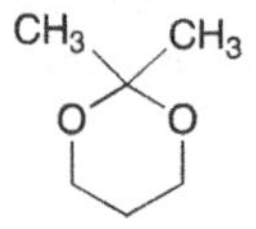 (six-membered ring containing two O atoms, with one ring carbon bearing two CH_3 groups)

Analysis

Recall the definitions to identify the functional groups:

- An ether has the general structure ROR.
- A hemiacetal has one C bonded to OH and OR.
- An acetal has one C bonded to two OR groups.

Solution

a. 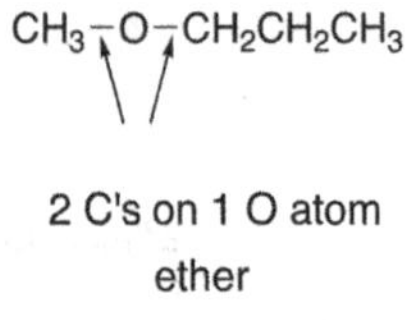 $CH_3-O-CH_2CH_2CH_3$

2 C's on 1 O atom

ether

b. 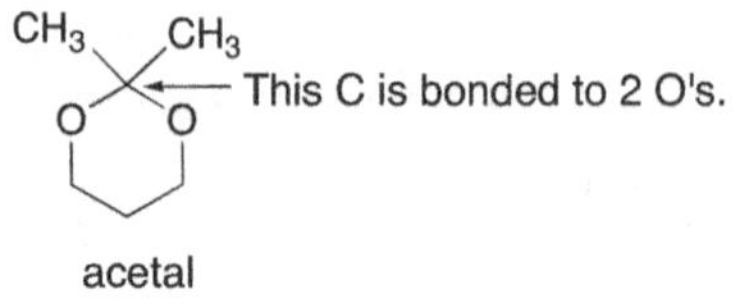 (ring structure) ← This C is bonded to 2 O's.

acetal

Note that the acetal in part (b) is part of a ring. It contains one ring carbon bonded to two oxygens, making it an acetal.

Example 16.8 What hydrolysis products are formed when the given acetal is treated with H_2SO_4 in H_2O?

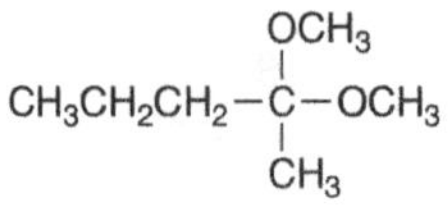

$CH_3CH_2CH_2-C(OCH_3)(CH_3)-OCH_3$

Analysis

To draw the products of hydrolysis:

- Locate the two C–OR bonds on the same carbon.
- Replace the two C–O single bonds with a carbonyl group (C=O).
- Each OR group then becomes a molecule of alcohol (ROH) product.

Solution

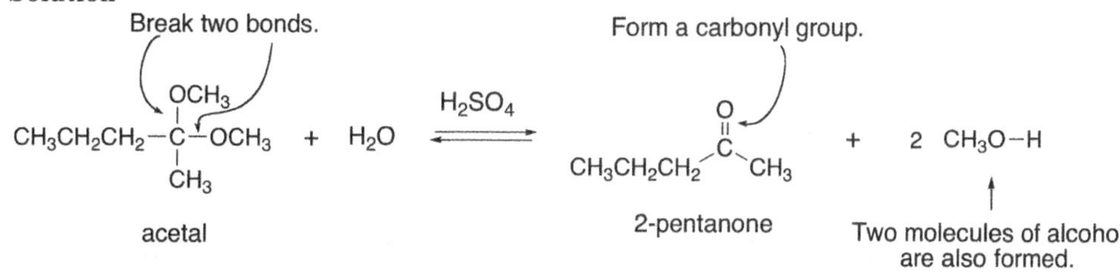

Self-Test

[1] Fill in the blank with one of the terms listed below.

Acetals (16.8)	Hydrolysis (16.8)	Reduction (16.6)
Aldehyde (16.1)	Ketone (16.1)	Tollens reagent (16.5)
Hemiacetal (16.8)	NADH (16.6)	

1. When bonds are cleaved by reaction with water, the reaction is a ______ reaction.
2. One reagent used to oxidize aldehydes selectively is the ______.
3. ______ are compounds that contain two OR groups (alkoxy groups) bonded to the same carbon.
4. An ______ has at least one H atom bonded to a carbonyl group.
5. A ______ contains an OH group (hydroxyl) and an OR group (alkoxy) bonded to the same carbon.
6. A ______ has two alkyl groups bonded to the carbonyl group.
7. ______ results in a decrease in the number of C–O bonds or an increase in the number of C–H bonds.
8. Biological systems use ______ as a reducing agent.

[2] Label each compound as a hemiacetal, acetal, ether, aldehyde, or ketone.

9. $CH_3CH_2CH_2-C(OH)(CH_3)-OCH_3$

10. (cyclohexane ring)–CHO

11. $CH_3CH_2-C(OCH_3)(CH_3)-OCH_3$

12. 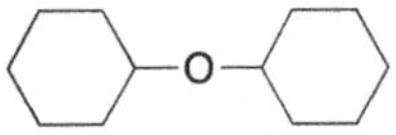 (cyclohexyl–O–cyclohexyl)

13. 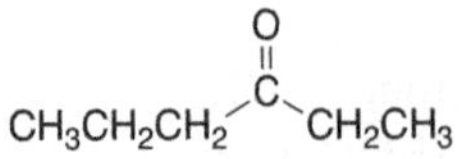 $CH_3CH_2CH_2C(=O)CH_2CH_3$

[3] Match the products with the correct reaction.

14. $CH_3CH_2CH_2CHO \xrightarrow{K_2Cr_2O_7}$

15. 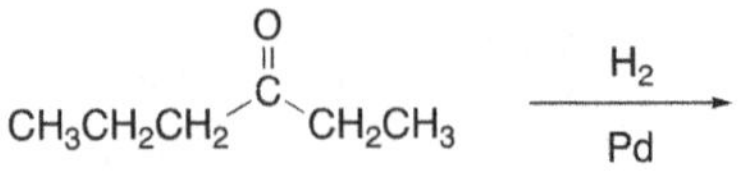 $CH_3CH_2CH_2C(=O)CH_2CH_3 \xrightarrow[Pd]{H_2}$

16. 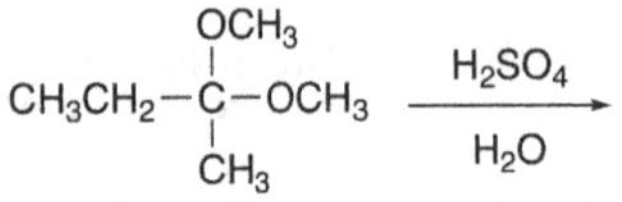 $CH_3CH_2C(OCH_3)(OCH_3)CH_3 \xrightarrow[H_2O]{H_2SO_4}$

17. 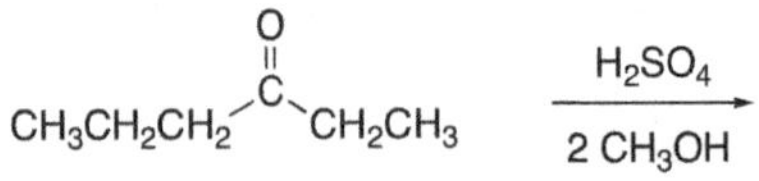 $CH_3CH_2CH_2C(=O)CH_2CH_3 \xrightarrow[2\ CH_3OH]{H_2SO_4}$

18. 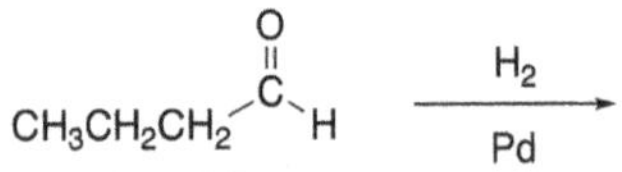 $CH_3CH_2CH_2C(=O)H \xrightarrow[Pd]{H_2}$

a. $CH_3CH_2CH_2CH_2OH$

b. $CH_3CH_2CH_2COOH$

c. 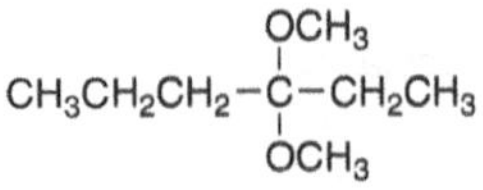 $CH_3CH_2CH_2-C(OCH_3)(OCH_3)-CH_2CH_3$

d. 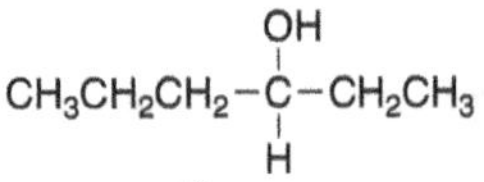 $CH_3CH_2CH_2-C(OH)(H)-CH_2CH_3$

e. 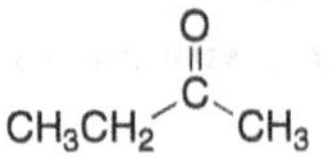 $CH_3CH_2C(=O)CH_3$ + 2 CH_3OH

[4] Which compound in each pair has the higher boiling point?

19. $CH_3(CH_2)_6CHO$ (**A**) or $CH_3(CH_2)_7OH$ (**B**)

20. 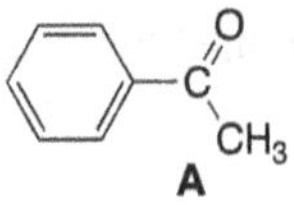 $C_6H_5C(=O)CH_3$ (**A**) or $C_6H_5CH(CH_3)_2$ (**B**)

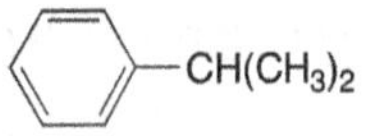

21. $CH_3(CH_2)_6CHO$ (**A**) or $CH_3(CH_2)_2CHO$ (**B**)

[5] Label each compound as water soluble or water insoluble.

22. $CH_3(CH_2)_7CHO$

23. 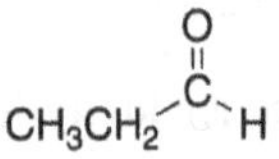 $CH_3CH_2C(=O)H$

24. $CH_3(CH_2)_2CHO$

25. 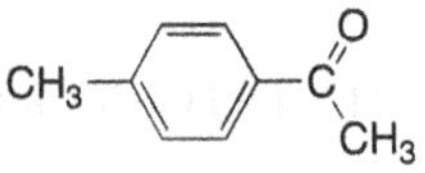 $CH_3-C_6H_4-C(=O)CH_3$

Answers to Self-Test

1. hydrolysis	6. ketone	11. acetal	16. e	21. **A**
2. Tollens reagent	7. Reduction	12. ether	17. c	22. insoluble
3. Acetals	8. NADH	13. ketone	18. a	23. soluble
4. aldehyde	9. hemiacetal	14. b	19. **B**	24. soluble
5. hemiacetal	10. aldehyde	15. d	20. **A**	25. insoluble

Solutions to In-Chapter Problems

16.1 An **aldehyde** has at least one H atom bonded to the carbonyl group.
A **ketone** has two alkyl groups bonded to the carbonyl group.

a. $CH_3CH_2C(=O)H$ aldehyde

b. $CH_3CH_2C(=O)CH_3$ ketone

c. $(CH_3)_3CC(=O)CH_3$ ketone

d. $(CH_3CH_2)_2CHC(=O)H$ aldehyde

16.2 Draw the constitutional isomers of molecular formula C_4H_8O and then label each compound using the definitions from Answer 16.1.

$CH_3CH_2C(=O)CH_3$ ketone

$CH_3CH_2CH_2C(=O)H$ aldehyde

$(CH_3)_2CHC(=O)H$ aldehyde

16.3 To name an aldehyde using the IUPAC system, use the steps in Example 16.1:

[1] Find the longest chain containing the CHO group, and change the *-e* ending of the parent alkane to the suffix ***-al.***

[2] Number the chain or ring to put the CHO group at C1, but omit this number from the name. Apply all of the other usual rules of nomenclature.

a. $(CH_3)_2CHCH_2CH_2CH_2CHO$

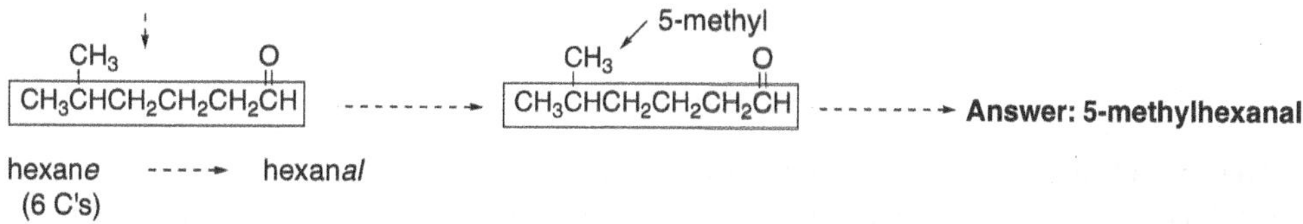

hexane ----→ hexan*al*
(6 C's)

b. $(CH_3)_3CC(CH_3)_2CH_2CHO$

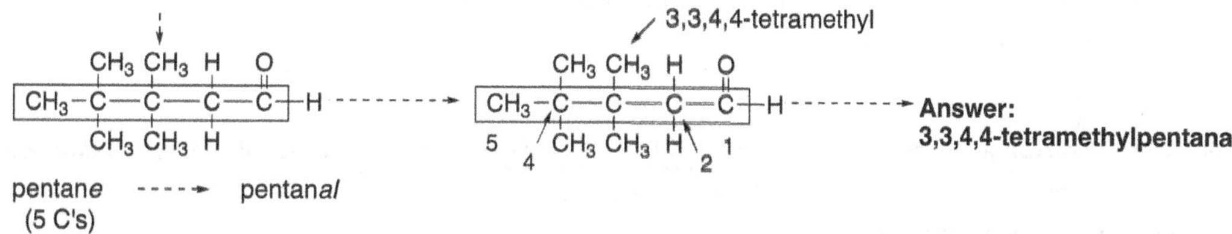

pentane ----→ pentan*al*
(5 C's)

c. $CH_3CH(CH_2CH_3)CH(CH_3)CH_2CH_2CH(CHO)CH_3$

re-draw

$CH_3CH_2CH(CH_3)CH(CH_3)CH_2CH_2CH(CH_3)CHO$ --------→ 2,5,6-trimethyl (numbered 6, 5, ..., 2, 1) --------→ **Answer: 2,5,6-trimethyloctanal**

octane ----→ octan*al*
(8 C's)

16.4 Work backwards from the name to draw each structure.

a. 2-chloro**propanal**

3 C chain with a CHO at C1 ----→

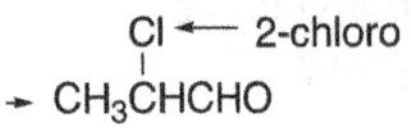

b. 3,4,5-triethyl**heptanal**

7 C chain with a CHO at C1 ----→

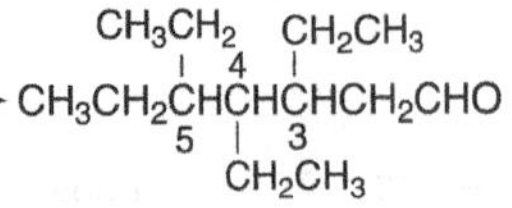

c. 3,6-diethyl**nonanal**

9 C chain with a CHO at C1 ----→

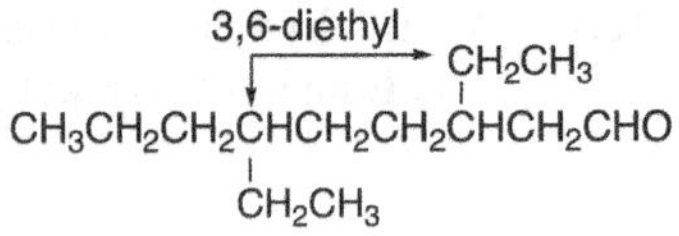

d. *o*-ethyl**benzaldehyde**

benzene ring with a CHO ----→

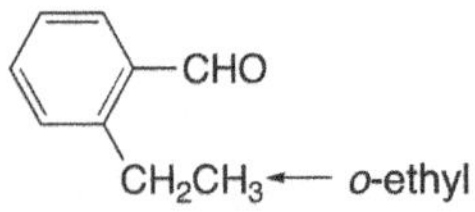

16.5 To name an aldehyde using the IUPAC system, use the steps in Example 16.1.

a.

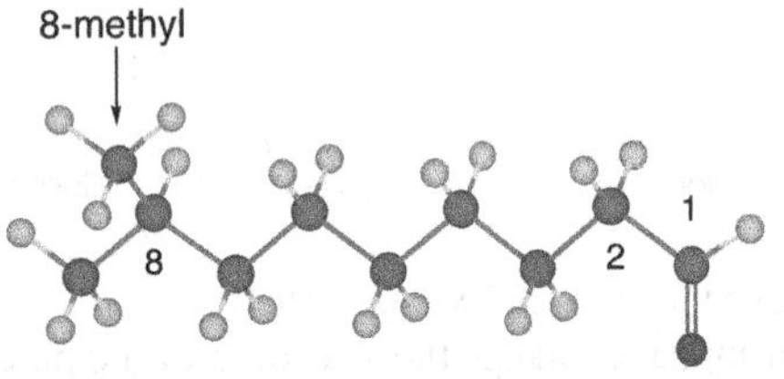

nonane (9 C's) ----→ nonanal ----→ **8-methylnonanal**

b.

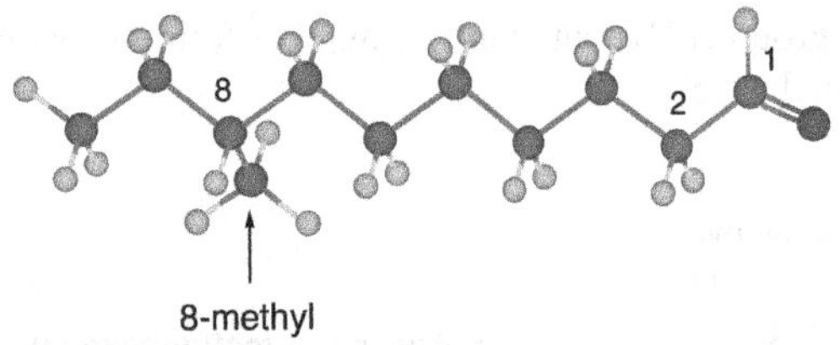

decane (10 C's) ----→ decanal ----→ **8-methyldecanal**

16.6 To name a ketone using IUPAC rules, use the steps in Example 16.2:

[1] Find the longest chain containing the carbonyl group, and change the ***-e*** ending of the parent alkane to the suffix ***-one.***

[2] Number the carbon chain to give the carbonyl carbon the lower number. Apply all of the other usual rules of nomenclature.

a.

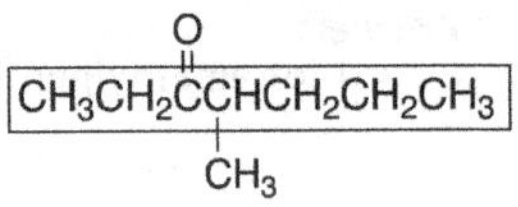

heptan*e* ----→ heptan*one*
(7 C's)

----→

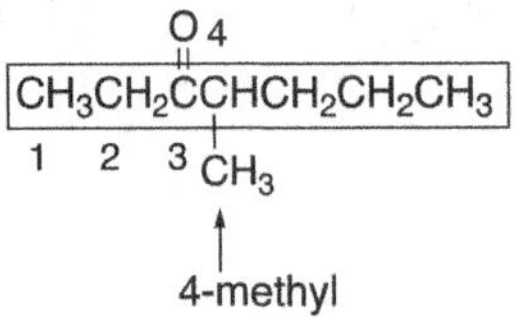

----→ **Answer: 4-methyl-3-heptanone**

b.

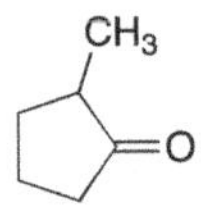

cyclopentan*e* ----→ cyclopentan*one*
(5 C's)

----→

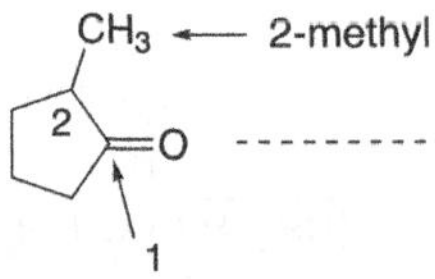

----→ **Answer: 2-methylcyclopentanone**

c. $CH_3C(CH_3)_2$—$C(=O)CH_2CH_2CH_2CH_3$ ----→ $CH_3C(CH_3)_2$—$C(=O)CH_2CH_2CH_2CH_3$ ----→ **Answer: 2,2-dimethyl-3-heptanone**

heptane ----→ heptan*one* (7 C's)

2,2-dimethyl

16.7 Work backwards from the name to draw each structure.

a. butyl ethyl ketone $CH_3CH_2CH_2CH_2C(=O)CH_2CH_3$ (butyl, ethyl)

c. *p*-ethyl**acetophenone**: *p*-ethyl → CH_3CH_2–C_6H_4–$C(=O)$–CH_3 (acetophenone)

b. 2-methyl-**3-pentanone**: 5 C chain with C=O at C3 ----→ $CH_3CH_2C(=O)CH(CH_3)CH_3$ ← 2-methyl

d. 2-propyl**cyclobutanone**: 4 C ring with C=O at C1 ----→ cyclobutanone with $CH_2CH_2CH_3$ ← 2-propyl

16.8 Aldehydes and ketones have *higher* boiling points than hydrocarbons of comparable size. Aldehydes and ketones have *lower* boiling points than alcohols of comparable size.

a. $CH_3CH_2C(=O)CH_2CH_3$ **A** — polar molecule, stronger forces, **higher boiling point**

$CH_3CH_2CH(CH_3)CH_2CH_3$ **B**

b. cyclohexanone **C**; cyclohexanol **D** — OH capable of hydrogen bonding, stronger forces, **higher boiling point**

16.9 Acetone will be soluble in water and organic solvents since it is a low molecular weight ketone (less than six carbons). Progesterone will be soluble only in organic solvents because it has many carbons and only two polar functional groups.

$CH_3C(=O)CH_3$ — small ketone, acetone

large molecule with two ketones, progesterone (ketone, ketone)

16.10 Compare the functional groups in each sunscreen. Dioxybenzone will most likely be washed off in water because it contains two hydroxyl groups and is the most water soluble.

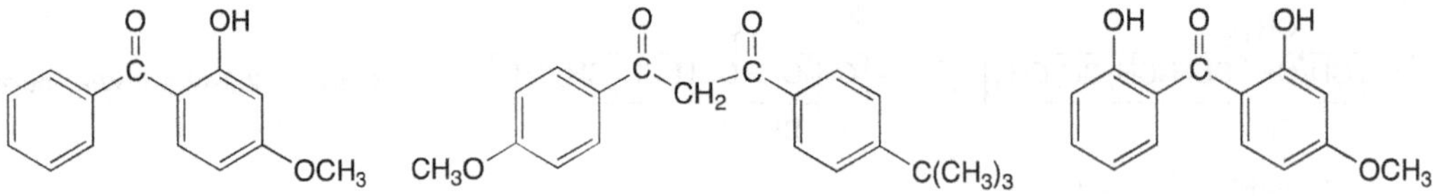

O OH
C
OCH3

oxybenzone
one hydroxyl group
one ketone
one ether

CH3O C CH2 C C(CH3)3

avobenzone
two ketones
one ether

OH O OH
C
OCH3

dioxybenzone
two hydroxyl groups
one ketone
one ether
most water soluble

16.11 Draw the product of each reaction using the guidelines in Example 16.3. Compounds that contain a C–H and C–O bond on the *same* carbon are oxidized with $K_2Cr_2O_7$.

- Aldehydes (RCHO) are oxidized to RCO_2H.
- Ketones (R_2CO) are *not* oxidized with $K_2Cr_2O_7$.

a. $CH_3CH_2CHO \xrightarrow{K_2Cr_2O_7} CH_3CH_2C(=O)OH$

b. $(CH_3CH_2)_2C{=}O \xrightarrow{K_2Cr_2O_7}$ No reaction

c. $CH_3C(CH_3){=}CHCH_2CH_2CH(CH_3)CH_2CHO \xrightarrow{K_2Cr_2O_7} CH_3C(CH_3){=}CHCH_2CH_2CH(CH_3)CH_2C(=O){-}OH$

16.12 Draw the product of each reaction using Example 16.4 as a guide. **Only aldehydes (RCHO) react with Tollens reagent, and they are oxidized to RCO_2H.** Ketones and alcohols are inert to oxidation.

a. $CH_3(CH_2)_6CHO \xrightarrow[NH_4OH]{Ag_2O} CH_3(CH_2)_6C(=O){-}OH$

b. 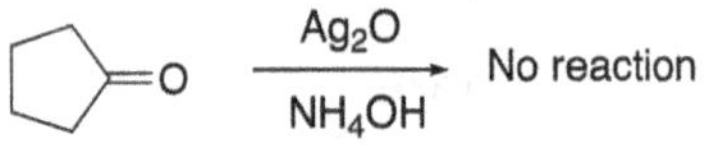 cyclopentanone (=O) $\xrightarrow[NH_4OH]{Ag_2O}$ No reaction

c. 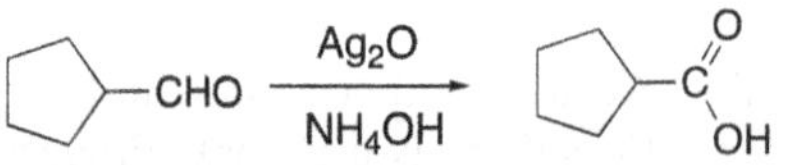 cyclopentyl–CHO $\xrightarrow[NH_4OH]{Ag_2O}$ cyclopentyl–C(=O)OH

d. 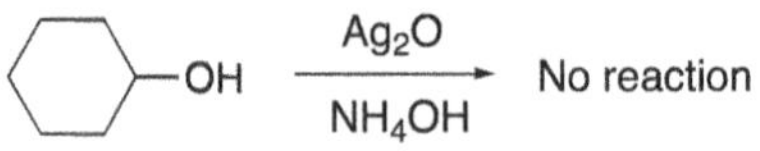 cyclohexyl–OH $\xrightarrow[NH_4OH]{Ag_2O}$ No reaction

16.13 3-Heptanone has molecular formula $C_7H_{14}O$.

a. $CH_3CH_2CH_2C(=O)CH_2CH_2CH_3$
4-heptanone
ketone
constitutional isomer that would not be oxidized by Tollens reagent

b. $CH_3CH_2CH_2CH_2CH_2CH_2CHO \xrightarrow{\text{Tollens reagent}} CH_3CH_2CH_2CH_2CH_2CH_2COOH$
heptanal
aldehyde
constitutional isomer that is oxidized by Tollens reagent

16.14 Draw the products of reduction using the steps in Example 16.5.

- Locate the C=O and mentally break one bond in the double bond.
- Mentally break the H–H bond of the reagent.
- Add one H atom to each atom of the C=O, forming new C–H and O–H single bonds.

a. 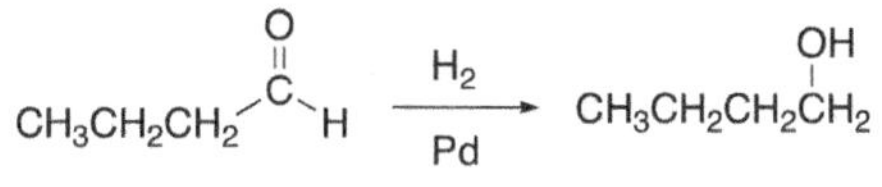 $CH_3CH_2CH_2CHO \xrightarrow[Pd]{H_2} CH_3CH_2CH_2CH_2OH$

c. 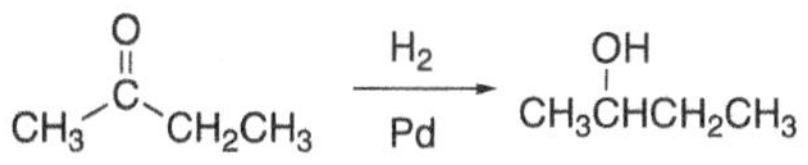 $CH_3COCH_2CH_3 \xrightarrow[Pd]{H_2} CH_3CH(OH)CH_2CH_3$

b. 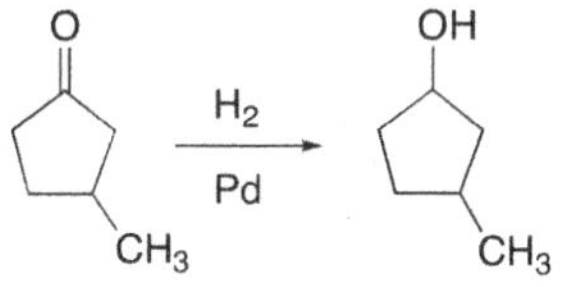 3-methylcyclopentanone $\xrightarrow[Pd]{H_2}$ 3-methylcyclopentanol

d. 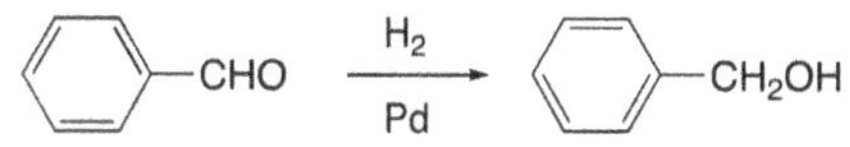 $C_6H_5CHO \xrightarrow[Pd]{H_2} C_6H_5CH_2OH$

16.15 Work backwards to determine what carbonyl compound is needed to prepare alcohol **A.**

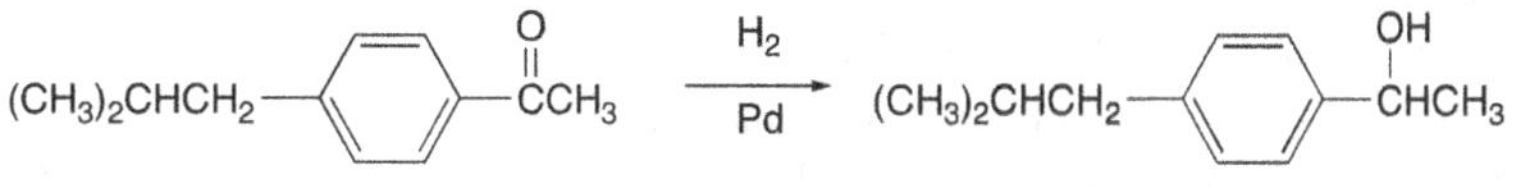

$(CH_3)_2CHCH_2-C_6H_4-COCH_3 \xrightarrow[Pd]{H_2} (CH_3)_2CHCH_2-C_6H_4-CH(OH)CH_3$

A

16.16

$\xrightarrow[Pd]{H_2}$

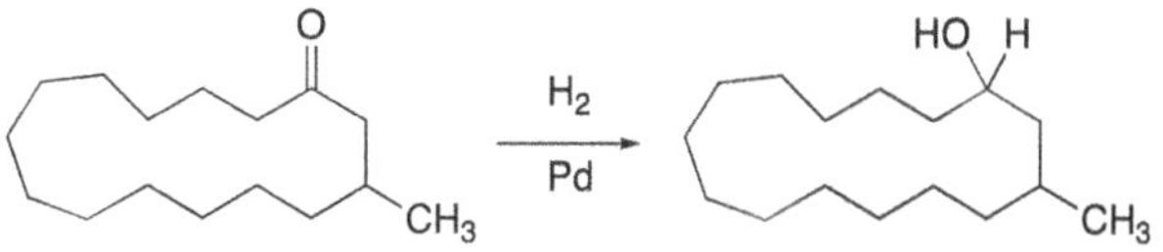

16.17 Recall that stereoisomers differ only in the three-dimensional arrangement of atoms in space, but all connectivity is identical. Constitutional isomers have the same molecular formula, but atoms are connected differently.

a. All-*trans*-retinal and 11-*cis*-retinal are stereoisomers, and differ only in the arrangement of groups around one double bond.
b. All-*trans*-retinal and vitamin A are not isomers. They have different molecular formulas.
c. Vitamin A and 11-*cis*-retinol are stereoisomers, and differ only in the arrangement of groups around one double bond.

16.18 The 11 trigonal planar carbons are labeled with (*).

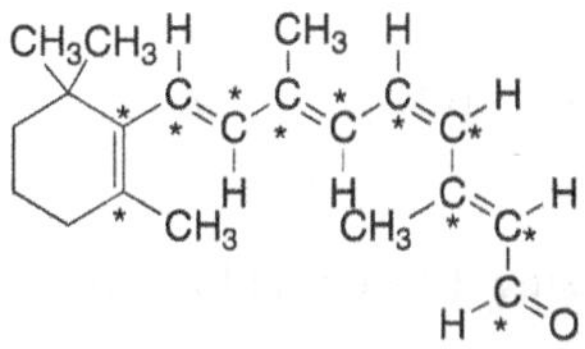

16.19 To form a hemiacetal and acetal from a carbonyl compound, use the steps in Example 16.6.

- Locate the C=O in the starting material.
- Break one C–O bond and add one equivalent of ROH across the double bond, placing the OR group on the carbonyl carbon. This forms the hemiacetal.
- Replace the OH group of the hemiacetal by OR to form the acetal.

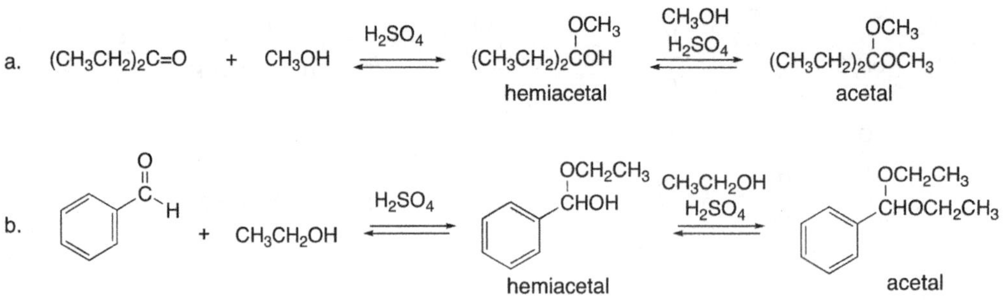

a. $(CH_3CH_2)_2C{=}O$ + CH_3OH $\xrightleftharpoons{H_2SO_4}$ $(CH_3CH_2)_2C(OCH_3)OH$ (hemiacetal) $\xrightleftharpoons[H_2SO_4]{CH_3OH}$ $(CH_3CH_2)_2C(OCH_3)OCH_3$ (acetal)

b. benzaldehyde (C_6H_5CHO) + CH_3CH_2OH $\xrightleftharpoons{H_2SO_4}$ $C_6H_5CH(OCH_2CH_3)OH$ (hemiacetal) $\xrightleftharpoons[H_2SO_4]{CH_3CH_2OH}$ $C_6H_5CH(OCH_2CH_3)OCH_2CH_3$ (acetal)

16.20 Recall the definitions from Example 16.7 to identify the functional groups:

- An ether has the general structure ROR.
- A hemiacetal has one C bonded to OH and OR.
- An acetal has one C bonded to two OR groups.

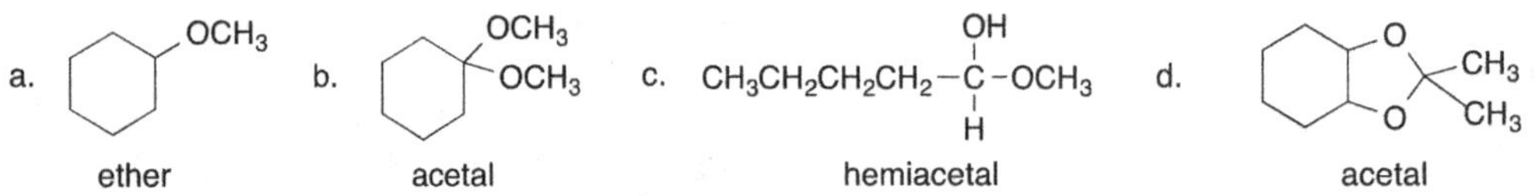

a. cyclohexyl–OCH_3: ether
b. cyclohexane C bonded to two OCH_3: acetal
c. $CH_3CH_2CH_2CH_2-CH(OH)-OCH_3$: hemiacetal
d. cyclic acetal with C bonded to two ring O and two CH_3: acetal

16.21

a. cyclic (1,3-dioxane): acetal
b. (1,4-dioxane): ether

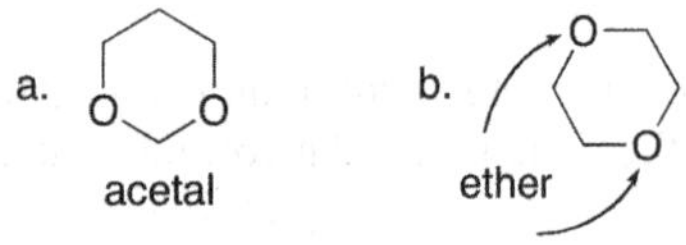

16.22 Acetal carbons are labeled with (*).

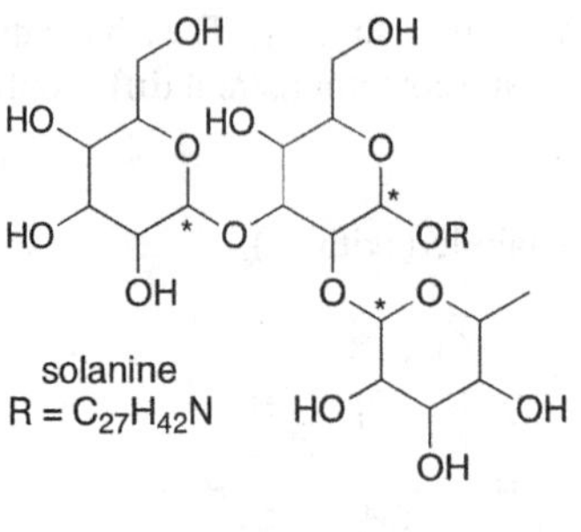

solanine
$R = C_{27}H_{42}N$

16.23 Draw the products of each reaction using the steps in Example 16.6.

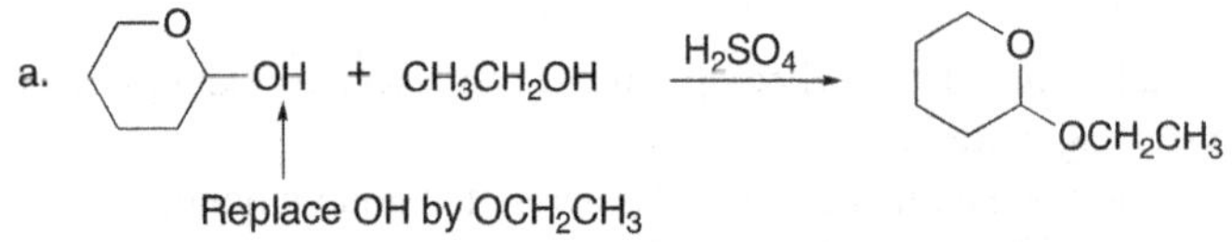

a. cyclic hemiacetal (ring O, C–OH) + CH_3CH_2OH $\xrightarrow{H_2SO_4}$ cyclic acetal (C–OCH_2CH_3)

Replace OH by OCH_2CH_3

b. [tetrahydrofuran-2-ol]–OH + [cyclohexanol]–OH $\xrightarrow{H_2SO_4}$ [tetrahydrofuranyl]–O–[cyclohexyl]

Replace OH by O–[cyclohexyl]

16.24 To draw the products of hydrolysis, use the steps in Example 16.8.

- Locate the two C–OR bonds on the same carbon.
- Replace the two C–O single bonds with a carbonyl group (C=O).
- Each OR group then becomes a molecule of alcohol (ROH) product.

a. $CH_3-C(OCH_3)_2-CH_2CH_2CH_3$ $\xrightleftharpoons[H_2SO_4]{H_2O}$ $CH_3COCH_2CH_2CH_3$ + 2 CH_3OH

c. [cyclohexyl]–$CH(OCH_3)_2$ $\xrightleftharpoons[H_2SO_4]{H_2O}$ [cyclohexyl]–CHO + 2 CH_3OH

b. 1,1-(CH_3CH_2O)(OCH_2CH_3)cyclohexane $\xrightleftharpoons[H_2SO_4]{H_2O}$ cyclohexanone + 2 CH_3CH_2OH

Solutions to Odd-Numbered End-of-Chapter Problems

16.25 Draw a structure to fit each description.

a. $CH_3CH_2CH_2CH(CH_2CH_3)CH_2CHO$ — aldehyde, $C_8H_{16}O$

b. $CH_3CH_2COCH(CH_3)CH_3$ — ketone, $C_6H_{12}O$

c. cyclopentanone — ketone, C_5H_8O

d. [cyclopentyl]–CHO — aldehyde, $C_6H_{10}O$

16.27 Compare C=O and C=C bonds.

a. Both are trigonal planar.
b. A C=O is polar, whereas a C=C is *not* polar.
c. Both functional groups undergo addition reactions.

16.29 To name the aldehyde and ketone, use the IUPAC rules in Examples 16.1 and 16.2.

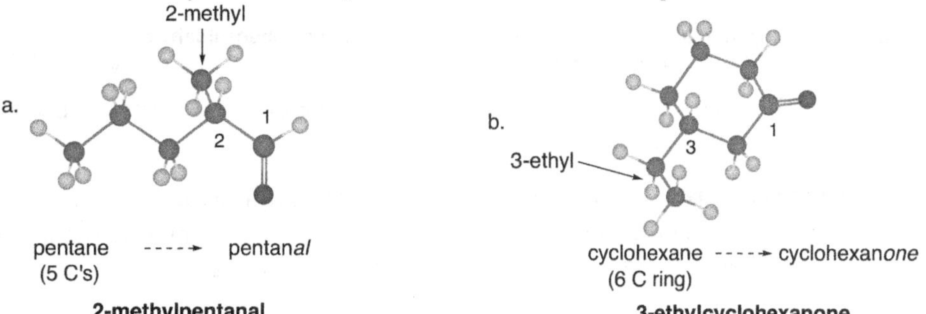

a. pentane (5 C's) ----→ pentan*al*

2-methylpentanal

b. cyclohexane (6 C ring) ----→ cyclohexan*one*

3-ethylcyclohexanone

16.31 To name an aldehyde using the IUPAC system, use the steps in Example 16.1:

[1] Find the longest chain containing the CHO group, and change the ***-e*** ending of the parent alkane to the suffix ***-al.***

[2] Number the chain or ring to put the CHO group at C1, but omit this number from the name. Apply all other usual rules of nomenclature.

a. $CH_3CH_2CH_2CH(CH_3)CH_2CHO$ → $CH_3CH_2CH_2CH(CH_3)CH_2CHO$ (CH₃ ← 3-methyl) → **Answer: 3-methylhexanal**

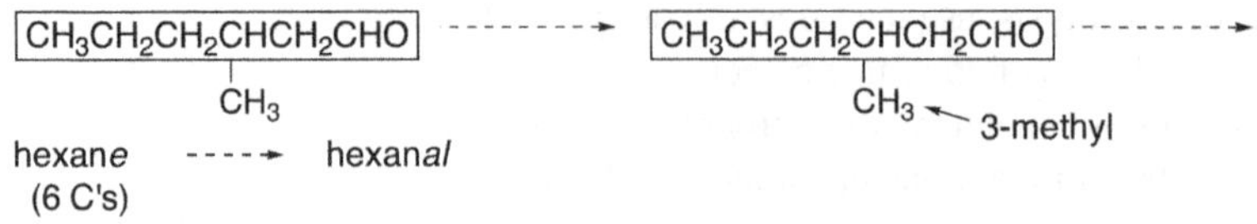

hexan*e* → hexan*al* (6 C's)

b. $CH_3CH_2CH(CH_3)CH_2CH(CH_3)CH_2CHO$ → numbered 5, 3, 1 (CH₃ groups ← 3,5-dimethyl) → **Answer: 3,5-dimethylheptanal**

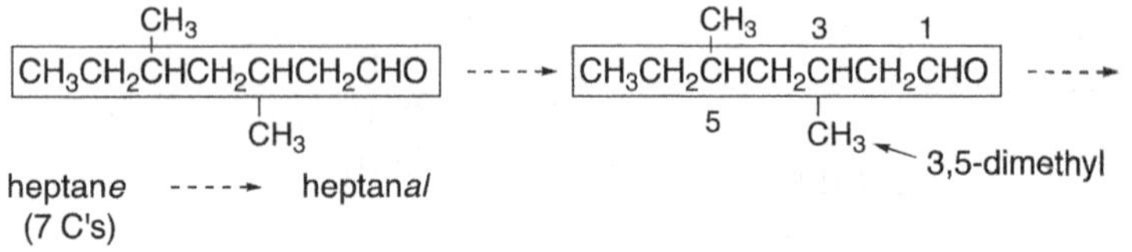

heptan*e* → heptan*al* (7 C's)

c. $CH_3CH_2CH_2CH(CH_2CHO)CH_2CH_2CH_3$ (O=C–H, H–C–H, H) → 3-propyl, C3 → **Answer: 3-propylhexanal**

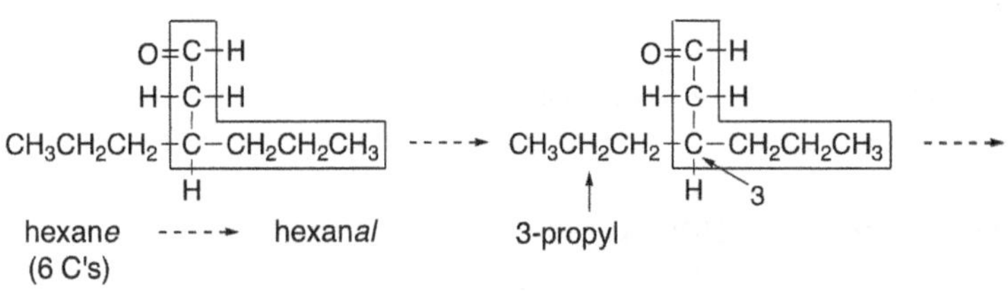

hexan*e* → hexan*al* (6 C's)

d. $CH_3CH_2C(CH_2CH_3)_2CH_2CH_2CH_2C(CH_3)_2CHO$ → 6 (CH₂CH₃), 2 (CH₃ ← 2,2-dimethyl), 1 → **Answer: 6,6-diethyl-2,2-dimethyloctanal**

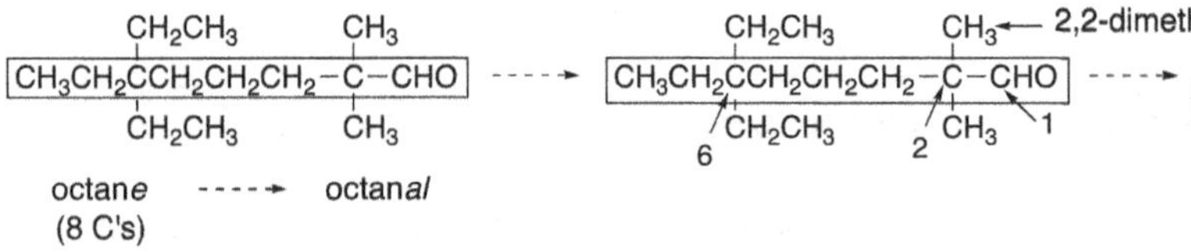

octan*e* → octan*al* (8 C's)

e. Cl–C₆H₄–CHO → Cl–C₆H₄–CHO (↑ *p*-chloro) → **Answer: *p*-chlorobenzaldehyde**

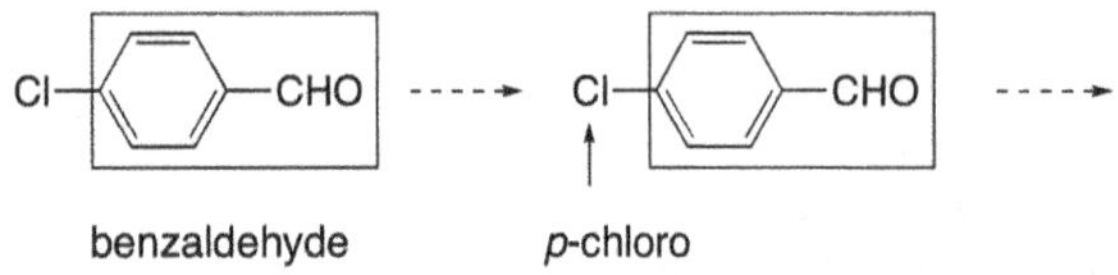

benzaldehyde

16.33 Work backwards to draw the structure.

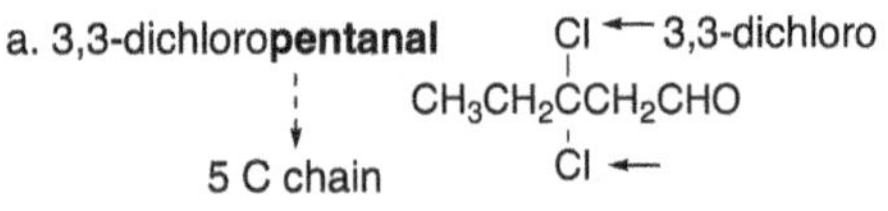

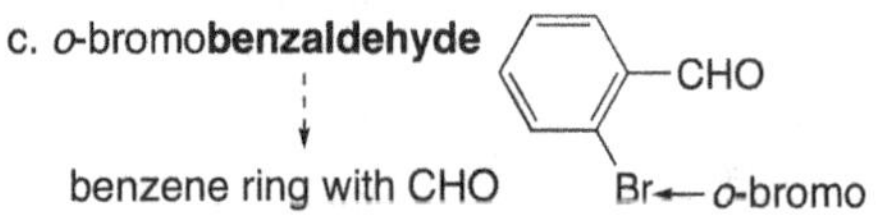

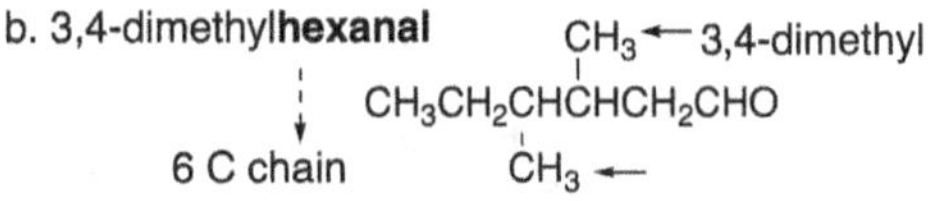

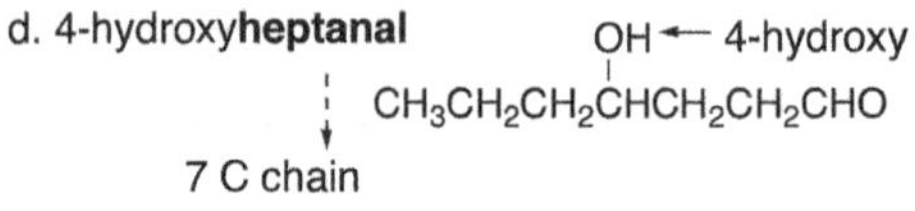

a. 3,3-dichloro**pentanal** → 5 C chain

$CH_3CH_2CCl_2CH_2CHO$ (Cl ← 3,3-dichloro)

b. 3,4-dimethyl**hexanal** → 6 C chain

$CH_3CH_2CH(CH_3)CH(CH_3)CH_2CHO$ (CH₃ ← 3,4-dimethyl)

c. *o*-bromo**benzaldehyde** → benzene ring with CHO

C₆H₄(CHO)Br (Br ← *o*-bromo)

d. 4-hydroxy**heptanal** → 7 C chain

$CH_3CH_2CH_2CH(OH)CH_2CH_2CHO$ (OH ← 4-hydroxy)

16.35 To name a ketone using IUPAC rules, use the steps in Example 16.2:

[1] Find the longest chain containing the carbonyl group, and change the ***-e*** ending of the parent alkane to the suffix ***-one.***

[2] Number the carbon chain to give the carbonyl carbon the lower number. Apply all of the other usual rules of nomenclature.

a. 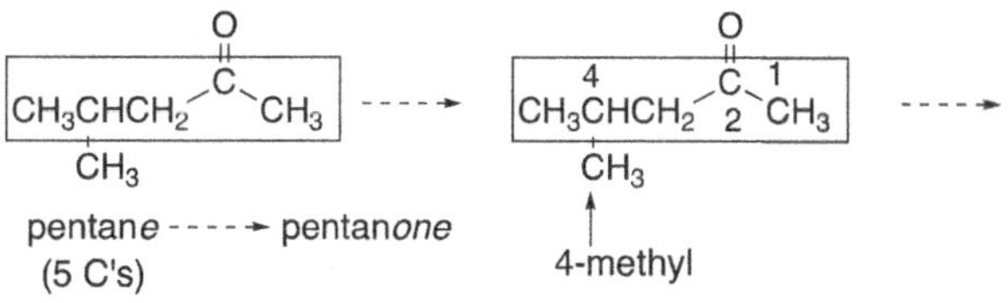 $CH_3CH(CH_3)CH_2C(=O)CH_3$ → numbered 4 ... 2, 1; 4-methyl

pentane → pentan*one* (5 C's)

Answer: 4-methyl-2-pentanone

b. 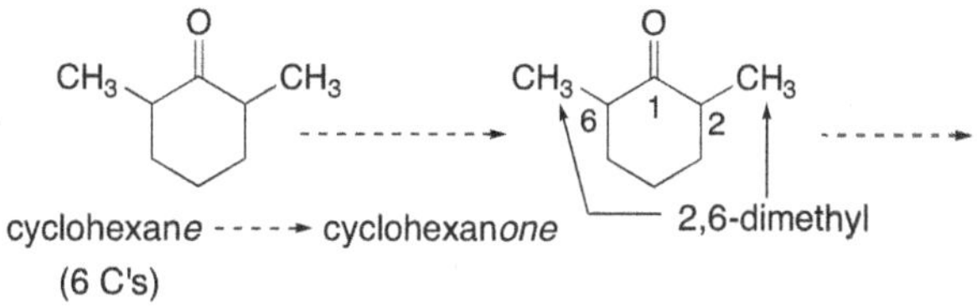 cyclohexane → cyclohexan*one* (6 C's); 2,6-dimethyl

Answer: 2,6-dimethylcyclohexanone

c. 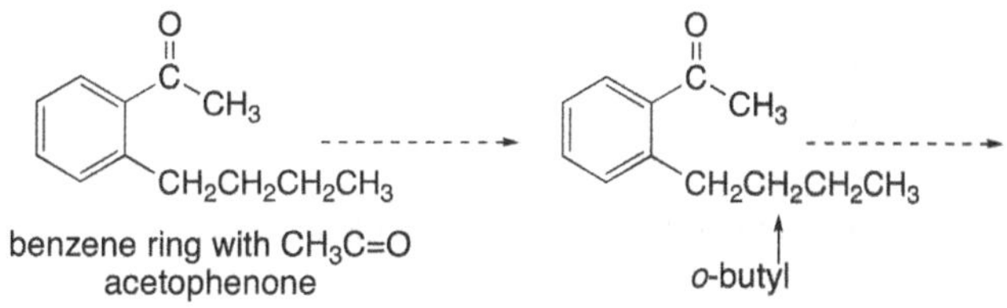 benzene ring with $CH_3C=O$ acetophenone; $CH_2CH_2CH_2CH_3$ → *o*-butyl

Answer: *o*-butylacetophenone

d. 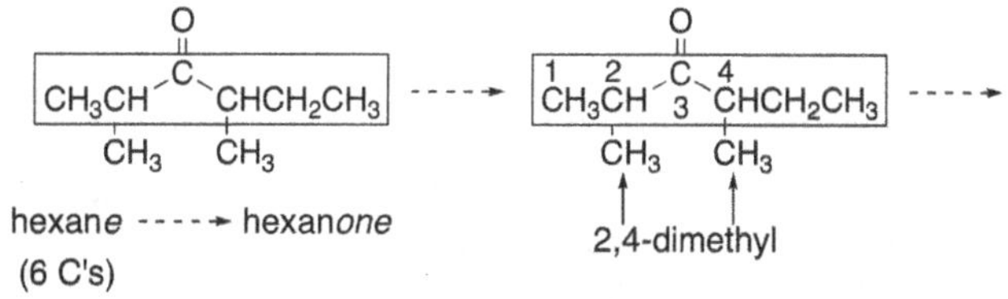 $CH_3CH(CH_3)C(=O)CH(CH_3)CH_2CH_3$ → numbered 1, 2, 3, 4; 2,4-dimethyl

hexane → hexan*one* (6 C's)

Answer: 2,4-dimethyl-3-hexanone

e. 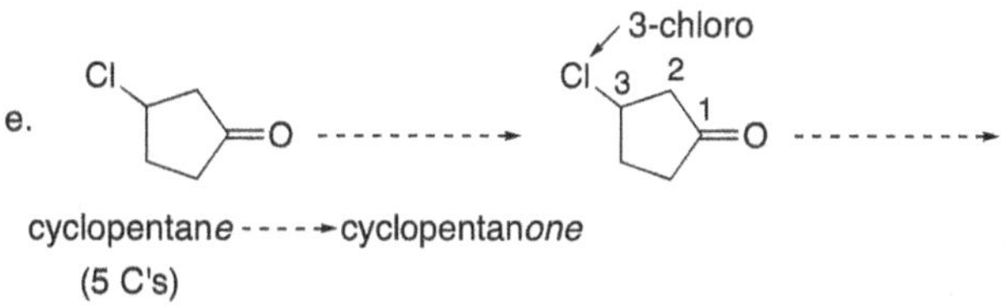 Cl, =O; numbered 1, 2, 3; 3-chloro

cyclopentane → cyclopentan*one* (5 C's)

Answer: 3-chlorocyclopentanone

16.37 Work backwards from the name to draw each structure.

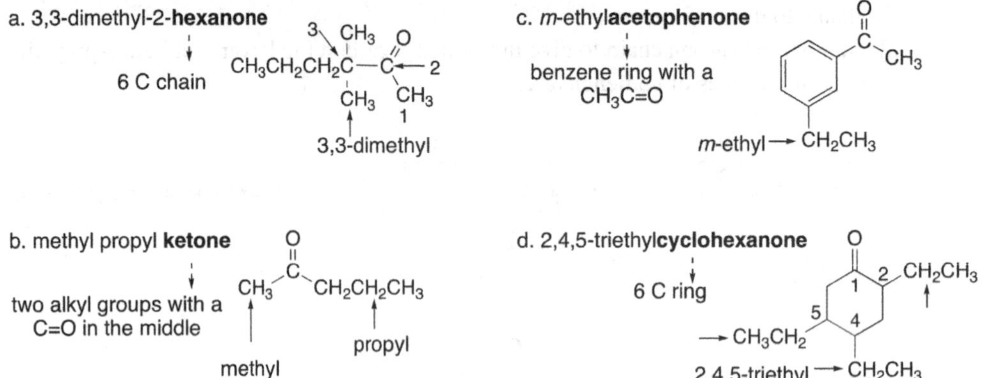

16.39 Draw the four aldehydes and then name them using the steps in Example 16.1.

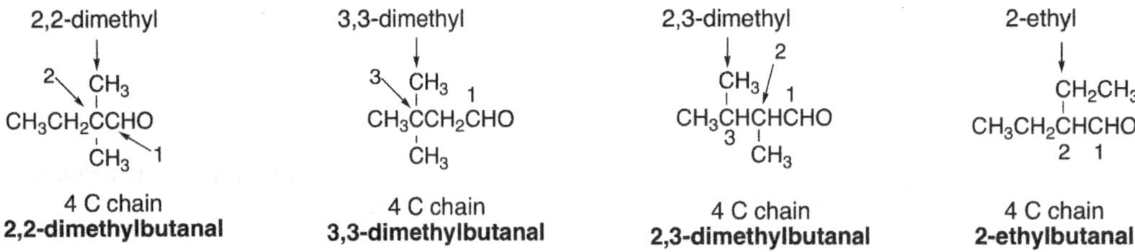

16.41 Draw the structure and correct each name.

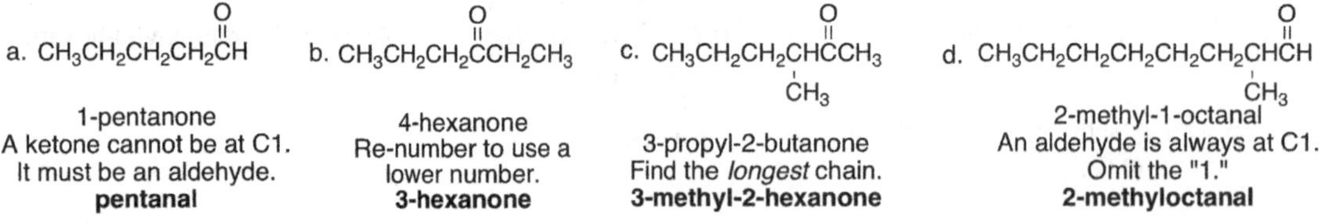

16.43 Draw benzaldehyde and then the hydrogen bond it can form with water.

hydrogen bond

16.45 Aldehydes and ketones have *higher* boiling points than hydrocarbons of comparable size. Aldehydes and ketones have *lower* boiling points than alcohols of comparable size.

a. $(CH_3)_3CCH_2CH_2CH_3$ or $(CH_3)_3CCH_2CHO$

hydrocarbon — aldehyde **higher boiling point**

b. cyclopentyl–$COCH_3$ or cyclopentyl–CH_2CH_2OH

ketone — alcohol **higher boiling point**

16.47 Aldehydes and ketones have *higher* melting points than hydrocarbons of comparable size. Aldehydes and ketones have *lower* melting points than alcohols of comparable size.

cyclopentyl–CH_3 (C), cyclopentanone (=O) (A), cyclopentyl–OH (B)

C → A → B

Increasing melting point

16.49 Low molecular weight aldehydes and ketones (less than six carbons) are water soluble.

a. benzene–CHO

7 C aldehyde
insoluble

b. $CH_3COCH_2CH_3$

4 C ketone
soluble

c. $CH_3CH_2CH_2CH_3$

hydrocarbon
insoluble

16.51 Draw the product of each reaction using the steps in Example 16.3. Compounds that contain a C–H and C–O bond on the *same* carbon are oxidized with $K_2Cr_2O_7$.

- Aldehydes (RCHO) are oxidized to RCO_2H.
- Ketones (R_2CO) are *not* oxidized with $K_2Cr_2O_7$.
- 1° Alcohols (RCH_2OH) are oxidized to RCO_2H (Section 14.5B).

a. $CH_3(CH_2)_4CHO \xrightarrow{K_2Cr_2O_7} CH_3(CH_2)_4COOH$

b. cyclopentyl–$CH_2CHO \xrightarrow{K_2Cr_2O_7}$ cyclopentyl–CH_2COOH

c. 2-ethylcyclopentanone (–CH_2CH_3) $\xrightarrow{K_2Cr_2O_7}$ No reaction

d. $CH_3(CH_2)_4CH_2OH \xrightarrow{K_2Cr_2O_7} CH_3(CH_2)_4COOH$

16.53 Draw the product of each reaction using Example 16.4 as a guide. **Only aldehydes (RCHO) react with Tollens reagent, and they are oxidized to RCO_2H.** Ketones and alcohols are inert to oxidation.

a. $CH_3(CH_2)_4CHO \xrightarrow[NH_4OH]{Ag_2O} CH_3(CH_2)_4COOH$

b. cyclopentyl–$CH_2CHO \xrightarrow[NH_4OH]{Ag_2O}$ cyclopentyl–CH_2COOH

c. 2-ethylcyclopentanone (–CH_2CH_3) $\xrightarrow[NH_4OH]{Ag_2O}$ No reaction

d. $CH_3(CH_2)_4CH_2OH \xrightarrow[NH_4OH]{Ag_2O}$ No reaction

16.55 Answer each question about erythrulose.

a, b. ketone → ; 1° → ; ← 1° ; erythrulose ; ↑ 2°

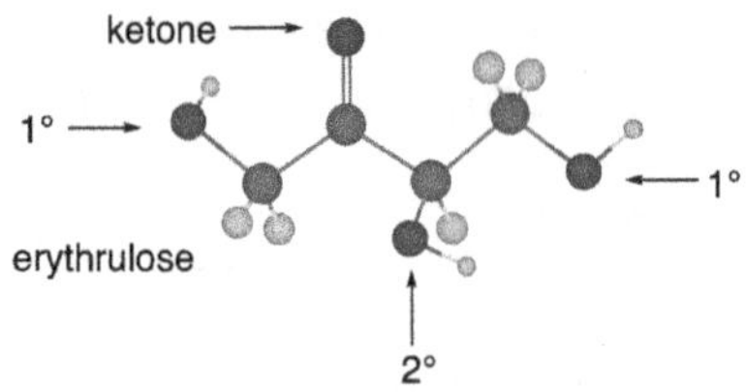

c. HOCH$_2$C(=O)CH(OH)CH$_2$OH $\xrightarrow{\text{Tollens reagent}}$ No reaction

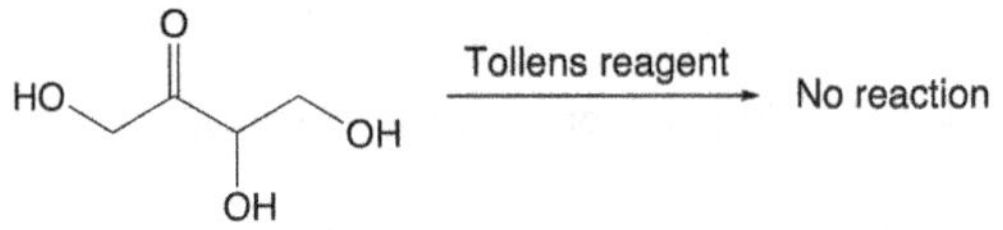

d. HOCH$_2$C(=O)CH(OH)CH$_2$OH $\xrightarrow{K_2Cr_2O_7}$ HOC(=O)C(=O)C(=O)C(=O)OH

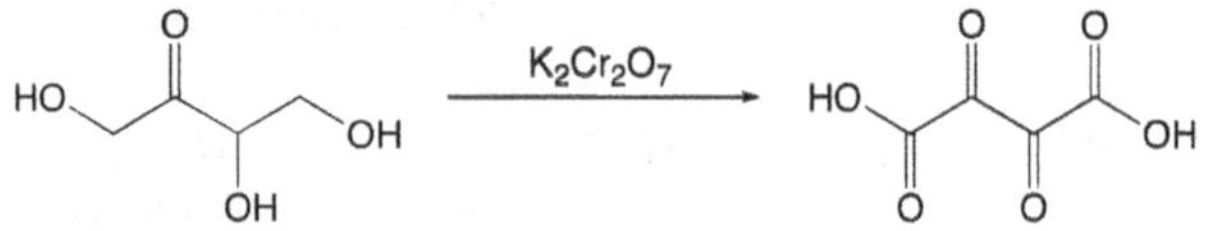

16.57 Work backwards to determine what aldehyde can be used to prepare each carboxylic acid.

a. $CH_3CH_2CH(CH_3)CH_2CO_2H \longleftarrow CH_3CH_2CH(CH_3)CH_2CHO$

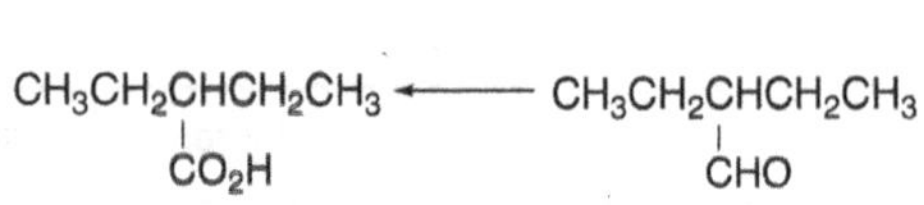

b. $CH_3-C_6H_4-CO_2H \longleftarrow CH_3-C_6H_4-CHO$

c. $CH_3CH_2CH(CO_2H)CH_2CH_3 \longleftarrow CH_3CH_2CH(CHO)CH_2CH_3$

16.59 Draw the products of reduction using the steps in Example 16.5.

- Locate the C=O and mentally break one bond in the double bond.
- Mentally break the H–H bond of the reagent.
- Add one H atom to each atom of the C=O, forming new C–H and O–H single bonds.

a. $CH_3CH_2-C_6H_4-CHO \xrightarrow[Pd]{H_2} CH_3CH_2-C_6H_4-CH_2OH$

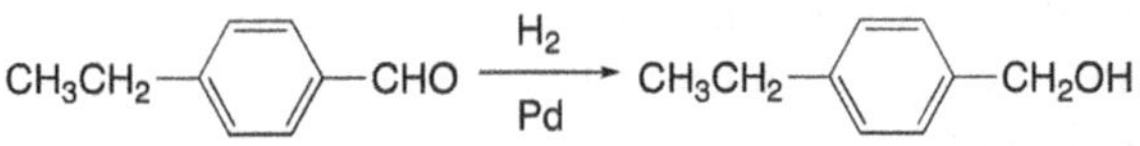

b. 4-methylcyclohexanone (CH_3, =O) $\xrightarrow[Pd]{H_2}$ 4-methylcyclohexanol (CH_3, OH)

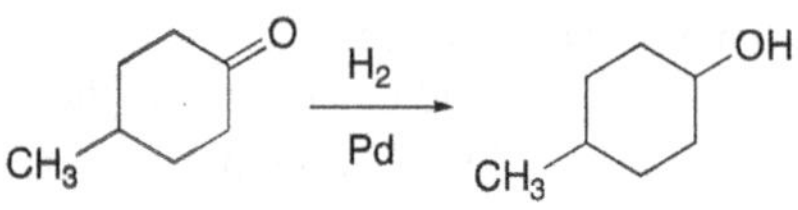

16.61

a, b: $CH_3CH_2-CH(CH_3)-(CH_2)_4CHO$ (↑ chirality center)

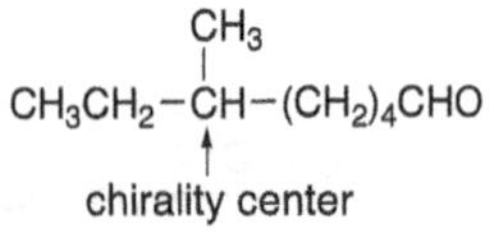

c. $CH_3CH_2-CH(CH_3)-(CH_2)_4CHO \xrightarrow[Pd]{H_2} CH_3CH_2-CH(CH_3)-(CH_2)_4CH_2OH$

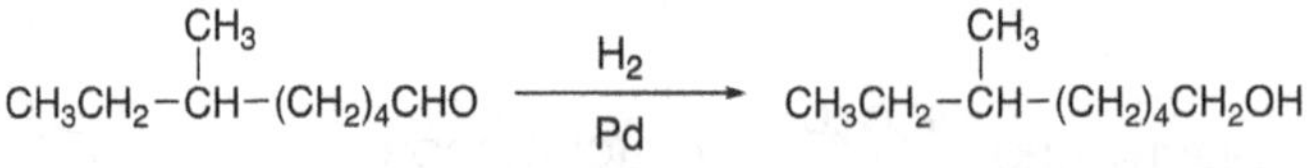

16.63 Work backwards to determine what carbonyl compound is needed to make each alcohol by a reduction reaction.

a. $CH_3CH_2CH_2CH_2CH_2OH \longleftarrow CH_3CH_2CH_2CH_2C(=O)H$

b. 3-methylcyclohexanol (OH, CH_3) ⟵ 3-methylcyclohexanone (O, CH_3)

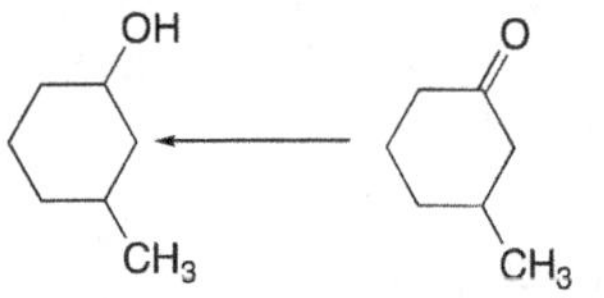

16.65 1-Methylcyclohexanol is a 3° alcohol and cannot be produced from the reduction of a carbonyl compound because only 1° or 2° alcohols can be formed in these reactions.

16.67 Recall the definitions from Example 16.7 to draw a compound of molecular formula $C_5H_{12}O_2$ that fits each description:

- An ether has the general structure ROR.
- A hemiacetal has one C bonded to OH and OR.
- An acetal has one C bonded to two OR groups.

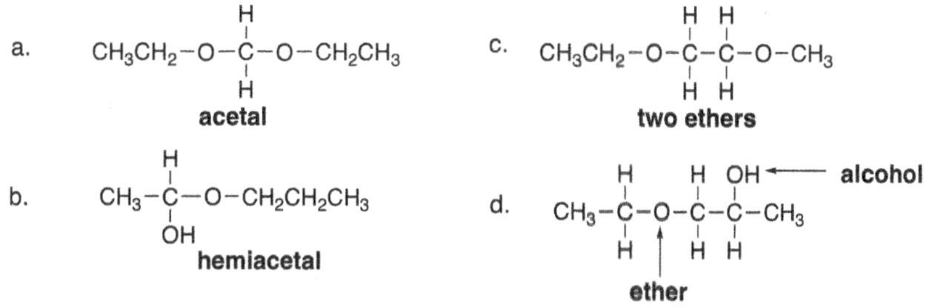

a. $CH_3CH_2-O-CH_2-O-CH_2CH_3$ **acetal**

b. $CH_3-CH(OH)-O-CH_2CH_2CH_3$ **hemiacetal**

c. $CH_3CH_2-O-CH_2-CH_2-O-CH_3$ **two ethers**

d. $CH_3-CH_2-O-CH_2-CH(OH)-CH_3$ **alcohol**, **ether**

16.69

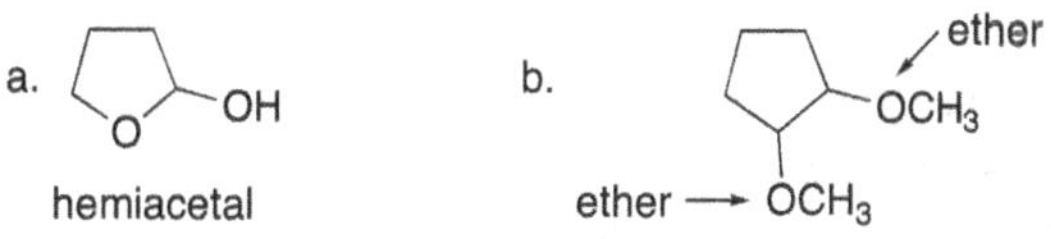

a. hemiacetal

b. ether, ether → OCH_3

16.71

a, b: alcohols, hemiacetal, amine

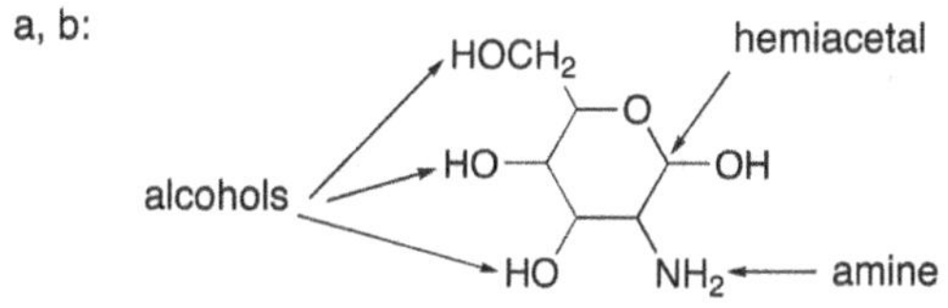

16.73 To form a hemiacetal and acetal from a carbonyl compound, use the steps in Example 16.6.

- Locate the C=O in the starting material.
- Break one C–O bond and add one equivalent of CH_3OH across the double bond, placing the OCH_3 group on the carbonyl carbon. This forms the hemiacetal.
- Replace the OH group of the hemiacetal by OCH_3 to form the acetal.

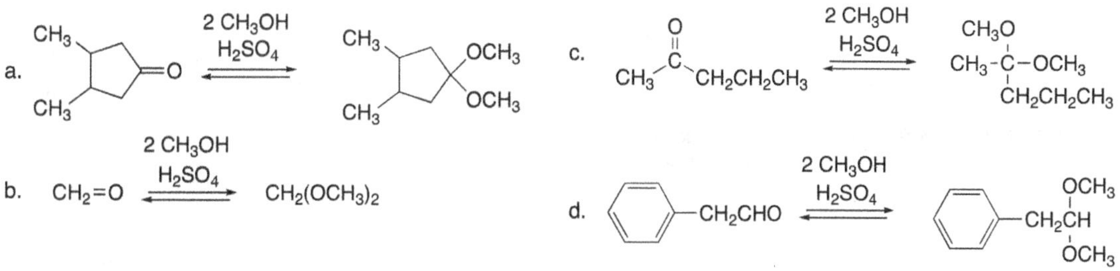

a. 2 CH_3OH, H_2SO_4

b. $CH_2=O$ $\xrightleftharpoons[]{2\ CH_3OH,\ H_2SO_4}$ $CH_2(OCH_3)_2$

c. $CH_3COCH_2CH_2CH_3$ $\xrightleftharpoons[]{2\ CH_3OH,\ H_2SO_4}$ $CH_3-C(OCH_3)_2-CH_2CH_2CH_3$

d. $C_6H_5-CH_2CHO$ $\xrightleftharpoons[]{2\ CH_3OH,\ H_2SO_4}$ $C_6H_5-CH_2CH(OCH_3)_2$

16.75 Draw the products of each reaction.

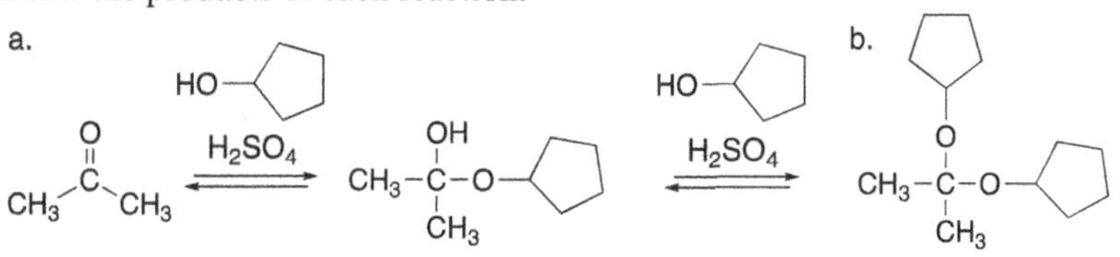

a. CH_3COCH_3 $\xrightleftharpoons[]{HO-C_5H_9,\ H_2SO_4}$ $CH_3-C(OH)(CH_3)-O-C_5H_9$

b. $\xrightleftharpoons[]{HO-C_5H_9,\ H_2SO_4}$ $CH_3-C(CH_3)(O-C_5H_9)-O-C_5H_9$

16.77 Answer each question.

a. [cyclic hemiacetal: five-membered ring with O, carbon bearing –OH] ← hemiacetal carbon

b. cyclic hemiacetal–OH ⇌ $HOCH_2CH_2CH_2CH{=}O$

c. cyclic hemiacetal–OH $\underset{}{\overset{CH_3OH,\ H_2SO_4}{\rightleftharpoons}}$ cyclic acetal–OCH_3

16.79 Draw the product of cyclization.

$HOCH_2CH_2CH_2CH(CH_3)CHO$ (C) ⇌ six-membered cyclic hemiacetal with OH and CH_3

16.81 To draw the products of hydrolysis, use the steps in Example 16.8.

- Locate the two C–OR bonds on the same carbon.
- Replace the two C–O single bonds with a carbonyl group (C=O).
- Each OR group then becomes a molecule of alcohol (ROH) product.

a. cyclohexane-1,1-diyl bis($OCH_2CH_2CH_3$) $\overset{H_2O,\ H_2SO_4}{\rightleftharpoons}$ cyclohexanone (=O) + 2 $HOCH_2CH_2CH_3$

b. $H{-}C(OCH_3)_2{-}CH_2CH_2CH_3$ $\overset{H_2O,\ H_2SO_4}{\rightleftharpoons}$ $H{-}C({=}O){-}CH_2CH_2CH_3$ + 2 $HOCH_3$

16.83 Answer each question about compound **A**.

a, b. CH_3–C₆H₄–C(=O)CH_3 (p-methyl) **p-methylacetophenone**

c. CH_3–C₆H₄–C(=O)CH_3 — seven trigonal planar C's

d. CH_3–C₆H₄–C(=O)CH_3 $\xrightarrow[\text{(acid)}]{2\ CH_3OH}$ CH_3–C₆H₄–C(OCH_3)$_2$–CH_3

16.85 Draw the products of each reaction.

a. $C_6H_5CHO \xrightarrow[Pd]{H_2} C_6H_5CH_2OH$

b. $C_6H_5CHO \xrightarrow{K_2Cr_2O_7} C_6H_5COOH$

c. $C_6H_5CHO \xrightarrow[NH_4OH]{Ag_2O} C_6H_5COOH$

d. $C_6H_5CHO \xrightarrow[H_2SO_4]{2\ CH_3OH} C_6H_5CH(OCH_3)_2$

e. $C_6H_5CHO \xrightarrow[H_2SO_4]{2\ CH_3CH_2OH} C_6H_5CH(OCH_2CH_3)_2$

f. $C_6H_5CH(OCH_2CH_3)_2 \xrightarrow[H_2SO_4]{H_2O} C_6H_5CHO + 2\ CH_3CH_2OH$

16.87 Draw the products of each reaction.

a. $CH_3CO(CH_2)_4CH_3 \xrightarrow[Pd]{H_2} CH_3CH(OH)(CH_2)_4CH_3$

b. $CH_3CO(CH_2)_4CH_3 \xrightarrow{K_2Cr_2O_7}$ No reaction

c. $CH_3CO(CH_2)_4CH_3 \xrightarrow[NH_4OH]{Ag_2O}$ No reaction

d. $CH_3CO(CH_2)_4CH_3 \xrightarrow[H_2SO_4]{2\ CH_3OH} CH_3C(OCH_3)_2(CH_2)_4CH_3$

e. $CH_3CO(CH_2)_4CH_3 \xrightarrow[H_2SO_4]{2\ CH_3CH_2OH} CH_3C(OCH_2CH_3)_2(CH_2)_4CH_3$

f. $CH_3C(OCH_2CH_3)_2(CH_2)_4CH_3 \xrightarrow[H_2SO_4]{H_2O} CH_3CO(CH_2)_4CH_3 + 2\ CH_3CH_2OH$

16.89 Draw the three constitutional isomers that can be converted to 1-pentanol. The starting material needs a C=O at C1 and a C=C.

$CH_2{=}CHCH_2CH_2CHO$
or
$CH_3CH{=}CHCH_2CHO$ $\xrightarrow[Pd]{H_2}$ $CH_3CH_2CH_2CH_2CH_2OH$
or
$CH_3CH_2CH{=}CHCHO$

16.91 Draw the products of each reaction.

a. [steroid with HO, two CH_3 groups and a ketone C=O on the five-membered ring] $\xrightarrow[Pd]{H_2}$ [same steroid with the C=O reduced to OH]

b. [Structure: steroid with HO, CH_3, CH_3, C=O] $\xrightarrow{K_2Cr_2O_7}$ [Structure: steroid with O=, CH_3, CH_3, C=O]

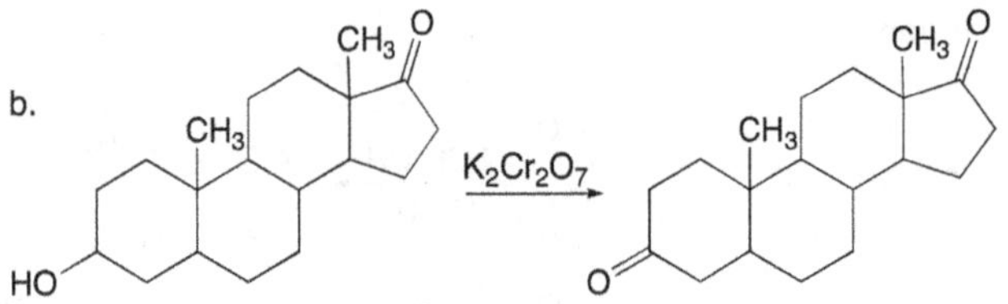

c. [Structure: steroid with HO, CH_3, CH_3, C=O] $\xrightarrow[H_2SO_4]{2\ CH_3OH}$ [Structure: steroid with HO, CH_3, CH_3, $C(OCH_3)_2$]

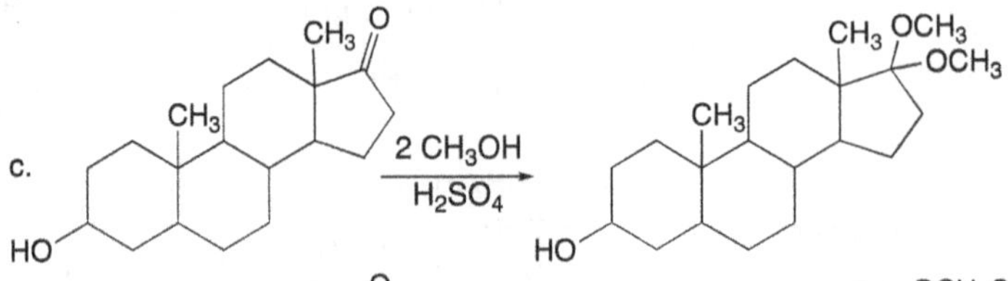

d. [Structure: steroid with HO, CH_3, CH_3, C=O] $\xrightarrow[H_2SO_4]{2\ CH_3CH_2OH}$ [Structure: steroid with HO, CH_3, CH_3, $C(OCH_2CH_3)_2$]

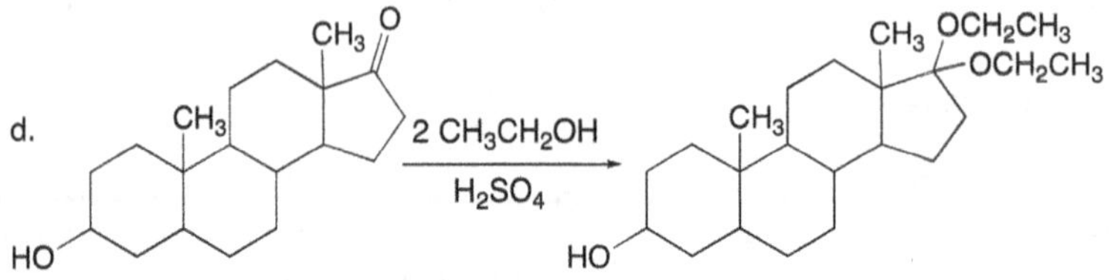

16.93 Draw the product of oxidation.

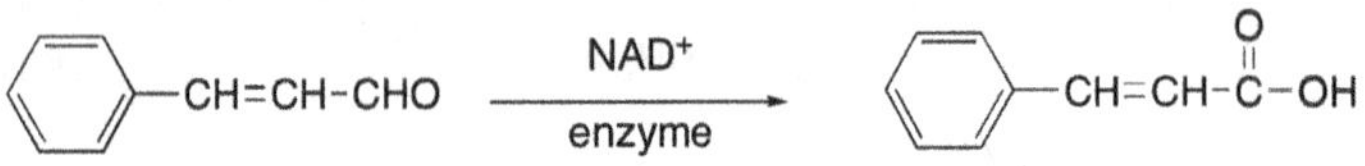

C_6H_5–CH=CH–CHO $\xrightarrow[\text{enzyme}]{NAD^+}$ C_6H_5–CH=CH–C(=O)–OH

16.95 Label each hemiacetal or alcohol.

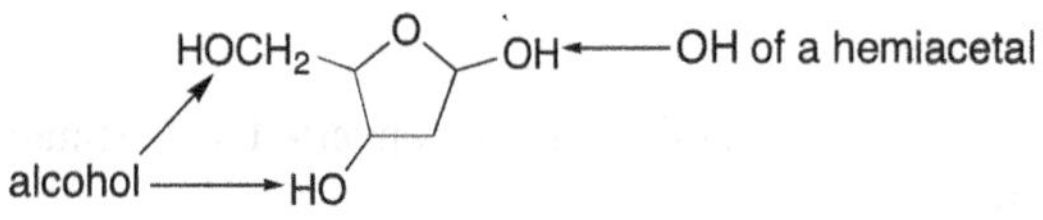

$HOCH_2$–[five-membered ring with O]–OH ← OH of a hemiacetal

alcohol → HO (ring OH); alcohol → $HOCH_2$

16.97 The main reaction that occurs in the rod cells in the retina is the conversion of 11-*cis*-retinal to its trans isomer. The cis double bond in 11-*cis*-retinal produces crowding, making the molecule unstable. Light energy converts this to the more stable trans isomer, and with this conversion an electrical impulse is generated in the optic nerve.

16.99 Identify the alcohol, acetal, hemiacetal, ether, and carboxylic acid functional groups.

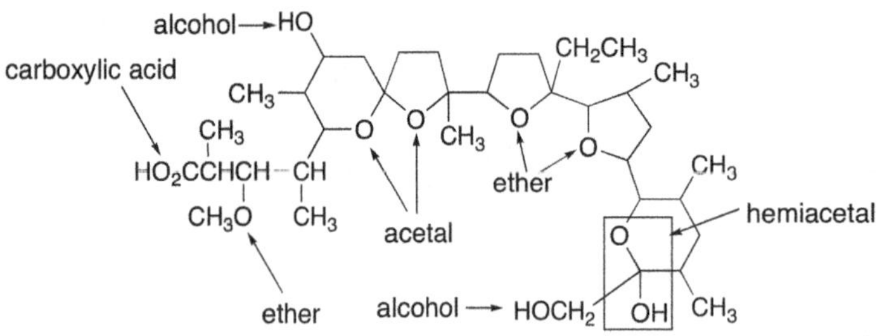

[Structure labels: alcohol → HO; carboxylic acid → HO_2CCHCH–CH (with CH_3, CH_3O, CH_3); CH_3; acetal → O; ether → CH_3O; ether → O; CH_2CH_3; CH_3; CH_3; hemiacetal → O / OH (boxed); alcohol → $HOCH_2$; CH_3]

Chapter 17 Carboxylic Acids, Esters, and Amides

Chapter Review

[1] What are the characteristics of carboxylic acids, esters, and amides?

- Carboxylic acids have the general structure RCOOH; esters have the general structure RCOOR'; amides have the general structure $RCONR'_2$, where R' = H or alkyl. (17.1)

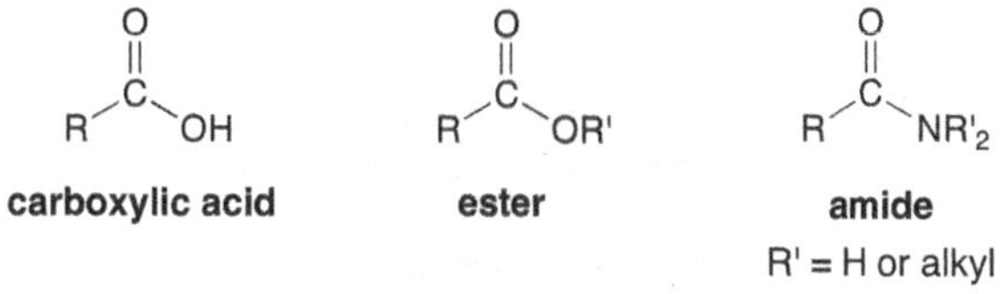

carboxylic acid **ester** **amide**
R' = H or alkyl

- The carbonyl carbon is trigonal planar, so bond angles are 120°. (17.1)

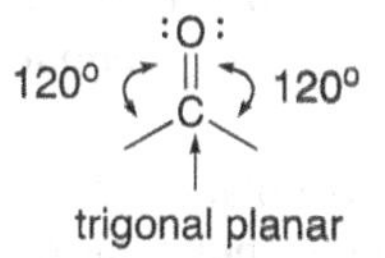

120° 120°
trigonal planar

- All acyl compounds have a polar C=O. RCO_2H, $RCONH_2$, and RCONHR' are capable of intermolecular hydrogen bonding. (17.3)

[2] How are carboxylic acids, esters, and amides named? (17.2)

- Carboxylic acids are identified by the suffix *-oic acid.*
- Esters are identified by the suffix *-ate*.
- Amides are identified by the suffix *-amide*.

[3] Give examples of useful carboxylic acids. (17.4)

- α-Hydroxy carboxylic acids are used in skin care products.

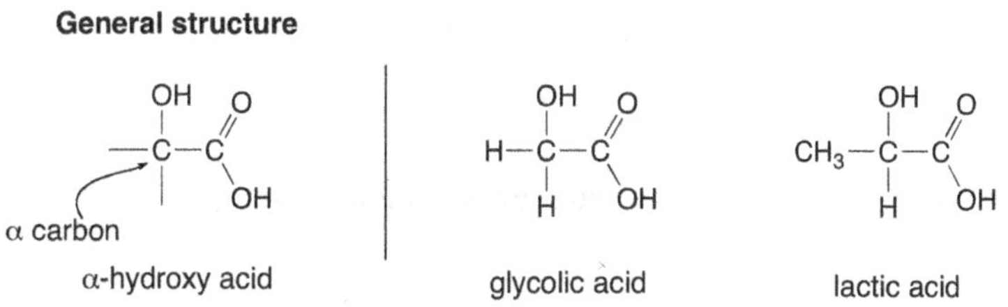

General structure

α carbon
α-hydroxy acid

glycolic acid

lactic acid

- Aspirin, ibuprofen, and naproxen are pain relievers and anti-inflammatory agents.
- Aspirin is an anti-inflammatory agent because it blocks the synthesis of prostaglandins from arachidonic acid.

[4] Give examples of useful esters and amides. (17.5)

- Some esters—ethyl butanoate, pentyl butanoate, and methyl salicylate—have characteristic odors and flavors.
- Benzocaine is the active ingredient in over-the-counter oral topical anesthetics.
- Acetaminophen is the active ingredient in Tylenol.

[5] What products are formed when carboxylic acids are treated with base? (17.6)

- Carboxylic acids react with bases to form carboxylate anions ($RCOO^-$).

$$RCOOH + Na^+\,OH^- \longrightarrow RCOO^-\,Na^+ + H_2O$$

- Carboxylate anions are water soluble and commonly used as preservatives.

[6] How does soap clean away dirt? (17.6C)

- Soaps are salts of carboxylic acids that have many carbon atoms in a long hydrocarbon chain. A soap is composed of an ionic head and a nonpolar hydrocarbon tail.

Structure of a soap

$$Na^+\ {}^-O-C(=O)-CH_2CH_2CH_2CH_2CH_2CH_2CH_2CH_2CH_2CH_2CH_2CH_2CH_2CH_2CH_2CH_2CH_3$$

ionic end — **polar head**; long, hydrocarbon chain — **nonpolar tail**

- Soap forms micelles in water with its polar heads on the surface and the hydrocarbon tails in the interior. Grease and dirt dissolve in the nonpolar tails, making it possible to wash them away with water.

[7] Discuss the acid–base chemistry of aspirin. (17.7)

- Aspirin remains in its neutral form in the stomach, and in this form it can cross a cell membrane and serve as an anti-inflammatory agent.
- A proton is removed from aspirin in the basic environment of the intestines to form an ionic carboxylate anion. This form is not absorbed.

[8] How are carboxylic acids converted to esters and amides? (17.8)

- Carboxylic acids are converted to esters by reaction with alcohols (R'OH) and acid (H_2SO_4).

$$RCOOH + R'OH \underset{}{\overset{H_2SO_4}{\rightleftharpoons}} \underset{\text{ester}}{RCOOR'} + H_2O$$

- Carboxylic acids are converted to amides by heating with ammonia (NH_3) or amines ($R'NH_2$ or R'_2NH).

$$RCOOH + R'_2NH \xrightarrow{\Delta} \underset{\text{amide}}{RCONR'_2} + H_2O$$

R' = H or alkyl

[9] What hydrolysis products are formed from esters and amides? (17.9)

- Esters are hydrolyzed to carboxylic acids (RCOOH) in the presence of an acid catalyst (H_2SO_4). Esters are converted to carboxylate anions ($RCOO^-$) with aqueous base (NaOH in H_2O).

$$RCOOR' + H_2O \xrightarrow{(H^+ \text{ or } OH^-)} RCOOH \text{ (in acid)} \text{ or } RCOO^- \text{ (in base)} + R'OH$$

- Amides are hydrolyzed to carboxylic acids (RCOOH) in the presence of an acid catalyst (HCl). Amides are converted to carboxylate anions ($RCOO^-$) with aqueous base (NaOH in H_2O).

$$RCONR'_2 + H_2O \xrightarrow{H^+} RCOOH + (R'_2NH_2)^+$$

R' = H or alkyl

$$RCONR'_2 + H_2O \xrightarrow{OH^-} RCOO^- + R'_2NH$$

[10] What are polyamides and polyesters and how are they formed? (17.10)

- Polyamides like nylon are polymers that contain many amide bonds. They are formed when a dicarboxylic acid is heated with a diamine.

Each NH_2 and COOH react.

$$H_2N{-}(CH_2)_6{-}NH_2 + HO{-}CO{-}(CH_2)_4{-}CO{-}OH + H_2N{-}(CH_2)_6{-}NH_2 + HO{-}CO{-}(CH_2)_4{-}CO{-}OH$$

loss of H_2O — loss of H_2O — loss of H_2O

$$\xrightarrow{\Delta}$$

$$\sim NH{-}(CH_2)_6{-}NH{-}CO{-}(CH_2)_4{-}CO{-}NH{-}(CH_2)_6{-}NH{-}CO{-}(CH_2)_4{-}CO \sim$$

new amide bonds

nylon 6,6

- Polyesters like PET are polymers that contain many ester bonds. They are formed when a dicarboxylic acid is treated with a diol in the presence of acid (H_2SO_4).

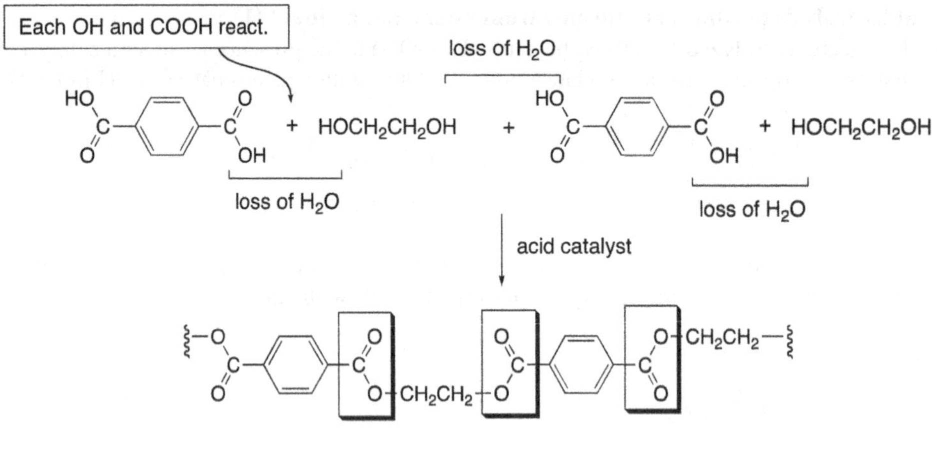

Each OH and COOH react.

loss of H_2O

$HO_2C-C_6H_4-CO_2H$ + $HOCH_2CH_2OH$ + $HO_2C-C_6H_4-CO_2H$ + $HOCH_2CH_2OH$

loss of H_2O loss of H_2O

acid catalyst

$-[O-CO-C_6H_4-CO-O-CH_2CH_2-O-CO-C_6H_4-CO-O-CH_2CH_2]-$

polyethylene terephthalate
PET

[11] How does penicillin act as an antibiotic? (17.11)

- The β-lactam of penicillin is more reactive than a regular amide and it reacts with an enzyme needed to synthesize the cell wall of a bacterium. Without a cell wall, the bacterium dies.

Problem Solving

[1] Nomenclature (17.2)

Example 17.1 Give the IUPAC name of the following carboxylic acid.

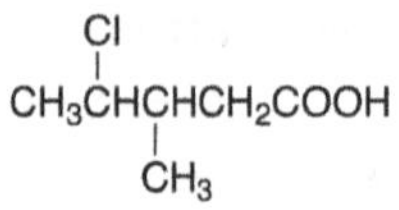

$CH_3CH(Cl)CH(CH_3)CH_2COOH$

Analysis and Solution

[1] Find and name the longest chain containing COOH:

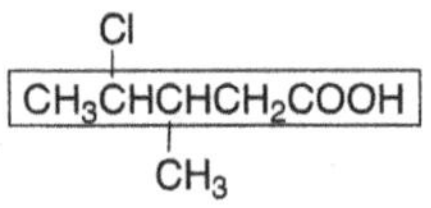

$CH_3CH(Cl)CH(CH_3)CH_2COOH$

pentan*e* ---→ pentan*oic acid*
(5 C's)

The COOH contributes one C to the longest chain.

[2] Number and name the substituents, making sure the COOH group is at C1:

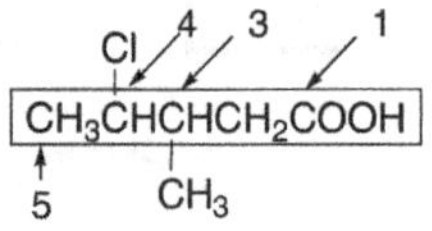

$CH_3CH(Cl)CH(CH_3)CH_2COOH$ (C5, C4, C3, C1)

one methyl substituent on C3
one chloro substituent on C4

Answer: 4-chloro-3-methylpentanoic acid

Example 17.2 Give the IUPAC name of the following ester.

$CH_3CH_2O{-}C(=O){-}CH_2CH_2CH_2CH_3$

Analysis and Solution

[1] Name the alkyl group on the O atom:

$CH_3CH_2O{-}C(=O){-}CH_2CH_2CH_2CH_3$

ethyl group

The word *ethyl* becomes the first part of the name.

[2] To name the acyl group, find and name the longest chain containing the carbonyl group, placing the C=O at C1:

$CH_3CH_2O{-}\boxed{C(=O){-}CH_2CH_2CH_2CH_3}$

pentano*ic acid* ---→ pentano*ate*
(5 C's)

Answer: ethyl pentanoate

---

[2] The Acidity of Carboxylic Acids (17.6)

Example 17.3 What products are formed when butanoic acid ($CH_3CH_2CH_2COOH$) reacts with potassium hydroxide (KOH)?

Analysis

In any acid–base reaction with a carboxylic acid:

- Remove a proton from the carboxyl group (COOH) and form the carboxylate anion ($RCOO^-$).
- Add a proton to the base. If the base has a (–1) charge to begin with, it becomes a neutral product when a proton (H^+) is added to it.
- Balance the charge of the carboxylate anion by drawing it as a salt with a metal cation.

Solution

$$CH_3CH_2CH_2C(=O){-}O{-}H \;+\; K^+\,OH^- \longrightarrow CH_3CH_2CH_2C(=O){-}O^-\,K^+ \;+\; H{-}O{-}H$$

butanoic acid; base

This proton is transferred from the acid to the base.

The carboxylate anion is formed as a potassium salt.

Thus, $CH_3CH_2CH_2COOH$ loses a proton to form $CH_3CH_2CH_2COO^-$, which is present in solution as its potassium salt, $CH_3CH_2CH_2COO^-K^+$. Hydroxide (OH^-) gains a proton to form H_2O.

Example 17.4 Give an acceptable name for each salt.

a. $CH_3CH(CH_3){-}C(=O){-}O^-\,K^+$

b. $CH_3CH_2CH_2{-}C(=O){-}O^-\,Na^+$

Analysis

Name the carboxylate salt by putting three parts together:

- the name of the metal cation
- the parent name that indicates the number of carbons in the parent chain
- the suffix, *-ate*

Solution

a. $CH_3CH(CH_3)COO^-\ K^+$

potassium cation

2-methyl substituent

parent + suffix
propano- -ate

- The first part of the name is the metal cation, potassium.
- The parent name is derived from the IUPAC name, propanoic acid. Change the ***-ic acid*** ending to ***-ate***; propanoic acid → propanoate.
- **Answer: potassium 2-methylpropanoate**

b. $CH_3CH_2CH_2COO^-\ Na^+$

sodium cation

parent + suffix
butano- -ate

- The first part of the name is the metal cation, sodium.
- The parent name is derived from the IUPAC name, butanoic acid. Change the ***-ic acid*** ending to ***-ate***; butanoic acid → butanoate.
- **Answer: sodium butanoate**

---

[3] The Conversion of Carboxylic Acids to Esters and Amides (17.8)

Example 17.5 What ester is formed when butanoic acid ($CH_3CH_2CH_2COOH$) is treated with methanol (CH_3OH) in the presence of H_2SO_4?

Analysis

To draw the products of any acyl substitution, arrange the carboxyl group of the carboxylic acid next to the functional group with which it reacts—the OH group of the alcohol in this case. Then replace the OH group of the carboxylic acid by the OR' group of the alcohol, forming a new C–O bond at the carbonyl carbon.

Remove OH and H to form H_2O.

$$RCOOH + H{-}OR' \longrightarrow RCOOR' + H{-}OH$$

carboxylic acid → ester (new C–O bond)

Solution

Replace the OH group of butanoic acid by the OCH_3 group of methanol to form the ester.

$$CH_3CH_2CH_2COOH + H{-}OCH_3 \xrightleftharpoons{H_2SO_4} CH_3CH_2CH_2COOCH_3 + H{-}OH$$

new C–O bond

OCH_3 replaces OH.

Example 17.6 What amide is formed when butanoic acid ($CH_3CH_2CH_2COOH$) is heated with propylamine ($CH_3CH_2CH_2NH_2$)?

Analysis

To draw the products of amide formation, arrange the carboxyl group of the carboxylic acid next to the H–N bond of the amine. Then replace the OH group of the carboxylic acid by the $NHCH_2CH_2CH_3$ group of the amine, forming a new C–N bond at the carbonyl carbon.

Solution

The reaction of RCOOH with an amine that has one alkyl group on the N atom ($R'NH_2$) forms a 2° amide (RCONHR').

$CH_3CH_2CH_2C(=O)OH$ + H–N(H)–$CH_2CH_2CH_3$ (propylamine) $\xrightarrow{\Delta}$ $CH_3CH_2CH_2C(=O)NHCH_2CH_2CH_3$ (2° amide) + H–OH

new C–N bond

$NHCH_2CH_2CH_3$ replaces OH.

---

[4] Hydrolysis of Esters and Amides (17.9)

Example 17.7 What products are formed when ethyl acetate ($CH_3CO_2CH_2CH_3$) is hydrolyzed with water in the presence of H_2SO_4?

Analysis

To draw the products of hydrolysis in acid, replace the OR' group of the ester by an OH group from water, forming a new C–O bond at the carbonyl carbon. A molecule of alcohol (R'OH) is also formed from the alkoxy group (OR') of the ester.

RC(=O)OR' (ester) + H–OH → RC(=O)OH + H–OR'

new C–O bond

OH replaces OR'.

Solution

Replace the OCH_2CH_3 group of ethyl acetate by the OH group of water to form acetic acid (CH_3CO_2H) and ethanol (CH_3CH_2OH).

$CH_3C(=O)OCH_2CH_3$ (ethyl acetate) + H–OH $\xrightleftharpoons{H_2SO_4}$ $CH_3C(=O)OH$ (acetic acid) + H–OCH_2CH_3 (ethanol)

OH replaces OCH_2CH_3.

Example 17.8 What products are formed when *N*-ethylacetamide ($CH_3CONHCH_2CH_3$) is hydrolyzed with water in the presence of NaOH?

Analysis

To draw the products of amide hydrolysis in base, replace the NHR' group of the amide by an oxygen anion (O^-), forming a new C–O bond at the carbonyl carbon. A molecule of amine ($R'NH_2$) is also formed from the nitrogen group (NHR') of the amide.

new C–O bond

amide + H–OH $\xrightarrow{NaOH}$ $RCO^-\ Na^+$ + H–N(H)–R'

O⁻ replaces NHR'.

Solution

Replace the $NHCH_2CH_3$ group of *N*-ethylacetamide by a negatively charged oxygen atom (O^-) to form sodium acetate ($CH_3CO_2^-\ Na^+$) and ethylamine ($CH_3CH_2NH_2$).

$CH_3CONHCH_2CH_3$ (*N*-ethylacetamide) + H–OH $\xrightarrow{NaOH}$ $CH_3CO_2^-\ Na^+$ + H–N(H)–CH_2CH_3

O^- replaces $NHCH_2CH_3$.

Self-Test

[1] Fill in the blank with one of the terms listed below.

Amides (17.1)
Carboxylic acids (17.10)
Condensation polymer (17.10)
Esters (17.1)
Hydrolysis (17.9)
α-Hydroxy acids (17.4)
Lactam (17.1)
Lactone (17.1)
Lipids (17.9)
Micelles (17.6)
Saponification (17.9)
Soaps (17.6)
Triacylglycerols (17.9)

1. A cyclic ester is called a ______.
2. When bonds are cleaved on reaction with water, this reaction is called ______.
3. ______ are carbonyl compounds that contain a nitrogen atom bonded to the carbonyl carbon.
4. ______ are spherical droplets having the ionic heads on the surface and the nonpolar tails packed together in the interior.
5. ______ contain a hydroxyl group on the α (alpha) carbon to the carboxyl group.
6. ______ contain three ester groups, each having a long carbon chain bonded to the carbonyl group.
7. ______ are organic compounds containing a carboxyl group.
8. ______ are salts of carboxylic acids that have many carbon atoms in a long hydrocarbon chain.
9. A cyclic amide is called a ______.
10. Basic hydrolysis of an ester is called ______.

11. ______ are carbonyl compounds that contain an alkoxy group (OR') bonded to the carbonyl carbon.
12. ______ are water-insoluble organic compounds found in biological systems.
13. A ______ is a polymer formed when monomers containing two functional groups come together with loss of a small molecule such as water.

[2] Label each compound as an amide, ester, ether, or carboxylic acid.

14. $CH_3CH_2{-}O{-}CH_2CH_2CH_3$

15. 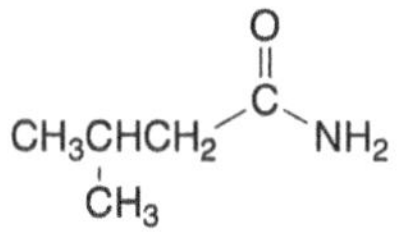 $CH_3CH(CH_3)CH_2C(=O)NH_2$

16. 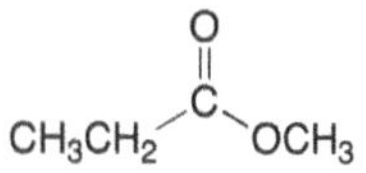 $CH_3CH_2C(=O)OCH_3$

17. 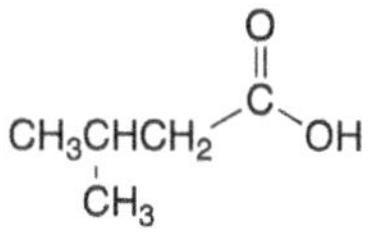 $CH_3CH(CH_3)CH_2C(=O)OH$

[3] Which compound in each pair has the higher boiling point?

18. A: CH_3CH_2OH B: $CH_3{-}C(=O)OH$

19. A: $CH_3{-}C(=O)NHCH_3$ B: $CH_3{-}C(=O)OCH_3$

20. A: $CH_3{-}C(=O)OCH_3$ B: $CH_3CH_2{-}C(=O)NH_2$

21. A: $H{-}C(=O)N(CH_3)_2$ B: $CH_3CH_2{-}C(=O)NH_2$

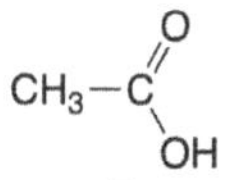

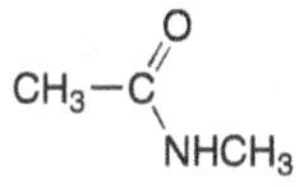

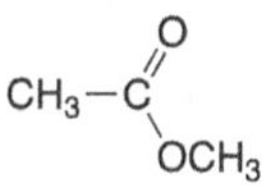

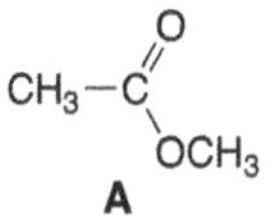

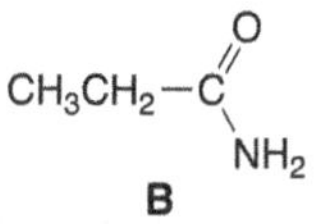

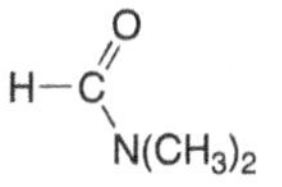

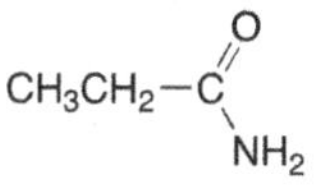

[4] Match the reaction with the products.

22. $CH_3CH(CH_3)CH_2C(=O)OH$ + $H{-}OCH_3$ $\xrightleftharpoons{H_2SO_4}$

23. $CH_3CH(CH_3)CH_2C(=O)OH$ + $H{-}NH{-}CH_2CH_3$ $\xrightarrow{\Delta}$

24. $CH_3CH(CH_3)CH_2C(=O)OCH_2CH_3$ + $H{-}OH$ $\xrightleftharpoons{H_2SO_4}$

25. $CH_3CH(CH_3)CH_2C(=O)NHCH_2CH_3$ + $H{-}OH$ $\xrightarrow{NaOH}$

a. $CH_3CH(CH_3)CH_2C(=O)O^-\ Na^+$ + $H{-}NH{-}CH_2CH_3$

b. $CH_3CH(CH_3)CH_2C(=O)OCH_3$ + $H{-}OH$

c. $CH_3CH(CH_3)CH_2C(=O)OH$ + $H{-}OCH_2CH_3$

d. $CH_3CH(CH_3)CH_2C(=O)NHCH_2CH_3$ + $H{-}OH$

Answers to Self-Test

1. lactone	6. Triacylglycerols	11. Esters	16. ester	21. **B**
2. hydrolysis	7. Carboxylic acids	12. Lipids	17. carboxylic acid	22. b
3. Amides	8. Soaps	13. condensation polymer	18. **B**	23. d
4. Micelles	9. lactam	14. ether	19. **A**	24. c
5. α-Hydroxy acids	10. saponification	15. amide	20. **B**	25. a

Solutions to In-Chapter Problems

17.1 **Carboxylic acids** are organic compounds containing a carboxyl group (COOH).
Esters are carbonyl compounds that contain an alkoxy group (OR') bonded to the carbonyl carbon.
Amides are carbonyl compounds that contain a nitrogen atom bonded to the carbonyl carbon.

a. $CH_3CH_2C(=O)OCH_2CH_3$ — ester

b. $CH_3C(=O)NH{-}CH_3$ — amide

c. $(CH_3)_3CC(=O)OH$ — carboxylic acid

d. $(CH_3)_2CHC(=O)N(CH_3)_2$ — amide

17.2 A **primary (1°) amide** contains one C–N bond. A 1° amide has the structure $RCONH_2$.
A **secondary (2°) amide** contains two C–N bonds. A 2° amide has the structure RCONHR'.
A **tertiary (3°) amide** contains three C–N bonds. A 3° amide has the structure $RCONR'_2$.

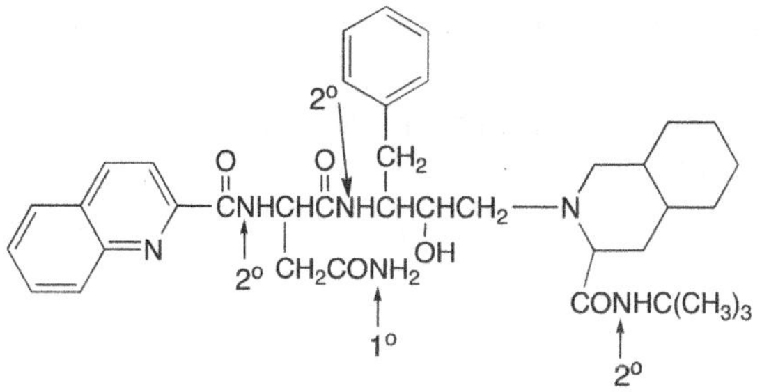

2° 2° 1° 2°

O O CH_2 CNHCHCNHCHCHCH$_2$—N N OH CH_2CONH_2 $CONHC(CH_3)_3$

17.3 Label the functional groups and the trigonal planar carbons in lisinopril.

a, b: The nine trigonal planar carbons are labeled (*).

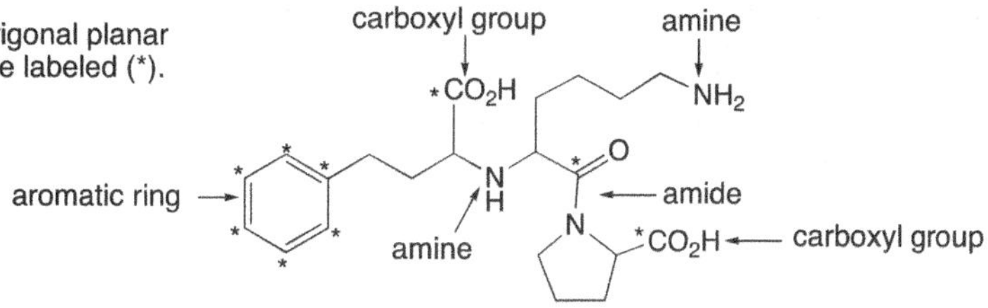

carboxyl group → *CO_2H
amine → NH_2
aromatic ring
amine (N–H)
amide
*CO_2H ← carboxyl group

17.4 Name the carboxylic acids as in Example 17.1.

a. $CH_3CH_2CH_2C(CH_3)_2CH_2COOH$ ------→ $CH_3CH_2CH_2C(CH_3)_2CH_2COOH$ (C3, C1) ------→ **Answer: 3,3-dimethylhexanoic acid**

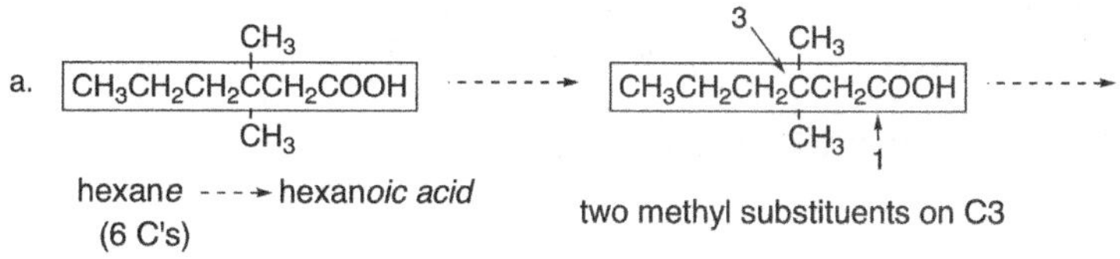

hexane ---→ hexan*oic acid* (6 C's)

two methyl substituents on C3

b. $CH_3CH(Cl)CH_2CH_2COOH$ ------→ $CH_3CH(Cl)CH_2CH_2COOH$ (C4, C1) ------→ **Answer: 4-chloropentanoic acid**

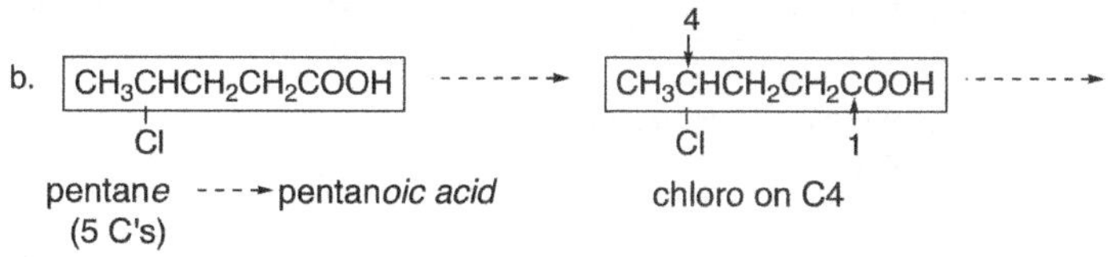

pentane ---→ pentan*oic acid* (5 C's)

chloro on C4

c. $CH_3CH_2CH(CH_2CH_3)CH_2CH(CH_2CH_3)COOH$ ------→ $CH_3CH_2CH(CH_2CH_3)CH_2CH(CH_2CH_3)COOH$ (C4, C2, C1) ------→ **Answer: 2,4-diethylhexanoic acid**

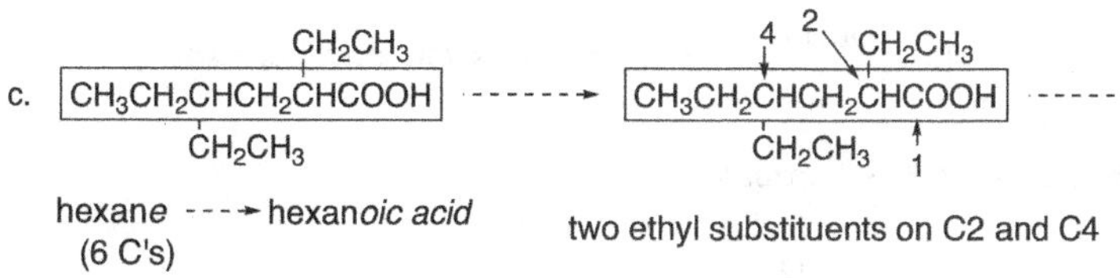

hexane ---→ hexan*oic acid* (6 C's)

two ethyl substituents on C2 and C4

17.5 Work backwards from the IUPAC name to draw the structure of each carboxylic acid.

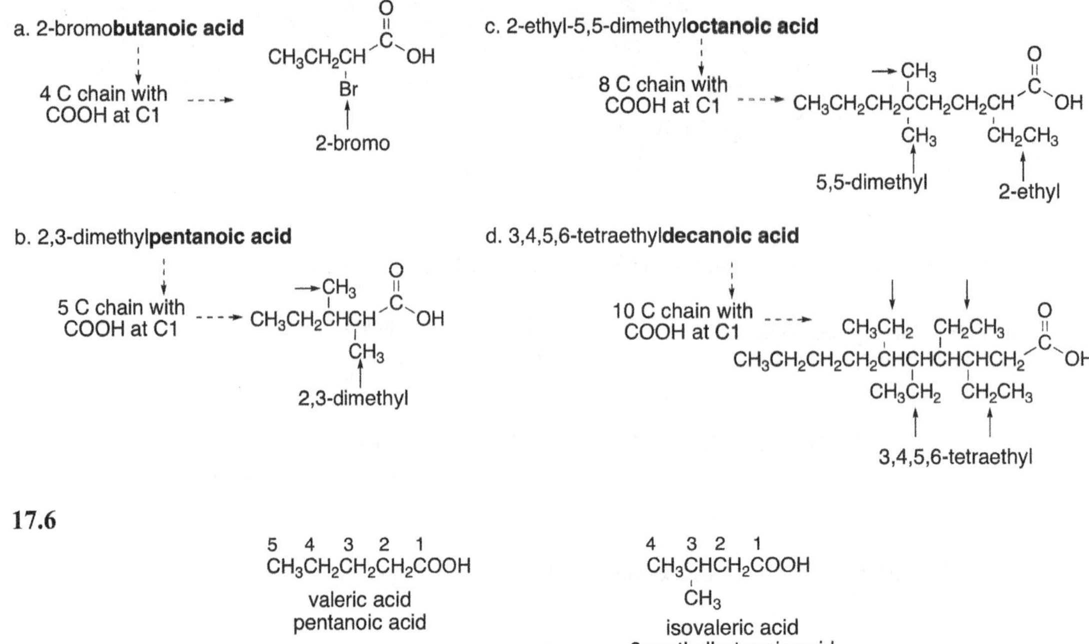

17.7 Use the steps in Example 17.2 to give the IUPAC name for each ester.

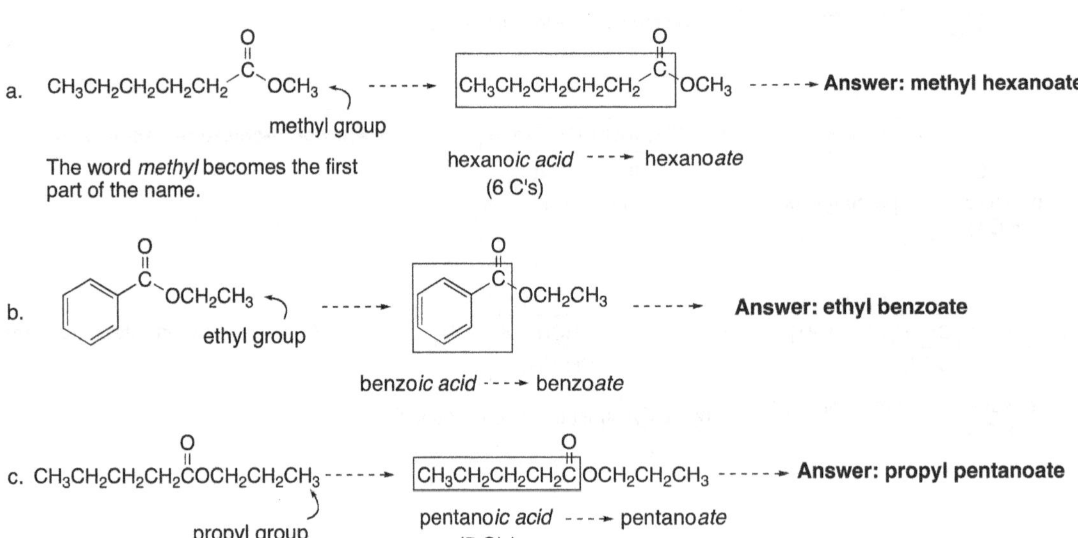

17.8 Work backwards to draw the structure from the IUPAC name.

a. propyl propanoate

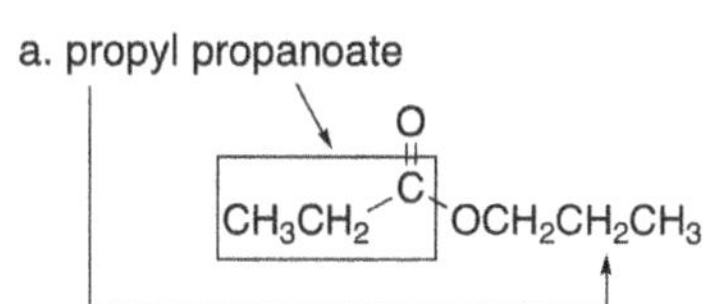

c. ethyl hexanoate

O

C

$CH_3CH_2CH_2CH_2CH_2$ OCH_2CH_3

b. butyl acetate

O

C

CH_3 $OCH_2CH_2CH_2CH_3$

d. methyl benzoate

O

C

OCH_3

17.9

O

C

H OCH_2CH_3

ethyl

derived from the 1 C carboxylic acid, formic acid

ethyl formate

17.10 To give each amide an IUPAC name, use the following steps:

Step [1] Name the alkyl group (or groups) bonded to the N atom of the amide. Use the prefix "*N*-" preceding the name of each alkyl group.

Step [2] Name the acyl group (RCO–) with the suffix *-amide*.

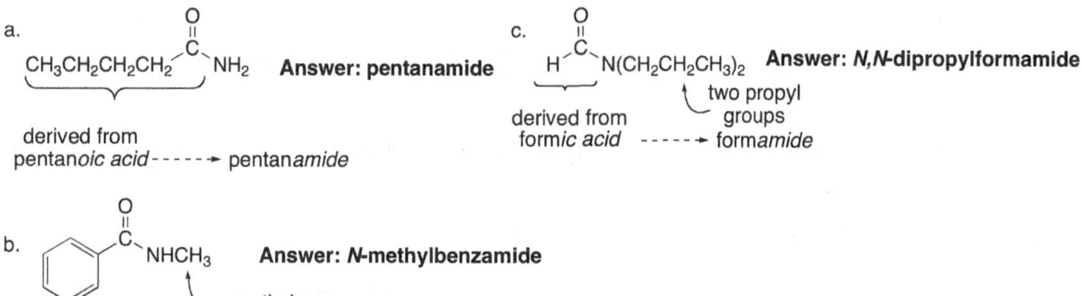

b.

O

C

$NHCH_3$

Answer: *N*-methylbenzamide

methyl group

derived from

benz*oic acid* → benz*amide*

17.11 Work backwards from the name to draw each amide.

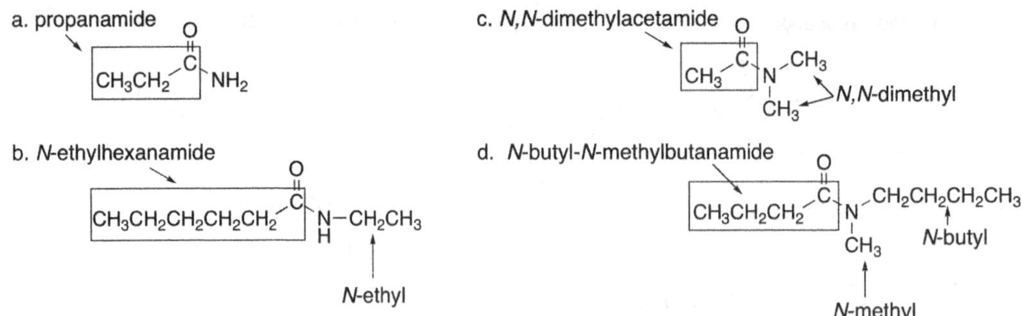

17.12 Use the following rules to determine which compound has a higher boiling point:

- Carboxylic acids have stronger intermolecular forces than esters, giving them higher boiling points and melting points when comparing compounds of comparable size.
- Carboxylic acids have higher boiling points and melting points than alcohols of comparable size.
- Primary (1°) and 2° amides have higher boiling points and melting points than esters and 3° amides of comparable size.

a. CH_3COOH (carboxylic acid, **higher boiling point**) or CH_3CH_2CHO (aldehyde)

b. $CH_3CH_2CH_2CH_2CONH_2$ (1° amide, **higher boiling point**) or $CH_3CH_2CO_2CH_2CH_3$ (ester)

17.13 α-Hydroxy acids contain a hydroxyl group on the α (alpha) carbon to the carboxyl group.

a. HO(O=)C–CH(OH)–CH(OH)–C(=O)OH — both hydroxyl groups bonded to α carbons

b. $CH_3CH(OH)CH_2$–C(=O)OH — no OH group on the α carbon

c. (2-hydroxyphenyl)–COOH — no OH group on the α carbon

17.14 Work backwards from the name to draw each ester.

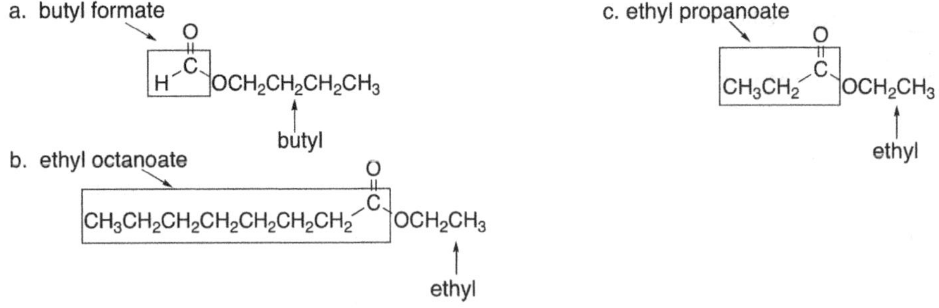

17.15 Draw the products of each acid–base reaction as in Example 17.3.

a. cyclohexane–COOH + NaOH ⟶ cyclohexane–CO_2^- Na^+ + H_2O

b. $CH_3CH_2CH_2COOH$ + Na_2CO_3 ⟶ $CH_3CH_2CH_2CO_2^-$ Na^+ + Na^+ HCO_3^-

17.16 Draw the products of each reaction.

a. benzene–COOH + NaOH ⟶ benzene–CO_2^- Na^+ + H_2O

b. benzene–COOH + Na_2CO_3 ⟶ benzene–CO_2^- Na^+ + Na^+ HCO_3^-

c. benzene–COOH + $NaHCO_3$ ⟶ benzene–CO_2^- Na^+ + H_2CO_3

17.17 Name each salt of a carboxylic acid using the steps in Example 17.4.

a. $CH_3CH_2CH_2CO_2^-$ Na^+ (sodium cation)

parent + suffix
butano- -ate

sodium butanoate

b. C_6H_5–COO^- Li^+ (lithium cation)

parent + suffix
benzo- -ate

lithium benzoate

17.18 Work backwards to draw the structure from the name.

sodium propanoate

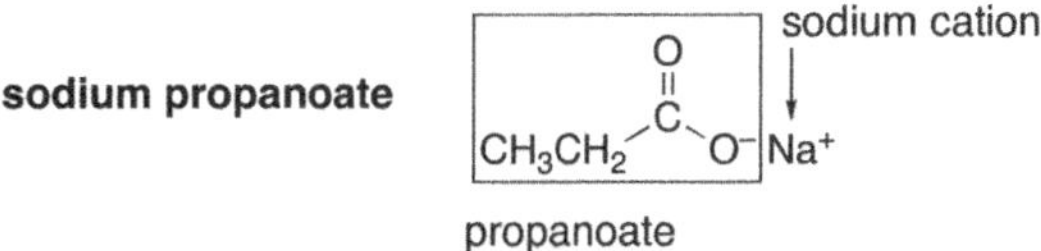

17.19 Soaps are salts of carboxylic acids that have many carbon atoms in a long hydrocarbon chain. A soap has two parts: a long hydrocarbon chain and an ionic end (polar head).

$CH_3CH_2CH_2CH_2CH_2CH_2CH_2CH_2CH_2CH_2CH_2CH_2CH_2CH_2CH_2CO_2^-$ K^+

potassium cation

hydrocarbon chain

carboxylate anion

17.20 Draw the acid–base reaction.

$$C_6H_4(OCOCH_3)COOH + NaOH \longrightarrow C_6H_4(OCOCH_3)COO^-\ Na^+ + H_2O$$

17.21 Answer each question about ibuprofen.

a. $(CH_3)_2CHCH_2-C_6H_4-CH(CH_3)COOH + NaOH \longrightarrow (CH_3)_2CHCH_2-C_6H_4-CH(CH_3)COO^-\ Na^+ + H_2O$

b. The neutral form of ibuprofen is present in the stomach because the stomach is acidic.
c. The ionized form of ibuprofen is present in the intestines because the intestines are basic.

17.22 Draw the products of each reaction as in Example 17.5.

a. $(CH_3)_2CHCOOH \xrightarrow[H_2SO_4]{CH_3CH_2OH} (CH_3)_2CHCOOCH_2CH_3$

b. $HCO_2H \xrightarrow[H_2SO_4]{CH_3CH_2OH} HCOOCH_2CH_3$

c. $CH_3(CH_2)_6COOH \xrightarrow[H_2SO_4]{CH_3CH_2OH} CH_3(CH_2)_6COOCH_2CH_3$

d. $C_6H_{11}-CO_2H \xrightarrow[H_2SO_4]{CH_3CH_2OH} C_6H_{11}-COOCH_2CH_3$

17.23 Identify **A** in the reaction.

2,5-dimethoxybenzyl alcohol (OCH_3, CH_2OH, OCH_3 on benzene) + $HOC(=O)CH_2CH(CH_3)_2$ $\xrightarrow{H_2SO_4}$ aryl–$CH_2O-C(=O)CH_2CH(CH_3)_2$ (with OCH_3 groups) **A** $\longrightarrow$ quinone–$CH_2O-C(=O)CH_2CH(CH_3)_2$ blattellaquinone

17.24 Work backwards to determine what carboxylic acid and alcohol are needed to prepare benzocaine.

$$H_2N-C_6H_4-CO_2H + CH_3CH_2OH \longrightarrow H_2N-C_6H_4-CO_2CH_2CH_3$$

benzocaine

17.25 Draw the products of each reaction as in Example 17.6.

a. $CH_3CH_2CH_2CH_2COOH \xrightarrow[\Delta]{NH_3} CH_3CH_2CH_2CH_2CONH_2$

b. $CH_3CH_2CH_2CH_2COOH \xrightarrow[\Delta]{(CH_3)_2NH} CH_3CH_2CH_2CH_2CON(CH_3)CH_3$

c. $CH_3CH_2CH_2CH_2COOH \xrightarrow[\Delta]{C_6H_{11}NH_2} CH_3CH_2CH_2CH_2CONH\text{-}C_6H_{11}$

d. $CH_3CH_2CH_2CH_2COOH \xrightarrow[\Delta]{CH_3NH_2} CH_3CH_2CH_2CH_2CONHCH_3$

17.26 Work backwards to determine what carboxylic acid and amine are needed to synthesize phenacetin.

$CH_3CH_2O\text{-}C_6H_4\text{-}NH_2 + HOCOCH_3 \longrightarrow CH_3CH_2O\text{-}C_6H_4\text{-}NHCOCH_3$

phenacetin

17.27 Draw the products of each reaction as in Example 17.7.

a. $CH_3(CH_2)_8COOCH_3 \xrightarrow[H_2SO_4]{H_2O} CH_3(CH_2)_8COOH + CH_3OH$

b. $CH_3CH(CH_3)CH_2COOCH_2CH_3 \xrightarrow[H_2SO_4]{H_2O} CH_3CH(CH_3)CH_2COOH + CH_3CH_2OH$

c. $C_6H_{11}\text{-}CO_2CH_2CH_2CH_3 \xrightarrow[H_2SO_4]{H_2O} C_6H_{11}\text{-}COOH + CH_3CH_2CH_2OH$

17.28 Basic hydrolysis of esters forms carboxylate anions and alcohols.

a. $CH_3(CH_2)_8COOCH_3 \xrightarrow[NaOH]{H_2O} CH_3(CH_2)_8COO^-\,Na^+ + CH_3OH$

b. $CH_3CH(CH_3)CH_2COOCH_2CH_3 \xrightarrow[NaOH]{H_2O} CH_3CH(CH_3)CH_2COO^-\,Na^+ + CH_3CH_2OH$

c. $C_6H_{11}CO_2CH_2CH_2CH_3 \xrightarrow[NaOH]{H_2O} C_6H_{11}CO^-\ Na^+ + CH_3CH_2CH_2OH$ (cyclohexyl–C(=O)O⁻ Na⁺)

17.29 Draw the products of hydrolysis of aspirin.

$$\text{aspirin (o-}CH_3C(=O)O\text{–}C_6H_4\text{–}COOH) \xrightarrow{H_2O} \text{o-}HO\text{–}C_6H_4\text{–}COOH + CH_3COH\ (CH_3C(=O)OH)$$

aspirin

17.30 Draw the products formed when each amide is treated with H_2O and H_2SO_4.

a. $CH_3(CH_2)_8C(=O)NH_2 \xrightarrow[H_2SO_4]{H_2O} CH_3(CH_2)_8C(=O)OH + NH_4^+\ HSO_4^-$

b. $CH_3CH(CH_3)CH_2C(=O)NHCH_3 \xrightarrow[H_2SO_4]{H_2O} CH_3CH(CH_3)CH_2C(=O)OH + (CH_3NH_3)^+\ HSO_4^-$

c. $C_6H_{11}CON(CH_2CH_2CH_3)_2 \xrightarrow[H_2SO_4]{H_2O} C_6H_{11}C(=O)OH + [(CH_3CH_2CH_2)_2NH_2]^+\ HSO_4^-$

17.31 Draw the products formed when each amide is treated with H_2O and NaOH as in Example 17.8.

a. $CH_3(CH_2)_8C(=O)NH_2 \xrightarrow[NaOH]{H_2O} CH_3(CH_2)_8C(=O)O^-\ Na^+ + NH_3$

b. $CH_3CH(CH_3)CH_2C(=O)NHCH_3 \xrightarrow[NaOH]{H_2O} CH_3CH(CH_3)CH_2C(=O)O^-\ Na^+ + CH_3NH_2$

c. $C_6H_{11}CON(CH_2CH_2CH_3)_2 \xrightarrow[NaOH]{H_2O} C_6H_{11}C(=O)O^-\ Na^+ + (CH_3CH_2CH_2)_2NH$

17.32 The triacylglycerol is hydrolyzed to glycerol and three carboxylic acids.

$$\begin{array}{l} CH_2\text{-}O\text{-}C(=O)\text{-}(CH_2)_{10}CH_3 \\ CH\text{-}O\text{-}C(=O)\text{-}(CH_2)_{12}CH_3 \\ CH_2\text{-}O\text{-}C(=O)\text{-}(CH_2)_{14}CH_3 \end{array} \xrightarrow[H_2SO_4]{H_2O} \begin{array}{l} CH_2\text{-}O\text{-}H \\ CH\text{-}O\text{-}H \\ CH_2\text{-}O\text{-}H \end{array} + \begin{array}{l} HO\text{-}C(=O)\text{-}(CH_2)_{10}CH_3 \\ + \\ HO\text{-}C(=O)\text{-}(CH_2)_{12}CH_3 \\ + \\ HO\text{-}C(=O)\text{-}(CH_2)_{14}CH_3 \end{array}$$

17.33 Work backwards to determine what two monomers are needed to prepare nylon 6,10.

$$H_2N\text{-}(CH_2)_6\text{-}NH_2 + HO\text{-}C(=O)\text{-}(CH_2)_8\text{-}C(=O)\text{-}OH$$

loss of H_2O

$$\downarrow$$

$$\text{-}N(H)\text{-}(CH_2)_6\text{-}N(H)\text{-}C(=O)\text{-}(CH_2)_8\text{-}\boxed{C(=O)\text{-}N(H)}\text{-}(CH_2)_6\text{-}N(H)\text{-}C(=O)\text{-}(CH_2)_8\text{-}C(=O)\text{-}$$

new amide bond

17.34 Draw the structure of Kodel. Since many bonds in the polymer backbone are part of a ring, the polymer is less flexible. This results in a stiffer fabric.

$$HOCH_2\text{-}C_6H_{10}\text{-}CH_2OH + HO\text{-}C(=O)\text{-}C_6H_4\text{-}C(=O)\text{-}OH$$

loss of H_2O

acid catalyst

$$\downarrow$$

$$\text{-}OCH_2\text{-}C_6H_{10}\text{-}CH_2\text{-}O\text{-}C(=O)\text{-}C_6H_4\text{-}C(=O)\text{-}O\text{-}CH_2\text{-}C_6H_{10}\text{-}CH_2\text{-}O\text{-}C(=O)\text{-}C_6H_4\text{-}C(=O)\text{-}$$

17.35 Polyesters can be converted back to their monomers by acid hydrolysis. The strong C–C bonds in polymers like polyethylene are not easily broken.

17.36

a. OH ← alcohol; enzyme; R–C(=O)–NH– (β-lactam ring with S, N, COOH); amide (β-lactam)

b. R–C(=O)–NH–; ester; O; HN; enzyme; COOH; amine

Solutions to Odd-Numbered End-of-Chapter Problems

17.37 Draw a structure that fits each description.

a. $CH_3C(CH_3)_2CH_2CH_2CH_2COOH$ — $C_8H_{16}O_2$ carboxylic acid

b. $CH_3CH_2CH_2CH_2COOCH_3$ — $C_6H_{12}O_2$ ester; OCH_3 ← methoxy group

c. cyclobutyl–$COOCH_3$ — $C_6H_{10}O_2$ ester

d. 2-methylcyclobutyl–COOH (CH_3) — $C_6H_{10}O_2$ carboxylic acid

17.39 Draw a structure with molecular formula $C_5H_{11}NO$ that fits each description.

a. $CH_3CH_2CH_2CH_2CONH_2$ — 1° amide

b. $CH_3CH_2CONH{-}CH_2CH_3$ — 2° amide

c. $CH_3CH_2CON(CH_3){-}CH_3$ — 3° amide

17.41 Draw the four esters with molecular formula $C_4H_8O_2$.

$CH_3CH_2CO{-}O{-}CH_3$ $\quad$ $CH_3CO{-}O{-}CH_2CH_3$ $\quad$ $HCOOCH_2CH_2CH_3$ $\quad$ $HCOOCH(CH_3)_2$

17.43 Name each compound using the rules in Examples 17.1 and 17.2.

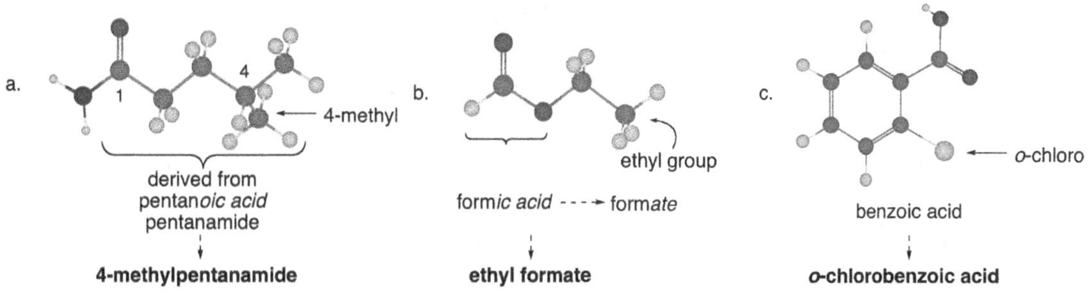

a. 4-methyl; derived from pentan*oic acid* pentanamide → **4-methylpentanamide**

b. ethyl group; form*ic acid* → form*ate* → **ethyl formate**

c. *o*-chloro; benzoic acid → ***o*-chlorobenzoic acid**

17.45 Name the carboxylic acids as in Example 17.1.

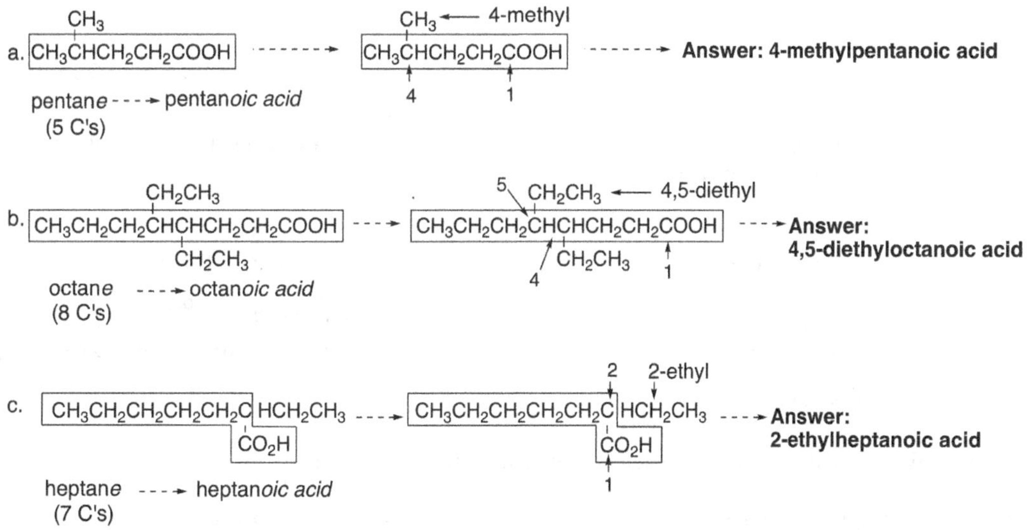

a. $CH_3CH(CH_3)CH_2CH_2COOH$ → CH_3 ← 4-methyl; C4 ... C1 → **Answer: 4-methylpentanoic acid**

pentan*e* → pentan*oic acid* (5 C's)

b. $CH_3CH_2CH_2CH(CH_2CH_3)CH(CH_2CH_3)CH_2CH_2COOH$ → 4,5-diethyl; C5, C4 ... C1 → **Answer: 4,5-diethyloctanoic acid**

octan*e* → octan*oic acid* (8 C's)

c. $CH_3CH_2CH_2CH_2CH_2CH(CO_2H)CH_2CH_3$ → C2, 2-ethyl; CO_2H = C1 → **Answer: 2-ethylheptanoic acid**

heptan*e* → heptan*oic acid* (7 C's)

17.47 Use the steps in Example 17.2 to give the IUPAC name for each ester.

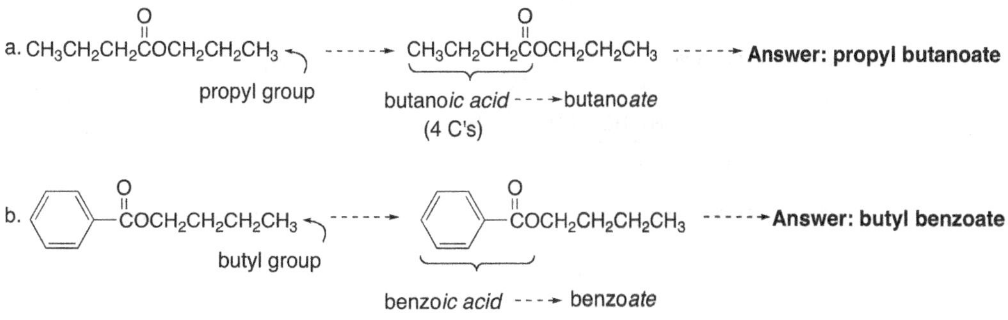

a. $CH_3CH_2CH_2COOCH_2CH_2CH_3$ (propyl group) → butano*ic acid* → butano*ate* (4 C's) → **Answer: propyl butanoate**

b. $C_6H_5COOCH_2CH_2CH_2CH_3$ (butyl group) → benzo*ic acid* → benzo*ate* → **Answer: butyl benzoate**

17.49 To give each amide an IUPAC name, use the following steps:

Step [1] Name the alkyl group (or groups) bonded to the N atom of the amide. Use the prefix "*N*-" preceding the name of each alkyl group.

Step [2] Name the acyl group (RCO–) with the suffix *-amide*.

a. $CH_3CH_2CH_2CH_2CH_2CONH_2$

derived from hexan*oic acid* -----→ hexan*amide*

Answer: hexanamide

b. $C_6H_5C(=O)N(CH_3)CH_2CH_3$ — ethyl group, methyl group

derived from benz*oic acid* -----→ benz*amide*

Answer: *N*-ethyl-*N*-methylbenzamide

17.51 Give the IUPAC name for each compound.

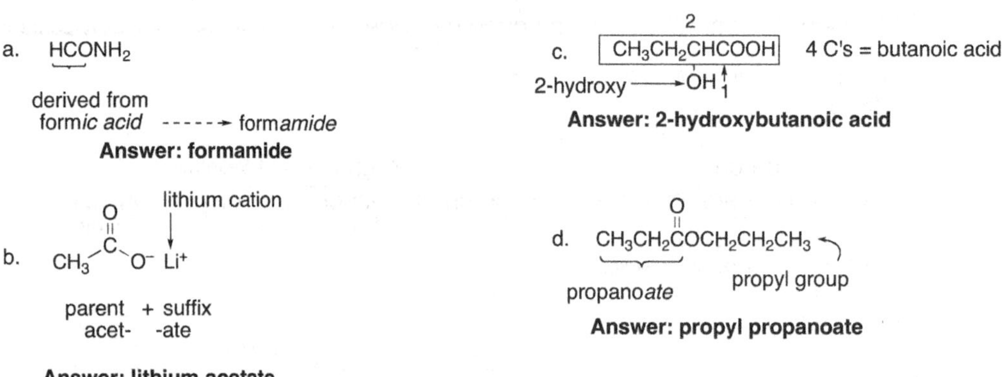

17.53 Work backwards to draw each structure from the given name.

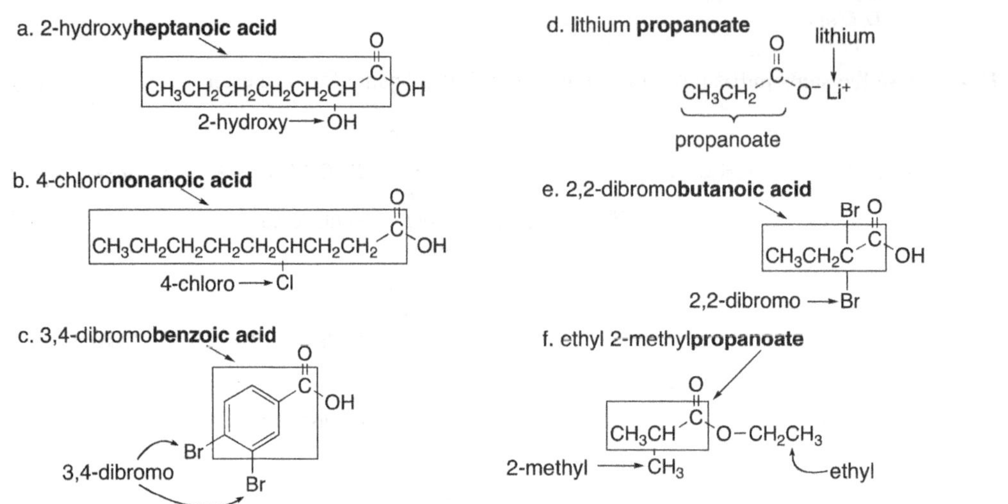

17.55 Work backwards to draw each structure from the given name.

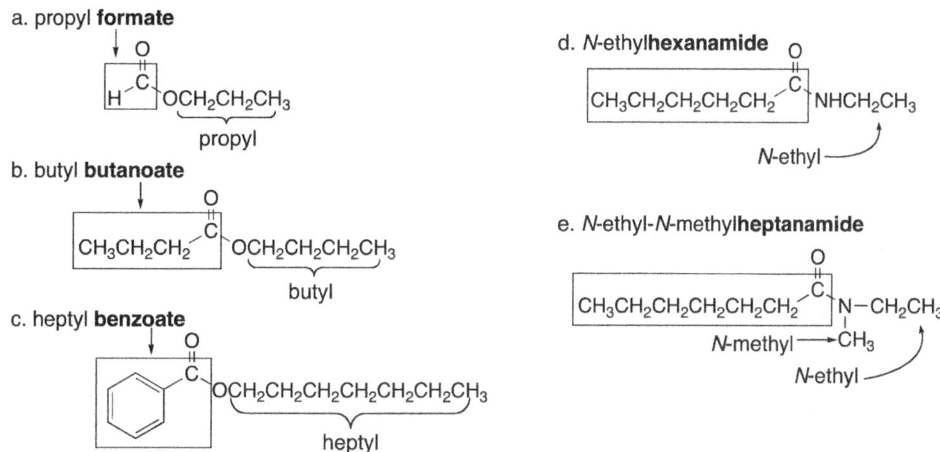

17.57 Draw the four carboxylic acids of molecular formula $C_5H_{10}O_2$ and give the IUPAC name of each one.

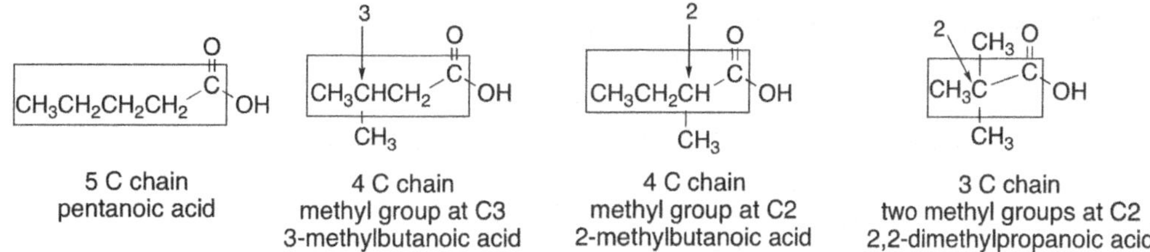

17.59 (a) HCO_2CH_3 can hydrogen bond to water because it has oxygen atoms with available lone pairs of electrons. (b) CH_3CH_2COOH can hydrogen bond to itself and to water because it has an OH group.

17.61 Use the following rules to rank the compounds in order of increasing boiling point:

- Carboxylic acids have stronger intermolecular forces than esters, giving them higher boiling points and melting points when comparing compounds of comparable size.
- Carboxylic acids have higher boiling points and melting points than alcohols of comparable size.
- Primary (1°) and 2° amides have higher boiling points and melting points than esters and 3° amides of comparable size.

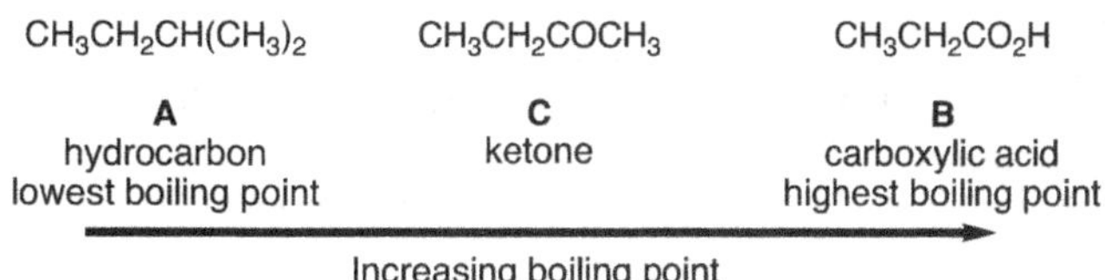

17.63 $CH_3CH_2CONH_2$ can form intermolecular hydrogen bonds, so it has stronger intermolecular forces than $CH_3CO_2CH_3$, which cannot.

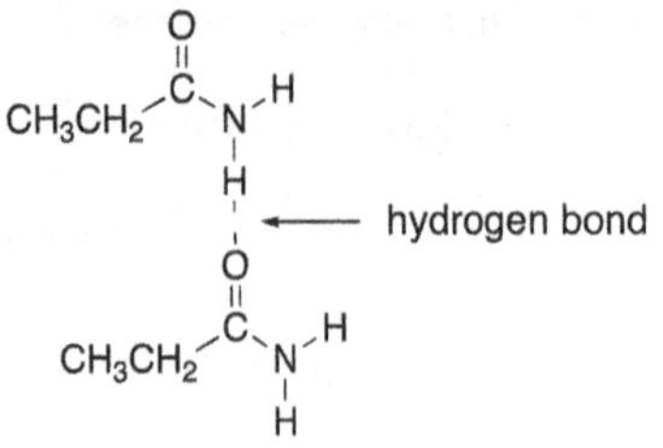

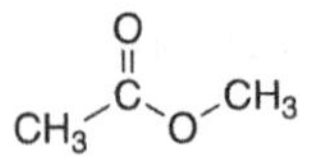

$CH_3CH_2C(=O)NH_2 \cdots$ ← hydrogen bond $\cdots CH_3CH_2C(=O)NH_2$

$CH_3C(=O)OCH_3$

no H bonded to the O atom

17.65 Draw the products of each acid–base reaction.

a. 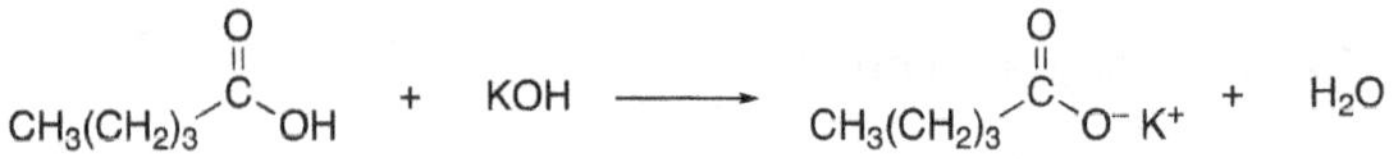 $CH_3(CH_2)_3COOH + KOH \longrightarrow CH_3(CH_2)_3COO^- K^+ + H_2O$

b. $(CH_3)_2CHCH_2CH_2COOH + Na_2CO_3 \longrightarrow (CH_3)_2CHCH_2CH_2CO_2^- Na^+ + NaHCO_3$

17.67 Draw the products of each reaction as in Example 17.5.

a. 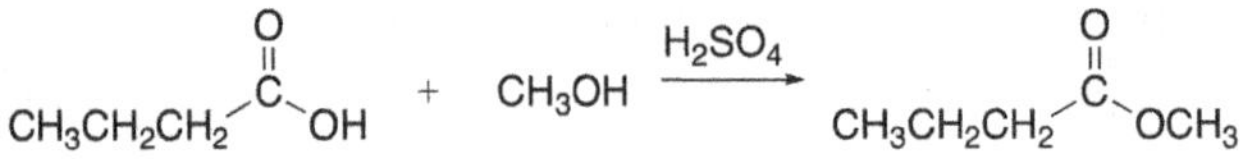 $CH_3CH_2CH_2COOH + CH_3OH \xrightarrow{H_2SO_4} CH_3CH_2CH_2COOCH_3$

b. 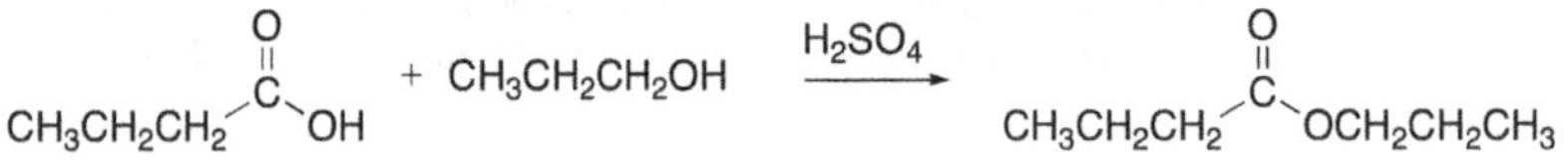 $CH_3CH_2CH_2COOH + CH_3CH_2CH_2OH \xrightarrow{H_2SO_4} CH_3CH_2CH_2COOCH_2CH_2CH_3$

c. 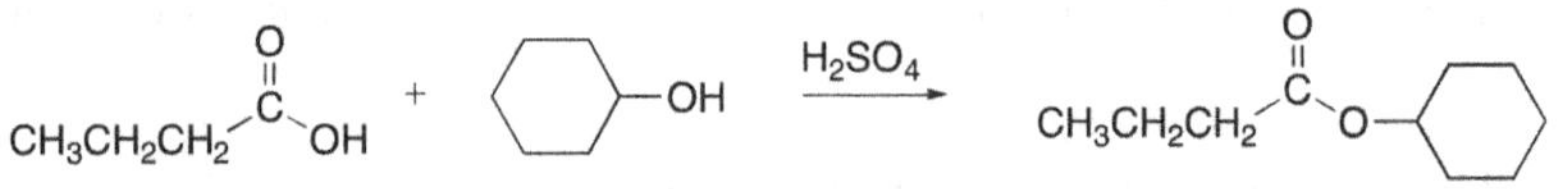 $CH_3CH_2CH_2COOH$ + cyclohexanol (C$_6$H$_{11}$–OH) $\xrightarrow{H_2SO_4} CH_3CH_2CH_2COO$–C$_6H_{11}$

d. 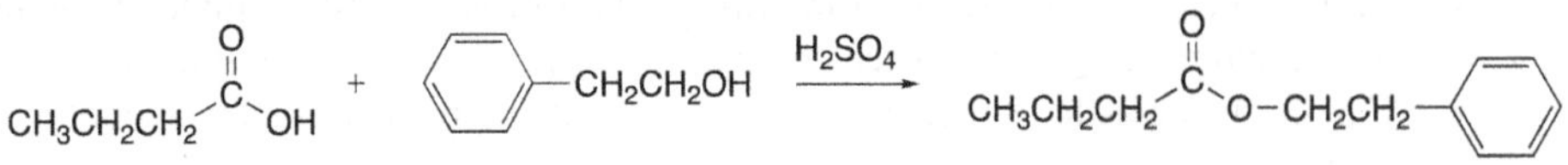 $CH_3CH_2CH_2COOH$ + C$_6$H$_5$–$CH_2CH_2OH \xrightarrow{H_2SO_4} CH_3CH_2CH_2COO$–$CH_2CH_2$–C$_6H_5$

17.69 Translate the ball-and-stick model to a line structure and work backwards to determine what carboxylic acid and alcohol are needed to synthesize methylparaben.

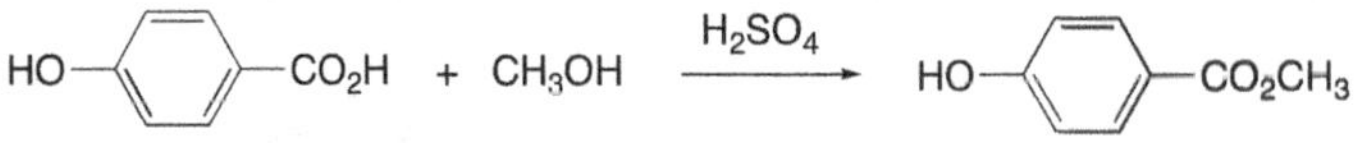

HO–C$_6$H$_4$–$CO_2H + CH_3OH \xrightarrow{H_2SO_4}$ HO–C$_6$H$_4$–CO_2CH_3

methylparaben

17.71 Draw the products of each reaction as in Example 17.6.

a. 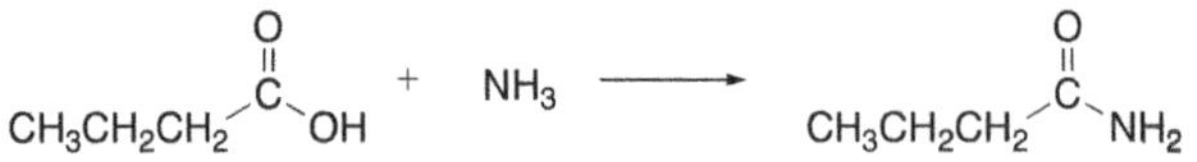 $CH_3CH_2CH_2COOH + NH_3 \longrightarrow CH_3CH_2CH_2CONH_2$

b. $CH_3CH_2CH_2C(=O)OH + CH_3CH_2NH_2 \longrightarrow CH_3CH_2CH_2C(=O)N(H)-CH_2CH_3$

c. $CH_3CH_2CH_2C(=O)OH + (CH_3CH_2)_2NH \longrightarrow CH_3CH_2CH_2C(=O)N(CH_2CH_3)-CH_2CH_3$

d. $CH_3CH_2CH_2C(=O)OH + CH_3NHCH_2CH_3 \longrightarrow CH_3CH_2CH_2C(=O)N(CH_2CH_3)-CH_3$

17.73 Work backwards to determine what carboxylic acid and alcohol are needed to synthesize each ester.

a. cyclohexyl–C(=O)OH + CH_3OH ⟶ cyclohexyl–CO_2CH_3

b. HO–C(=O)–CH_3 + cyclohexyl–OH ⟶ cyclohexyl–O–C(=O)–CH_3

17.75 Work backwards to determine what carboxylic acid and amine are needed to make acetaminophen.

HO–C_6H_4–NH_2 + CH_3–C(=O)–OH ⟶ HO–C_6H_4–N(H)–C(=O)–CH_3

This bond must be formed.

acetaminophen

17.77 Draw the products of each reaction as in Example 17.7.

a. $CH_3CH_2CH_2C(=O)OCH(CH_3)_2 \xrightarrow[H_2SO_4]{H_2O} CH_3CH_2CH_2C(=O)OH + HOCH(CH_3)_2$

b. decalinyl–O–C(=O)–CH_3 $\xrightarrow[H_2SO_4]{H_2O}$ decalinyl–OH + HO–C(=O)–CH_3

c. cyclohexyl–CH_2CH_2–O–C(=O)–H $\xrightarrow[H_2SO_4]{H_2O}$ cyclohexyl–CH_2CH_2–OH + HO–C(=O)–H

17.79 Draw the products when each amide is treated with H_2O and HCl.

a. 2-methoxybenzamide (ring with $CONH_2$ and OCH_3) $\xrightarrow[HCl]{H_2O}$ 2-methoxybenzoic acid (ring with $COOH$ and OCH_3) + NH_4^+ Cl^-

b. $(CH_3)_3CCON(CH_3)_2$ $\xrightarrow[HCl]{H_2O}$ $(CH_3)_3C\text{-}COOH$ + $[(CH_3)_2NH_2]^+$ Cl^-

c. cyclohexyl–NH–C(=O)CH$_3$ $\xrightarrow[HCl]{H_2O}$ [cyclohexyl–NH_3]$^+$ Cl^- + HO–C(=O)CH$_3$

17.81 Draw the products of ester hydrolysis.

$C_6H_5\text{-}CH_2\text{-}C(=O)OCH_2CH_3$ $\xrightarrow[H_2SO_4]{H_2O}$ $C_6H_5\text{-}CH_2\text{-}C(=O)OH$ + CH_3CH_2OH

17.83 Work backwards to determine what two monomers are needed to prepare the polyamide.

HO–C(=O)–C$_6$H$_4$–C(=O)–OH + $NH_2-(CH_2)_4-NH_2$

loss of H_2O

–C(=O)–C$_6$H$_4$–C(=O)–N(H)–$(CH_2)_4$–N(H)–C(=O)–C$_6$H$_4$–C(=O)–N(H)–$(CH_2)_4$–N(H)–

new amide bond

17.85 Draw the structure of the polyester PTT.

HOOC–C$_6$H$_4$–COOH + $HOCH_2CH_2CH_2OH$

H_2SO_4

–O–C(=O)–C$_6$H$_4$–C(=O)–$OCH_2CH_2CH_2O$–C(=O)–C$_6$H$_4$–C(=O)–O–$CH_2CH_2CH_2$–

PTT

17.87 Saponification is the hydrolysis of an ester with strong base, which forms a metal salt of a carboxylate anion, $RCOO^-M^+$. Esterification forms a new ester, RCOOR', from a carboxylic acid and an alcohol.

17.89 Draw the products of each reaction.

a. $CH_3COOH + CH_3OH \xrightleftharpoons{H_2SO_4} CH_3COOCH_3 + H_2O$

b. $(CH_3)_2CHOCOCH_3 + H_2O \xrightleftharpoons{H_2SO_4} (CH_3)_2CHOH + HOCOCH_3$

c. $(CH_3)_2CHCH_2OCOCH(CH_3)_2 + H_2O \xrightarrow{NaOH} Na^{+}\,{}^{-}OCOCH(CH_3)_2 + (CH_3)_2CHCH_2OH$

d. $CH_3(CH_2)_4COOH + NaOH \longrightarrow CH_3(CH_2)_4COO^-\,Na^+ + H_2O$

17.91 Answer each question.

a. $CH_3CH(CH_3)CH_2CH_2CH_2COOH$ → **5-methylhexanoic acid**

5-methyl → CH_3 (on C5); 6 C's hexanoic acid (C1 = carbonyl carbon)

b. $CH_3CH_2CH(CH_3)CH_2CH_2COOH$

isomer

c. $CH_3CH_2CH_2CH_2CH_2COOCH_3$

isomer

d. $CH_3CH(CH_3)CH_2CH_2CH_2COOH \xrightarrow{NaOH} CH_3CH(CH_3)CH_2CH_2CH_2COO^-\,Na^+ + H_2O$

e. **A** is insoluble in H_2O, but soluble in an organic solvent.

f. $CH_3CH(CH_3)CH_2CH_2CH_2COOH \xrightarrow[H_2SO_4]{CH_3CH_2OH} CH_3CH(CH_3)CH_2CH_2CH_2COOCH_2CH_3 + H_2O$

g. 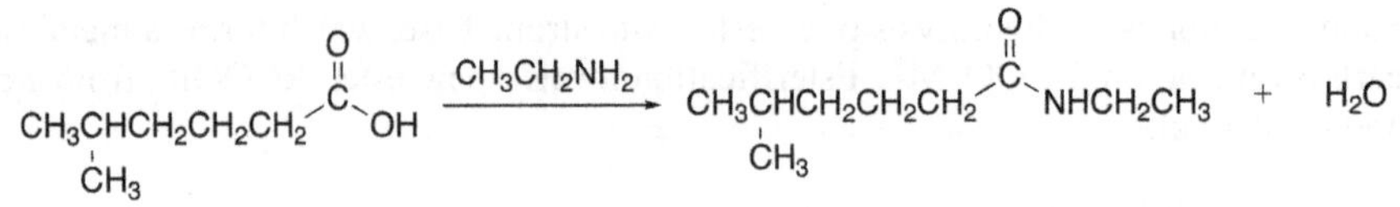 $CH_3CHCH_2CH_2CH_2C(=O)OH$ (with CH_3 on the second carbon) $\xrightarrow{CH_3CH_2NH_2}$ $CH_3CHCH_2CH_2CH_2C(=O)NHCH_2CH_3$ (with CH_3 on the second carbon) + H_2O

17.93 A soap contains both a long chain hydrocarbon and a carboxylic acid salt.

a. $CH_3CO_2^-\ Na^+$ — short chain carboxylic acid

b. $CH_3(CH_2)_{14}CO_2^-\ Na^+$ — sodium cation (Na^+); long chain carboxylate anion. This is a soap because it contains both a long chain and a carboxylic acid salt.

c. $CH_3(CH_2)_{12}COOH$ — no salt

17.95 Answer each question about naproxen.

a. 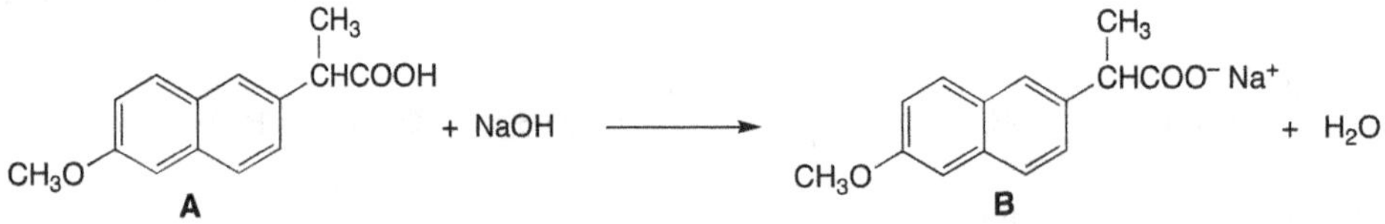 A (naphthalene ring bearing CH_3O and $CHCOOH$ with CH_3) + NaOH ⟶ B (naphthalene ring bearing CH_3O and $CHCOO^-\ Na^+$ with CH_3) + H_2O

b. In the stomach, naproxen exists as the neutral carboxylic acid (**A**).
c. In the intestines, naproxen exists as the ionized carboxylate anion (**B**).

17.97 Aspirin acts as an anti-inflammatory agent by inhibiting the production of prostaglandins, which are part of the inflammatory cascade of reactions.

17.99 Soap is able to dissolve nonpolar hydrocarbons in water because the nonpolar end of the soap binds to dirt and hydrocarbons, whereas the polar end then hydrogen bonds with water, thereby rendering the dirt soluble in water.

Structure of a soap

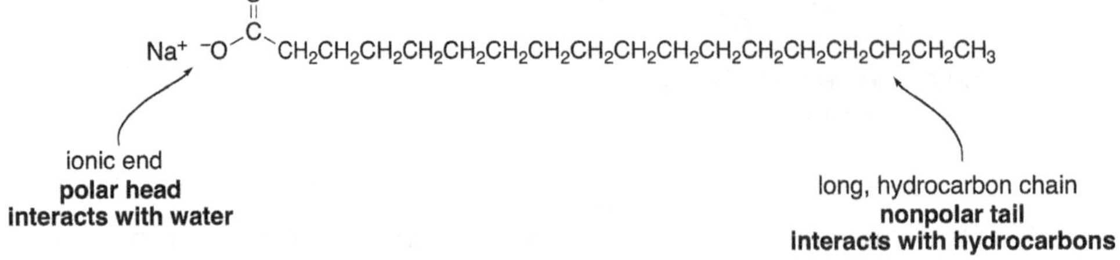

$Na^+\ {}^-O-C(=O)-CH_2CH_2CH_2CH_2CH_2CH_2CH_2CH_2CH_2CH_2CH_2CH_2CH_2CH_2CH_2CH_2CH_3$

ionic end
polar head
interacts with water

long, hydrocarbon chain
nonpolar tail
interacts with hydrocarbons

17.101 Draw the products of the hydrolysis of aspartame by breaking the amide and ester bonds.

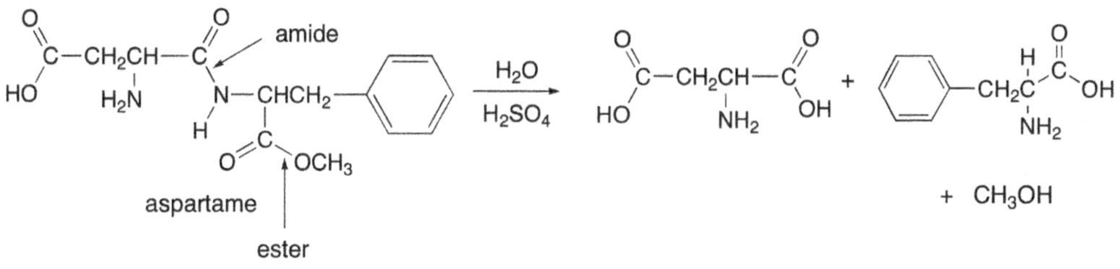

aspartame

amide

ester

$\xrightarrow[H_2SO_4]{H_2O}$

+ CH_3OH

17.103 Draw the products of an intramolecular reaction. A molecule of water is also formed in each reaction.

a.

b.

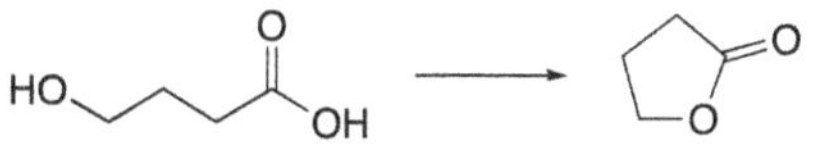

Chapter 18 Amines and Neurotransmitters

Chapter Review

[1] What are the characteristics of amines? (18.1, 18.3)

- Amines are organic nitrogen compounds, formed by replacing one or more hydrogen atoms of NH_3 by alkyl groups.
- Primary (1°) amines have one C–N bond; 2° amines have two C–N bonds; 3° amines have three C–N bonds.

R–N̈(H)–H	R–N̈(R)–H	R–N̈(R)–R
1° amine	2° amine	3° amine

- An amine has a lone pair of electrons on the N atom, and is trigonal pyramidal in shape.
- Amines contain polar C–N and N–H bonds. Primary (1°) and 2° amines can hydrogen bond. Intermolecular hydrogen bonds between N and H are weaker than those between O and H.
- Tertiary (3°) amines have lower boiling points than 1° and 2° amines of comparable size, because 3° amines cannot hydrogen bond.

[2] How are amines named? (18.2)

- Primary (1°) amines are identified by the suffix *-amine*.

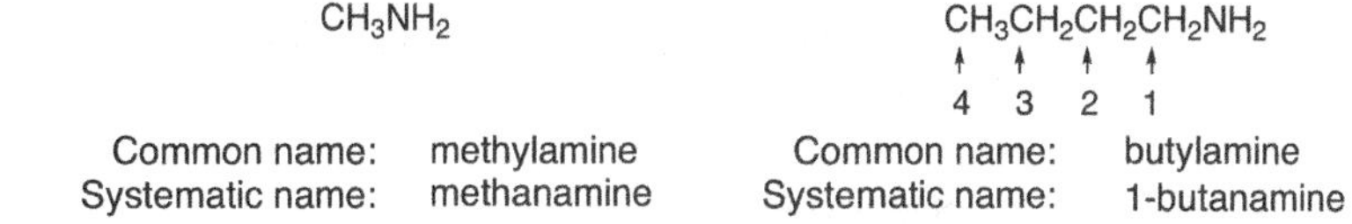

- Secondary (2°) and 3° amines with identical alkyl groups are named by adding the prefix *di-* or *tri-* to the name of the 1° amine.
- Secondary (2°) and 3° amines with different alkyl groups are named as *N*-substituted 1° amines.

CH_3CH_2–N(CH_2CH_3)–CH_2CH_3	$CH_3CH_2CH_2$–N(H)–$CH_2CH_2CH_3$	CH_3CHCH_3–N(H)–CH_3
triethylamine	dipropylamine	*N*-methyl-2-propanamine

[3] What are alkaloids? Give examples of common alkaloids. (18.4, 18.5)

- Alkaloids are naturally occurring amines isolated from plant sources.
- Examples of alkaloids include caffeine (from coffee and tea), nicotine (from tobacco), morphine and codeine (from the opium poppy), quinine (from the Cinchona tree), and atropine (from the deadly nightshade plant).

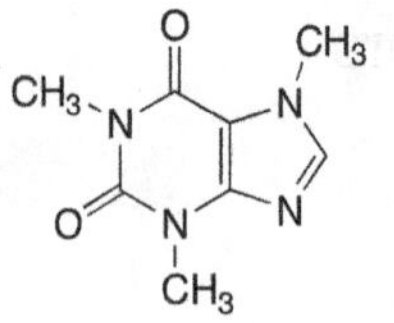

caffeine

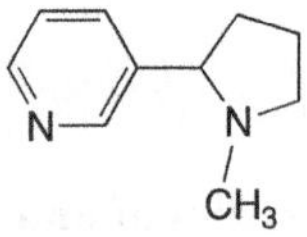

nicotine

[4] What products are formed when an amine is treated with acid? (18.6)

- Amines act as proton acceptors in water and acid. For example, the reaction of RNH_2 with HCl forms the water-soluble ammonium salt RNH_3^+ Cl^-.

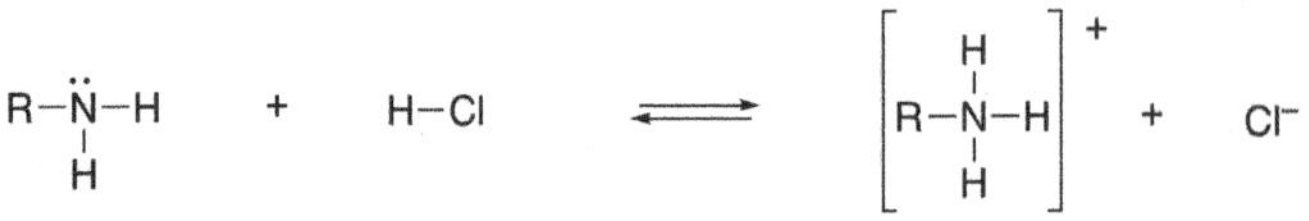

$$R{-}\ddot{N}H_2 + H{-}Cl \rightleftharpoons [R{-}NH_3]^+ + Cl^-$$

[5] What are the characteristics of ammonium salts and how are they named? (18.6, 18.7)

- An ammonium salt consists of a positively charged ammonium ion and an anion.
- An ammonium salt is named by changing the suffix *-amine* of the parent amine to the suffix *-ammonium,* followed by the name of the anion.

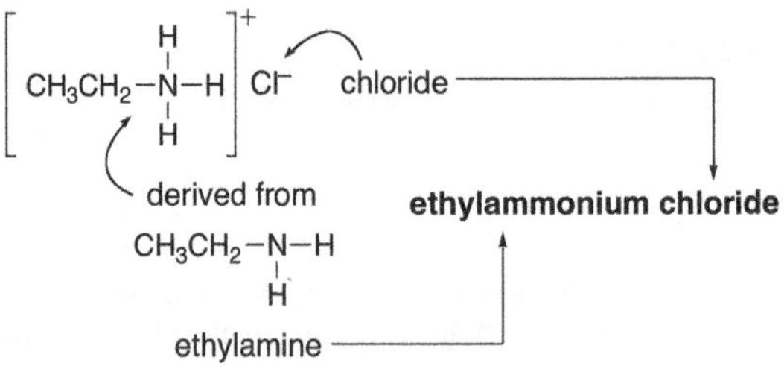

$[CH_3CH_2{-}NH_3]^+$ Cl^- chloride → **ethylammonium chloride**

derived from $CH_3CH_2{-}NH_2$ ethylamine

- Ammonium salts are water-soluble solids.
- Water-insoluble amine drugs are sold as their ammonium salts to increase their solubility in the aqueous environment of the blood.

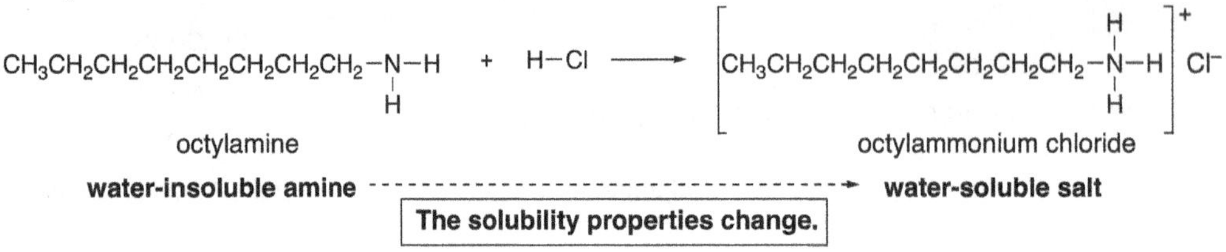

$$CH_3CH_2CH_2CH_2CH_2CH_2CH_2CH_2{-}NH_2 + H{-}Cl \longrightarrow [CH_3CH_2CH_2CH_2CH_2CH_2CH_2CH_2{-}NH_3]^+\ Cl^-$$

octylamine — octylammonium chloride

water-insoluble amine → **water-soluble salt**

The solubility properties change.

[6] What are neurotransmitters and how do they differ from hormones? (18.8, 18.9)

- A neurotransmitter is a chemical messenger that transmits a nerve impulse from a neuron to another cell.
- A hormone is a compound produced by an endocrine gland that travels through the bloodstream to a target tissue or organ.

[7] What roles do dopamine and serotonin play in the body? (18.8)

- Dopamine affects movement, emotions, and pleasure. Too little dopamine causes Parkinson's disease. Too much dopamine causes schizophrenia. Dopamine plays a role in addiction.

- Serotonin is important in mood, sleep, perception, and temperature regulation. A deficiency of serotonin causes depression. SSRIs are antidepressants that effectively increase the concentration of serotonin.

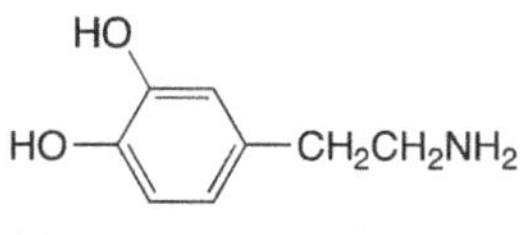

dopamine

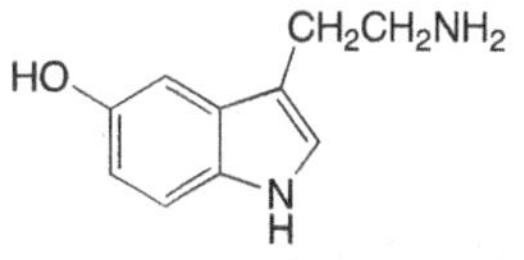

serotonin

[8] Give examples of important derivatives of 2-phenylethylamine. (18.9)

- Derivatives of 2-phenylethylamine contain a benzene ring bonded to a two-carbon chain that is bonded to a nitrogen atom.

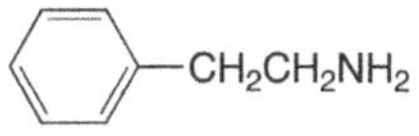

2-phenylethylamine

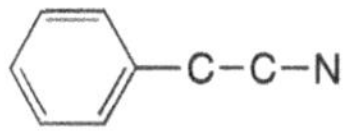

common structural feature

- Examples of 2-phenylethylamine derivatives include epinephrine, norepinephrine, amphetamine, and methamphetamine. Albuterol and salmeterol are also derivatives of 2-phenylethylamine that are used to treat asthma.

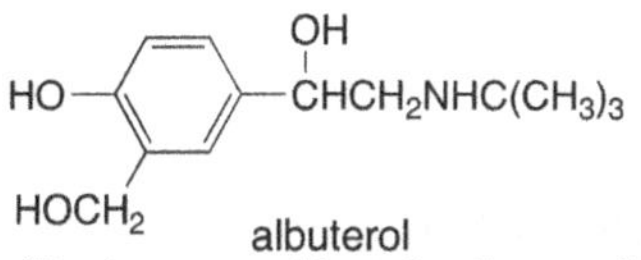

albuterol
(Trade names: Ventolin, Proventil)

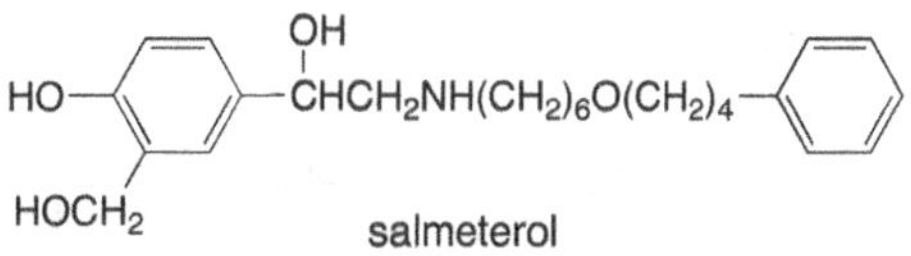

salmeterol
(Trade name: Serevent)

[9] What is histamine, and how do antihistamines and anti-ulcer drugs work? (18.10)

- Histamine is an amine with a wide range of physiological effects. Histamine dilates capillaries, is responsible for the runny nose and watery eyes of allergies, and stimulates the secretion of stomach acid.

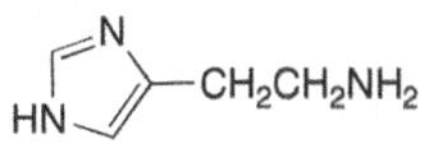

histamine

- Antihistamines bind to the H1 histamine receptor and inhibit vasodilation, so they are used to treat the symptoms of colds and allergies.
- Anti-ulcer drugs bind to the H2 histamine receptor and reduce the production of stomach acid.

Problem Solving

[1] Structure and Bonding (18.1)

Example 18.1 Classify each amine in the following compounds as 1°, 2°, or 3°.

a. $CH_3C(CH_3)(CH_2CH_3)NHCH_3$ b. $CH_3CH_2CH(CH_3)CH_2NH_2$

Analysis
To determine whether an amine is 1°, 2°, or 3°, count the number of carbons bonded to the nitrogen atom. A 1° amine has one C–N bond, and so forth.

Solution
Draw out the structure or add H's to the skeletal structure to clearly see how many C–N bonds the amine contains.

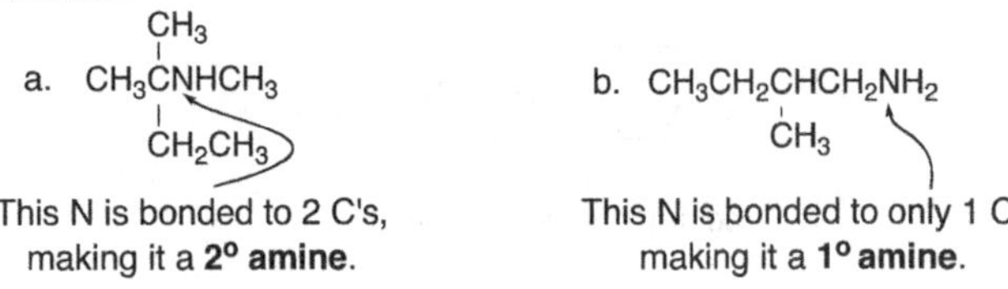

[2] Nomenclature (18.2)

Example 18.2 Give a systematic name for each amine.

a. cyclohexyl–N($CH_2CH_2CH_3$)(CH_2CH_3)   b. $CH_3CH_2C(Cl)(CH_2CH_3)CH_2CH_2CH_2NH_2$

Analysis and Solution

a. For a 3° amine, one alkyl group on N is the principal R group and the others are substituents.

[1] Name the ring bonded to the N:

6 C's in the ring → cyclohexanamine

[2] Name the substituents:

a propyl and ethyl group on N

- 2 N's are needed, one for each alkyl group.
- Alphabetize the *e* of **ethyl** before the *p* of **propyl**.

Answer: *N*-ethyl-*N*-propylcyclohexanamine

b. [1] For a 1° amine, find and name the longest chain containing the amine nitrogen:

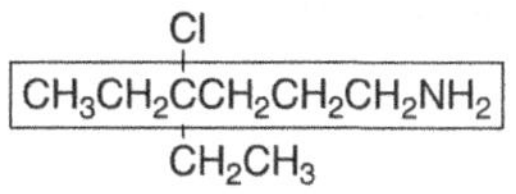

$$CH_3CH_2C(Cl)(CH_2CH_3)CH_2CH_2CH_2NH_2$$

hexane ---→ hexan*amine*
(6 C's)

[2] Number the carbon skeleton:

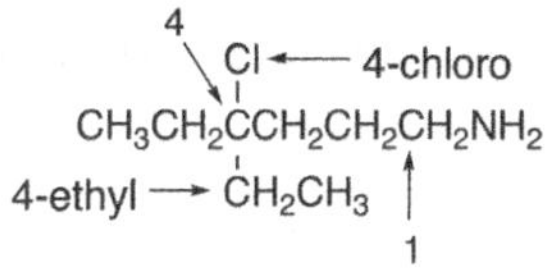

4 — Cl ← 4-chloro
$CH_3CH_2CCH_2CH_2CH_2NH_2$
4-ethyl → CH_2CH_3
1

You must use a number to show the location of the NH_2 group.

Answer: 4-chloro-4-ethyl-1-hexanamine

---

[3] Physical Properties (18.3)

Example 18.3 Which compound in each pair has the higher boiling point?
a. $CH_3CH_2N(CH_3)_2$ or $CH_3CH_2CH_2CH_2NH_2$
b. $CH_3OCH_2CH_2CH_3$ or $CH_3CH_2CH_2CH_2NH_2$

Analysis
Keep in mind the general rule: For compounds of comparable size, **the stronger the intermolecular forces, the higher the boiling point.** Compounds that can hydrogen bond have higher boiling points than compounds that are polar but cannot hydrogen bond. Polar compounds have higher boiling points than nonpolar compounds.

Solution
a. The 1° amine ($CH_3CH_2CH_2CH_2NH_2$) has N–H bonds, so intermolecular hydrogen bonding is possible. The 3° amine [$CH_3CH_2N(CH_3)_2$] has only C–H bonds, so there is no possibility of intermolecular hydrogen bonding. $CH_3CH_2CH_2CH_2NH_2$ has a higher boiling point because it has stronger intermolecular forces.

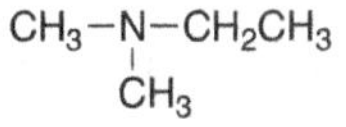

$CH_3-N(CH_3)-CH_2CH_3$
a 3° amine with only C–H bonds

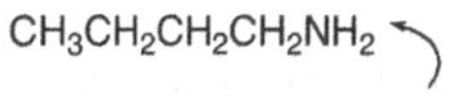

$CH_3CH_2CH_2CH_2NH_2$
a 1° amine with N–H bonds
intermolecular hydrogen bonding
higher boiling point

b. The 1° amine ($CH_3CH_2CH_2CH_2NH_2$) has N–H bonds, so intermolecular hydrogen bonding is possible. The ether ($CH_3OCH_2CH_2CH_3$) has only C–H bonds, so there is no possibility of intermolecular hydrogen bonding. $CH_3CH_2CH_2CH_2NH_2$ has a higher boiling point because it has stronger intermolecular forces.

$CH_3-O-CH_2CH_2CH_3$
an ether with only C–H bonds

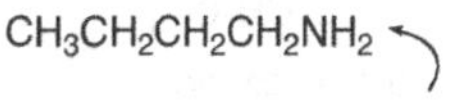

$CH_3CH_2CH_2CH_2NH_2$
a 1° amine with two N–H bonds
intermolecular hydrogen bonding
higher boiling point

---

[4] Amines as Bases (18.6)

Example 18.4 What products are formed when $CH_3CH_2NHCH_3$ reacts with HCl?

Analysis

In any acid–base reaction with an amine:

- Locate the N atom of the amine and add a proton to it. Since the amine nitrogen is neutral to begin with, adding a proton gives it a (+1) charge.
- Remove a proton from the acid (HCl) and form its conjugate base (Cl^-).

Solution

Transfer a proton from the acid to the base. Use the lone pair on the N atom to form the new bond to the proton of the acid.

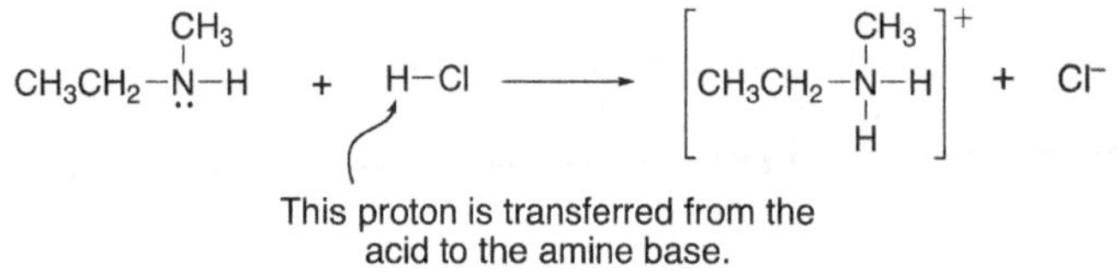

$$CH_3CH_2{-}\ddot{N}(CH_3){-}H + H{-}Cl \longrightarrow \left[CH_3CH_2{-}N(CH_3)(H){-}H\right]^+ + Cl^-$$

This proton is transferred from the acid to the amine base.

Thus, HCl loses a proton to form Cl^-, and the N atom of the amine gains a proton to form an ammonium cation.

Example 18.5 Name each ammonium salt.

a. $[(CH_3CH_2)_3NH]^+\ CH_3COO^-$ b. $(CH_3NH_3)^+\ Br^-$

Analysis

To name an ammonium salt, draw out the four groups bonded to the N atom. Remove one hydrogen from the N atom to draw the structure of the parent amine. Then put the following two parts of the name together:

- Name the ammonium ion by changing the suffix *-amine* of the parent amine to the suffix *-ammonium*.
- Add the name of the anion.

Solution

a. 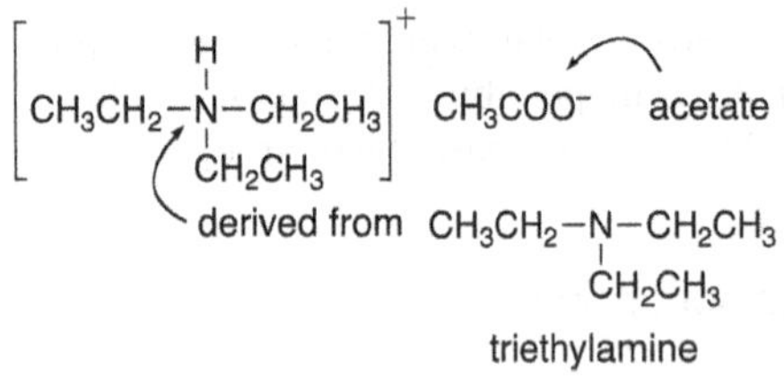 $\left[CH_3CH_2{-}N(H)(CH_2CH_3){-}CH_2CH_3\right]^+\ CH_3COO^-$ acetate

derived from $CH_3CH_2{-}N(CH_2CH_3){-}CH_2CH_3$

triethylamine

- Change the name triethyl*amine* to triethyl*ammonium*.
- Add the name of the anion, acetate.
- Answer: **triethylammonium acetate**

b. 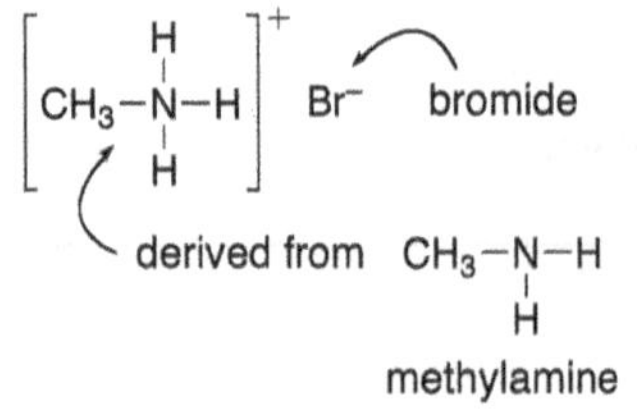 $\left[CH_3{-}N(H)(H){-}H\right]^+\ Br^-$ bromide

derived from $CH_3{-}N(H){-}H$

methylamine

- Change the name methyl*amine* to methyl*ammonium*.
- Add the name of the anion, bromide.
- Answer: **methylammonium bromide**

Self-Test

[1] Fill in the blank with one of the terms listed below.

Amines (18.1)
Ammonium salts (18.6)
Antihistamines (18.10)
Anti-ulcer drugs (18.10)
Axon (18.8)
Heterocycle (18.1)
Hormone (18.9)
Neuron (18.8)
Neurotransmitter (18.8)
Postsynaptic neuron (18.8)
Presynaptic neuron (18.8)
Quaternary ammonium salts (18.1)
SSRIs (18.8)
Synapse (18.8)

1. ______ are water-soluble solids.
2. ______ act by inhibiting the reuptake of serotonin by the presynaptic neuron, effectively increasing the concentration of serotonin.
3. A ______ is a compound produced by an endocrine gland, which then travels through the bloodstream to a target tissue or organ.
4. ______ are organic nitrogen compounds that are considered to be derivatives of NH_3.
5. A ______ consists of many short filaments called dendrites connected to a cell body.
6. A ______ is a cell that releases a neurotransmitter.
7. ______ are ammonium salts with nitrogen atoms bonded to *four* alkyl groups.
8. A ______ is a chemical messenger that transmits nerve impulses from one neuron to another.
9. A ______ is a cell that contains the receptors that bind a neurotransmitter.
10. Dendrites are separated from each other by a small gap called a ______.
11. A ______ is a ring that contains a heteroatom such as N, O, or S.
12. ______ bind to the H2 histamine receptor, reducing acid secretion in the stomach.
13. ______ bind to the H1 histamine receptor and inhibit vasodilation. They are used to treat the symptoms of the common cold and environmental allergies.
14. A long stem called an ______ protrudes from the cell body of a neuron.

[2] Label each compound as a 1°, 2°, or 3° amine.

15. CH_3NH_2
16. 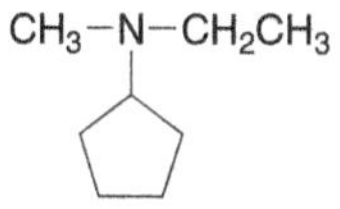 $CH_3-N-CH_2CH_3$ (N bonded to cyclopentyl)
17. 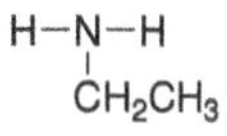 $H-N-H$ (N bonded to CH_2CH_3)
18. 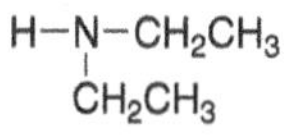 $H-N-CH_2CH_3$ (N bonded to CH_2CH_3)

[3] Pick the compound in each pair with the higher boiling point.

19. $CH_3CH_2NH_2$ (**A**) or CH_3CH_2OH (**B**)

20. 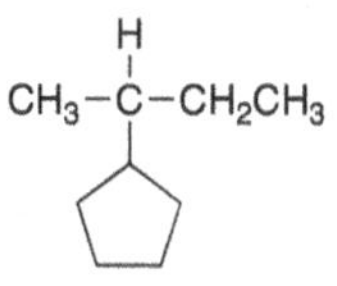 $CH_3-\overset{H}{C}-CH_2CH_3$ (C bonded to cyclopentyl) (**A**) or $CH_3-N-CH_2CH_3$ (N bonded to cyclopentyl) (**B**)

21. CH_3OCH_3 (**A**) or $H-N-CH_3$ (N bonded to CH_3) (**B**)

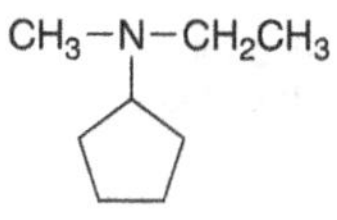

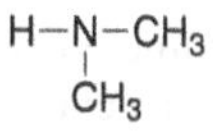

[4] Match the reactants and products.

22. 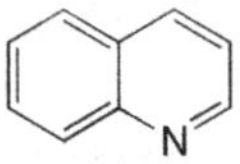 quinoline + HCl ⟶

23. $CH_3CH_2CHCH_2CH_3$ (with $NHCH_3$ on C3) + H_2O ⟶

24. 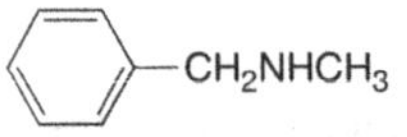 C_6H_5–CH_2NHCH_3 + HCl ⟶

25. 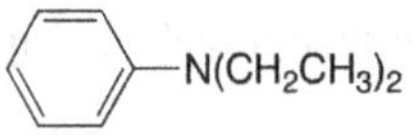 C_6H_5–$N(CH_2CH_3)_2$ + HCl ⟶

a. 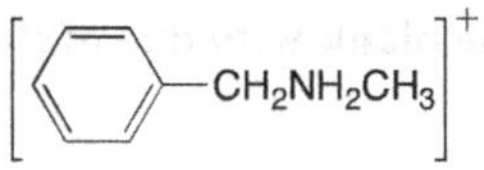 $[C_6H_5\text{–}CH_2NH_2CH_3]^+ + Cl^-$

b. 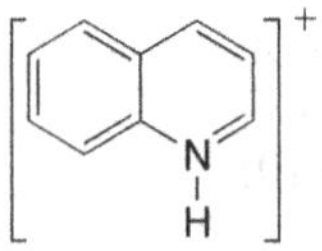 [quinoline with N–H]$^+$ + Cl^-

c. $[CH_3CH_2CHCH_2CH_3$ (with NH_2CH_3 on C3)$]^+ + OH^-$

d. 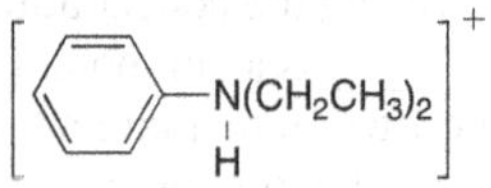 $[C_6H_5\text{–}N(CH_2CH_3)_2$ with H on N$]^+ + Cl^-$

Answers to Self-Test

1. Ammonium salts
2. SSRIs
3. hormone
4. Amines
5. neuron
6. presynaptic neuron
7. Quaternary ammonium salts
8. neurotransmitter
9. postsynaptic neuron
10. synapse
11. heterocycle
12. Anti-ulcer drugs
13. Antihistamines
14. axon
15. 1°
16. 3°
17. 1°
18. 2°
19. **B**
20. **B**
21. **B**
22. b
23. c
24. a
25. d

Solutions to In-Chapter Problems

18.1 To determine whether an amine is 1°, 2°, or 3°, count the number of carbons bonded to the nitrogen atom as in Example 18.1. A 1° amine has one C–N bond, and so forth.

a. 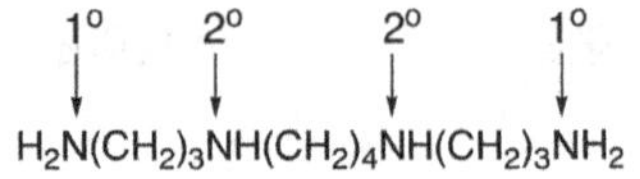 $H_2N(CH_2)_3NH(CH_2)_4NH(CH_2)_3NH_2$ — labels from left to right: 1°, 2°, 2°, 1°

b. 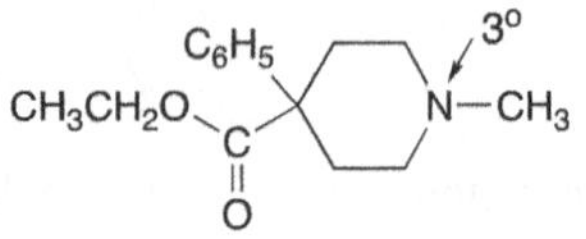 CH_3CH_2O–C(=O)– on a piperidine ring carbon bearing C_6H_5; ring N–CH_3 labeled 3°

18.2 To determine whether an amine is 1°, 2°, or 3°, count the number of carbons bonded to the nitrogen. To classify an alcohol as 1°, 2°, or 3°, count the number of carbons bonded to the carbon with the OH group.

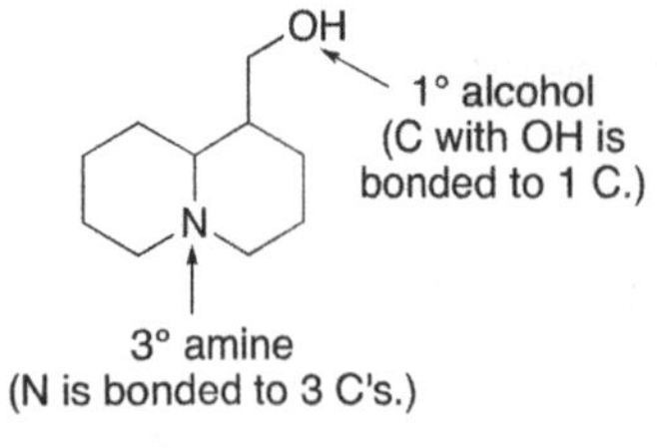

OH: 1° alcohol (C with OH is bonded to 1 C.)

N: 3° amine (N is bonded to 3 C's.)

18.3 Label the amine and hydroxyl group in scopolamine, a drug used to treat motion sickness, as 1°, 2°, or 3° as in Example 18.1.

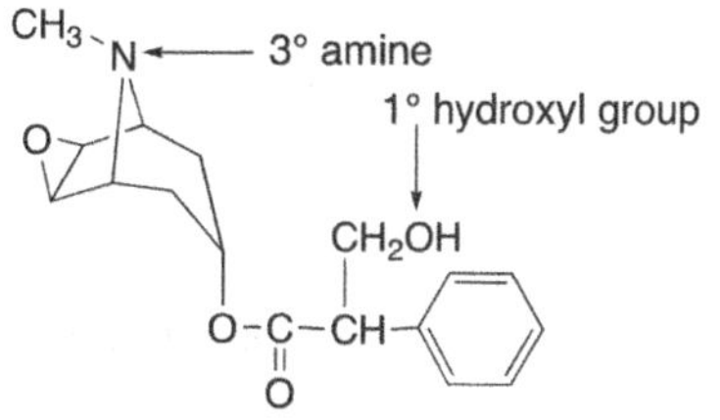

CH_3–N ← 3° amine

1° hydroxyl group → CH_2OH

O–C(=O)–CH–(benzene ring)

scopolamine

18.4 Methamphetamine is a 2° amine. Give the molecular shape around each atom by counting groups.

a. 2° amine

b. (benzene ring)–$CH_2CH(CH_3)NHCH_3$

(1) 3 groups trigonal planar

(2) 4 groups tetrahedral

(3) 3 atoms 1 lone pair trigonal pyramidal

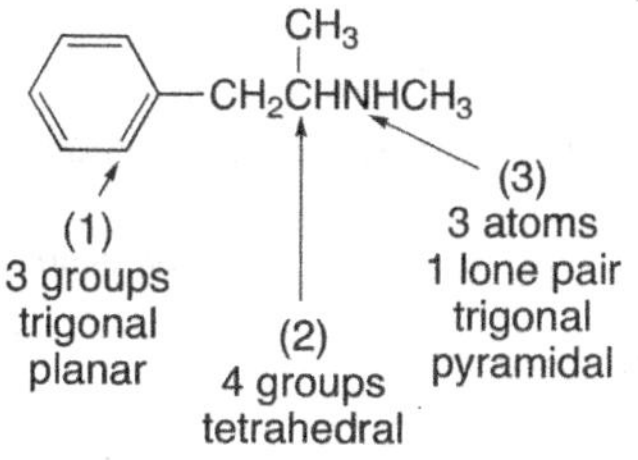

18.5 Name each amine as in Example 18.2.

a. $CH_3CH_2CH(NH_2)CH_3$ - - - → $CH_3CH_2CH(NH_2)CH_3$ (C2 and C1 numbered) - - - → **Answer: 2-butanamine**

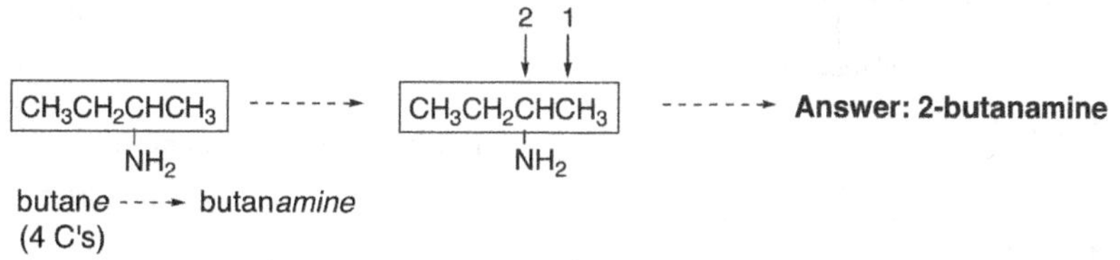

butane - - → butan*amine*
(4 C's)

b. $CH_3CH_2CH_2NHCH_3$ - - - → $CH_3CH_2CH_2NHCH_3$ (C1 numbered) - - - → **Answer: *N*-methyl-1-propanamine**

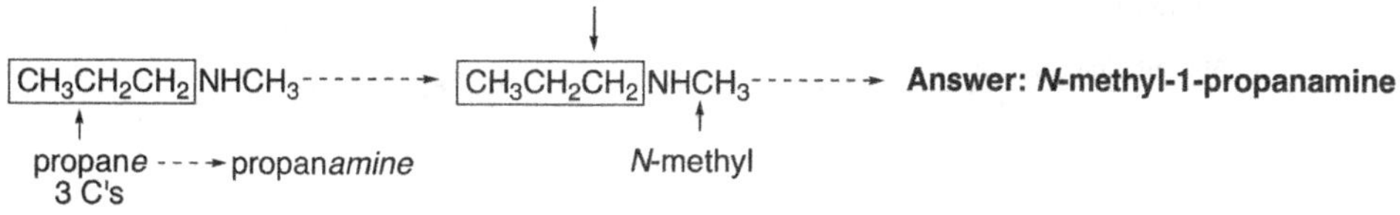

propane - - → propan*amine*
3 C's

N-methyl

c. (cyclohexane ring)–$N(CH_3)_2$ - - - → (cyclohexane ring)–$N(CH_3)_2$ - - - → **Answer: *N,N*-dimethylcyclohexanamine**

6 C's in the ring
cyclohexan*amine*

N,N-dimethyl

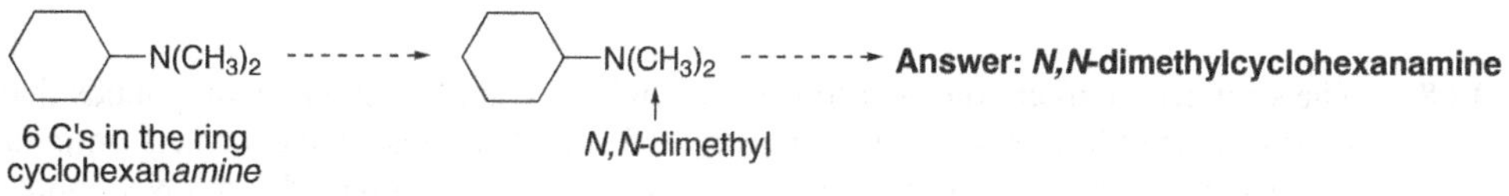

d. $(CH_3CH_2CH_2CH_2)_2NH$ - - - → **Answer: dibutylamine**

2 butyl groups
(4 C's)

18.6 Work backwards to draw the structure corresponding to each name.

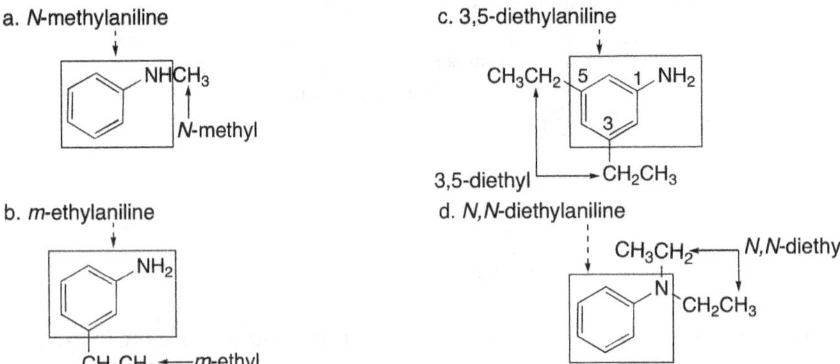

18.7 Work backwards to draw the structure corresponding to each name.

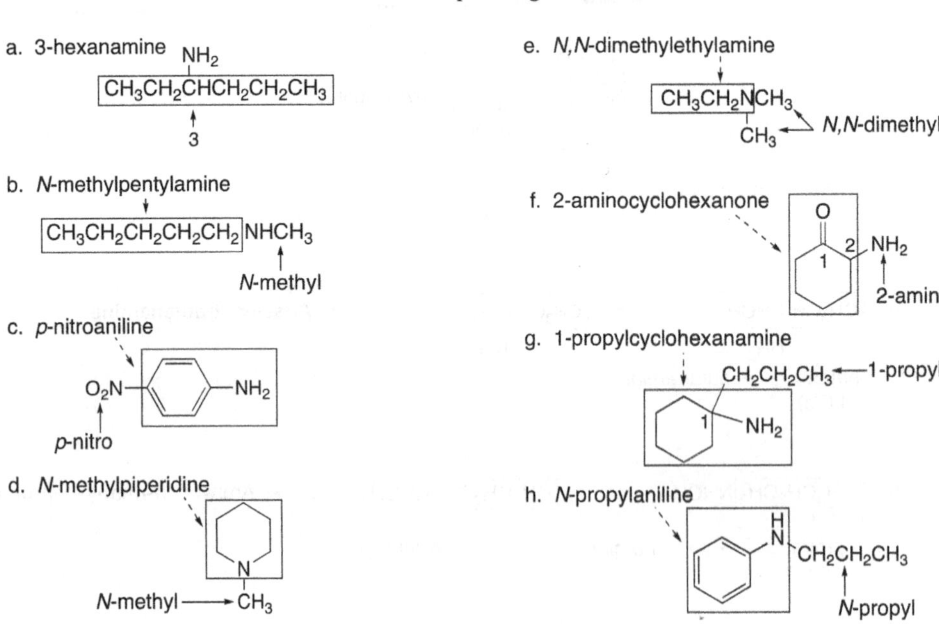

18.8 The stronger the intermolecular forces, the higher the boiling point. Compounds that can hydrogen bond have stronger intermolecular forces than those that cannot. Compounds with OH bonds have stronger hydrogen bonding interactions than compounds with NH bonds.

a. $CH_3COCH_2CH_3$ (drawn as $CH_3\overset{O}{\overset{\|}{C}}CH_2CH_3$) or $(CH_3)_2CHCH_2NH_2$
hydrogen bonding possible
higher boiling point

b. $(CH_3)_2CHCH_2NH_2$ or $(CH_3)_2CHCH_2OH$
OH bond for stronger hydrogen bonds
higher boiling point

18.9 Caffeine is soluble in the organic solvent CH_2Cl_2 because caffeine is organic and "like dissolves like."

18.10 Nicotine has two heterocycles: pyridine and pyrrolidine.

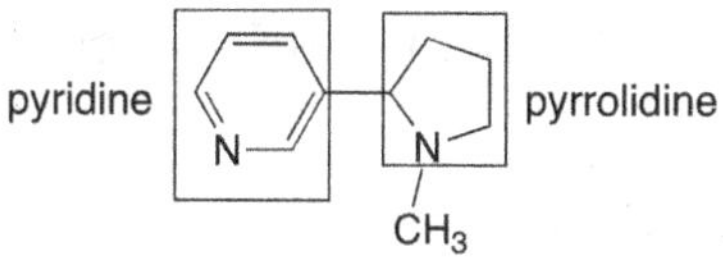

pyridine | pyrrolidine | N | N | CH_3

18.11 Morphine contains a 3° amine.

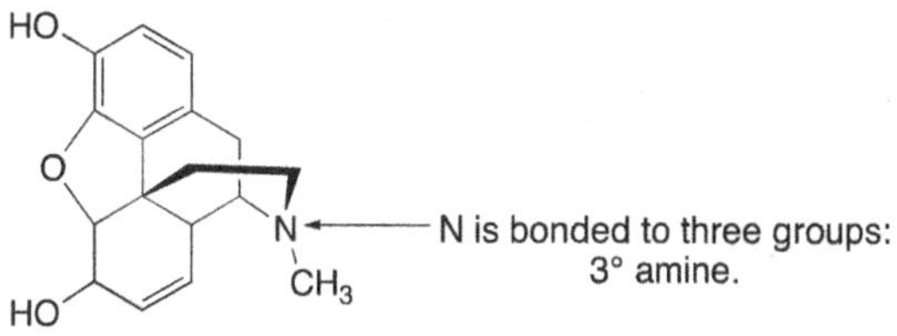

HO | O | HO | N | CH_3 | N is bonded to three groups: 3° amine.

18.12 Identify the functional groups in heroin.

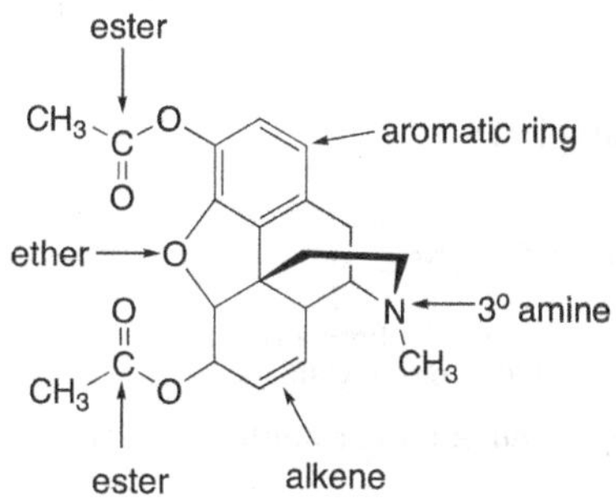

ester | aromatic ring | ether | 3° amine | ester | alkene

18.13 Quinine has 11 trigonal planar carbon atoms [part (a), labeled with *] and nine tetrahedral carbon atoms [part (b), all other C's].

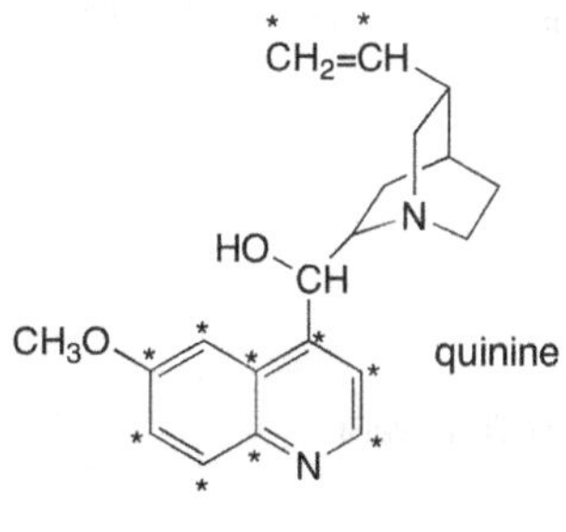

$CH_2{=}CH$ | HO | CH | N | CH_3O | N | quinine

18.14 Draw the products when the amines are treated with HCl as in Example 18.4.

a. $CH_3CH_2NH_2 \xrightarrow{HCl} (CH_3CH_2NH_3)^+ + Cl^-$

b. $(CH_3CH_2)_2NH \xrightarrow{HCl} [(CH_3CH_2)_2NH_2]^+ + Cl^-$

c. $(CH_3CH_2)_3N \xrightarrow{HCl} [(CH_3CH_2)_3NH]^+ + Cl^-$

18.15 Draw the products of each reaction.

a. $CH_3CH_2CH_2CH_2{-}NH_2$ + HCl ⟶ $(CH_3CH_2CH_2CH_2{-}NH_3)^+$ + Cl^-

b. $(CH_3)_2NH$ + C_6H_5COOH ⟶ $[(CH_3)_2NH_2]^+$ + $C_6H_5COO^-$

c. piperidine (ring with N–H) + H_2O ⟶ [piperidinium, ring N with H H]$^+$ + OH^-

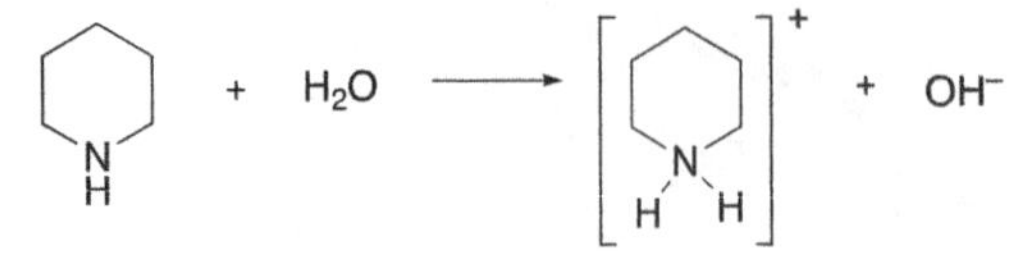

18.16 Draw the products of the reaction.

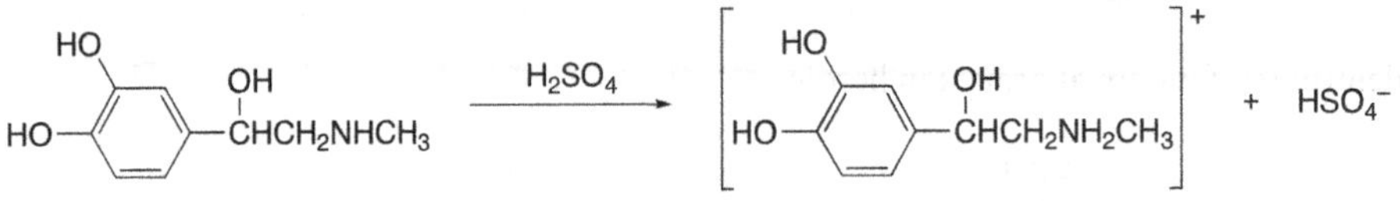

HO, HO–(benzene ring)–CH(OH)CH₂NHCH₃ $\xrightarrow{H_2SO_4}$ [HO, HO–(benzene ring)–CH(OH)CH₂NH₂CH₃]$^+$ + HSO_4^-

18.17 Name each ammonium salt as in Example 18.5.

a. $(CH_3NH_3)^+$ Cl^- — chloride; derived from methylamine — **methylammonium chloride**

b. $[(CH_3CH_2CH_2)_2NH_2]^+$ Br^- — bromide; derived from dipropylamine — **dipropylammonium bromide**

c. $[(CH_3)_2NHCH_2CH_3]^+$ CH_3COO^- — acetate; derived from ethyldimethylamine — **ethyldimethylammonium acetate**

18.18 Ammonium salts are water-soluble solids. A water-insoluble amine can be converted to a water-soluble ammonium salt by treatment with acid.

a. $(CH_3CH_2)_3N$ — 3° amine with 6 C's — **water insoluble**

b. $[(CH_3CH_2)_3NH]^+$ Br^- — ammonium salt — **water soluble**

c. $CH_3CH_2NH_2$ — 1° amine with 2 C's — **water soluble**

d. [piperidinium, ring N with H H]$^+$ Cl^- — ammonium salt — **water soluble**

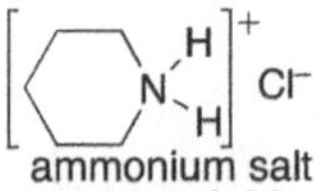

18.19 Draw the products of each reaction.

a. $[(CH_3CH_2)_3NH]^+$ Br^- $\xrightarrow{NaOH}$ $(CH_3CH_2)_3N$ + H_2O + NaBr

b. $(CH_3CH_2NH_3)^+$ HSO_4^- $\xrightarrow{NaOH}$ $CH_3CH_2NH_2$ + H_2O + $NaHSO_4$

c. [piperidinium, ring N with H H]$^+$ Cl^- $\xrightarrow{NaOH}$ piperidine (ring NH) + H_2O + NaCl

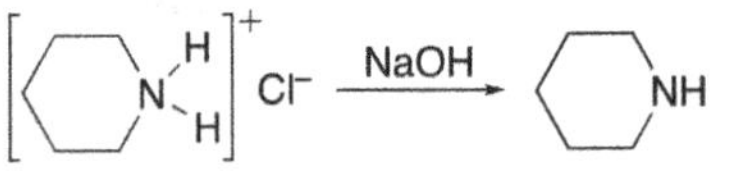

18.20 Draw the structure of each amine from which the ammonium salt is derived.

HO
OH
CHCH$_2$–N–CH$_3$
H
phenylephrine

O
CCH$_2$CH$_3$
C–CH$_2$–CH–N–CH$_3$
CH$_3$ CH$_3$
methadone

18.21 A quaternary ammonium salt like Bitrex has four R groups bonded to N, so there is no proton available on N that can be removed by a base to form an amine.

18.22 The chirality center in each compound is labeled.

HO–CH$_2$C*HNH$_2$ (COOH)
tyrosine

HO, HO–CH$_2$C*HNH$_2$ (COOH)
L-dopa

* = chirality center

HO, HO–C*HCH$_2$NH$_2$ (OH)
norepinephrine

HO, HO–CH$_2$CHNH$_2$ (H)
dopamine
achiral

18.23 COOH must be removed and OH must be added.

The COOH group is removed, and the OH group is added.

CH$_2$CHNH$_2$
COOH
N
H
tryptophan

HO
CH$_2$CH$_2$NH$_2$
N
H
serotonin

18.24 Serotonin, bufotenin, and psilocin have the same ring system with a two-carbon chain containing a N atom bonded to it. They all contain an OH group bonded to the six-membered ring.

1°
HO
CH$_2$CH$_2$NH$_2$
N–H ← 2°
serotonin

3°
HO
CH$_2$CH$_2$N(CH$_3$)$_2$
N–H ← 2°
bufotenin

3°
OH
CH$_2$CH$_2$N(CH$_3$)$_2$
N–H ← 2°
psilocin

18.25 To convert phenylephrine to methamphetamine requires the addition of a methyl group and removal of two OH groups.

18.26 The atoms of 2-phenylethylamine are in bold in each compound.

a. CH_3O, CH_3O, CH_3O — **$CH_2CH_2NH_2$**

mescaline

b. $(CH_3CH_2)_2N$–C(=O)–, N–CH_3, N–H

LSD

18.27 Classify each alcohol and amine as 1°, 2°, or 3°. OH groups not labeled are phenols.

2° alcohol → OH; HO–, –$CHCH_2NHC(CH_3)_3$; 1° alcohol → $HOCH_2$; 2° amine

albuterol
(Trade names: Ventolin, Proventil)

2° alcohol → OH; HO–, –$CHCH_2NH(CH_2)_6O(CH_2)_4$–; 1° alcohol → $HOCH_2$; 2° amine

salmeterol
(Trade name: Serevent)

18.28 Draw the complete structure for cimetidine.

Solutions to Odd-Numbered End-of-Chapter Problems

18.29 To determine whether an amine or amide is 1°, 2°, or 3°, count the number of carbons bonded to the nitrogen atom as in Example 18.1.

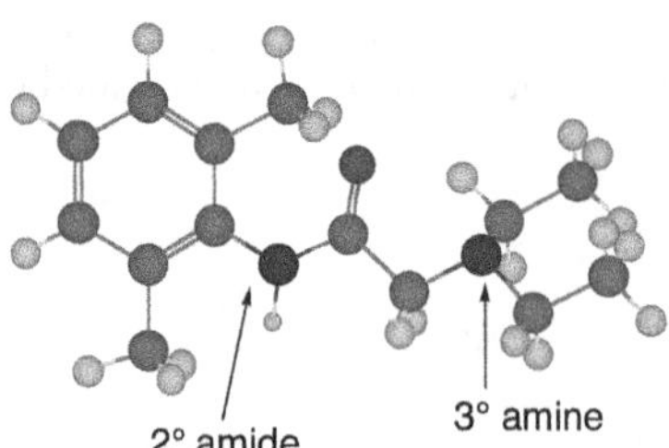

18.31 Convert the ball-and-stick model to a line structure and classify the N-containing functional groups.

2° amide

3° amine

3° amine

18.33 To determine whether an amine is 1°, 2°, or 3°, count the number of carbons bonded to the nitrogen atom as in Example 18.1.

a. $CH_3CH(CH_3)CH_2CH_2NHCH_3$ — 2°

b. N–CH_3 (piperidine ring) — 3°

18.35

a. a compound that contains a 1° amine and a 1° alcohol

$HOCH_2CH_2CH_2CH_2NH_2$

C

b. a compound that contains a 2° amine and a 2° alcohol

$CH_3NHCH_2CH(OH)CH_3$

D

c. a compound that contains a 1° amine and a 3° alcohol

$CH_3C(CH_3)(OH)CH_2NH_2$

B

d. a compound that contains a 3° amine and a 1° alcohol

$(CH_3)_2NCH_2CH_2OH$

A

18.37

$CH_3CH_2CH_2\overset{*}{C}HCH_3$ (with NH_2 on the starred carbon)

$CH_3CH_2\overset{*}{C}HCH_3$ (with CH_2NH_2 on the starred carbon)

$CH_3\overset{*}{C}HCH(CH_3)_2$ (with NH_2 on the starred carbon)

* = chirality center

18.39 Name each amine as in Example 18.2.

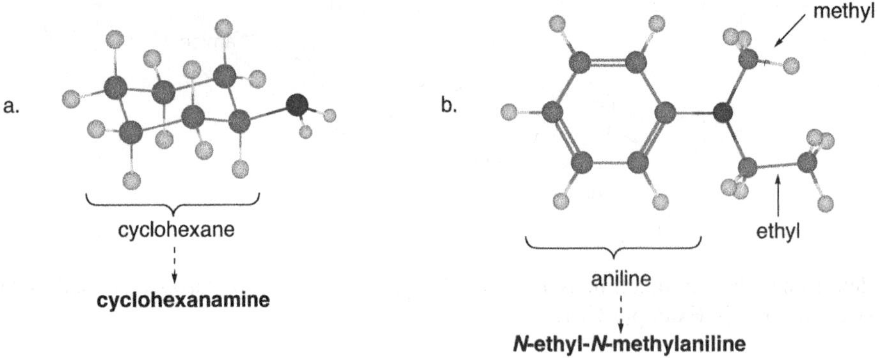

18.41 Name each amine as in Example 18.2.

a. $$CH_3CH_2-\overset{\displaystyle H}{\overset{|}{N}}-CH_2CH_3$$ ----→ **Answer: diethylamine**

(2 ethyl groups)

b. $$\boxed{CH_3CH_2\overset{NH_2}{\overset{|}{C}}HCH_2CH_2CH_2CH_2CH_3}$$ --→ $$\boxed{\underset{1}{C}H_3\underset{2}{C}H_2\underset{3}{\overset{NH_2}{\overset{|}{C}}}HCH_2CH_2CH_2CH_2CH_3}$$ --→ **Answer: 3-octanamine**

octane ---→ octan*amine*
(8 C's)

c. $$\boxed{CH_3CH_2CH_2\overset{NHCH_3}{\overset{|}{C}}HCH_2CH_3}$$ --→ $$\boxed{CH_3CH_2CH_2\underset{3}{\overset{NHCH_3}{\overset{|}{C}}}H\underset{2}{C}H_2\underset{1}{C}H_3}$$ (NHCH$_3$ ← *N*-methyl) --→ **Answer: *N*-methyl-3-hexanamine**

hexane ---→ hexan*amine*
(6 C's)

d. $$\boxed{CH_3\underset{NHCH_2CH_3}{\underset{|}{C}}HCH_2CH_2CH_2CH_2CH_3}$$ --→ $$\boxed{\overset{1}{C}H_3\overset{2}{\underset{NHCH_2CH_3}{\underset{|}{C}}}HCH_2CH_2CH_2CH_2CH_3}$$ (NHCH$_2$CH$_3$ ← *N*-ethyl) --→ **Answer: *N*-ethyl-2-heptanamine**

heptane ---→ heptan*amine*
(7 C's)

18.43 Work backwards to draw the structure corresponding to each name.

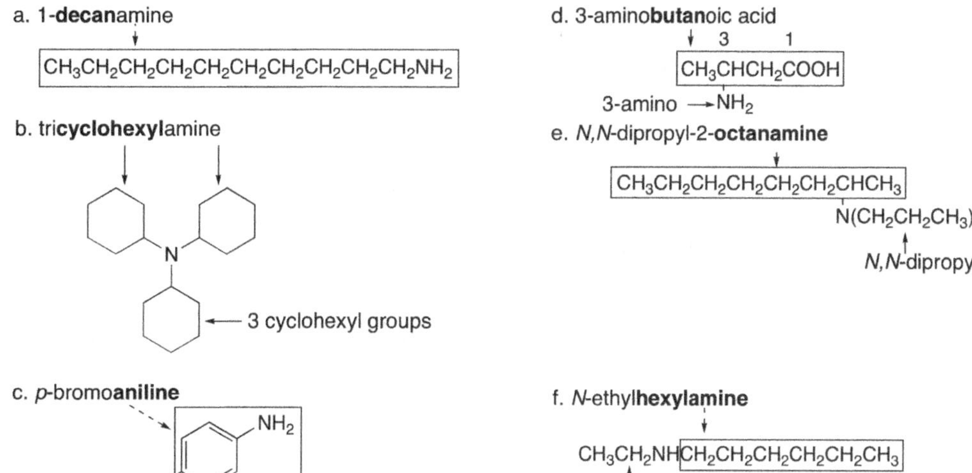

18.45 Work backwards from the name to draw each ammonium salt.

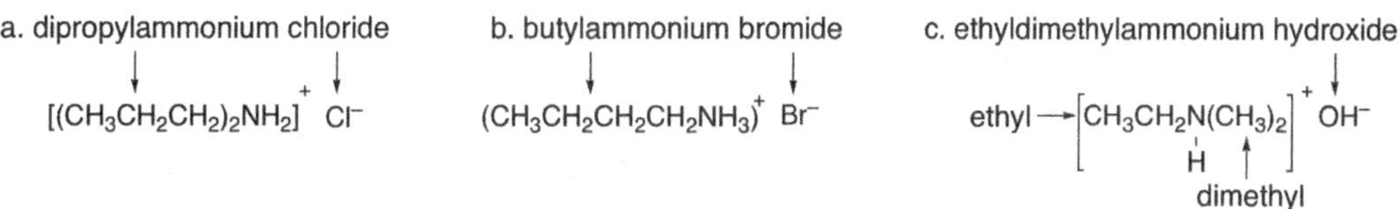

18.47 Draw and name the four amines of molecular formula C_3H_9N.

$(CH_3)_3N$	$CH_3CH_2CH_2NH_2$	$CH_3CH_2NHCH_3$	$CH_3CH(CH_3)NH_2$
trimethylamine	1-propanamine	*N*-methylethanamine	2-propanamine

18.49 Pyridine is capable of hydrogen bonding with water, so it is more water soluble than benzene. Pyridine has a higher boiling point than benzene because it contains polar C–N bonds, while benzene is a nonpolar hydrocarbon.

18.51

a. $CH_3CH_2CH_2NH_2$ or $CH_3CH_2CH_2CH_2CH_2CH_2NH_2$
larger surface area
higher boiling point

b. $CH_3(CH_2)_7N(CH_3)_2$ or $CH_3(CH_2)_9NH_2$
NH bond for hydrogen bonding
higher boiling point

18.53 The hydrogen bond is drawn as a dashed line.

$$\begin{array}{l} \text{CH}_3\text{CH}_2\text{–N(H)–CH}_2\text{CH}_3 \cdots \text{H–N(CH}_2\text{CH}_3)_2 \end{array}$$

18.55 Primary amines can hydrogen bond to each other, whereas 3° amines cannot. Therefore, 1° amines have higher boiling points than 3° amines of similar size. Since all amines contain nitrogen atoms, any amine can hydrogen bond to water. Therefore, 1° and 3° amines have similar solubility properties.

18.57 Draw the reaction of each amine with water by transferring a proton from H_2O to the amine.

a. $CH_3CH_2NH_2 + H_2O \longrightarrow (CH_3CH_2NH_3)^+ + OH^-$

b. $(CH_3CH_2)_2NH + H_2O \longrightarrow [(CH_3CH_2)_2NH_2]^+ + OH^-$

c. $(CH_3CH_2)_3N + H_2O \longrightarrow [(CH_3CH_2)_3NH]^+ + OH^-$

18.59 Draw the products of each reaction.

a. $CH_3CH_2CH_2N(CH_3)_2 + HCl \longrightarrow [CH_3CH_2CH_2NH(CH_3)_2]^+ + Cl^-$

b. $CH_3CH_2CH(NH_2)CH_2CH_3 + H_2SO_4 \longrightarrow [CH_3CH_2CH(NH_3)CH_2CH_3]^+ + HSO_4^-$

c. pyridine + HBr ⟶ [pyridinium (N–H)]$^+$ + Br^-

d. $CH_3CH_2-\overset{+}{N}H(CH_2CH_3)-CH_2CH_3 + NaOH \longrightarrow CH_3CH_2-N(CH_2CH_3)-CH_2CH_3 + H_2O + Na^+$

18.61 Draw the products of each reaction.

a. 2-propylpiperidine (N–H, $CH_2CH_2CH_3$) $\xrightarrow{HCl}$ [2-propylpiperidinium (NH_2, $CH_2CH_2CH_3$)]$^+$ Cl^-

b. 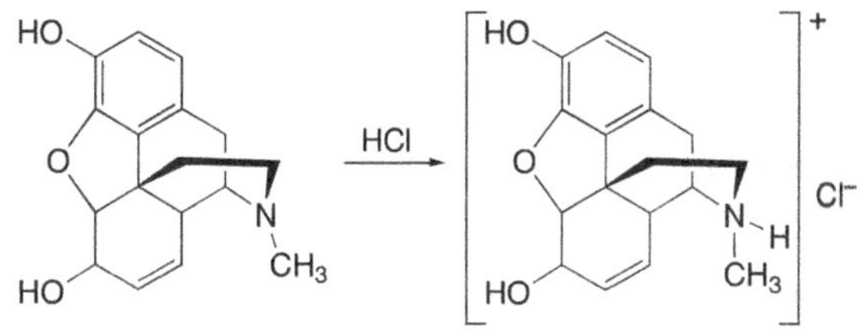 [Chemical structure: morphine + HCl → protonated morphine ammonium ion with $N-H$, CH_3, HO groups; Cl^-]

18.63 The heterocycle in both coniine and morphine is piperidine (labeled in bold).

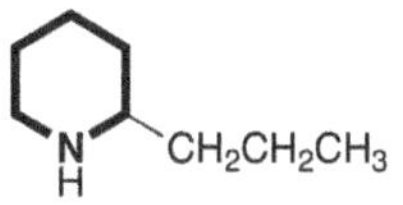

[Chemical structures: coniine (piperidine ring with $CH_2CH_2CH_3$); morphine (HO, O, N, CH_3)]

coniine

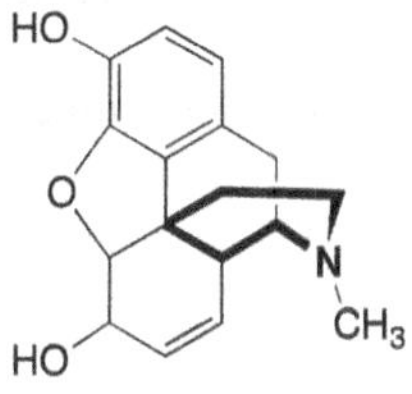

morphine

18.65 Caffeine is a mild stimulant, imparting a feeling of alertness after consumption. It also increases heart rate, dilates airways, and stimulates the secretion of stomach acid. These effects are observed because caffeine increases glucose production, making an individual feel energetic.

18.67 An alkaloid solution is slightly basic because its amine pulls off a proton from water, forming OH^- and an ammonium ion.

18.69 Dopamine affects brain processes that control movement, emotions, and pleasure. Normal dopamine levels give an individual a pleasurable, satisfied feeling. Increased levels result in an intense "high." Drugs such as heroin, cocaine, and alcohol increase dopamine levels. When there is too little dopamine in the brain, an individual loses control of fine motor skills and Parkinson's disease results.

18.71 Serotonin plays an important role in mood, sleep, perception, and temperature regulation. We get sleepy after eating a turkey dinner on Thanksgiving because the unusually high level of tryptophan in turkey is converted to serotonin. A deficiency of serotonin causes depression.

18.73 Dopamine [part (a)] and norepinephrine [part (b)] are derived from tyrosine, serotonin [part (c)] is derived from tryptophan, and histamine [part (d)] is derived from histidine.

18.75 The atoms of 2-phenylethylamine are labeled in bold in the compounds below.

a. 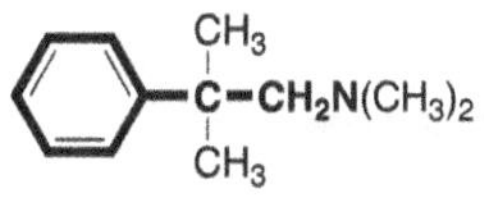 [Structure: benzene ring–$C(CH_3)_2$–$CH_2N(CH_3)_2$]

b. 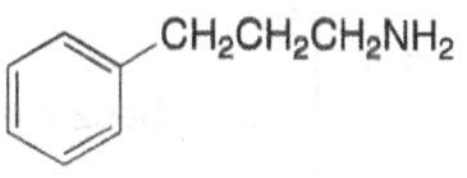 [Structure: benzene ring–$CH_2CH_2CH_2NH_2$]

not a derivative of 2-phenylethylamine

c. 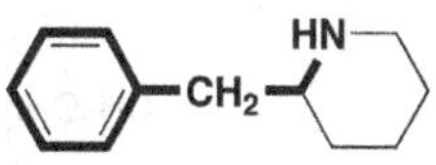 [Structure: benzene ring–CH_2–piperidine (HN)]

18.77 The atoms of 2-phenylethylamine are labeled in bold.

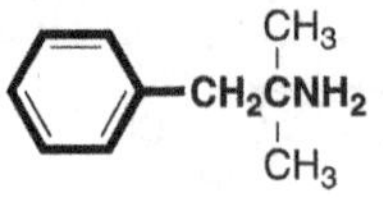

C_6H_5**$-CH_2C(CH_3)_2NH_2$**

18.79 Chlorpheniramine is an example of an antihistamine. Antihistamines bind to the H1 histamine receptor, but they evoke a different response than histamine. An antihistamine like chlorpheniramine or diphenhydramine, for example, inhibits vasodilation, so it is used to treat the symptoms of the common cold and environmental allergies.

18.81 Answer each question about benzphetamine.

a. and b.

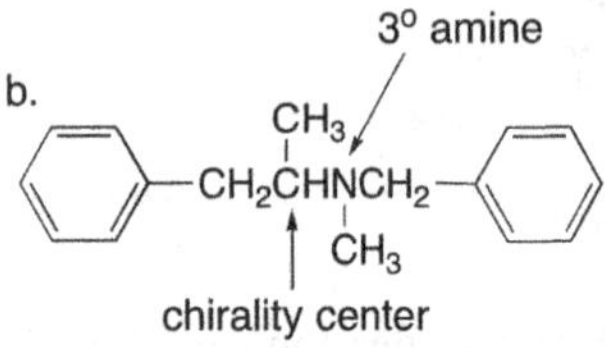

$C_6H_5-CH_2CH(CH_3)N(CH_3)CH_2-C_6H_5$

3° amine

chirality center

c.

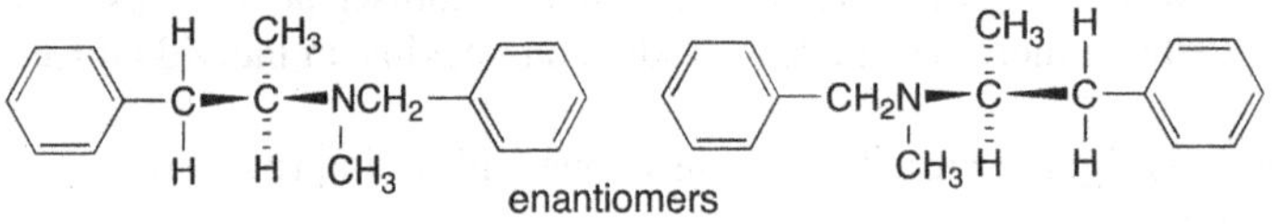

H CH₃ / C—C—NCH₂ / H H CH₃ and CH₃ H / CH₂N—C—C / CH₃ H H

enantiomers

d.

CH_3, $CH_2CH(CH_3)NH_2$ and $CH_2C_6H_5$ on benzene ring

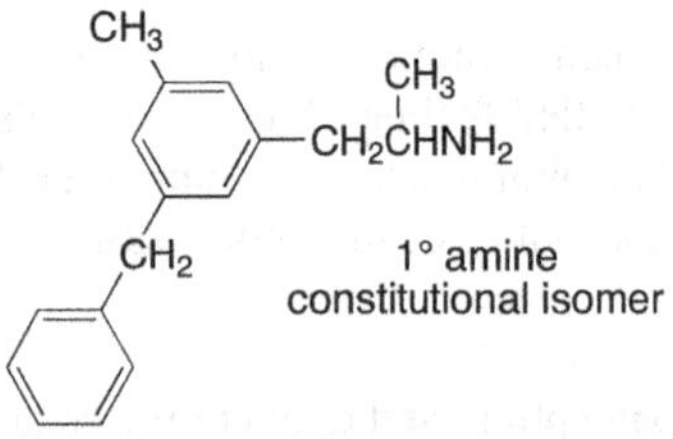

1° amine
constitutional isomer

e.

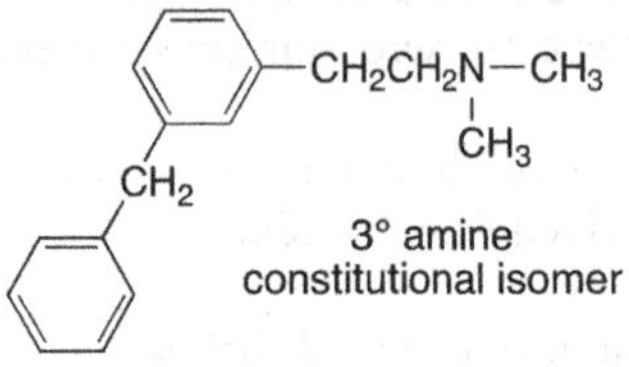

$CH_2CH_2N(CH_3)-CH_3$ and $CH_2C_6H_5$ on benzene ring

3° amine
constitutional isomer

f. $[C_6H_5-CH_2CH(CH_3)NH(CH_3)CH_2-C_6H_5]^+$ Cl^- benzphetamine hydrochloride

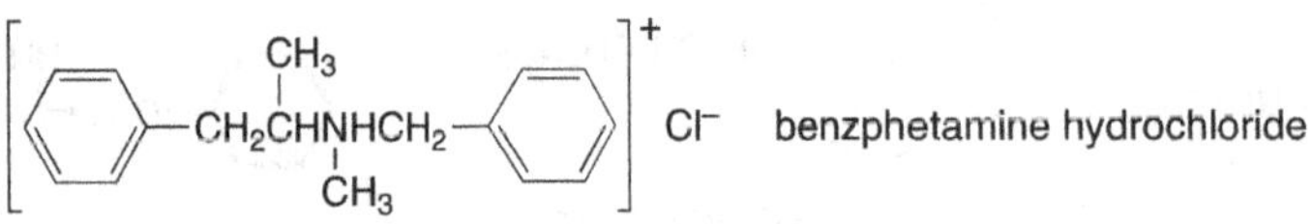

g. $C_6H_5-CH_2CH(CH_3)N(CH_3)CH_2-C_6H_5$ + CH_3COOH ⟶ $[C_6H_5-CH_2CH(CH_3)NH(CH_3)CH_2-C_6H_5]^+$ + CH_3COO^-

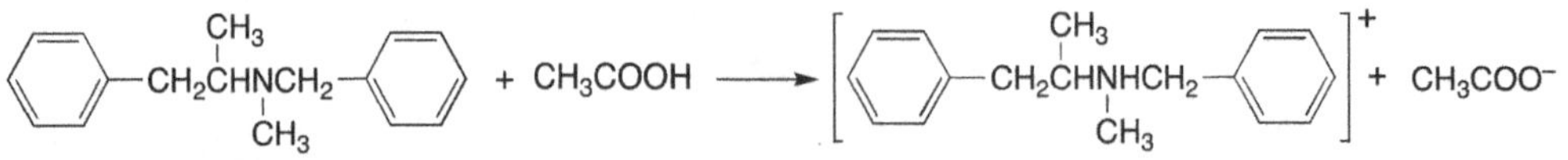

18.83 Answer the questions about Ritalin.

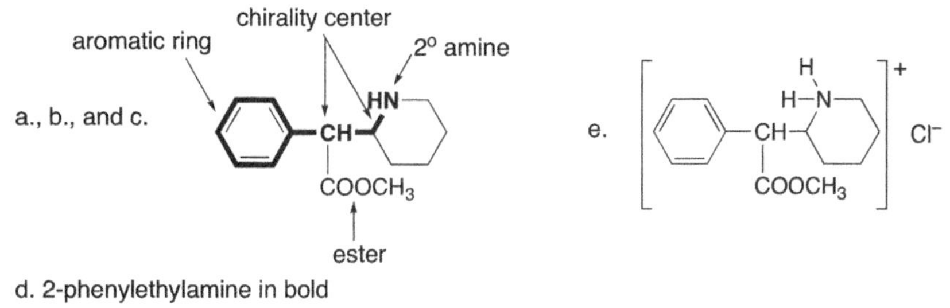

18.85 Work backwards to draw the structure of the amine and carboxylic acid that are used to form Chlortrimeton, an ammonium salt.

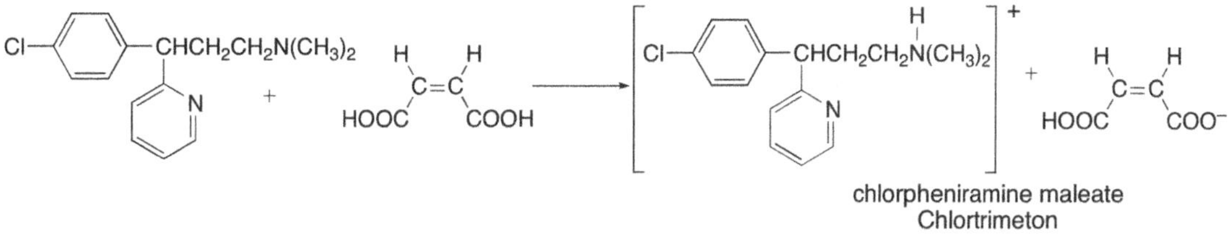

18.87 A vasodilator dilates blood vessels, whereas a bronchodilator dilates airways in the lungs. Histamine is a vasodilator and albuterol is a bronchodilator.

18.89 Albuterol will exist in the ionic form in the stomach (lower pH) and in the neutral form in the intestines.

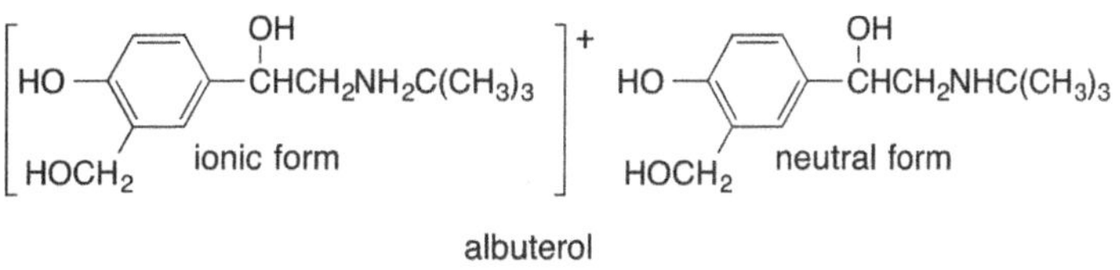

18.91 Heroin has two esters, which can be made from the two OH groups in morphine. Add acetic acid (CH_3COOH) and H_2SO_4 to morphine to make the two esters in heroin.

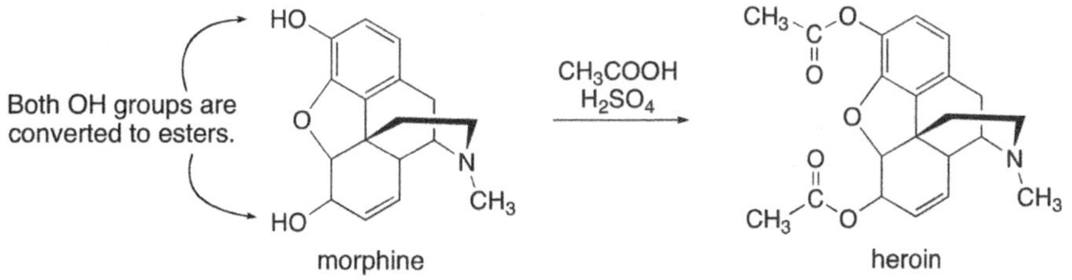

Chapter 19 Lipids

Chapter Review

[1] What are the general characteristics of lipids? (19.1)

- Lipids are biomolecules that contain many nonpolar C–C and C–H bonds, making them soluble in organic solvents and insoluble in water.
- Hydrolyzable lipids, including waxes, triacylglycerols, and phospholipids, can be converted to smaller molecules on reaction with water.

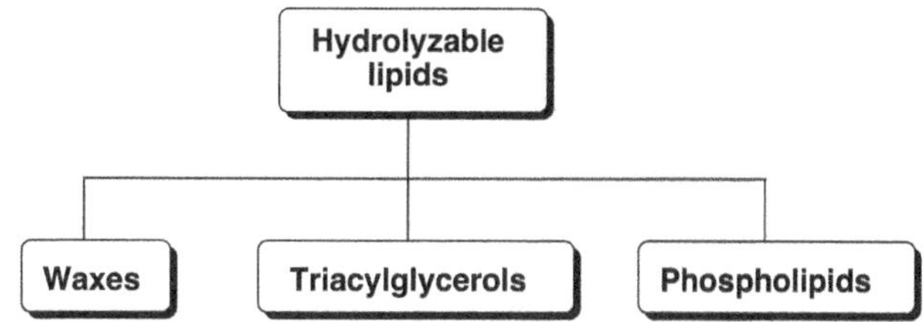

- Nonhydrolyzable lipids, including steroids, fat-soluble vitamins, and eicosanoids, cannot be cleaved into smaller units by hydrolysis.

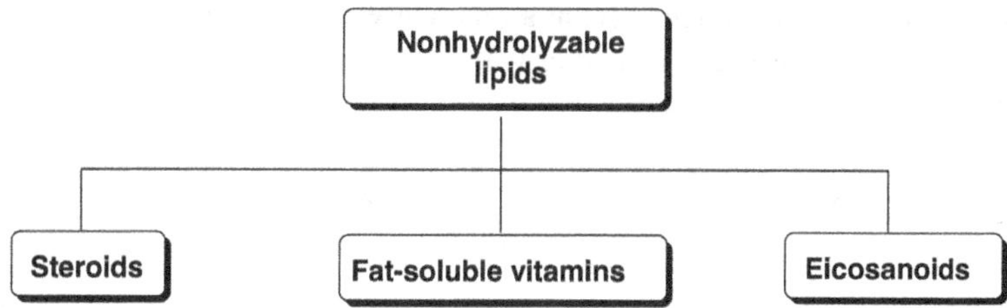

[2] How are fatty acids classified and what is the relationship between their melting points and the number of double bonds they contain? (19.2)

- Fatty acids are saturated if they contain no carbon–carbon double bonds and unsaturated if they contain one or more double bonds. Naturally occurring unsaturated fatty acids generally contain cis double bonds.
- As the number of double bonds in the fatty acid increases, its melting point decreases.

[3] What are waxes? (19.3)

- A wax is an ester (RCOOR') formed from a fatty acid (RCOOH) and a high molecular weight alcohol (R'OH). Since waxes contain many nonpolar C–C and C–H bonds, they are hydrophobic.
- Waxes (RCOOR') are hydrolyzed to fatty acids (RCOOH) and alcohols (R'OH).

Wax General structure

RCOOR' (R and R': long chains of C's)

Formation

RCOOH (fatty acid) + H–OR' (alcohol) → RCOOR' (wax) + H_2O

[4] What are triacylglycerols, and how do the triacylglycerols in a fat and oil differ? (19.4)

- Triacylglycerols, or triglycerides, are triesters formed from glycerol and three molecules of fatty acids. A monounsaturated triacylglycerol contains one carbon–carbon double bond, whereas polyunsaturated triacylglycerols have more than one carbon–carbon double bond.

$$\underset{\text{glycerol}}{\begin{array}{l} CH_2\text{-}OH \\ | \\ CH\text{—}OH \\ | \\ CH_2\text{-}OH \end{array}} + \underset{\text{fatty acids}}{\begin{array}{l} HO\text{—}\overset{O}{\overset{\|}{C}}\text{—}R \\ HO\text{—}\overset{O}{\overset{\|}{C}}\text{—}R' \\ HO\text{—}\overset{O}{\overset{\|}{C}}\text{—}R'' \end{array}} \longrightarrow \underset{\text{triacylglycerol}}{\begin{array}{l} CH_2\text{-}O\text{—}\overset{O}{\overset{\|}{C}}\text{—}R \\ CH\text{—}O\text{—}\overset{O}{\overset{\|}{C}}\text{—}R' \\ CH_2\text{-}O\text{—}\overset{O}{\overset{\|}{C}}\text{—}R'' \end{array}}$$

R groups have 11–19 C's.

- Fats are triacylglycerols derived from fatty acids having few double bonds, making them solids at room temperature. Fats are generally obtained from animal sources.
- Oils are triacylglycerols derived from fatty acids having a larger number of double bonds, making them liquids at room temperature. Oils are generally obtained from plant sources.

[5] What hydrolysis products are formed from a triacylglycerol? (19.5)

- Triacylglycerols are hydrolyzed in acid or with enzymes (in biological systems) to form glycerol and three molecules of fatty acids.

$$\underset{\text{triacylglycerol}}{\begin{array}{l} CH_2\text{-}O\text{—}\overset{O}{\overset{\|}{C}}\text{—}R \\ CH\text{—}O\text{—}\overset{O}{\overset{\|}{C}}\text{—}R' \\ CH_2\text{-}O\text{—}\overset{O}{\overset{\|}{C}}\text{—}R'' \end{array}} + 3\,H_2O \xrightarrow[\text{or lipase}]{H_2SO_4} \underset{\text{glycerol}}{\begin{array}{l} CH_2\text{-}OH \\ | \\ CH\text{—}OH \\ | \\ CH_2\text{-}OH \end{array}} + \underbrace{HO\text{—}\overset{O}{\overset{\|}{C}}\text{—}R + HO\text{—}\overset{O}{\overset{\|}{C}}\text{—}R' + HO\text{—}\overset{O}{\overset{\|}{C}}\text{—}R''}_{\text{three fatty acids}}$$

- Base hydrolysis of a triacylglycerol forms glycerol and sodium salts of fatty acids—soaps.

$$\underset{\text{triacylglycerol}}{\begin{array}{l} CH_2\text{-}O\text{—}\overset{O}{\overset{\|}{C}}\text{—}R \\ CH\text{—}O\text{—}\overset{O}{\overset{\|}{C}}\text{—}R' \\ CH_2\text{-}O\text{—}\overset{O}{\overset{\|}{C}}\text{—}R'' \end{array}} + 3\ NaOH \xrightarrow{H_2O} \underset{\text{glycerol}}{\begin{array}{l} CH_2\text{-}OH \\ | \\ CH\text{—}OH \\ | \\ CH_2\text{-}OH \end{array}} + \underbrace{Na^{+}\,{}^{-}O\text{—}\overset{O}{\overset{\|}{C}}\text{—}R + Na^{+}\,{}^{-}O\text{—}\overset{O}{\overset{\|}{C}}\text{—}R' + Na^{+}\,{}^{-}O\text{—}\overset{O}{\overset{\|}{C}}\text{—}R''}_{\text{soaps}}$$

[6] What are the major types of phospholipids? (19.6)

- All phospholipids contain a phosphorus atom, and have a polar (ionic) head and two nonpolar tails. Phosphoacylglycerols are derived from glycerol, two molecules of fatty acids, phosphate, and an alcohol (either ethanolamine or choline). Sphingomyelins are derived from sphingosine, a fatty acid (that forms an amide), phosphate, and an alcohol (either ethanolamine or choline).

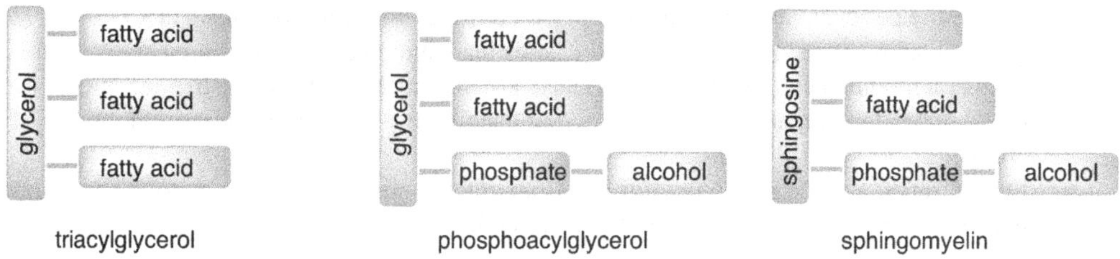

[7] Describe the structure of the cell membrane. How do molecules and ions cross the cell membrane? (19.7)

- The main component of the cell membrane is phospholipids, arranged in a lipid bilayer with the ionic heads oriented towards the outside of the bilayer, and the nonpolar tails on the interior.
- Small molecules like O_2 and CO_2 diffuse through the membrane from the side of higher concentration to the side of lower concentration. Larger polar molecules and some ions (Cl^-, HCO_3^-, and glucose) travel through channels created by integral membrane proteins (facilitated diffusion). Some cations (Na^+, K^+, and Ca^{2+}) must travel against the concentration gradient, a process called active transport, which requires energy input.

[8] What are the main structural features of steroids? What is the relationship between the steroid cholesterol and cardiovascular disease? (19.8)

- Steroids are tetracyclic lipids that contain three six-membered rings and one five-membered ring.

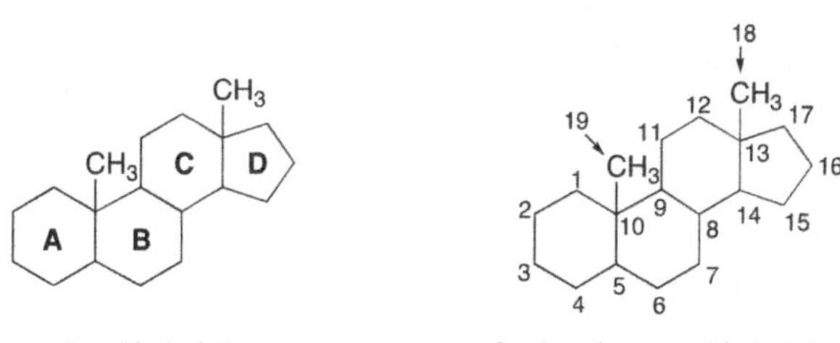

- Because cholesterol is insoluble in the aqueous medium of the blood, it is transported through the bloodstream in water-soluble particles called lipoproteins. Low-density lipoprotein particles (LDLs) transport cholesterol from the liver to the tissues. If the blood cholesterol level is high, it forms plaque on the walls of arteries, increasing the risk of heart attack and stroke. High-density lipoprotein particles (HDLs) transport cholesterol from the tissues to the liver, where it is metabolized or eliminated.

[9] What is a hormone? Give examples of steroid hormones. (19.9)

- A hormone is a molecule that is synthesized in one part of an organism, and elicits a response at a different site. Steroid hormones include estrogens and progestins (female sex hormones), androgens (male sex hormones), and adrenal cortical steroids such as cortisone, which are synthesized in the adrenal gland.

[10] Which vitamins are fat soluble? (19.10)

- Fat-soluble vitamins are lipids required in small quantities for normal cell function, and which cannot be synthesized in the body. Vitamins A, D, E, and K are fat soluble.

[11] What are the general characteristics of prostaglandins and leukotrienes? (19.11)

- Prostaglandins are a group of carboxylic acids that contain a five-membered ring and are derived from arachidonic acid. Prostaglandins cause inflammation, decrease gastric secretions, inhibit blood platelet aggregation, stimulate uterine contractions, and relax the smooth muscle of the uterus.
- Leukotrienes, acyclic molecules derived from arachidonic acid, contribute to the asthmatic response by constricting smooth muscles in the lungs.

$CH_3(CH_2)_4(CH{=}CHCH_2)_4(CH_2)_2COOH$

arachidonic acid
(four cis double bonds)

a fatty acid

HO, HO, $CH_2CH{=}CH(CH_2)_3COOH$, $CH{=}CHCH(CH_2)_4CH_3$, OH

$PGF_{2\alpha}$
a prostaglandin

H, H, H, H, H, H, H, H, CH_2, C_5H_{11}, $CHCH(OH)(CH_2)_3COOH$, $S{-}CH_2$, $CHCONHCH_2COOH$, $NHCOCH_2CH_2CHCOOH$, NH_2

LTC_4
a leukotriene

Problem Solving

[1] Introduction to Lipids (19.1)

Example 19.1 Which compounds are likely to be lipids?

a. $HO_2CCH_2{-}C(CH_2OH)(CH_3){-}CH_3$

b. (steroid structure with HO, CH_3, CH_3, CH_3, $CH(CH_2)_3CH(CH_3)_2$)

Analysis
Lipids contain many nonpolar C–C and C–H bonds and few polar bonds.

Solution

a. Compound (a) is unlikely to be a lipid because it contains a polar OH group and a carboxyl group (COOH), and has only six carbon atoms.
b. Compound (b) is likely to be a lipid because it contains many nonpolar C–C and C–H bonds, and only one OH group.

---

[2] Fatty Acids (19.2)

Example 19.2 Draw the skeletal structure for palmitoleic acid (Table 19.2), and label the hydrophobic and hydrophilic portions.

Analysis

- Skeletal structures have a carbon at the intersection of two lines and at the end of every line. The double bond must have the cis arrangement in an unsaturated fatty acid.
- The nonpolar C–C and C–H bonds comprise the hydrophobic portion of a molecule and the polar bonds comprise the hydrophilic portion.

Solution

The skeletal structure for palmitoleic acid is drawn below. The hydrophilic portion is the COOH group and the hydrophobic portion is the rest of the molecule.

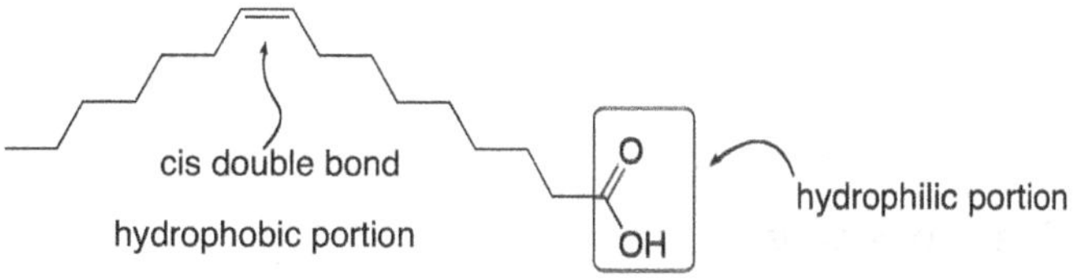

cis double bond
hydrophobic portion
O
OH
hydrophilic portion

---

[3] Waxes (19.3)

Example 19.3 Draw the structure of a wax formed from an 18-carbon straight-chain alcohol and the fatty acid $CH_3(CH_2)_{10}COOH$.

Analysis

To draw the wax, arrange the carboxyl group of the fatty acid (RCOOH) next to the OH group of the alcohol (R'OH) with which it reacts. Then, replace the OH group of the fatty acid with the OR' group of the alcohol, forming an ester RCOOR' with a new C–O bond at the carbonyl carbon.

Solution

Draw the structures of the fatty acid and the alcohol, and replace the OH group of the 12-carbon acid with the $O(CH_2)_{17}CH_3$ group of the alcohol.

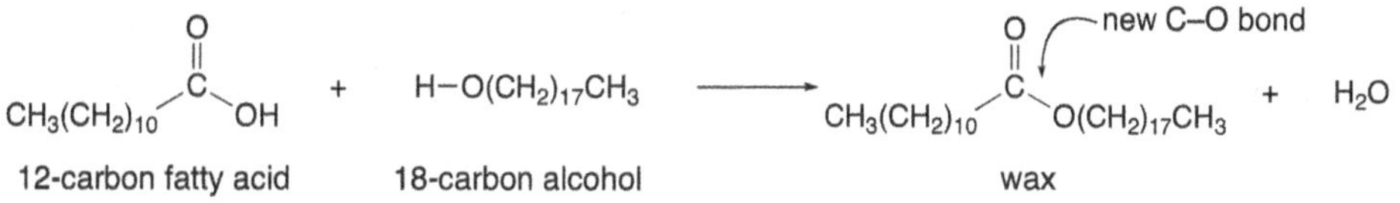

$$CH_3(CH_2)_{10}C(=O)OH + H{-}O(CH_2)_{17}CH_3 \longrightarrow CH_3(CH_2)_{10}C(=O)O(CH_2)_{17}CH_3 + H_2O$$

12-carbon fatty acid — 18-carbon alcohol — wax (new C–O bond)

---

[4] Triacylglycerols–Fats and Oils (19.4)

Example 19.4 Draw the structure of a triacylglycerol formed from glycerol, one molecule of palmitic acid, and two molecules of linoleic acid. Bond the palmitic acid to the 2° OH group (OH on the middle carbon atom) of glycerol.

Analysis

To draw the triacylglycerol, arrange each OH group of glycerol next to the carboxyl group of a fatty acid. Then join each O atom of glycerol to a carbonyl carbon of a fatty acid, thus forming three new C–O bonds. Structures of the fatty acids are presented in Table 19.2.

Solution

Form three new ester bonds (RCOOR') from the OH groups of glycerol and the three fatty acids (RCOOH).

$$\begin{array}{l} CH_2\text{-}OH \\ | \\ CH\text{-}OH \\ | \\ CH_2\text{-}OH \\ \text{glycerol} \\ 2^\circ\ OH \end{array} + \begin{array}{l} HO\text{-}C(=O)\text{-}(CH_2)_7CH{=}CHCH_2CH{=}CH(CH_2)_4CH_3 \\ HO\text{-}C(=O)\text{-}(CH_2)_{14}CH_3 \quad \text{palmitic acid} \\ HO\text{-}C(=O)\text{-}(CH_2)_7CH{=}CHCH_2CH{=}CH(CH_2)_4CH_3 \quad \text{linoleic acid} \end{array} \longrightarrow \begin{array}{l} CH_2\text{-}O\text{-}C(=O)\text{-}(CH_2)_7CH{=}CHCH_2CH{=}CH(CH_2)_4CH_3 \\ CH\text{-}O\text{-}C(=O)\text{-}(CH_2)_{14}CH_3 \\ CH_2\text{-}O\text{-}C(=O)\text{-}(CH_2)_7CH{=}CHCH_2CH{=}CH(CH_2)_4CH_3 \\ \text{triacylglycerol} \end{array} + 3\ H_2O$$

---

[5] Hydrolysis of Triacylglycerols (19.5)

Example 19.5 Draw the products formed when the given triacylglycerol is hydrolyzed with water in the presence of sulfuric acid.

$$\begin{array}{l} CH_2\text{-}O\text{-}C(=O)\text{-}(CH_2)_{14}CH_3 \\ CH\text{-}O\text{-}C(=O)\text{-}(CH_2)_{10}CH_3 \\ CH_2\text{-}O\text{-}C(=O)\text{-}(CH_2)_7CH{=}CH(CH_2)_7CH_3 \end{array}$$

Analysis

To draw the products of ester hydrolysis, cleave the three C–O single bonds at the carbonyl carbons to form glycerol and three fatty acids (RCOOH).

Solution

Bonds broken during hydrolysis

$$\begin{array}{l} CH_2\text{-}O\text{-}C(=O)\text{-}(CH_2)_{14}CH_3 \\ CH\text{-}O\text{-}C(=O)\text{-}(CH_2)_{10}CH_3 \\ CH_2\text{-}O\text{-}C(=O)\text{-}(CH_2)_7CH{=}CH(CH_2)_7CH_3 \end{array} \xrightarrow[H_2SO_4]{3\ H\text{-}OH} \begin{array}{l} CH_2\text{-}OH \\ | \\ CH\text{-}OH \\ | \\ CH_2\text{-}OH \\ \text{glycerol} \end{array} + \begin{array}{l} HO\text{-}C(=O)\text{-}(CH_2)_{14}CH_3 \quad \text{palmitic acid} \\ HO\text{-}C(=O)\text{-}(CH_2)_{10}CH_3 \quad \text{lauric acid} \\ HO\text{-}C(=O)\text{-}(CH_2)_7CH{=}CH(CH_2)_7CH_3 \quad \text{oleic acid} \end{array}$$

---

[6] Phospholipids (19.6)

Example 19.6 Draw the structure of a cephalin formed from two molecules of myristic acid.

Analysis

Substitute the 14-carbon saturated fatty acid myristic acid for the R and R' groups in the general structure of a cephalin molecule. In a cephalin, $-CH_2CH_2NH_3^+$ forms part of the phosphodiester.

Solution

$$\begin{array}{l} CH_2\text{-}O\text{-}\overset{\overset{O}{\|}}{C}\text{-}R \\ | \\ CH\text{-}O\text{-}\overset{\overset{O}{\|}}{C}\text{-}R' \\ | \\ CH_2\text{-}O\text{-}\overset{\overset{O}{\|}}{\underset{\underset{O^-}{|}}{P}}\text{-}O\text{-}CH_2CH_2\overset{+}{N}H_3 \end{array}$$

R and R' ← from myristic acid

general structure

$$\begin{array}{l} CH_2\text{-}O\text{-}\overset{\overset{O}{\|}}{C}\text{-}(CH_2)_{12}CH_3 \\ | \\ CH\text{-}O\text{-}\overset{\overset{O}{\|}}{C}\text{-}(CH_2)_{12}CH_3 \\ | \\ CH_2\text{-}O\text{-}\overset{\overset{O}{\|}}{\underset{\underset{O^-}{|}}{P}}\text{-}O\text{-}CH_2CH_2\overset{+}{N}H_3 \end{array}$$

Answer

Self-Test

[1] Fill in the blank with one of the terms listed below.

Cell membrane (19.7)
Fats (19.4)
Fat-soluble vitamins (19.10)
Fatty acids (19.2)
Hormone (19.9)
Lipids (19.1)
Oils (19.4)
Phospholipids (19.6)
Saturated fatty acids (19.2)
Soaps (19.5)
Triacylglycerols (19.4)
Unsaturated fatty acids (19.2)
Waxes (19.3)

1. ______ are triesters formed from glycerol and three molecules of fatty acids.
2. ______ are carboxylic acids that have no double bonds in their long hydrocarbon chains.
3. ______ are liquids at room temperature and derived from fatty acids having one or more carbon–carbon double bonds.
4. ______ are biomolecules that are soluble in organic solvents but insoluble in water.
5. ______ are esters formed from a fatty acid and a high molecular weight alcohol.
6. ______ are metal salts of fatty acids.
7. ______ are carboxylic acids with long chains of 12–20 carbon atoms.
8. The vitamins A, D, E, and K are a group of lipids called ______.
9. ______ are solids at room temperature and are derived from fatty acids having few carbon–carbon double bonds.
10. A ______ is a molecule that is synthesized in one part of an organism, which then elicits a response at a different site.
11. ______ are carboxylic acids that have one or more carbon–carbon double bonds in their long hydrocarbon chains.
12. The cytoplasm is the aqueous medium inside the cell, separated from water outside the cell by the ______.
13. ______ are lipids that contain a phosphorus atom.

[2] Which molecule in each pair has a higher melting point?

14. $CH_3(CH_2)_{14}COOH$ **A** $CH_3(CH_2)_4CH{=}CHCH_2CH{=}CH(CH_2)_5COOH$ **B**

15. $CH_3(CH_2)_2CH{=}CH(CH_2)_8COOH$ **A** $CH_3(CH_2)_2CH{=}CHCH_2CH{=}CH(CH_2)_3COOH$ **B**

16. $CH_3CH{=}CHCH_2CH{=}CH(CH_2)_7COOH$ **A** $CH_3(CH_2)_6CH{=}CH(CH_2)_4COOH$ **B**

[3] Choose the method of transport across a cell membrane for each species.

a. simple diffusion b. facilitated transport c. active transport

17. HCO_3^- 18. CO_2 19. Na^+ 20. glucose

[4] Match the vitamin with the description.

a. vitamin A b. vitamin D c. vitamin E d. vitamin K

21. This vitamin is synthesized in the body from cholesterol.
22. This vitamin is converted to 11-*cis*-retinal, the light-sensitive compound responsible for vision in all vertebrates.
23. A deficiency of this vitamin leads to excessive and sometimes fatal bleeding because of inadequate blood clotting.

[5] Match the hormone with a description of its actions.

a. estradiol and estrone b. progesterone c. testosterone and androsterone

24. This hormone is often called the "pregnancy hormone" because it prepares the uterus for the implantation of a fertilized egg.
25. These hormones control the development of secondary sex characteristics in males.

Answers to Self-Test

1. Triacylglycerols	6. Soaps	11. Unsaturated fatty acids	16. **B**	21. b
2. Saturated fatty acids	7. Fatty acids	12. cell membrane	17. b	22. a
3. Oils	8. fat-soluble vitamins	13. Phospholipids	18. a	23. d
4. Lipids	9. Fats	14. **A**	19. c	24. b
5. Waxes	10. hormone	15. **A**	20. b	25. c

Solutions to In-Chapter Problems

19.1 Recall from Example 19.1 that lipids contain many nonpolar C–C and C–H bonds and few polar bonds.

a. 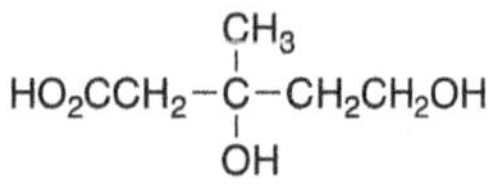 $HO_2CCH_2{-}C(CH_3)(OH){-}CH_2CH_2OH$

two polar OH groups and a COOH
six carbons
not a lipid

b. 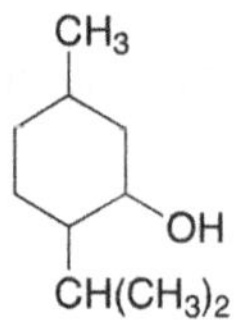 CH_3, OH, $CH(CH_3)_2$

one polar OH group
10 carbons
lipid

c. 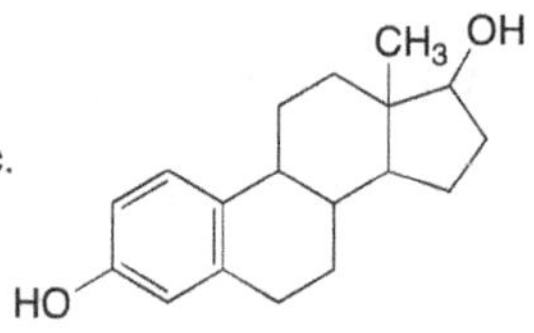 CH_3, OH, HO

two polar OH groups
18 carbons
lipid

19.2 Since lipids contain many nonpolar C–C bonds, they are soluble in nonpolar and weakly polar organic solvents. Therefore, lipids are likely to be soluble in (a) CH_2Cl_2 and (c) $CH_3CH_2CH_2CH_2CH_3$. The 5% aqueous NaCl solution (b) is not a solution in which lipids are soluble because it is a polar solvent.

19.3 Answer the questions as in Example 19.2.

- Skeletal structures have a carbon at the intersection of two lines and at the end of every line. The double bond must have the cis arrangement in an unsaturated fatty acid.
- The nonpolar C–C and C–H bonds comprise the hydrophobic portion of a molecule and the polar bonds comprise the hydrophilic portion.
- For the same number of carbons, increasing the number of double bonds decreases the melting point of a fatty acid.

a. and b.

$CH_3(CH_2)_{16}COOH$ =

A

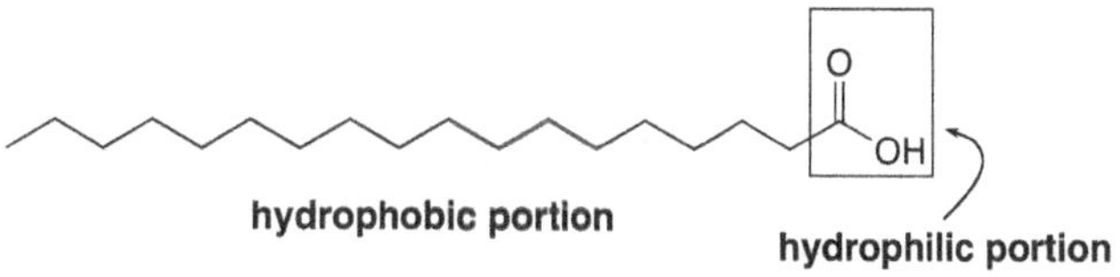

O OH

hydrophobic portion **hydrophilic portion**

$CH_3(CH_2)_4CH{=}CHCH_2CH{=}CH(CH_2)_7COOH$ =

B

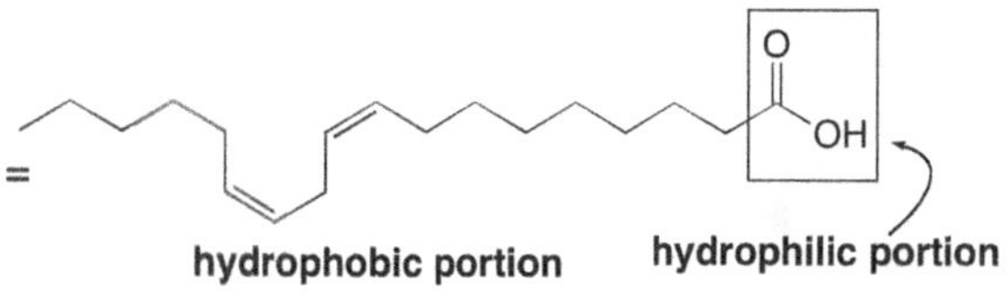

O OH

hydrophobic portion **hydrophilic portion**

c. **A** will have the higher melting point because the molecules can pack together better.

19.4 In omega-*n* acids, *n* is the carbon at which the first double bond occurs in the carbon chain, beginning at the end of the chain that contains the CH_3 group.

a. 1 3 COOH

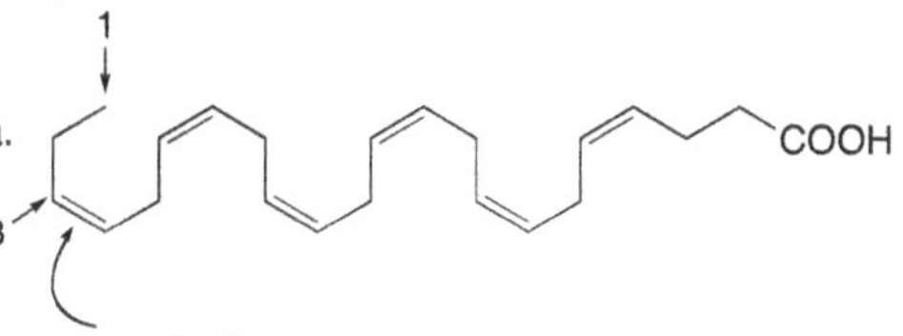

b. first C=C at C3 → an **omega-3 acid**

19.5 In omega-*n* acids, *n* is the carbon at which the first double bond occurs in the carbon chain, beginning at the end of the chain that contains the CH_3 group.

first C=C at C9 → an **omega-9 acid**

a. $CH_3CH_2CH_2CH_2CH_2CH_2CH_2CH_2CH{=}CHCH_2CH_2CH_2CH_2CH_2CH_2CH_2COOH$

1 2 3 4 5 6 7 8 9

first C=C at C6 → an **omega-6 acid**

b. 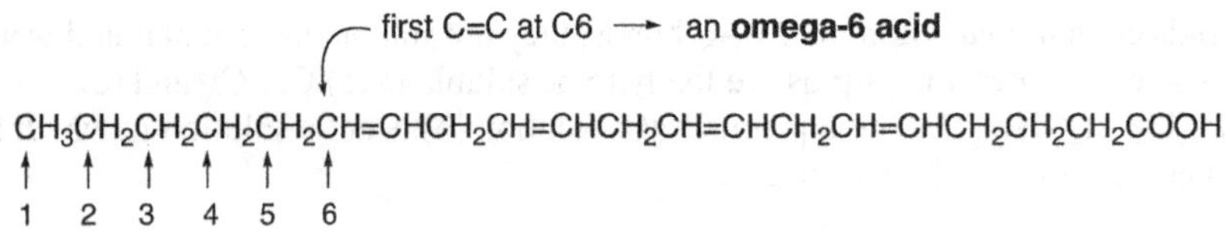 $CH_3CH_2CH_2CH_2CH_2CH{=}CHCH_2CH{=}CHCH_2CH{=}CHCH_2CH{=}CHCH_2CH_2CH_2COOH$

1 2 3 4 5 6

19.6 Use the steps in Example 19.3 to draw each wax.

a.  $CH_3(CH_2)_{16}COOH + H{-}O(CH_2)_9CH_3 \longrightarrow CH_3(CH_2)_{16}C(=O)O(CH_2)_9CH_3 + H_2O$

b. 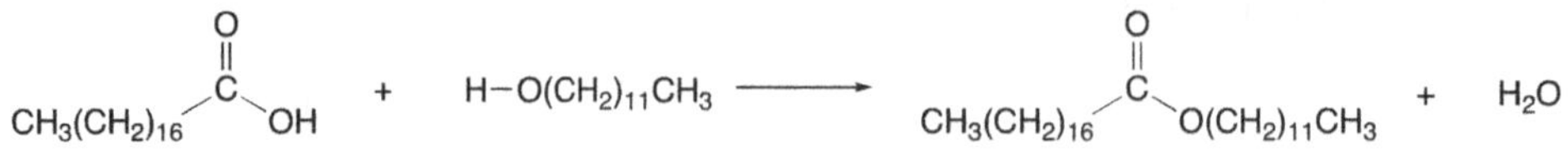 $CH_3(CH_2)_{16}COOH + H{-}O(CH_2)_{11}CH_3 \longrightarrow CH_3(CH_2)_{16}C(=O)O(CH_2)_{11}CH_3 + H_2O$

c. $CH_3(CH_2)_{16}COOH + H{-}O(CH_2)_{29}CH_3 \longrightarrow CH_3(CH_2)_{16}C(=O)O(CH_2)_{29}CH_3 + H_2O$

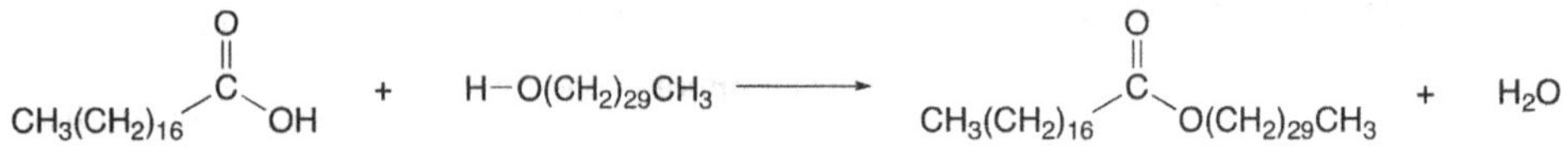

19.7 Draw the structure of the wax formed from eicosenoic acid.

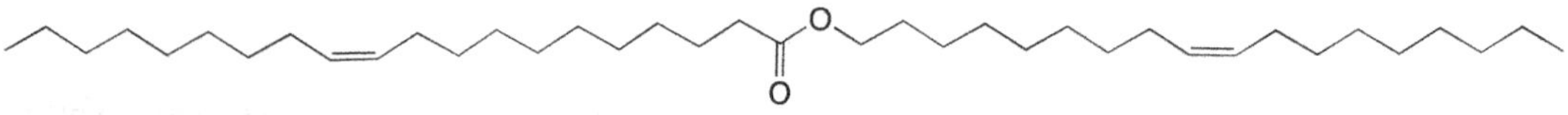

O

O

19.8 Beeswax is hydrophobic because it contains very long hydrocarbon chains. It will therefore be insoluble in a very polar solvent like water, only slightly soluble in the less polar solvent ethanol, and very soluble in the weakly polar solvent chloroform.

19.9 Draw the products of the hydrolysis of cetyl laurate.

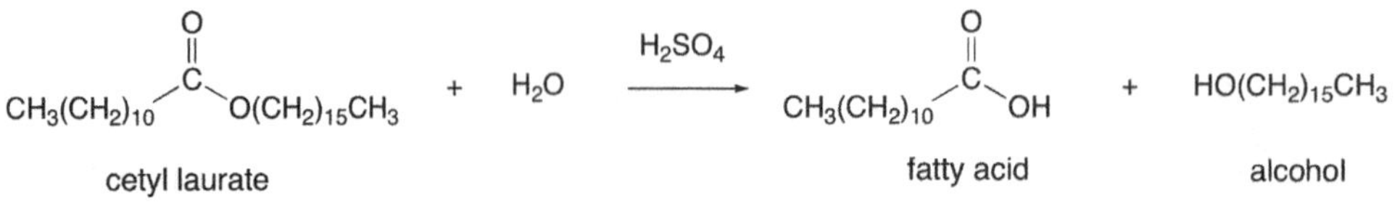

$$CH_3(CH_2)_{10}C(=O)O(CH_2)_{15}CH_3 + H_2O \xrightarrow{H_2SO_4} CH_3(CH_2)_{10}C(=O)OH + HO(CH_2)_{15}CH_3$$

cetyl laurate — fatty acid — alcohol

19.10 Draw the structure of each triacylglycerol as in Example 19.4.

a. 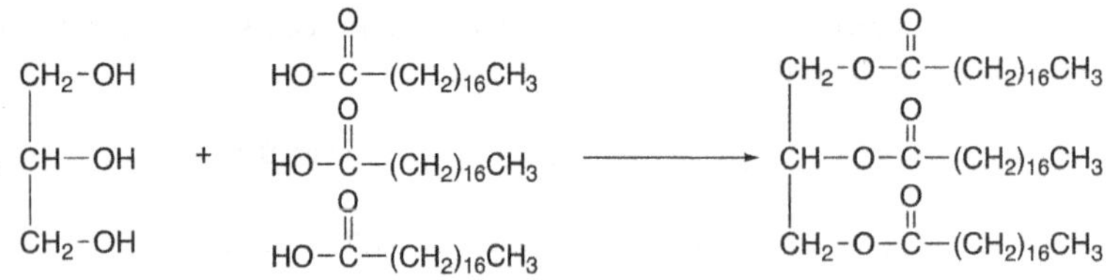 $CH_2{-}OH$, $CH{-}OH$, $CH_2{-}OH$ + 3 $HO{-}C(=O){-}(CH_2)_{16}CH_3$ ⟶ $CH_2{-}O{-}C(=O){-}(CH_2)_{16}CH_3$, $CH{-}O{-}C(=O){-}(CH_2)_{16}CH_3$, $CH_2{-}O{-}C(=O){-}(CH_2)_{16}CH_3$

b. 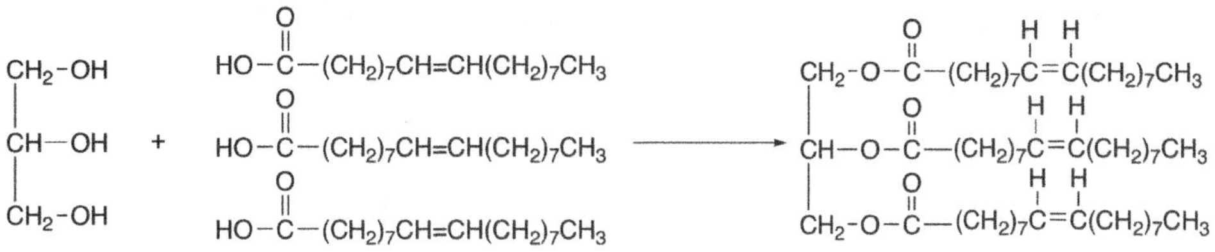 Glycerol + 3 oleic acid → triacylglycerol

$$\begin{matrix} CH_2-OH \\ | \\ CH-OH \\ | \\ CH_2-OH \end{matrix} \quad + \quad \begin{matrix} HO-C(=O)-(CH_2)_7CH=CH(CH_2)_7CH_3 \\ HO-C(=O)-(CH_2)_7CH=CH(CH_2)_7CH_3 \\ HO-C(=O)-(CH_2)_7CH=CH(CH_2)_7CH_3 \end{matrix} \longrightarrow \begin{matrix} CH_2-O-C(=O)-(CH_2)_7C(H)=C(H)(CH_2)_7CH_3 \\ | \\ CH-O-C(=O)-(CH_2)_7C(H)=C(H)(CH_2)_7CH_3 \\ | \\ CH_2-O-C(=O)-(CH_2)_7C(H)=C(H)(CH_2)_7CH_3 \end{matrix}$$

19.11

$$\begin{matrix} CH_2-O-C(=O)-(CH_2)_{16}CH_3 & \leftarrow \text{from stearic acid} \\ | & \\ CH-O-C(=O)-(CH_2)_7CH=CH(CH_2)_7CH_3 & \leftarrow \text{from oleic acid} \\ | & \\ CH_2-O-C(=O)-(CH_2)_{16}CH_3 & \leftarrow \text{from stearic acid} \end{matrix}$$

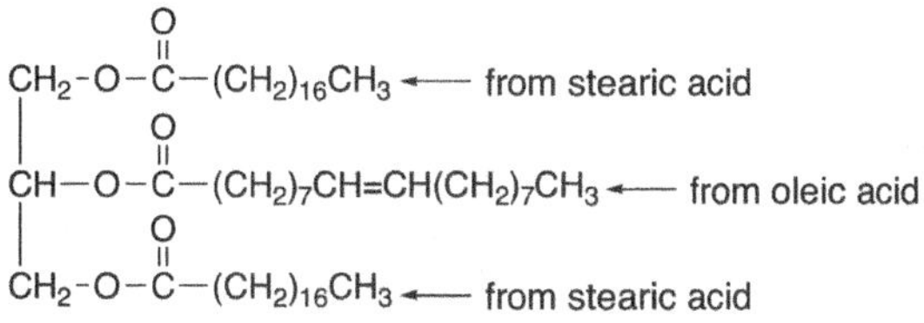

19.12 Draw a triacylglycerol to fit each description.

a. 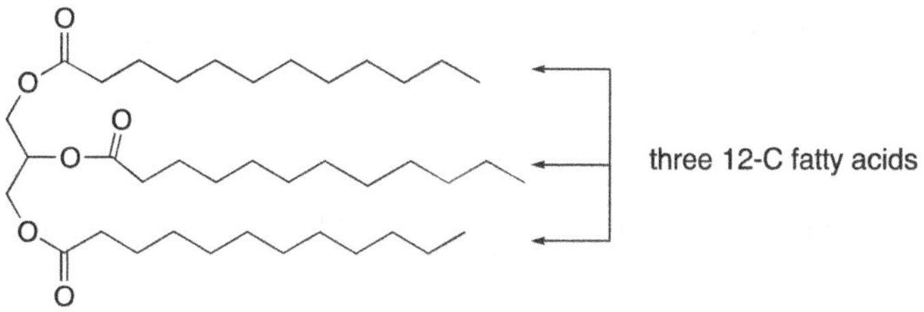 three 12-C fatty acids

b. 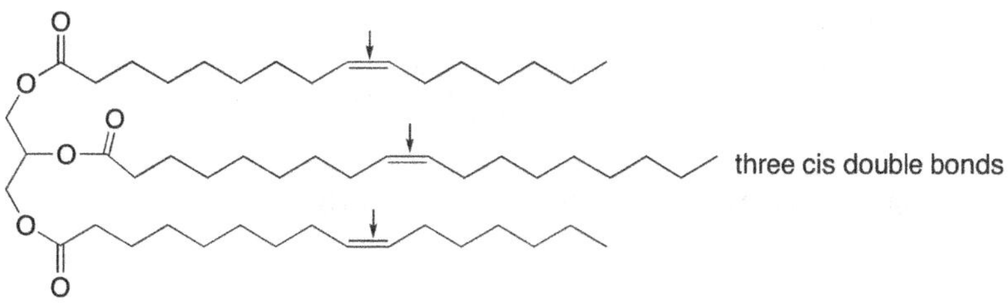 three cis double bonds

c. trans double bonds

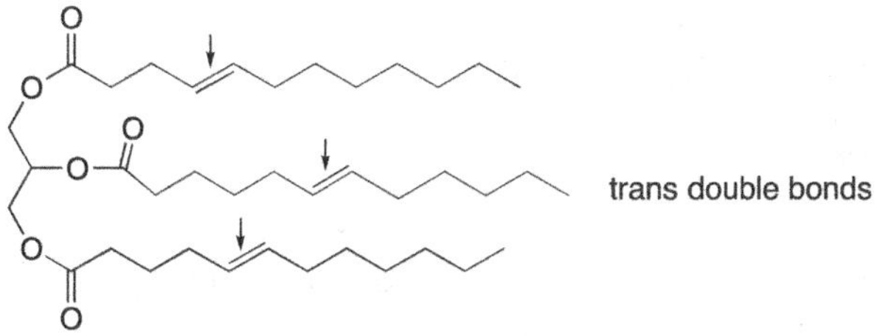

19.13 Draw the products of the hydrolysis of each triacylglycerol as in Example 19.5.

a.

$$\begin{array}{l} CH_2\text{-}O\text{-}\overset{O}{\overset{\|}{C}}\text{-}(CH_2)_{12}CH_3 \\ CH\text{-}O\text{-}\overset{O}{\overset{\|}{C}}\text{-}(CH_2)_{12}CH_3 \\ CH_2\text{-}O\text{-}\overset{O}{\overset{\|}{C}}\text{-}(CH_2)_{12}CH_3 \end{array} \xrightarrow[H_2SO_4]{3\ H\text{-}OH} \begin{array}{l} CH_2\text{-}OH \\ CH\text{-}OH \\ CH_2\text{-}OH \end{array} + 3\ HO\text{-}\overset{O}{\overset{\|}{C}}\text{-}(CH_2)_{12}CH_3$$

b.

$$\begin{array}{l} CH_2\text{-}O\text{-}\overset{O}{\overset{\|}{C}}\text{-}(CH_2)_{12}CH_3 \\ CH\text{-}O\text{-}\overset{O}{\overset{\|}{C}}\text{-}(CH_2)_7CH{=}CH(CH_2)_5CH_3 \\ CH_2\text{-}O\text{-}\overset{O}{\overset{\|}{C}}\text{-}(CH_2)_7CH{=}CH(CH_2)_7CH_3 \end{array} \xrightarrow[H_2SO_4]{3\ H\text{-}OH} \begin{array}{l} CH_2\text{-}OH \\ CH\text{-}OH \\ CH_2\text{-}OH \end{array} + \begin{array}{l} HO\text{-}\overset{O}{\overset{\|}{C}}\text{-}(CH_2)_{12}CH_3 \\ HO\text{-}\overset{O}{\overset{\|}{C}}\text{-}(CH_2)_7CH{=}CH(CH_2)_5CH_3 \\ HO\text{-}\overset{O}{\overset{\|}{C}}\text{-}(CH_2)_7CH{=}CH(CH_2)_7CH_3 \end{array}$$

19.14 Balance the equation for the combustion of tristearin.

$$2\ \begin{array}{l} CH_2\text{-}O\text{-}\overset{O}{\overset{\|}{C}}\text{-}(CH_2)_{16}CH_3 \\ CH\text{-}O\text{-}\overset{O}{\overset{\|}{C}}\text{-}(CH_2)_{16}CH_3 \\ CH_2\text{-}O\text{-}\overset{O}{\overset{\|}{C}}\text{-}(CH_2)_{16}CH_3 \end{array} + 163\ O_2 \xrightarrow[\text{enzymes}]{} 114\ CO_2 + 110\ H_2O$$

tristearin
$C_{57}H_{110}O_6$

total of 338 O atoms on each side
220 H atoms
114 C atoms

19.15 Draw the soap prepared by saponification of each triacylglycerol.

a.

$$\begin{array}{l} CH_2\text{-}O\text{-}\overset{O}{\overset{\|}{C}}\text{-}(CH_2)_{12}CH_3 \\ CH\text{-}O\text{-}\overset{O}{\overset{\|}{C}}\text{-}(CH_2)_{12}CH_3 \\ CH_2\text{-}O\text{-}\overset{O}{\overset{\|}{C}}\text{-}(CH_2)_{12}CH_3 \end{array} \xrightarrow[H_2O]{NaOH} 3\ Na^+\ {}^-O\text{-}\overset{O}{\overset{\|}{C}}\text{-}(CH_2)_{12}CH_3$$

b.

$$\begin{array}{l} CH_2\text{-}O\text{-}\overset{O}{\overset{\|}{C}}\text{-}(CH_2)_{12}CH_3 \\ CH\text{-}O\text{-}\overset{O}{\overset{\|}{C}}\text{-}(CH_2)_7CH{=}CH(CH_2)_5CH_3 \\ CH_2\text{-}O\text{-}\overset{O}{\overset{\|}{C}}\text{-}(CH_2)_7CH{=}CH(CH_2)_7CH_3 \end{array} \xrightarrow[H_2O]{NaOH} \begin{array}{l} Na^+\ {}^-O\text{-}\overset{O}{\overset{\|}{C}}\text{-}(CH_2)_{12}CH_3 \\ Na^+\ {}^-O\text{-}\overset{O}{\overset{\|}{C}}\text{-}(CH_2)_7CH{=}CH(CH_2)_5CH_3 \\ Na^+\ {}^-O\text{-}\overset{O}{\overset{\|}{C}}\text{-}(CH_2)_7CH{=}CH(CH_2)_7CH_3 \end{array}$$

19.16 Draw the structure of the two cephalins as in Example 19.6.

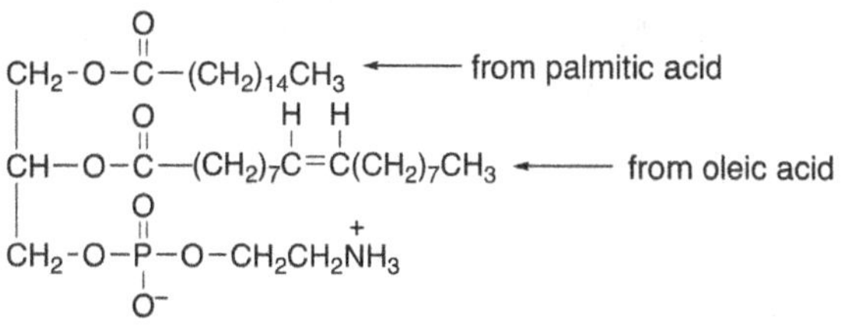

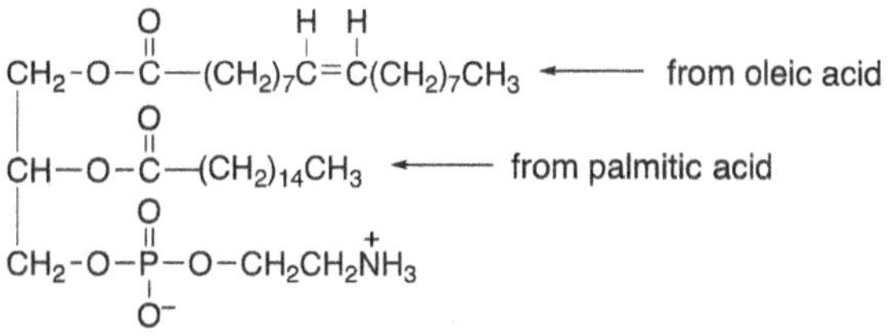

$CH_2\text{-}O\text{-}C(=O)\text{-}(CH_2)_{14}CH_3$ ← from palmitic acid
$CH\text{-}O\text{-}C(=O)\text{-}(CH_2)_7CH{=}CH(CH_2)_7CH_3$ ← from oleic acid
$CH_2\text{-}O\text{-}P(=O)(O^-)\text{-}O\text{-}CH_2CH_2\overset{+}{N}H_3$

$CH_2\text{-}O\text{-}C(=O)\text{-}(CH_2)_7CH{=}CH(CH_2)_7CH_3$ ← from oleic acid
$CH\text{-}O\text{-}C(=O)\text{-}(CH_2)_{14}CH_3$ ← from palmitic acid
$CH_2\text{-}O\text{-}P(=O)(O^-)\text{-}O\text{-}CH_2CH_2\overset{+}{N}H_3$

19.17 A **triacylglycerol** has glycerol as the backbone, and three nonpolar side chains formed from esters with fatty acids.
A **phosphoacylglycerol** has glycerol as the backbone, a phosphodiester on a terminal carbon, and two nonpolar side chains formed from esters with fatty acids.
A **sphingomyelin** has sphingosine as the backbone, one amide, and a phosphodiester on a terminal carbon.

a. phosphoacylglycerol, cephalin

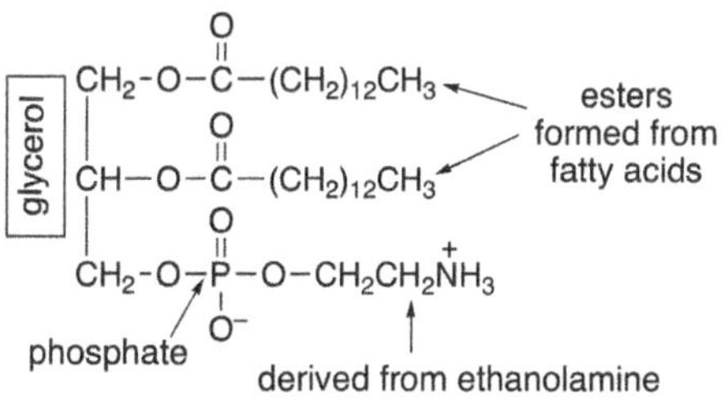

glycerol:
$CH_2\text{-}O\text{-}C(=O)\text{-}(CH_2)_{12}CH_3$
$CH\text{-}O\text{-}C(=O)\text{-}(CH_2)_{12}CH_3$ — esters formed from fatty acids
$CH_2\text{-}O\text{-}P(=O)(O^-)\text{-}O\text{-}CH_2CH_2\overset{+}{N}H_3$
phosphate; derived from ethanolamine

c. sphingomyelin

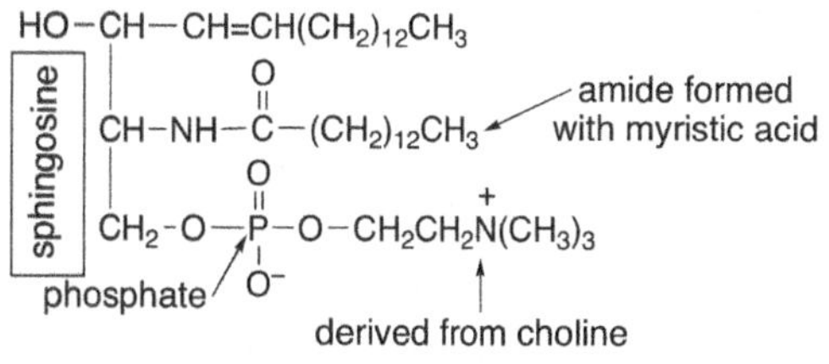

sphingosine:
$HO\text{-}CH\text{-}CH{=}CH(CH_2)_{12}CH_3$
$CH\text{-}NH\text{-}C(=O)\text{-}(CH_2)_{12}CH_3$ ← amide formed with myristic acid
$CH_2\text{-}O\text{-}P(=O)(O^-)\text{-}O\text{-}CH_2CH_2\overset{+}{N}(CH_3)_3$
phosphate; derived from choline

b. triacylglycerol

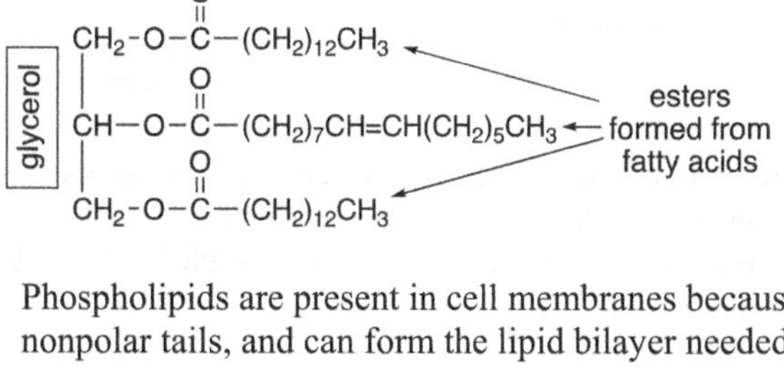

glycerol:
$CH_2\text{-}O\text{-}C(=O)\text{-}(CH_2)_{12}CH_3$
$CH\text{-}O\text{-}C(=O)\text{-}(CH_2)_7CH{=}CH(CH_2)_5CH_3$ — esters formed from fatty acids
$CH_2\text{-}O\text{-}C(=O)\text{-}(CH_2)_{12}CH_3$

19.18 Phospholipids are present in cell membranes because they have an ionic polar head and two nonpolar tails, and can form the lipid bilayer needed for cell membrane function.
Triacylglycerols are basically nonpolar compounds, so they have no polar head to attract water on the outside of a membrane.

19.19 Membrane **A** is formed from the phospholipids linoleic and oleic acids, so it will be more fluid or pliable because it contains unsaturated fatty acids. Membrane **B** is formed from the saturated fatty acids stearic and palmitic acids, which have no double bonds and therefore pack very tightly, making it more rigid.

19.20 Ions don't diffuse readily through the interior of the cell membrane because it is hydrophobic and ions are insoluble in a nonpolar medium.

19.21 Answer each question about cholesterol.

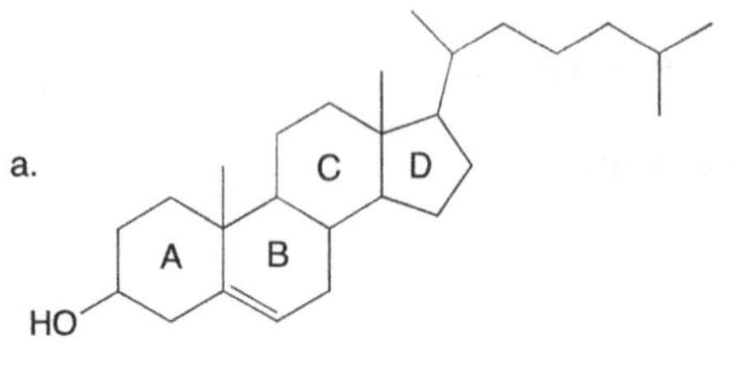

a. A B C D HO

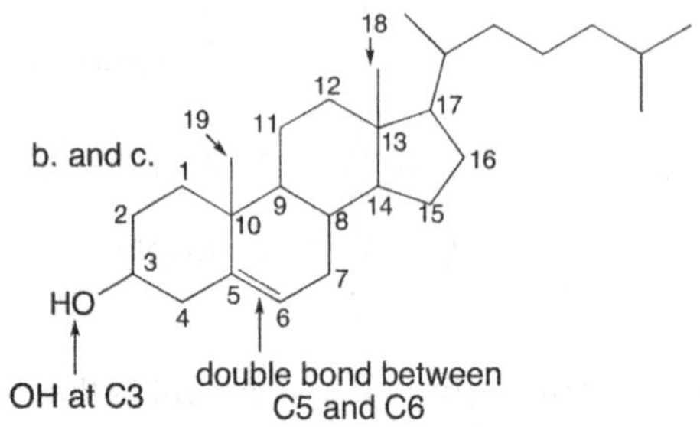

b. and c. 1 2 3 4 5 6 7 8 9 10 11 12 13 14 15 16 17 18 19 HO

OH at C3

double bond between C5 and C6

d. Cholesterol contains one polar C–O bond and one polar O–H bond from the polar OH group. The large hydrocarbon skeleton with nonpolar C–C and C–H bonds makes it water insoluble.

19.22 Triacylglycerols would be found in the interior of a lipoprotein particle, because this is the hydrophobic region of lipoproteins.

19.23 Label the functional groups.

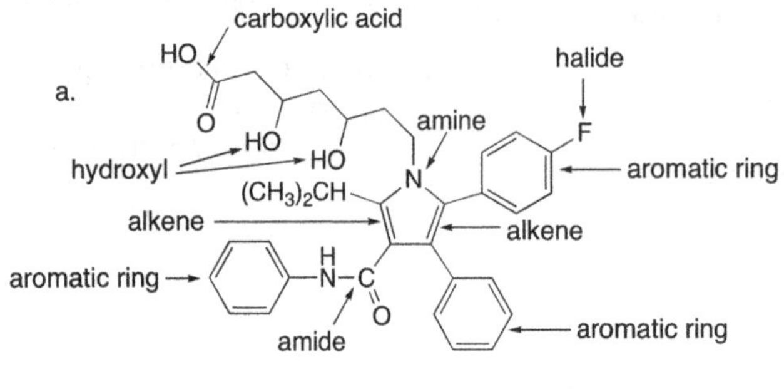

a. carboxylic acid, HO, O, halide, F, amine, HO, HO, hydroxyl, $(CH_3)_2CH$, N, aromatic ring, alkene, alkene, aromatic ring, H–N–C=O, amide, aromatic ring

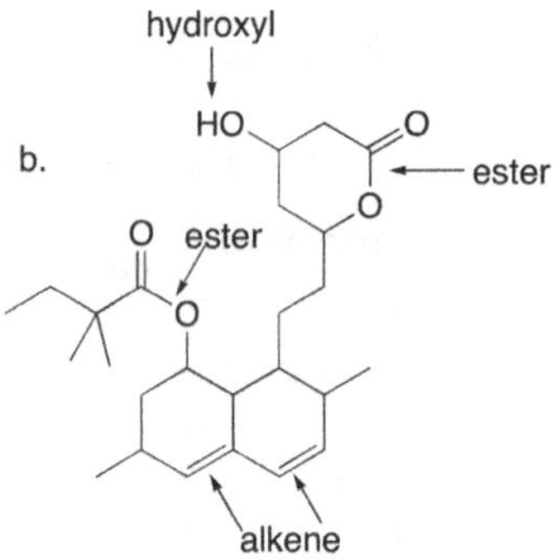

b. hydroxyl, HO, O, ester, O, ester, O, O, alkene

19.24 a. Estrone has a phenol (a benzene ring with a hydroxyl group), whereas progesterone has a ketone and C=C in ring A. Progesterone also has a methyl group bonded to C10.

b. Estrone has a ketone at C17, whereas progesterone has a C–C bond, which is attached to a ketone.

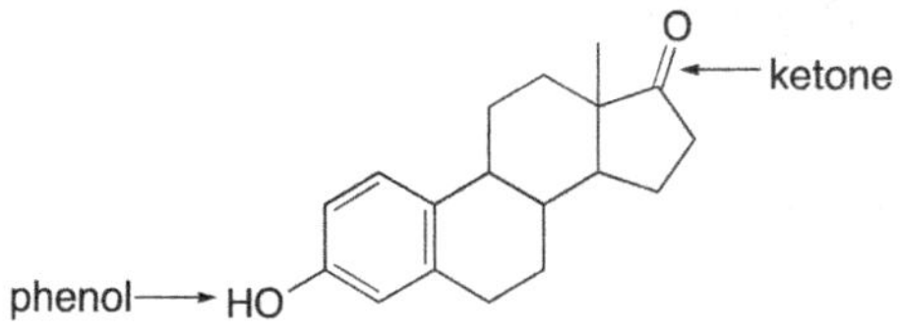

19.25 Testosterone has a methyl group at C10 that nandrolone lacks.

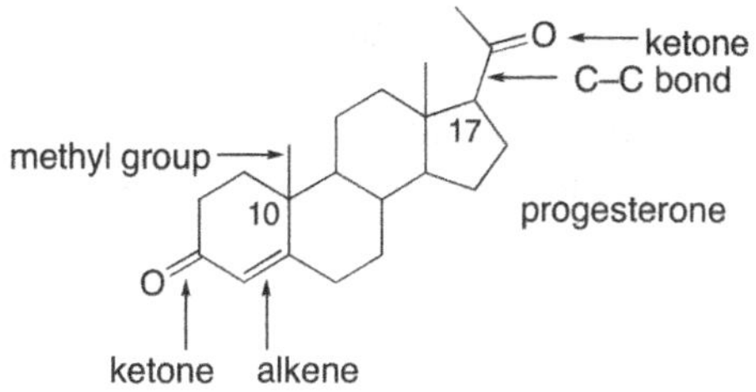

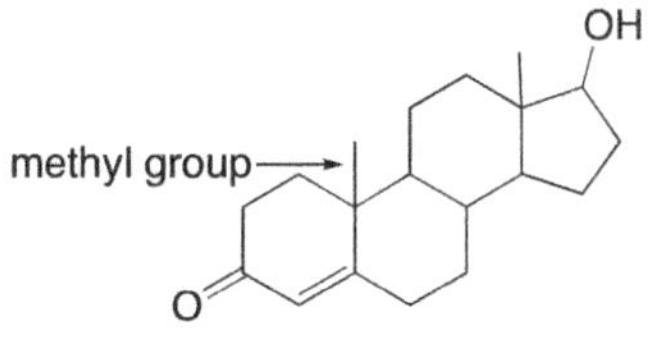

19.26 Water-soluble vitamins are excreted in the urine whereas fat-soluble vitamins are stored in the body. When a person ingests a large quantity of a water-soluble vitamin, much of it is excreted in the urine. When a person ingests a large quantity of a fat-soluble vitamin, it can be retained in the fat in the body, potentially building up to toxic levels.

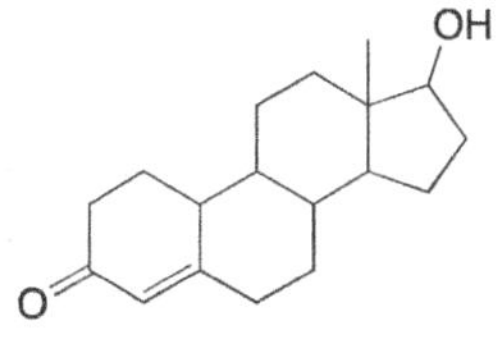

19.27 Label the functional groups and draw the skeletal structure for PGE_2.

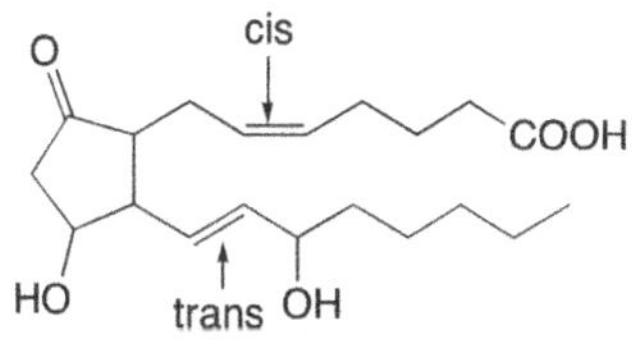

19.28 Label the functional groups and the double bonds as cis or trans in LTC_4.

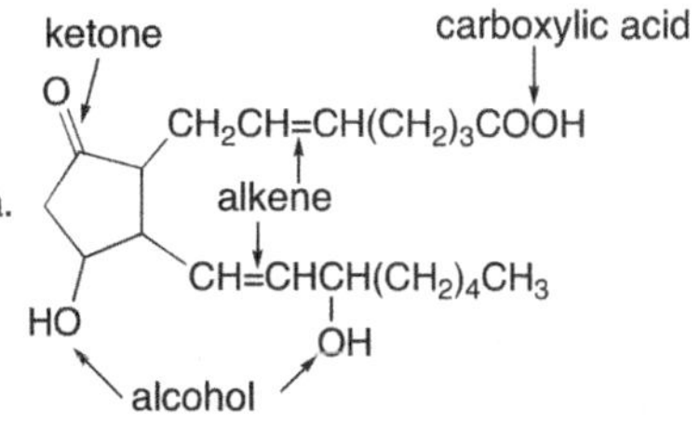

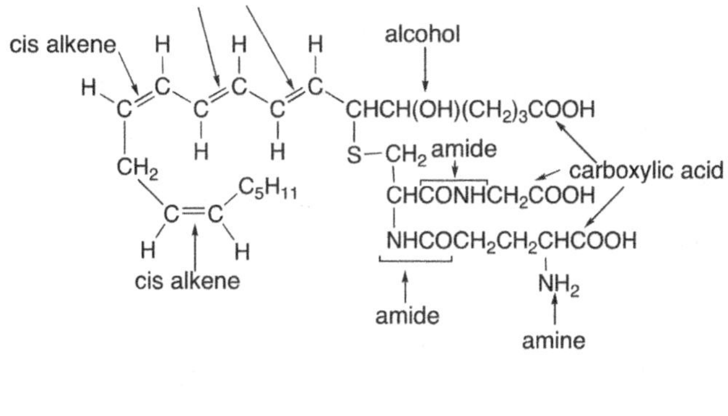

Solutions to Odd-Numbered End-of-Chapter Problems

19.29 **Hydrolyzable lipids** include waxes, triacylglycerols, and phospholipids.
Nonhydrolyzable lipids include steroids, fat-soluble vitamins, and eicosanoids.

a. prostaglandin—nonhydrolyzable
b. triacylglycerol—hydrolyzable
c. leukotriene—nonhydrolyzable
d. vitamin A—nonhydrolyzable
e. phosphoacylglycerol—hydrolyzable
f. lecithin—hydrolyzable
g. cholesterol—nonhydrolyzable

19.31 A wax is hydrophobic. As a result, it is soluble in (b) CH_2Cl_2 and (c) $CH_3CH_2OCH_2CH_3$, both organic solvents, but insoluble in (a) water, which is polar.

19.33 For the same number of carbons, increasing the number of double bonds decreases the melting point of a fatty acid. In order of increasing melting point:

$CH_3(CH_2)_5CH{=}CH(CH_2)_7COOH$, $CH_3(CH_2)_7CH{=}CH(CH_2)_7COOH$, $CH_3(CH_2)_{16}COOH$

19.35 For the same number of carbons, increasing the number of double bonds decreases the melting point of a fatty acid. In order of increasing melting point:

$$\underset{\textbf{B}}{CH_3CH_2CH{=}CH(CH_2)_7CO_2H} < \underset{\textbf{A}}{CH_3(CH_2)_{10}CO_2H} < \underset{\textbf{C}}{CH_3(CH_2)_{12}CO_2H}$$

19.37 Answer each question about 7,10,13,16,19-docosapentaenoic acid.
a. and b.

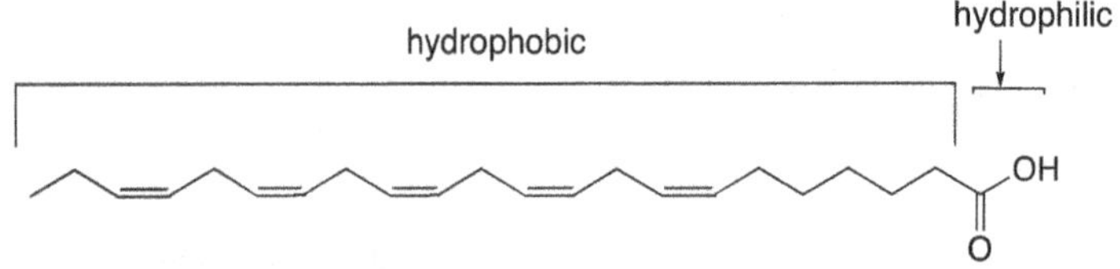

c. The melting point of the cis isomer would be lower than the melting point of the trans isomer.
d. This fatty acid will be a liquid at room temperature.
e. 7,10,13,16,19-Docosapentaenoic acid is an omega-3 fatty acid because there is a double bond at the third C from the left (numbered from the CH_3 group).

19.39

a.  cis CO_2H

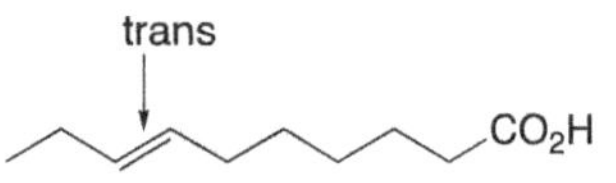

b. omega-3 acid

c. trans CO_2H

d. 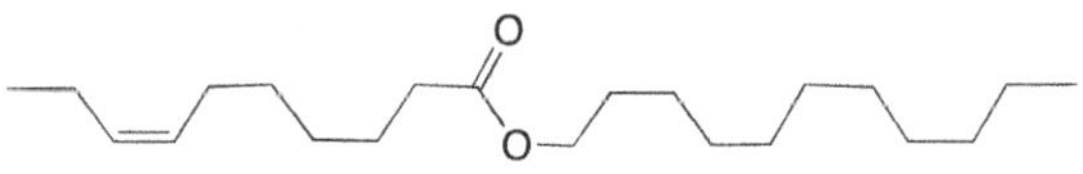 O O

19.41 Draw the structure of each wax.

a.  $CH_3(CH_2)_{14}C(=O)OH + H{-}O(CH_2)_{21}CH_3 \longrightarrow CH_3(CH_2)_{14}C(=O)O(CH_2)_{21}CH_3 + H_2O$

b. $CH_3(CH_2)_{14}C(=O)OH$ + $H{-}O(CH_2)_{11}CH_3$ ⟶ $CH_3(CH_2)_{14}C(=O)O(CH_2)_{11}CH_3$ + H_2O

c. $CH_3(CH_2)_{14}C(=O)OH$ + $H{-}O(CH_2)_9CH_3$ ⟶ $CH_3(CH_2)_{14}C(=O)O(CH_2)_9CH_3$ + H_2O

19.43 Draw the products of the hydrolysis of each wax.

a. $CH_3(CH_2)_{16}C(=O)O(CH_2)_{17}CH_3$ + H_2O $\xrightarrow{H_2SO_4}$ $CH_3(CH_2)_{16}C(=O)OH$ + $HO(CH_2)_{17}CH_3$

b. $CH_3(CH_2)_{12}C(=O)O(CH_2)_{25}CH_3$ + H_2O $\xrightarrow{H_2SO_4}$ $CH_3(CH_2)_{12}C(=O)OH$ + $HO(CH_2)_{25}CH_3$

c. $CH_3(CH_2)_{14}C(=O)O(CH_2)_{27}CH_3$ + H_2O $\xrightarrow{H_2SO_4}$ $CH_3(CH_2)_{14}C(=O)OH$ + $HO(CH_2)_{27}CH_3$

19.45 a. **B,** macadamia nuts, is high in unsaturated triacylglycerols.
b. **C,** cheese, is high in saturated fat.
c. **A,** margarine, most likely contains trans fat.

19.47 Draw a triacylglycerol that fits each description.

a.
$CH_2{-}O{-}C(=O){-}(CH_2)_{10}CH_3$ ← lauric acid
$CH{-}O{-}C(=O){-}(CH_2)_{12}CH_3$ ← myristic acid
$CH_2{-}O{-}C(=O){-}(CH_2)_7CH{=}CHCH_2CH{=}CH(CH_2)_4CH_3$ ← linoleic acid

triacylglycerol

b.
$CH_2{-}O{-}C(=O){-}$[skeletal chain]
$CH{-}O{-}C(=O){-}$[skeletal chain]
$CH_2{-}O{-}C(=O){-}(CH_2)_7CH{=}CHCH_2CH{=}CH(CH_2)_4CH_3$ (two cis double bonds)

unsaturated triacylglycerol

c.
$CH_2{-}O{-}C(=O){-}(CH_2)_{12}CH_3$
$CH{-}O{-}C(=O){-}(CH_2)_{12}CH_3$
$CH_2{-}O{-}C(=O){-}(CH_2)_{12}CH_3$

saturated triacylglycerol

d.
$CH_2{-}O{-}C(=O){-}(CH_2)_{12}CH_3$
$CH{-}O{-}C(=O){-}(CH_2)_{12}CH_3$
$CH_2{-}O{-}C(=O){-}(CH_2)_7CH{=}CH(CH_2)_7CH_3$ (one double bond)

monounsaturated triacylglycerol

19.49 Draw the triacylglycerol that fits the description.

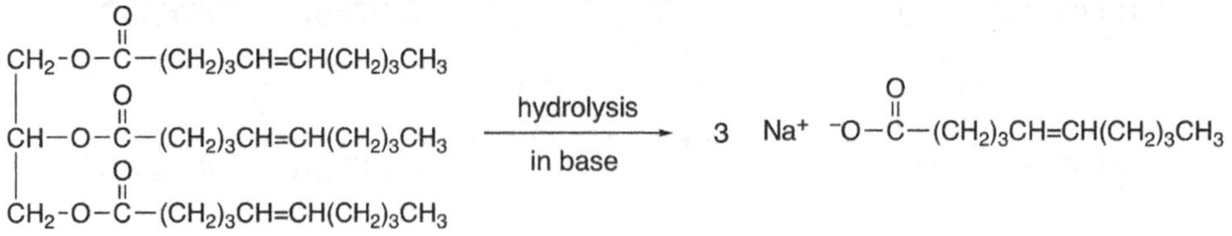

$$\begin{array}{l} CH_2\text{-}O\text{-}\overset{O}{\overset{\|}{C}}\text{-}(CH_2)_3CH{=}CH(CH_2)_3CH_3 \\ CH\text{-}O\text{-}\overset{O}{\overset{\|}{C}}\text{-}(CH_2)_3CH{=}CH(CH_2)_3CH_3 \\ CH_2\text{-}O\text{-}\overset{O}{\overset{\|}{C}}\text{-}(CH_2)_3CH{=}CH(CH_2)_3CH_3 \end{array} \xrightarrow[\text{in base}]{\text{hydrolysis}} 3\ \ Na^+\ {}^-O\text{-}\overset{O}{\overset{\|}{C}}\text{-}(CH_2)_3CH{=}CH(CH_2)_3CH_3$$

19.51 Answer each question.

Compound	**a. General structure**	**b. Example**	**c. Water soluble (Y/N)**	**d. Hexane soluble (Y/N)**
[1] Fatty acid	RCOOH	$CH_3(CH_2)_{10}CO_2H$	N	Y
[2] Soap	$RCOO^-Na^+$	$CH_3(CH_2)_{10}CO_2^-\ Na^+$	Y	N
[3] Wax	RCOOR'	$CH_3(CH_2)_6CO_2(CH_2)_7CH_3$	N	Y
[4] Triacylglycerol	$CH_2\text{-}O\text{-}\overset{O}{\overset{\|}{C}}\text{-}R$ / $CH\text{-}O\text{-}\overset{O}{\overset{\|}{C}}\text{-}R'$ / $CH_2\text{-}O\text{-}\overset{O}{\overset{\|}{C}}\text{-}R''$	$CH_2\text{-}O\text{-}\overset{O}{\overset{\|}{C}}\text{-}(CH_2)_{12}CH_3$ / $CH\text{-}O\text{-}\overset{O}{\overset{\|}{C}}\text{-}(CH_2)_{12}CH_3$ / $CH_2\text{-}O\text{-}\overset{O}{\overset{\|}{C}}\text{-}(CH_2)_{12}CH_3$	N	Y

19.53 Answer each question about the triacylglycerol.

a.

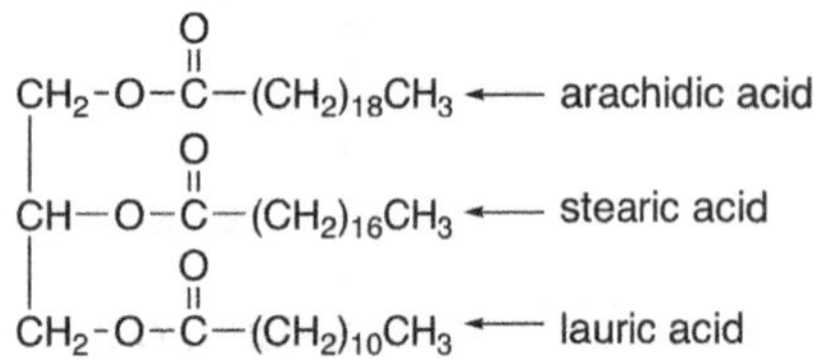

$$\begin{array}{ll} CH_2\text{-}O\text{-}\overset{O}{\overset{\|}{C}}\text{-}(CH_2)_{18}CH_3 & \leftarrow \text{arachidic acid} \\ CH\text{-}O\text{-}\overset{O}{\overset{\|}{C}}\text{-}(CH_2)_{16}CH_3 & \leftarrow \text{stearic acid} \\ CH_2\text{-}O\text{-}\overset{O}{\overset{\|}{C}}\text{-}(CH_2)_{10}CH_3 & \leftarrow \text{lauric acid} \end{array}$$

e.

$$\begin{array}{l} CH_2\text{-}OH \\ CH\text{-}OH \\ CH_2\text{-}OH \end{array} \quad + \quad \begin{array}{l} HOOC(CH_2)_{18}CH_3 \\ HOOC(CH_2)_{16}CH_3 \\ HOOC(CH_2)_{10}CH_3 \end{array}$$

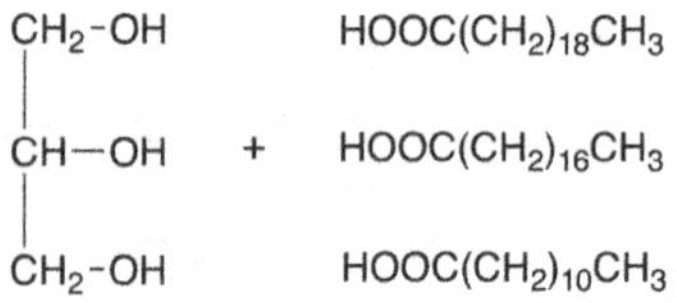

b. It would be a solid at room temperature because it is formed from saturated fatty acids.
c. The long hydrocarbon chains are hydrophobic.
d. The ester linkages are hydrophilic.

19.55

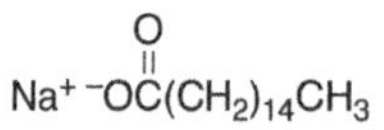

$$Na^+\ {}^-O\overset{O}{\overset{\|}{C}}(CH_2)_{14}CH_3$$

sodium palmitate

19.57 Draw the products of triacylglycerol hydrolysis.

a.

$$\begin{array}{l} CH_2\text{-}O\text{-}\overset{O}{\overset{\|}{C}}\text{-}(CH_2)_{14}CH_3 \\ CH\text{-}O\text{-}\overset{O}{\overset{\|}{C}}\text{-}(CH_2)_{14}CH_3 \\ CH_2\text{-}O\text{-}\overset{O}{\overset{\|}{C}}\text{-}(CH_2)_{16}CH_3 \end{array} \xrightarrow[H_2SO_4]{H_2O} \begin{array}{l} CH_2\text{-}OH \\ CH\text{-}OH \\ CH_2\text{-}OH \end{array} + \begin{array}{l} 2\ HO\overset{O}{\overset{\|}{C}}\text{-}(CH_2)_{14}CH_3 \\ HO\overset{O}{\overset{\|}{C}}\text{-}(CH_2)_{16}CH_3 \end{array}$$

$$\begin{array}{l} CH_2\text{-}O\text{-}\overset{O}{\overset{\|}{C}}\text{-}(CH_2)_{14}CH_3 \\ CH\text{-}O\text{-}\overset{O}{\overset{\|}{C}}\text{-}(CH_2)_{14}CH_3 \\ CH_2\text{-}O\text{-}\overset{O}{\overset{\|}{C}}\text{-}(CH_2)_{16}CH_3 \end{array} \xrightarrow[NaOH]{H_2O} \begin{array}{l} CH_2\text{-}OH \\ CH\text{-}OH \\ CH_2\text{-}OH \end{array} + \begin{array}{l} 2\ Na^{+}\ {}^{-}O\overset{O}{\overset{\|}{C}}\text{-}(CH_2)_{14}CH_3 \\ Na^{+}\ {}^{-}O\overset{O}{\overset{\|}{C}}\text{-}(CH_2)_{16}CH_3 \end{array}$$

b.

$$\begin{array}{l} CH_2\text{-}O\text{-}\overset{O}{\overset{\|}{C}}\text{-}(CH_2)_{14}CH_3 \\ CH\text{-}O\text{-}\overset{O}{\overset{\|}{C}}\text{-}(CH_2)_7CH{=}CH(CH_2)_7CH_3 \\ CH_2\text{-}O\text{-}\overset{O}{\overset{\|}{C}}\text{-}(CH_2)_7CH{=}CH(CH_2)_5CH_3 \end{array} \xrightarrow[H_2SO_4]{H_2O} \begin{array}{l} CH_2\text{-}OH \\ CH\text{-}OH \\ CH_2\text{-}OH \end{array} + \begin{array}{l} HO\overset{O}{\overset{\|}{C}}\text{-}(CH_2)_{14}CH_3 \\ HO\overset{O}{\overset{\|}{C}}\text{-}(CH_2)_7CH{=}CH(CH_2)_7CH_3 \\ HO\overset{O}{\overset{\|}{C}}\text{-}(CH_2)_7CH{=}CH(CH_2)_5CH_3 \end{array}$$

$$\begin{array}{l} CH_2\text{-}O\text{-}\overset{O}{\overset{\|}{C}}\text{-}(CH_2)_{14}CH_3 \\ CH\text{-}O\text{-}\overset{O}{\overset{\|}{C}}\text{-}(CH_2)_7CH{=}CH(CH_2)_7CH_3 \\ CH_2\text{-}O\text{-}\overset{O}{\overset{\|}{C}}\text{-}(CH_2)_7CH{=}CH(CH_2)_5CH_3 \end{array} \xrightarrow[NaOH]{H_2O} \begin{array}{l} CH_2\text{-}OH \\ CH\text{-}OH \\ CH_2\text{-}OH \end{array} + \begin{array}{l} Na^{+}\ {}^{-}O\overset{O}{\overset{\|}{C}}\text{-}(CH_2)_{14}CH_3 \\ Na^{+}\ {}^{-}O\overset{O}{\overset{\|}{C}}\text{-}(CH_2)_7CH{=}CH(CH_2)_7CH_3 \\ Na^{+}\ {}^{-}O\overset{O}{\overset{\|}{C}}\text{-}(CH_2)_7CH{=}CH(CH_2)_5CH_3 \end{array}$$

19.59 Phospholipids are lipids that contain a phosphorus atom. Sphingomyelins (c) contain phosphorus. Triacylglycerols (a), leukotrienes (b), and fatty acids (d) do not.

19.61 Draw a phospholipid that fits each description.

a.

$$\begin{array}{l} CH_2\text{-}O\text{-}\overset{O}{\overset{\|}{C}}\text{-}(CH_2)_7CH{=}CH(CH_2)_5CH_3 \leftarrow \text{from palmitoleic acid} \\ CH\text{-}O\text{-}\overset{O}{\overset{\|}{C}}\text{-}(CH_2)_7CH{=}CH(CH_2)_5CH_3 \leftarrow \text{from palmitoleic acid} \\ CH_2\text{-}O\text{-}\overset{O}{\overset{\|}{P}}\text{-}O\text{-}CH_2CH_2\overset{+}{N}H_3 \\ \qquad\quad\ \ |\\ \qquad\quad\ O^{-} \end{array}$$

cephalin

b.

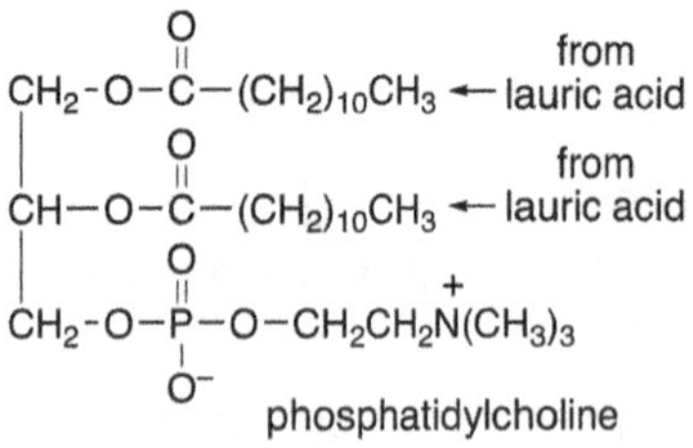

```
        O
        ||                     from
CH2-O-C-(CH2)10CH3  ← lauric acid
 |      O
 |      ||                     from
CH-O-C-(CH2)10CH3  ← lauric acid
 |      O
 |      ||                 +
CH2-O-P-O-CH2CH2N(CH3)3
        |
        O−
```

phosphatidylcholine

c.

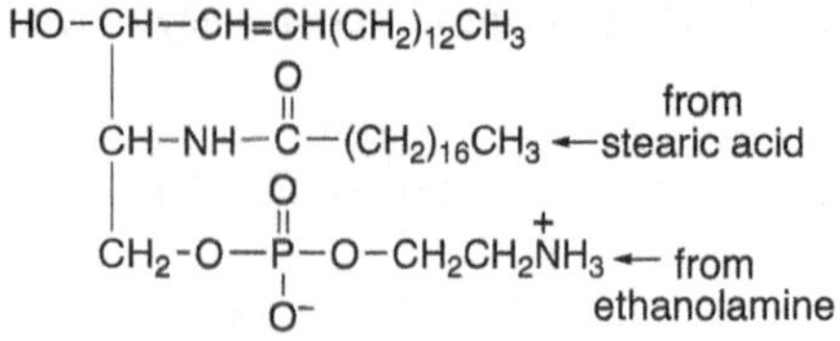

```
HO-CH-CH=CH(CH2)12CH3
    |   O
    |   ||                       from
   CH-NH-C-(CH2)16CH3  ← stearic acid
    |   O
    |   ||               +
   CH2-O-P-O-CH2CH2NH3  ← from
         |                 ethanolamine
         O−
```

sphingomyelin

19.63 Diffusion is the movement of small molecules through a membrane along a concentration gradient. Facilitated transport is the transport of molecules through channels in the cell membrane. O_2 and CO_2 move by diffusion, whereas glucose and Cl^- move by facilitated transport.

19.65 Draw the anabolic steroid 4-androstene-3,17-dione.

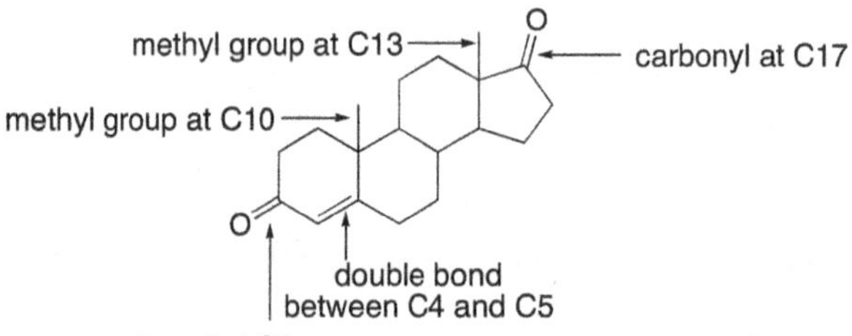

methyl group at C13

carbonyl at C17

methyl group at C10

double bond between C4 and C5

carbonyl at C3

19.67 Cholesterol is insoluble in the aqueous medium of the bloodstream. By being bound to a lipoprotein particle, however, it can be transported in the aqueous solution of the blood.

19.69 Low-density lipoproteins (LDLs) transport cholesterol from the liver to the tissues where it is incorporated in cell membranes. High-density lipoproteins (HDLs) transport cholesterol from the tissues back to the liver. When LDLs supply more cholesterol than is needed, LDLs deposit cholesterol on the walls of arteries, forming plaque. Atherosclerosis is a disease that results from the buildup of these fatty deposits, restricting the flow of blood, increasing blood pressure, and increasing the likelihood of a heart attack or stroke. As a result, LDL cholesterol is often called "bad" cholesterol. Since HDL cholesterol transports cholesterol back to the liver and removes it from the bloodstream, high levels of HDLs reduce the risk of heart disease and stroke, and it is called "good" cholesterol.

19.71 Answer each question about estrone and testosterone.

a.

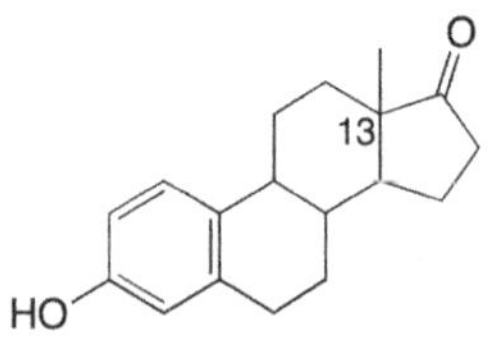

O 13 HO

estrone

OH 13 O

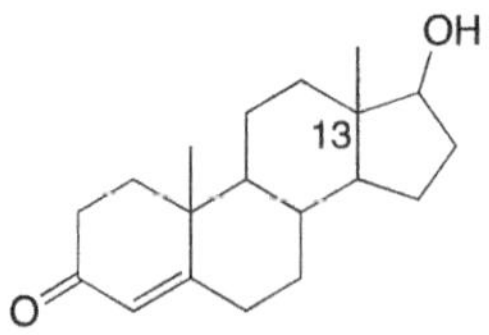

testosterone

b. The estrogen (left) and androgen (right) both contain the four rings of the steroid skeleton. Both contain a methyl group bonded to C13.

c. The estrogen has an aromatic A ring and a hydroxyl group on this ring. The androgen has a carbonyl on the A ring but does not contain an aromatic ring. The androgen also contains a C=C in the A ring and an additional CH_3 group at C10. The D rings are also different. The estrogen contains a carbonyl at C17, whereas the androgen has an OH group.

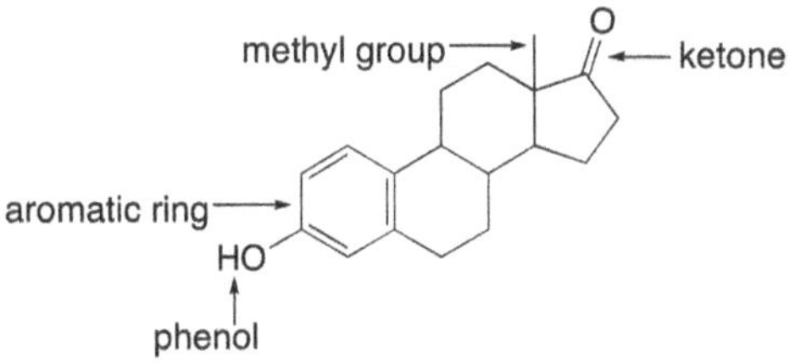

methyl group → ketone; aromatic ring; HO; phenol

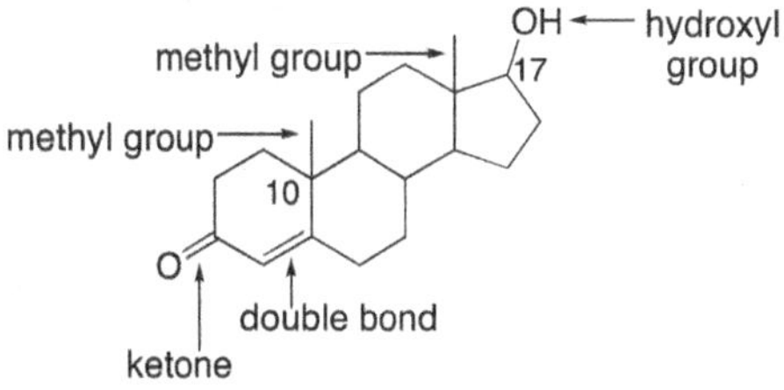

methyl group; methyl group; OH ← hydroxyl group; 17; 10; O; ketone; double bond

d. Estrogens, synthesized in the ovaries, control the menstrual cycle and secondary sexual characteristics of females. Androgens, synthesized in the testes, control the development of male secondary sexual characteristics.

19.73 Prostaglandins and leukotrienes are two types of eicosanoids, a group of biologically active compounds containing 20 carbon atoms derived from the fatty acid arachidonic acid. Prostaglandins are a group of carboxylic acids that contain a five-membered ring. Leukotrienes do not contain a ring. They both mediate biological activity at the site where they are formed. Prostaglandins mediate inflammation and uterine contractions, whereas leukotrienes stimulate smooth muscle contraction in the lungs, leading to the narrowing of airways in individuals with asthma.

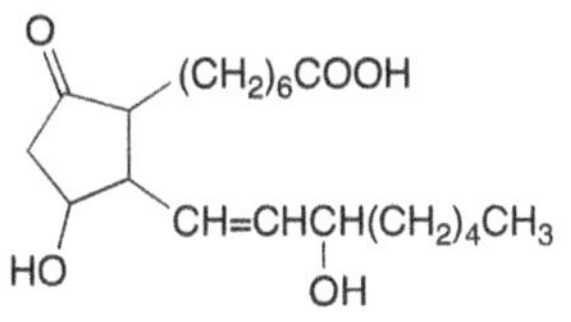

$(CH_2)_6COOH$; $CH{=}CHCH(CH_2)_4CH_3$; OH; HO

a prostaglandin
PGE_1
five-membered ring

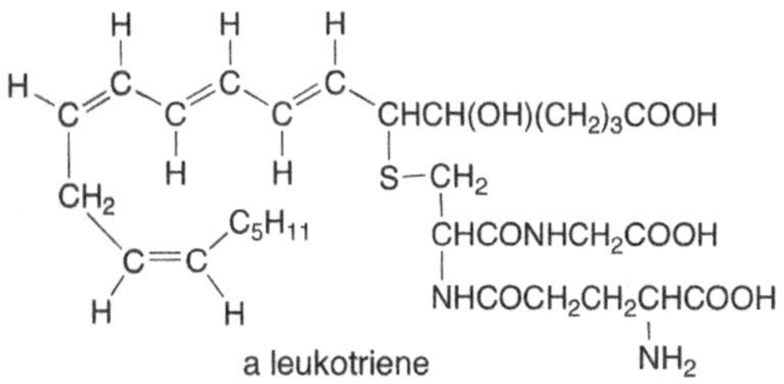

$CHCH(OH)(CH_2)_3COOH$; $S{-}CH_2$; $CHCONHCH_2COOH$; $NHCOCH_2CH_2CHCOOH$; NH_2; CH_2; C_5H_{11}

a leukotriene
LTC_4
no ring

19.75 All prostaglandins contain a five-membered ring and a carboxyl group (COOH).

19.77 Aspirin and celecoxib are both anti-inflammatory medicines. Aspirin inhibits the activity of both the COX-1 and COX-2 enzymes, whereas celecoxib inhibits the activity of COX-2 only.

19.79 Answer each question about vitamins A and D.

	Vitamin A	**Vitamin D**
a.	10 Tetrahedral carbons	21 Tetrahedral carbons
b.	10 Trigonal planar carbons	Six trigonal planar carbons
e.	Required for normal vision	Regulates calcium and phosphorus metabolism
f.	Deficiency causes night blindness.	Deficiency causes rickets and skeletal deformities.
g.	Found in liver, kidney, oily fish, and dairy	Found in milk and breakfast cereals

c. and d.

vitamin A

alkene alkene alkene alcohol

CH_3 H CH_3 H CH_3

$CH_2O{-}H$

H H H H

CH_3 CH_3

polar C–O,O–H

vitamin D

alkene

CH_3 $CH(CH_3)(CH_2)_3CH(CH_3)_2$

H C–C H

alcohol

H–O– $=CH_2$

polar C–O,O–H alkene

19.81 Draw an example of each type of lipid.

a. (skeletal structure with OH and =O) a monounsaturated fatty acid

b. $CH_3(CH_2)_{12}COO(CH_2)_{15}CH_3$
a wax that contains a total of 30 carbons

c.
$CH_2{-}O{-}C(=O){-}(CH_2)_{12}CH_3$
$CH{-}O{-}C(=O){-}(CH_2)_{12}CH_3$ a saturated triacylglycerol
$CH_2{-}O{-}C(=O){-}(CH_2)_{12}CH_3$

d. $HO{-}CH{-}CH{=}CH(CH_2)_{12}CH_3$
$CH{-}NH{-}C(=O){-}(CH_2)_{12}CH_3$
$CH_2{-}O{-}P(=O)(O^-){-}O{-}CH_2CH_2\overset{+}{N}H_3$
a sphingomyelin derived from ethanolamine

19.83 a. prostaglandin—[5] carboxyl group, [6] five-membered ring
b. androgen—[6] five-membered ring, [7] steroid nucleus
c. triacylglycerol—[1] glycerol, [2] fatty acid, [4] ester
d. wax—[2] fatty acid, [4] ester

19.85 Decide which terms describe each block diagram.

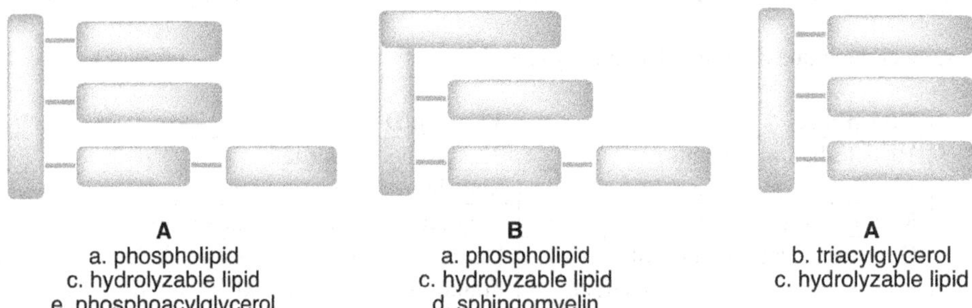

A	B	A
a. phospholipid c. hydrolyzable lipid e. phosphoacylglycerol	a. phospholipid c. hydrolyzable lipid d. sphingomyelin	b. triacylglycerol c. hydrolyzable lipid

19.87 **A** represents a soap, and **B** represents a phosphoacylglycerol. Both compounds have ionic polar heads and nonpolar tails from many C–C and C–H bonds derived from fatty acids.

19.89 Coconut oil is a liquid at room temperature because the hydrocarbon chains of lauric acid have only 12 carbons in them, making them short enough that the triacylglycerol remains a liquid.

19.91 Vegetable oils are composed of triacylglycerols, whereas motor oil, derived from petroleum, is mostly alkanes and other long chain hydrocarbons.

19.93 Humans cannot survive on a completely fat-free diet. Certain fatty acids and fat-soluble vitamins are required in the diet.

19.95 Saturated fats should be avoided in the diet because they are more likely to lead to atherosclerosis and heart disease.

19.97 Animals that live in colder climates have triacylglycerols with more unsaturated fatty acid side chains because the unsaturated triacylglycerols have lower melting points. They remain liquid at the lower temperatures of their climate, allowing the cells to remain more fluid with less rigid cell membranes than saturated triacylglycerols would allow for.

19.99 Recall that increasing the number of C–O bonds is an oxidation, whereas increasing the number of C–H bonds is a reduction.

a.

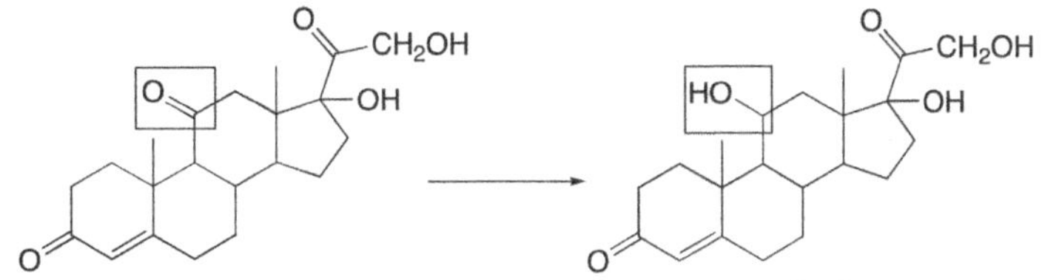

O, CH_2OH, OH, O, O

cortisone

HO, O, CH_2OH, OH, O

cortisol

C=O reduced to CH–OH
reduction

b.

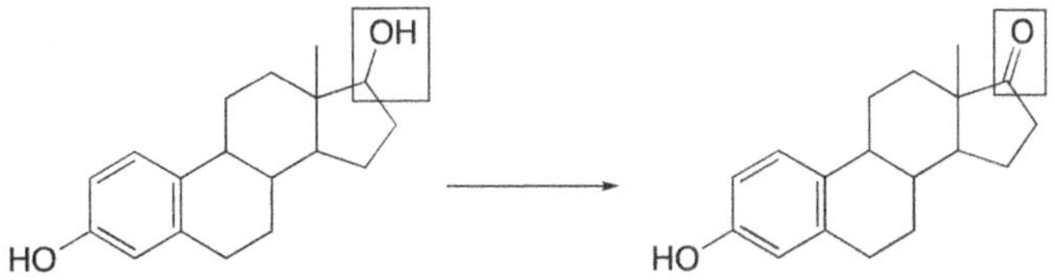

OH, HO

estradiol

O, HO

estrone

CH–OH oxidized to C=O
oxidation

c. PGE$_2$: cyclopentanone ring with HO, side chains $CH_2CH{=}CH(CH_2)_3COOH$ and $CH{=}CHCH(OH)(CH_2)_4CH_3$ → PGE$_1$: side chains $(CH_2)_6COOH$ and $CH{=}CHCH(OH)(CH_2)_4CH_3$

PGE$_2$ PGE$_1$

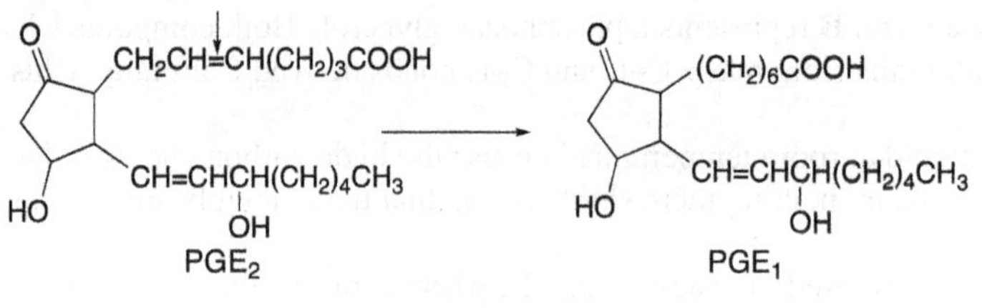

C=C reduced to CH_2CH_2
reduction

Chapter 20 Carbohydrates

Chapter Review

[1] Identify the three major types of carbohydrates. (20.1)

- Monosaccharides, which cannot be hydrolyzed to simpler compounds, have three to six carbons with a carbonyl group at either the terminal carbon or the carbon adjacent to it. Generally, all other carbons have OH groups bonded to them.
- Disaccharides are composed of two monosaccharides.
- Polysaccharides are composed of three or more monosaccharides.

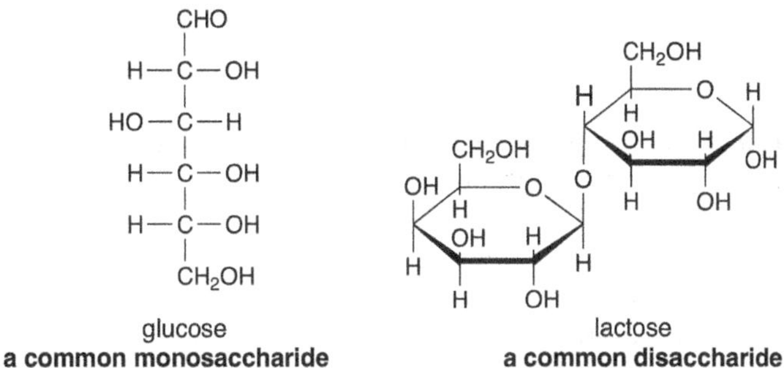

glucose
a common monosaccharide

lactose
a common disaccharide

[2] What are the major structural features of monosaccharides? (20.2)

- Monosaccharides with a carbonyl group at C1 are called aldoses and those with a carbonyl at C2 are called ketoses. Generally, all other carbons have OH groups bonded to them. The terms triose, tetrose, and so forth are used to indicate the number of carbons in the chain.

3–6 C's
C1
C2
C3
C4
C=O
OH on all (or most) other C's
monosaccharide
carbonyl at C1 ---→ aldehyde ---→ **aldose**
carbonyl at C2 ---→ ketone ---→ **ketose**
CHO
H–C–OH
CH_2OH
glyceraldehyde
CH_2OH
C=O
CH_2OH
dihydroxyacetone

- The acyclic form of monosaccharides is drawn with Fischer projection formulas. A D sugar has the OH group of the chirality center farthest from the carbonyl on the right side. An L sugar has the OH group of the chirality center farthest from the carbonyl on the left side.

Two enantiomers of glyceraldehyde

CHO
H—OH
CH_2OH
Fischer projection formula
chirality center
CHO
H–C–OH
CH_2OH
D-glyceraldehyde
CHO
HO–C–H
CH_2OH
chirality center
L-glyceraldehyde

[3] How are the cyclic forms of monosaccharides drawn? (20.3)

- In aldohexoses, the OH group on C5 reacts with the aldehyde carbonyl to give two cyclic hemiacetals called anomers. The acetal carbon is called the anomeric carbon. The α anomer has the OH group on the new chirality center drawn down for a D sugar and the β anomer has the OH group drawn up.

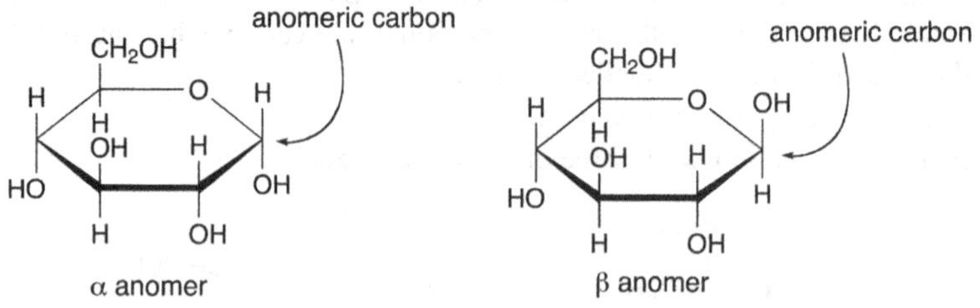

anomeric carbon

α anomer

anomeric carbon

β anomer

[4] What reduction and oxidation products are formed from monosaccharides? (20.4)

- Monosaccharides are reduced to alditols with H_2 and Pd.

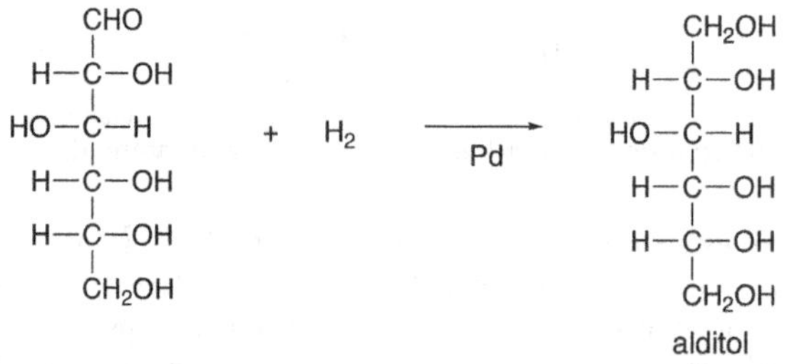

CHO, H–C–OH, HO–C–H, H–C–OH, H–C–OH, CH_2OH + H_2 —Pd→ CH_2OH, H–C–OH, HO–C–H, H–C–OH, H–C–OH, CH_2OH

alditol

- Monosaccharides are oxidized to aldonic acids with Benedict's reagent. Sugars that are oxidized with Benedict's reagent are called reducing sugars.

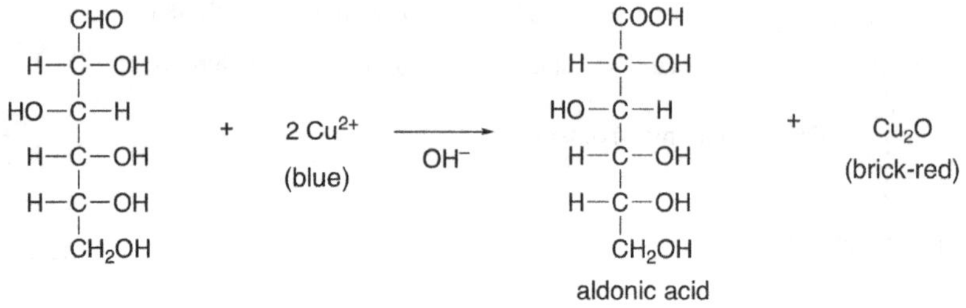

CHO, H–C–OH, HO–C–H, H–C–OH, H–C–OH, CH_2OH + $2\ Cu^{2+}$ (blue) —OH^-→ COOH, H–C–OH, HO–C–H, H–C–OH, H–C–OH, CH_2OH + Cu_2O (brick-red)

aldonic acid

[5] What are the major structural features of disaccharides? (20.5)

- Disaccharides contain two monosaccharide units joined by an acetal C–O bond called a glycosidic linkage. An α glycoside has the glycosidic linkage oriented down and a β glycoside has the glycosidic linkage oriented up.
- Disaccharides are hydrolyzed to two monosaccharides by the cleavage of the glycosidic C–O bond with water.

α glycoside

disaccharide + H_2O ⟶

- Lactose, the principal disaccharide in milk, and sucrose (table sugar) are common disaccharides.

[6] What are the differences in the polysaccharides cellulose, starch, and glycogen? (20.6)

- Cellulose, starch, and glycogen are all polymers of the monosaccharide glucose.
- Cellulose is an unbranched polymer composed of repeating glucose units joined in 1→4-β-glycosidic linkages. Cellulose forms long chains that stack in three-dimensional sheets. The human digestive system does not contain the needed enzyme to metabolize cellulose.
- There are two forms of starch—amylose, which is an unbranched polymer, and amylopectin, which is a branched polysaccharide polymer. Both forms contain 1→4-α-glycosidic linkages, and the polymer winds in a helical arrangement. Starch is digestible because the human digestive system has the needed amylase enzyme to catalyze hydrolysis.
- Glycogen resembles amylopectin but is more extensively branched. Glycogen is the major form in which polysaccharides are stored in animals.

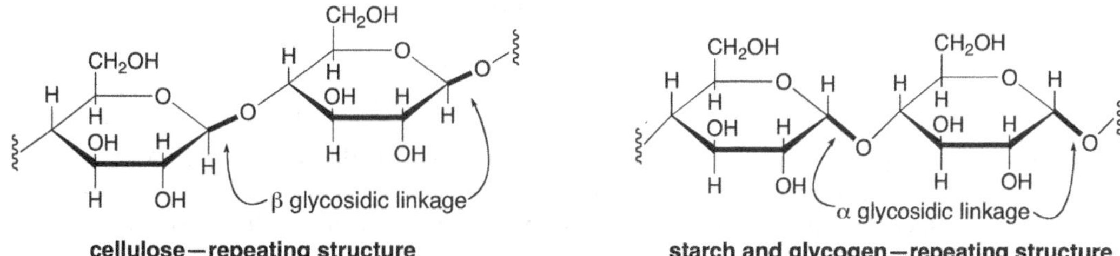

cellulose—repeating structure **starch and glycogen—repeating structure**

[7] Give examples of some carbohydrate derivatives that contain amino groups, amides, or carboxylate anions. (20.7)

- Glycosaminoglycans are a group of unbranched carbohydrates derived from amino sugars (such as D-glucosamine) and glucuronate units. Examples include hyaluronate, which forms a gel-like matrix in joints and the vitreous humor of the eye; chondroitin, which is a component of cartilage and tendons; and heparin, an anticoagulant.

D-glucosamine — *N*-acetyl-D-glucosamine — D-glucuronate

- Chitin, a polymer of *N*-acetyl-D-glucosamine, forms the hard exoskeleton of crabs, lobsters, and shrimp.

[8] What role do carbohydrates play in determining blood type? (20.8)

- Human blood type—A, B, AB, or O—is determined by three or four monosaccharides attached to a membrane protein on the surface of red blood cells. There are three different carbohydrate sequences, one for each of the A, B, and O blood types. Blood type AB contains the sequences for both blood type A and blood type B. Since the blood of an individual may contain antibodies to another blood type, blood type must be known before giving a transfusion.

Problem Solving

[1] Monosaccharides (20.2)

Example 20.1 Classify each monosaccharide by the type of carbonyl group and the number of carbons in the chain.

a.
CH_2OH
C=O
H—C—OH
H—C—OH
HO—C—H
HO—C—H
CH_2OH

b.
CHO
H—C—OH
H—C—OH
HO—C—H
HO—C—H
CH_2OH

Analysis

Identify the type of carbonyl group to label the monosaccharide as an aldose or ketose. An aldose has the C=O at C1, so that a hydrogen atom is bonded to the carbonyl carbon. A ketose has two carbons bonded to the carbonyl carbon. Count the number of carbons in the chain to determine the suffix, *-triose*, *-tetrose*, and so forth.

Solution

a.
CH_2OH
C=O (ketone)
H—C—OH
H—C—OH
HO—C—H
HO—C—H
CH_2OH

7 C's in the chain

Answer: ketoheptose

b.
CHO (aldehyde)
H—C—OH
H—C—OH
HO—C—H
HO—C—H
CH_2OH

6 C's in the chain

Answer: aldohexose

Example 20.2 Consider the given ketohexose. (a) Label all chirality centers. (b) Classify the ketohexose as a D or L monosaccharide. (c) Draw the enantiomer.

```
       CH2OH
        |
        C=O
        |
    H—C—OH
        |
   HO—C—H
        |
   HO—C—H
        |
       CH2OH
```

Analysis

- A chirality center has four different groups around a carbon atom.
- The labels D and L are determined by the position of the OH group on the chirality center farthest from the carbonyl group: a D sugar has the OH group on the right and an L sugar has the OH group on the left.
- To draw an enantiomer, draw the mirror image so that each group is a reflection of the group in the original compound.

Solution

a. The three carbons that contain a H and OH group in the ketohexose are chirality centers.

```
       CH2OH
        |
        C=O
        |
    H—C*—OH
        |
   HO—C*—H
        |
   HO—C*—H
        |
       CH2OH
```

* = chirality center

b. The ketohexose is an L sugar because the OH group on the chirality center farthest from the carbonyl is on the left.

```
       CH2OH
        |
        C=O
        |
    H—C—OH
        |
   HO—C—H
        |
 → HO—C—H
        |
       CH2OH
```

L sugar

c. The enantiomer of the L-ketohexose has all three OH groups on the chirality centers on the opposite side of the carbon chain.

```
       CH2OH
        |
        C=O
        |
   HO—C—H
        |
    H—C—OH
        |
    H—C—OH
        |
       CH2OH
```

enantiomer

---

[2] The Cyclic Forms of Monosaccharides (20.3)

Example 20.3 Draw the β anomer of the D monosaccharide.

```
        CHO
        |
    H—C—OH
        |
    H—C—OH
        |
   HO—C—H
        |
    H—C—OH
        |
       CH2OH
```

Analysis and Solution

[1] Draw a hexagon with an O atom in the upper right corner. Add the CH_2OH above the ring on the first carbon to the left of the O atom.

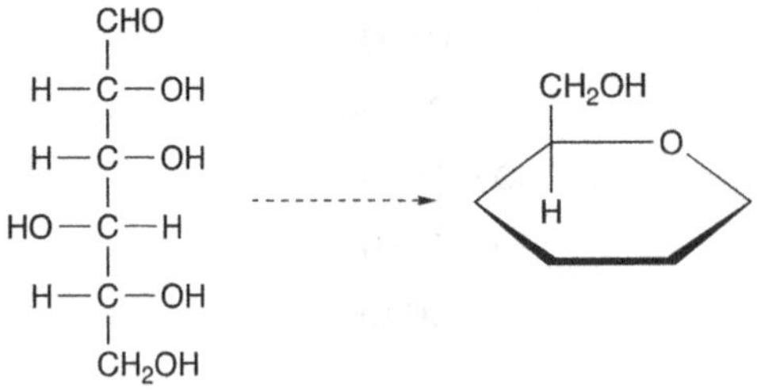

[2] Draw the anomeric carbon on the first carbon clockwise from the O atom.

- The β anomer has the OH group drawn up.

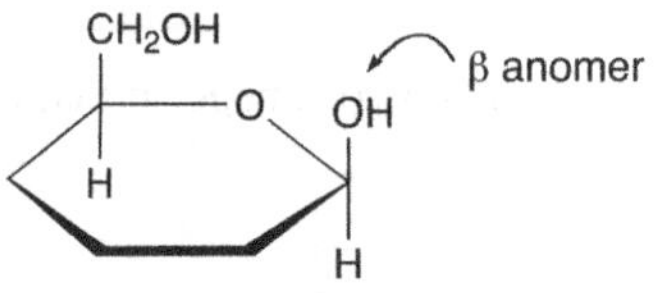

[3] Add the OH groups and H atoms to the three remaining carbons (C2–C4).

- Groups on the *right* side in the acyclic form are drawn *down*, below the six-membered ring, whereas groups on the *left* side in the acyclic form are drawn *up*, above the six-membered ring.

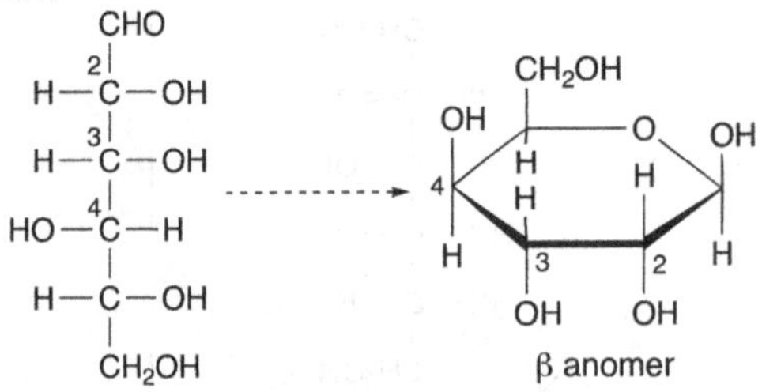

β anomer

[3] Reduction and Oxidation of Monosaccharides (20.4)

Example 20.4 Draw the structure of the compound formed when the following monosaccharide is treated with H_2 in the presence of a Pd catalyst?

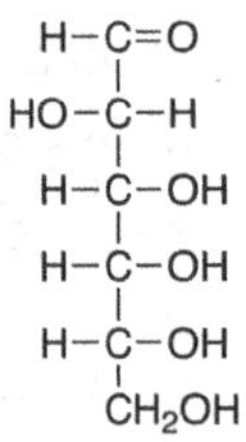

Analysis

- Locate the C=O and mentally break one bond in the double bond.
- Mentally break the H–H bond of the reagent and add one H atom to each atom of the C=O, forming new C–H and O–H bonds.

Solution

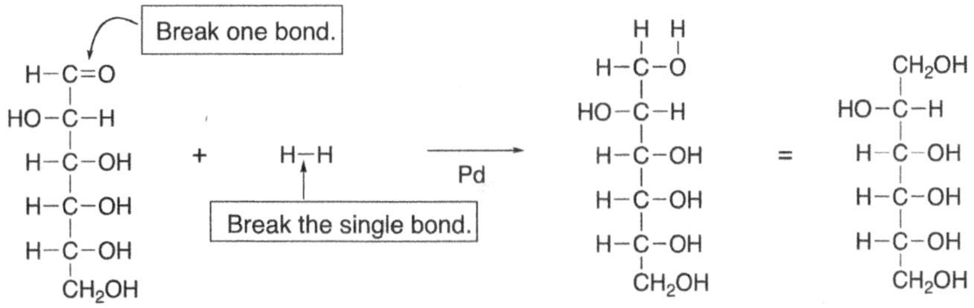

Break one bond.

H–C=O / HO–C–H / H–C–OH / H–C–OH / H–C–OH / CH_2OH + H–H $\xrightarrow{Pd}$ (H H) H–C–O / HO–C–H / H–C–OH / H–C–OH / H–C–OH / CH_2OH = CH_2OH / HO–C–H / H–C–OH / H–C–OH / H–C–OH / CH_2OH

Break the single bond.

Example 20.5 Draw the product formed when each monosaccharide is oxidized with Benedict's reagent.

a. CHO / H–C–OH / HO–C–H / HO–C–H / CH_2OH

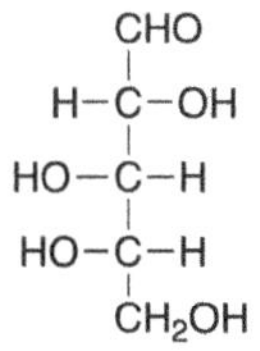

b. CH_2OH / C=O / H–C–OH / H–C–OH / CH_2OH

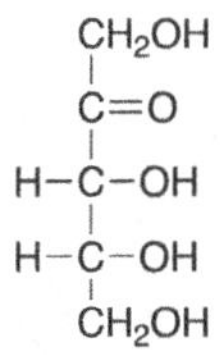

Analysis

To draw the oxidation product of an aldose, convert the CHO group to COOH. To oxidize a ketose, first rearrange the ketose to an aldose by moving the carbonyl group to C1 to give an aldehyde. Then convert the CHO group to COOH.

Solution

The monosaccharide in part (a) is an aldose, so its oxidation product can be drawn directly. The monosaccharide in part (b) is a ketose, so it undergoes rearrangement prior to oxidation.

a. aldehyde CHO / H–C–OH / HO–C–H / HO–C–H / CH_2OH $\xrightarrow{\text{oxidation}}$ COOH / H–C–OH / HO–C–H / HO–C–H / CH_2OH

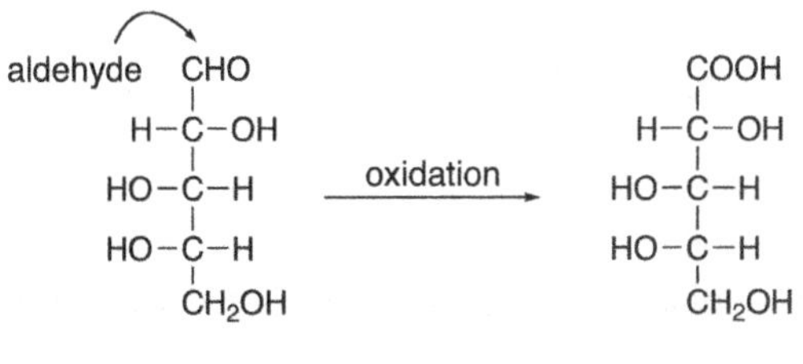

b. ketone CH_2OH / C=O / H–C–OH / H–C–OH / CH_2OH $\xrightarrow{\text{rearrangement}}$ CHO / H–C–OH / H–C–OH / H–C–OH / CH_2OH $\xrightarrow{\text{oxidation}}$ COOH / H–C–OH / H–C–OH / H–C–OH / CH_2OH

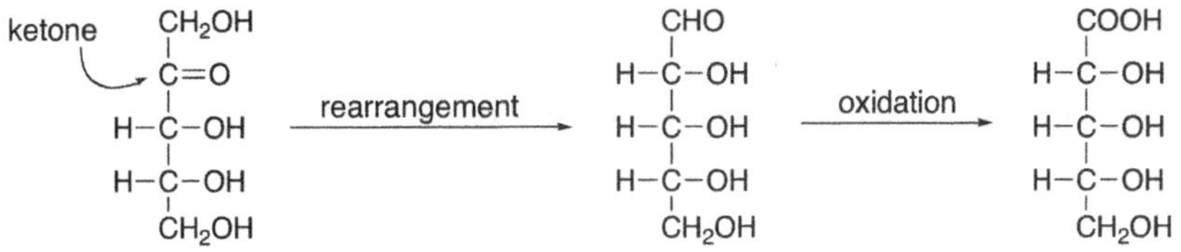

[4] Disaccharides (20.5)

Example 20.6 (a) Locate the glycosidic linkage in the disaccharide. (b) Number the carbon atoms in both rings. (c) Classify the glycosidic linkage as α or β, and use numbers to designate its location.

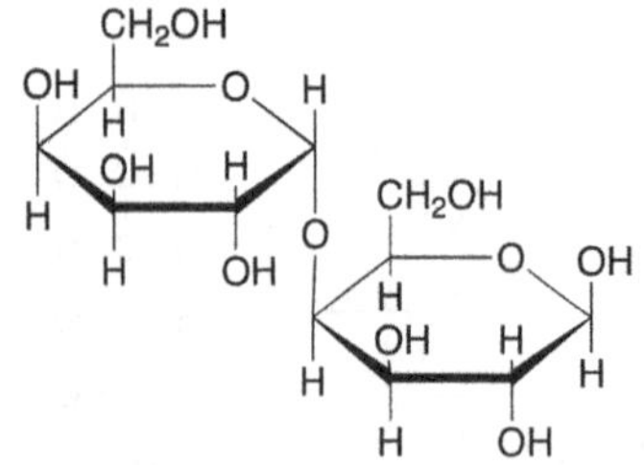

Analysis and Solution

a and b. The glycosidic linkage is the acetal C–O bond that joins the two monosaccharides. Each ring is numbered beginning at the anomeric carbon, the carbon bonded to two oxygen atoms.

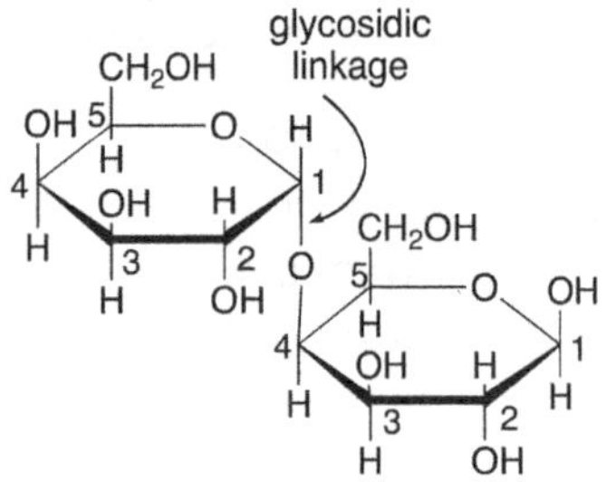

c. The disaccharide has an α glycosidic linkage because the C–O bond is drawn down. The glycosidic linkage joins C1 of one ring to C4 of the other ring, so it is called a 1→4-α-glycosidic linkage.

Example 20.7 Draw the products formed when the following disaccharide is hydrolyzed with water.

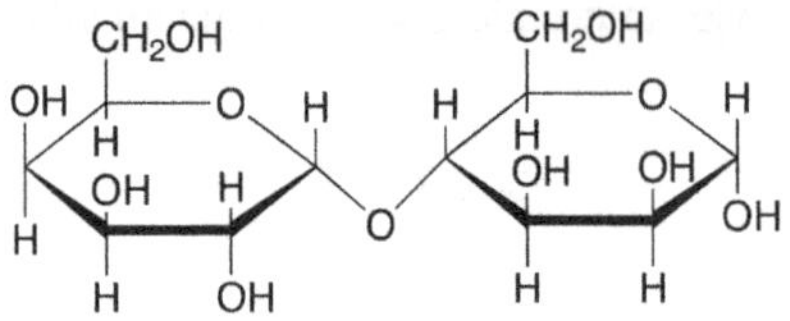

Analysis

Locate the glycosidic linkage, the acetal C–O bond that joins the two monosaccharides. Cleave the C–O bond by adding the elements of H_2O across the bond.

Solution

Two monosaccharides are formed by cleaving the glycosidic linkage.

glycosidic linkage

$+ \; H{-}OH \longrightarrow$

Add OH onto the ring.

Add H to the O atom.

Self-Test

[1] Fill in the blank with one of the terms listed below.

Aldonic acid (20.4)	Glycosidic linkage (20.5)	Nonreducing sugars (20.4)
Anomeric carbon (20.3)	Haworth projections (20.3)	Polysaccharides (20.1)
Carbohydrates (20.1)	Hexose (20.2)	Reducing sugars (20.4)
Disaccharides (20.1)	Monosaccharides (20.1)	Triose (20.2)

1. ______ are composed of two monosaccharides joined together.
2. The new C–O bond that joins the rings of two monosaccharides together is called a ______.
3. Flat, six-membered rings used to represent the cyclic hemiacetals of glucose and other sugars are called ______.
4. ______, commonly referred to as sugars and starches, are polyhydroxy aldehydes and ketones, or compounds that can be hydrolyzed to them.
5. A ______ is a monosaccharide that has six carbons.
6. In the cyclic form of monosaccharides, the carbon atom that is part of the hemiacetal is a new chirality center called the ______.
7. A ______ is a monosaccharide that has three carbons.
8. Carbohydrates that are oxidized with Benedict's reagent are called ______.
9. ______ or simple sugars are the simplest carbohydrates.
10. Carbohydrates that do not react with Benedict's reagent are called ______.
11. ______ have three or more monosaccharides joined together.
12. The aldehyde carbonyl of an aldose is easily oxidized with a variety of reagents to form a carboxyl group, yielding an ______.

[2] Label each monosaccharide as an aldose or a ketose.

a. aldose b. ketose

13. CHO / H—C—OH / H—C—OH / CH_2OH

14. CH_2OH / C=O / H—C—OH / HO—C—H / CH_2OH

15. CHO / H—C—OH / HO—C—H / HO—C—H / H—C—OH / CH_2OH

[3] Label each monosaccharide as an L or D sugar.

a. L sugar b. D sugar

16.
```
     CHO
     |
 H — C — OH
     |
HO — C — H
     |
     CH2OH
```

17.
```
     CH2OH
     |
     C=O
     |
 H — C — OH
     |
HO — C — H
     |
     CH2OH
```

18.
```
     CHO
     |
 H — C — OH
     |
HO — C — H
     |
HO — C — H
     |
 H — C — OH
     |
     CH2OH
```

[4] Label each cyclic monosaccharide as an α or β anomer.

a. α anomer b. β anomer

19.
```
        CH2OH
  OH  ——————— O     H
     H              
     H        H       
  H                   OH
       OH     OH
```

20.
```
        CH2OH
  H   ——————— O     H
     H              
     OH       OH      
  HO                  OH
       H      H
```

21.
```
        CH2OH
  OH  ——————— O     OH
     H              
     OH       H       
  H                   H
       H      OH
```

[5] Pick the product of each reaction.

a.
```
     COOH
     |
 H — C — OH
     |
HO — C — H
     |
 H — C — OH
     |
     CH2OH
```

b.
```
     CH2OH
     |
HO — C — H
     |
HO — C — H
     |
 H — C — OH
     |
 H — C — OH
     |
     CH2OH
```

c.
```
     CH2OH
     |
 H — C — OH
     |
HO — C — H
     |
 H — C — OH
     |
     CH2OH
```

d.
```
     COOH
     |
HO — C — H
     |
HO — C — H
     |
 H — C — OH
     |
 H — C — OH
     |
     CH2OH
```

22.
```
     CHO
     |
HO — C — H
     |
HO — C — H       2 Cu2+
     |          ———————→
 H — C — OH       OH−
     |
 H — C — OH
     |
     CH2OH
```

24.
```
     CHO
     |
HO — C — H
     |
HO — C — H         H2
     |          ———————→
 H — C — OH         Pd
     |
 H — C — OH
     |
     CH2OH
```

23.
```
     CH2OH
     |
     C=O
     |             2 Cu2+
HO — C — H       ———————→
     |              OH−
 H — C — OH
     |
     CH2OH
```

25.
```
 H — C=O
     |
 H — C — OH
     |              H2
HO — C — H       ———————→
     |              Pd
 H — C — OH
     |
     CH2OH
```

Answers to Self-Test

1. Disaccharides	6. anomeric carbon	11. Polysaccharides	16. a	21. b
2. glycosidic linkage	7. triose	12. aldonic acid	17. a	22. d
3. Haworth projections	8. reducing sugars	13. a	18. b	23. a
4. Carbohydrates	9. Monosaccharides	14. b	19. a	24. b
5. hexose	10. nonreducing sugars	15. a	20. a	25. c

Solutions to In-Chapter Problems

20.1 Label the hemiacetal carbon and the hydroxyl groups in lactose. Recall that a 1° hydroxyl group is bonded to a carbon bonded to one other carbon. A 2° hydroxyl group is bonded to a carbon bonded to two other carbons.

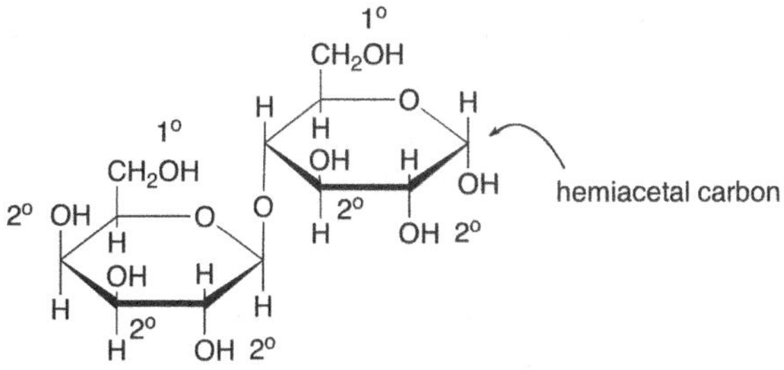

1°
CH_2OH
H O H
1° H
OH H
CH_2OH OH hemiacetal carbon
2° OH O O 2°
H H OH 2°
OH H
H H
2°
H OH 2°

20.2 Classify each monosaccharide by the type of carbonyl group and the number of carbons in the chain as in Example 20.1.

a. 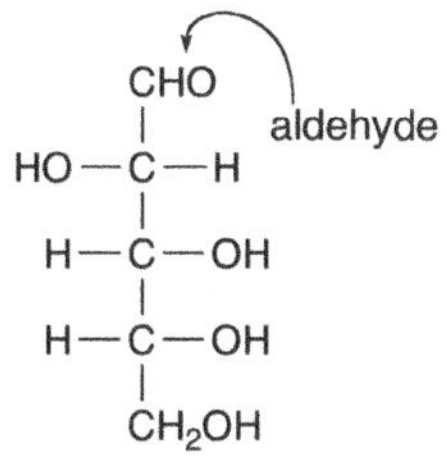
CHO ← aldehyde
HO—C—H
H—C—OH
H—C—OH
CH_2OH
arabinose
5 C's in the chain
aldopentose

b. 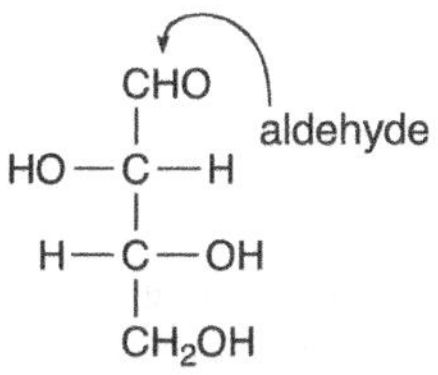
CHO ← aldehyde
HO—C—H
H—C—OH
CH_2OH
threose
4 C's in the chain
aldotetrose

c. 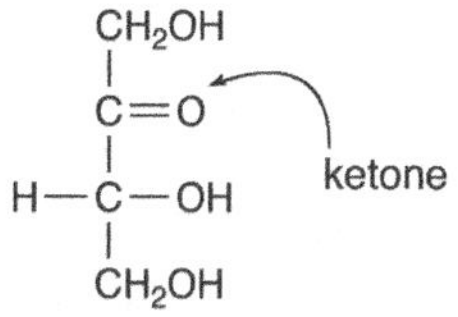
CH_2OH
C=O ← ketone
H—C—OH
CH_2OH
erythrulose
4 C's in the chain
ketotetrose

20.3

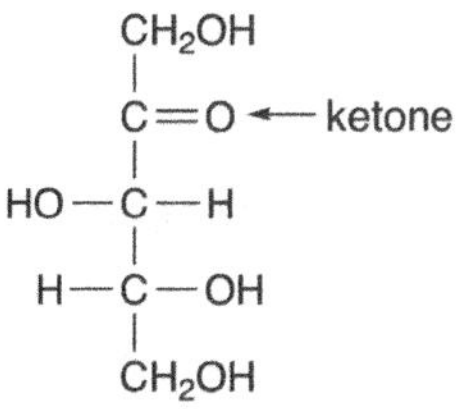

CH_2OH
C=O ← ketone
HO—C—H
H—C—OH
CH_2OH
5 C's in the chain
ketopentose

20.4 Draw the structure of each monosaccharide.

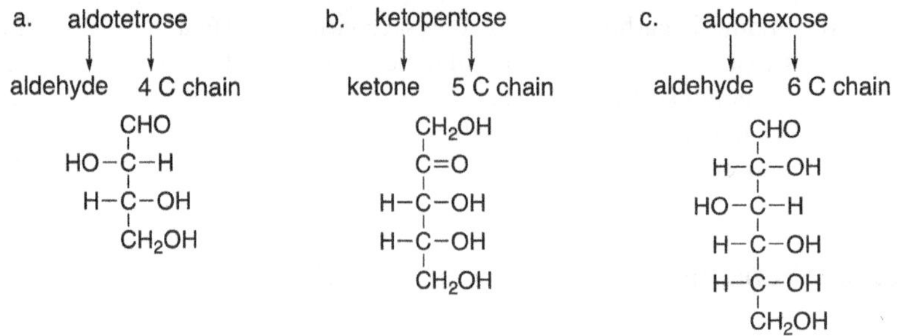

20.5 Water solubility increases as the number of polar groups increases. Hexane is a nonpolar hydrocarbon, insoluble in water. 1-Decanol is a long-chain alcohol and is slightly soluble in water due to the polar OH group. Glucose, with multiple hydroxyl groups, is water soluble.

hexane < 1-decanol < glucose

Increasing water solubility

20.6 Recall that in a Fischer projection:

- A carbon atom is located at the intersection of the two lines of the cross.
- The horizontal bonds come forward, on wedges.
- The vertical bonds go back, on dashed lines.

CHO / HO–C–H / CH_2OH → CHO / HO–┼–H / CH_2OH

L-glyceraldehyde — Fischer projection formula

20.7 Use the definitions in Example 20.2 to [1] label all chirality centers; [2] classify the monosaccharide as D or L; [3] draw the enantiomer.

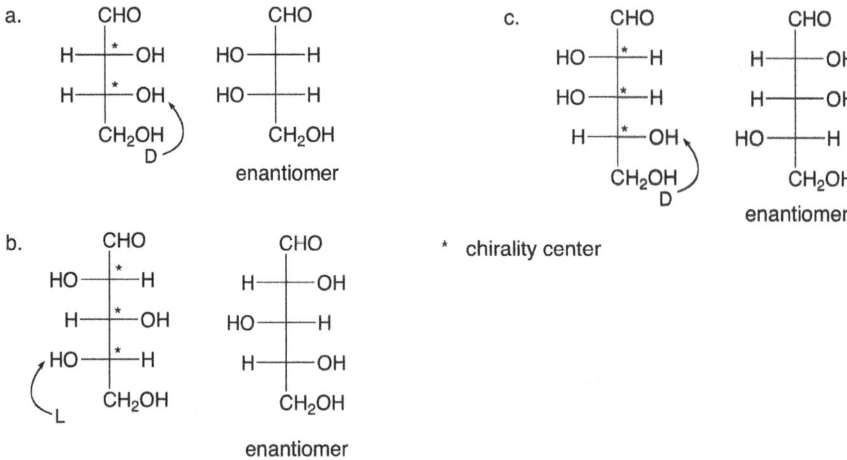

20.8 Answer each question about the monosaccharide.

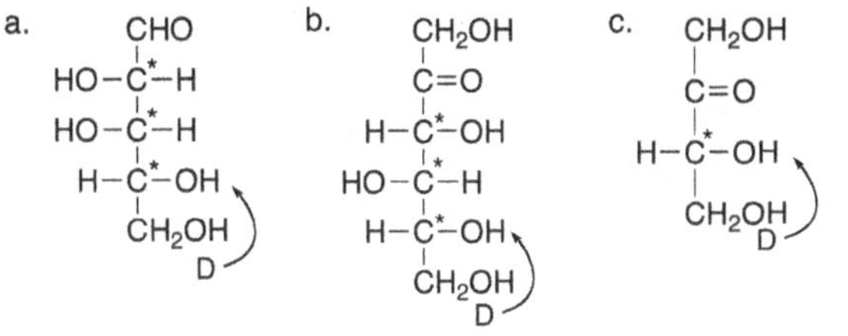

a. 5 C's, aldehyde, aldopentose

b. 6 C's, ketone, ketohexose

c. 4 C's, ketone, ketotetrose

* chirality center

20.9 D-Glucose and D-fructose are constitutional isomers (same molecular formula, different connectivity) because glucose has an aldehyde and fructose has a ketone. D-Galactose and D-fructose are also constitutional isomers because galactose has an aldehyde and fructose has a ketone.

20.10 The hemiacetal carbon is the carbon bonded to an OH and an OR group.

- The α anomer has the OH group drawn down, below the ring.
- The β anomer has the OH group drawn up, above the ring.

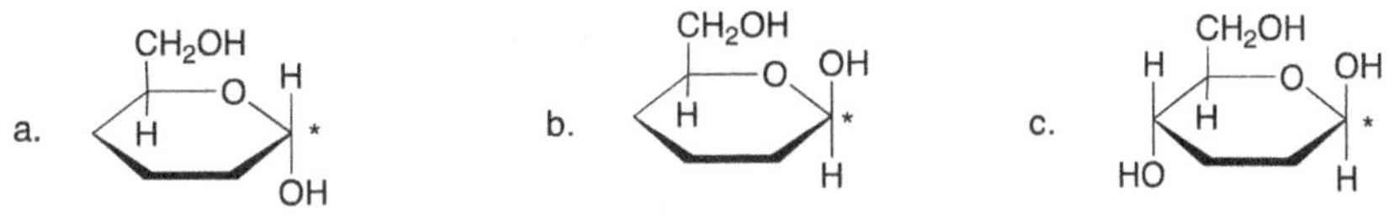

a. α anomer b. β anomer c. β anomer

* hemiacetal C

20.11 Convert each monosaccharide to the indicated anomer as in Example 20.3.

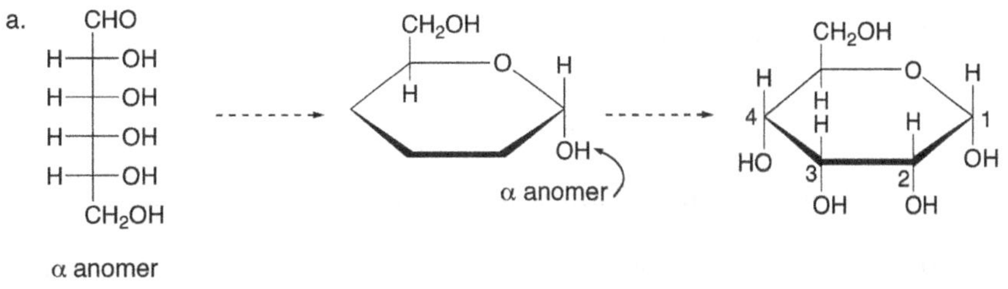

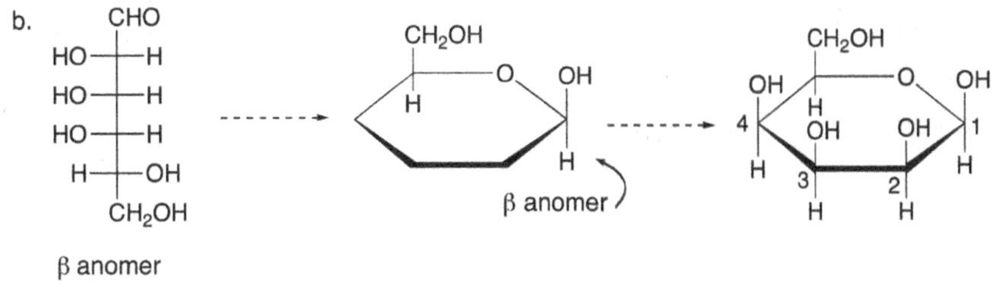

a. α anomer

b. β anomer

c.

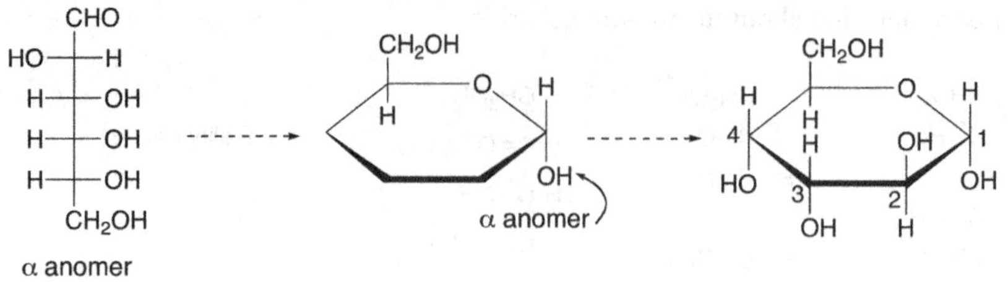

CHO
HO—H
H—OH
H—OH
H—OH
CH_2OH
α anomer

α anomer

20.12 The anomeric carbon is the carbon bonded to an OH and an OR group.

- The α anomer has the OH group drawn down, below the ring.
- The β anomer has the OH group drawn up, above the ring.

a. α anomer

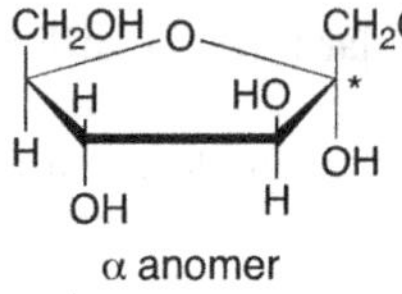

b. α anomer

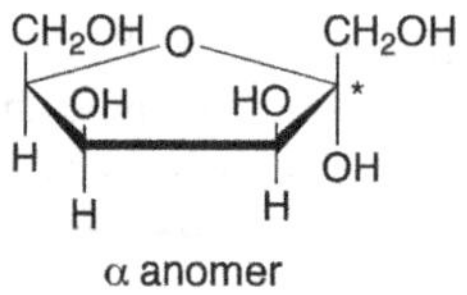

c. β anomer

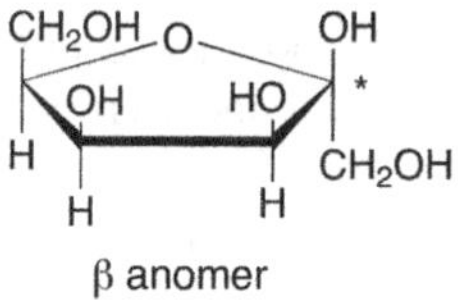

* anomeric C

20.13 Draw the products of each reduction reaction as in Example 20.4.

a. CHO, H—C—OH, H—C—OH, H—C—OH, CH_2OH $\xrightarrow[Pd]{H_2}$ CH_2OH, H—C—OH, H—C—OH, H—C—OH, CH_2OH

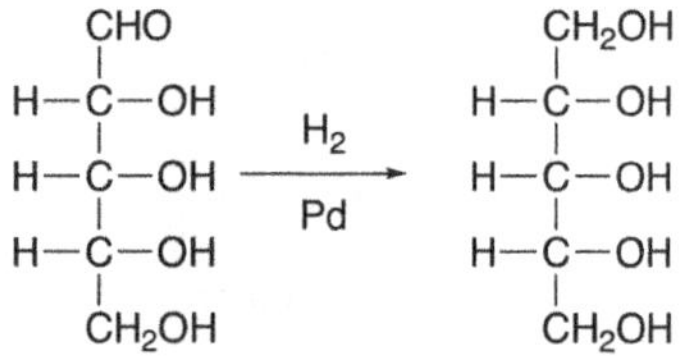

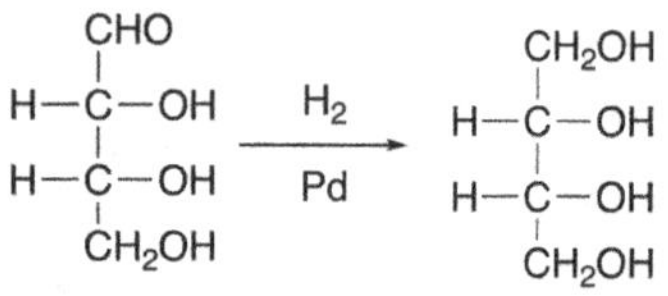

b. CHO, H—C—OH, HO—C—H, HO—C—H, H—C—OH, CH_2OH $\xrightarrow[Pd]{H_2}$ CH_2OH, H—C—OH, HO—C—H, HO—C—H, H—C—OH, CH_2OH

c. CHO, H—C—OH, H—C—OH, CH_2OH $\xrightarrow[Pd]{H_2}$ CH_2OH, H—C—OH, H—C—OH, CH_2OH

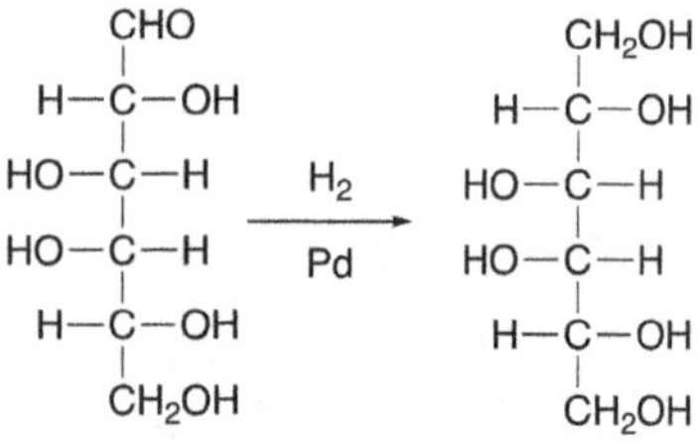

20.14 Draw the products of the oxidation of each monosaccharide as in Example 20.5.

a. CHO, H—C—OH, H—C—OH, H—C—OH, CH_2OH $\xrightarrow[OH^-]{2\ Cu^{2+}}$ COOH, H—C—OH, H—C—OH, H—C—OH, CH_2OH

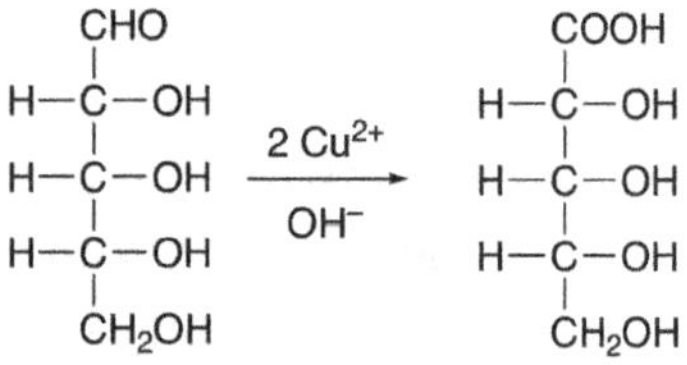

b. CHO, H—C—OH, H—C—OH, CH_2OH $\xrightarrow[OH^-]{2\ Cu^{2+}}$ COOH, H—C—OH, H—C—OH, CH_2OH

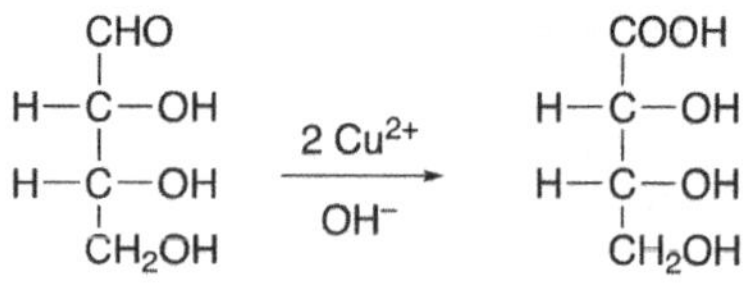

c.

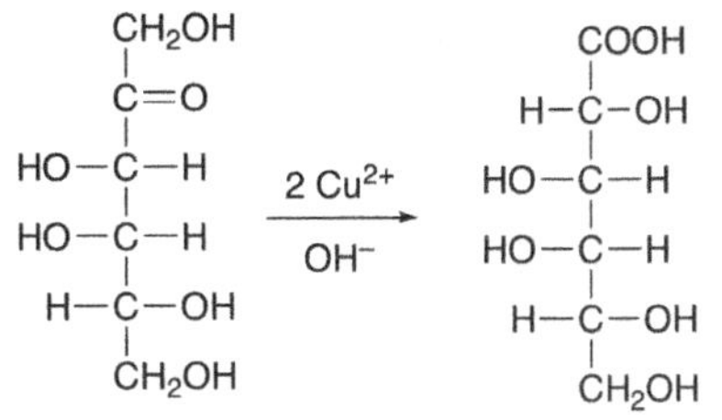

```
    CH2OH                      COOH
     |                          |
     C=O                      H-C-OH
     |                          |
  HO-C-H     2 Cu2+          HO-C-H
     |       ------->           |
  HO-C-H      OH-            HO-C-H
     |                          |
   H-C-OH                     H-C-OH
     |                          |
    CH2OH                      CH2OH
```

20.15 a. The compound has OH groups on all carbons, making it an alditol.
b. The compound contains a COOH group, making it an aldonic acid.

20.16 Draw the products of each reaction.

a.

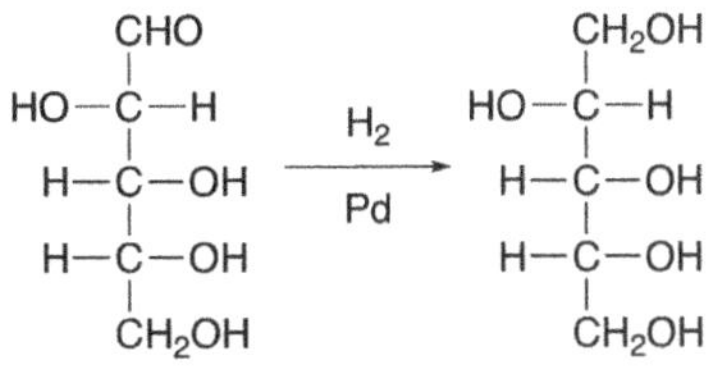

```
     CHO                  CH2OH
     |                     |
  HO-C-H      H2        HO-C-H
     |       ---->         |
   H-C-OH     Pd         H-C-OH
     |                     |
   H-C-OH                H-C-OH
     |                     |
    CH2OH                 CH2OH
```

b.

```
     CHO                    COOH
     |                      |
  HO-C-H     2 Cu2+      HO-C-H
     |       ------->       |
   H-C-OH     OH-         H-C-OH
     |                      |
   H-C-OH                 H-C-OH
     |                      |
    CH2OH                  CH2OH
```

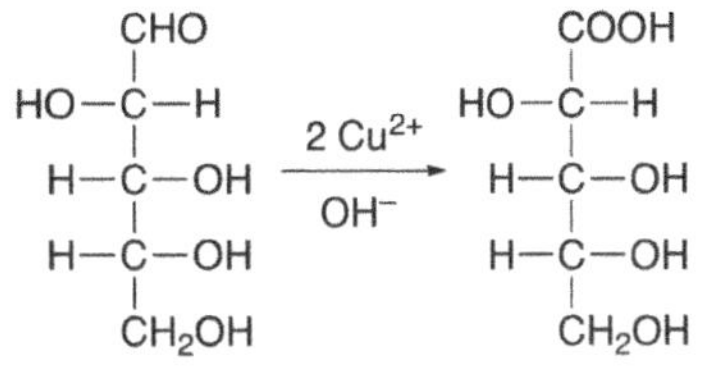

20.17 Answer each question about the disaccharide as in Example 20.6.

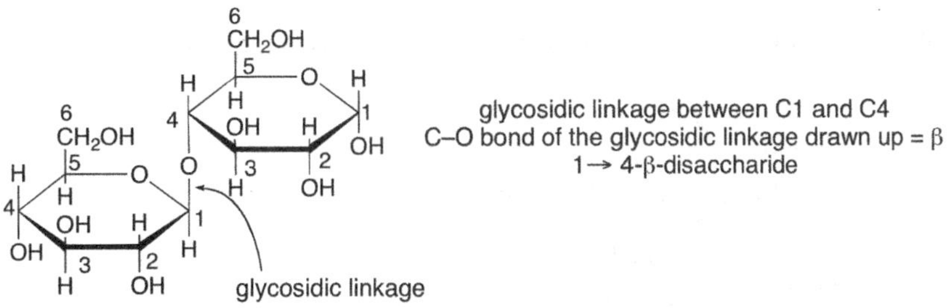

glycosidic linkage

glycosidic linkage between C1 and C4
C–O bond of the glycosidic linkage drawn up = β
1→ 4-β-disaccharide

20.18 Draw the products of the hydrolysis of cellobiose as in Example 20.7.

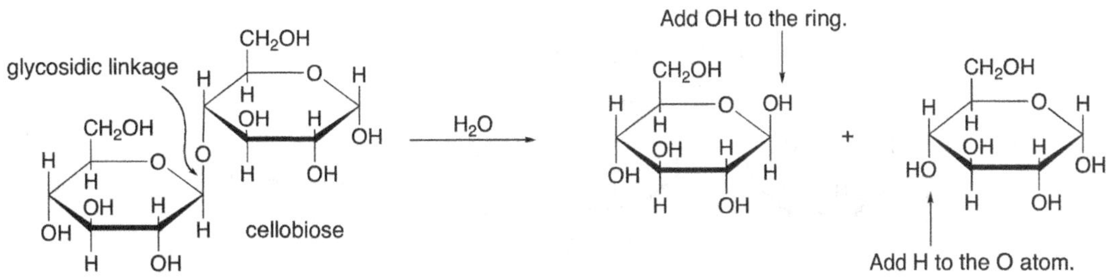

glycosidic linkage

cellobiose

H_2O

Add OH to the ring.

Add H to the O atom.

20.19 Label the acetal (C bonded to two OR groups) and hemiacetal (C bonded to an OH and an OR) groups in lactose.

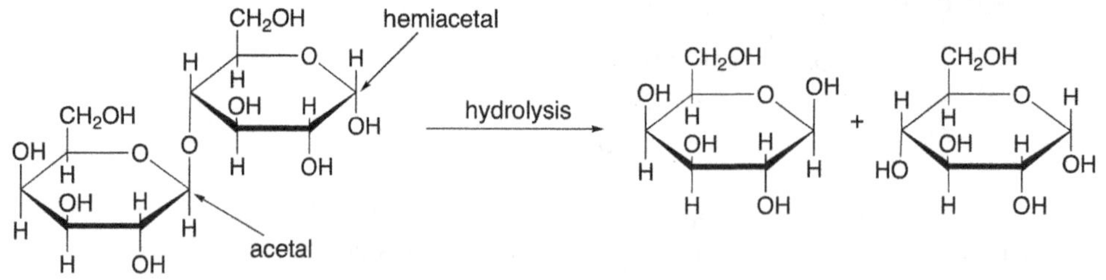

hemiacetal
hydrolysis
acetal

20.20 Locate the three acetals in rebaudioside A.

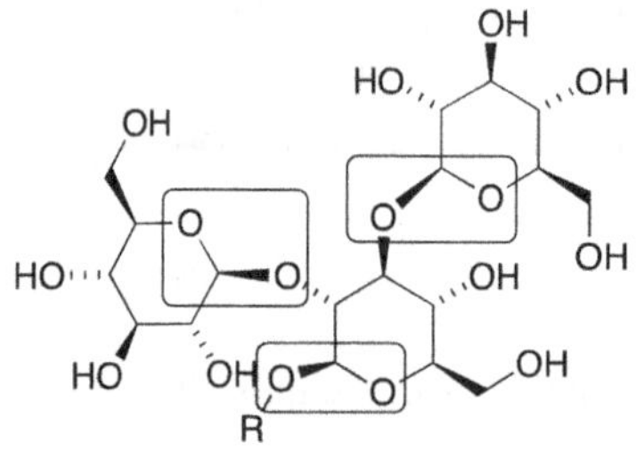

20.21 Label all the acetal carbons in sucralose, and classify the alkyl halides as 1°, 2°, or 3°.

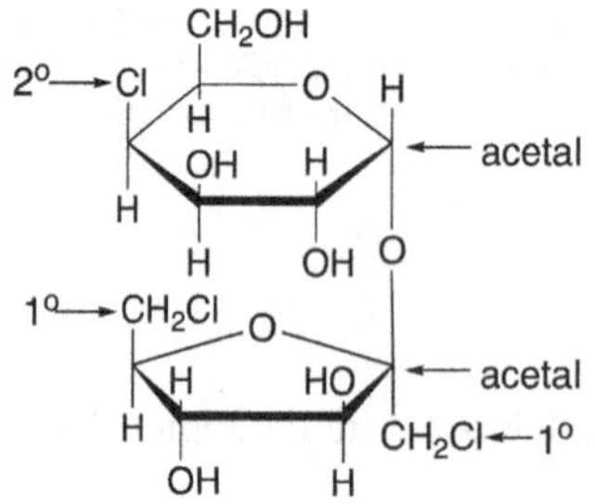

2° → Cl
acetal
1° → CH_2Cl
acetal
CH_2Cl ← 1°

20.22 Label all the acetal carbons in amylopectin.

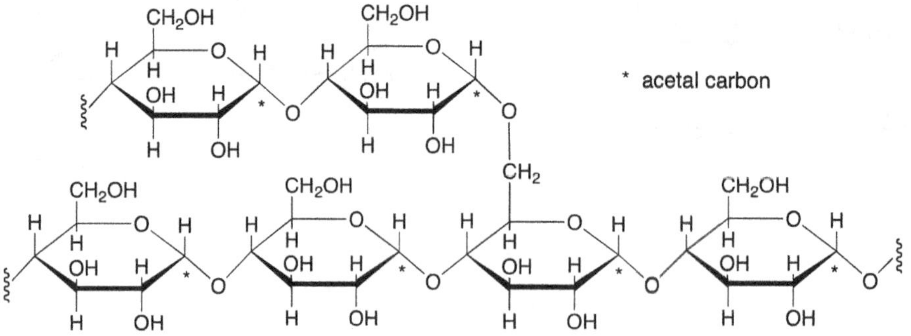

* acetal carbon

20.23 Cellulose is water insoluble because the many OH groups are already hydrogen bonded to other OH groups in the interior of the three-dimensional structure. The OH groups are therefore less available for hydrogen bonding to water.

20.24 Classify the glycosidic linkages in chondroitin and heparin as α or β. Recall that β glycosidic linkages have a C–O bond *above* the plane of the six-membered ring, whereas α glycosidic linkages have a C–O bond *below* the plane of the six-membered ring.

1→3-β-linkage

chondroitin

1→4-α-linkage

heparin

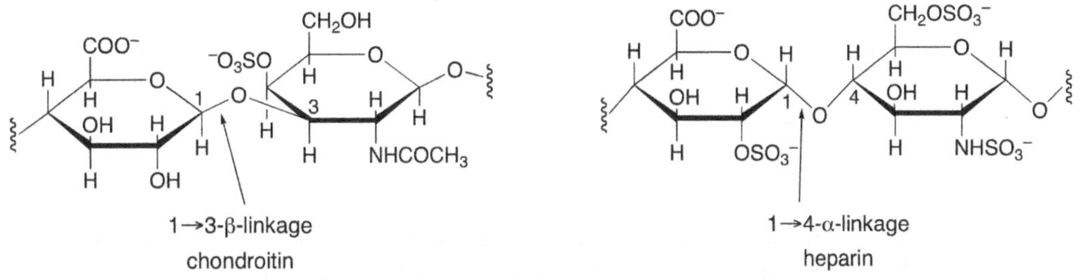

20.25 Change the β glycosidic linkages to α glycosidic linkages in chitin.

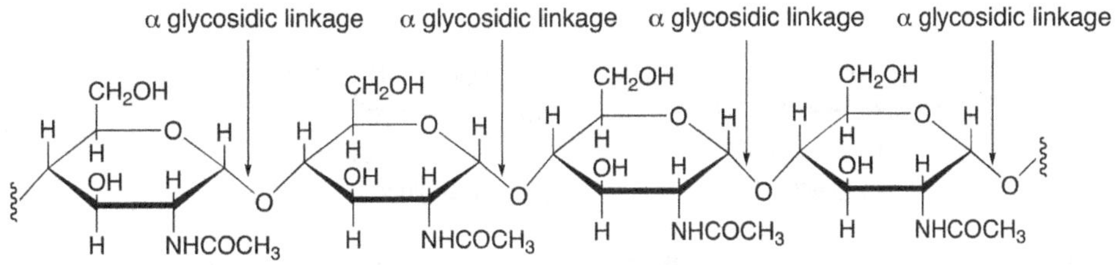

α glycosidic linkage α glycosidic linkage α glycosidic linkage α glycosidic linkage

20.26 *N*-Acetyl-D-glucosamine and *N*-acetyl-D-galactosamine are stereoisomers because they differ only in the configuration at C4.

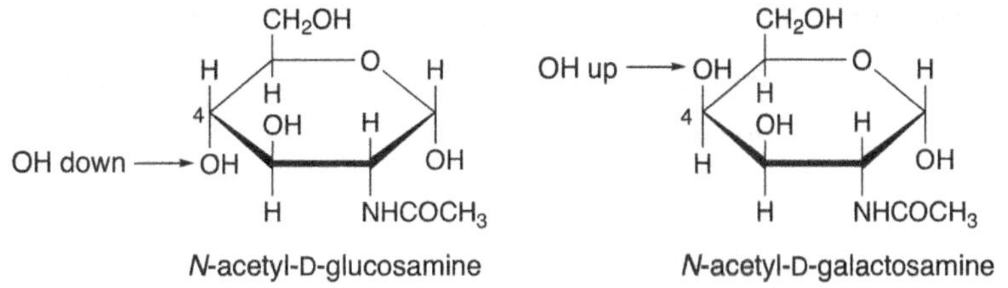

OH down → OH

OH up → OH

N-acetyl-D-glucosamine

N-acetyl-D-galactosamine

Solutions to Odd-Numbered End-of-Chapter Problems

20.27 Draw the structure of each compound.

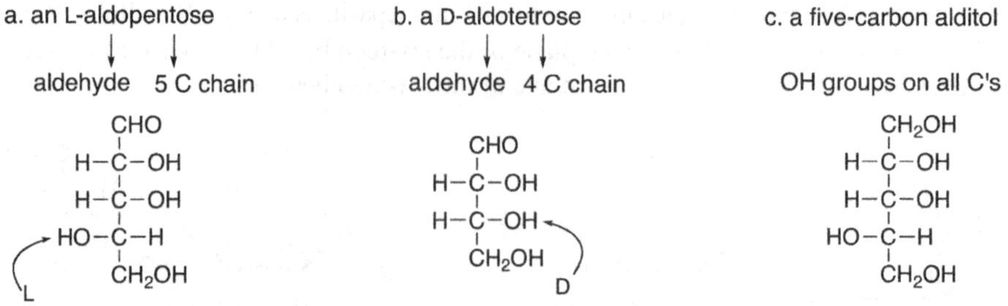
a. an L-aldopentose
aldehyde
5 C chain
L
b. a D-aldotetrose
aldehyde
4 C chain
D
c. a five-carbon alditol
OH groups on all C's

20.29 α-D-Glucose and β-D-glucose are not enantiomers because they differ in the orientation of only one OH at C1. Enantiomers are mirror images of each other, so they differ at *all* chirality centers.

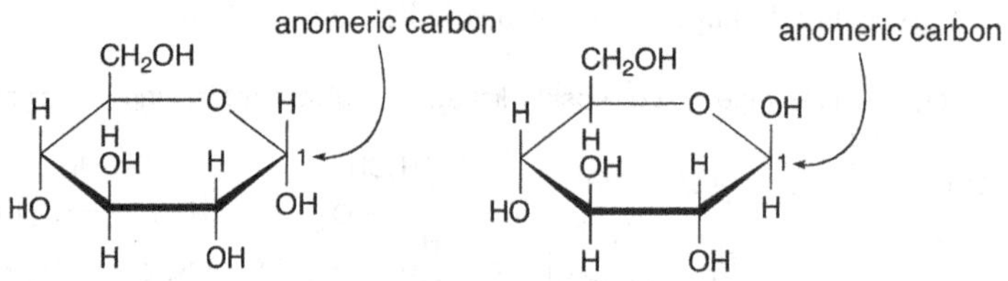
anomeric carbon
anomeric carbon

α-D-glucose β-D-glucose

20.31 Classify each monosaccharide by the type of carbonyl group and the number of carbons in the chain as in Example 20.1.

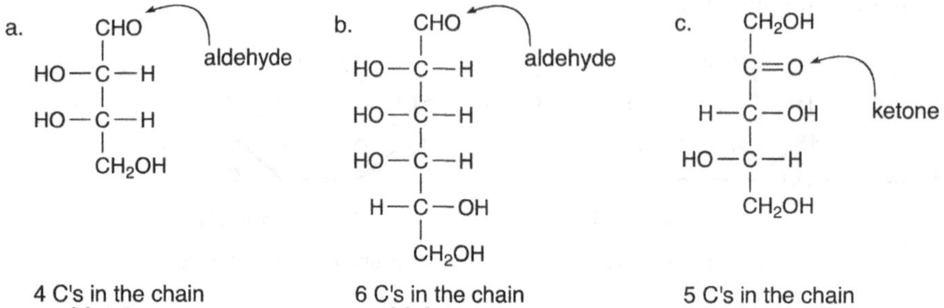
a.
aldehyde
b.
aldehyde
c.
ketone

4 C's in the chain
aldotetrose

6 C's in the chain
aldohexose

5 C's in the chain
ketopentose

20.33 The carbon skeleton for each compound is drawn from the ball-and-stick model, but the 3-D arrangement around chirality centers is not shown.

CHO
H–C–OH
H–C–OH
CH_2OH
A

a. aldotetrose
b. two chirality centers

CHO
H–C–OH
H–C–OH
H–C–OH
H–C–OH
CH_2OH
B

a. aldohexose
b. four chirality centers

20.35 Draw the Fischer projection and then classify the monosaccharide as in Example 20.1.

a. CHO (aldehyde)
H—+—OH
CH_2OH

b. 3 C's in the chain = aldotriose

20.37 Answer each question about the monosaccharides.

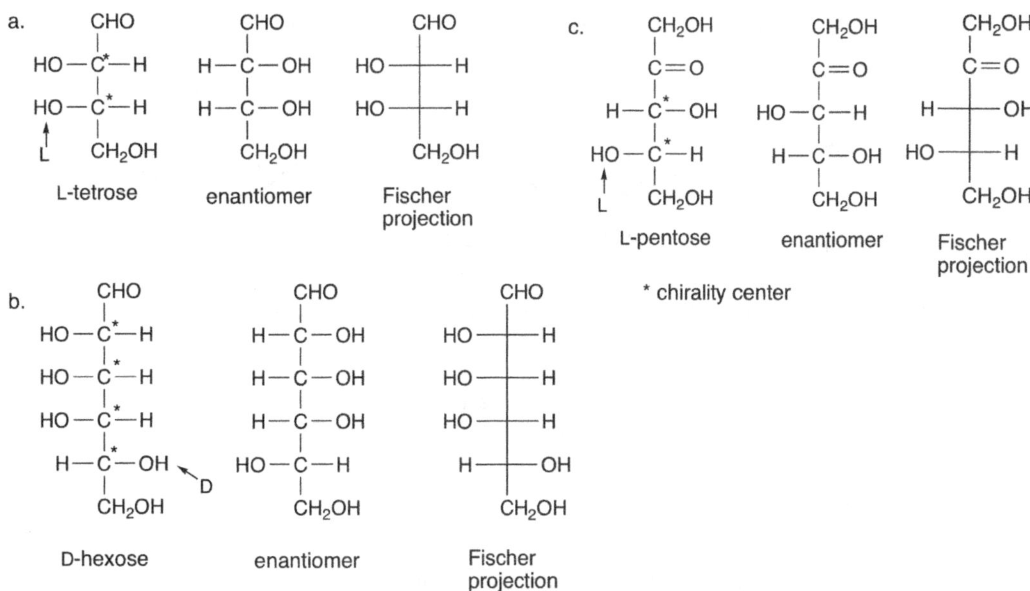

20.39 Answer each question about monosaccharides **A, B,** and **C.**

a. **A** and **B** are stereoisomers with a different three-dimensional arrangement at the indicated chirality centers.

CHO
H—C—OH
H—C—OH ←
CH_2OH
A

CHO
H—C—OH
HO—C—H ←
CH_2OH
B

b. **B** and **C** (or **A** and **C**) are constitutional isomers.

CHO / H—C—OH / HO—C—H / CH_2OH — **B** aldehyde

CH_2OH / C=O / H—C—OH / CH_2OH — **C** ketone

CHO / H—C—OH / H—C—OH / CH_2OH — **A** aldehyde

CH_2OH / C=O / H—C—OH / CH_2OH — **C** ketone

c. Enantiomer of **B:**

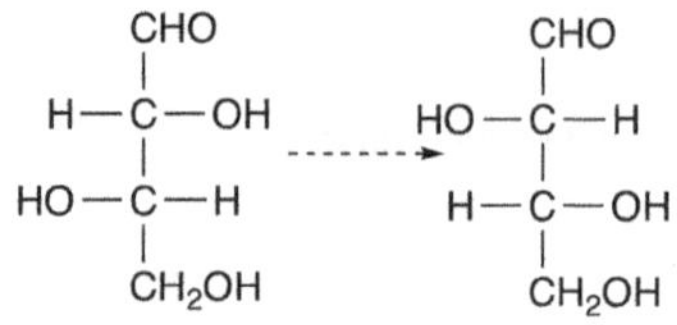

CHO / H—C—OH / HO—C—H / CH_2OH — **B** ⟶ CHO / HO—C—H / H—C—OH / CH_2OH

d. Fischer projection of **A:**

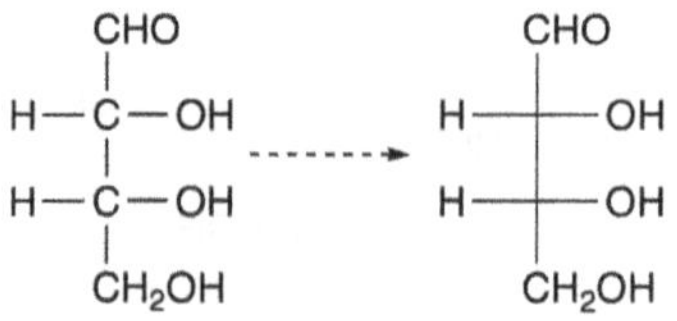

CHO / H—C—OH / H—C—OH / CH_2OH — **A** ⟶ CHO / H—+—OH / H—+—OH / CH_2OH

20.41 Convert the monosaccharide to both anomers as in Example 20.3.

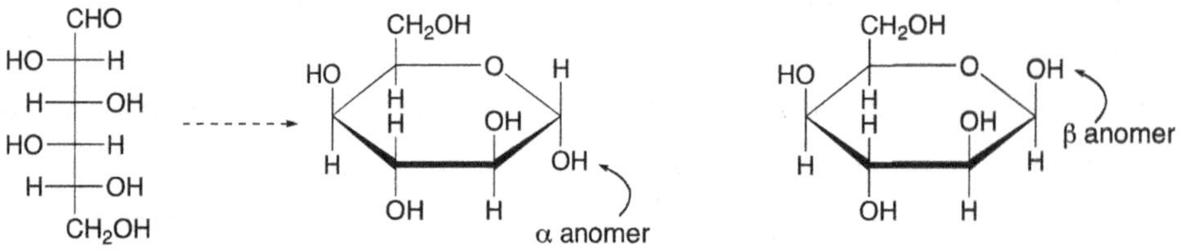

CHO / HO—+—H / H—+—OH / HO—+—H / H—+—OH / CH_2OH ⟶ α anomer; β anomer

20.43 Answer each question about the cyclic monosaccharide, drawn in a skeletal structure.

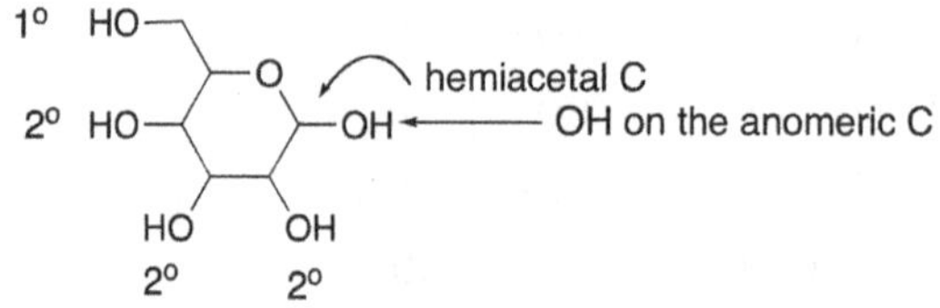

1° HO; 2° HO; HO 2°; OH 2°; hemiacetal C; OH ⟵ OH on the anomeric C

20.45 Answer each question about the cyclic monosaccharide.

a. and b. OH ⟵ β anomer (OH group up); ⟵ hemiacetal carbon

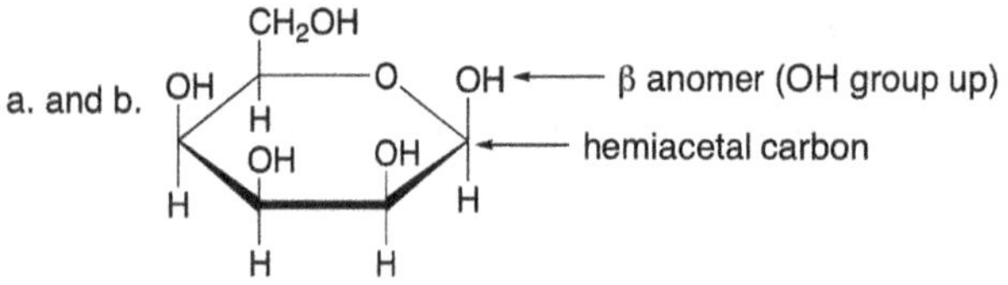

c. α anomer

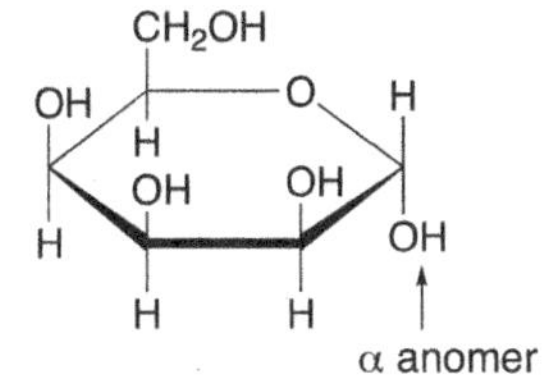

d. stereoisomer: different arrangement

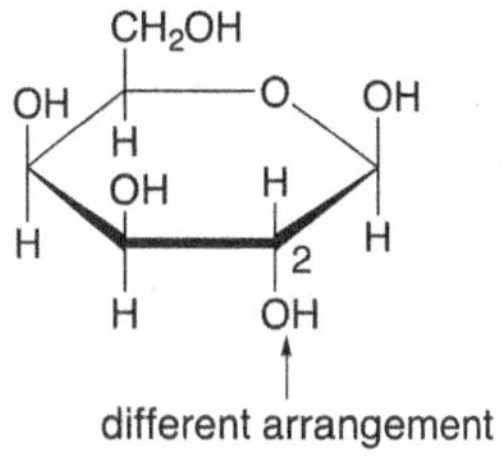

20.47 Answer the questions for each cyclic monosaccharide.

a. β anomer; hemiacetal anomeric carbon

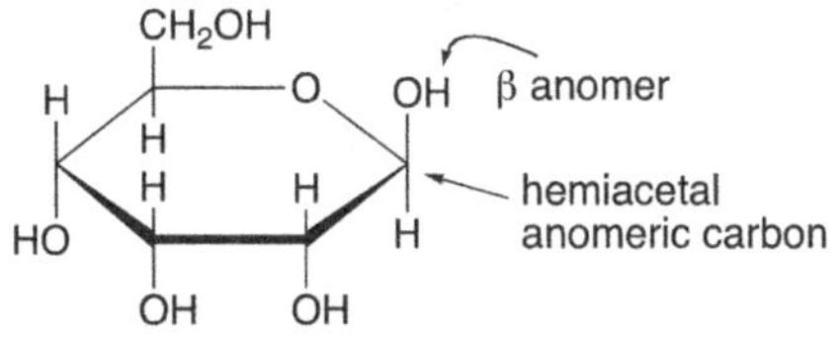

b. hemiacetal anomeric carbon; α anomer

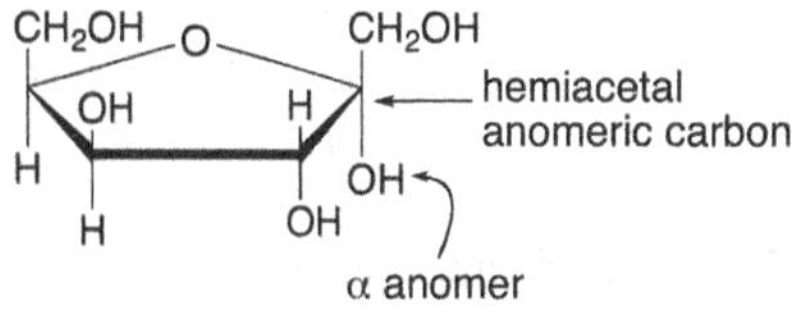

20.49 The hemiacetal carbon is the carbon bonded to an OH and an OR group.

- The α anomer has the OH group drawn down, below the ring.
- The β anomer has the OH group drawn up, above the ring.

a. β anomer; * hemiacetal C

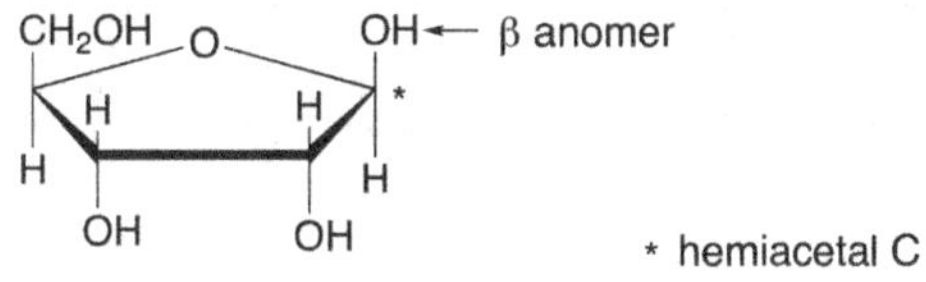

b. α anomer

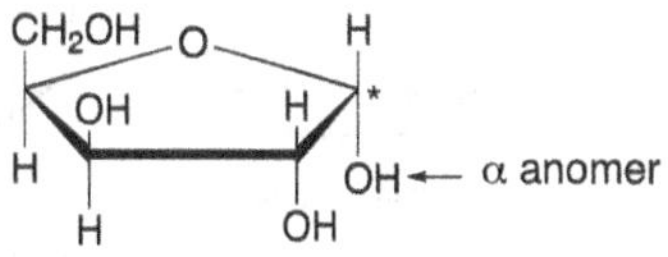

20.51 a. The compound contains a COOH, making it an **aldonic acid.**
b. The compound contains OH groups on each carbon, making it an **alditol.**
c. The compound contains an aldehyde, making it a **monosaccharide.**

20.53 Draw the products of each reaction.

a. CHO / HO–C–H / H–C–OH / CH_2OH $\xrightarrow[Pd]{H_2}$ CH_2OH / HO–C–H / H–C–OH / CH_2OH

CHO / HO–C–H / H–C–OH / CH_2OH $\xrightarrow[OH^-]{2\ Cu^{2+}}$ COOH / HO–C–H / H–C–OH / CH_2OH

b. CHO / H–C–OH / HO–C–H / HO–C–H / CH_2OH $\xrightarrow[Pd]{H_2}$ CH_2OH / H–C–OH / HO–C–H / HO–C–H / CH_2OH

CHO / H–C–OH / HO–C–H / HO–C–H / CH_2OH $\xrightarrow[OH^-]{2\ Cu^{2+}}$ COOH / H–C–OH / HO–C–H / HO–C–H / CH_2OH

c.

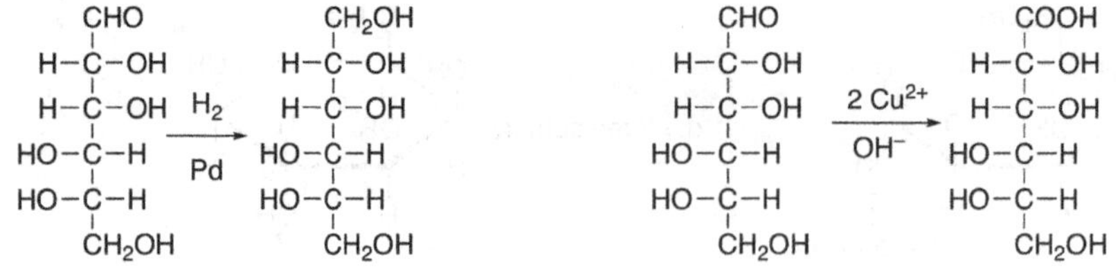

CHO / H–C–OH / H–C–OH / HO–C–H / HO–C–H / CH_2OH $\xrightarrow[Pd]{H_2}$ CH_2OH / H–C–OH / H–C–OH / HO–C–H / HO–C–H / CH_2OH

CHO / H–C–OH / H–C–OH / HO–C–H / HO–C–H / CH_2OH $\xrightarrow[OH^-]{2\ Cu^{2+}}$ COOH / H–C–OH / H–C–OH / HO–C–H / HO–C–H / CH_2OH

20.55 Draw the products of the oxidation of each monosaccharide as in Example 20.5.

a. CHO / H–C–OH / H–C–OH / HO–C–H / CH_2OH $\xrightarrow[OH^-]{2\ Cu^{2+}}$ COOH / H–C–OH / H–C–OH / HO–C–H / CH_2OH

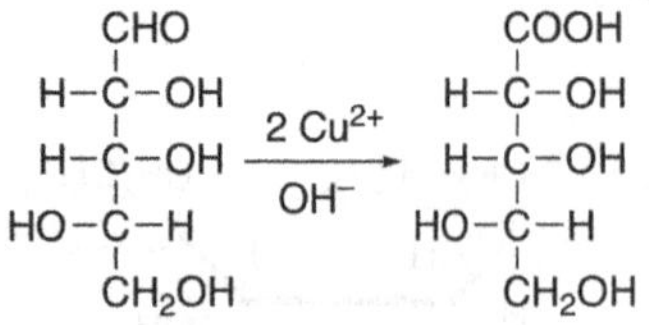

b. CH_2OH / C=O / H–C–OH / H–C–OH / CH_2OH $\xrightarrow[OH^-]{2\ Cu^{2+}}$ COOH / H–C–OH / H–C–OH / H–C–OH / CH_2OH

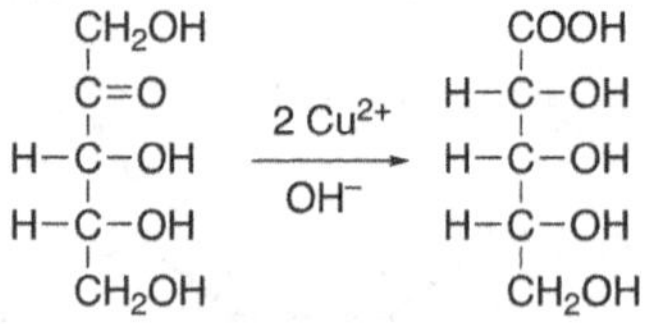

20.57 Reducing sugars are carbohydrates that are oxidized with Benedict's reagent (Cu^{2+}). Glucose, an aldohexose, is a reducing sugar.

20.59 Draw the products of the hydrolysis of the disaccharides as in Example 20.7.

a.

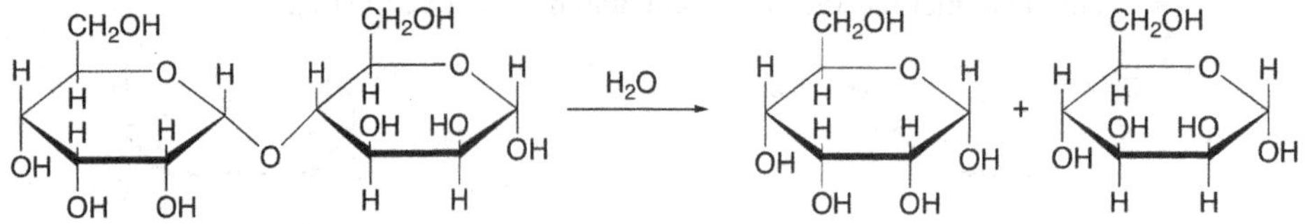

b.

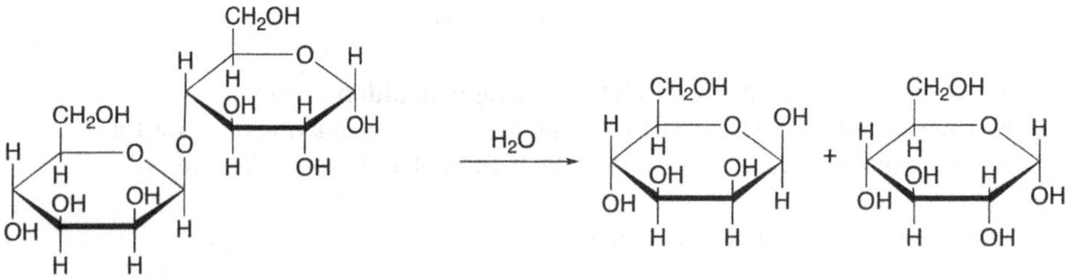

20.61 a. Lactose is a disaccharide, so **A,** with two monosaccharide units, represents lactose.
b. Cellulose is an unbranched polysaccharide, represented by **C.**
c. Amylopectin is a branched polysaccharide, represented by **B.**

20.63 Draw the disaccharide with an α glycosidic linkage.

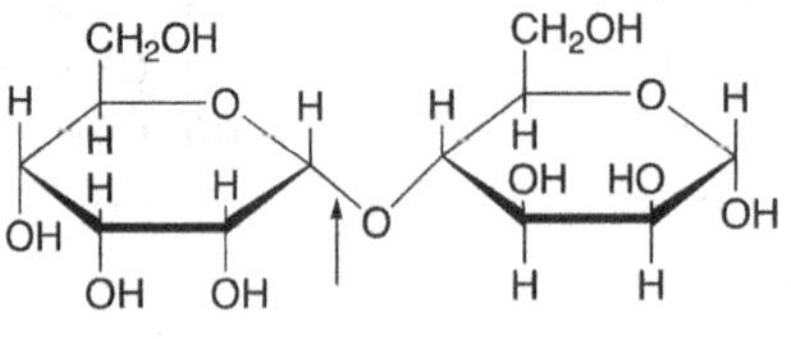

α glycosidic linkage

20.65 Answer each question about isomaltose.

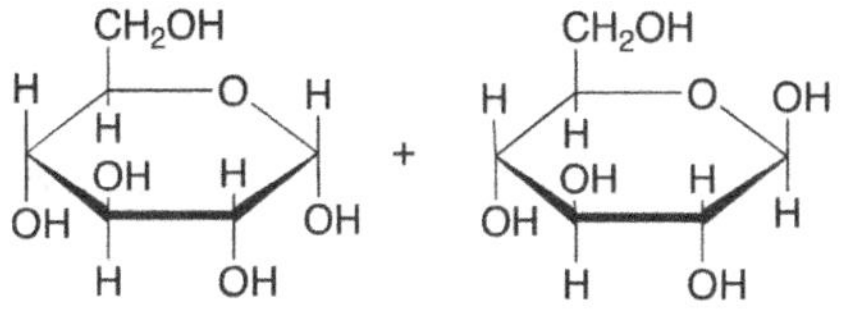

a. and b.

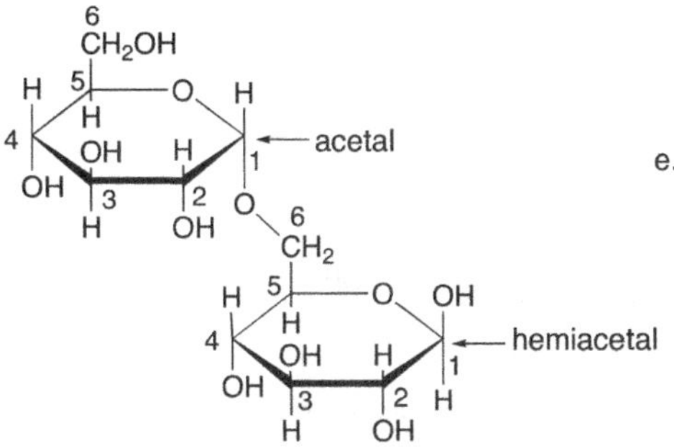

c. 1→6-α-glycosidic linkage

d. β anomer (hemiacetal OH group up)

20.67 Draw the disaccharide formed from two galactose units joined by a 1→4-β-glycosidic linkage.

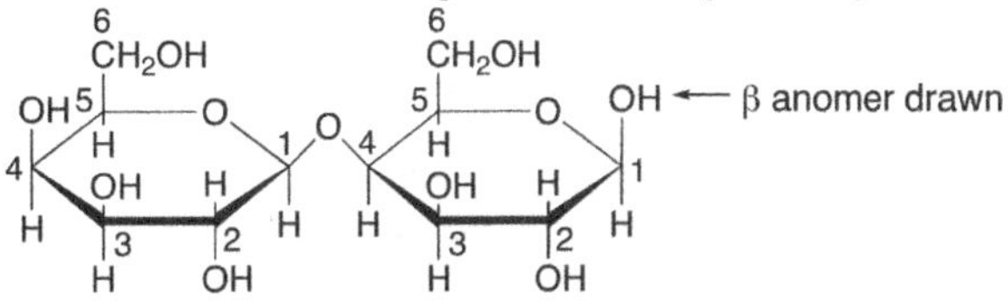

20.69 Cellulose and amylose are both composed of repeating glucose units. In cellulose the glucose units are joined by a 1→4-β-glycosidic linkage, whereas in amylose they are joined by a 1→4-α-glycosidic linkage. This leads to very different three-dimensional shapes, with cellulose forming sheets and amylose forming helices.

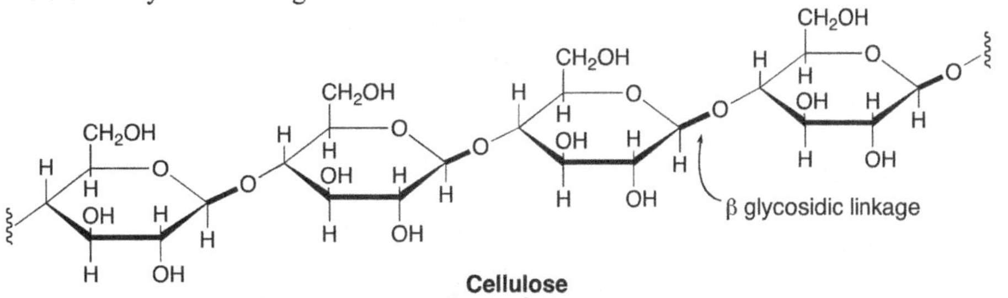

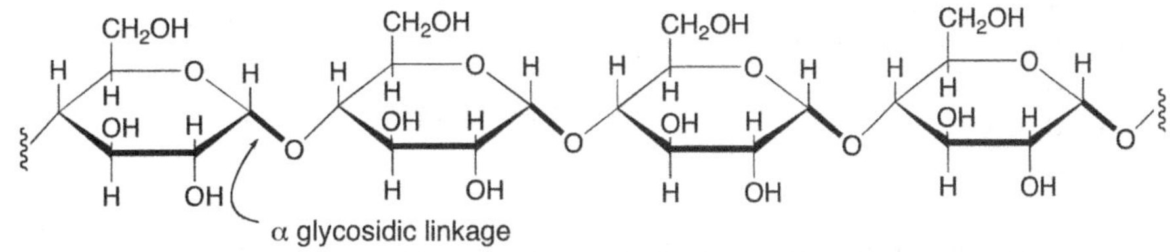

20.71 Draw a short segment of a polysaccharide that contains three galactose units joined together in 1→4-α-glycosidic linkages.

CH₂OH, O, H, OH, 1, 4, 1→4-α-glycosidic linkage

20.73 a. The major polysaccharide in pasta is starch.
b. The major polysaccharide in grass is cellulose.
c. Ice cream contains the disaccharide lactose.

20.75 Answer each question about the monosaccharide.

a. and b. CHO; HO–C–H; HO–C–H; HO–C–H; H–C–OH; CH_2OH — D sugar; 6 C's; aldehyde; **aldohexose**

c. CHO; H–C–OH; H–C–OH; H–C–OH; HO–C–H; CH_2OH — enantiomer

d. CHO; HO–C*–H; HO–C*–H; HO–C*–H; H–C*–OH; CH_2OH — * chirality center

e. CH_2OH, OH, O, H, H, OH, HO, H, OH, H, H — α anomer

f. CHO; HO–C–H; HO–C–H; HO–C–H; H–C–OH; CH_2OH —($2\ Cu^{2+}$, OH^-)→ COOH; HO–C–H; HO–C–H; HO–C–H; H–C–OH; CH_2OH

g. CHO; HO–C–H; HO–C–H; HO–C–H; H–C–OH; CH_2OH —(H_2, Pd)→ CH_2OH; HO–C–H; HO–C–H; HO–C–H; H–C–OH; CH_2OH

h. a reducing sugar

20.77 Lactose intolerance results from a lack of the enzyme lactase. It results in abdominal cramping and diarrhea. Galactosemia results from the inability to metabolize galactose. As a result, galactose accumulates in the liver, causing cirrhosis, and in the brain, leading to mental retardation.

20.79 Fructose is a naturally occurring sugar with more perceived sweetness per gram than sucrose, so fructose provides the same amount of sweetness in fewer grams and fewer calories. Sucralose is a synthetic sweetener; that is, it is not naturally occurring and is therefore labeled artificial.

20.81 An individual with type A blood can receive only blood types A and O, because he or she will produce antibodies and an immune response to B or AB blood. He or she can donate to individuals with either type A or AB blood, however, because the type A polysaccharides are common to both and no immune response will be generated.

20.83 The long sheets of polysaccharides in chitin are similar to cellulose in that they have β glycosidic linkages and are not digestible by humans.

20.85 Hyaluronate is a glycosaminoglycan found in the extracellular fluid that lubricates joints and the vitreous humor of the eye. Chondroitin is a component of cartilage and tendons. Heparin is a glycosaminoglycan stored in the mast cells of the liver and other organs and prevents blood clotting.

20.87 Answer the questions about fructose.

fructose		product 1		product 2
CH_2OH		CH_2OH		CH_2OH
C=O		HO–C–H		H–C–OH
HO–C–H	reduction →	HO–C–H	+	HO–C–H
H–C–OH		H–C–OH		H–C–OH
H–C–OH		H–C–OH		H–C–OH
CH_2OH		CH_2OH		CH_2OH

fructose

stereoisomers
They differ in configuration at a single carbon.

Chapter 21 Amino Acids, Proteins, and Enzymes

Chapter Review

[1] What are the main structural features of an amino acid? (21.2)

- Amino acids contain an amino group (NH_2) on the α carbon to the carboxyl group (COOH). Because they contain both an acid and a base, amino acids exist in their neutral form as zwitterions having the general structure $^+H_3NCH(R)COO^-$. Because they are salts, amino acids are water soluble and have high melting points.

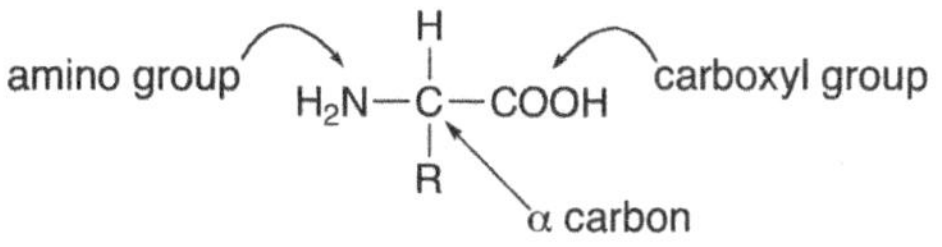

amino group — $H_2N{-}CH(R){-}COOH$ — carboxyl group; α carbon

α-amino acid

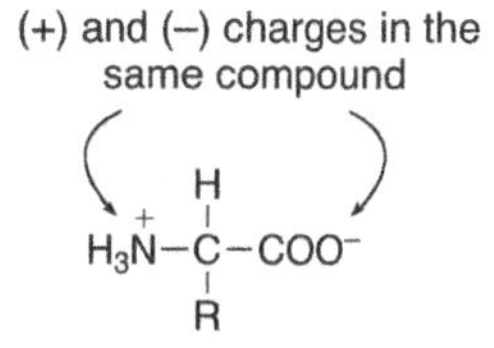

(+) and (–) charges in the same compound — $H_3\overset{+}{N}{-}CH(R){-}COO^-$

zwitterion

- All amino acids except glycine (R = H) have a chirality center on the α carbon. The L-amino acids are naturally occurring.
- Amino acids are subclassified as neutral, acidic, or basic by the functional groups present in the R group, as shown in Table 21.2.

[2] Describe the acid–base properties of amino acids. (21.3)

- Neutral, uncharged amino acids exist as zwitterions containing an ammonium cation ($-NH_3^+$) and a carboxylate anion ($-COO^-$).
- When strong acid is added, the carboxylate anion gains a proton and the amino acid has a net +1 charge. When strong base is added, the ammonium cation loses a proton and the amino acid has a net –1 charge.

$$H_3\overset{+}{N}{-}CH(CH_3){-}COOH \xleftarrow{H^+} H_3\overset{+}{N}{-}CH(CH_3){-}COO^- \xrightarrow{OH^-} H_2N{-}CH(CH_3){-}COO^-$$

overall +1 charge, pH ≤ 2 — pH ≈ 6 — overall –1 charge, pH ≥ 10

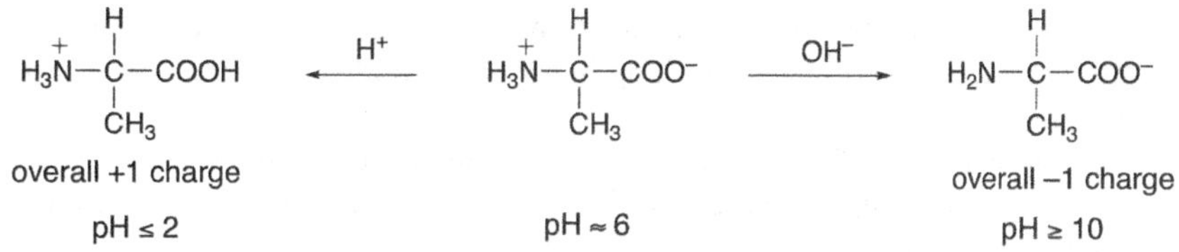

[3] What are the main structural features of peptides? (21.4)

- Peptides contain amino acids, called amino acid residues, joined together by amide (peptide) bonds. The amino acid that contains the free $-NH_3^+$ group on the α carbon is called the N-terminal amino acid, and the amino acid that contains the free $-COO^-$ group on the α carbon is called the C-terminal amino acid.
- Peptides are written from left to right, from the N-terminal to C-terminal end, using the one- or three-letter abbreviations for the amino acids listed in Table 21.2.

$$H_3\overset{+}{N}-\underset{\underset{CH_3}{|}}{CH}-\overset{\overset{O}{\|}}{C}-\underset{\underset{H}{|}}{N}-\underset{\underset{CH_2OH}{|}}{CH}-\overset{\overset{O}{\|}}{C}-O^-$$

N-terminal amino acid C-terminal amino acid

alanylserine
Ala–Ser

[4] Give examples of simple biologically active peptides. (21.5)

- Enkephalins are pentapeptides that act as sedatives and pain killers by binding to pain receptors.
- Oxytocin is a nonapeptide hormone that stimulates the contraction of uterine muscles and initiates the flow of milk in nursing mothers.
- Vasopressin is a nonapeptide hormone that serves as an antidiuretic; that is, vasopressin causes the kidneys to retain water.

[5] What are the general characteristics of the primary, secondary, tertiary, and quaternary structures of proteins? (21.6)

- The primary structure of a protein is the particular sequence of amino acids joined together by amide bonds.
- The two most common types of secondary structure are the α-helix and the β-pleated sheet. Both structures are stabilized by hydrogen bonds between the N–H and C=O groups.
- The tertiary structure is the three-dimensional shape adopted by the entire peptide chain. London dispersion forces stabilize hydrophobic interactions between nonpolar amino acids. Hydrogen bonding and ionic interactions occur between polar or charged amino acid residues. Disulfide bonds are covalent sulfur–sulfur bonds that occur between cysteine residues in different parts of the peptide chain.
- When a protein contains more than one polypeptide chain, the quaternary structure describes the shape of the protein complex formed by two or more chains.

[6] What are the basic features of fibrous proteins like α-keratin and collagen? (21.7)

- Fibrous proteins are composed of long linear polypeptide chains that serve structural roles and are water insoluble.
- α-Keratin in hair is a fibrous protein composed almost exclusively of α-helix units that wind together to form a superhelix. Disulfide bonds between chains make the resulting bundles of protein chains strong.
- Collagen, found in connective tissue, is composed of a superhelix formed from three elongated left-handed helices.

[7] What are the basic features of globular proteins like hemoglobin and myoglobin? (21.7)

- Globular proteins have compact shapes and are folded to place polar amino acids on the outside to make them water soluble. Hemoglobin and myoglobin are both conjugated proteins composed of a protein unit and a heme molecule. The Fe^{2+} ion of the heme binds oxygen. While myoglobin has a single polypeptide chain, hemoglobin contains four peptide chains that form a single protein molecule.

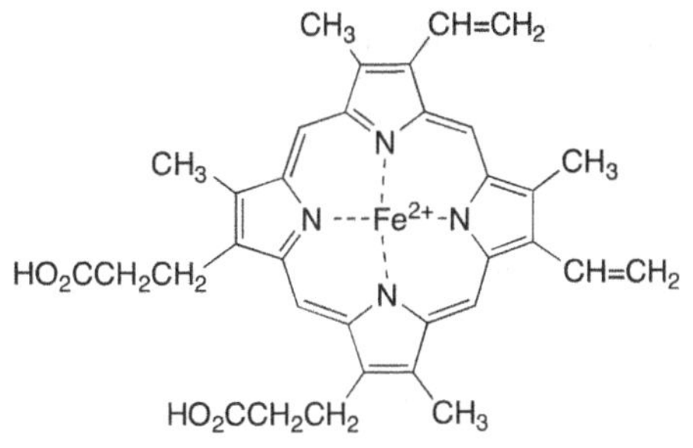

heme

[8] What products are formed when a protein is hydrolyzed? (21.8)

- Hydrolysis breaks up the primary structure of a protein to form the amino acids that comprise it. All the amide bonds are broken by the addition of water, forming a carboxylate anion ($-COO^-$) in one amino acid and an ammonium cation ($-NH_3^+$) in the other.

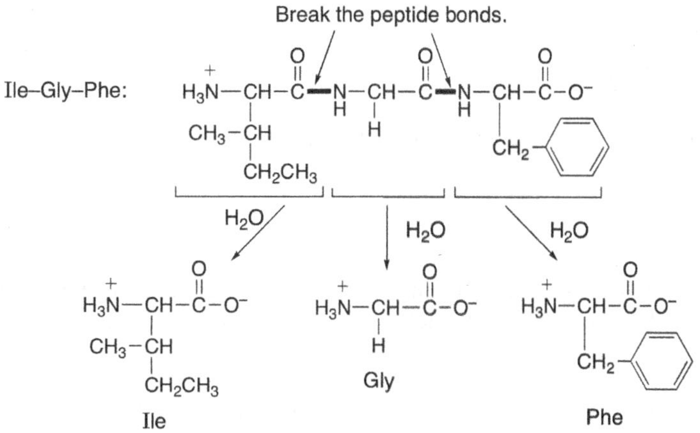

Break the peptide bonds.

Ile–Gly–Phe: $H_3\overset{+}{N}-CH(CH(CH_3)CH_2CH_3)-CO-NH-CH_2-CO-NH-CH(CH_2C_6H_5)-COO^-$

$\xrightarrow{H_2O}$ Ile + Gly + Phe

[9] What is denaturation? (21.8)

- Denaturation is a process that alters the shape of a protein by disrupting the secondary, tertiary, or quaternary structure. High temperature, acid, base, and agitation can denature a protein. Compact water-soluble proteins uncoil and become less water soluble.

[10] What are the main structural features of enzymes and how are enzymes classified? (21.9, 21.10)

- Enzymes are biological catalysts that greatly increase the rate of biological reactions and are highly specific for a substrate or a type of substrate. An enzyme binds a substrate at its active site, forming an enzyme–substrate complex by either the lock-and-key model or the induced-fit model.
- Enzymes are classified into six categories by the type of reaction they catalyze: oxidoreductases, transferases, hydrolases, isomerases, lyases, and ligases (Table 21.3).

[11] What are the three types of enzyme inhibitors? (21.10)

- Enzyme inhibitors cause an enzyme to lose activity. Irreversible inhibition occurs when an inhibitor covalently binds the enzyme and permanently destroys its activity. Competitive reversible inhibition occurs when the inhibitor is structurally similar to the substrate and competes with it for the active site. Noncompetitive reversible inhibition occurs when an inhibitor binds to a location other than the active site, altering the shape of the active site, and thereby rendering it inaccessible to the substrate.

[12] How are enzymes used in medicine? (21.11)

- Measuring blood enzyme levels is used to diagnose heart attacks and diseases that cause higher-than-normal concentrations of certain enzymes to enter the blood.
- Drugs that inhibit the action of an enzyme can be used to kill bacteria. ACE inhibitors are used to treat high blood pressure. HIV protease inhibitors are used to treat HIV by binding to an enzyme needed by the virus to replicate itself.

Problem Solving

[1] Amino Acids (21.2)

Example 21.1 Draw the Fischer projection for each amino acid.

a. L-serine b. D-methionine

Analysis

To draw an amino acid in a Fischer projection, place the $-COO^-$ group at the top and the R group at the bottom. The L isomer has the $-NH_3^+$ on the left side and the D isomer has the $-NH_3^+$ on the right side.

Solution

a. For serine, R = CH_2OH

COO^-
$H_3\overset{+}{N}$—|—H
CH_2OH

$\overset{+}{N}H_3$ on left
L isomer

b. For methionine, R = $CH_2CH_2SCH_3$

COO^-
H—|—$\overset{+}{N}H_3$
$CH_2CH_2SCH_3$

$\overset{+}{N}H_3$ on right
D isomer

[2] Acid–Base Behavior of Amino Acids (21.3)

Example 21.2 Draw the structure of the amino acid leucine at each pH: (a) 6; (b) 2; (c) 11.

Analysis

A neutral amino acid exists in its zwitterionic form (no net charge) at its isoelectric point, which is pH ≈ 6. The zwitterionic forms of neutral amino acids appear in Table 21.2. At low pH (≤ 2), the carboxylate anion is protonated and the amino acid has a net positive (+1) charge. At high pH (≥ 10), the ammonium cation loses a proton and the amino acid has a net negative (–1) charge.

Solution

a. At pH = 6, the neutral, zwitterionic form of leucine predominates.

$$H_3\overset{+}{N}-\underset{CH_2CH(CH_3)_2}{\overset{H}{C}}-COO^-$$

neutral
pH = 6

b. At pH = 2, leucine has a net +1 charge.

$$H_3\overset{+}{N}-\underset{CH_2CH(CH_3)_2}{\overset{H}{C}}-COOH$$

+1 charge
pH = 2

c. At pH = 11, leucine has a net –1 charge.

$$H_2N-\underset{CH_2CH(CH_3)_2}{\overset{H}{C}}-COO^-$$

–1 charge
pH = 11

[3] Peptides (21.4)

Example 21.3 Identify the individual amino acids used to form the following tripeptide. What is the name of the tripeptide?

$$H_3\overset{+}{N}-\underset{CH_2OH}{CH}-\overset{O}{\overset{\|}{C}}-\underset{H}{N}-\underset{H}{CH}-\overset{O}{\overset{\|}{C}}-\underset{H}{N}-\underset{CH_2C(=O)NH_2}{CH}-\overset{O}{\overset{\|}{C}}-O^-$$

Analysis

- Locate the amide bonds in the peptide. To draw the structures of the individual amino acids, break the amide bonds by adding water. Add O^- to the carbonyl carbon to form a carboxylate anion ($-COO^-$) and 2 H's to the N atom to form an ammonium cation ($-NH_3^+$).
- Identify the amino acids by comparing with Table 21.2.
- To name the peptide: [1] Name the C-terminal amino acid. [2] Name the other amino acids as substituents by changing the *-ine* (or *-ic acid*) ending to the suffix *-yl*. Place the names of the substituent amino acids in order from left to right.

Solution

To determine the amino acids that form the peptide, work backwards. Break the amide bonds (in bold) that join the amino acids together. This forms serine, glycine, and asparagine.

$$H_3\overset{+}{N}-\underset{CH_2OH}{CH}-\overset{O}{\overset{\|}{C}}\boldsymbol{-}\underset{H}{N}-\underset{H}{CH}-\overset{O}{\overset{\|}{C}}\boldsymbol{-}\underset{H}{N}-\underset{CH_2C(=O)NH_2}{CH}-\overset{O}{\overset{\|}{C}}-O^-$$

$$H_3\overset{+}{N}-\underset{CH_2OH}{CH}-\overset{O}{\overset{\|}{C}}-O^- \quad H_3\overset{+}{N}-\underset{H}{CH}-\overset{O}{\overset{\|}{C}}-O^- \quad H_3\overset{+}{N}-\underset{CH_2C(=O)NH_2}{CH}-\overset{O}{\overset{\|}{C}}-O^-$$

serine — glycine — asparagine (C-terminal amino acid)

The tripeptide is named as a derivative of the C-terminal amino acid, asparagine, with serine and glycine as substituents; thus, the tripeptide is named **serylglycylasparagine.**

---

[4] Protein Hydrolysis and Denaturation (21.8)

Example 21.4 Draw the structures of the amino acids formed by the hydrolysis of the peptide below.

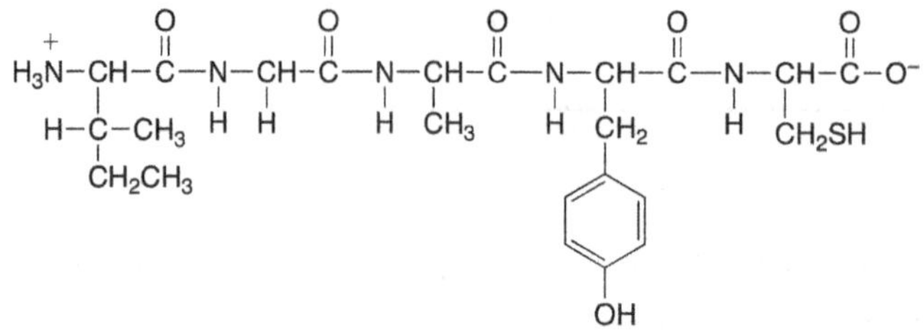

$$H_3\overset{+}{N}-CH(CH(CH_3)CH_2CH_3)-C(=O)-NH-CH_2-C(=O)-NH-CH(CH_3)-C(=O)-NH-CH(CH_2C_6H_4OH)-C(=O)-NH-CH(CH_2SH)-C(=O)-O^-$$

Analysis

Locate each amide bond in the protein or peptide backbone. To draw the hydrolysis products, break each amide bond by adding the elements of H_2O to form a carboxylate anion ($-COO^-$) in one amino acid and an ammonium cation ($-NH_3^+$) in the other.

Solution

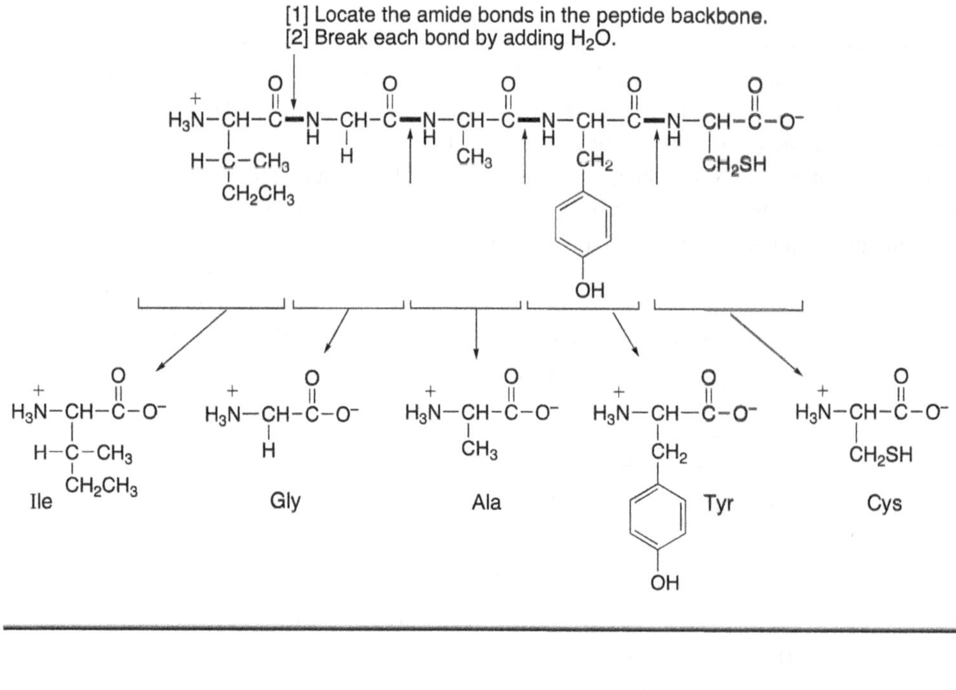

[1] Locate the amide bonds in the peptide backbone.
[2] Break each bond by adding H_2O.

$$H_3\overset{+}{N}-CH(CH(CH_3)CH_2CH_3)-C(=O)\,|\,NH-CH_2-C(=O)\,|\,NH-CH(CH_3)-C(=O)\,|\,NH-CH(CH_2C_6H_4OH)-C(=O)\,|\,NH-CH(CH_2SH)-C(=O)-O^-$$

Ile: $H_3\overset{+}{N}-CH(CH(CH_3)CH_2CH_3)-C(=O)-O^-$

Gly: $H_3\overset{+}{N}-CH(H)-C(=O)-O^-$

Ala: $H_3\overset{+}{N}-CH(CH_3)-C(=O)-O^-$

Tyr: $H_3\overset{+}{N}-CH(CH_2C_6H_4OH)-C(=O)-O^-$

Cys: $H_3\overset{+}{N}-CH(CH_2SH)-C(=O)-O^-$

Self-Test

[1] Fill in the blank with one of the terms listed below.

Cofactor (21.9)
C-Terminal amino acid (21.4)
Dipeptide (21.4)
Enzymes (21.9)
α-Helix (21.6)
Inhibitor (21.10)
Noncompetitive inhibitor (21.10)
N-Terminal amino acid (21.4)
Peptide bonds (21.4)
β-Pleated sheet (21.6)
Primary structure (21.6)
Secondary structure (21.6)
Tertiary structure (21.6)
Tripeptide (21.4)
Zwitterion (21.2)

1. The ______ of a protein is the particular sequence of amino acids that is joined together by peptide bonds.
2. The amide bonds in peptides and proteins are called ______.
3. A ______ contains both a positive and a negative charge.
4. A ______ is a metal ion or a nonprotein organic molecule needed for an enzyme-catalyzed reaction to occur.
5. ______ are proteins that serve as biological catalysts for reactions in all living organisms.
6. The amino acid in a peptide with the free $-NH_3^+$ group on the α carbon is called the ______.
7. The three-dimensional shape adopted by the entire peptide chain is called its ______.
8. The ______ forms when two or more peptide chains, called strands, line up side-by-side.
9. A ______ has two amino acids joined together by one amide bond.
10. An ______ is a molecule that causes an enzyme to lose activity.
11. The ______ forms when a peptide chain twists into a right-handed or clockwise spiral.
12. The amino acid in a peptide with the free $-COO^-$ group on the α carbon is called the ______.
13. A ______ has three amino acids joined together by two amide bonds.
14. The three-dimensional arrangement of localized regions of a protein is called its ______.
15. A ______ binds to an enzyme but does not bind at the active site.

[2] Choose the term for each label in the figures below.

a. enzyme–substrate complex　　b. substrate　　c. active site　　d. products

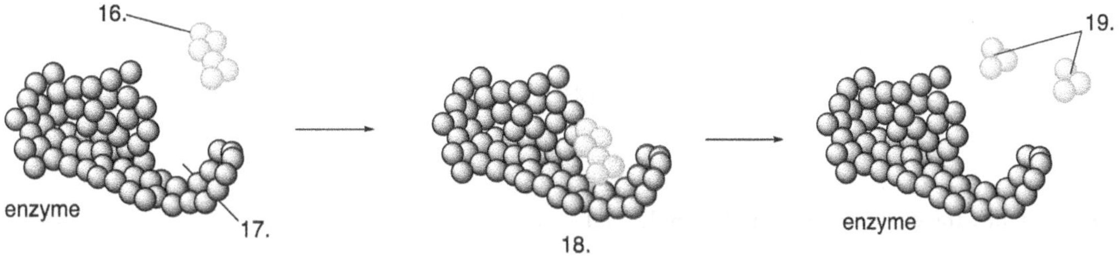

16.
enzyme
17.
18.
19.
enzyme

[3] Pick the appropriate abbreviation for each amino acid.

a. Gln　　b. Val　　c. Phe　　d. Pro

20. $H_3\overset{+}{N}-C(H)(CH_2C_6H_5)-COO^-$

21. $H_3\overset{+}{N}-C(H)(CH_2CH_2CONH_2)-COO^-$

22. $H_3\overset{+}{N}-C(H)(CH(CH_3)_2)-COO^-$

23. (pyrrolidine ring with $-COO^-$, ring $\overset{+}{N}$ bearing H, H)

[4] Label each amino acid as an L or D isomer.

a. L isomer b. D isomer

24. Fischer projection: COO^- (top); $H_3\overset{+}{N}$ (left); H (right); $CH_2CH_2CONH_2$ (bottom)

25. Fischer projection: COO^- (top); H (left); $\overset{+}{N}H_3$ (right); $CH(CH_3)_2$ (bottom)

Answers to Self-Test

1. primary structure	6. N-terminal amino acid	11. α-helix	16. b	21. a
2. peptide bonds	7. tertiary structure	12. C-terminal amino acid	17. c	22. b
3. zwitterion	8. β-pleated sheet	13. tripeptide	18. a	23. d
4. cofactor	9. dipeptide	14. secondary structure	19. d	24. a
5. Enzymes	10. inhibitor	15. noncompetitive inhibitor	20. c	25. b

Solutions to In-Chapter Problems

21.1 Identify the other functional groups in each amino acid.

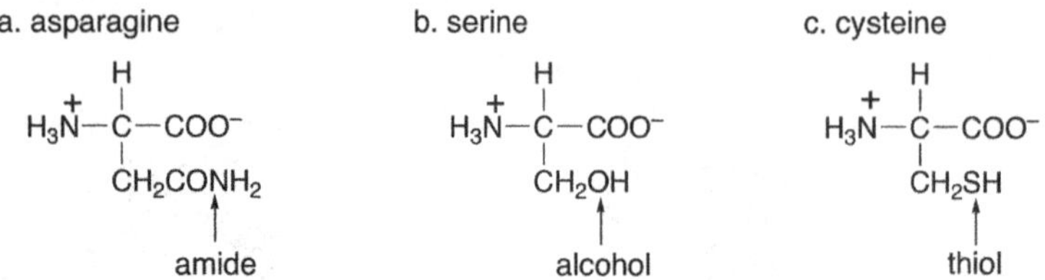

21.2 Draw the enantiomers as in Example 21.1.

a. Fischer projection: COO^- (top); $H_3\overset{+}{N}$ (left); H (right); $CH_2C_6H_5$ (bottom) — $\overset{+}{N}H_3$ on left, L

Fischer projection: COO^- (top); H (left); $\overset{+}{N}H_3$ (right); $CH_2C_6H_5$ (bottom) — $\overset{+}{N}H_3$ on right, D

phenylalanine

b. Fischer projection: COO^- (top); $H_3\overset{+}{N}$ (left); H (right); $CH_2CH_2SCH_3$ (bottom) — $\overset{+}{N}H_3$ on left, L

Fischer projection: COO^- (top); H (left); $\overset{+}{N}H_3$ (right); $CH_2CH_2SCH_3$ (bottom) — $\overset{+}{N}H_3$ on right, D

methionine

21.3 Refer to Table 21.2 to determine the names of the following amino acids. Naturally occurring amino acids are L-amino acids.

a. Fischer projection: COO^- on top, $H_3\overset{+}{N}$ left, H right, CH_2OH bottom — L-serine, naturally occurring

b. Fischer projection: COO^- on top, H left, $\overset{+}{N}H_3$ right, $CH_2CH_2COO^-$ bottom — D-glutamic acid

c. Fischer projection: COO^- on top, $H_3\overset{+}{N}$ left, H right, CH_2COO^- bottom — L-aspartic acid, naturally occurring

21.4 Draw the structure of valine at each pH as in Example 21.2.

a. $H_3\overset{+}{N}-CH(CH(CH_3)_2)-COO^-$ — pH = 6, predominant at p*I*

b. $H_3\overset{+}{N}-CH(CH(CH_3)_2)-COOH$ — pH = 2

c. $H_2N-CH(CH(CH_3)_2)-COO^-$ — pH = 11

21.5 The amino acid is serine.

$H_3\overset{+}{N}-CH(CH_2OH)-COO^-$ — neutral form

$H_3\overset{+}{N}-CH(CH_2OH)-COOH$ — positively charged form

$H_2N-CH(CH_2OH)-COO^-$ — negatively charged form

21.6 The amino acid with the free $-NH_3^+$ group on the α carbon is called the N-terminal amino acid. The amino acid with the free $-COO^-$ group on the α carbon is called the C-terminal amino acid.

a. $H_3\overset{+}{N}-CH(CH_2CH(CH_3)_2)-C(=O)-NH-CH(CH_3)-C(=O)-O^-$

N-terminal: leucine; C-terminal: alanine

c. $H_3\overset{+}{N}-CH(CH_2C_6H_5)-C(=O)-NH-CH(CH_3)-C(=O)-NH-CH(CH_2CH_2CONH_2)-C(=O)-O^-$

N-terminal: phenylalanine; C-terminal: glutamine

b. Arg–His–Asn–Tyr

N-terminal: arginine; C-terminal: tyrosine

d. Val–Thr–Pro–Phe

N-terminal: valine; C-terminal: phenylalanine

21.7 Peptides are named from left to right as substituents of the C-terminal amino acid.

a. and b. alanylglycylleucylmethionine

N-terminal amino acid alanine; C-terminal amino acid methionine

c. Ala–Gly–Leu–Met

21.8 Draw the structure of each dipeptide by joining the $-COO^-$ and $-NH_3^+$ groups of the two different amino acids.

a. Gly–Phe

$$H_3\overset{+}{N}-CH_2-\overset{O}{\overset{\|}{C}}-NH-CH(CH_2C_6H_5)-COO^-$$

b. Gln–Ile

$$H_3\overset{+}{N}-CH(CH_2CH_2CONH_2)-\overset{O}{\overset{\|}{C}}-NH-CH(CH(CH_3)CH_2CH_3)-COO^-$$

c. Leu–Cys

$$H_3\overset{+}{N}-CH(CH_2CH(CH_3)_2)-\overset{O}{\overset{\|}{C}}-NH-CH(CH_2SH)-COO^-$$

21.9 Draw the two dipeptides formed from leucine and asparagine.

a.

$$H_3\overset{+}{N}-CH(CH_2CH(CH_3)_2)-\overset{O}{\overset{\|}{C}}-NH-CH(CH_2CONH_2)-COO^-$$

$$H_3\overset{+}{N}-CH(CH_2CONH_2)-\overset{O}{\overset{\|}{C}}-NH-CH(CH_2CH(CH_3)_2)-COO^-$$

b. N-terminal: leucine
C-terminal: asparagine

N-terminal: asparagine
C-terminal: leucine

c. Leu–Asn

Asn–Leu

21.10 Identify the amino acids in the dipeptides as in Example 21.3. The amide bonds joining the two amino acids are drawn in **bold.**

a.

$$H_3\overset{+}{N}-CH(CH_3)-\overset{O}{\overset{\|}{C}}\boldsymbol{-}NH-CH(CH(CH_3)CH_2CH_3)-\overset{O}{\overset{\|}{C}}-O^-$$

$$H_3\overset{+}{N}-CH(CH_3)-\overset{O}{\overset{\|}{C}}-O^- \qquad H_3\overset{+}{N}-CH(CH(CH_3)CH_2CH_3)-\overset{O}{\overset{\|}{C}}-O^-$$

alanine and isoleucine
Ala–Ile

b.

$$H_3\overset{+}{N}-CH(CH_2C_6H_4OH)-\overset{O}{\overset{\|}{C}}\boldsymbol{-}NH-CH(CH(CH_3)_2)-\overset{O}{\overset{\|}{C}}-O^-$$

$$H_3\overset{+}{N}-CH(CH_2C_6H_4OH)-\overset{O}{\overset{\|}{C}}-O^- \qquad H_3\overset{+}{N}-CH(CH(CH_3)_2)-\overset{O}{\overset{\|}{C}}-O^-$$

tyrosine and valine
Tyr–Val

21.11 Name the dipeptide and indicate the peptide bond as in Example 21.3.

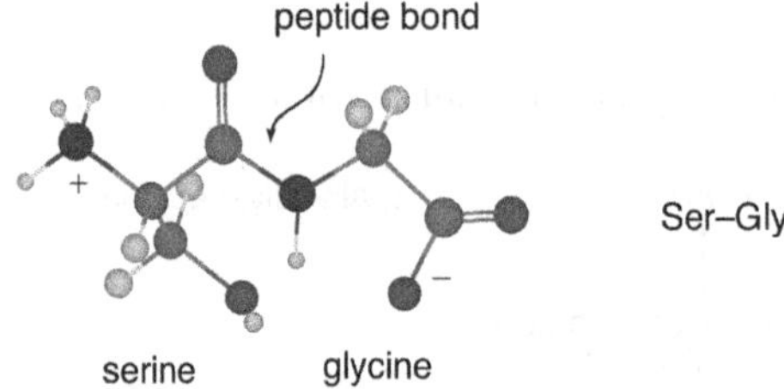

Ser–Gly

21.12 When cysteine is oxidized, a disulfide bond forms.

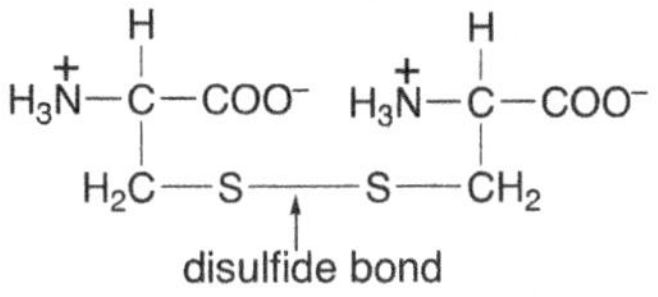

$$H_3\overset{+}{N}-\underset{H_2C-S}{\overset{H}{C}}-COO^- \quad H_3\overset{+}{N}-\underset{S-CH_2}{\overset{H}{C}}-COO^-$$

disulfide bond

21.13 Yes, two different proteins can be composed of the same amino acids because the amino acids can be ordered differently. Examples include Ala–Gly–Ile–Trp and Gly–Trp–Ala–Ile.

21.14 Draw each pair of amino acids.

a. Ser: $H_3\overset{+}{N}-CH(CH_2O-H)-COO^-$; Tyr: $H_3\overset{+}{N}-CH(CH_2-C_6H_4-\ddot{O}-H)-COO^-$

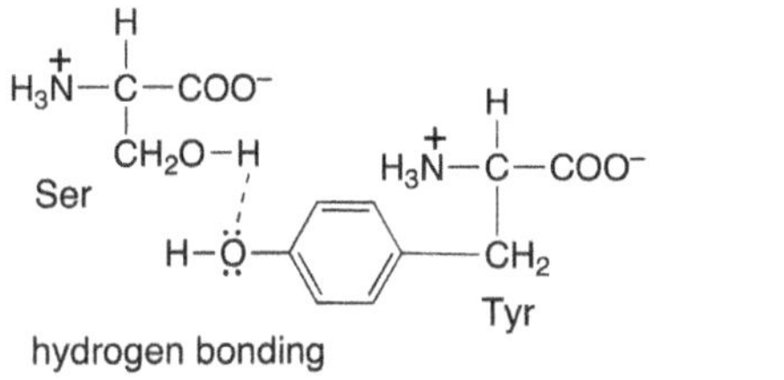

hydrogen bonding

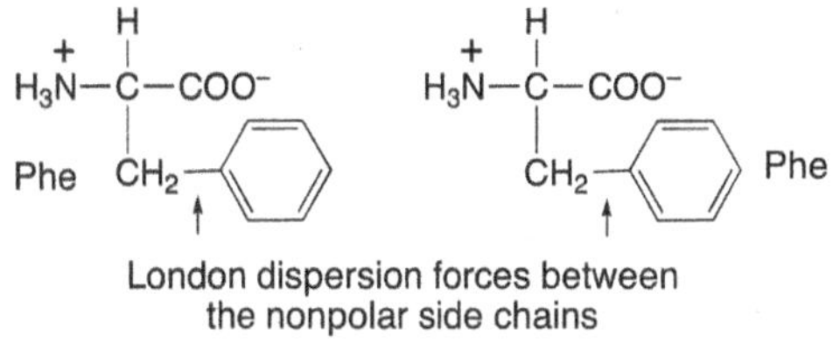

b. Val: $H_3\overset{+}{N}-CH(CH(CH_3)_2)-COO^-$; Leu: $H_3\overset{+}{N}-CH(CH_2CH(CH_3)_2)-COO^-$

London dispersion forces between the nonpolar side chains

c. Phe: $H_3\overset{+}{N}-CH(CH_2-C_6H_5)-COO^-$; Phe: $H_3\overset{+}{N}-CH(CH_2-C_6H_5)-COO^-$

London dispersion forces between the nonpolar side chains

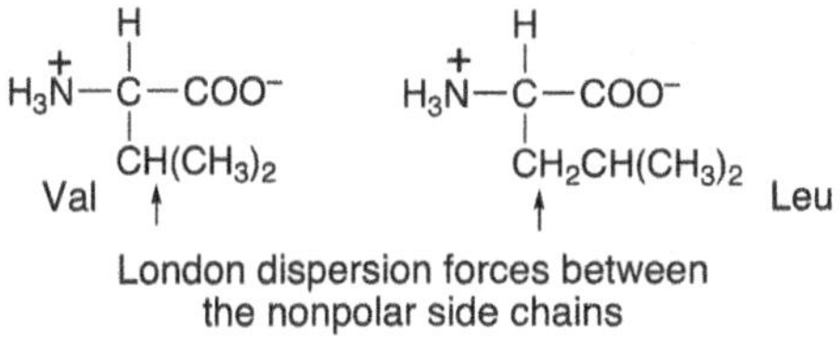

21.15 Glycine has no large side chain, which makes it possible for the β sheets to stack well together.

21.16 Hemoglobin is more water soluble than keratin because it is a globular protein that folds with its polar groups on its exterior, thus making it possible to form hydrogen bonds with water. Keratin is water insoluble because it has more nonpolar groups on its exterior, and this does *not* allow for hydrogen bonding with water.

21.17 Draw the products formed by the hydrolysis of each tripeptide as in Example 21.4. The products of hydrolysis are the individual amino acids.

a. Ala–Leu–Gly

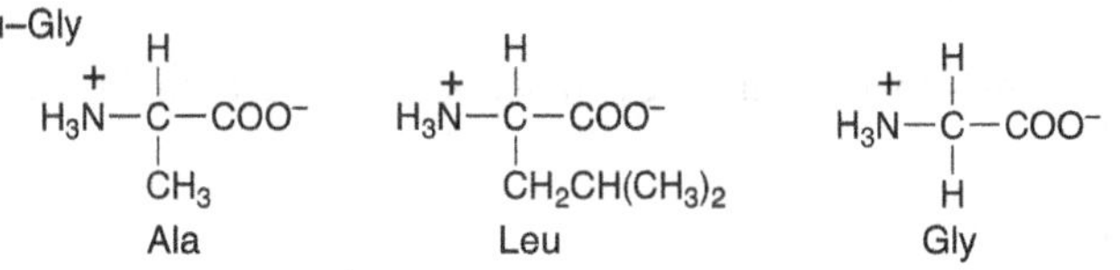

Ala: $H_3\overset{+}{N}-CH(CH_3)-COO^-$; Leu: $H_3\overset{+}{N}-CH(CH_2CH(CH_3)_2)-COO^-$; Gly: $H_3\overset{+}{N}-CH(H)-COO^-$

b. Ser–Thr–Phe

Ser: $H_3\overset{+}{N}-CH(CH_2OH)-COO^-$; Thr: $H_3\overset{+}{N}-CH(CH(OH)CH_3)-COO^-$; Phe: $H_3\overset{+}{N}-CH(CH_2-C_6H_5)-COO^-$

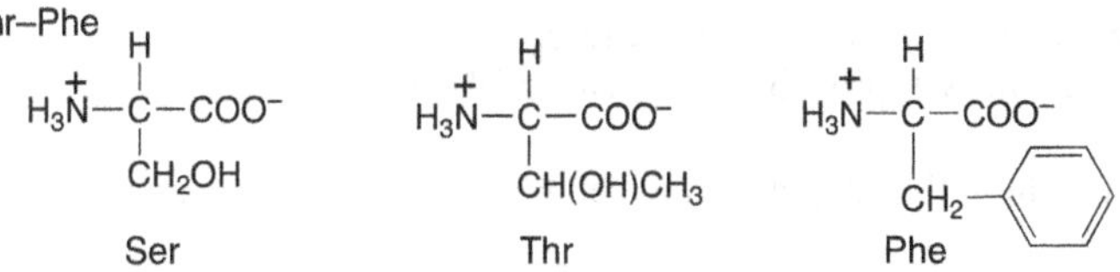

c. Leu–Tyr–Asn

$H_3\overset{+}{N}-CH(CH_2CH(CH_3)_2)-COO^-$ Leu

$H_3\overset{+}{N}-CH(CH_2-C_6H_4-OH)-COO^-$ Tyr

$H_3\overset{+}{N}-CH(CH_2CONH_2)-COO^-$ Asn

21.18 Heating collagen disrupts the hydrogen bonding and other intermolecular forces. The superhelix unwinds, making it less ordered, turning it into the jelly-like substance gelatin.

21.19

a. $^-O_2C-CH_2-CH_2-CO_2^- \longrightarrow {}^-O_2C(H)C=C(H)CO_2^-$

2 H's removed
dehydrogenase

b. $^-O_2C-CH(OH)-CH_2CO_2^- \longrightarrow {}^-O_2C-C(=O)-CH_2CO_2^-$

2 H's removed
dehydrogenase

21.20 A cis alkene is isomerized to a trans alkene so the enzyme is an isomerase.

$HO_2C(H)C=C(H)CO_2H \longrightarrow HO_2C(H)C=C(H)CO_2H$ (cis → trans)

21.21 a. Because a phosphate is transferred to glycerol, the enzyme can be called **glycerol kinase.**
b. Water is added to fumerate, so the enzyme can be called **fumarate synthase.**
c. Because the starting material is glucose 6-phosphate and the reaction is an isomerization, the enzyme can be called **glucose 6-phosphate isomerase.**

21.22 a. Aspartate transaminase catalyzes the transfer of an NH_2 from aspartate to another compound.
b. Glyceraldehyde 3-phosphate dehydrogenase catalyzes the addition (or removal) of 2 H's from glyceraldehyde 3-phosphate.

21.23 a. The reaction is faster at 35 °C rather than 25 °C.
b. The enzyme loses activity at > 40 °C, so the reaction is faster at 35 °C.

21.24 The optimum pH is the pH at which enzyme activity is the highest. In this case, activity is highest at about pH 8.

21.25 a. Lowering the pH decreases the rate.
b. Raising the pH decreases the rate.
c. Decreasing the temperature decreases the rate.
d. At 50 °C the enzyme will probably be considerably denatured, so the rate will be decreased.

21.26 If **Y** increases the rate of the enzyme-catalyzed reaction, then **Y** is a positive regulator.

21.27 A competitive inhibitor has a shape and structure similar to the substrate, so it competes with the substrate for binding to the active site.

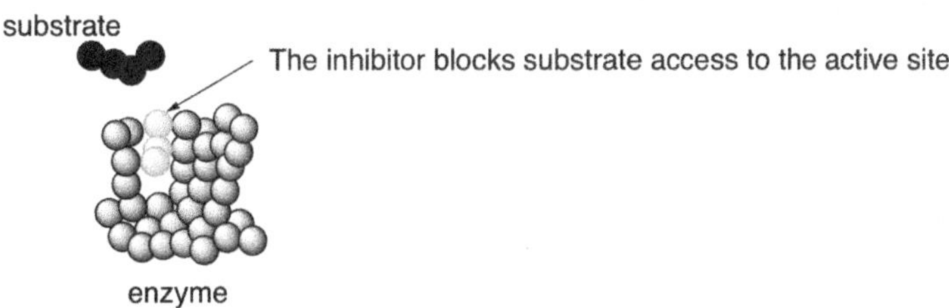

21.28 Sarin is an irreversible inhibitor because it forms a covalent bond at the enzyme's active site.

21.29 Fibrin and thrombin circulate as inactive zymogens (fibrinogen and prothrombin), so that the blood does not clot unnecessarily. They are activated as required at a bleeding point to form a clot.

21.30 Both captopril and enalapril are derived from the amino acid proline and contain an amide and a carboxylate anion. Captopril also contains a thiol (SH), whereas enalapril contains three additional functional groups—an amine, a carboxylate anion, and a benzene ring.

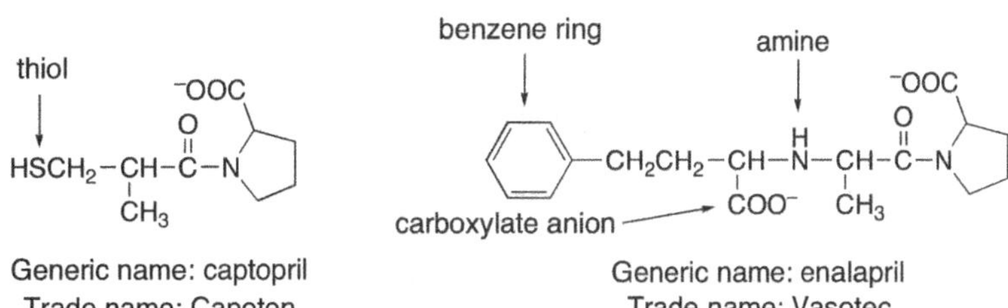

Solutions to Odd-Numbered End-of-Chapter Problems

21.31 Alanine is an ionic salt with extremely strong electrostatic forces, leading to its high melting point and making it a solid at room temperature. Pyruvic acid is a neutral polar molecule with weaker intermolecular forces, so it is a liquid at room temperature.

21.33 Draw an amino acid to fit each requirement.

a. $H_3\overset{+}{N}-CH(CH_2OH)-COO^-$ — CH_2OH: 1° alcohol

b. $H_3\overset{+}{N}-CH(CH_2CONH_2)-COO^-$ — CH_2CONH_2: amide

c. $H_3\overset{+}{N}-CH(CH_2C_6H_5)-COO^-$ — benzene ring: aromatic ring

d. neutral amino acid

$H_3\overset{+}{N}-CH(CH(CH_3)_2)-COO^-$ — $CH(CH_3)_2$: 3° carbon

21.35 Answer each question about the amino acids.

	[1] L Enantiomer	[2] Classification	[3] Three-letter Symbol	[4] One-letter Symbol
a.	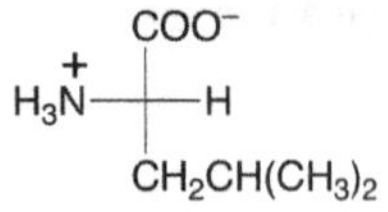 COO^-; $H_3\overset{+}{N}$—C—H; $CH_2CH(CH_3)_2$; leucine	neutral	Leu	L
b.	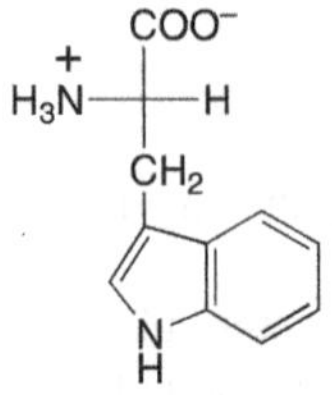 COO^-; $H_3\overset{+}{N}$—C—H; CH_2–(indol-3-yl, N–H); tryptophan	neutral	Trp	W
c.	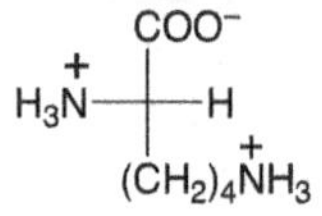 COO^-; $H_3\overset{+}{N}$—C—H; $(CH_2)_4\overset{+}{N}H_3$; lysine	basic	Lys	K
d.	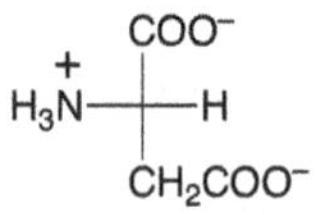 COO^-; $H_3\overset{+}{N}$—C—H; CH_2COO^-; aspartic acid	acidic	Asp	D

21.37 Draw and label the enantiomers of each amino acid.

a. methionine

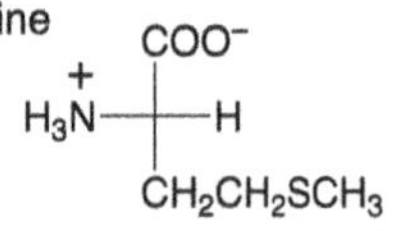

L: COO^- (top); $H_3\overset{+}{N}$—C—H; $CH_2CH_2SCH_3$ (bottom)

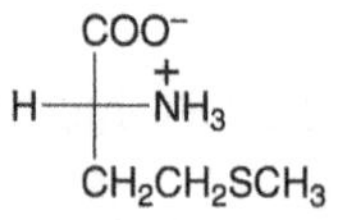

D: COO^- (top); H—C—$\overset{+}{N}H_3$; $CH_2CH_2SCH_3$ (bottom)

b. asparagine

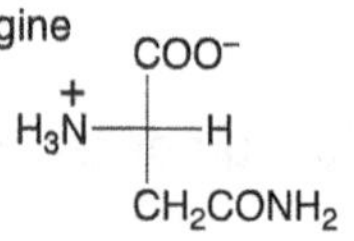

L: COO^- (top); $H_3\overset{+}{N}$—C—H; CH_2CONH_2 (bottom)

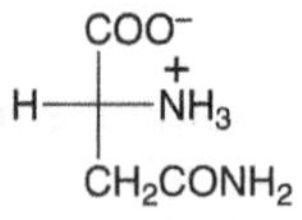

D: COO^- (top); H—C—$\overset{+}{N}H_3$; CH_2CONH_2 (bottom)

21.39 Answer each question about the amino acids.

	[1] Amino Acid	[2] Three-letter Symbol	[3] One-letter Symbol	[4] Classification
a.	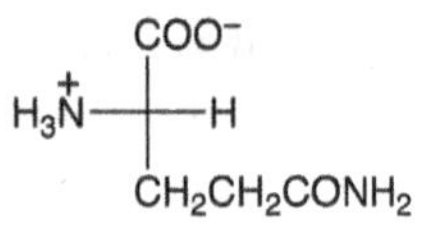 COO^-; $H_3\overset{+}{N}$—C—H; $CH_2CH_2CONH_2$; glutamine	Gln	Q	neutral
b.	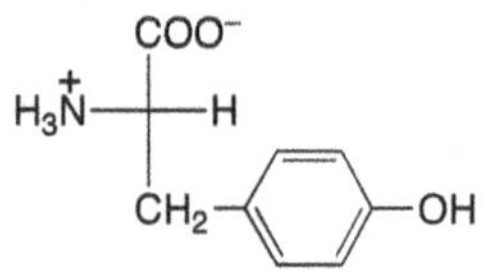 COO^-; $H_3\overset{+}{N}$—C—H; CH_2–C_6H_4–OH; tyrosine	Tyr	Y	neutral

21.41

a, b, c. COO^-

$H_3\overset{+}{N}-C-H$

$CH_2CH_2SCH_3$

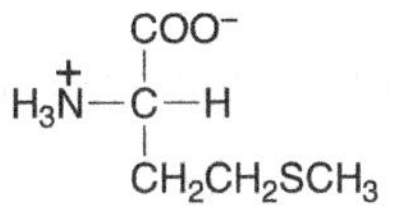

Met

M

methionine

neutral form present at the isoelectric point

21.43 Draw the structure of leucine at each pH as in Example 21.2.

a. 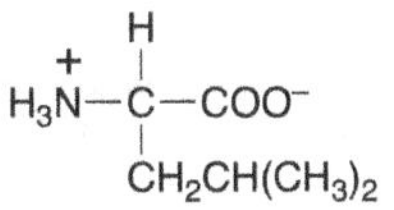 $H_3\overset{+}{N}-C(H)-COO^-$ with $CH_2CH(CH_3)_2$

pH = 6

predominant form at p*I*

b. 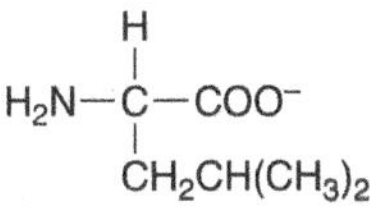 $H_2N-C(H)-COO^-$ with $CH_2CH(CH_3)_2$

pH = 10

c. 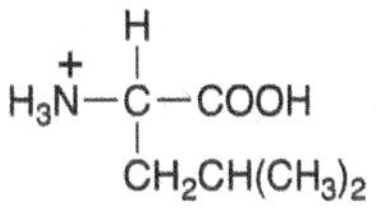 $H_3\overset{+}{N}-C(H)-COOH$ with $CH_2CH(CH_3)_2$

pH = 2

21.45 Locate the peptide bond and name the dipeptide as in Example 21.3.

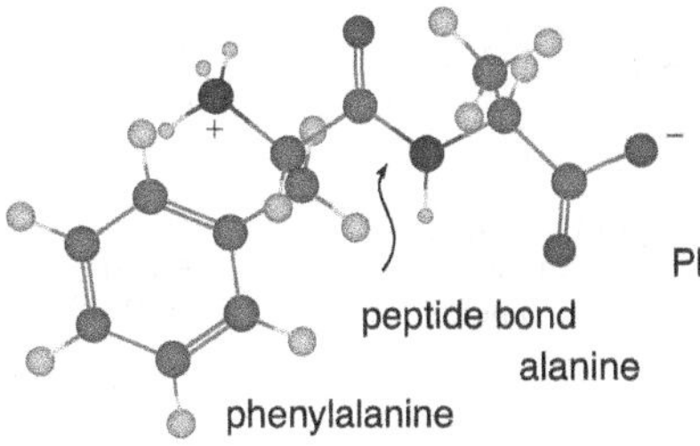

peptide bond

alanine

phenylalanine

Phe–Ala

21.47 The amino acid with the free $–NH_3^+$ group on the α carbon is called the N-terminal amino acid. The amino acid with the free $–COO^-$ group on the α carbon is called the C-terminal amino acid.

a, b, c.

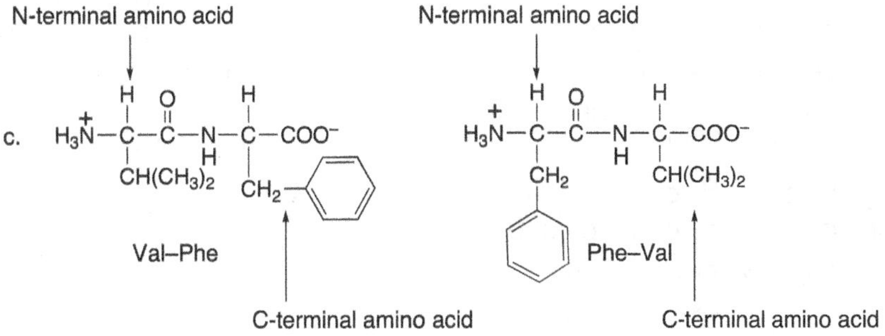

N-terminal amino acid

$H_3\overset{+}{N}-C(H)(CH(CH_3)_2)-C(=O)-N(H)-C(H)(CH_2C_6H_5)-COO^-$

Val–Phe

C-terminal amino acid

N-terminal amino acid

$H_3\overset{+}{N}-C(H)(CH_2C_6H_5)-C(=O)-N(H)-C(H)(CH(CH_3)_2)-COO^-$

Phe–Val

C-terminal amino acid

21.49 Answer each question about the tripeptides.

a. leucylvalyltryptophan

[1] $H_3\overset{+}{N}$–C(H)(CH$_2$CH(CH$_3$)$_2$)–C(=O)–*–N(H)–C(H)(CH(CH$_3$)$_2$)–C(=O)–*–N(H)–C(H)(CH$_2$–indolyl)–COO$^-$

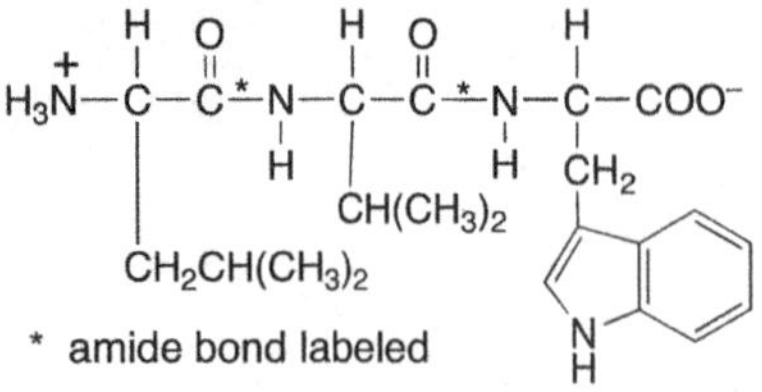

[2] * amide bond labeled

[3] N-terminal: leucine
C-terminal: tryptophan

[4] Leu–Val–Trp

b. phenylalanylserylthreonine

[1] $H_3\overset{+}{N}$–C(H)(CH$_2$–C$_6$H$_5$)–C(=O)–*–N(H)–C(H)(CH$_2$OH)–C(=O)–*–N(H)–C(H)(CH(OH)CH$_3$)–COO$^-$

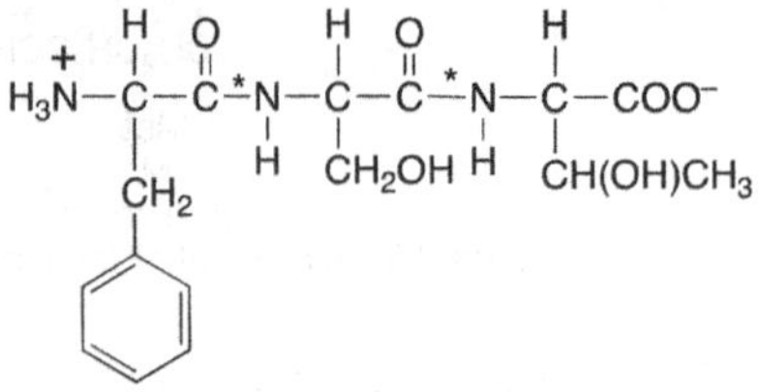

[2] * amide bond labeled

[3] N-terminal: phenylalanine
C-terminal: threonine

[4] Phe–Ser–Thr

21.51 Answer each question about the tripeptide.

a. $H_3\overset{+}{N}$–CH(CH(CH$_3$)$_2$)–C(=O)–N(H)–CH(H)–C(=O)–N(H)–CH(CH$_2$–C$_6$H$_5$)–C(=O)–O$^-$

valine
N-terminal
amino acid

glycine

phenylalanine
C-terminal
amino acid

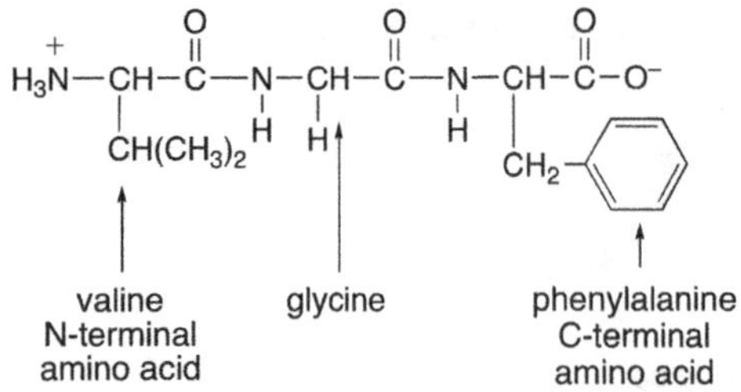

Val–Gly–Phe

b. $H_3\overset{+}{N}$–CH(CH$_2$CH(CH$_3$)$_2$)–C(=O)–N(H)–CH(CH$_2$–C$_6$H$_4$–OH)–C(=O)–N(H)–CH(CH$_2$CH$_2$SCH$_3$)–C(=O)–O$^-$

leucine
N-terminal
amino acid

tyrosine

methionine
C-terminal
amino acid

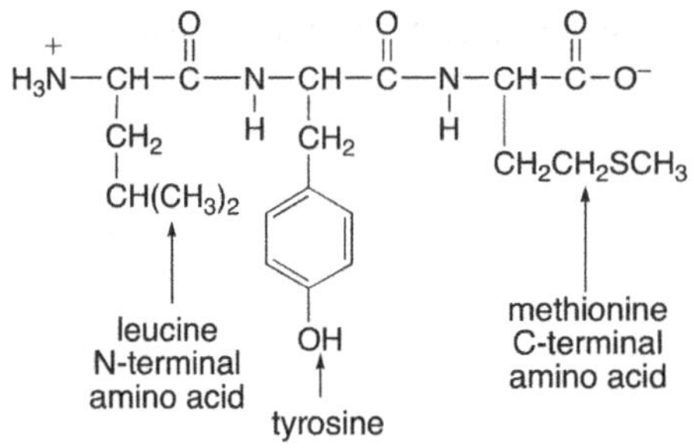

Leu–Tyr–Met

21.53 Draw the structure of the three tripeptides.

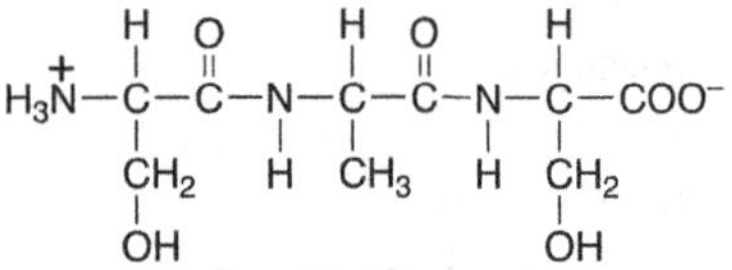

$H_3\overset{+}{N}$–C(H)(CH$_2$OH)–C(=O)–N(H)–C(H)(CH$_3$)–C(=O)–N(H)–C(H)(CH$_2$OH)–COO$^-$

Ser–Ala–Ser

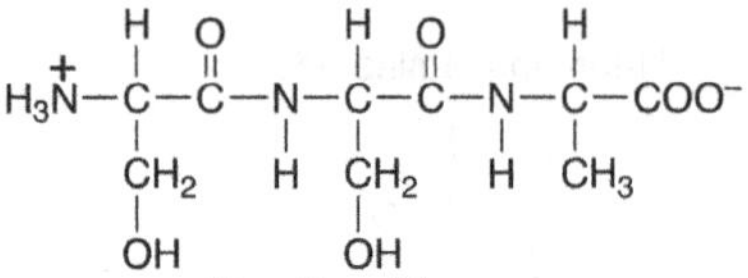

$H_3\overset{+}{N}$–C(H)(CH$_2$OH)–C(=O)–N(H)–C(H)(CH$_2$OH)–C(=O)–N(H)–C(H)(CH$_3$)–COO$^-$

Ser–Ser–Ala

$H_3\overset{+}{N}$–C(H)(CH$_3$)–C(=O)–N(H)–C(H)(CH$_2$OH)–C(=O)–N(H)–C(H)(CH$_2$OH)–COO$^-$

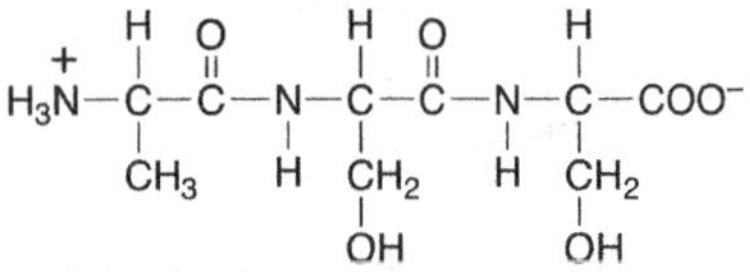

Ala–Ser–Ser

21.55 Draw the products formed by the hydrolysis of the tripeptide as in Example 21.4.

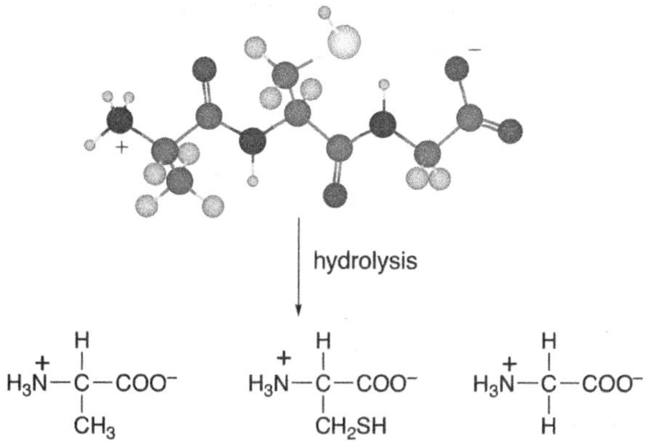

21.57 Draw the products formed by the hydrolysis of each tripeptide as in Example 21.4.

a.

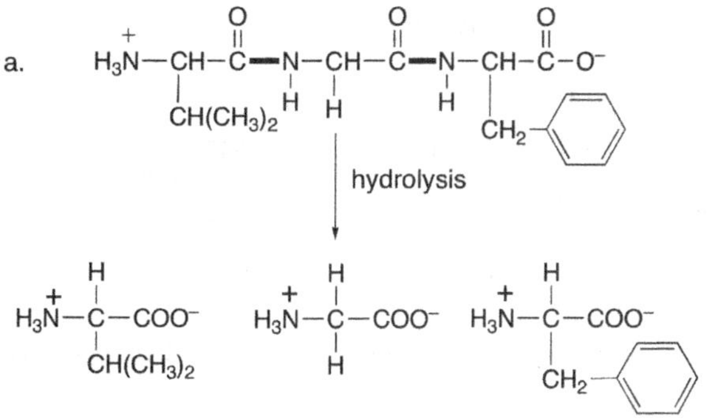

b.

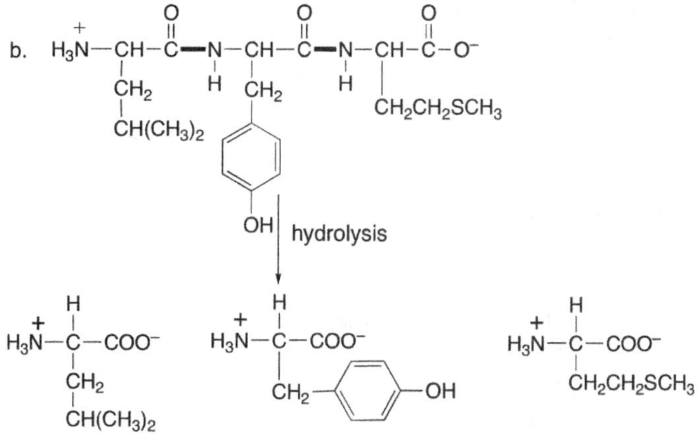

21.59 Draw the products of the hydrolysis of bradykinin.

Arg–Pro–Pro–Gly–Phe–Ser–Pro–Phe–Arg

bradykinin

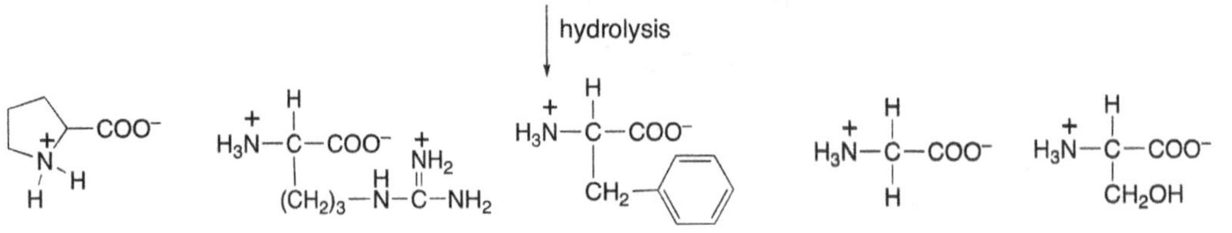

hydrolysis

Pro 3 moles | Arg 2 moles | Phe 2 moles | Gly 1 mole | Ser 1 mole

21.61 Draw the products of the hydrolysis of the peptide with chymotrypsin.

Hydrolysis occurs here.

Gly–Tyr–Gly–Ala–Phe–Val

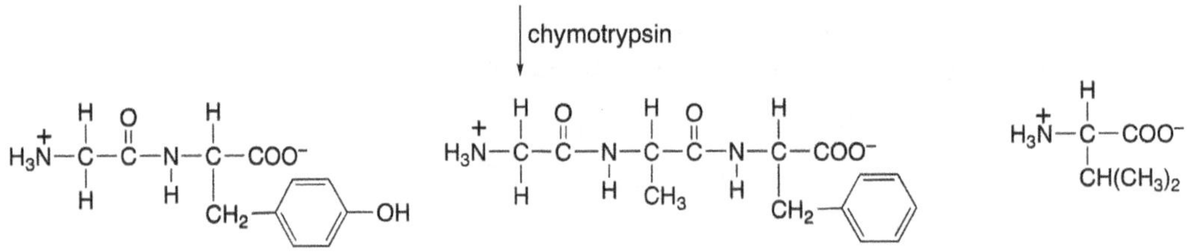

chymotrypsin

Gly–Tyr | Gly–Ala–Phe | Val

21.63 Use Figure 21.10 to determine which types of intermolecular forces occur between each amino acid pair.

a. isoleucine and valine: London dispersion forces of the nonpolar side chains

b. threonine and phenylalanine: London dispersion forces. Since the side chain of phenylalanine has no O or N atom, no hydrogen bonding is possible.

c. Lys and Glu: electrostatic attraction of the charged side chains

d. Arg and Asp: electrostatic attraction of the charged side chains

21.65 Draw the structure.

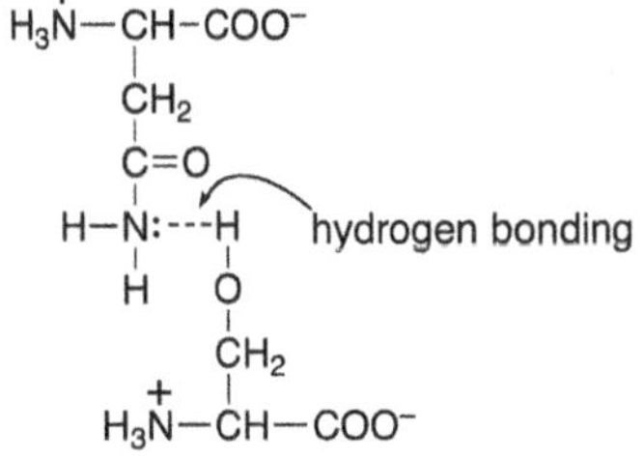

hydrogen bonding

21.67 Use Figure 21.6 to label the regions of secondary structure.

The α-helix is the corkscrew-shaped region.

The β-pleated sheets are drawn as flat ribbons with gentle curves.

The areas of random coil are shown as thin string-like lines.

21.69 Compare keratin and hemoglobin.

	a. Secondary Structure	**b. H_2O Solubility**	**c. Function**	**d. Location**
Hemoglobin	globular with much α-helix	soluble	carries oxygen to tissues	blood
Keratin	α-helix	insoluble	gives strength to tissues	nails, hair

21.71 When a protein is heated and denatured, the primary structure is unaffected. The 2°, 3°, and 4° structures may be altered by unfolding.

21.73 a. Insulin is a hormone that controls glucose levels.
b. Myoglobin stores oxygen in muscle.
c. α-Keratin forms hard tissues such as hair and nails.
d. Chymotrypsin is a protease that hydrolyzes peptide bonds.
e. Oxytocin is a hormone that stimulates uterine contractions and induces the release of breast milk.

21.75 a. **B** represents the enzyme–substrate complex.
b. **C** represents competitive inhibition.
c. **A** represents noncompetitive inhibition.

21.77

a.  $^-O_2CCH_2-C(CO_2^-)(OH)-CH(H)-CO_2^- \longrightarrow {}^-O_2CCH_2-C(CO_2^-)=C(H)-CO_2^-$

(OH H boxed in reactant)

H_2O is lost.
dehydrase

b. 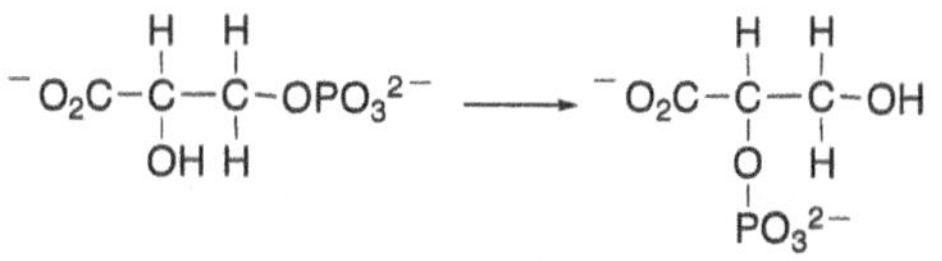 $^-O_2C-CH(OH)-CH_2-OPO_3^{2-} \longrightarrow {}^-O_2C-CH(OPO_3^{2-})-CH_2-OH$

The reactant and product are isomers, so the enzyme is an isomerase.

21.79

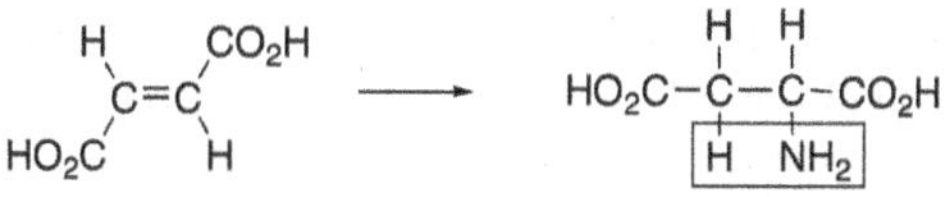

$HO_2C(H)C=C(H)CO_2H \longrightarrow HO_2C-CH(H)-CH(NH_2)-CO_2H$

(H NH₂ boxed in product)

NH_3 is added, so the enzyme is a synthase.

21.81 a. Citrate decarboxylase catalyzes the loss of CO_2 from citrate.
b. Oxalate reductase catalyzes the reduction of oxalate.
c. Serine transaminase catalyzes the transfer of an NH_2 group from serine to another compound.

21.83 a. The optimum pH for enzyme **A** is about pH 6.
b. The optimum pH for enzyme **B** is about pH 10.
c. At pH 4, enzyme **A** has higher activity.
d. At pH 9, enzyme **B** has higher activity.

21.85 In positive allosteric control, a regulator makes the active site more able to bind the substrate, and the rate of reaction increases. In negative allosteric control, a regulator makes the active site less able to bind the substrate, and the rate of reaction decreases.

21.87 Captopril inhibits the angiotensin-converting enzyme, blocking the conversion of angiotensinogen to angiotensin. This reduces the concentration of angiotensin, which in turn lowers blood pressure.

21.89 A noncompetitive inhibitor binds to the enzyme but does not bind at the active site.

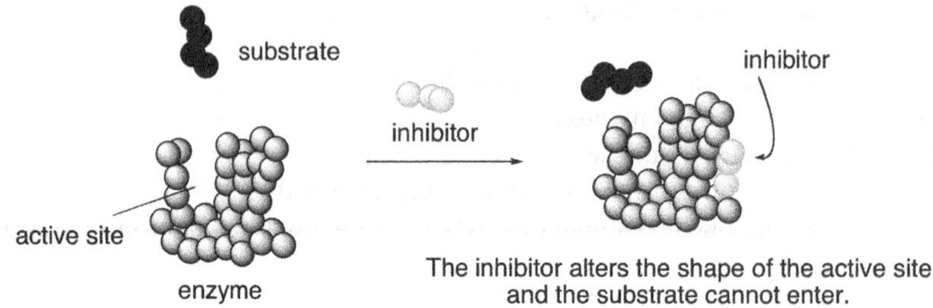

The inhibitor alters the shape of the active site and the substrate cannot enter.

21.91 The α-keratin in nails has more cysteine residues to form disulfide bonds. The larger the number of disulfide bonds, the harder the substance. Therefore, fingernails are harder than skin.

21.93 Humans cannot synthesize the amino acids methionine and lysine, so they must be obtained in the diet. Diets that include animal products readily supply all of the needed amino acids, but plant sources generally do not have sufficient amounts of all the essential amino acids. Grains—wheat, rice, and corn—are low in lysine, and legumes—beans, peas, and peanuts—are low in methionine, but a combination of these foods provides all of the needed amino acids.

21.95 Cauterization denatures the proteins in a wound.

21.97 When surgical instruments are heated with steam in an autoclave, the high temperature causes any bacterial enzymes to be denatured and inactive, so the instruments are sterilized.

21.99 Penicillin inhibits the formation of the bacterial cell wall by irreversibly binding to an enzyme needed for its construction. It does not affect humans because human cells have a cell membrane, instead of a rigid cell wall. Sulfanilamide inhibits the production of folic acid and therefore reproduction in bacteria, but humans do not synthesize folic acid (they must ingest it instead), so it does not affect humans.

21.101 Both aspartic acid and glutamic acid have two carboxylic acid groups. At low pH they have a +1 charge with both acid groups protonated, but at a high pH both acid groups are ionized, leading to a net charge of –2.

$$H_3\overset{+}{N}-\underset{CH_2COOH}{\overset{H}{\overset{|}{\underset{|}{C}}}}-COOH \qquad\qquad NH_2-\underset{CH_2COO^-}{\overset{H}{\overset{|}{\underset{|}{C}}}}-COO^-$$

Asp at low pH
+1 charge

Asp at high pH
–2 charge

Chapter 22 Nucleic Acids and Protein Synthesis

Chapter Review

[1] What are the main structural features of nucleosides and nucleotides? (22.1)

- A nucleoside contains a monosaccharide joined to a nitrogen-containing base at the anomeric carbon.
- A nucleotide contains a monosaccharide joined to a nitrogen-containing base, and a phosphate bonded to the 5'-OH group of the monosaccharide.
- The monosaccharide is either ribose or 2-deoxyribose, and the bases are abbreviated as A, G, C, T, and U.

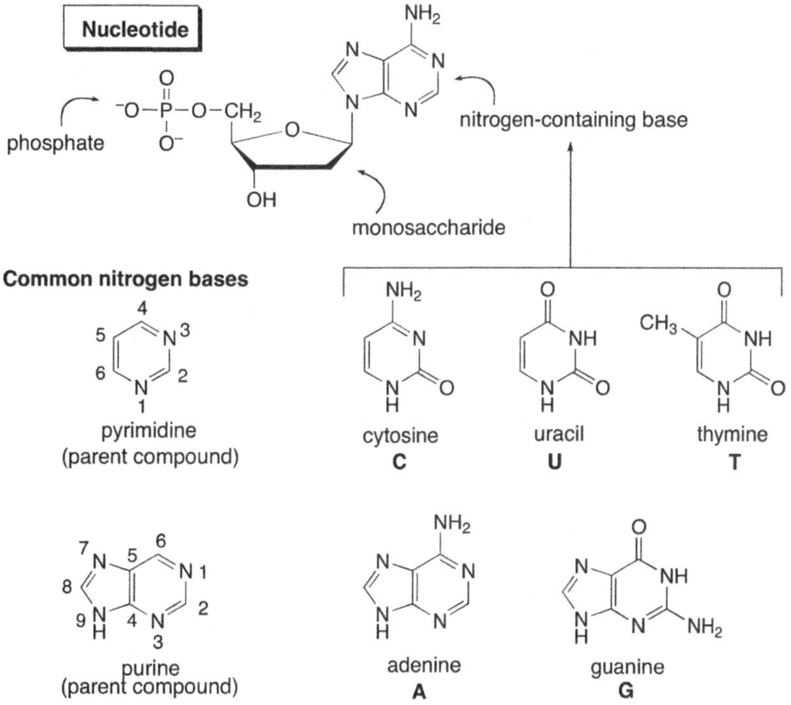

Nucleotide

phosphate — monosaccharide — nitrogen-containing base

Common nitrogen bases

pyrimidine (parent compound) — cytosine **C** — uracil **U** — thymine **T**

purine (parent compound) — adenine **A** — guanine **G**

[2] How do the nucleic acids DNA and RNA differ in structure? (22.2)

- DNA is a polymer of deoxyribonucleotides where the sugar is 2-deoxyribose and the bases are A, G, C, and T. DNA is double stranded.
- RNA is a polymer of ribonucleotides where the sugar is ribose and the bases are A, G, C, and U. RNA is single stranded.

[3] Describe the basic features of the DNA double helix. (22.3)

- DNA consists of two polynucleotide strands that wind into a right-handed double helix. The sugar–phosphate backbone lies on the outside of the helix and the bases lie on the inside. The two strands run in opposite directions; that is, one strand runs from the 5' end to the 3' end and the other runs from the 3' end to the 5' end. The double helix is stabilized by hydrogen bonding between complementary base pairs; A pairs with T, and C pairs with G.

[4] Outline the main steps in the replication of DNA. (22.4)

- The replication of DNA is semiconservative; an original DNA molecule forms two DNA molecules, each of which has one strand from the parent DNA and one new strand.
- In replication, DNA unwinds and the enzyme DNA polymerase catalyzes replication on both strands. The identity of the bases on the template strand determines the order of the bases on the new strand, with A pairing with T and C pairing with G. Replication occurs from the 3' end to the 5' end of each strand. One strand, the leading strand, grows continuously, while the other strand, the lagging strand, is synthesized in fragments and then joined together with a DNA ligase enzyme.

[5] List the three types of RNA molecules and describe their functions. (22.5)

- Ribosomal RNA (rRNA), which consists of one large subunit and one small subunit, provides the site where proteins are assembled.
- Messenger RNA (mRNA) contains the sequence of nucleotides that determines the amino acid sequence in a protein. mRNA is transcribed from DNA such that each DNA gene corresponds to a different mRNA molecule.
- Transfer RNA (tRNA) contains an anticodon that identifies the amino acid that it carries on its acceptor stem and delivers that amino acid to a growing polypeptide.

[6] What is transcription? (22.6)

- Transcription is the synthesis of mRNA from DNA. The DNA helix unwinds and RNA polymerase catalyzes RNA synthesis from the 3' to 5' end of the template strand, forming mRNA with complementary bases.

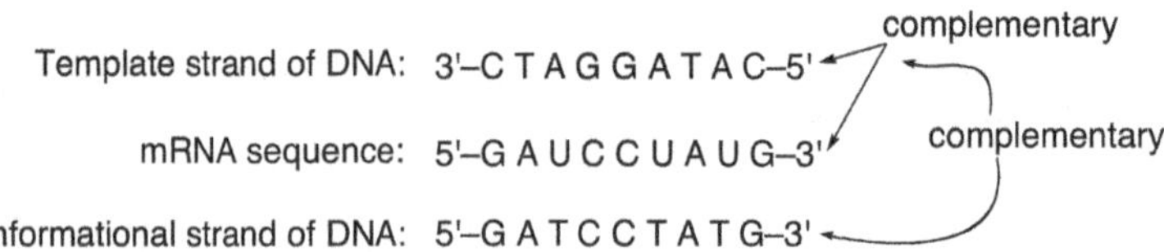

[7] What are the main features of the genetic code? (22.7)

- mRNAs contain sequences of three bases called codons that code for individual amino acids. There are 61 codons that correspond to the 20 amino acids, as well as three stop codons that signal the end of protein synthesis.

[8] How are proteins synthesized by the process of translation? (22.8)

- Translation begins with initiation, the binding of the ribosomal subunits to mRNA and the arrival of the first tRNA with an amino acid. During elongation, tRNAs bring individual amino acids to the ribosome one after another, and new peptide bonds are formed. Termination occurs when a stop codon is reached.

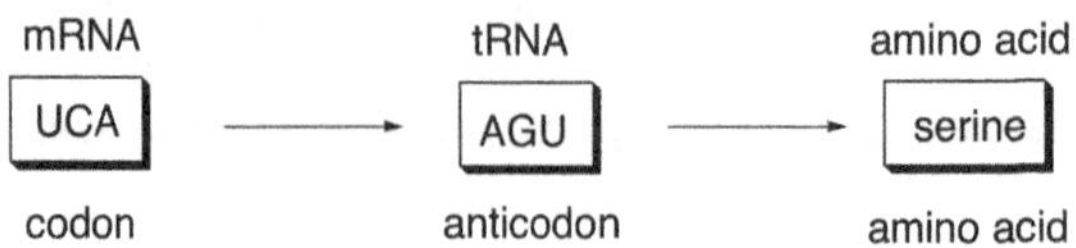

[9] What is a mutation and how are mutations related to genetic diseases? (22.9)

- Mutations are changes in the nucleotide sequence in a DNA molecule. A mutation that causes an inherited condition may result in a genetic disease.
    - A point mutation results in the substitution of one nucleotide for another.

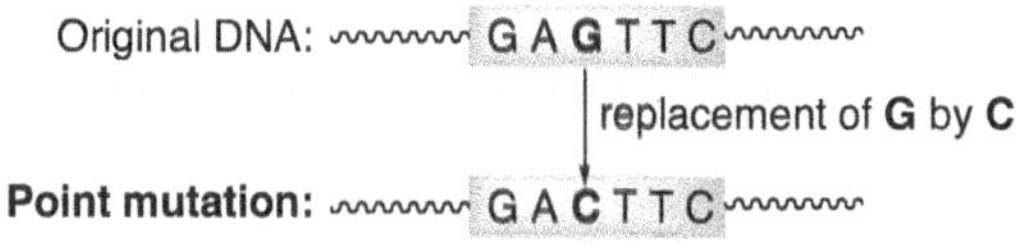
Original DNA: GAGTTC
replacement of G by C
Point mutation: GACTTC

- Deletion and insertion mutations result in the loss or addition of nucleotides, respectively.

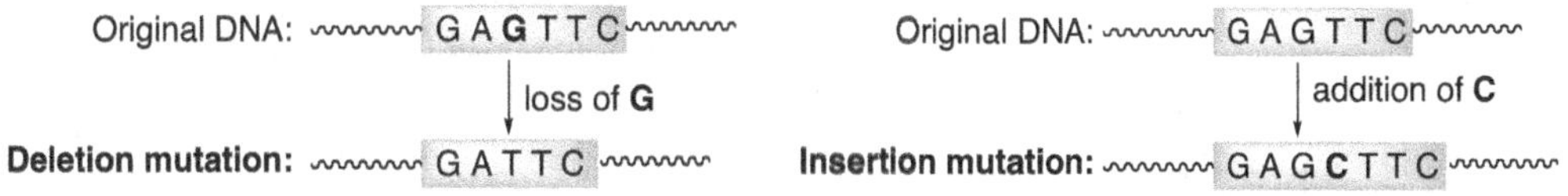
Original DNA: GAGTTC
loss of G
Deletion mutation: GATTC
Original DNA: GAGTTC
addition of C
Insertion mutation: GAGCTTC

[10] What are the principal features of three techniques that use DNA in the laboratory—recombinant DNA, the polymerase chain reaction (PCR), and DNA fingerprinting? (22.10)

- Recombinant DNA is synthetic DNA formed when a segment of DNA from one source is inserted into the DNA of another source. When inserted into a bacterium, recombinant DNA can be used to prepare large quantities of useful proteins.
- The polymerase chain reaction (PCR) is used to amplify a portion of a DNA molecule to produce millions of copies of a single gene.
- DNA fingerprinting is used to identify an individual by cutting DNA with restriction endonucleases to give a unique set of fragments.

[11] What are the main characteristics of viruses? (22.11)

- A virus is an infectious agent that contains either DNA or RNA within a protein coat. When the virus invades a host cell, it uses the biochemical machinery of the host to replicate. A retrovirus contains RNA and a reverse transcriptase that allow the RNA to synthesize viral DNA, which then transcribes RNA that directs the synthesis of viral proteins.

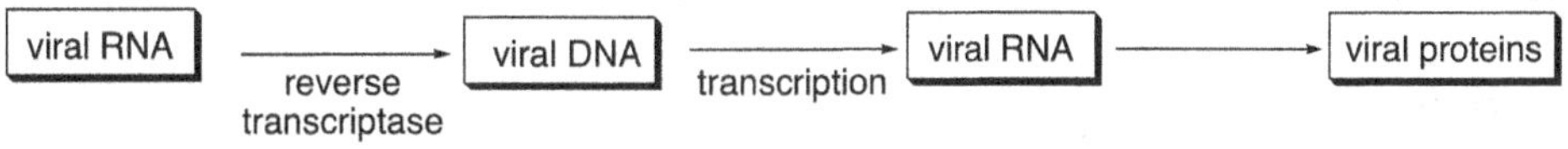
viral RNA
reverse transcriptase
viral DNA
transcription
viral RNA
viral proteins

Problem Solving

[1] Nucleosides and Nucleotides (22.1)

Example 22.1 Identify the base and monosaccharide used to form the following nucleoside, and then name it.

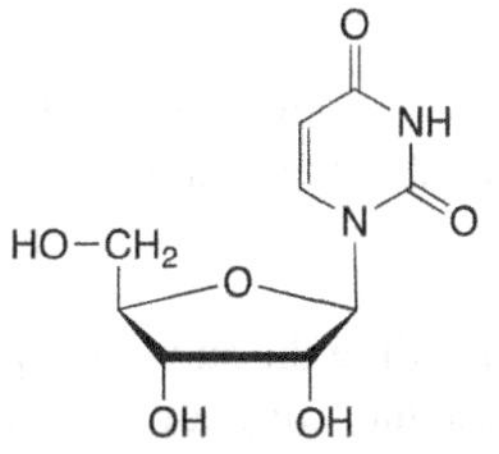
O
NH
N
O
HO−CH2
O
OH
OH

Analysis

- The sugar portion of the nucleoside contains the five-membered ring. If there is an OH group at C2', the sugar is ribose, but if there is no OH group at C2', the sugar is deoxyribose.
- The base is joined to the five-membered ring as an *N*-glycoside. A pyrimidine base has one ring, and is derived from cytosine, uracil, or thymine. A purine base has two rings and is derived from either adenine or guanine.
- Nucleosides derived from pyrimidines end in the suffix *-idine*. Nucleosides derived from purines end in the suffix *-osine*.

Solution

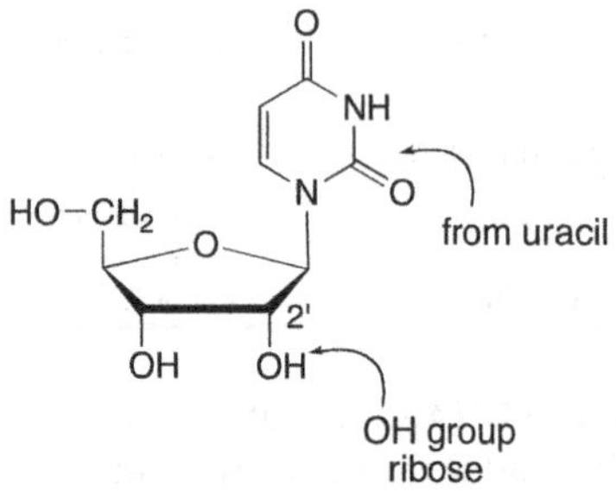

from uracil

OH group ribose

The sugar has an OH at C2', so it is derived from ribose. The base is uracil. To name the ribonucleoside, change the suffix of the base to *-idine*; thus, uracil → uridine.

Example 22.2 Draw the structure of the nucleotide dTMP.

Analysis

Translate the abbreviation to a name; dTMP is deoxythymidine 5'-monophosphate. First, draw the sugar. Since there is a *deoxy* prefix in the name, dTMP is a deoxyribonucleotide and the sugar is deoxyribose. Then draw the base, in this case thymine, bonded to C1' of the sugar ring. Finally, add the phosphate. dTMP has one phosphate group bonded to the 5'-OH of the nucleoside.

Solution

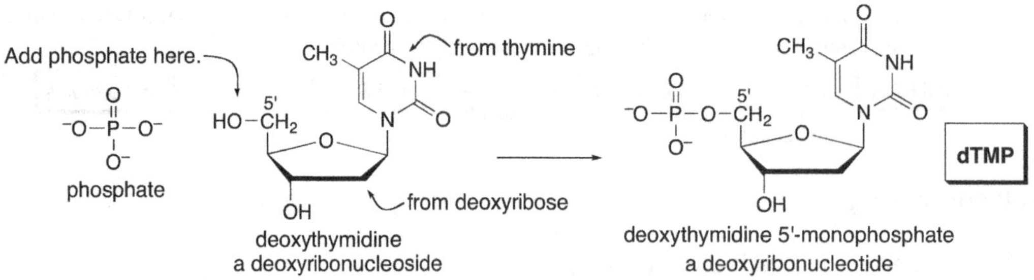

Add phosphate here.

phosphate

from thymine

from deoxyribose

deoxythymidine
a deoxyribonucleoside

dTMP

deoxythymidine 5'-monophosphate
a deoxyribonucleotide

---

[2] Nucleic Acids (22.2)

Example 22.3 (a) Draw the structure of a dinucleotide formed by joining the 3'-OH group of GMP to the 5'-phosphate in CMP. (b) Label the 5' and 3' ends. (c) Name the dinucleotide.

Analysis

Draw the structure of each nucleotide, including the sugar, the phosphate bonded to C5', and the base at C1'. In this case the sugar is ribose because the names of the mononucleotides do not contain the prefix *deoxy*. Bond the 3'-OH group to the 5'-phosphate to form the phosphodiester bond. The name of the dinucleotide begins with the nucleotide that contains the free phosphate at the 5' end.

Solution

a. and b.

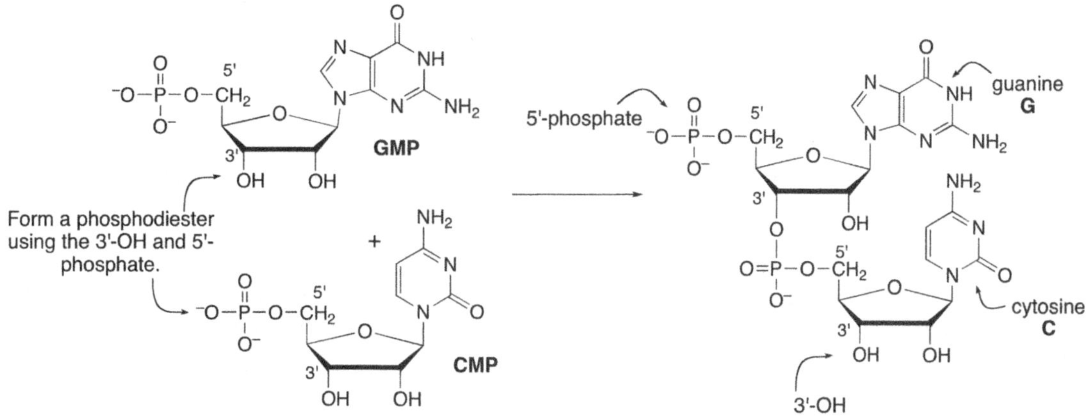

c. Polynucleotides are named beginning at the 5' end, so this dinucleotide is named GC.

---

[3] The DNA Double Helix (22.3)

Example 22.4 Write the sequence of the complementary strand of the following portion of a DNA molecule: 5'–AGGTATC–3'.

Analysis

The complementary strand runs in the opposite direction, from the 3' end to the 5' end. Use base pairing to determine the corresponding sequence on the complementary strand: A pairs with T and C pairs with G.

Solution

Original strand: 5'–A G G T A T C–3'

↓ ↓ ↓ ↓ ↓ ↓ ↓

Complementary strand: 3'–T C C A T A G–5'

---

[4] Replication (22.4)

Example 22.5 What is the sequence of a newly synthesized DNA segment if the template strand has the sequence 3'–CACGTC–5'?

Analysis

The newly synthesized strand runs in the opposite direction, from the 5' end to the 3' end in this example. Use base pairing to determine the corresponding sequence on the new strand: A pairs with T and C pairs with G.

Solution

Template strand: 3'–C A C G T C–5'

New strand: 5'–G T G C A G–3'

---

[5] Transcription (22.6)

Example 22.6 If a portion of the template strand of a DNA molecule has the sequence 3'–TACACATTC–5', what is the sequence of the mRNA molecule produced from this template? What is the sequence of the informational strand of this segment of the DNA molecule?

Analysis
mRNA has a base sequence that is complementary to the template from which it is prepared. mRNA has a base sequence that is identical to the informational strand of DNA, except that it contains the base U instead of T.

Solution

Template strand of DNA: 3'–T A C A C A T T C–5'

mRNA sequence: 5'–A U G U G U A A G–3'

Informational strand of DNA: 5'–A T G T G T A A G–3'

(complementary; complementary)

---

[6] The Genetic Code (22.7)

Example 22.7 Derive the amino acid sequence that is coded for by the following mRNA sequence.

5' UUG CAG AGA GCA CCG GAG 3'

Analysis
Use Table 22.3 to identify the codons that correspond to each amino acid. Codons are written from the 5' end to the 3' end of an mRNA molecule and correspond to a peptide written from the N-terminal end to the C-terminal end.

Solution

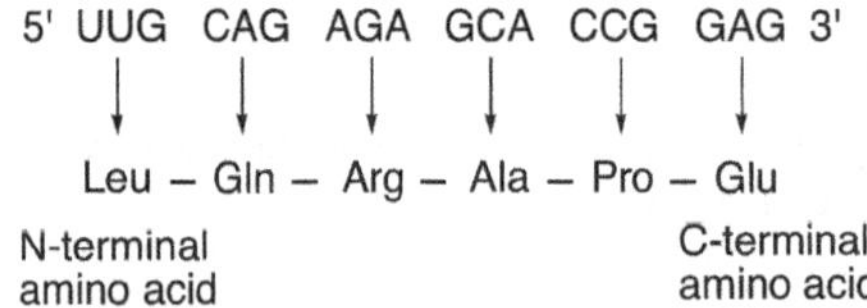

[7] Translation and Protein Synthesis (22.8)

Example 22.8 What sequence of amino acids would be formed from the following mRNA sequence: 5' ACA UAU CAG ACC UGG 3'? List the anticodons contained in each of the needed tRNA molecules.

Analysis
Use Table 22.3 to determine the amino acid that is coded for by each codon. The anticodons contain complementary bases to the codons: A pairs with U, and C pairs with G.

Solution

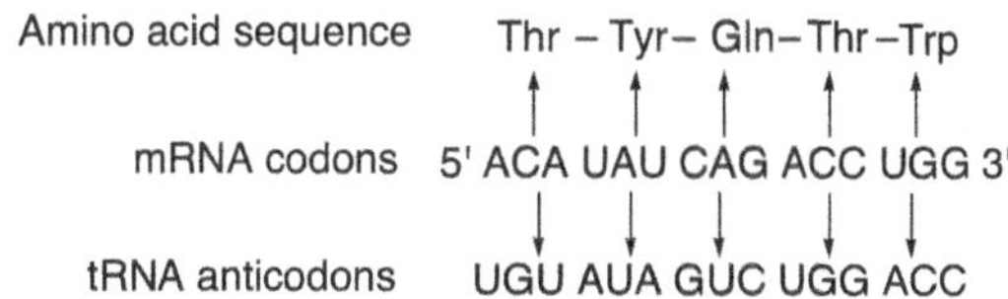

Example 22.9 What polypeptide would be synthesized from the following template strand of DNA: 3' GCG AGT GCT ACG 5'?

Analysis
To determine what polypeptide is synthesized from a DNA template, two steps are needed. First, use the DNA sequence to determine the transcribed mRNA sequence: C pairs with G, T pairs with A, and A (on DNA) pairs with U (on mRNA). Then, use the codons in Table 22.3 to determine what amino acids are coded for by a given codon in mRNA.

Solution

DNA template strand ⟶ 3' GCG AGT GCT ACG 5'

mRNA ⟶ 5' CGC UCA CGA UGC 3'

Polypeptide ⟶ Arg– Ser– Arg– Cys

---

[8] Mutations and Genetic Diseases (22.9)

Example 22.10 (a) What dipeptide is produced from the following segment of DNA: CAGATG?
(b) What happens to the dipeptide when a point mutation occurs and the DNA segment contains the sequence CTGATG instead?

Analysis
Transcribe the DNA sequence to an mRNA sequence with complementary base pairs. Then use Table 22.3 to determine what amino acids are coded for by each codon.

Solution
a. GUC codes for valine and UAC codes for tyrosine, so the dipeptide Val–Tyr results.

b. GAC codes for aspartic acid, so the point mutation results in the synthesis of the dipeptide Asp–Tyr.

Self-Test

[1] Fill in the blank with one of the terms listed below.

Codon (22.7)
Deoxyribonucleic acid (22.1, 22.3)
Gene (22.1)
Lagging strand (22.4)
Leading strand (22.4)
Messenger RNA (22.5)
Nucleic acids (22.1, 22.2)
Nucleoside (22.1)
Nucleotides (22.1)
Ribonucleic acids (22.1, 22.5)
Ribosomal RNA (22.5)
Transcription (22.6)
Transfer RNA (22.5)
Virus (22.11)

1. ______ stores the genetic information of an organism and transmits that information from one generation to another.
2. During DNA replication, the ______ grows continuously.
3. ______ are polymers of nucleotides.
4. Nucleic acids are unbranched polymers composed of repeating monomers called ______.
5. ______ interprets the genetic information in mRNA and brings specific amino acids to the site of protein synthesis in the ribosome.
6. A ______ is an infectious agent consisting of a DNA or RNA molecule that is contained within a protein coating.
7. ______ is the carrier of information from DNA (in the cell nucleus) to the ribosomes (in the cytoplasm).
8. A sequence of three nucleotides (a triplet) codes for a specific amino acid. Each triplet is called a ______.
9. ______ is the synthesis of messenger RNA from DNA.
10. During DNA replication, the ______ is synthesized in small fragments, which are then joined together by a DNA ligase enzyme.
11. A ______ is formed by joining the anomeric carbon of a monosaccharide with a nitrogen atom of a base.
12. A ______ is a portion of a DNA molecule responsible for the synthesis of a single protein.
13. ______ translate the genetic information contained in DNA into the proteins needed for all cellular functions.
14. ______ is found in the ribosomes in the cytoplasm of the cell.

[2] Decide if the statement is describing RNA or DNA.

a. RNA b. DNA

15. A polymer of deoxyribonucleotides
16. The monosaccharide is ribose.
17. The bases are A, G, C, and U.
18. The bases are A, G, C, and T.

[3] What amino acid is coded for by each codon?

19. AAC
20. GCU
21. UGG
22. CGU

[4] Label each structure as a purine or pyrimidine.

a. purine b. pyrimidine

23.

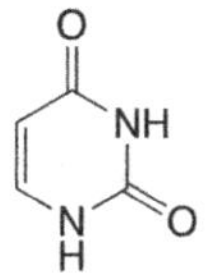

24.

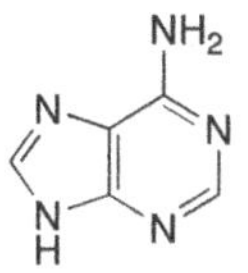

25.

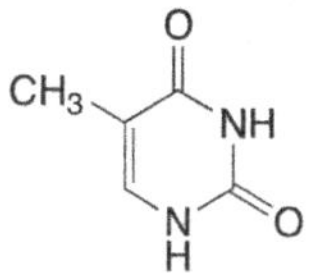

Answers to Self-Test

1. Deoxyribonucleic acid	6. virus	11. nucleoside	16. a	21. Trp
2. leading strand	7. Messenger RNA	12. gene	17. a	22. Arg
3. Nucleic acids	8. codon	13. Ribonucleic acids	18. b	23. b
4. nucleotides	9. Transcription	14. Ribosomal RNA	19. Asn	24. a
5. Transfer RNA	10. lagging strand	15. b	20. Ala	25. b

Solutions to In-Chapter Problems

22.1 Name each nucleoside as in Example 22.1.

a. 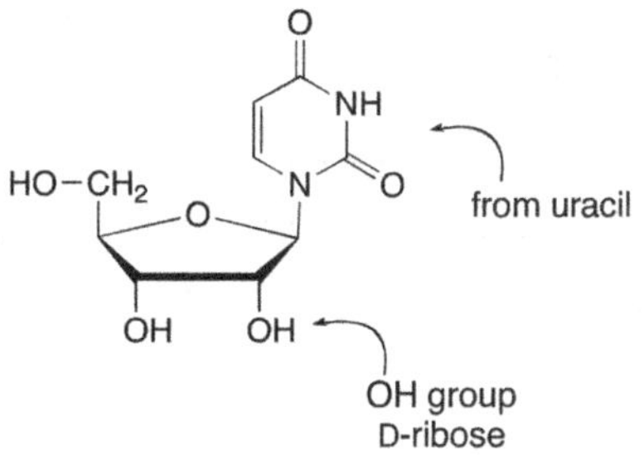 from uracil; OH group D-ribose

uridine

b. 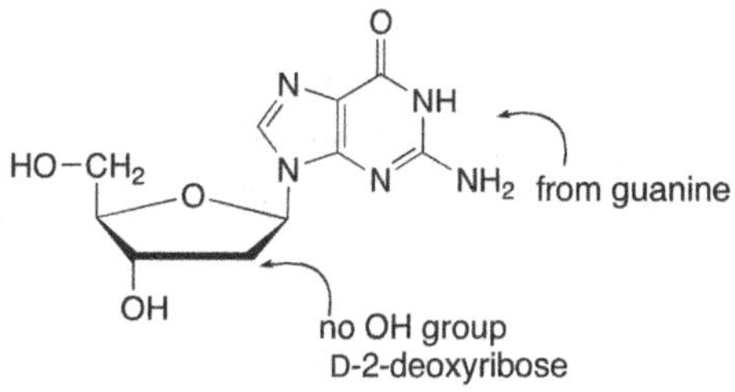 from guanine; no OH group D-2-deoxyribose

deoxyguanosine

22.2 Draw the structure of guanosine.

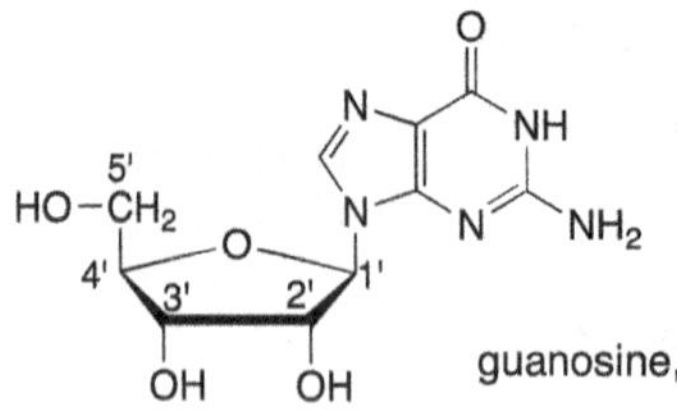

guanosine, a ribonucleoside

22.3

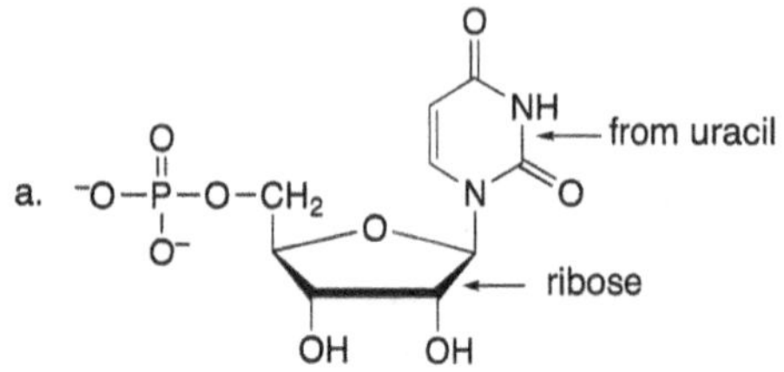

a. uridine 5'-monophosphate, UMP

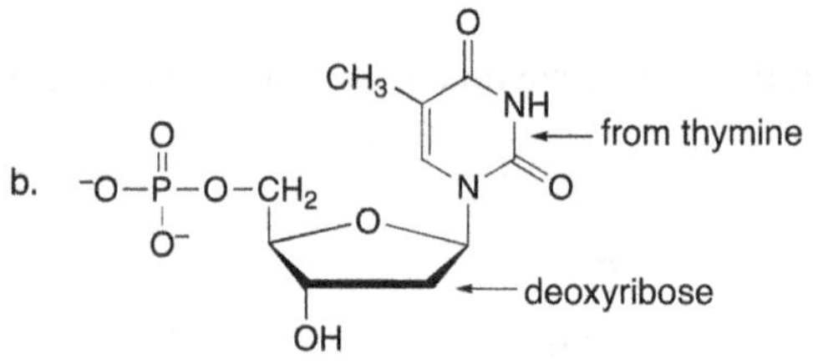

b. deoxythymidine 5'-monophosphate, dTMP

22.4 Answer each question about the nucleic acids.

a. the sugar ribose: RNA
b. the sugar deoxyribose: DNA
c. the base T: DNA
d. the base U: RNA
e. the nucleotide GMP: RNA
f. the nucleotide dCMP: DNA

22.5 a. False. A nucleotide has a phosphate unit and deoxycytidine has no phosphate.
b. True. Deoxycytidine contains a base and a monosaccharide, making it a nucleoside.
c. False. Deoxycytidine has no phosphate.
d. True. Cytosine, the base in deoxycytidine contains a single ring, making it a pyrimidine.

22.6 Draw the structure of each nucleotide as in Example 22.2.

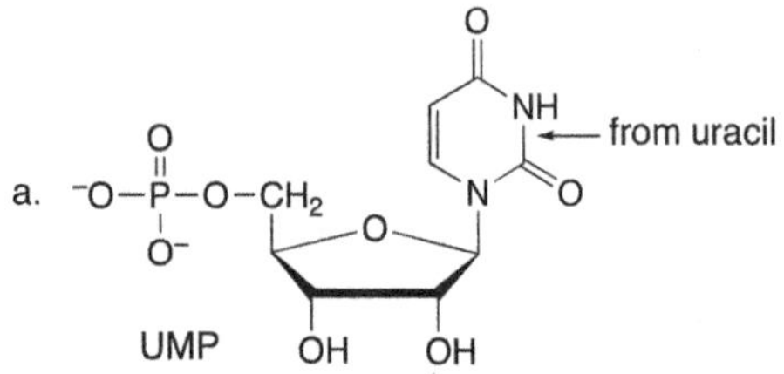

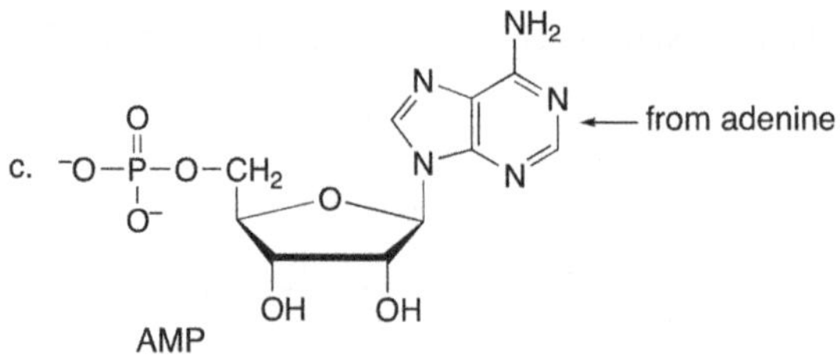

a. UMP

c. AMP

b. dTMP

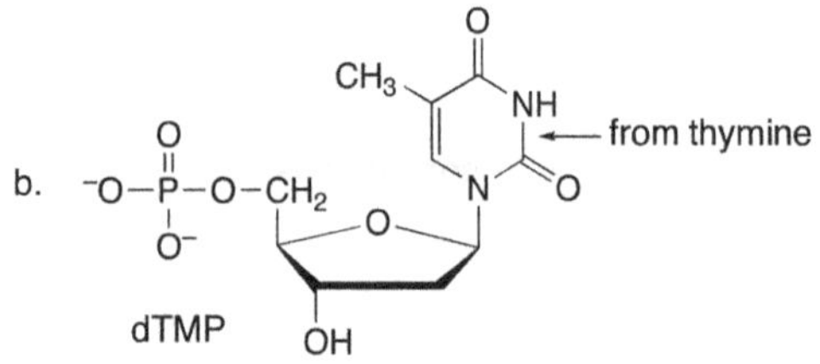

22.7 Write out the name for each abbreviation using Table 22.1. Recall that the first capital letter indicates the base, and the second indicates the number of phosphate groups (mono-, di-, or tri- = M, D, or T, respectively).

a. GTP: guanosine 5'-triphosphate
b. dCDP: deoxycytidine 5'-diphosphate
c. dTTP: deoxythymidine 5'-triphosphate
d. UDP: uridine 5'-diphosphate

22.8 Draw the structure of a dinucleotide formed by joining the 3'-OH group of dTMP to the 5'-phosphate in dGMP as in Example 22.3.

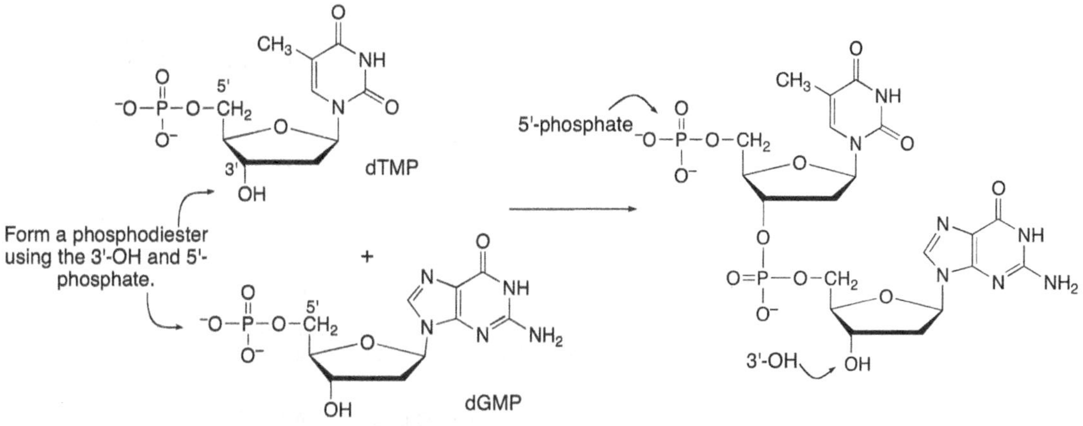

Form a phosphodiester using the 3'-OH and 5'-phosphate.

dTMP + dGMP → 5'-phosphate … 3'-OH

22.9 Draw the structure of each polynucleotide. In (a), the monosaccharide must be ribose because the dinucleotide contains the base U. In (b), the monosaccharide must be deoxyribose because the trinucleotide contains the base T.

a. C, U

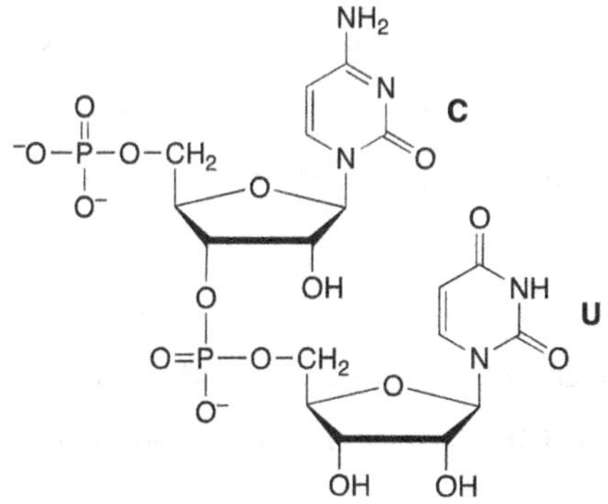

b. T, A, G

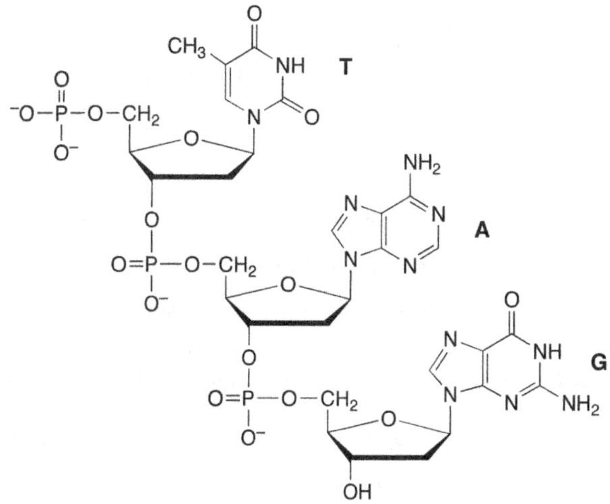

22.10 a. True.

b. False. There are five phosphodiesters linking the six nucleotides together.

c. False. The nucleotide at the 5' end contains the base adenine.

d. True.

e. True.

f. False. RNA does not contain the base T.

22.11 Write the complementary strand for each of the following strands of DNA as in Example 22.4.

a. 5'–A A A C G T C C–3'
Complementary strand: 3'–T T T G C A G G–5'

b. 5'–T A T A C G C C–3'
Complementary strand: 3'–A T A T G C G G–5'

c. 5'–A T T G C A C C C G C–3'
Complementary strand: 3'–T A A C G T G G G C G–5'

d. 5'–C A C T T G A T C G G–3'
Complementary strand: 3'–G T G A A C T A G C C–5'

22.12 Draw the sequence of a newly synthesized DNA segment as in Example 22.5.

a. 3'–A G A G T C T C–5'
New strand: 5'–T C T C A G A G–3'

b. 5'–A T T G C T C–3'
New strand: 3'–T A A C G A G–5'

c. 3'–A T C C T G T A C–5'
New strand: 5'–T A G G A C A T G–3'

d. 5'–G G C C A T A C T C–3'
New strand: 3'–C C G G T A T G A G–5'

22.13 Draw the mRNA and the informational strand of DNA as in Example 22.6.

a. 3'–TGCCTAACG–5' Template strand of DNA
[1] 5'–ACGGAUUGC–3' mRNA sequence
[2] 5'–ACGGATTGC–3' Informational strand

b. 3'–GACTCC–5' Template strand of DNA
[1] 5'–CUGAGG–3' mRNA sequence
[2] 5'–CTGAGG–3' Informational strand

c. 3'–TTAACGCGA–5' Template strand of DNA
[1] 5'–AAUUGCGCU–3' mRNA sequence
[2] 5'–AATTGCGCT–3' Informational strand

d. 3'–CAGTGACCGTAC–5' Template strand of DNA
[1] 5'–GUCACUGGCAUG–3' mRNA sequence
[2] 5'–GTCACTGGCATG–3' Informational strand

22.14 Work backwards to draw the template strand of DNA from which each mRNA was synthesized.

a. 5'–U G G G G C A U U–3'
3'–A C C C C G T A A–5'

b. 5'–G U A C C U–3'
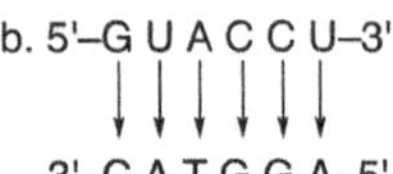
3'–C A T G G A–5'

c. 5'–C C G A C G A U G–3'
3'–G G C T G C T A C–5'

d. 5'–G U A G U C A C G–3'
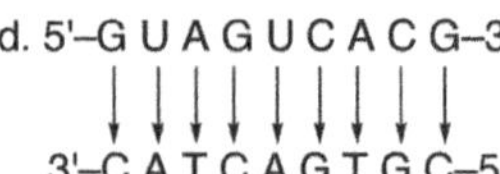
3'–C A T C A G T G C–5'

22.15 Use Table 22.3 to determine what amino acid is coded for by each codon.

a. GCC: Ala
b. AAU: Asn
c. CUA: Leu
d. AGC: Ser
e. CAA: Gln
f. AAA: Lys

22.16 Use Table 22.3 to determine what codons code for each amino acid.

a. glycine: GGU, GGC, GGA, GGG
b. isoleucine: AUU, AUC, AUA
c. lysine: AAA, AAG
d. glutamic acid: GAA, GAG

22.17 Derive the amino acid sequence that is coded for by each mRNA sequence as in Example 22.7.

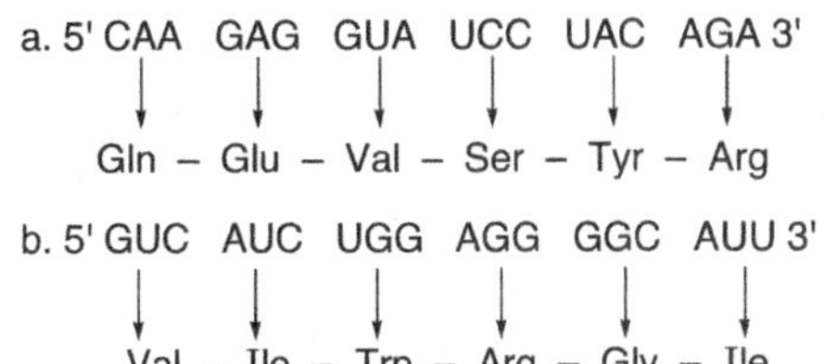

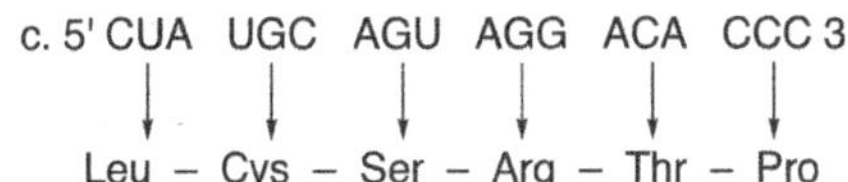

22.18 Use Table 22.3 to determine possible codons for each amino acid sequence.

a. Met–Arg–His–Phe
5'–AUG CGC CAU UUU–3'
b. Gly–Ala–Glu–Gln
5'–GGC GCC GAA CAA–3'
c. Gln–Asn–Gly–Ile–Val
5'–CAA AAU GGA AUU GUG–3'
d. Thr–His–Asp–Cys–Trp
5'–ACU CAU GAU UGC UGG–3'

22.19 Answer each question about the mRNA sequence.

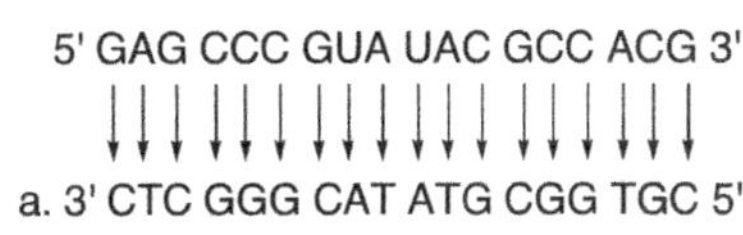

DNA template strand

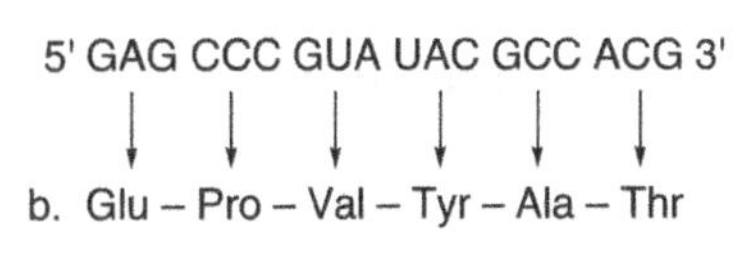

peptide

22.20 The anticodon has three nucleotides that are complementary to the codon, and is found on the tRNA. Use Table 22.3 to determine the amino acid represented by each codon.

a. CGG
anticodon: GCC
amino acid: Arg
b. GGG
anticodon: CCC
amino acid: Gly
c. UCC
anticodon: AGG
amino acid: Ser
d. AUA
anticodon: UAU
amino acid: Ile
e. CCU
anticodon: GGA
amino acid: Pro
f. GCC
anticodon: CGG
amino acid: Ala

22.21 The anticodon has three nucleotides that are complementary to the codon, and is found on a tRNA molecule, as in Example 22.8. Use Table 22.3 to determine the amino acid represented by each codon.

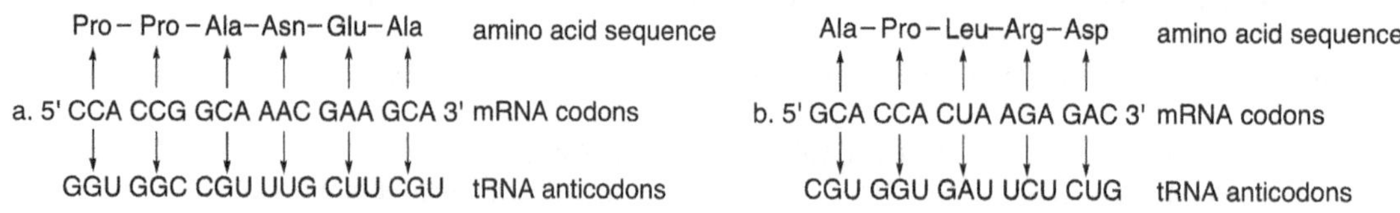

22.22 Draw the polypeptide synthesized from each template DNA, as in Example 22.8.

a. 3' TCT CAT CGT AAT GAT TCG 5' ⟵ DNA template strand ⟶ b. 3' GCT CCT AAA TAA CAC TTA 5'

5' AGA GUA GCA UUA CUA AGC 3' ⟵ mRNA ⟶ 5' CGA GGA UUU AUU GUG AAU 3'

Arg – Val – Ala – Leu – Leu – Ser ⟵ Polypeptide ⟶ Arg – Gly – Phe – Ile – Val – Asn

22.23 Draw the dipeptide and explain the effect of the mutations as in Example 22.10.

a. AAC TGA ⟶ UUG ACU ⟶ Leu – Thr
DNA mRNA dipeptide

b. [1] AAC GGA ⟶ UUG CCU ⟶ [1] Leu – Pro
[2] ATC TGA ⟶ UAG ACU ⟶ [2] Stop codon, no dipeptide is formed.
[3] AAT TGA ⟶ UUA ACU ⟶ [3] Leu – Thr
DNA mRNA dipeptide

22.24 Compare the peptides made from the amino acids.

DNA: TAT GCA CTT —deletion⟶ TAA CTT
mRNA: AUA CGU GAA AUU GAA
Ile – Arg – Glu —loss of Arg⟶ Ile – Glu

22.25 Draw the cut segment of DNA.

5' ~~~~ C C A   sticky ends   A G C T T G G A T T ~~~~ 3'
3' ~~~~ G G T T C G A   A C C T A A ~~~~ 5'

22.26 The antibiotics sulfanilamide and penicillin act upon enzymes found only in bacteria; therefore, they affect bacterial growth and reproduction. Viruses use the starting materials and biochemical processes of the host to reproduce, so antibiotics are ineffective in treating viral infections.

Solutions to Odd-Numbered End-of-Chapter Problems

22.27 a. The polynucleotide is double stranded: DNA.
b. The polynucleotide may contain adenine: DNA and RNA.
c. The polynucleotide may contain dGMP: DNA.
d. The polynucleotide is a polymer of ribonucleotides: RNA.

22.29

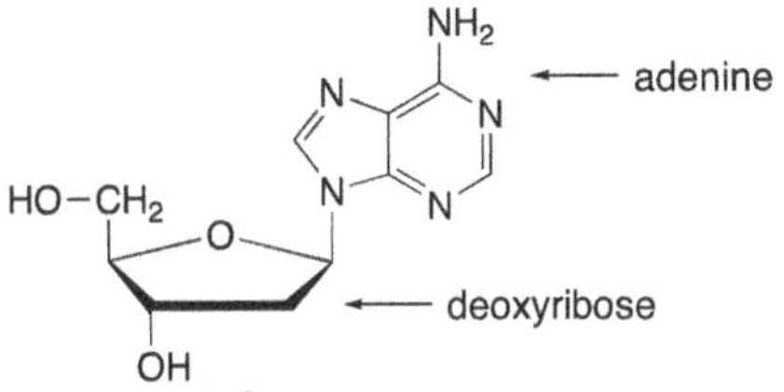

NH_2 ← adenine
HO–CH_2 ← deoxyribose
OH
A

a. deoxyadenosine
b. nucleoside
c. The nucleoside is in DNA only, because the sugar is deoxyribose.

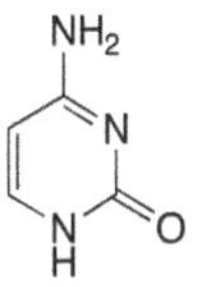

NH_2
B

a. cytosine
b. base
c. The base is found in both DNA and RNA.

22.31 Name each nucleoside as in Example 22.1.

a. NH_2 from cytosine
HO–CH_2
OH
no OH group
D-2-deoxyribose

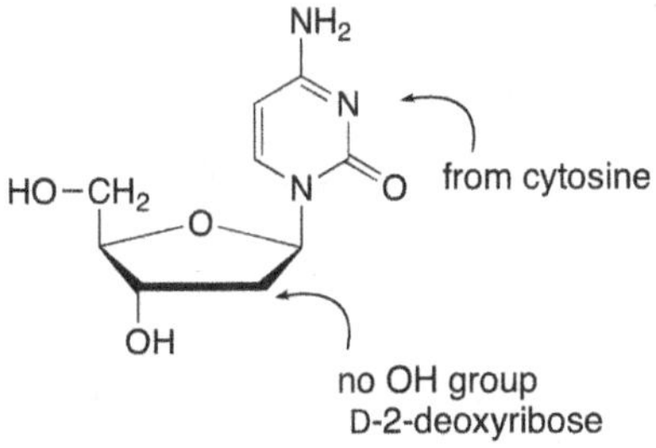

deoxycytidine

b. from guanine
monophosphate
OH OH
OH group
D-ribose

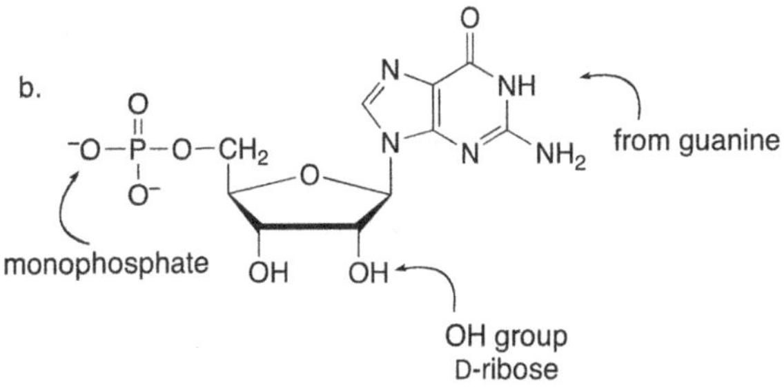

guanosine 5'-monophosphate

22.33

a. CH_3 ← thymine
$^{-}O-P-O-CH_2$
← deoxyribose
OH

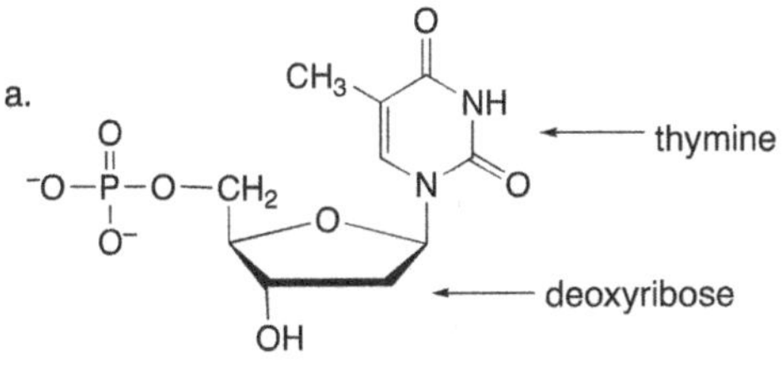

b. deoxythymidine 5'-monophosphate, dTMP

22.35 Draw each structure.

a. 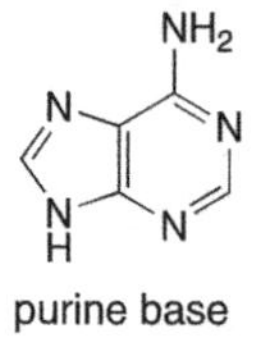
NH2
purine base

b. 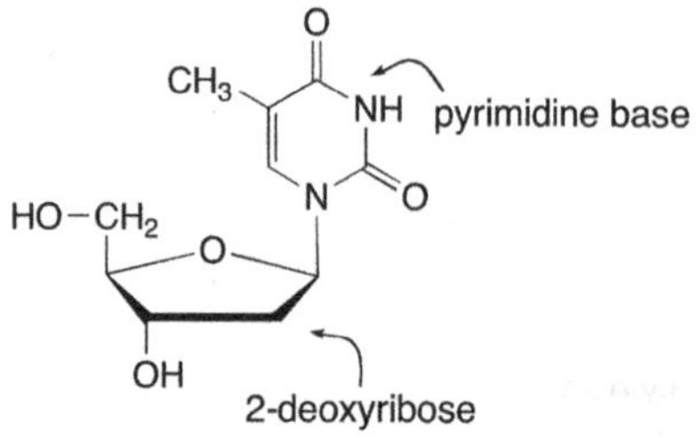
HO–CH2
CH3
NH pyrimidine base
OH
2-deoxyribose

c. 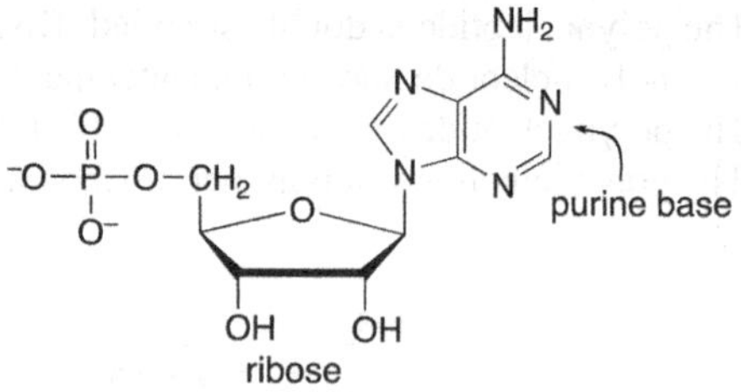
$^{-}O-P(=O)(O^{-})-O-CH_2$
NH2
purine base
OH OH
ribose

d. 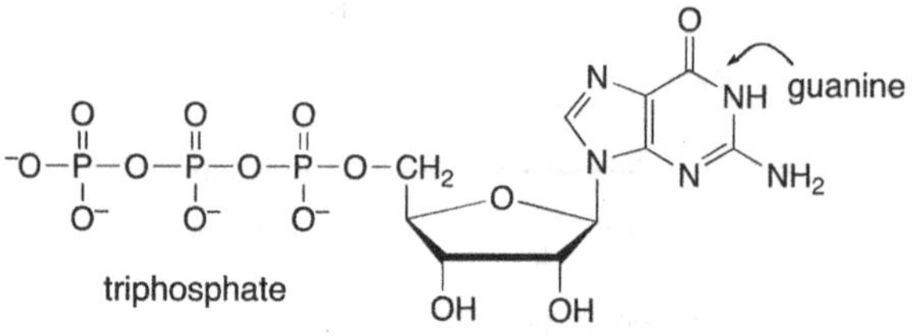
$^{-}O-P(=O)(O^{-})-O-P(=O)(O^{-})-O-P(=O)(O^{-})-O-CH_2$
NH guanine
NH2
triphosphate
OH OH

22.37 Draw the structure of each nucleoside or nucleotide.

a. 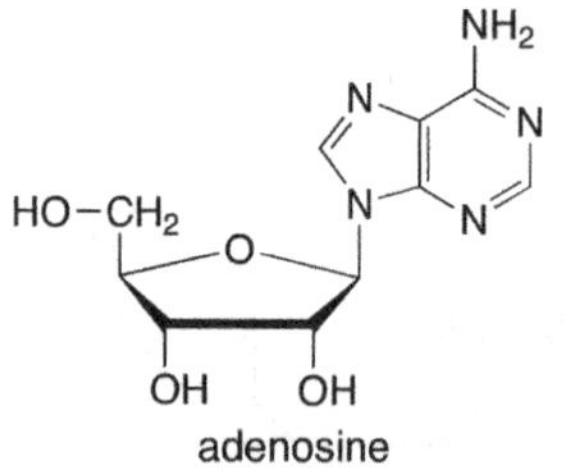
HO–CH2
NH2
OH OH
adenosine

b. 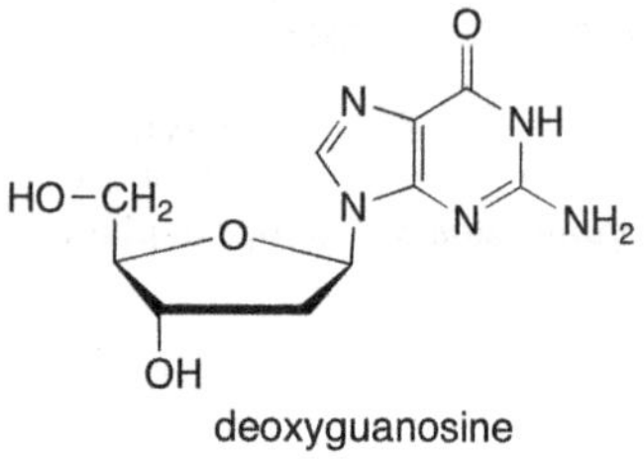
HO–CH2
NH
NH2
OH
deoxyguanosine

c. 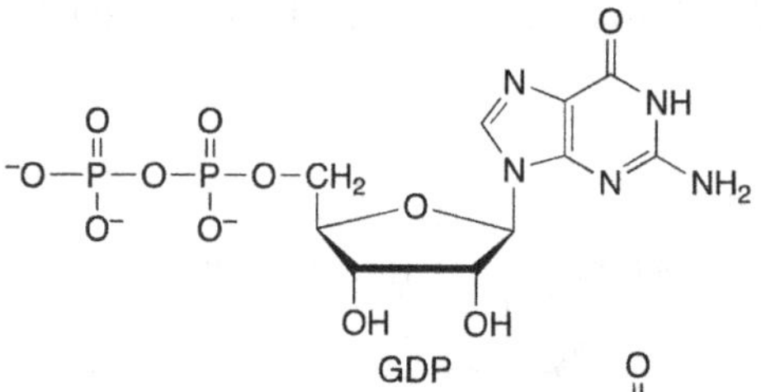
$^{-}O-P(=O)(O^{-})-O-P(=O)(O^{-})-O-CH_2$
NH
NH2
OH OH
GDP

d. 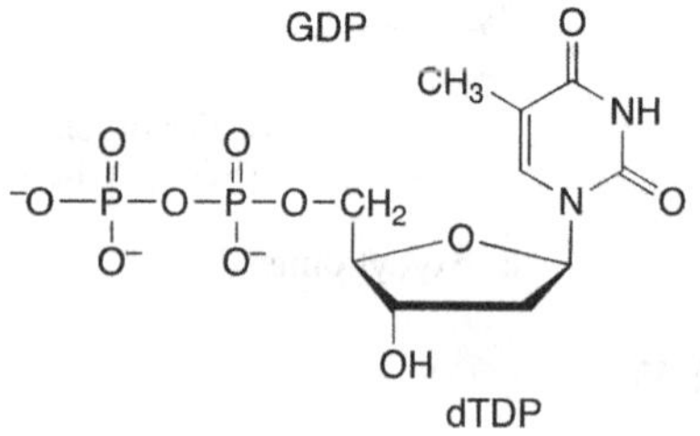
$^{-}O-P(=O)(O^{-})-O-P(=O)(O^{-})-O-CH_2$
CH3
NH
OH
dTDP

22.39

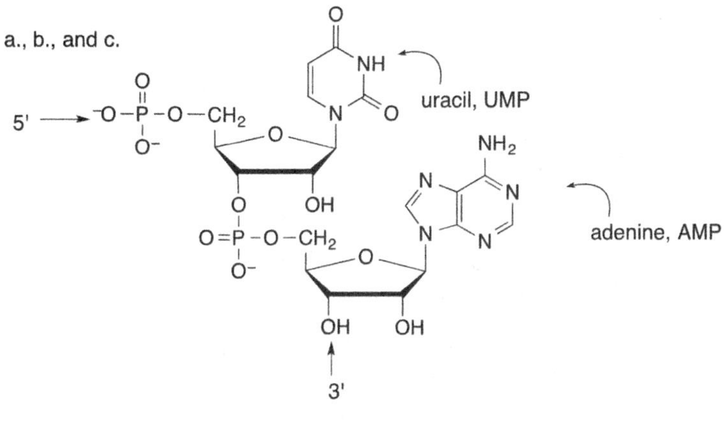

a., b., and c.

5' → uracil, UMP

adenine, AMP

3'

d. This dinucleotide is a ribonucleotide because the sugar rings contain an OH group on C2'.
e. UA

22.41 Draw the two possible dinucleotides.

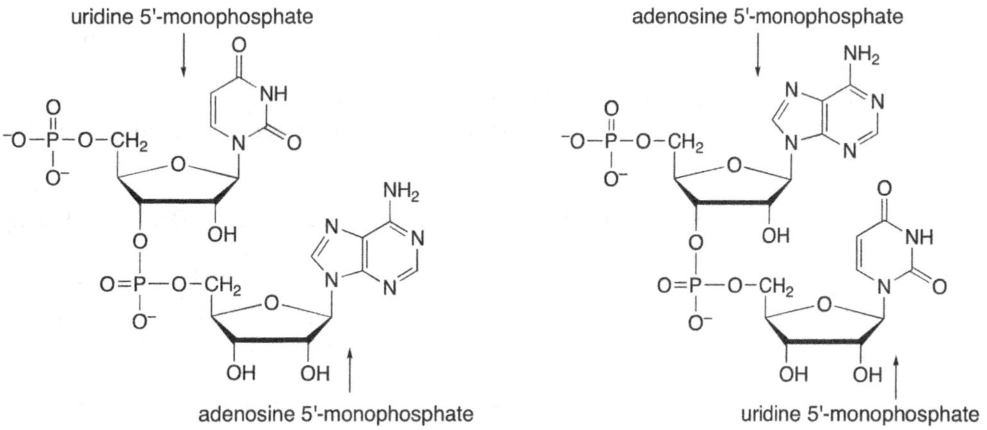

uridine 5'-monophosphate

adenosine 5'-monophosphate

adenosine 5'-monophosphate

uridine 5'-monophosphate

22.43 Draw the structure of each dinucleotide.

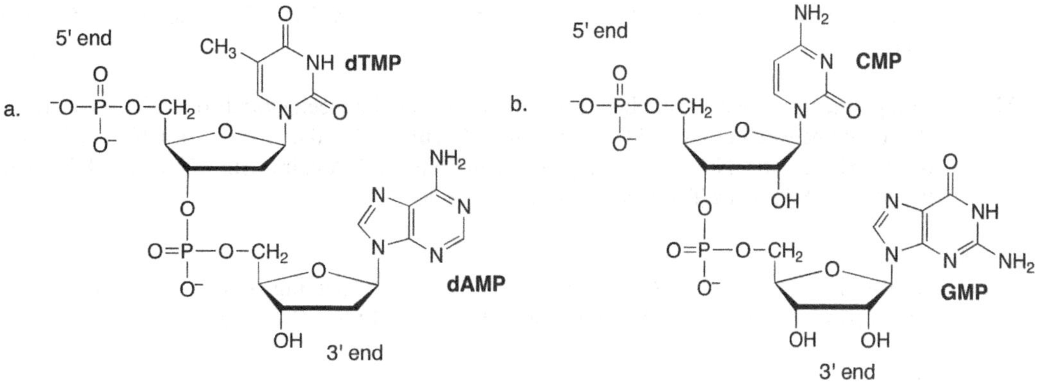

a. 5' end **dTMP** **dAMP** 3' end

b. 5' end **CMP** **GMP** 3' end

22.45 Draw the deoxyribonucleotide.

5' end
T
G
A
3' end

22.47 a. The DNA double helix has 2-deoxyribose as the only sugar. The sugar–phosphate groups are on the outside of the helix.
b. The 5' end has a phosphate and the 3' end has an OH group.
c. Hydrogen bonding occurs in the interior of the helix between base pairs: A pairs with T and G pairs with C.

22.49 Write the complementary strand for each of the following strands of DNA as in Example 22.4.

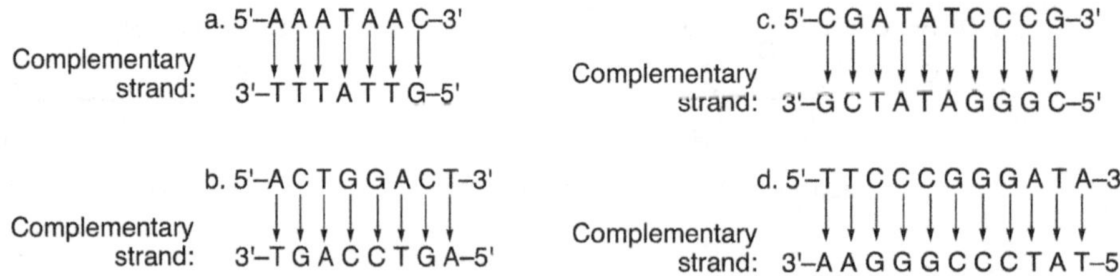

22.51 If 27% of the bases in DNA are A, the same percentage must be T because they are complementary base pairs. The remaining 46% are equally divided between G and C (23% each).

22.53 Complementary base pairing between G and C occurs between strands of DNA, two sites on RNA, and between DNA and RNA, whereas A pairs with either T or U. In DNA, A–T base pairs occur, whereas in RNA, A–U base pairs occur. When DNA pairs with RNA, A in DNA pairs with U in RNA, while T in DNA pairs with A in RNA.

22.55

A: 3' end
B: template strand
C: 5' end
D: leading strand
E: lagging strand
F: template strand
G: 5' end

22.57 Draw the sequence of a newly synthesized DNA segment as in Example 22.5.

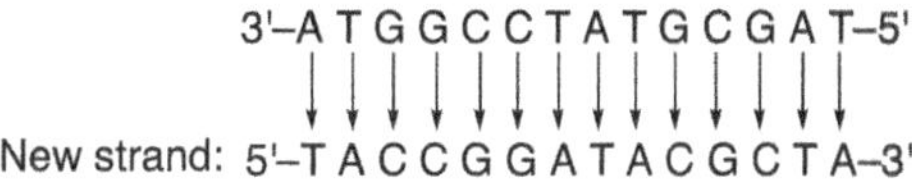

22.59 Replication occurs in only one direction, from the 3' end to the 5' end of the template strand, but the strands of DNA run in opposite directions. The leading strand can be formed continuously from the 3' end as it is unwound. The lagging strand runs in the opposite direction, so it must be formed in segments that are then joined by DNA ligase.

22.61 Draw the mRNA sequence transcribed from each sequence of DNA as in Example 22.6.

a. DNA: 5'–AAATAAC–3'
mRNA: 3'–UUUAUUG–5'

b. DNA: 5'–ACTGGACT–3'
mRNA: 3'–UGACCUGA–5'

c. DNA: 5'–CGATATCCCG–3'
mRNA: 3'–GCUAUAGGGC–5'

d. DNA: 5'–TTCCCGGGATA–3'
mRNA: 3'–AAGGGCCCUAU–5'

22.63 Draw the mRNA and the informational strand of DNA as in Example 22.6.

a. 3'–ATGGCTTA–5' Template strand of DNA
[1] 5'–UACCGAAU–3' mRNA sequence
[2] 5'–TACCGAAT–3' Informational strand of DNA

b. 3'–CGGCGCTTA–5' Template strand of DNA
[1] 5'–GCCGCGAAU–3' mRNA sequence
[2] 5'–GCCGCGAAT–3' Informational strand of DNA

c. 3'–GGTATACCG–5' Template strand of DNA
[1] 5'–CCAUAUGGC–3' mRNA sequence
[2] 5'–CCATATGGC–3' Informational strand of DNA

d. 3'–TAGGCCGTA–5' Template strand of DNA
[1] 5'–AUCCGGCAU–3' mRNA sequence
[2] 5'–ATCCGGCAT–3' Informational strand of DNA

22.65 The anticodon has the three nucleotides that are complementary to the codon, and is found on the tRNA. Use Table 22.3 to determine the amino acid represented by each codon.

a. CUG
anticodon: GAC
amino acid: Leu

b. UUU
anticodon: AAA
amino acid: Phe

c. AAG
anticodon: UUC
amino acid: Lys

d. GCA
anticodon: CGU
amino acid: Ala

22.67 Fill in the missing information.

mRNA: G C G
anticodon: C G C
amino acid: alanine

22.69 Derive the amino acid sequence that is coded for by each mRNA sequence as in Example 22.7.

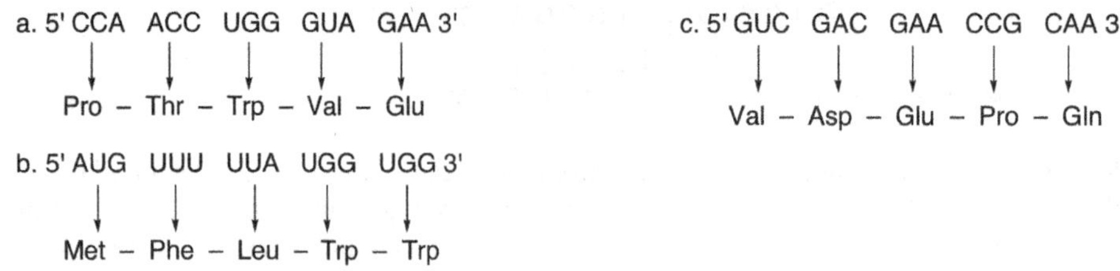

22.71 Work backwards to derive the mRNA sequence from the amino acid sequence.

a. Ile – Met – Lys – Ser – Tyr → AUU AUG AAA AGU UAU

b. Pro – Gln – Glu – Asp – Phe → CCU CAA GAA GAU UUU

c. Thr – Ser – Asn – Arg → ACU AGU AAU CGU

22.73 Work backwards to write the sequence of the DNA template strand from which the mRNA was synthesized, and then give the peptide synthesized by the mRNA.

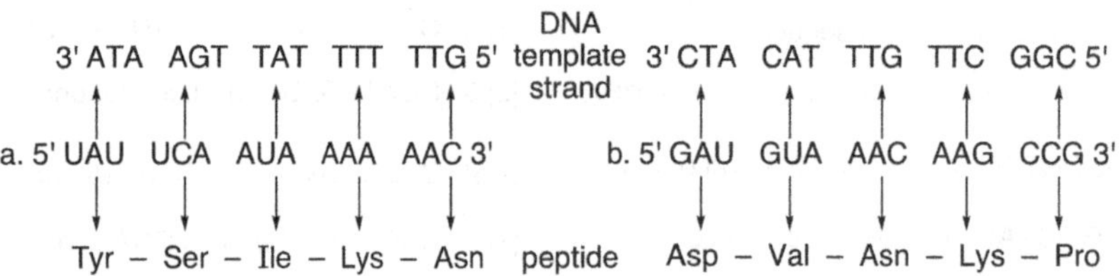

22.75 Answer each question about the mRNA sequence.

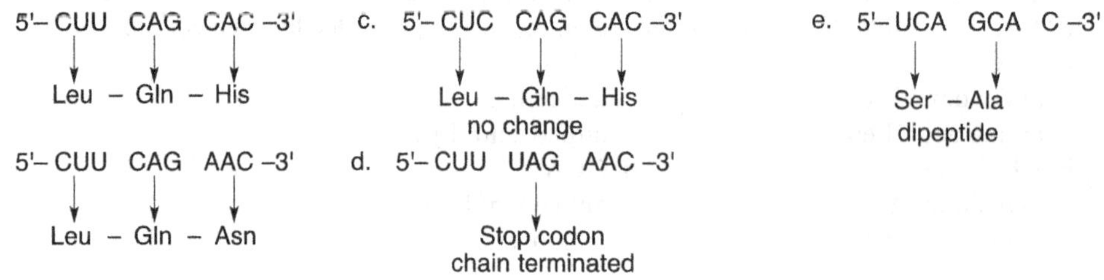

22.77 Answer each question about the DNA sequence.

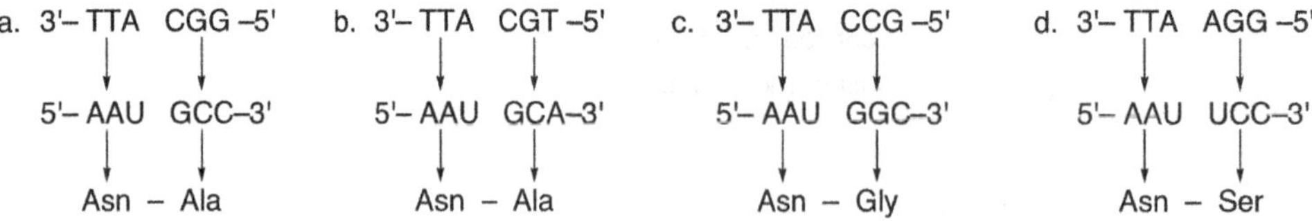

22.79 Draw the cut segment of DNA.

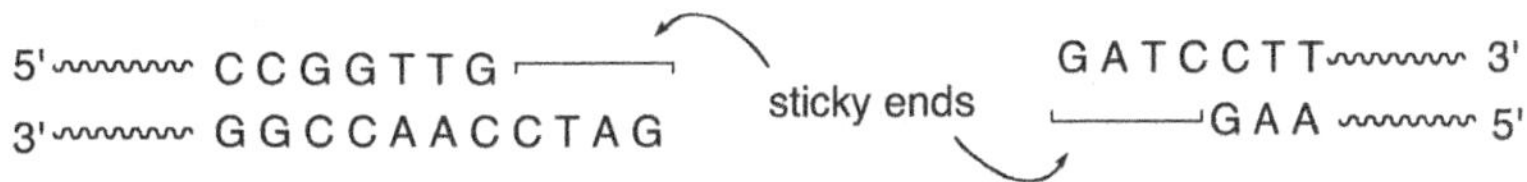

22.81 Circular plasmid DNA is isolated from bacteria and cleaved at a specific site with a restriction endonuclease. This forms "sticky ends" of DNA that are then combined with another DNA source, perhaps human, with complementary "sticky ends" to the plasmid DNA. When these segments of DNA come together in the presence of a DNA ligase enzyme, circular DNA is re-formed. This recombinant DNA can then be re-inserted into bacteria.

22.83 The DNA fragments can help determine genetic relationships. A line is drawn through each fragment found in the parents. Ovals are drawn around the fragments in the children that are not found in either parent.

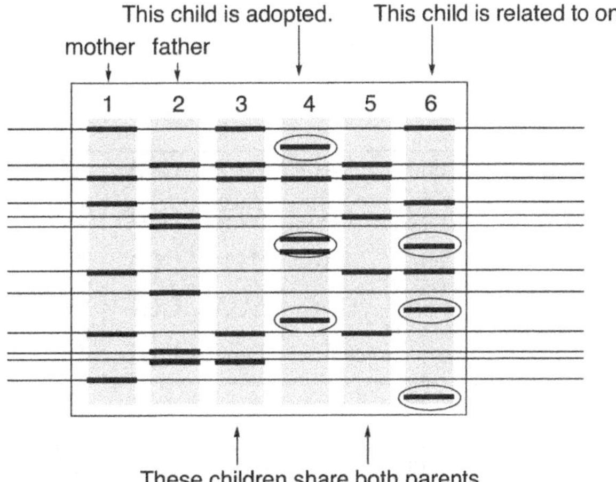

a. Lanes 3 and 5 represent DNA of children that share both parents because they have DNA fragments from both parents (lines drawn through each fragment).
b. Lane 4 represents DNA from an adopted child because the DNA fragments have little relationship to the parental DNA fragments (four of five fragments are not found in either parent).

22.85 Lane 3 contains DNA of the father. Only lane 3 contains a band that is present in the DNA of each of the children.

22.87 a. Lamivudine resembles the nucleoside cytidine.
b. Lamivudine inhibits reverse transcription because it is a nucleoside analogue that gets incorporated into a DNA chain, but because it does not contain a 3' hydroxyl group, synthesis is terminated.

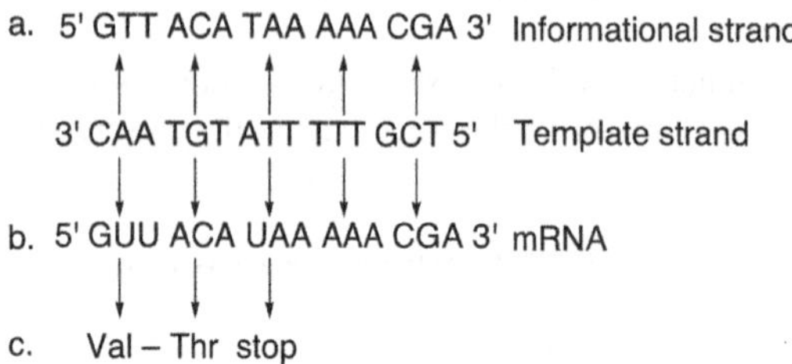

NH2
N
HO–CH2
O
N
O
from cytosine
3'
OH
cytidine

22.89 Answer each question about the DNA sequence.

a. 5' GTT ACA TAA AAA CGA 3' Informational strand

3' CAA TGT ATT TTT GCT 5' Template strand

b. 5' GUU ACA UAA AAA CGA 3' mRNA

c. Val – Thr stop

22.91

DNA informational strand:	5' end	AAC	TAT	CCA	ACG	AAG	ATG	3' end
DNA template strand:	3' end	TTG	ATA	GGT	TGC	TTC	TAC	5' end
mRNA codons:	5' end	AAC	UAU	CCA	ACG	AAG	AUG	3' end
tRNA anticodons:		UUG	AUA	GGU	UGC	UUC	UAC	
Polypeptide:		Asn	Tyr	Pro	Thr	Lys	Met	

22.93

DNA informational strand:	5' end	AAC	GTA	TCA	ACT	CAC	ATG	3' end
DNA template strand:	3' end	TTG	CAT	AGT	TGA	GTG	TAC	5' end
mRNA codons:	5' end	AAC	GUA	UCA	ACU	CAC	AUG	3' end
tRNA anticodons:		UUG	CAU	AGU	UGA	GUG	UAC	
Polypeptide:		Asn	Val	Ser	Thr	His	Met	

22.95 To determine the number of bases in the gene, multiply the number of amino acids by three, because each amino acid is coded for by a codon of three bases.

$$325 \times 3 = 975 \text{ bases}$$

22.97 Work backwards from the amino acid sequence to determine a base pair sequence that could code for met-enkephalin.

ATA	CCA	CCA	AAA	TAC	DNA
↑	↑	↑	↑	↑	
UAU	GGU	GGU	UUU	AUG	mRNA
↑	↑	↑	↑	↑	
Tyr –	Gly –	Gly –	Phe –	Met	peptide

22.99 Use Table 21.2 to identify each amino acid in the peptide, and then use Table 22.3 to determine the DNA sequence.

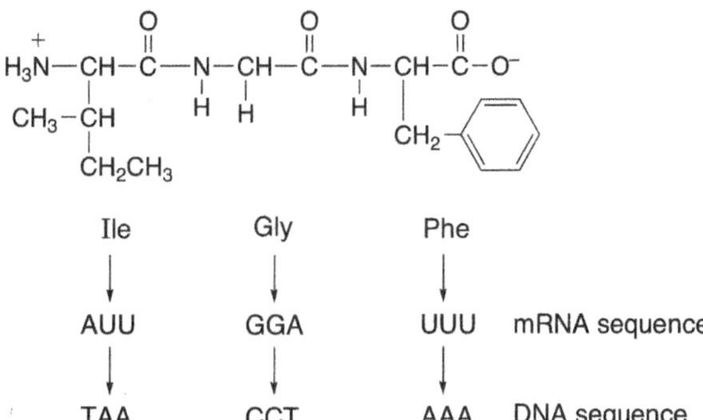

Chapter 23 Metabolism and Energy Production

Chapter Review

[1] What is metabolism and where is energy produced in cells? (23.1)

- Metabolism is the sum of all the chemical reactions that take place in an organism. Catabolic reactions break down large molecules and release energy, while anabolic reactions synthesize larger molecules and require energy.
- Energy is produced in the mitochondria, sausage-shaped organelles that contain an outer and inner cell membrane. Energy is produced in the matrix, the area surrounded by the inner membrane.

[2] What are the four stages of metabolism? (23.2)

- Metabolism begins with digestion in stage [1], in which large molecules—polysaccharides, proteins, and triacylglycerols—are hydrolyzed to smaller molecules—monosaccharides, amino acids, fatty acids, and glycerol.
- In stage [2], biomolecules are degraded into two-carbon acetyl units.

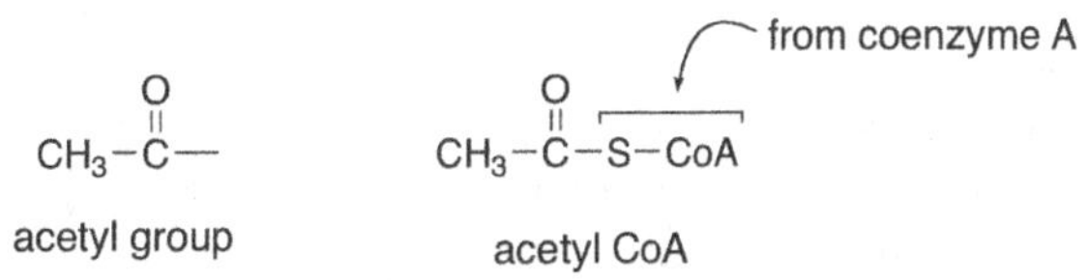

$CH_3-C(=O)-$ acetyl group

$CH_3-C(=O)-S-CoA$ acetyl CoA (S–CoA: from coenzyme A)

- The citric acid cycle comprises stage [3]. The citric acid cycle converts two carbon atoms to two molecules of CO_2, and forms reduced coenzymes NADH and $FADH_2$ that carry electrons to the electron transport chain.
- In stage [4], the electron transport chain and oxidative phosphorylation produce ATP, and oxygen is converted to water.

[3] What is ATP and how do coupled reactions with ATP drive energetically unfavorable reactions? (23.3)

- ATP is the primary energy-carrying molecule in metabolic pathways. The hydrolysis of ATP cleaves one phosphate group and releases 7.3 kcal/mol of energy.
- The hydrolysis of ATP provides the energy to drive a reaction that requires energy. A pair of reactions of this sort is said to be coupled.

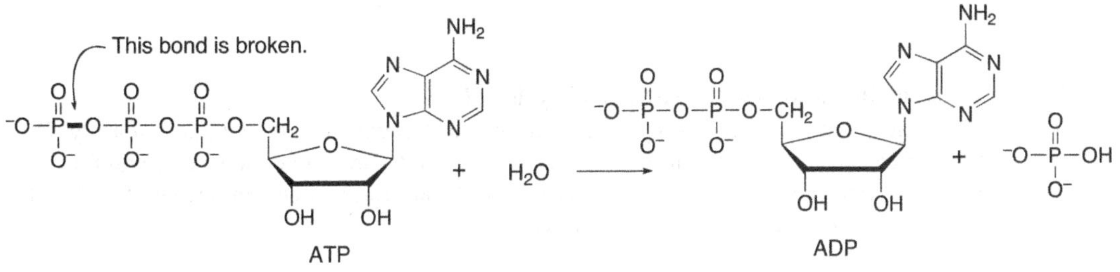

ATP + H_2O → ADP + $^{-}O-P(=O)(O^{-})-OH$

(This bond is broken.)

[4] List the main coenzymes in metabolism and describe their roles. (23.4)

- Nicotinamide adenine dinucleotide (NAD^+) is a biological oxidizing agent that accepts electrons and protons, thus generating its reduced form, NADH. NADH is a reducing agent that donates electrons and protons, thus re-forming NAD^+.

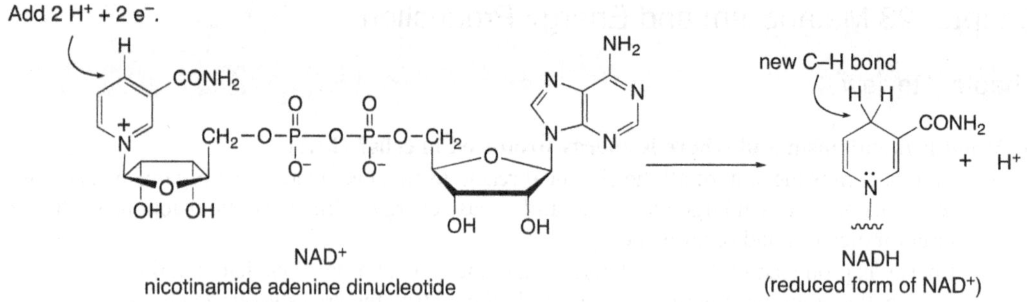

Add 2 H^+ + 2 e^-.

NAD$^+$
nicotinamide adenine dinucleotide

new C–H bond

NADH
(reduced form of NAD$^+$)

- Flavin adenine dinucleotide (FAD) is a biological oxidizing agent that accepts electrons and protons, thus yielding its reduced form, $FADH_2$. $FADH_2$ is a reducing agent that donates electrons and protons, thus re-forming FAD.

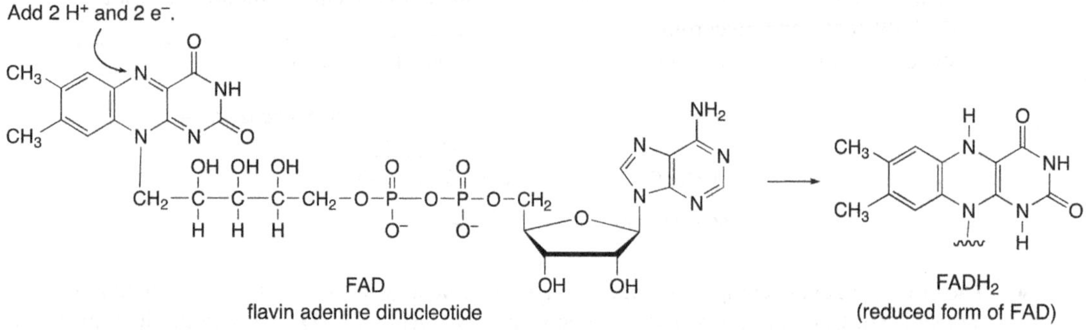

Add 2 H^+ and 2 e^-.

FAD
flavin adenine dinucleotide

$FADH_2$
(reduced form of FAD)

- Coenzyme A reacts with acetyl groups (CH_3CO–) to form a high-energy thioester that delivers two-carbon acetyl groups to other substrates.

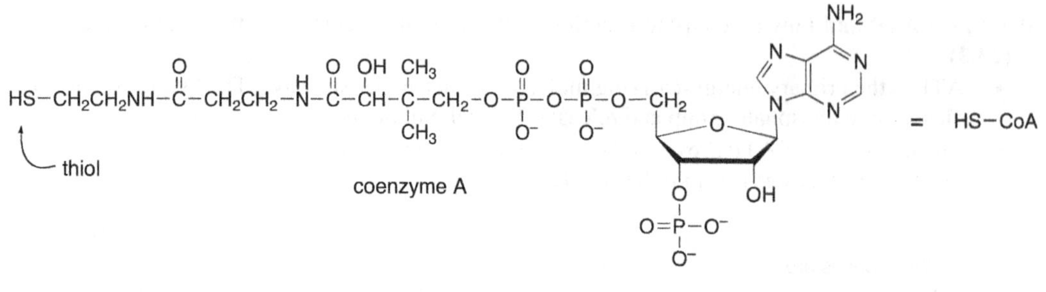

thiol

coenzyme A

= HS–CoA

[5] What are the main features of the citric acid cycle? (23.5)

- The citric acid cycle is an eight-step cyclic pathway that begins with the addition of acetyl CoA to a four-carbon substrate. In the citric acid cycle, two carbons are converted to CO_2 and four molecules of reduced coenzymes (3 NADH + $FADH_2$) are formed. One molecule of GTP, a high-energy nucleoside triphosphate, is also formed.

$$CH_3-\overset{\overset{\displaystyle O}{||}}{C}-SCoA + 2\,H_2O + 3\,NAD^+ + FAD + GDP + HPO_4^{2-}$$

overall reaction

$$2\,CO_2 + HSCoA + 3\,NADH + 3\,H^+ + FADH_2 + GTP$$

exhaled gas — The coenzyme re-enters the cycle. — The reduced coenzymes enter the electron transport chain. — energy source

[6] What are the main components of the electron transport chain and oxidative phosphorylation? (23.6)

- The electron transport chain is a multistep process that takes place in the inner membrane of mitochondria. Electrons from reduced coenzymes enter the chain and are passed from one molecule to another in a series of redox reactions, releasing energy along the way. At the end of the chain, electrons and protons react with inhaled oxygen to form water.
- H^+ ions are pumped across the inner membrane of the mitochondrion, forming a high concentration of H^+ ions in the intermembrane space, thus creating a potential energy gradient. When the H^+ ions travel through the channel in the ATP synthase enzyme, this energy is used to convert ADP to ATP—a process called oxidative phosphorylation.

[7] Why do compounds such as cyanide act as poisons when they disrupt the electron transport chain? (23.7)

- Since all catabolic pathways converge at the electron transport chain, these steps are needed to produce energy for normal cellular processes. Compounds that disrupt a single step can halt ATP synthesis, so that an organism cannot survive.
- Cyanide from HCN irreversibly binds to the Fe^{3+} ion of the enzyme cytochrome oxidase and, as a result, Fe^{3+} cannot be reduced to Fe^{2+} and water cannot be formed from oxygen. Since the electron transport chain is disrupted, energy is not generated for oxidative phosphorylation, ATP is not synthesized, and cell death often follows.

Problem Solving

[1] ATP and Energy Production (23.3)

Example 23.1 The phosphorylation of glycerol to glycerol 3-phosphate requires 2.2 kcal/mol of energy. This unfavorable reaction can be driven by the hydrolysis of ATP to ADP. (a) Write the equation for the coupled reaction. (b) How much energy is released in the coupled reaction?

$$\text{glycerol} + HPO_4^{2-} \longrightarrow \text{glycerol 3-phosphate} + H_2O$$

Analysis

Write the equation for the hydrolysis of ATP to ADP, which releases 7.3 kcal/mol of energy (Section 23.3A). Add together the substances in this equation and the given equation to give the net equation—that is, the equation for the coupled reaction. To determine the overall energy change, add together the energy changes for each step.

Solution

a. The coupled equation shows the overall reaction that combines the phosphorylation of glycerol and the hydrolysis of ATP.

Cross out compounds that appear on both sides of the reaction arrows.

							Energy change
Phosphorylation: [1]	glycerol	+	HPO_4^{2-} (crossed out)	→	glycerol 3-phosphate	+ H_2O (crossed out)	+2.2 kcal/mol
Hydrolysis: [2]	ATP	+	H_2O (crossed out)	→	ADP	+ HPO_4^{2-} (crossed out)	–7.3 kcal/mol
Coupled reaction:	glycerol	+	ATP	→	glycerol 3-phosphate	+ ADP	

b. The overall energy is the sum of the energy changes for each step:
+2.2 kcal/mol + (–7.3) kcal/mol = –5.1 kcal/mol.

---

[2] Coenzymes in Metabolism (23.4)

Example 23.2 Label the reaction as an oxidation or reduction, and give the reagent, NAD^+ or NADH, that would be used to carry out the reaction.

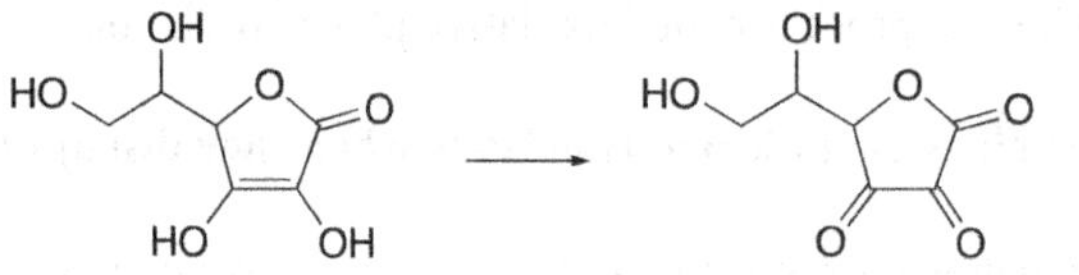

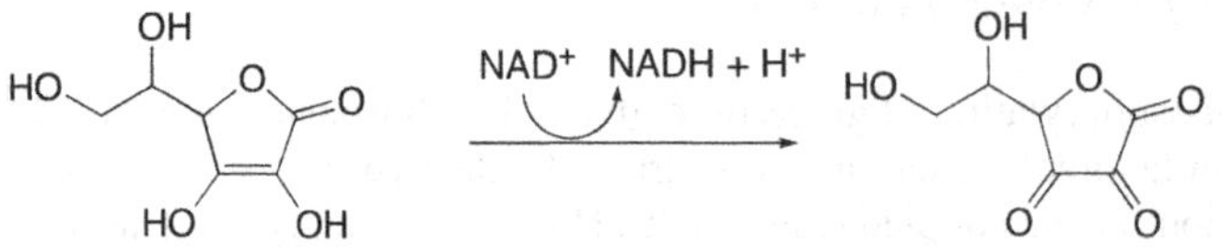

ascorbic acid → dehydroascorbic acid

Analysis

Count the number of C–O bonds in the starting material and product. Oxidation increases the number of C–O bonds and reduction decreases the number of C–O bonds. NAD^+ is a coenzyme used for an oxidation, and NADH is a coenzyme used for a reduction.

Solution

The conversion of ascorbic acid to dehydroascorbic acid is an oxidation, because the product has more C–O bonds than the reactant. To carry out the oxidation, the oxidizing agent NAD^+ could be used.

ascorbic acid —(NAD^+ → NADH + H^+)→ dehydroascorbic acid

---

[3] The Citric Acid Cycle (23.5)

Example 23.3 (a) Using curved arrow symbolism, write out the reaction of isocitrate with NAD^+ to form α-ketoglutarate and carbon dioxide. (b) Classify the reaction as an oxidation, reduction, or decarboxylation.

Analysis

Use Figure 23.7 to draw the structures for isocitrate and α-ketoglutarate. Draw the organic reactant and product on the horizontal arrow and the oxidizing reagent NAD^+, which is converted to NADH, on the curved arrow. Oxidation reactions result in a loss of electrons, a loss of hydrogen, or a gain of oxygen. Reduction reactions result in a gain of electrons, a gain of hydrogen, or a loss of oxygen. A decarboxylation results in the loss of CO_2.

Solution

a. Equation:

$$\underset{\text{isocitrate}}{{}^-O_2C-CH_2-CH(CO_2^-)-CH(OH)-CO_2^-} \xrightarrow{NAD^+ \;\to\; NADH + H^+} \underset{\alpha\text{-ketoglutarate}}{{}^-O_2C-CH_2-CH_2-C(=O)-CO_2^-} + CO_2$$

b. Two processes occur in this reaction. An alcohol with one C–O bond is converted to a carbonyl group with two C–O bonds. This reaction is an oxidation because the number of C–O bonds increases. Also, CO_2 is removed from the starting material, so decarboxylation occurs as well.

Self-Test

[1] Fill in the blank with one of the terms listed below.

Acetyl CoA (23.4)	Catabolism (23.1)	Electron transport chain (23.6)
Adenosine 5'-diphosphate (ADP, 23.3)	Citric acid cycle (23.5)	Metabolism (23.1)
Adenosine 5'-triphosphate (ATP, 23.3)	Coenzyme A (23.4)	Oxidative phosphorylation (23.6)
Anabolism (23.1)	Coupled reactions (23.3)	

1. ______ is formed by adding three phosphates to the 5'-OH group of a nucleoside composed of the sugar ribose and the base adenine.
2. The ______ is a multistep process that relies on four enzyme systems, called complexes I, II, III, and IV, as well as mobile electron carriers.
3. ______ differs from other coenzymes in this chapter because it is not an oxidizing or reducing agent.
4. The ______ is a cyclic metabolic pathway that begins with the addition of acetyl CoA to a four-carbon substrate and ends when the same four-carbon compound is formed as a product eight steps later.
5. ______ is formed by adding two phosphates to the 5'-OH group of a nucleoside composed of the sugar ribose and the base adenine.
6. ______ are pairs of reactions that occur together. The energy released by one reaction provides the energy to drive the other reaction.
7. ______ is a process in which energy from the oxidation of reduced coenzymes is used to transfer a phosphate group.
8. When an acetyl group is bonded to coenzyme A, the product is called ______.
9. ______ is the breakdown of large molecules into smaller ones with the release of energy.
10. ______ is the sum of all the chemical reactions that take place in an organism.

11. ______ is the synthesis of large molecules from smaller ones, and energy is generally absorbed in the process.

[2] Fill in the blank with one of the terms listed below.

a. organelles
b. matrix
c. mitochondria
d. cytoplasm
e. intermembrane space

12. The ______ is the region of the cell between the cell membrane and the nucleus.
13. ______ are specialized structures, each of which has a specific function.
14. ______ are small, sausage-shaped organelles in which energy production takes place.
15. The area between the outer membrane and inner membrane of a mitochondrion is called the ______.
16. Energy production occurs within the ______, the area surrounded by the inner membrane of the mitochondrion.

[3] Classify each molecule as an oxidizing agent, a reducing agent, or neither.

a. oxidizing agent
b. reducing agent
c. neither

17. NADH
18. FAD
19. ADP
20. A coenzyme that gains hydrogen atoms
21. A coenzyme that loses hydrogen atoms

[4] Fill in the table with the terms listed.

a. monosaccharides
b. glycolysis
c. triacylglycerols
d. amino acid catabolism
e. protein
f. fat

Food	Basic Unit	Components	Catabolic Pathway
Olive oil	22.	Fatty acids and glycerol	Fatty acid oxidation
Pasta	Carbohydrates	23.	24.
Chicken breast	Protein	Amino acids	25.

Answers to Self-Test

1. Adenosine 5'-triphosphate (ATP)
2. electron transport chain
3. Coenzyme A
4. citric acid cycle
5. Adenosine 5'-diphosphate (ADP)
6. Coupled reactions
7. Oxidative phosphorylation
8. acetyl CoA
9. Catabolism
10. Metabolism
11. Anabolism
12. d
13. a
14. c
15. e
16. b
17. b
18. a
19. c
20. a
21. b
22. c
23. a
24. b
25. d

Solutions to In-Chapter Problems

23.1 By funneling all catabolic pathways into a single common pathway, it is possible to use the same reactions and the same enzyme systems to metabolize all types of biomolecules.

23.2 Hydrolysis of GTP to GDP is similar to the hydrolysis of ATP to ADP shown in Figure 23.4.

$$\text{GTP} + H_2O \longrightarrow \text{GDP} + HPO_4^{2-}$$

23.3 Write the equation for the coupled reaction and determine how much energy is released as in Example 23.1.

Cross out compounds that appear on both sides of the reaction arrows.

a.

[1] glucose + ~~HPO_4^{2-}~~ ⟶ glucose 1-phosphate + ~~H_2O~~

[2] ATP + ~~H_2O~~ ⟶ ADP + ~~HPO_4^{2-}~~

Coupled reaction: glucose + ATP ⟶ glucose 1-phosphate + ADP

b. Overall energy change: +5.0 kcal/mol + (–7.3) kcal/mol = –2.3 kcal/mol
(Reaction [1]) (Reaction [2])

23.4 Write out each reaction using curved arrow symbolism.

a. fructose —(ATP → ADP)⟶ fructose 6-phosphate

b. glucose —(ATP → ADP)⟶ glucose 6-phosphate

23.5 Calculate the energy change in the reaction.

–10.3 kcal/mol of energy released

creatine phosphate + ADP ⟶ creatine + ATP

+7.3 kcal/mol of energy absorbed

Overall energy change: –10.3 kcal/mol + 7.3 kcal/mol = –3.0 kcal/mol of energy released

23.6 Label each reaction as an oxidation or reduction and give the reagent that would be used to carry out the reaction as in Example 23.2.

a. $H_2C{=}O$ —(NADH + H^+ → NAD^+)⟶ CH_3OH

one fewer C–O bond
reduction

b. $CH_3{-}CH(OH){-}COO^-$ —(NAD^+ → NADH + H^+)⟶ $CH_3{-}C({=}O){-}COO^-$

one more C–O bond
oxidation

23.7 Riboflavin is water soluble because it has numerous polar groups (OH and NH) that can hydrogen bond to water.

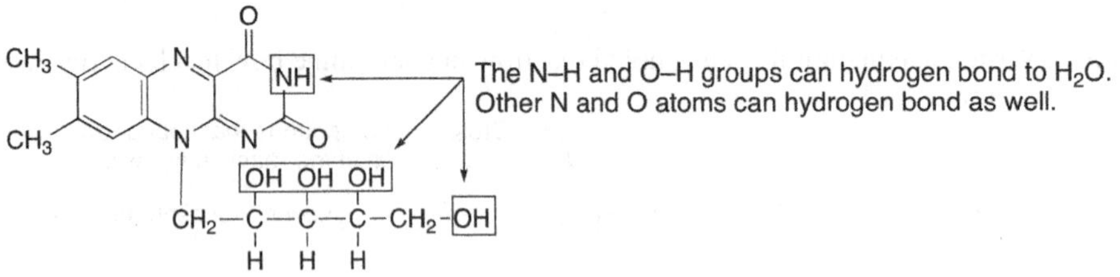

The N–H and O–H groups can hydrogen bond to H_2O. Other N and O atoms can hydrogen bond as well.

riboflavin
vitamin B_2

23.8 The reaction is an oxidation because the product has two fewer C–H bonds than the reactant. FAD, an oxidizing agent, would be used in the reaction.

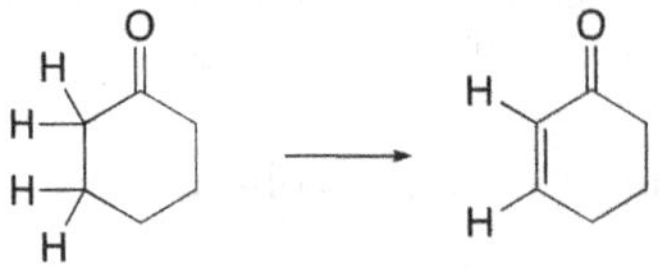

23.9 The many polar groups in vitamin B_5 make it water soluble.

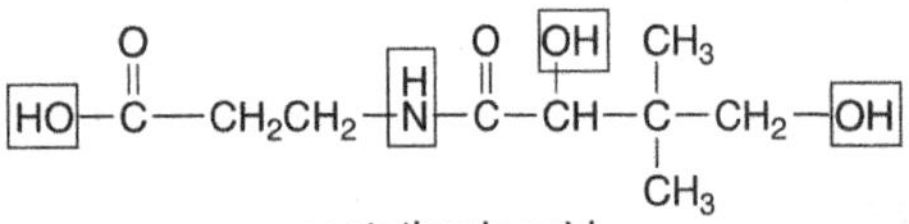

All N–H and O–H bonds can hydrogen bond.

pantothenic acid
vitamin B_5

23.10 Draw the products of hydrolysis of the thioester $CH_3CH_2CH_2COSCoA$.

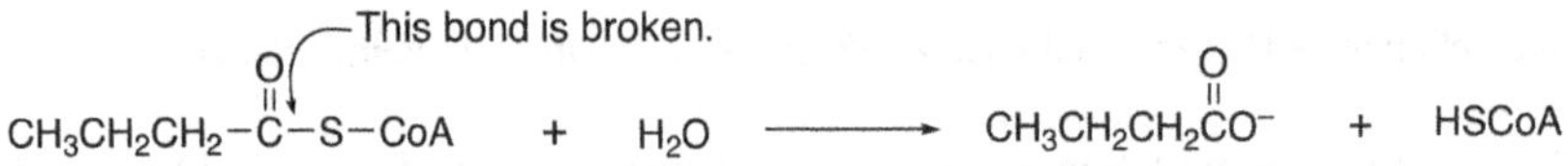

This bond is broken.

$$CH_3CH_2CH_2\overset{O}{\overset{\|}{C}}{-}S{-}CoA + H_2O \longrightarrow CH_3CH_2CH_2\overset{O}{\overset{\|}{C}}O^- + HSCoA$$

23.11 Write out the reaction with curved arrow symbolism, and classify the reaction as an oxidation, reduction, or decarboxylation as in Example 23.3.

a. 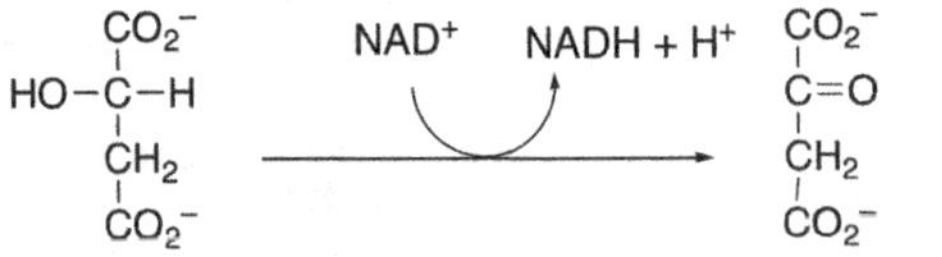 malate (CO_2^-–C(OH)H–CH_2–CO_2^-) $\xrightarrow{NAD^+ \;\; NADH + H^+}$ oxaloacetate (CO_2^-–C=O–CH_2–CO_2^-)

malate oxaloacetate

b. This reaction is an oxidation, because an alcohol with one C–O bond is converted to a carbonyl group with two C–O bonds.

23.12 Succinate dehydrogenase is so named because the elements of hydrogen (H_2) are removed from succinate, and dehydrogenase enzymes remove hydrogen atoms.

23.13 Because malate is the substrate and two hydrogens are removed in the reaction, the enzyme can be called malate dehydrogenase.

23.14 If NADH and $FADH_2$ were not oxidized in the electron transport chain, the citric acid cycle and aerobic metabolism would cease, because NAD^+ and FAD are required in the reactions of the citric acid cycle.

23.15 a. The conversion of Fe^{3+} to Fe^{2+} is a reduction reaction because Fe^{3+} gains an electron.
b. Fe^{3+} is an oxidizing agent because it gains electrons, causing another species to be oxidized.

23.16 The pH would be lower in the intermembrane space because the H^+ concentration is higher in the intermembrane space than in the matrix.

Solutions to Odd-Numbered End-of-Chapter Problems

23.17 Mitochondria contain an outer membrane and an inner membrane with many folds. The area between these two membranes is called the intermembrane space. Energy production occurs within the matrix, the area surrounded by the inner membrane.

23.19 The steps of catabolism in order: hydrolysis of starch, the conversion of glucose to acetyl CoA, the citric acid cycle, the electron transport chain, oxidative phosphorylation.

23.21 a. cleavage of a protein with chymotrypsin: stage [1]
b. oxidation of a fatty acid to acetyl CoA: stage [2]
c. oxidation of malate to oxaloacetate with NAD^+: stage [3]
d. conversion of ADP to ATP with ATP synthase: stage [4]
e. hydrolysis of starch to glucose with amylase: stage [1]

23.23 Coupled reactions are pairs of reactions that occur together. The energy released by one reaction provides the energy to drive the other reaction.

23.25

$$CH_3-\overset{O}{\overset{\|}{C}}-O-\overset{O}{\overset{\|}{P}}(O^-)-O^- \xrightarrow{H_2O} CH_3\overset{O}{\overset{\|}{C}}O^- + HPO_4^{2-}$$

This bond is broken. (the C–O bond)

acetyl phosphate

23.27 Write the equation for the coupled reaction and determine how much energy is released as in Example 23.1.

a.

	Energy change
succinate + HSCoA ⟶ succinyl CoA + ~~H_2O~~	+9.4 kcal/mol
ATP + ~~H_2O~~ ⟶ ADP + HPO_4^{2-}	−7.3 kcal/mol

Coupled reaction: ATP + succinate + HSCoA ⟶ succinyl CoA + ADP + HPO_4^{2-}

b. The reaction is not energetically favored because an input of energy is required (+2.1 kcal/mol).

Overall energy change: +9.4 kcal/mol + (−7.3) kcal/mol = +2.1 kcal/mol

23.29 Answer each question about the phosphorylation of glucose.

a. glucose + HPO_4^{2-} ⟶ glucose 1-phosphate + H_2O

b. Coupled reaction: glucose + ATP ⟶ glucose 1-phosphate + ADP

c. Overall energy change: +5.0 kcal/mol + (–7.3) kcal/mol = –2.3 kcal/mol

phosphorylation (+5.0 kcal/mol); ATP hydrolysis (–7.3 kcal/mol)

23.31 Answer each question about the phosphorylation of fructose 6-phosphate.

a. The energy change for the reverse reaction is –3.9 kcal/mol.

b.

		Energy change
[1]	fructose 6-phosphate + HPO_4^{2-} ⟶ fructose 1,6-bisphosphate + H_2O	+3.9 kcal/mol
[2]	ATP + H_2O ⟶ ADP + HPO_4^{2-}	–7.3 kcal/mol

(HPO_4^{2-} and H_2O are crossed out.)

Coupled reaction: fructose 6-phosphate + ATP ⟶ fructose 1,6-bisphosphate + ADP

c. Overall energy change: +3.9 kcal/mol + (–7.3) kcal/mol = –3.4 kcal/mol
(Reaction [1]) (Reaction [2])

23.33 Answer each question about the reaction.

a. Coupled reaction: 1,3-bisphosphoglycerate + ADP ⟶ 3-phosphoglycerate + ATP

b. –11.8 kcal/mol + 7.3 kcal/mol = –4.5 kcal/mol

c. Yes, this is an energetically favorable coupled reaction that could be used to synthesize ATP from ADP.

23.35 Draw the structure of GTP and the hydrolysis product.

a. GTP

b. GDP

removal of one phosphate group

23.37 An oxidizing agent causes an oxidation reaction to occur, so the oxidizing agent is reduced. A reducing agent causes a reduction reaction to occur, so the reducing agent is oxidized.

a. $FADH_2$: reducing agent b. ATP: neither c. NAD^+: oxidizing agent

23.39 When a substrate is oxidized, NAD^+ is reduced to NADH because it gains a hydrogen. NAD^+ is an oxidizing agent.

23.41 Label the reaction as an oxidation or reduction and give the reagent that would be used to carry out the reaction as in Example 23.2.

NAD⁺ → NADH + H⁺

cyclohexanol (–OH) → cyclohexanone (=O)

one more C–O bond
oxidation

23.43 Find the step(s) in the citric acid cycle with each characteristic. Use Figure 23.7 as a guide.

a. The reaction generates NADH: steps [3], [4], and [8].
b. CO_2 is removed: steps [3] and [4].
c. The reaction utilizes FAD: step [6].
d. The reaction forms a new carbon–carbon single bond: step [1].

23.45 Draw the structural formula of **A.** Then find the structure in the citric acid cycle to answer each question.

a. $^-O_2C(H)C{=}C(H)CO_2^-$ (trans)
A
fumarate

b. immediate precursor to **A:**
succinate

c. formed from **A:**
malate

23.47 Look at Figure 23.7 to find intermediates in the citric acid cycle with each characteristic.

a. CO_2^-–CH_2–$C^*H(CO_2^-)$–$C^*H(OH)$–CO_2^-
isocitrate
* = chirality center

b. malate: CO_2^-–[HO]C–H–CH_2–CO_2^-; isocitrate: CO_2^-–CH_2–H–C–CO_2^-–[HO]C–H–CO_2^-
malate isocitrate
2° alcohols

23.49 Answer each question about the conversion of isocitrate to α-ketoglutarate.

isocitrate → [3a] (a. oxidation; c. NAD⁺ here) → oxalosuccinate → [3b] (b. decarboxylation) → α-ketoglutarate

d. The elements of H_2 (hydrogen) are removed in converting isocitrate to oxalosuccinate, so the enzyme that catalyzes this step is called isocitrate dehydrogenase.

23.51 A hydration reaction (addition of H_2O) occurs in step [7] of the citric acid cycle.

$$\underset{\text{fumarate}}{\begin{array}{c} CO_2^- \\ | \\ CH \\ \| \\ HC \\ | \\ CO_2^- \end{array}} + H_2O \longrightarrow \underset{\text{malate}}{\begin{array}{c} CO_2^- \\ | \\ HO{-}C{-}H \\ | \\ CH_2 \\ | \\ CO_2^- \end{array}}$$

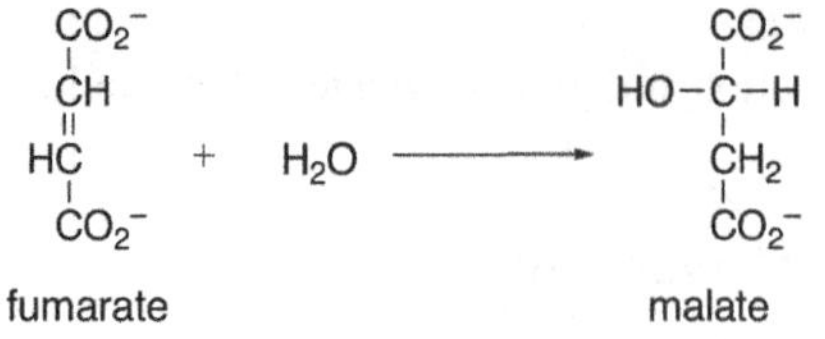

23.53 Aerobic respiration means that a reaction requires oxygen. O_2 is converted to H_2O in the final stage of electron transport.

23.55
a. $FADH_2$ donates electrons to the electron transport chain.
b. ADP is a substrate for the formation of ATP.
c. ATP synthase catalyzes the formation of ATP from ADP.
d. The inner mitochondrial membrane contains the four complexes for the electron transport chain. ATP synthase is also embedded in the membrane and contains the H^+ ion channel that allows H^+ to return to the matrix.

23.57 The final products of the electron transport chain are NAD^+, FAD, and H_2O.

23.59 $FADH_2$ enters the electron transport chain at complex II, whereas NADH enters at complex I, so they result in different amounts of ATP formation.

23.61
a. Pepsin catalyzes the hydrolysis of proteins to amino acids in the stomach.
b. Succinate dehydrogenase catalyzes the conversion of succinate to fumarate in step [6] of the citric acid cycle.
c. ATP synthase catalyzes the oxidative phosphorylation of ADP to ATP.

23.63 Each two-carbon fragment of acetyl CoA that enters the citric acid cycle produces three molecules of NADH and one molecule of $FADH_2$. These then enter the electron transport chain and serve as reducing agents. For each NADH that enters the electron transport chain, there are 2.5 ATPs produced, and for each $FADH_2$, there are 1.5 ATPs produced. In addition, there is one GTP (an ATP equivalent) produced in the citric acid cycle.

In total:			
	3 NADH × 2.5 ATP/NADH	=	7.5 ATPs
	1 $FADH_2$ × 1.5 ATP/$FADH_2$	=	1.5 ATPs
	1 GTP (ATP equivalent)	=	1 ATP
			10 ATPs

23.65 ATP has three phosphate groups, whereas ADP has two.

23.67 Answer each question about glucose 1-phosphate.

a. [Structure: glucose ring with CH_2OH, ring O, C1 bearing $O{-}P(=O)(O^-){-}O^-$; labels H, OH, HO, H, OH; C1 indicated]

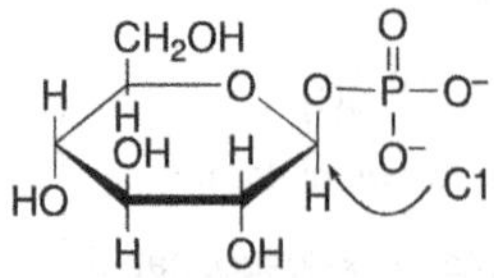

glucose 1-phosphate

b. 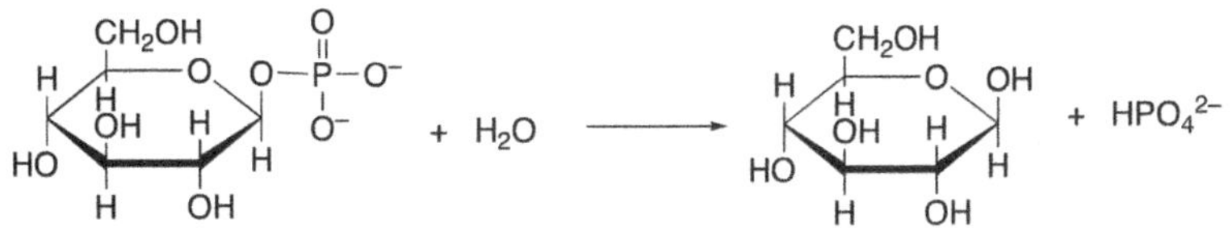 glucose 1-phosphate + H_2O ⟶ + HPO_4^{2-}

23.69 The citric acid cycle generates the reducing agents NADH and $FADH_2$ that enter the electron transport chain. In the process, NAD^+ and FAD, coenzymes needed for oxidation reactions in the citric acid cycle, are also produced.

23.71 There would be more mitochondria in the heart because it has far more metabolic needs than bone.

23.73 Creatine phosphate is a high-energy phosphate that generates energy upon hydrolysis.

23.75 Use conversion factors to solve the problem.

$$500.\ \text{Cal} \times \frac{1\ \text{kcal}}{1\ \text{Cal}} \times \frac{1\ \text{mol ATP}}{7.3\ \text{kcal}} = \mathbf{68\ moles}$$

$$68\ \text{mol} \times \frac{6.02 \times 10^{23}\ \text{molecules}}{1\ \text{mol}} = \mathbf{4.1 \times 10^{25}\ molecules\ of\ ATP}$$

Chapter 24 Carbohydrate, Lipid, and Protein Metabolism

Chapter Review

[1] What are the main elements that provide clues to the outcome of a biochemical reaction? (24.2)

- To understand the course of a biochemical reaction, examine the functional groups that are added or removed, the reagents (coenzymes or other materials), and the enzyme. The name of an enzyme is often a clue as to the type of reaction.

[2] What are the main aspects of glycolysis? (24.3)

- Glycolysis is a linear, 10-step pathway that converts glucose to two three-carbon pyruvate molecules. In the energy-investment phase, steps [1]–[5], the energy from two ATP molecules is used for phosphorylation and two three-carbon products are formed. In the energy-generating phase, steps [6]–[10], two pyruvate molecules ($CH_3COCO_2^-$), 2 NADHs, and 4 ATPs are generated.
- The net result of glycolysis, considering both phases, is 2 $CH_3COCO_2^-$, 2 NADHs, and 2 ATPs.

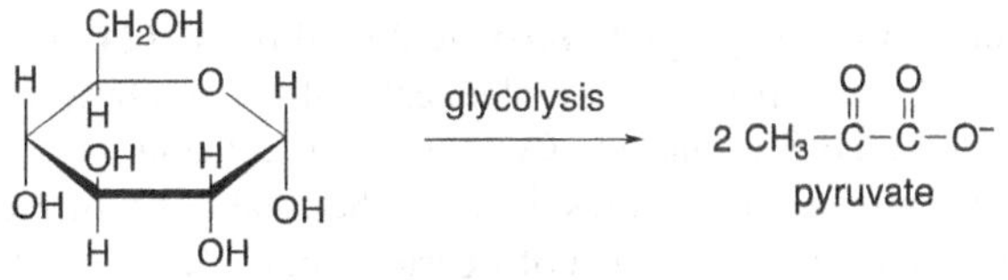

$$\text{glucose} \xrightarrow{\text{glycolysis}} 2\ CH_3-\overset{O}{\overset{\|}{C}}-\overset{O}{\overset{\|}{C}}-O^-\ \text{(pyruvate)}$$

[3] What are the major pathways for pyruvate metabolism? (24.4)

- When oxygen is plentiful, pyruvate is converted to acetyl CoA, which can enter the citric acid cycle.
- When the oxygen level is low, the anaerobic metabolism of pyruvate forms lactate and NAD^+.
- In yeast and other microorganisms, pyruvate is converted to ethanol and CO_2 by fermentation.

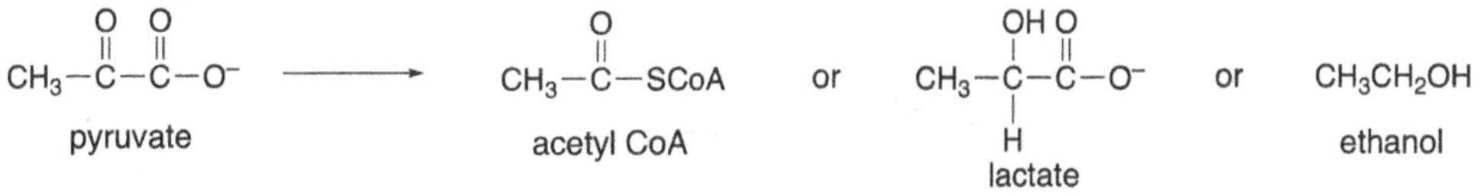

$$\underset{\text{pyruvate}}{CH_3-\overset{O}{\overset{\|}{C}}-\overset{O}{\overset{\|}{C}}-O^-} \longrightarrow \underset{\text{acetyl CoA}}{CH_3-\overset{O}{\overset{\|}{C}}-SCoA} \ \text{or}\ \underset{\text{lactate}}{CH_3-\overset{OH}{\overset{|}{\underset{H}{\underset{|}{C}}}}-\overset{O}{\overset{\|}{C}}-O^-} \ \text{or}\ \underset{\text{ethanol}}{CH_3CH_2OH}$$

[4] How much ATP is formed by the complete catabolism of glucose? (24.5)

- To calculate the amount of ATP formed in the catabolism of glucose, we must take into account the ATP yield from glycolysis, the oxidation of two molecules of pyruvate to two molecules of acetyl CoA, the citric acid cycle, and oxidative phosphorylation.
- As shown in Figure 24.6, the complete catabolism of glucose forms six CO_2 molecules and 32 molecules of ATP.

[5] What are the main features of gluconeogenesis? (24.6)

- Gluconeogenesis is the synthesis of glucose from noncarbohydrate sources—lactate, amino acids, or glycerol. Gluconeogenesis converts two molecules of pyruvate to glucose. Conceptually, gluconeogenesis is the reverse of glycolysis, but three steps in gluconeogenesis require different enzymes.

$$2\ CH_3-\overset{\overset{\displaystyle O}{\|}}{C}-CO_2^- \xrightarrow{\text{gluconeogenesis}} C_6H_{12}O_6$$

pyruvate → glucose

- Gluconeogenesis occurs when the body has depleted its supplies of glucose and stored glycogen, and takes place during sustained physical exercise and fasting.

[6] Describe the main features of the β-oxidation of fatty acids. (24.7)

- β-Oxidation is a spiral metabolic pathway that sequentially cleaves two-carbon acetyl CoA units from an acyl CoA derived from a fatty acid. Each cycle of β-oxidation consists of a four-step sequence that forms one molecule each of acetyl CoA, NADH, and $FADH_2$.

$$CH_3(CH_2)_{16}-\overset{\overset{\displaystyle O}{\|}}{C}-OH \longrightarrow CH_3(CH_2)_{16}-\overset{\overset{\displaystyle O}{\|}}{C}-SCoA \longrightarrow 9\ CH_3-\overset{\overset{\displaystyle O}{\|}}{C}-SCoA$$

fatty acid → thioester → acetyl CoA

[7] How much ATP is formed from complete fatty acid oxidation? (24.7)

- To determine the ATP yield from the complete catabolism of a fatty acid, we must consider the ATP used up in the synthesis of the acyl CoA, the ATP generated from coenzymes produced during β-oxidation, and the ATP that results from the catabolism of each acetyl CoA.
- As an example, the complete catabolism of stearic acid, $C_{18}H_{36}O_2$, yields 120 ATPs.

[8] What are ketone bodies and how do they play a role in metabolism? (24.8)

- Ketone bodies—acetoacetate, β-hydroxybutyrate, and acetone—are formed when acetyl CoA levels exceed the capacity of the citric acid cycle. Ketone bodies can be re-converted to acetyl CoA and metabolized for energy. When the level of ketone bodies is high, the pH of the blood can be lowered, causing ketoacidosis.

$$2\ CH_3-\overset{\overset{\displaystyle O}{\|}}{C}-SCoA \xrightarrow[\text{several steps}]{} CH_3-\overset{\overset{\displaystyle O}{\|}}{C}-CH_2-\overset{\overset{\displaystyle O}{\|}}{C}-O^- \xrightarrow{NADH + H^+ \rightarrow NAD^+} CH_3-\underset{H}{\overset{OH}{C}}-CH_2-\overset{\overset{\displaystyle O}{\|}}{C}-O^-$$

acetoacetate → β-hydroxybutyrate

acetoacetate → CO_2 +

$$CH_3-\overset{\overset{\displaystyle O}{\|}}{C}-CH_3$$

acetone

[9] What are the main features of amino acid catabolism? (24.9)

- The catabolism of amino acids involves two parts. First, the amino group is removed by transamination followed by oxidative deamination. The NH_4^+ ion formed enters the urea cycle where it is converted to urea and eliminated in urine. The carbon skeletons of the amino acids are catabolized by a variety of pathways to yield pyruvate, acetyl CoA, or an intermediate in the citric acid cycle.

$$R{-}\overset{\overset{+}{NH_3}}{\underset{H}{C}}{-}CO_2^- \xrightarrow[\text{[2] oxidative deamination}]{\text{[1] transamination}} R{-}\overset{O}{\overset{\|}{C}}{-}CO_2^- + NH_4^+$$

amino acid → α-keto acid

Problem Solving

[1] Understanding Biochemical Reactions (24.2)

Example 24.1 Analyze the following reaction by considering the functional groups that change and the name of the enzyme.

$$\text{3-phosphoglycerate } ({}^{2-}O_3P{-}O{-}CH_2{-}CH(OH){-}C(=O){-}O^-) \xrightarrow[\text{phosphoglycerate kinase}]{\text{ATP} \rightarrow \text{ADP}} \text{1,3-bisphosphoglycerate } ({}^{2-}O_3P{-}O{-}CH_2{-}CH(OH){-}C(=O){-}O{-}PO_3^{2-})$$

Analysis

- Consider the functional groups that are added or removed as a clue to the type of reaction.
- Classify the reagent (when one is given) as to the type of reaction it undergoes.
- Use the name of the enzyme as a clue to the reaction type.

Solution

A phosphate group is added to 3-phosphoglycerate in the reaction, meaning that a **phosphorylation** has occurred. Since kinase enzymes catalyze the addition or removal of phosphates, the enzyme used for the reaction is phosphoglycerate kinase. One phosphate bond is broken in ATP, giving ADP as one of the by-products.

Example 24.2 Analyze the following reaction by considering the functional groups that change, the coenzyme utilized, and the name of the enzyme.

$$CH_3{-}\overset{O}{\overset{\|}{C}}{-}CO_2^- + CO_2 \xrightarrow[\text{pyruvate carboxylase}]{\text{ATP} \rightarrow \text{ADP} + HPO_4^{2-}} {}^-O_2CCH_2{-}\overset{O}{\overset{\|}{C}}{-}CO_2^-$$

Analysis

- Consider the functional groups that are added or removed as a clue to the type of reaction.
- Classify the reagent—that is, any coenzyme or other reactant—as to the type of reaction it undergoes.
- Use the name of the enzyme as a clue to the reaction type.

Solution

In this reaction, a carboxylate ($-COO^-$) is added to the reactant pyruvate from CO_2. Since carboxylase enzymes catalyze the addition of a carboxylate, the enzyme is called pyruvate carboxylase. ATP is the coenzyme used, and hydrolysis of one phosphate group provides the energy for the reaction, forming ADP and HPO_4^{2-}.

---

[2] The Catabolism of Triacylglycerols (24.7)

Example 24.3 Consider oleic acid, $C_{17}H_{33}CO_2H$. (a) How many molecules of acetyl CoA are formed from the complete β-oxidation of oleic acid? (b) How many cycles of β-oxidation are needed for the complete catabolism of oleic acid?

Analysis
The number of carbons in the fatty acid determines the number of molecules of acetyl CoA formed and the number of times β-oxidation occurs.

- The number of molecules of acetyl CoA equals one-half the number of carbons in the original fatty acid.
- Because the final turn of the cycle forms *two* molecules of acetyl CoA, the number of cycles is one fewer than the number of acetyl CoA molecules formed.

Solution
Oleic acid has 18 carbons, so it forms nine molecules of acetyl CoA from eight cycles of β-oxidation.

---

[3] Amino Acid Metabolism (24.9)

Example 24.4 What products are formed in the following transamination reaction?

$$CH_3CH_2CH(CH_3)-\overset{\overset{+}{N}H_3}{\underset{H}{C}}-CO_2^- + {}^-O_2CCH_2CH_2-\overset{O}{\overset{\|}{C}}-CO_2^- \xrightarrow[\text{transaminase}]{}$$

α-ketoglutarate

Analysis
To draw the products of transamination, convert the C–H and C–NH_3^+ groups of the amino acid to a C=O, forming an α-keto acid.

Solution

$$CH_3CH_2CH(CH_3)-\overset{\overset{+}{N}H_3}{\underset{H}{C}}-CO_2^- + {}^-O_2CCH_2CH_2-\overset{O}{\overset{\|}{C}}-CO_2^-$$

$$\downarrow \text{transaminase}$$

$$CH_3CH_2CH(CH_3)-\overset{O}{\overset{\|}{C}}-CO_2^- + {}^-O_2CCH_2CH_2-\overset{\overset{+}{N}H_3}{\underset{H}{C}}-CO_2^-$$

new C–NH_3^+ bond

The amino group is removed from the amino acid.

Example 24.5 What final products are formed when the following amino acid is subjected to transamination followed by oxidative deamination?

$$(CH_3)_2CH{-}\overset{\overset{+}{N}H_3}{\underset{H}{C}}{-}CO_2^-$$

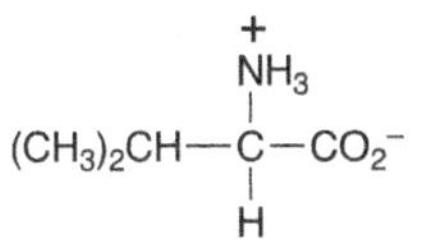

Analysis

To draw the organic product, replace the C–H and C–NH_3^+ on the α carbon of the amino acid by C=O. NH_4^+ is formed from the amino group.

Solution

$$\underset{\alpha\text{ carbon}}{(CH_3)_2CH{-}\overset{\overset{+}{N}H_3}{\underset{H}{C}}{-}CO_2^-} \xrightarrow[\text{[2] oxidative deamination}]{\text{[1] transamination}} (CH_3)_2CH{-}\overset{O}{\overset{\|}{C}}{-}CO_2^- \; + \; NH_4^+$$

Self-Test

[1] Fill in the blank with one of the terms listed below.

Cori cycle (24.6)	Ketogenesis (24.8)	β-Oxidation (24.7)
Fermentation (24.4)	Ketogenic amino acids (24.9)	Oxidative deamination (24.9)
Glucogenic amino acids (24.9)	Ketone bodies (24.8)	Transamination (24.9)
Gluconeogenesis (24.6)	Ketosis (24.8)	
Glycolysis (24.3)	Kinase (24.2)	

1. ______ is an anabolic process that synthesizes glucose from pyruvate.
2. A ______ catalyzes the transfer of a phosphate group from one substrate to another.
3. Cycling compounds from the muscle to the liver and back to the muscle is called the ______.
4. ______ is a process in which two-carbon acetyl CoA units are sequentially cleaved from a fatty acid.
5. Under some circumstances ketone bodies accumulate, a condition called ______.
6. ______ are catabolized to pyruvate or an intermediate in the citric acid cycle, so these compounds can be used to synthesize glucose.
7. ______ is the anaerobic conversion of glucose to ethanol and CO_2.
8. ______ is the synthesis of ketone bodies from acetyl CoA.
9. ______ is the transfer of an amino group from an amino acid to an α-keto acid, usually α-ketoglutarate.
10. In ______, the C–H and C–NH_3^+ bonds on the α carbon of glutamate are converted to C=O and an ammonium ion, NH_4^+.
11. ______ is a catabolic process that converts glucose to pyruvate.
12. Acetoacetate, β-hydroxybutyrate, and acetone are all ______.

13. ______ are converted to acetyl CoA, or the related thioester acetoacetyl CoA, $CH_3COCH_2COSCoA$. These catabolic products cannot be used to synthesize glucose, but they can be converted to ketone bodies and yield energy by this pathway.

[2] Match the type of enzyme with the action.

a. carboxylase b. decarboxylase c. dehydrogenase

14. Addition of a carboxylate ($–COO^-$)
15. Removal of two hydrogen atoms
16. Removal of carbon dioxide (CO_2)

[3] Pick the reaction conditions under which pyruvate is converted to each of the following:

a. acetyl CoA b. lactate c. ethanol

17. fermentation
18. aerobic conditions
19. anaerobic conditions

[4] Match the enzyme with the reaction it catalyzes.

a. hexokinase
b. alcohol dehydrogenase
c. pyruvate decarboxylase
d. phosphofructokinase
e. phosphohexose isomerase

20. fructose → fructose 6-phosphate
21. pyruvate → acetaldehyde
22. glucose → glucose 6-phosphate
23. acetaldehyde → ethanol
24. glucose 6-phosphate → fructose 6-phosphate

Answers to Self-Test

1. Gluconeogenesis	6. Glucogenic amino acids	11. Glycolysis	16. b	21. c
2. kinase	7. Fermentation	12. ketone bodies	17. c	22. a
3. Cori cycle	8. Ketogenesis	13. Ketogenic amino acids	18. a	23. b
4. β-Oxidation	9. Transamination	14. a	19. b	24. e
5. ketosis	10. oxidative deamination	15. c	20. d	

Solutions to In-Chapter Problems

24.1 Analyze the reaction as in Example 24.1.

$$^-O-\overset{O}{\overset{\|}{P}}(O^-)-O-CH_2-\overset{O}{\overset{\|}{C}}-CH_2OH \xrightarrow{\text{triose phosphate isomerase}} {}^-O-\overset{O}{\overset{\|}{P}}(O^-)-O-CH_2-\overset{OH}{CH}-\overset{O}{\overset{\|}{C}}-H$$

dihydroxyacetone phosphate → glyceraldehyde 3-phosphate

The reactant and product are constitutional isomers that differ in the location of a carbonyl and hydroxyl group. Thus, an isomerase enzyme converts one isomer to another.

24.2 Analyze the reaction as in Example 24.2.

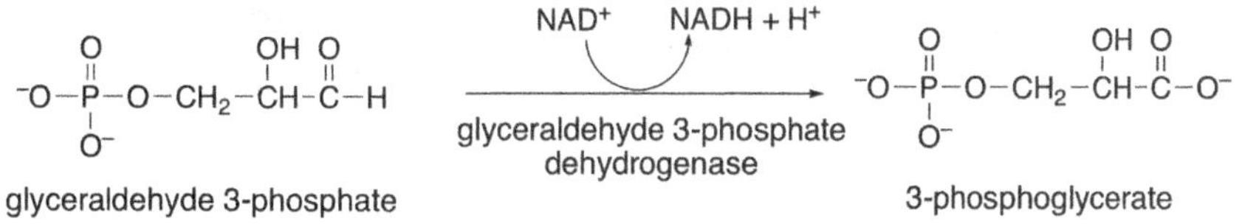

$$^{-}O-\overset{\overset{O}{\|}}{\underset{\underset{O^-}{|}}{P}}-O-CH_2-\overset{OH}{\overset{|}{CH}}-\overset{O}{\overset{\|}{C}}-H \xrightarrow[\text{glyceraldehyde 3-phosphate dehydrogenase}]{NAD^+ \quad NADH + H^+} {}^{-}O-\overset{\overset{O}{\|}}{\underset{\underset{O^-}{|}}{P}}-O-CH_2-\overset{OH}{\overset{|}{CH}}-\overset{O}{\overset{\|}{C}}-O^-$$

glyceraldehyde 3-phosphate → 3-phosphoglycerate

The aldehyde (CHO) in the reactant is oxidized to a carboxylate anion (COO^-). NAD^+ is the oxidizing agent, which gets reduced to NADH. A dehydrogenase enzyme removes hydrogen from the reactant in the oxidation.

24.3 Compare glucose 6-phosphate (**A**) and fructose 6-phosphate (**B**).

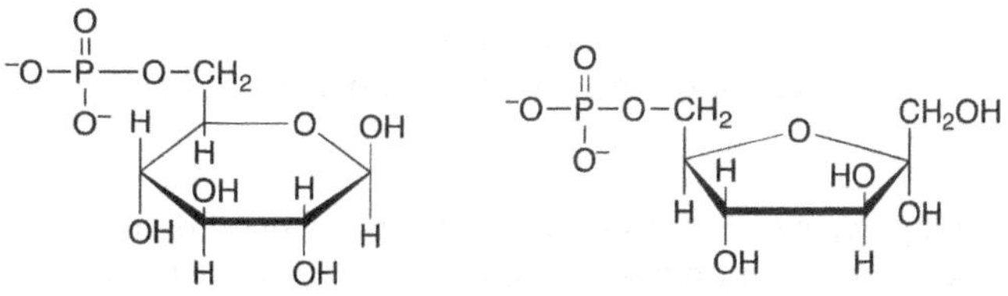

glucose 6-phosphate
A

fructose 6-phosphate
B

same molecular formula
different arrangement of atoms
constitutional isomers

24.4 The energy of the coupled reaction (–3.4 kcal/mol) represents the sum of the energies of the phosphorylation of fructose 6-phosphate and the hydrolysis of ATP (–7.3 kcal/mol).

phosphorylation + (–7.3 kcal/mol) = –3.4 kcal/mol
phosphorylation = +3.9 kcal/mol

Thus, the phosphorylation of fructose 6-phosphate requires 3.9 kcal/mol.

24.5 Use Figures 24.2, 24.3, and 24.4 to determine which steps utilize kinases, and write the species that is phosphorylated.

step [1]: glucose; step [3]: fructose 6-phosphate; step [7]: ADP; step [10]: ADP

24.6 Use Figures 24.2, 24.3, and 24.4 to determine which steps involve the conversion of one constitutional isomer to another.

step [2]: phosphohexose isomerase
step [5]: triose phosphate isomerase
step [8]: phosphoglycerate mutase

24.7 Citrate is one intermediate in the citric acid cycle, so the rate of glycolysis decreases when there are high levels of citrate.

24.8 Compare galactose and mannose.

a.

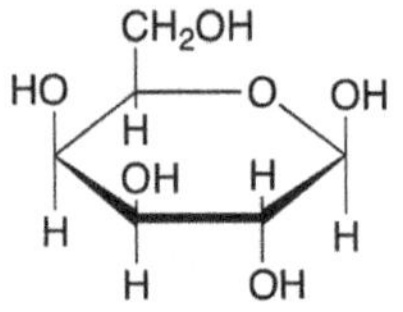

galactose

same molecular formula
same connectivity of atoms
different 3-D arrangement
stereoisomers

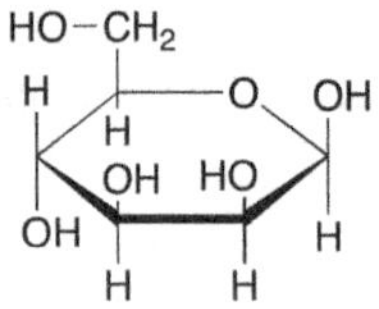

mannose

b.

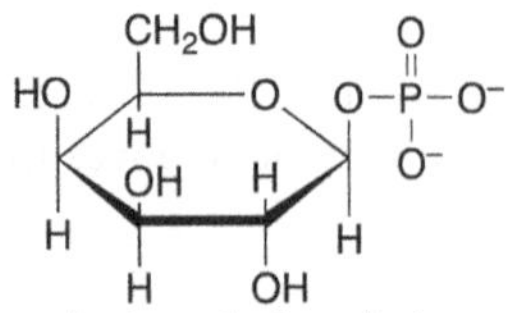

galactose 1-phosphate

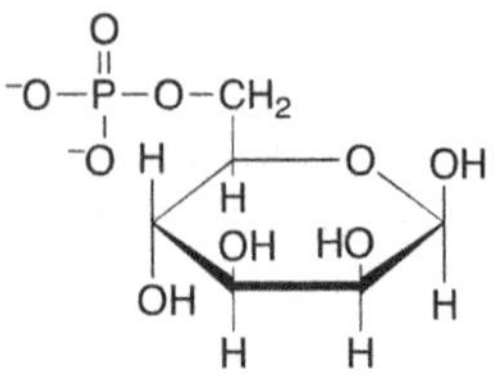

mannose 6-phosphate

c. same molecular formula
different connectivity of atoms
constitutional isomers

24.9 NADH serves as the reducing agent for the conversion of pyruvate to both lactate and ethanol. The NAD^+ formed in the process can then be used as the oxidant in step [6] of glycolysis, making the process anaerobic.

24.10 a. The conversion of pyruvate to acetyl CoA and the conversion of pyruvate to ethanol both create a two-carbon compound from the three-carbon pyruvate with loss of CO_2.
b. The conversions are different in that the production of acetyl CoA generates NADH, whereas the production of ethanol consumes NADH.

24.11 a. The conversions of pyruvate to lactate and pyruvate to ethanol both use NADH to generate a reduced compound.
b. The conversions are different in that ethanol is a two-carbon product formed by decarboxylation, whereas lactate is a three-carbon compound formed by the addition of hydrogen.

24.12 Use Figure 24.6 to determine how much ATP is produced in each transformation.

a. glucose $\rightarrow$ 2 pyruvate: 7 ATP
b. pyruvate $\rightarrow$ acetyl CoA: 2.5 ATP per pyruvate
c. glucose $\rightarrow$ 2 acetyl CoA: 12 ATP
d. 2 acetyl CoA $\rightarrow$ 4 CO_2: 20 ATP

24.13 The conversion of pyruvate to acetyl CoA and steps [3] and [4] of the citric acid cycle form CO_2.

24.14 The aerobic oxidation of glucose to CO_2 is much more efficient than the anaerobic conversion of glucose to lactate. Thirty-two (32) ATPs are generated with aerobic metabolism, whereas only two ATPs are generated with anaerobic metabolism to lactate.

24.15 Since gluconeogenesis is formally the reverse of glycolysis, use Figures 24.2, 24.3, and 24.4 to determine the starting material and product for each step in gluconeogenesis.

	Starting Material	Product
a.	pyruvate	phosphoenolpyruvate
b.	2-phosphoglycerate	3-phosphoglycerate
c.	glucose 6-phosphate	glucose

24.16 The reaction involves the transfer of a phosphate group from ATP to glycerol by glycerol kinase. No oxidation or reduction occurs.

$$\underset{\text{glycerol}}{\begin{array}{c} CH_2-OH \\ | \\ H-C-OH \\ | \\ CH_2-OH \end{array}} \xrightarrow[\text{glycerol kinase}]{\text{ATP} \quad \text{ADP}} \underset{\text{glycerol 3-phosphate}}{\begin{array}{c} CH_2-OH \\ | \\ H-C-OH \\ | \\ CH_2-O-\overset{O}{\overset{\|}{P}}-O^- \\ \quad\quad\quad | \\ \quad\quad\quad O^- \end{array}}$$

24.17 For each fatty acid, determine the number of molecules of acetyl CoA formed from the complete catabolism and the number of cycles of β-oxidation needed for complete oxidation as in Example 24.3.

a. arachidic acid ($C_{20}H_{40}O_2$): 20 carbons, [1] 10 molecules of acetyl CoA, [2] 9 cycles
b. palmitoleic acid ($C_{16}H_{30}O_2$): 16 carbons, [1] 8 molecules of acetyl CoA, [2] 7 cycles

24.18 Calculate the number of molecules of ATP released on the catabolism of palmitic acid ($C_{16}H_{32}O_2$).
[1] Determine the amount of ATP required to synthesize the acyl CoA from the fatty acid.
[2] Add up the ATP generated from the coenzymes produced during β-oxidation.
[3] Determine the amount of ATP that results from each acetyl CoA, and add the results for steps [1]–[3].

[1] 2 ATPs are required for the conversion of $C_{16}H_{32}O_2$ to $C_{15}H_{31}COSCoA$ (–2 ATPs).
[2] 7 cycles of β-oxidation:

7 NADH	×	2.5 ATP/NADH	=	17.5 ATPs
7 $FADH_2$	×	1.5 ATP/$FADH_2$	=	10.5 ATPs
		From reduced coenzymes:		28 ATPs

[3] From acetyl CoA:
8 acetyl CoA × 10 ATP/acetyl CoA = 80 ATPs

Total steps [1]–[3]:
(–2) + 28 + 80 = 106 ATP molecules from palmitic acid
Answer

24.19 Calculate the number of molecules of ATP released on the catabolism of arachidic acid using the steps in Answer 24.18.

[1] 2 ATPs are required for the conversion of $C_{20}H_{40}O_2$ to $C_{19}H_{39}COSCoA$ (–2 ATPs).
[2] 9 cycles of β-oxidation:

$$9\ \text{NADH} \times 2.5\ \text{ATP/NADH} = 22.5\ \text{ATPs}$$
$$9\ \text{FADH}_2 \times 1.5\ \text{ATP/FADH}_2 = 13.5\ \text{ATPs}$$
$$\text{From reduced coenzymes: } 36\ \text{ATPs}$$

[3] From acetyl CoA:

$$10\ \text{acetyl CoA} \times 10\ \text{ATP/acetyl CoA} = 100\ \text{ATPs}$$

Total steps [1]–[3]:

$$(-2) + 36 + 100 = 134\ \text{ATP molecules from arachidic acid}$$

Answer

24.20 Calculate the number of moles of ATP formed per gram of palmitic acid.

$$\frac{106\ \text{mol ATP}}{1\ \text{mol palmitic acid}} \times \frac{1\ \text{mol}}{256\ \text{g}} = \frac{0.414\ \text{mol ATP}}{1\ \text{g palmitic acid}}$$

24.21 Compare the number of ATPs generated per carbon atom.

a. $\frac{32\ \text{ATP}}{6\ \text{C's in glucose}} = 5.3\ \text{ATP/C in glucose}$

b. $\frac{120\ \text{ATP}}{18\ \text{C's in stearic acid}} = 6.7\ \text{ATP/C in stearic acid}$ ← more energy per carbon

c. Because stearic acid produces more ATP per carbon atom than glucose, lipids are more effective energy-storing molecules than carbohydrates.

24.22 All three compounds contain at least one carbonyl group.

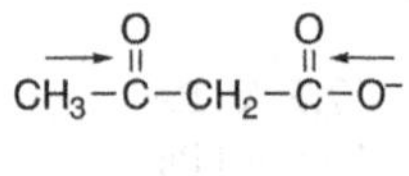

$CH_3-C(=O)-CH_2-C(=O)-O^-$ acetoacetate

$CH_3-C(=O)-CH_3$ acetone

$CH_3-CH(OH)-CH_2-C(=O)-O^-$ β-hydroxybutyrate

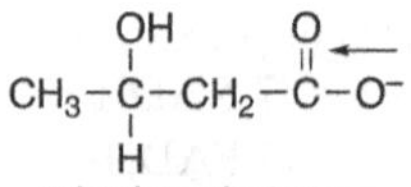

24.23 β-Hydroxybutyrate contains an alcohol and a carboxylate anion, but no ketone.

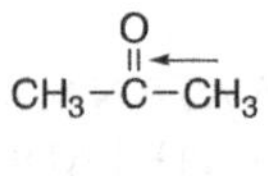

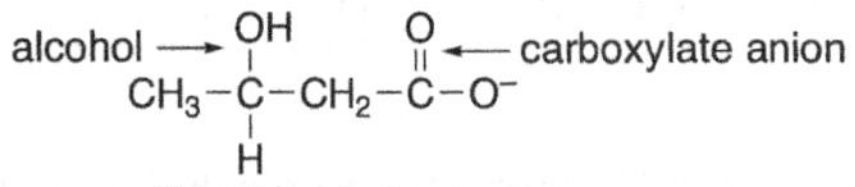

alcohol → OH; carboxylate anion → $C(=O)O^-$

$CH_3-CH(OH)-CH_2-C(=O)-O^-$

β-hydroxybutyrate

24.24 When acetoacetate is converted to acetone, a CO_2 group is removed. Therefore, the enzyme is a decarboxylase.

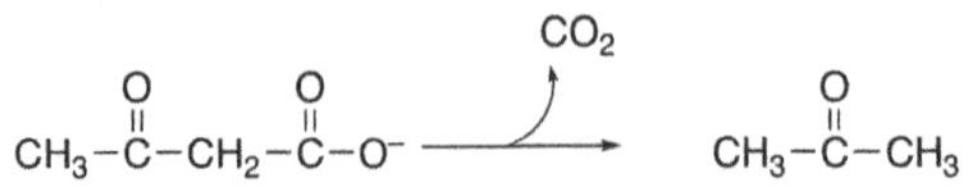

$$CH_3-C(=O)-CH_2-C(=O)-O^- \xrightarrow{-CO_2} CH_3-C(=O)-CH_3$$

acetoacetate → acetone

24.25 An increased concentration of acetyl CoA increases the production of ketone bodies, because acetyl CoA is the reactant that starts ketogenesis.

24.26 Draw the products when each amino acid undergoes transamination as in Example 24.4.

a. 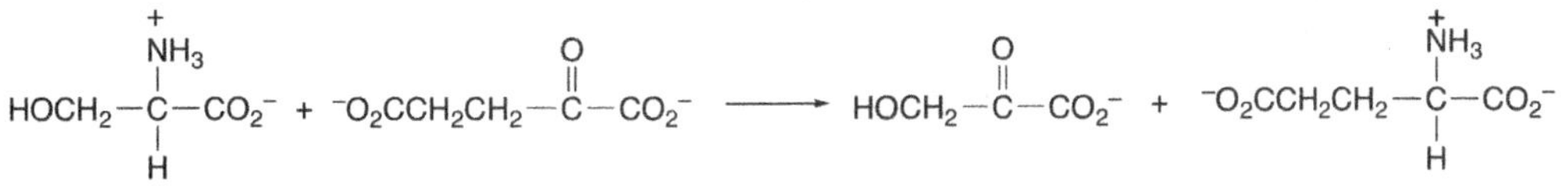 $HOCH_2-C(H)(\overset{+}{N}H_3)-CO_2^- + {}^-O_2CCH_2CH_2-C(=O)-CO_2^- \longrightarrow HOCH_2-C(=O)-CO_2^- + {}^-O_2CCH_2CH_2-C(H)(\overset{+}{N}H_3)-CO_2^-$

serine

b. 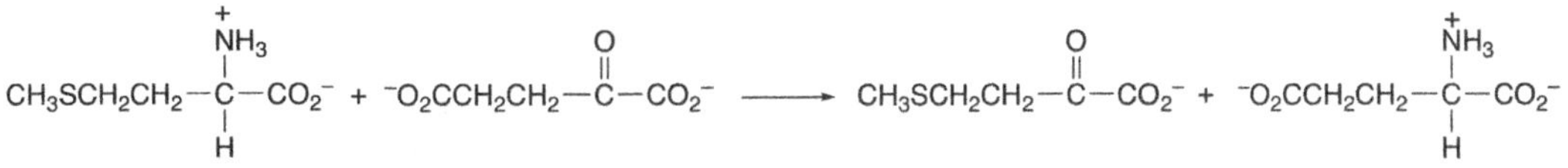 $CH_3SCH_2CH_2-C(H)(\overset{+}{N}H_3)-CO_2^- + {}^-O_2CCH_2CH_2-C(=O)-CO_2^- \longrightarrow CH_3SCH_2CH_2-C(=O)-CO_2^- + {}^-O_2CCH_2CH_2-C(H)(\overset{+}{N}H_3)-CO_2^-$

methionine

c. 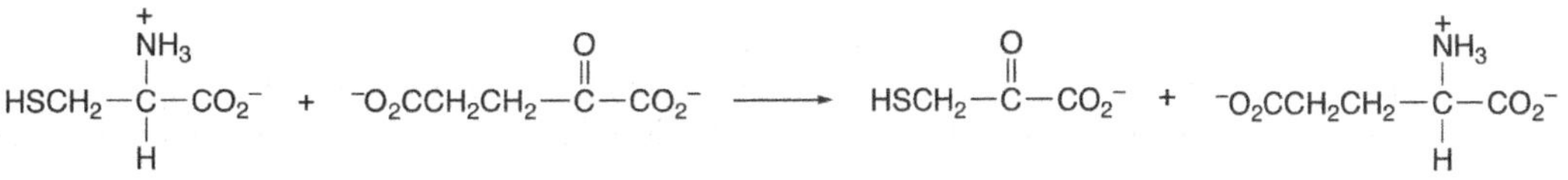 $HSCH_2-C(H)(\overset{+}{N}H_3)-CO_2^- + {}^-O_2CCH_2CH_2-C(=O)-CO_2^- \longrightarrow HSCH_2-C(=O)-CO_2^- + {}^-O_2CCH_2CH_2-C(H)(\overset{+}{N}H_3)-CO_2^-$

cysteine

24.27 Draw the products of transamination and oxidative deamination as in Example 24.5.

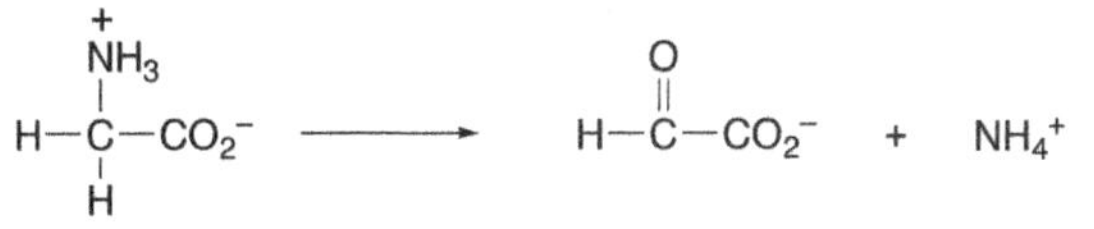

a. 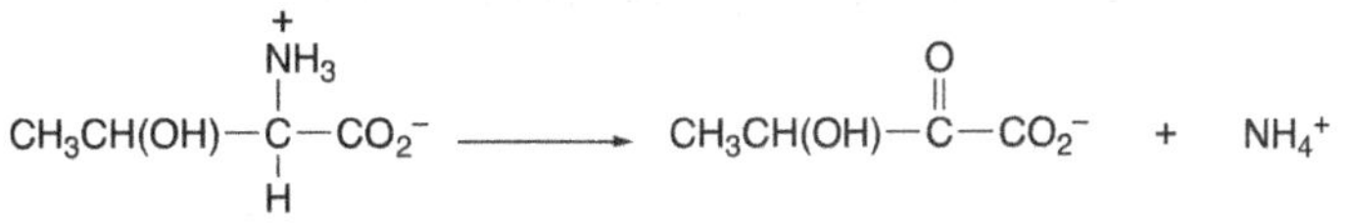 $CH_3CH(OH)-C(H)(\overset{+}{N}H_3)-CO_2^- \longrightarrow CH_3CH(OH)-C(=O)-CO_2^- + NH_4^+$

b. $H-C(H)(\overset{+}{N}H_3)-CO_2^- \longrightarrow H-C(=O)-CO_2^- + NH_4^+$

c. $CH_3CH_2CH(CH_3)-C(H)(\overset{+}{N}H_3)-CO_2^- \longrightarrow CH_3CH_2CH(CH_3)-C(=O)-CO_2^- + NH_4^+$

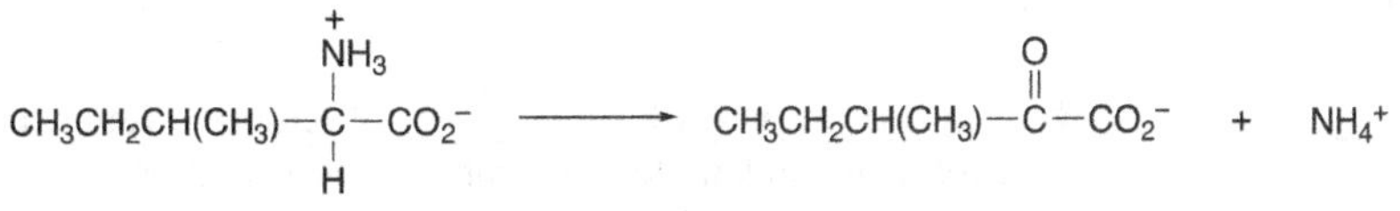

24.28 Use Figure 24.10 to determine what metabolic intermediates are produced from each amino acid.

a. cysteine: pyruvate
b. aspartic acid: oxaloacetate
c. valine: succinyl CoA
d. threonine: acetyl CoA, pyruvate, and succinyl CoA

Solutions to Odd-Numbered End-of-Chapter Problems

24.29 Analyze the reactions as in Examples 24.1 and 24.2.

a. $CH_3-CH(OH)-CH_2-C(=O)-SCoA \xrightarrow[\beta\text{-hydroxyacyl CoA dehydrogenase}]{NAD^+ \rightarrow NADH + H^+} CH_3-C(=O)-CH_2-C(=O)-SCoA$

Oxidation occurs with loss of hydrogen by a dehydrogenase enzyme. The coenzyme NAD^+ serves as an oxidant.

b. $^-O_2CCH_2-C(=O)-CO_2^- \xrightarrow[\text{phosphoenolpyruvate carboxykinase}]{GTP \rightarrow GDP} CH_2=C(O-PO_3^{2-})-C(=O)-O^- + CO_2$

Transfer of phosphate from GTP to the reactant occurs, forming GDP. Since a kinase enzyme catalyzes the transfer of a phosphate group, the enzyme used here is phosphoenolpyruvate carboxykinase. One carboxylate ($-COO^-$) is also lost, breaking a C–C bond and forming CO_2.

24.31 This reaction converts one isomer to another because there is a change in position of the phosphate; thus, the enzyme is called a "mutase" (phosphoglucomutase).

24.33 Convert **A** to a structural formula, and then answer each question about glycolysis.

a. $CH_2=C(O-PO_3^{2-})-C(=O)-O^-$ (phosphate drawn as $^-O-P(=O)(O^-)-O-$)

A
phosphoenolpyruvate

b. immediate precursor to **A**: 2-phosphoglycerate
c. formed from **A**: pyruvate

24.35 Compare the energy-investment phase and the energy-generating phase of glycolysis.

	Energy-Investment Phase:	**Energy-Generating Phase:**
a. reactant/final product	glucose, glyceraldehyde 3-phosphate	glyceraldehyde 3-phosphate, pyruvate
b. ATP used/formed	2 ATP used	4 ATP/glucose produced
c. coenzymes used or formed	no coenzymes used or formed	2 NAD^+ used and 2 NADH produced per glucose

24.37

a. steps that form ATP: 7, 10
b. steps that use ATP: 1, 3
c. step that forms a reduced coenzyme: 6
d. step that breaks a C–C bond: 4

24.39 Six molecules of CO_2 are formed during the catabolism of glucose. The conversion of pyruvate to acetyl CoA forms one CO_2 for each pyruvate. The conversions of isocitrate to α-ketoglutarate and α-ketoglutarate to succinyl CoA in the citric acid cycle form one CO_2 for each acetyl CoA.

2 pyruvate to 2 acetyl CoA	2 CO_2 released
2 isocitrate to 2 α-ketoglutarate	2 CO_2 released
2 α-ketoglutarate to 2 succinyl CoA	2 CO_2 released
	6 CO_2 total

24.41

a. $$\underset{\text{glucose}}{C_6H_{12}O_6} + 2\,NAD^+ \xrightarrow{2\,ADP \;\to\; 2\,ATP} \underset{\text{pyruvate}}{2\,CH_3COCO_2^-} + 2\,NADH + 2\,H^+$$

b. $$\underset{\text{glucose}}{C_6H_{12}O_6} \xrightarrow{2\,ADP \;\to\; 2\,ATP} \underset{\text{ethanol}}{2\,CH_3CH_2OH} + 2\,CO_2$$

24.43

$$\underset{\text{pyruvate}}{CH_3{-}\overset{O}{\overset{\|}{C}}{-}\overset{O}{\overset{\|}{C}}{-}O^-} \xrightarrow{NADH + H^+ \;\to\; NAD^+} \underset{\text{lactate}}{CH_3{-}\underset{H}{\underset{|}{\overset{OH}{\overset{|}{C}}}}{-}\overset{O}{\overset{\|}{C}}{-}O^-}$$

24.45

a. Under aerobic conditions, pyruvate carbons become CO_2.
   Under anaerobic conditions, pyruvate carbons become lactate.
b. Under aerobic conditions, NAD^+ and FAD are used and NADH and $FADH_2$ are formed.
   Under anaerobic conditions, NADH is used in converting pyruvate to lactate and NAD^+ is formed.

24.47 Glycolysis takes glucose and breaks it into smaller units, producing pyruvate. Gluconeogenesis is the synthesis of glucose from lactate, amino acids, and glycerol.

24.49 The rate of glycolysis is affected by each of the following conditions.

a. low ATP concentration: increased
b. low ADP concentration: decreased
c. high carbohydrate diet: increased
d. low carbohydrate diet: decreased

24.51 In the Cori cycle, glucose in muscle is catabolized to pyruvate and then converted to lactate, which is transported to the liver. In the liver, lactate is converted to glucose by gluconeogenesis and then transported back to the muscle.

24.53 The following metabolic products are formed from pyruvate under each condition.

a. anaerobic conditions in the body: lactate
b. anaerobic conditions in yeast: ethanol
c. aerobic conditions: CO_2

24.55 Use Figure 24.6 to fill in the boxes.

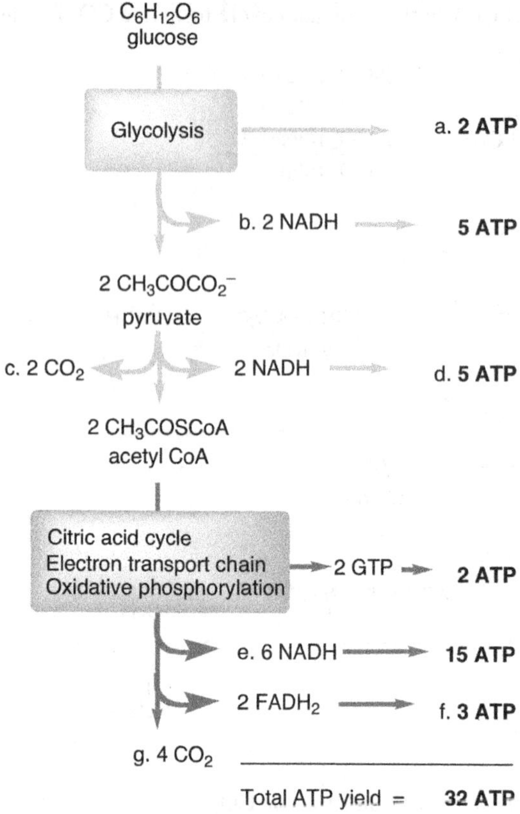

24.57 In fermentation, the six carbons in glucose become two carbons in two molecules of CO_2 and four carbons in two molecules of CH_3CH_2OH.

24.59 When a fatty acid is converted to an acyl CoA, two phosphate bonds of ATP are hydrolyzed, forming AMP. Therefore, 2 ATP equivalents are used.

24.61 For myristic acid, determine the number of molecules of acetyl CoA formed and the number of cycles of β-oxidation needed as in Example 24.3.

myristic acid, $C_{13}H_{27}CO_2H$: 14 carbons, [1] 7 molecules of acetyl CoA, [2] 6 cycles

24.63 For hexanoic acid, the compound depicted in the ball-and-stick model, determine the number of molecules of acetyl CoA formed and the number of cycles of β-oxidation needed as in Example 24.3.

hexanoic acid, $CH_3(CH_2)_4CO_2H$: 6 carbons: (a) 3 molecules of acetyl CoA; (b) 2 cycles

24.65 Calculate the number of molecules of ATP formed from myristic acid using the steps in Answer 24.18.

[1] 2 ATPs are required for the conversion of $CH_3(CH_2)_{12}CO_2H$ to $CH_3(CH_2)_{12}COSCoA$ (–2 ATPs).

[2] 6 cycles of β-oxidation:

6 NADH	×	2.5 ATP/NADH	=	15 ATPs
6 $FADH_2$	×	1.5 ATP/$FADH_2$	=	9 ATPs
		From reduced coenzymes:		24 ATPs

[3] From acetyl CoA:

7 acetyl CoA × 10 ATP/acetyl CoA = 70 ATPs

Total steps [1]–[3]:

(–2) + 24 + 70 = 92 ATP molecules from myristic acid
Answer

24.67

$C_{15}H_{31}CO_2H$

↓

$CH_3(CH_2)_{12}-CH_2-CH_2-\overset{O}{\overset{\|}{C}}-SCoA$
C_{16} acyl CoA

→ (FAD → $FADH_2$; acyl CoA dehydrogenase) [1] **Oxidation**

$CH_3(CH_2)_{12}-CH{=}CH-\overset{O}{\overset{\|}{C}}-SCoA$

↓ (H_2O; enoyl CoA hydratase) [2] **Hydration**

$CH_3(CH_2)_{12}-\overset{OH}{\overset{|}{C}H}-\overset{H}{\overset{|}{C}H}-\overset{O}{\overset{\|}{C}}-SCoA$

↓ (NAD^+ → NADH + H^+; β-hydroxyacyl CoA dehydrogenase) [3] **Oxidation**

$CH_3(CH_2)_{12}-\overset{O}{\overset{\|}{C}}-CH_2-\overset{O}{\overset{\|}{C}}-SCoA$

↓ (HSCoA; acyl CoA acyltransferase) [4] **Cleavage**

$CH_3(CH_2)_{12}-\overset{O}{\overset{\|}{C}}-SCoA$ + $CH_3-\overset{O}{\overset{\|}{C}}-SCoA$

C_{14} acyl CoA; 2 C's

24.69 Answer each question about decanoic acid.

a. $CH_3(CH_2)_6-\underset{\beta}{CH_2}-\underset{\alpha}{CH_2}-\overset{O}{\overset{||}{C}}-OH$

b. $CH_3(CH_2)_6-\underset{\beta}{CH_2}-\underset{\alpha}{CH_2}-\overset{O}{\overset{||}{C}}-SCoA$

c. Five acetyl CoA molecules are formed, since there are 10 carbons.

d. Four cycles of β-oxidation are needed for complete oxidation.

e. [1] –2 ATPs for conversion to acyl CoA
[2] 4 cycles of β-oxidation:
4 NADH x 2.5 ATP/NADH = 10 ATPs
4 $FADH_2$ x 1.5 ATP/$FADH_2$ = 6 ATPs
From reduced coenzymes: 16 ATPs
[3] 5 acetyl CoA x 10 ATP/acetyl CoA = 50 ATPs

Total steps [1]–[3]
(–2) + 16 + 50 = **64 molecules of ATP**

24.71 Calculate the number of moles of ATP formed per gram of decanoic acid.

$$\frac{64 \text{ mol ATP}}{1 \text{ mol decanoic acid}} \times \frac{1 \text{ mol}}{172 \text{ g}} = \frac{0.37 \text{ mol ATP}}{1 \text{ g decanoic acid}}$$

24.73 **A** and **C** are ketone bodies. **B** is not a ketone body formed from acetyl CoA.

24.75 In poorly controlled diabetes, glucose cannot be metabolized due to a lack of insulin. Therefore, fatty acids are used for metabolism and ketone bodies are formed to a greater extent.

24.77 Ketogenic amino acids are converted to acetyl CoA or a related thioester and can be converted to ketone bodies. Glucogenic amino acids are catabolized to pyruvate or another intermediate in the citric acid cycle and can be converted to glucose.

24.79 Draw the products when each amino acid undergoes transamination as in Example 24.4.

a. $H-\overset{\overset{+}{N}H_3}{\underset{H}{C}}-CO_2^- \longrightarrow H-\overset{O}{\overset{||}{C}}-CO_2^-$

b. $C_6H_5-CH_2-\overset{\overset{+}{N}H_3}{\underset{H}{C}}-CO_2^- \longrightarrow C_6H_5-CH_2-\overset{O}{\overset{||}{C}}-CO_2^-$

24.81 Draw the products of transamination as in Example 24.4.

a. $CH_3-\overset{\overset{+}{N}H_3}{\underset{H}{C}}-CO_2^- + {}^-O_2CCH_2-\overset{O}{\overset{||}{C}}-CO_2^- \longrightarrow CH_3-\overset{O}{\overset{||}{C}}-CO_2^- + {}^-O_2CCH_2-\overset{\overset{+}{N}H_3}{\underset{H}{C}}-CO_2^-$

b. $(CH_3)_2CHCH_2-\overset{\overset{+}{N}H_3}{\underset{H}{C}}-CO_2^- + {}^-O_2CCH_2-\overset{O}{\overset{||}{C}}-CO_2^- \longrightarrow (CH_3)_2CHCH_2-\overset{O}{\overset{||}{C}}-CO_2^- + {}^-O_2CCH_2-\overset{\overset{+}{N}H_3}{\underset{H}{C}}-CO_2^-$

24.83 Draw the products of oxidative deamination.

a. $(CH_3)_2CHCH_2-\overset{\overset{+}{NH_3}}{\underset{H}{C}}-CO_2^-$ $\longrightarrow$ $(CH_3)_2CHCH_2-\overset{O}{\overset{\|}{C}}-CO_2^-$ + NH_4^+

b. $C_6H_5-CH_2-\overset{\overset{+}{NH_3}}{\underset{H}{C}}-CO_2^-$ $\longrightarrow$ $C_6H_5-CH_2-\overset{O}{\overset{\|}{C}}-CO_2^-$ + NH_4^+

24.85 Use Figure 24.10 to determine what metabolic intermediates are produced from each amino acid.

a. phenylalanine: acetoacetyl CoA and fumarate
b. glutamic acid: α-ketoglutarate
c. asparagine: oxaloacetate
d. glycine: pyruvate

24.87 Yes, amino acids can be both ketogenic and glucogenic. Phenylalanine, for example, can be degraded by both glucogenic and ketogenic routes and is therefore both ketogenic and glucogenic.

24.89 A spiral pathway involves the same set of reactions repeated on increasingly smaller substrates; example: β-oxidation of fatty acids. A cyclic pathway is a series of reactions that regenerates the first reactant; example: citric acid cycle.

24.91 Glucose is metabolized anaerobically by yeast to form CO_2 and ethanol. The CO_2 causes bread to rise.

24.93 Pyruvate is metabolized anaerobically to lactate in the cornea, because the blood supply (and therefore oxygen) is very limited.

24.95 The Atkins diet calls for the ingestion of only protein and fat. The diet has an absence of carbohydrates to be metabolized, so ketone bodies are formed.

24.97 Compare the reactant and product to determine what type of reaction has occurred.

a. $CH_3-\overset{O}{\overset{\|}{C}}-CH_2OH$ $\xrightarrow{\text{kinase}}$ $CH_3-\overset{O}{\overset{\|}{C}}-CH_2OPO_3^{2-}$
addition of phosphate

b. $CH_3-\overset{O}{\overset{\|}{C}}-CH_2OH$ $\xrightarrow{\text{dehydrogenase}}$ $CH_3-\overset{O}{\overset{\|}{C}}-CHO$
loss of H_2

c. $CH_3-\overset{O}{\overset{\|}{C}}-CH_2OH$ $\xrightarrow{\text{carboxylase}}$ $HO_2CCH_2-\overset{O}{\overset{\|}{C}}-CH_2OH$
addition of CO_2

24.99 Calculate the number of ATP molecules from the catabolism of one glycerol molecule.

$$\text{glycerol} \xrightarrow{[1]} \text{glycerol 3-phosphate} \xrightarrow{[2]} \text{dihydroxyacetone phosphate} \xrightarrow{[3]} \text{pyruvate} \xrightarrow{[4]} \text{acetyl CoA} \xrightarrow{[5]} CO_2$$

[1] 1 ATP required for conversion of glycerol to glycerol 3-phosphate	–1 ATP
[2] glycerol 3-phosphate → dihydroxyacetone phosphate: 1 NADH → 2.5 ATPs	2.5 ATPs
[3] dihydroxyacetone phosphate → pyruvate: 2 ATPs + 1 NADH → 2.5 ATPs =	4.5 ATPs
[4] pyruvate → acetyl CoA: 1 NADH → 2.5 ATPs	2.5 ATPs
[5] acetyl CoA → CO_2: 1 acetyl CoA × 10 ATP/acetyl CoA =	10 ATPs
Total steps [1]–[5]:	**18.5 ATPs**

24.101 Calculate the amount of ATP generated during catabolism of a gram of glucose compared to a gram of arachidic acid as in Section 24.7.

$$\frac{32 \text{ mol ATP}}{1 \text{ mol glucose}} \times \frac{1 \text{ mol glucose}}{180.2 \text{ g glucose}} = \frac{0.18 \text{ mol ATP}}{1 \text{ g glucose}} \quad \text{(Section 24.7)}$$

Answer 24.19 →

$$\frac{134 \text{ mol ATP}}{1 \text{ mol arachidic acid}} \times \frac{1 \text{ mol arachidic acid}}{312.5 \text{ g arachidic acid}} = \frac{0.429 \text{ mol ATP}}{1 \text{ g arachidic acid}}$$

One gram of arachidic acid produces more ATP than one gram of glucose (0.429 mol/g compared to 0.18 mol/g). This result supports the fact that lipids are more efficient at storing energy than carbohydrates.

Chapter 25 Body Fluids

Solutions to In-Chapter Problems

25.1 Besides K^+ and HPO_4^{2-}, other prevalent ions in intracellular fluids are Mg^{2+}, SO_4^{2-}, and protein anions.

25.2 Multiply the volume of body fluid (42 L) by each value in Figure 25.1 to answer each part.

a. 42 L x 0.64 = 27 L of intracellular fluids

b. 42 L x 0.25 = 11 L of interstitial fluids

c. 42 L x 0.08 = 3 L of plasma

25.3

	Erythrocytes	Leukocytes	Platelets
a. Prevalence	5×10^6 per µL most prevalent	7×10^3 per µL least prevalent	3.5×10^5 per µL
b. Function	Transport O_2 and CO_2	Destroy bacteria and other foreign species	Contain machinery to initiate blood clot formation
c. Presence of a nucleus	No	Yes	No
d. Life span	120 days	3–4 days	5–9 days
e. Organ where synthesized	Bone marrow	Bone marrow	Bone marrow and spleen

25.4 Whole blood contains all the substances (a)–(e), whereas plasma contains albumin (a), water (c), and electrolytes (d).

25.5 Blood serum contains no fibrinogen, whereas blood plasma contains fibrinogen.

25.6 **Anemia** is caused when too few erythrocytes are produced or too many are lost or destroyed. **Leukemia** is caused by an overproduction of leukocytes that do not mature and are unable to destroy invading foreign substances. **Hemophilia** is caused when an individual lacks one or more clotting factors, so blood does not clot.

25.7 Fibrinogen is an inactive zymogen that is converted to fibrin by the enzyme thrombin. Fibrin cannot be stored in blood because it would cause the blood to clot.

25.8 The blood in veins is dark red, because hemoglobin is complexed with fewer oxygen molecules.

25.9 Since O_2 flows from the capillary to **B, B** must represent the tissues, which have a lower partial pressure of O_2 than the lungs.

25.10 Because CO_2 flows from the tissues where it has a pressure of 46 mm Hg to the capillary, the pressure in the capillary must be less; thus, **A** = 38 mm Hg. Because O_2 flows from the capillary to the tissues where its pressure is 40 mm Hg, the pressure of O_2 in the capillary must be greater; thus, **B** = 97 mm Hg.

25.11

a. O_2 and CO_2 always flow from a region of higher partial pressure to a region of lower partial pressure by diffusion.
b. Carbaminohemoglobin is formed when an NH_2 group of hemoglobin reacts with CO_2. When carbaminohemoglobin reaches the lungs, it is re-converted to hemoglobin and CO_2, which is exhaled.
c. Alveoli are small air sacs in the lungs where gas exchange (O_2 and CO_2) occurs between the air and the capillaries.
d. Carbonic anhydrase is the enzyme that converts CO_2 to carbonic acid. When carbonic acid reaches the lungs it is re-converted to CO_2 and exhaled.

25.12

a. $-COO^- + H_3O^+ \longrightarrow -COOH + H_2O$

b. $-NH_3^+ + OH^- \longrightarrow -NH_2 + H_2O$

25.13 A heavy smoker cannot expel as much CO_2. When $[CO_2]$ increases, $[H_3O^+]$ increases and pH decreases. Thus a pH less than 7.35 (a) may result.

25.14 When $[HCO_3^-]$ decreases, the equilibrium shifts to form more HCO_3^- and H_3O^+. Increasing $[H_3O^+]$ decreases pH.

25.15

a. Morphine is more polar because it has OH groups and fewer nonpolar C–C and C–H bonds.
b. The less polar heroin molecules can cross the blood–brain barrier more readily.
c. Heroin is more potent because it is less polar and can enter the brain more readily.

25.16 Xe is a nonpolar atom that can readily penetrate the brain by crossing the blood–brain barrier.

25.17 Only blood cells and large proteins remain in the blood. Thus the filtrate contains (a) K^+, (c) glucose, (d) water, and (e) waste products.

25.18 Creatine is water soluble because it is ionic. Creatinine is a small polar molecule with NH bonds capable of intermolecular hydrogen bonding with water, making it water soluble.

25.19 A diuretic increases the output of urine.

25.20 With no vasopressin, less urine is retained and urine output increases.

Solutions to Odd-Numbered End-of-Chapter Problems

25.21 The three principal body fluids are intracellular fluid (64%), interstitial fluid (25%), and plasma (8%).

25.23 Graph (a) corresponds to intracellular fluid because the K^+ concentration is high and the Cl^- concentration is low. Graph (b) corresponds to interstitial fluid because the K^+ concentration is low and the Cl^- concentration is high.

25.25

a. **A** represents platelets, **B** represents leukocytes, and **C** represents erythrocytes.
b. **C** is present in the highest concentration.
c. **B** is present in the lowest concentration.

25.27

a. Erythrocytes and platelets have no nucleus.
b. Erythrocytes survive for the longest time period.
c. Leukocytes are key elements of the immune system.
d. Erythrocytes carry hemoglobin.
e. Platelets contain the machinery for clot formation.

25.29

a. Globulins are present in whole blood [1], blood plasma [2], and blood serum [3].
b. Fibrinogen is present in whole blood [1] and blood plasma [2].
c. Erythrocytes are present in whole blood [1].
d. Urea is present in whole blood [1], blood plasma [2], and blood serum [3].

25.31

a. Veins [2] carry blood from the tissues to the heart.
b. Arteries [1] contain blood that is rich in oxygen.
c. Capillaries [3] contain a thin wall that is permeable to water and small molecules.

25.33

a. Vitamin K is needed for the synthesis of certain clotting factors.
b. Fibrinogen is the soluble zymogen that is converted to insoluble fibrin, which forms a mesh around a group of blood cells in a clot.
c. Platelets contain the machinery for clot formation. They collect at a site of injury in a blood vessel.
d. Thrombin is the protein that converts fibrinogen to fibrin.

25.35 O_2 complexes with the Fe^{2+} of hemoglobin, whereas CO_2 reacts with an NH_2 group of a peptide chain of hemoglobin.

25.37 There are three mechanisms for CO_2 transport from the tissues to the lungs.
[1] CO_2 can be transported as a gas dissolved in blood plasma.
[2] CO_2 reacts with hemoglobin to form carbaminohemoglobin, which is re-converted to hemoglobin and CO_2 and exhaled in the lungs.
[3] CO_2 is converted to carbonic acid by carbonic anhydrase, and then re-converted to CO_2 in the lungs.

25.39 O_2 always diffuses from a region of higher concentration to a region of lower concentration. O_2 flows from the air (160 mm Hg) to the alveoli and bloodstream (100 mm Hg) to the tissues (40 mm Hg).

25.41 The three main buffer systems in the body are:

[1] Carbonic acid (H_2CO_3)/bicarbonate (HCO_3^-) buffer in the blood. CO_2 and H_2O react to form H_2CO_3, so [CO_2] is directly related to pH. Increasing [CO_2] increases [H_2CO_3] and [H_3O^+], and pH decreases. Decreasing [CO_2] decreases [H_3O^+], and pH increases.

[2] Dihydrogen phosphate ($H_2PO_4^-$)/hydrogen phosphate (HPO_4^{2-}) buffer in intracellular fluids. HPO_4^{2-} reacts with excess acid, and $H_2PO_4^-$ reacts with excess base.

[3] Proteins in intracellular fluids and the blood. Carboxylate anions ($-COO^-$) react with excess acid and ammonium cations ($-NH_3^+$) react with excess base.

25.43

a. Acidosis (pH < 7.35) results from any of the conditions listed in the first column of Table 25.1: decreased respiratory rate, advanced congestive heart failure, advanced lung disease, starvation, diabetes in poor control, excessive exercise, or kidney failure.
b. Alkalosis (pH > 7.45) results from any of the conditions listed in the second column of Table 25.1: hyperventilation, excessive vomiting with loss of stomach acid, antacid overdose, or excess aldosterone.

25.45 When more CO_2 is expelled in hyperventilation, [CO_2] decreases, [H_3O^+] decreases, and pH increases.

25.47 Compound (b) is more likely to be a general anesthetic, because it is a neutral molecule. Compound (a) is ionic, so it cannot cross the blood–brain barrier.

25.49 Dopamine can leave the brain but it cannot enter the brain by crossing the blood–brain barrier.

25.51 Only blood cells and large proteins remain in the blood. Thus, the filtrate contains (a) sodium, (b) sulfate, and (e) urea. The capillaries retain (c) leukocytes and (d) platelets.

25.53 Aldosterone increases the sodium concentration in the blood, more fluid is retained, and urine output decreases.